Chemometrics in Food Control Based on Elemental Analysis

Piotr Szefer

Chemometrics in Food Control Based on Elemental Analysis

Springer

Piotr Szefer
Department of Bromatology
Medical University of Gdansk
Gdansk, Pomorskie, Poland

ISBN 978-3-031-99686-3 ISBN 978-3-031-99687-0 (eBook)
https://doi.org/10.1007/978-3-031-99687-0

This Springer imprint is published by the registered company Springer Nature Switzerland AG
The registered company address is: Gewerbestrasse 11, 6330 Cham, Switzerland

This book is dedicated to my grandson Filip

Preface

The book provides a comprehensive review of the characterization and control of all food groups, i.e., meat, fish, seafood and their products; milk and dairy products; honey and its products; cereals and grains; olive oils and oilseeds; vegetables; legume seeds and spices; mushrooms; fruits, including fruit juices; tea; coffee, cocoa, sugar and their products; mineral and tap water; wines; and other alcoholic beverages.

Various issues are raised, including in-depth discussions related to classifying and controlling the rich assortment of animal and plant foods of origin. This aim was achieved using state-of-the-art analytical methods coupled with modern chemometric techniques, including multivariate models.

Often, topics that are strictly suited to discussion in the section "Geographical differentiation" or "Intervarietal differentiation" for particular food groups go beyond these research areas. In such circumstances, this issue has been discussed in other subchapters of the book, e.g., "Interregional differentiation", "Interspecies differentiation", or "Differentiation of other parameters".

The most important aspects concerning the analytical and chemometric workshop are presented in tabular form in ELECTRONIC SUPPLEMENTARY MATERIAL 1. The detailed classification of of samples with respect to their geographical origin and animal/botanical characteristics is presented in ELECTRONIC SUPPLEMENTARY MATERIAL 2.

It has been demonstrated that the other main topics concern traceability, i.e., the possibility of verifying and establishing the interregional and interspecies/brand to distinguish a foodstuff; hence, the use of a combination of modern analytical and chemometric tools is not only recommended but also essentially needed. Since food is characterized by a complex matrix, the strategies concerned multivariate elemental profile including organic compound analysis.

This book will be useful for food chemists, quality control professionals, nutrition and public health educators, analytical chemists, food researchers needing to use chemical techniques in practice, food manufacturers, environmental scientists and students, and practising professions in food chemistry and food safety.

Gdańsk

Piotr Szefer
Autumn 2024

Acknowledgements

I would like to thank the *Springer Nature* team for their help in the production of this book, and I express particular gratitude to Gabriela Prostko and Sabrina Reba, Associate Editors from Springer Group, for their constructive cooperation, help and understanding.

My special thanks go to Professor Małgorzata Grembecka, the Head of the Department of Bromatology of the Medical University of Gdańsk. She was fully and exceptionally understanding, enabling me to work effectively on the preparation of the book project.

I give particular thanks to my wife, Krystyna, for her patience and understanding during the preparation of this book.

I am also grateful to various publishers and authors for permission to use copyrighted figures from previously published papers.

Finally, I thank Maksymilian Biniakiewicz, the Head of Printing House of the University of Gdańsk, who professionally processed drawings originating from previously published articles in accordance with Springer's instructions.

Competing Interests The author has no competing interests to declare that are relevant to the content of this manuscript.

Contents

About the Author

Piotr Szefer has been employed since 1972 in the Department of Analytical Chemistry, Medical University of Gdańsk (MUG). He was awarded Full Professorship in January 2000. His research is focused on analytical chemistry, environmental analytics, trace analysis, food and marine chemistry and bioanalytics. From 1996 to 2002, he was the Dean of the Faculty of Pharmacy at MUG. He was the Head of the Department of Food Sciences at MUG. He has 5 written books published by Elsevier Science B.V., CRC Press–Taylor & Francis, Marine Board, European Science Foundation and Maralte BV. Piotr Szefer was also Guest Editor of two special issues of journals, *Applied Geochemistry* (Pergamon–Elsevier Science Ltd., 1998) and the *Marine Pollution Bulletin* (Pergamon–Elsevier, 2004). He has published ca. 50 book chapters, including 20 international chapters (by, e.g., Springer, Elsevier, Taylor & Francis and Nova Science Publishers, Inc.), approximately 370 papers and over 300 symposial abstracts. He has been a member of more than 30 international and national research consortiums, expert panels, working groups and scientific associations and organizations (e.g., WG MB-ESF, SCOR and WGMS-ICES) as well as a member of 20 editorial boards of international journals (those published by, e.g., Elsevier, Springer-Versita, Taylor & Francis and Bentham). He has reviewed more than 2000 manuscripts for approximately 120 different journals, mainly those with international reputations. He has published with approximately 100 Polish scientists and approximately 150 scientists from approximately 50 countries. He has visited 16 countries as a visiting professor or research scientist. He has received (as a host scientist) different researchers from 17 countries in his department. According to Google Scholar, his h-index is 50; his i10-index is 122; and he has 8053 citations. Piotr Szefer has recently been ranked among the top 2% of researchers in the world. The rank was prepared by Stanford University in cooperation with Elsevier and SciTech Strategies. According to the portal Research.com, he has also been ranked among the top 1% of the best representatives in his field of science.

Acronyms and Abbreviations

^{13}C-EA/LC-IRMS	liquid chromatography coupled to isotope ratio mass spectrometry;
^{1}H HR-MAS NMR	^{1}H high resolution magic angle spinning spectroscopy;
1stHPYT	first HarvestPlus Yield Trial
2-FAC	2-furoic acid
2-FAL	2-furaldehyde
2T2D	2-trace 2-dimensional correlation analysis
5-HMF	(5-hydroxymethyl)furfural
5-MFA	5-methylfuraldehyde
AA	antioxidant activity
AA	total free amino acids
AAEA	ascorbic acid equivalent antioxidant capacity
AAS	atomic absorption spectroscopy
AAS–AMA	atomic absorption spectrometer–advanced mercury analyser
ABM	aceto balsamico di modena IGP
ACHE	acetylcholinesterase
ACY	total anthocyanins
AES	atomic emission spectrometer
AFS	atomic fluorescence spectroscopy
AHC	agglomerative hierarchical clustering
AMA	advanced mercury analyser
AMF	arbuscular mycorrhizal fungi
AMMI	additive main effects and multiplicative interactions analysis
AMS	accelerator mass spectrometry
ANCOVA	analysis of covariance
ANN	artificial neural network
ANOVA	analysis of variance
AOAC	Association of Official Analytical Chemists

AOC	Appellation d'Origine Controlôée
AOP	antioxidant potential
AOX	acyl-CoA oxidase
ASCP	derivative anodic chronopotentiometric stripping analysis
ATR-FTIR	attenuated total reflectance–Fourier transform infrared spectroscopy
AUC	area under the curve
AUC-ROC	area under the ROC CURVE
AVA	Agri-Food and Veterinary Authority
AVQ-ANN	adaptive vector quantization-artificial neural network
AW	water activity
BDA	Bayes discrimination analysis
BFR	brominated flame retardants
BGR	geoscience and natural resources
BP-ANN	back-propagation feed-forward–artificial neural network
BP-NN	back propagation artificial neural networks
BTS	butyltins
BW	weighted regression coefficients
C&RT (Cart)	classification and regression trees
CA	cluster analysis
CAC	catalytically active compounds
CAF	caffeine
CAP	canonical analysis of principal components
CAPC	canonical analysis of principal coordinates
CAR	conditional autoregressive model
CART	classification and regression trees
CART	regression trees
CAT	catalase
CAT	catechins
CBE	charge balance error
CBO	Certified Brand of Origin
CBOQ	Certified Brand of Origin and Quality
CC	canonical correlation
CCA	canonical correlation analysis
CCC	cophenetic correlation coefficient
CDA (DA)	canonical discriminant analysis
CDF	cumulative distribution function
CDFs	canonical discriminant functions
CE	capillary electrophoresis
CEC	cation exchange capacity
CE-DAD	capillary electrophoretic–diode-array detector
CEN	European Committee for Standardization

CE-UV capillary electrophoresis with UV-detection method
CF canonical function
CFA continuous-flow analysis
CF-EA-IRMS continuous flow elemental analyser coupled to elemental analyser isotope ratio-mass spectrometry
CF-EA-IRMS isotopic ratio mass spectrometry in continuous flow mode
CF-IRMS continuous flow isotope ratio mass spectrometry
CFP chemical fertilizers and pesticides
CF-TC-IRMS continuous flow total combustion–elemental analyser–isotope ratio mass spectrometry
CHROMA C^*_{ab}
CIE Commission Internationale de l'Eclairage
CIE L*a*b* colour space L*—Lightness axis; a*—Red to green axis; b*—Yellow to blue axis; C*—Saturation; ho—Hue angle/tint
CIMMYT International Maize and Wheat Improvement Centre
CLA hierarchical cluster analysis
CLas Candidatus *Liberibacter asiaticus*;
COD *coefficient of determination*
COI cytochrome oxidase I
Colour values a* and b* value
ComDim common components and specific weights analysis
COX cytochrome-c-oxidase
CP cold-pressed extra-virgin olive oil
CPA cold plasma ashing
CP-ANN counterpropagation artificial neural networks
CRM certified reference material
CRT Classification and Regression Tree
CS citrate synthase
CSI Calcite Saturation Index
CSIA compound specific isotope analysis
CSIA-IRMS compound-specific isotope analysis through gas chromatography coupled to isotope-ratio mass spectrometry
CSLM confocal scanning laser microscopy
CV cross validation method
CV cross validation
CV K-fold cross validation
CV potential curves

CV	cross validation
CVA	canonical variate analysis
CV–AAS	cold-vapour atomic absorption spectroscopy
CV-ANOVA	cross validation-analysis of variance
CVR	classification via regression
CWT	continuous wavelet transform
CWT-RF	random forest based on continuous wavelet transform
D.O.Ca.	Denominación de Origen Calificada
d_{20}	relative density
DA	discriminant analysis
DAN	deep adaptation networks
DANN	domain adaptive neural network
DBT	dibutyltin
DC	divergence criterion
DD-SIMCA	data-driven soft independent modelling of class analogy
DE	diatomaceous earth
DE	dry extract
DEHP	di(2-ethylexyl)phthalate
DFA	discriminant function analysis
DGC	density gradient centrifugation
DIC	dissolved inorganic carbon
DL-PCBS	dioxin-like PCBS
DM	dry matter
DMA	direct mercury analysis
DOC	controlled designation of origin (Denominazione di Origine Controllata)
DOP	Denominazione d'Origine Protetta—in English PDO (Protected Designation of Origin)
DOs	denominations of origin
DPASV	differential pulse anodic stripping voltammetry
DP-LIBS	double-pulse lips
DPLS	discriminant partial least square
DPPH-RSA	2,2-diphenyl-1-picrylhydrazyl free radical scavenging activity
DPR KOREA	Democratic People's Republic of Korea
DRC-CP-MS	dynamic reaction cell for inductively coupled plasma mass spectrometry
DT	decision tree
DTA	decision tree algorithm
DTS	decision trees
E	redox potential
EA-IRMS	elemental analyser –isotope ratio-mass spectrometry

EC	electrical conductivity
EC	electrolytic conductivity
EDGC	Euclidean distances from the group centroids
EDXRF	energy dispersive x-ray fluorescence
EF	elemental fingerprints
EFA	elemental fingerprint analysis
EFDS	elliptical Fourier descriptors
EFSA	European Food Safety Authority
EGS Geochemistry Expert Group	EuroGeoSurveys Geochemistry Expert Group
EICPSF-MS	element inductively coupled plasma sector field mass spectrometry
ELM	extreme learning machine
EMMA	end-member mixing analysis
EN	electronic nose
ENAA	epithermal neutron activation analysis; CP-ANN–counterpropagation artificial neural network
ERT	extremely randomized trees
ESI-MS/MS	electrospray ionization with tandem mass spectrometry
ET	electronic tongue
ET AAS	electrothermal atomic absorption spectroscopy
ETS	electronic tongues
EV	Eigenvalue
EVOO	extra-virgin olive oil
EXT	aqueous extract
FA	factor analysis
FA	fatty acids
FAA	fatty acid analysis
FAAS	flame atomic absorption spectroscopy
FAC	fatty acid composition
FAES	flame atomic emission spectrometry
FAMD	factor analysis of mixed data
FAMES	fatty acid methyl esters
FAP	fatty acid profile
FAS	fatty acids
FCF	Fulton's condition factor
FDA	factorial discriminant analysis
FDA	Fisher discrimination analysis
FDHAC	fuzzy divisive hierarchical associative clustering
FEAS	Food, Agricultural, and Environmental Sciences
FES	flame emission spectroscopy
FF	flesh firmness

FFP	fresh fruit pulp
FHDC	Fenghuangdancong tea
FIA	Fibonacci index analysis
FIA	flow injection analysis
FP	flame photometry
FPA	flavour profile analysis
FPCA	fuzzy principal component analysis
FPCA-LDA	fuzzy principal component analysis combined with linear discriminant analysis
FR	fixed residua
FRA	factorial regression model
FS	fluorescence spectroscopy
FS -LIBS	femtosecond libs
FS-FAAS	fast sequential flame atomic-absorption spectrometry
FSRA	forward stepwise regression analysis
fs-LIBS (FS-LIBS)	femtosecond laser-induced breakdown spectroscopy
FTIR	Fourier transform infrared
FW	fruit weight
GA	genetic algorithm
GA	gross alpha activity
GB	gross beta activity
GC	gas chromatography
GC/C/IRMS	gas chromatography via a combustion interface–isotope-ratio mass spectrometry
GC-C-IRMS	gas chromatography–combustion–isotope ratio mass spectrometry
GC-EI-IRMS	gas chromatography via electron ionization–isotope ratio-mass spectrometry
GC-FID	gas chromatography–flame-ionization detector
GC-IRMS	gas chromatography–isotope ratio mass spectrometry
GC-MS	gas chromatography–mass spectrometry
GC-MS-MS	gas chromatography–tandem mass spectrometry
GC-P-IRMS	gas chromatography–pyrolysis–isotope ratio mass spectrometry
GC-QIT-MS	gas chromatography–quadrupole ion trap–mass spectrometry
GDA	general discriminant analysis
GEMAS	Geochemical Mapping of Agricultural and Grazing Land Soil of Europe
GF-AAS	graphite-furnace atomic absorption spectroscopy

GFEC	grain zinc concentration
GIS	Geographic Information System
GLC-FID	gas–liquid chromatography–flame ionization detector
GLM	general linear model
GLM	generalized linear model
GLM PROCEDURE	general linear model
GLM-ANOVA	general linear model— analysis of variance
GLP	Green Lungs of Poland
Gly	glycine
GM	genetic modified
GMP	good manufacturing practice
GNIP	Global Network of Isotopes in Precipitation
GNN	genetic neural network
GOXH2O2 and GPXCHP	selenium dependent and sum of SE-dependent and SE-independent glutathione peroxidases
GPA	generalized procrustes analysis
GPA-MLR	generalized procrustes analysis–multiple linear regression
GPC	grain protein content
GR	glutathione reductase
GRS	gamma-ray spectroscopy
GSFR	green stem and foliar retention
GSH	level ID total glutatione
GSI	Gonadosomatic Index
GST	glutathione s-transferases
GZnC	grain zinc concentration
HAFS	hydride atomic fluorescence spectrometry
HC	hierarchical clustering
HCA	hierarchical cluster analysis
HCD	hierarchical cluster dendrogram
HCPC	hierarchical clustering on principal components
HC-PLSR	hierarchical cluster-based partial least squares regression
HFCS	fructose corn syrup
HG-AAS	hydrogen-generation atomic absorption spectroscopy
HG-AFS	hydride atomic fluorescence spectrometry
HI	Hazard Index
HICS	high-income countries
HIPC	high performance ion chromatography
HIS	hepatosomatic index
HLB	Huanglongbing most serious bacterial disease of citrus

HM or HMS	heavy metal or heavy metals
HMF	5-(hydroxymethyl) furfural
HMF	5-(hydroxymethyl)-2-furaldehyde
HPAEC/PAD	high pH anion-exchange chromatography with pulsed amperometric detection
HPAEC/PAD	high-performance anion-exchange chromatography with pulsed amperometric detection
HPGC	high pressure gas chromatography
HPIC	high performance ion chromatography
HPIC-CD	high pressure ion chromatography–conductivity detector
HPLC	high performance liquid chromatography
HPLC-DAD	high-performance liquid chromatography with diode-array detection
HPLC-DAD-MS	high performance liquid chromatography with diode-array detection and mass spectrometry detection
HPLC-DAD/UV/VIS	high-performance liquid chromatography with diode-array vs UV–Vis detection
HPLC/ESI-MS/MS	high performance liquid chromatography/ electrospray ionization tandem mass spectrometry
HPLC-DRI	high performance liquid chromatography-differential refractive index
HPLC-ELSD	high-performance liquid chromatography with evaporative light scattering detection
HPLC-FD	high-performance liquid chromatography with fluorescence detection
HPLC-HHPN-AFS	high-performance liquid chromatography-atomic fluorescence spectroscopy with evaporative light scattering detection
HPLC-MRM	high-performance liquid chromatography with multiple reaction monitoring detector
HPLC–PDA	high-performance liquid chromatography–photodiode array detection
HPLC-RID	high-performance liquid chromatography with refractive index detection
HPLC-UV-VIS/PDA	high-performance liquid chromatography with UV–Vis vs diode-array detectors
HR CS FAAS	high-resolution continuum-source flame atomic absorption spectrophotometry
HR-CS-AAS	high resolution continuum source atomic absorption spectrometry
HRI	Health Risk Index

HR-ICP-MS	high-resolution inductively-coupled plasma-mass spectrometry
HR-ICP-MS (ICP-HRMS)	high-resolution inductively-coupled plasma-mass spectrometry
HR-MAS NMR	^{1}H high resolution magic angle spinning–nuclear magnetic resonance spectroscopy
HSD	honestly significant difference
HSD	honestly significantly different Tukey's test for unequal N analysis
HS-SPME-GC/MS	head space-solid phase micro extraction followed by gas chromatography coupled with mass spectrometry
HS-SPME-GC-MS	headspace solid-phase microextraction coupled to gas chromatography–mass spectrometry
HT	hierarchical tree models
HTP-IRMS	high-temperature pyrolysis–isotope ratio mass spectrometry
HTS	high throughput sequencing
HTST	high temperature short time
Hue	h_{ab}
IAA	indispensible amino acid
IC	ion chromatography
ICA	independent component analysis
ICAP	induced coupled argon plasma spectrometry
IC-CD	ion chromatography with conductivity detection
ICP-AES	inductively coupled plasma atomic emission spectroscopy
ICP-MS	inductively-coupled plasma-mass spectrometry
ICP-MS WITH ORS	inductively-coupled plasma-mass spectrometry equipped with an octopole reaction system collision/reaction cell technology
ICP-MS-DRC	inductively coupled plasma mass spectrometer with dynamic reaction cell technology
ICP-OES	inductively-coupled plasma-optical emission spectroscopy
ICPS	inductively-coupled plasma spectrometry
ICP-SFMS	inductively coupled plasma sector field mass spectrometry
ICP-TOF-MS	inductively coupled plasma time-of-flight mass spectroscopy
IEC	ion exchange chromatography
IGP	Indication of Geographical Protection

IHC	International Honey Commission
IMA	Indian Musa accessions
IOVW	International Organisation of Vine and Wine
IPLS	interval partial least-squares
IRIS	isotopic ratio infrared spectroscopy
IRMS	isotope ratio-mass spectrometry
ISCIRA	Internal Standard Stable Carbon Isotope Ratio Analysis;
ISE	ion selective electrode
IST	independent-sample t test
Jackknife LOO	jackknife leave-one-out procedure
JY	juice yield
KANN	Kohonen artificial neural network
k0-INAA	K 0-standardization of instrumental neutron activation analysis
KDE	kernel density estimates
Kf-CV	k-fold cross validation
KMC	kinetic Monte Carlo
K-means	k-means classification
KMO	Kaiser-Meyer-Olkin measure of sampling adequacy
KMO	Kaiser-Meyer-Olkin test
kNN (KNNA)	k-nearest neighbour analysis (k-nearest neighbours algorithm)
k-NN (KNN)	k-nearest neighbours
KNN	nearest neighbour methods
KS	Kennard-Stone algorithm
LAD	least absolute deviation
LA-ICP-MS	laser ablation inductively coupled plasma mass spectrometry
LC-IRMS	liquid chromatography-isotope ratio-mass spectrometry
LC-MS/MS	liquid chromatography with tandem mass spectrometry
LC-PDA	liquid chromatography with photodiode array detector
LC-TQ-ESI-MS	liquid chromatography vs triple quadrupole mass spectrometry equipped with electrospray ionization source
LD	linear discriminant function
LDA	linear discriminant analysis
LDA	linear discriminant function analysis
LDA and QDA	linear and quadratic discriminant analyses
LDF	linear discriminant function
Leu	leucine

LIBS	laser-induced breakdown spectroscopy
LIBS GDT	LIBS graphical development tool computer program
LIGHTNESS	L*
LLE-CGC	liquid-liquid extraction–capillary gas chromatography
LMICS	low middle income countries
LOD	limit of detection
LOI	loss-on-ignition method
LOO CV	leave-one-out cross-validation
LOQ	limit of quantification
LPCA	linear principal component analysis
LPP	low-polymeric polyphenols
LPR	local polynomial regression
LR	linear regression
LR	logistic regression
LRA	linear regression analysis
LSD	Fisher's least significant difference
LSD	least significant difference
LVQ	learning vector quantization
LVQ-ANN	learning vector quantization–artificial neural network
LVQNN	learning vector quantization neural network
LVs	latent values
LWIA	liquid water isotope analyser
MAD	median absolute deviation
MAL	maximum allowable limit
MANOVA	multivariate analysis of variance
MAS	UV-Vis molecular absorption spectrometry
MBC	Mozzarella di Bufala Campana
MBT	monobutyltin
MCA	multivariable correlation analysis
MCC	multielement composition content
MCD	minimum covariance determinant
MCDA	multi-criteria decision analysis also known as multi-criteria decision-making (MCDM)
MC-ICP-MS	multicollector–inductively coupled plasma–mass spectrometry
MCLS	promulgated maximum contaminant levels
MC-TIMS	multi-collector thermal ionization mass spectrometry
MD	Mahalanobis distance
MDA	multidimensional analysis
MDM	multi-criteria decision-making analysis
MDS	multidimensional scaling

MEA	multi-element analysis
MEC	multi-element composition
MeHg	methylmercury
MEP	multi-element profile
Met	methionine
MGO	methylglyoxal
MHC	multivariate hierarchical clustering
MICRO-CT	X-ray micro-computed tomography
MIRS	mid-infrared spectroscopy
MK MODEL	mechanistic Munich–Kohn model
ML	machine learning algorithm
MLF	multi-layer feed-forward
MLP	multilayer perceptron
MLP-ANN	multi-layer perceptron–artificial neural network
MLP-BP	multi-layered perceptron neural networks with backpropagation algorithm
MLR	multiple linear regression
MLRA	multiple linear regression analysis
MLs	maximum levels
MP-AES	microwave plasma atomic emission spectrometry
MPI	Metal Pollution Index
MPLS	multi-way partial least squares
MR	multiple regression
MRA	multiple regression analyses
MRL	maximum residue level
MRM	multiple-reaction-monitoring
MRT	Duncan's multiple range test
MS	mass spectrometry
MSC	multiplicative scatter correction
MSD	squared Mahalanobis distance
MUFA	monounsaturated fatty acid
MVE PEEL	*S. melongena* var. *esculentum*
MVEF	MVE flesh
MVS PEEL	*Solanum melongena* var. *serpentinum*
MVSP	MVS peel
MW2D	moving-window two-dimensional correlation
N	force/load
NAA	neutron activation analysis
NB	naïve Bayes
NB MLA	NB naïve Bayes machine-learning algorithm
NCS	non-centrifugal cane sugar
NCV	nested cross-validation
NE-LIBS	nanoparticle-enhanced LIBS

NER%	non error rate
NFE	nitrogen free extract
NIPALS	non-linear iterative partial least squares
NIR	near-infrared
NIRS	near-infrared spectroscopy
NIRTS	near-infrared transmittance (NIT) spectroscopy
NITS	near-infrared transmittance spectroscopy
NLRA	nominal logistic regression analysis
NMDS	non-metric multidimensional scaling analysis
nMDS	non-metric multidimensional scaling ordination
NMR (^{1}H NMR)	nuclear magnetic resonance
NN	neural network
NNB	neural network bagging
NNS	nearest neighbours search
non-GM	non-genetically modified
NPC	normal probability curve
ns-LIBS (NS-LIBS)	nanosecond laser-induced breakdown spectroscopy
OCC	overall correct classification rate
OCPLS	one-class partial least squares
OCRF	one class random forest
OD	optical density
OECD	Organization for Economic Co-Operation and Development
OIV	International Organization of Vine and Wine
OLS	ordinary least squares technique
One-at-a-time CV	one-at-a-time cross-validation
OPA	orthogonal projection analysis
OPLS- DA	orthogonal partial least squares-discriminant analysis
OPLS-DA	orthogonal projection to latent structure–discriminant analysis
OT	organic tin
OTC	organotin compounds
OTS	organotins
PA	path coefficient analysis;
PA	procrustes analysis
PAD	pulsed amperometric detection
PAM	partitioning around medoids algorithm
PC	grain protein content
PCA	principal component analysis
PCA-LDA	a hybrid PCA-LDA model
PCA-RF	principle component analysis-based on random forest

PCDD/FS	polychlorinated diphenyldioxins/furans
PCR	polymerase chain reaction
PCR	principal component regression
PCR-RAPD	polymerase chain reaction–randomly amplified polymorphic DNA
PCR-RAPD MARKERS	polymerase chain reaction-randomly amplified polymorphic DNA markers
PCVG	principle components variable grouping
PDA	pairwise discriminant analysis
PDO	protected designation of origin
PE	polyethylene
PERMANOVA	permutational multivariate analysis of variance
PET	polyethylene terephthalate
PF	potential functions
PFA	prognostic factor analysis
PGES	platinum group elements
PGI	protected geographical indication
PHE	phenylalanine
PIXE	particle (proton) induced X-ray emission
PLS	partial least squares regression
PLS-DA	partial least squares regression–discriminant analysis
PLSR	partial least square regression
PLSRA	partial least squares regression analysis
PLS-UVE	partial least squares regression-uninformative variable elimination
PNN	probabilistic neural network analysis
PNN	probabilistic neural network
PoCu	potential curves
POL	total polyphenols
POND	aquaculture production method with many aquatic species cultured in ponds
POV	peroxide value
PR	pattern recognition
PRIMER	Plymouth routines in multivariate ecological research
PRO	proanthocyanidins
PRO	proline
PROMETHEE	multicriteria decision making method
PRP	pentagonal radar plot
PTES	potentially toxic elements
PTWI	provisional tolerable weekly intake
PUFA	polyunsaturated fatty acid
PWT	Pingwu Fuzhuan brick tea
PXRF	portable X-ray fluorescence analysers

Q res	Q-residual
QDA	quadratic discriminant analysis
QDA	quadratic discriminant function analysis
QDF	quadratic discriminant function
Q-ICP-MS	quadrupole inductively coupled plasma mass spectrometry
Q-ICP-MS	single quadrupole inductively coupled plasma mass spectrometry
Q-ICP-MS (ICP-Q-MS)	single quadrupole inductively coupled plasma mass spectrometry
QR	quadratic regression
QTL	quantitative trait loci
R	dry residue
R2	coefficients of determination
RA	regression analysis
RAS	recirculating aquaculture system
RBF	radial basis function
Rc2 VALUE	calibration coefficient of determination
R-CA	R-cluster analysis
RDA	recommended daily allowance
RDA	redundancy analysis
RDA	regularized discriminant analysis
RDF	radial distribution function
REE	rare earth elements
REML	restricted maximum likelihood
RF	random forest classification
RF	random forests; Spearman
RF MLA	random forest machine-learning algorithm
RF	random forest model
RFA	random forest algorithm
RFC	random forest classifier
RFD	reference dose
RFI	random forest importance
RG	rice grain
RGP	rice grain pellet
RI	refractive index
RILs	recombinant inbred lines
RMFA	R-mode factor analysis
RMLR	ridge multinomial logistic regression
RMSE	root mean square error
RMSEC	root mean square errors of calibration
RMSECV	root mean square errors of cross-validation
RMSEP	root mean square error of prediction
ROC	receiver operating characteristic analysis
RPA	radar plot analysis

RPD	residual predictive deviation
RP-HPLC-PDA	reverse-phase high performance liquid chromatography photodiode-array detection
RPP	rice powder pellet
RPPBA	rice powder pellet with boric acid
RRMSEP	relative root-mean square error of prediction
RRS	resonance Raman spectroscopy
RS	Raman spectroscopy
RSA	radical-scavenging activity
RSD	relative standard deviation
RT-qPCR	real-time quantitative-PCR
RTs	regression trees
SAI	safe and adequate daily intake
SAS	statistical analysis software
SAS	surface active substances
SCIRA	stable carbon isotope ratio analysis
SD	second derivative
SDA	stepwise discriminant analysis
SEC-UV	size-exclusion liquid chromatography coupled on-line to UV
SECV	standard errors of cross validation
SEM	scanning electron microscopy
SER	serine
SFAs	saturated fatty acids
SIA	stable isotope analysis
SIC	stable isotope composition
SIMCA	soft independent modelling by class analogy
SIMPER	similarity percentage
SIR	stable isotope ratio
SIR	syringaldehyde
SIRA	stable isotope ratio analysis
SIRC	stable isotopic-ratio composition
SLDA	stepwise linear discriminant analysis
SLR	simple linear regression
SLST	solid-liquid-solid-transformation
SMLR	sparse multinomial logistic regression
SMR	stepwise multiple regression
SNIF-NMR	site-specific natural isotopic fractionation by nuclear magnetic resonance
SNK	Student-Newman-Keuls or Newman-Keuls
SNP	single nucleotide polymorphism
SNV	standard normal variate
SOAN	self-organizing with adaptive neighbourhood network
SOD	superoxide dismutase

SOM	Kohonen self-organizing map
SOMS	Kohonen's self-organizing maps
SPA	successive projection algorithm
SPA-LDA	successive projections algorithm
SPB	stepwise Bayesian procedure
SPCA	sparse version of PCA
SPLOM	scatter plot matrix
sPLS-DA	sparse partial least squares discriminant analysis
SPSS	statistical package for the social sciences
SR	selectivity ratio
SR	skin resistance
SREG MODEL	site regression model
SS-HR-CS ET AAS	solid sampling continuous source high resolution electrothermal atomic absorption spectrometry
SSR	simple sequence repeats
SST	soluble solids
STEM-EDS	scanning transmission electron microscopy energy dispersive spectroscopy
Stepwise CDA	stepwise canonical discriminant analysis
STI	Stress Tolerance Index
Strip Charts of ME	strip charts of multielements
Strip Charts of SIR	strip charts of stable isotope ratios
SVD	singular value decomposition
SVM	support vector machine
SVS	stepwise variable selection
T	temperature
TA	total monomeric anthocyanins
TA	titratable acidity
TA	total acidity
TA	total alkalinity
TAA	total amino acids
TARLS	tolerable average residue levels
TBT	tributyltin
TCD	thermal conductivity detector
TCD	total colour difference
TDE	total dry extract
TDI	tolerable daily intake
TDS	total dissolved solids
TE	test set
TEF	trace element fingerprint
TF	total flavonoids
TF	transfer factor
TFA	trans fatty acid
TFC	total flavonoid content

TH	total hardness
THQ	target hazard quotient
TIC	total inorganic carbon
TIC	total ion chromatography
TIMS	thermal ionization mass spectrometry
TKW	thousand kernel weight
TOC	total organic carbon
TOSC	total oxyradical scavenging capacity toward peroxyl
TP	total polyphenols;
TPC	total phenolic compounds
TPC	total polyphenolic compounds contents
TPC	total polyphenol content
TPHT	triphenyltin
TR	training set
TS	total solid content
TS	traditional statistics
TS	test set
TSG	traditional specialty guaranteed
t-SNE	t-distributed stochastic neighbour embedding
TSS	total soluble solids
TTA	total titratable acidity
TTC	total tannin content
TUIL	tolerable upper intake levels
Tukey's HSD	honestly significantly different—unequal N Tukey test
TVC	total caloric value
TXRF	total reflection X-ray fluorescence
TYR	tyrosine
UFA	unsaturated fatty acid
UHPLC	ultra-high-performance liquid chromatography
U-HPLC-DAD	ultrahigh performance liquid chromatography-diode array detector
UHPLC-DAD-MS/MS	ultrahigh performance liquid chromatography/diode array detector with tandem mass spectrometry
UHPLC-LTQ ORBITRAP MS	ultrahigh-performance liquid chromatography coupled with electrospray ionization hybrid linear trap quadrupole Orbitrap mass spectrometry
UHT	ultrahigh temperature
UNEQ	unequal dispersed classes
UPGMA	unweighted pair group method with arithmetic mean

UPLC-MS/MS	ultra-performance liquid chromatography-tandem mass spectrometry
UPMGA	unweighted pair group method with an arithmetic mean
USDA	United States Department of Agriculture
USFA	unsaturated fatty acid
USN WITH CFN	ultrasonic nebulizer with a cross-flow nebulizer
UTC-SIR-MS	stable isotope ratio mass spectrometer with universal triple collector
UTL	upper threshold limit
UVE	uninformative variable elimination
VA	validation test
VA	volatile acidity
VES	variable employment strategies
VIP	variable importance for the projection
VIP	variable importance in projection
VOCs	volatile organic compounds
VOO	virgin olive oil
VPDB	Vienna Peedee Belemnite Scale
VP-SEM/EDS	variable pressure scanning electron microscope equipped with energy-dispersive X-ray spectrometry
VQBCP	vector quantization-based classification procedure
VQBCP-ANN	vector quantization-based classification procedure-artificial neural networks;
VSCA	volatile short-chain acids
WD-XRF	wavelength dispersive x-ray spectrometer
WS-CRDS	wavelength-scanned cavity ring-down spectroscopy
XHLJ	Westlake Xihu Longjing green tea
XPS	X-ray photoelectron spectroscopy
XRD	X-ray diffraction analysis
XRF	X-ray fluorescence spectroscopy or X-ray fluorescence
XRM	Micro-computed tomography (microCT) or X-ray microscopy
δ^2H Py-CSIA	δ^2H composition of specific compounds released by direct analytical pyrolysis;
ΔE	variation of colour
Δp	weight loss

Chapter 1
Key Issues Constituting the Book Content

Contents

1.1 General Food Characteristics

1.1.1 Nutrient-Specific Food Groups

1.1.1.1 Meat and Its Products

Meat is considered one of the most nutritious and favoured food components available to the masses, helping to fulfil most bodily requirements. It has played a crucial role in human evolution and is an essential component of a well-balanced diet. Meat is a good source of proteins, zinc, iron, selenium and phosphorus, followed by vitamin A and B-complex vitamins. The average content of meat protein is approximately 23%. The individual values vary from higher to lower values depending on the type of meat source. Meat fat and its fatty acid profile are the leading topics of numerous debates in relation to its consumption, but its moderate consumption is

P. Szefer, *Chemometrics in Food Control Based on Elemental Analysis*, https://doi.org/10.1007/978-3-031-99687-0_1

always advised to lead a healthy life. The fat content of animal carcasses is estimated to range between 8% and 20%. Meat quality is affected by the animal breed type, how the animal is reared, including the feeding source (grains, pasture and grass), the genetic predisposition of the animal, post-mortem techniques and nutritional status during production. These factors affect the fat, lean and connective tissue components of meat and, for this reason, influence meat quality. Meats, i.e. chicken, pork, lamb and beef, are great sources of protein, essential amino acids, fats and fatty acids; carbohydrates and minerals, such as iron, zinc, iodine; vitamins (especially B_{12}) and essential fatty acids. These nutritional ingredients are conducive to human health; therefore, it has been recommended that meat and poultry be eaten every week as part of a well-adjusted diet (Ahmad et al. 2018).

Notably, the bioavailability of elements in meat is much greater than that in plant sources. The levels of both macro- and micro elements in animal-based foods are generally lower than in plant-based foods. Factors such as age, sex, type of muscle and cut of meat, and the cooking process strongly influence the elemental profile of meat. The concentrations and distributions of toxic elements, i.e. arsenic, cadmium, lead and mercury, are normalized in relation to the European Union and international regulations that establish the maximum levels of these elements in various meat products and their tolerable intakes. Since the bioavailability of these elements strongly influences their toxicity, special attention should be given to the analytical aspects of toxic elements, preferably via speciation analysis (Ortega-Barrales and Fernández-de Córdova 2015).

According to Cobos and Díaz (2014), healthier and functional meat products deserve special attention because they include low-sodium, low-fat, lipid-modified and nitrite-reduced meat products as well as meat and meat products with bioactive compounds.

1.1.1.2 Fish Flesh and Other Seafood (Mussels and Crustaceans)

Knowledge of the chemical profile of seafood is essential for assessing the quality of the raw material, storage stability and application of technological processes (Gokoglu and Yerlikaya 2015).

Fish is composed of 70–84% water, 15–24% protein, 0.1–22% fat, 0.1–1% carbohydrate and mineral elements, which are responsible for the nutritional value, functional aspects and sensory characteristics of the flesh. The muscle proteins of seafood are nutritionally valuable and easily digestible, with a well-balanced amino acid composition. Moreover, fish is a rich source of polyunsaturated fatty acids (omega-3 fatty acids) with a relatively high unsaturated lipid content. Compared with other protein sources, fish is low in calories (Gokoglu and Yerlikaya 2015). Specifically, fish protein consists of 20–30% sarcoplasmic proteins, which are composed mainly of albumins in addition to haemoproteins; 66–77% myofibrillar proteins, such as myosin, actin, actomyosin and troponin and 3–5% stromal proteins, including collagens (Kim 2015). Seafood provides a number of vitamins, such as A, D, E and B_{12}, as well as biophilic elements, such as selenium and iodine, which are very important in the diet (Gokoglu and Yerlikaya 2015). Moreover, macro elements, such as sodium, potassium, calcium, magnesium, phosphorus, sulphur and chlorine, as well as micro elements, are abundant in fish (Klein and Hurlbut 1985).

Seafood is a good source of mineral elements. In addition to the macro elements mentioned above, there are minor and trace elements. In terms of the benefits of minor elements, the contents of zinc, iron, manganese, copper, selenium and iodine, depending on their speciation, also significantly contribute to seafood consumption. Edible crustaceans (shrimp, lobster and crayfish) contain, on average, 78–83% moisture, 15–19% protein, 0.5–2.0% fat, 1.3–2.1% mineral elements and 23–41% edible portions. Analogical data for molluscs (oysters, scallops and mussels) show 80–83% moisture, 9–16% protein, 0.1–1.3% fat, 1.4–2.0% mineral elements and 10–44% edible portions (Durazzo et al. 2022). The distribution of macro- and micronutrients in fish and seafood, as well as their role in metabolism, growth and health, is key to the health value of fish and seafood in the human diet. For example, selenium, an essential element, has implications for cardiovascular and neurological diseases; it also plays a role in protection against cancer and neutralizing the toxicological effects of mercury, which is a persistent pollutant that is biomagnified in high-trophic-level predatory fish and seafood (Durazzo et al. 2022).

The bioaccumulation of organic and inorganic pollutants in marine fish and edible molluscs and crustaceans is briefly discussed within the context of European Union (EU) food quality standards. The health risks associated with the human consumption of seafood have been assessed mainly by the EFSA, an independent European agency funded by the EU. The potential health risks arising from the consumption of "frutti di mare" are examined from the standpoint of estimated PTWIs for toxic pollutants (Szefer 2013).

This issue deserves special attention due to the exposure of human populations to potentially toxic metals through fish and seafood consumption. For example, the most important case studies have been reported for mercury in fish and shellfish species; cadmium in squids; copper and zinc in bivalves, molluscs and crustaceans; arsenic and selenium in seafood and calcium, magnesium, sodium and potassium in fish and shellfish. In this context, analysis of the corresponding trends in the accumulation of mineral elements, detoxification processes, toxicity levels and extraordinary exposure levels within various ecosystems should be taken into account. Therefore, analytical methods, including alternative processes of mineralization, sample treatments and the assortment of analytical instrumentation, appear to be indispensable tools for determining the contents of essential and toxic elements in fish and seafood. The exposure of fish and seafood to toxic elements is associated with the emergence of serious risks and diseases that can occur in human populations. In this context, the main historical events have been reported, considering the most important effects of these elements on health (Marcovecchio et al. 2015).

1.1.1.3 Milk and Dairy Products

On average, cow milk contains 87% water, 3.9% fat, 4.9% lactose, 3.5% protein, 0.7% mineral elements and vitamins and other minor constituents. On a dry basis, raw whole milk contains 29.36% fat, 26.98% protein (22.22% casein, 4.76% whey protein), 38.1% lactose and 5.56% ash. The composition of non-fat solids in skim milk is 52.15% lactose, 38.71% protein (31.18% casein, 7.53% whey protein), 1.08% fat and 8.06% ash. The physical properties of milk include fluid flow, mixing and churning, emulsification and homogenization and heat transfer processes such as pasteurization, sterilization, evaporation, dehydration, chilling and freezing (Kailasapathy 2016; Magan et al. 2021; Goyal et al. 2023; Singh et al. 2023).

Milk and dairy products are important sources of mineral elements, mainly Ca, P, Mg, Na, K and Zn. They are present in the form of inorganic ions or are components of organic molecules, mainly proteins and fats. The interphase distribution of macro- and micro elements in milk between the soluble and micellar phases affects both milk's nutritional value and the stability of the colloidal dispersion. The balance between Ca and P may be disturbed by the differing features of the technical processes, such as pH, temperature, pressure and concentration, to which milk can be exposed. When a mild process is used, the elemental balance is only slightly affected and can return to its original state. The application of stricter technological processing conditions usually leads to irreversible changes, resulting in the formation of a new mineral element system. Milk products such as yogurt and cheese with relatively high levels of mineral elements are desirable from a nutritional standpoint; therefore, dairy products have been fortified with major and minor elements. This contributed to the development of appropriate methods that meet the requirements of both research and routine control procedures (de la Fuente and Juárez 2015).

Investigations of the elemental composition of bird eggs are scarce despite their importance in human nutrition. Eggs are a main source of Na, K, P and Ca in the human diet, but can also be abundant in Mn, Fe, Zn and Cu, depending on their origin. The Se content in eggs has been improved by changing the hen diet. According to Armenta and de la Guardia (2015b), additional studies are needed to investigate the possibilities offered by ICP–MS for maximum quantification of numerous elements in eggs produced in different countries. In the near future, speciation studies could also be performed to determine the Se, As and Hg forms in eggs produced from farms using roxarsone, an arsenic animal feed additive commonly used to promote growth and control disease.

1.1.1.4 Honey and Its Products

Honey contains 80–85% carbohydrates, 15–17% water, 0.3% protein, 0.2% ash and small amounts of amino acids, phenols, pigments and vitamins (Bogdanov et al. 2007). In addition, minor concentrations of mineral components, organic acids, enzymes, pollen, fragrance and flavour compounds are present (da Silva et al. 2016; Herrero-Latorre et al. 2017; Bobis et al. 2020). Honey contains various contents of mineral elements, with K as the main element. Other elements include Cl, S, Ca, Na, P, Mg, Si, Fe, Mn and Cu. Compared with lighter honey, darker honey is characterized by a greater content of mineral elements. The trace element levels in honey are dependent mainly on the botanical origin of the honey, and they constitute 3.68% of the honey mass (Tafere 2021).

Honey mainly contains sugars such as glucose and fructose. It has been reported that heat and crystallization can affect the colour of honey. The USDA classifies honey into seven colour categories: water white, extra white, white, extra light amber, light amber, amber and dark amber. Compared with darker honey, light-coloured honey generally has a milder flavour. Its composition is particularly variable, depending on its botanical and geographical origins. Herrero-Latorre et al. (2017) reported the use of hives as a passive sampling tool while considering honeybees and honey as environmental bioindicators for micro elements, macro elements and radionuclides.

1.1.1.5 Cereal and Its Products

Ruiz-de-Cenzano et al. (2015) reviewed recently published papers that report different methodologies for determining the mineral elements in cereals and pulses. The authors presented a rich and diverse set of data on the concentrations of major, minor, trace and ultratrace elements in barley, maize, millet, oat, rice, rye, wheat and other cereal products, together with beans, chickpeas, lentils, peas, peanuts and soybeans from different geographical origins, as well as the potential contributions of cereals and pulses to the daily intake of mineral elements. Importantly, essential micronutrients present in cereals include vitamins, mineral elements, antioxidants and phytochemicals. However, processing may change the levels of mineral elements in grains and affect their bioavailability. Arbitrary ingredients can be added to improve processing or to produce innovative bakery products, which often have increased nutritional value. Compared with all other cereals, wheat (*Triticum aestivum*) is the most important crop for bread production because it is better for baking than all other cereals (Pena 2002).

There has been an increasing trend in recent years to consume breads containing wholegrain, multigrain or other functional ingredients, which are becoming increasingly important in the bakery industry and on the market. According to Collar (2015), new ingredients and the development of new or emerging technologies are vital challenges in cereal investigations to establish baked products with high nutritional value. Studies on the influence of these processes and ingredient innovations on the elemental profile, availability and fortification of cereal products require thorough research.

1.1.1.6 Olive Oils and Seed Oils

It has been reported that the quantification of major, minor and trace elements in edible oils and fats is a subject of interest from technological and toxicological viewpoints, although this issue is of less interest with regard to edible oils than with regard to other foods. Moreover, the occurrence of mineral elements in this food matrix does not seem interesting enough from a nutritional point of view, e.g. owing to their low levels. There is generally no information available pertaining to speciation analysis of elemental concentrations in edible oils. However, according to Knochen and Morales (2015), it can be expected that speciation data will be available, as has occurred with other foods.

1.1.1.7 Vegetables, Legume Seeds and Fruits

Vegetables and fruits, because of their relatively high fibre levels and healthful phytochemicals, have a protective effect against a variety of chronic diseases, i.e. cardiac disease, type 2 diabetes, obesity and several types of cancer. On the one hand, such food types are rich in dietary macro- and micronutrients, but on the other hand, they may also be contaminated with toxic elements such as As, Cd, Pb and Hg. Various environmental, geochemical and meteorological factors directly affect the elemental composition of vegetables and fruits. On the basis of publicly accessible national databases (US FDA, West African, Tanzanian, Danish, Finnish, Japanese, Malaysian and Australian), it has been possible to detect elements that commonly occur in plant foods (Hoo Fung et al. 2015). The most common macronutrients and minor elements are K, Mg, Ca, P, Zn, Fe and Cu. No significant variation was detected in the case of macronutrient levels in more common vegetables and fruits, in contrast to micro elements, which are distinguished by elevated levels of Mn and Cr in both raw and processed plant foods. Furthermore, parameters such as contaminated soils and irrigation are responsible for the accumulation of elements in vegetables and fruits.

1.1.1.8 Spices, Salt and Vinegar

Gonzálvez et al. (2015) determined and characterized the mineral elemental composition of spices, e.g. vanilla, oregano, pepper, cinnamon, cumin, ginger and turmeric; some herbs, such as thyme, rosemary, mint, parsley, tea, lemon balm, sage, linden, lavender, blackberry and rose and garlic and onion. The elemental profiles of salt and vinegar were also analyzed. On the one hand, the importance of salt (except NaCl), vinegar and spices as sources of mineral elements in the human diet has been reduced compared to that previously thought. On the other hand, controlling the presence of toxic elements in this food group is very important because of the global market of these products and their high added value. For example, garlic and onion are clear examples of the significance of spices in human foods, although another case can be cited in this regard, i.e. herbal products, which are very commonly employed in regional and national cooking practices. The most important impact of the analytical control of the elemental profile of spices pertains to their provenance and production control to protect the quality of raw and end products. According to Gonzálvez et al. (2015), environmentally friendly and accurate methods, including speciation analysis of mineral elements, are needed to assess the mineral composition of spices in the context of their use in industrial food processes. Moreover, it has been recommended to use available methodologies with improvements in sensitivity and selectivity to determine the identification of the fingerprint of special products and to avoid, in the global market, the introduction of improper production practices in contaminated areas.

1.1.1.9 Mushrooms

Mushrooms, as attractive functional foods, are a source of essential bioactive compounds. Among the various wild edible species, *Armillaria mellea*, *Calocybe gambosa*, *Clitocybe odora* and *Coprinus comatus*, which contain chemical constituents such as phenolic compounds, macronutrients, sugars, fatty acids, tocopherols and ascorbic acid, are appreciated in gastronomy. According to Vaz et al. (2011), the four species of mushrooms could be useful in well-balanced diets as a source of bioactive compounds.

In addition to their use as functional foods, wild culinary mushrooms are widely used as medicaments because of their antibacterial, antifungal, antiviral, antioxidant and anticancer activities (Tirkey et al. 2023).

1.1.1.10 Tea, Coffee and Their Infusions

The consumption of infusions of tea and coffee provides various levels of essential and toxic elements to humans. Their contents depend on numerous factors, including the technique of infusion preparation. For example, the mass/volume ratio, type of equipment and infusion time may significantly vary depending on the number of cultures consuming these infusions. Although the characteristics of *C. sinensis* infusions are frequently reported, such information is scarce for coffee. According to Vázquez and Vélez (2015), increasing the number of samples of this product, including a broad assortment of varieties, geographical origin and the effect of infusion preparation methods, is desirable. Moreover, future studies should unify the way in which element concentrations are expressed. On the basis of daily recommended intakes, the consumption of these infusions can provide many variable levels of mineral elements with significant contents of Mn, Mg, Al and F. To assess the risks or benefits related to these products more adequately, it would be advisable to consider not only the bulk content of mineral elements but also their speciation and bioavailability. This approach provides key factors that are very helpful for accurate risk evaluation, as is performed with other foodstuffs (Vázquez and Vélez 2015).

1.1.1.11 Cocoa, Sugar and Its Products

Sugar and honey are composed of numerous mineral elements at variable levels. Major elements such as Ca, K, Mg and Na that are present in cocoa, sugar and its products are distinguished by their endogenous origin and can indirectly originate from soil substrata and sugar-rich and melliferous plants (Pohl et al. 2015). However, most external sources of mineral elements in sugar and honey can be specified; examples include environmental pollution and the technology and processing that are used for both food types. These sources may be responsible for increases in the contents of some minor elements, e.g. Al, Cu, Fe, Mn and Zn, as well as trace elements such as As, Cd, Co, Cr, Hg, Ni, Pb and Sn (Pohl et al. 2015). With respect to the elemental profile, the questions of quality control and quality assurance of sugar and honey are particularly important since they are common in cooking and the production of various beverages and food products. In the latter case, sugar and honey are commonly used as sweeteners; hence, the determination of the elemental distribution patterns of these products can enable quality and safety assessments. Moreover, the elemental composition of sugar and honey appears to be a fingerprint of their geographical origin and of the methods of their pretreatment and processing. Therefore, it could possibly be used for accurate classification and for distinguishing different brands of sugar and honey, or as a confirmation of the authenticity of these products. Knowledge of elemental contents can also be valuable in relation to the processing of sugar and honey, and facilitating difficult decision-making to improve their high quality. Monitoring trace elemental contents may also be useful in tracking changes in cumulative environmental pollution in a studied area (Pohl et al. 2015).

1.1.1.12 Mineral and Drinking Water

Water covers approximately 70% of the Earth's surface, of which 97% is salt water. It has been reported that natural mineral water is a source of major and minor elements that can compensate for nutritional deficiencies in humans; hence, water, a calorie-free drink, provides humans with the necessary mineral elements. Any increase or decrease in the current WHO guideline values for mineral elements in drinking water requires control of their concentrations (Dutta and Das 2015). Recent epidemiological studies suggested the need for reassessment of the safe upper limits of some elements in drinking water. This is the case with Se, which has a narrow range of acceptable concentrations; in this range, Se has nutritional value. Beyond this range, Se has toxic effects. Given the different biological activities of different chemical species, the selenium provided to humans through drinking water may be responsible for toxicity at much lower levels than previously thought. Therefore, health risks may occur at exposures below the current EU and WHO upper limits and guidelines. For Ca and Mg in drinking water, these nutritive macro elements are desirable for cardiovascular disease (CVD). It has been reported that a lower heart disease risk is observed in areas with high levels of Ca and Mg, which are the main components of water hardness. It has been assumed that the contribution of hard water, especially that rich in Mg, is responsible for a decreased risk of CVD, although international regulations on either the minimum or maximum limits for Ca, Mg or water hardness are lacking (Dutta and Das 2015; Vinceti et al. 2013; Anne 2011).

1.1.1.13 Wine and Other Alcoholic Beverages

Studies of wine samples for contents of macro- and micro elements have been used, for example, to determine legal limits for export purposes; to monitor the content of trace contaminants and macro elements, e.g. Ca and K; to control authenticity purposes; to trace and indicate potential changes in colour and oxidation and to promote wine health benefits (Cozzolino 2015). Elemental profiles have been studied extensively by the winemaking industry. The guarantee of wine authenticity or regionality on the basis of multi-element profiles and/or stable isotopic analysis, including multivariate analysis, is recommended as the most useful approach in the modern winemaking industry (Galani-Nikolakaki and Kallithrakas 2007; Cozzolino 2015).

Recently, the implementation of new analytical techniques such as X-rays has opened new and promising development prospects for the effective analysis of elements in grapes and wine (Cozzolino 2015).

1.1.2 Authenticity of Food

The authenticity of foods and beverages is an issue of key significance for all participants in the food chain. Because manufacturers are responsible for ensuring the quality of foodstuffs, food authorities are obliged to control the safety and traceability of foods. The increased awareness of clients in recent years has resulted in demands regarding safe, honest and accordingly labelled foods, etc. However, when healthier, safer, autochthonous and environmentally friendly foods are more expensive, consumers can be increasingly vulnerable to adulteration and fraud. In this context, the concept of food authenticity encompasses various issues pertaining to adulteration and to mislabelling various parameters such as origin (e.g. geographical origin/cultivar/breed), production methods (e.g. organic/conventional production) and processing techniques (e.g. conservation/stabilization). In particular, traceability issues related to geographical provenance are among the key food authenticity problems in Europe, especially given that the European quality policy has strict rules to identify and preserve the trademark of products corresponding to the assigned origin and/or to certain processing methods. Among the geographical determinants, the protected designation of origin (PDO) pertains to the localization of the geographic region in which foodstuffs are produced, whereas the protected geographical indication (PGI) identifies areas where at least one of the three production processes, i.e. production, transformation or elaboration, has been carried out. Another indicator, named the traditional specialties guaranteed (TSG), preserves traditional production methods. The application of state-of-the-art and cost-effective methods has become an urgent need in the food industry to guarantee the authenticity and traceability of foodstuffs (Agriopoulou et al. 2023; González-Domínguez 2020, 2022; González-Domínguez et al. 2022; Visciano and Schirone 2021; Sammut et al. 2021; Szefer and Nriagu 2007; The Council of the European Union 2012).

1.1.3 Issue of Counterfeit Food

The food industry and national administrations are facing serious challenges worldwide regarding counterfeit food. Counterfeit food has attracted particular attention in the EU, mainly because of the troublesome cases that have occurred, e.g. the presence of horse meat in internationally available and commercially recognizable meatballs. Since this incident, further cases have exacerbated the problem of counterfeit food in recent years. A wide range of these food fraud incidents deceive the consumer, e.g. counterfeiting alcoholic beverages, swapping species and adulterating meat products and counterfeiting a product's country of origin, resulting in harm to public health. This situation creates many problems conducive to causing harm to consumers and may cause reputational damage to a company or food sector, and loss of profits and market share. Moreover, other consequences of food adulteration may lead to a fall or collapse in the share price, cause a loss of access to markets and even lead to political and economy-wide consequences. According to Spink (2019), increasing sophistication and dangers of food fraud have occurred over the past 35 years. To fight against fraudsters operating within food supply chains, many collaborative international networks, such as the EU Food Integrity Project, the Interpol Intellectual Property Crime College, INTERPOL/Europol Operation Opson and the EU Food Fraud Network, have been engaged. This extensive coordinated action has led to the recognition of complexity and the challenge of detecting and deterring food fraud. Owing to the acquired knowledge and many years of experience, it has become possible to expand the current broad, harmonized and thorough regulatory and standard certification requirements for food manufacturers, suppliers and retailers. Thus, there has been a need for a more focused and systematic procedure, e.g. to conduct vulnerability assessments and organize and implement a food fraud prevention strategy. The overall food fraud prevention strategy begins by implementing basic requirements for food, and its development involves the use of criminology root cause analysis for final resource allocation decision-making on the basis of the COSO principle of enterprise risk management (ERM) (Spink 2019).

A brief overview by Hellberg et al. (2021) presents key issues pertaining to the history of food fraud and current challenges and vulnerabilities that the food industry is facing. In addition, great importance is attached to requirements for conformity with regulatory and industry standards to alleviate the tendency towards food fraud, with a focus on the Global Food Safety Initiative (GFSI) benchmarking requirements. There are various strategies and tools for overcoming food fraud, including appropriate analytical detection methods. Comprehensive information on the adulteration of certain food groups was provided by Hellberg et al. (2021). It refers to fraud in the food service and retail sectors regarding meat and poultry products, seafood, coffee and tea, fats and oils, herbs and spices, nonalcoholic ready-to-drink products, wine and other alcoholic beverages, dairy products, grains and cereals, honey, organic foods, dietary supplements and probiotic products. There are currently many recorded accidents involving the most common counterfeit foods, i.e. seafood, dairy products, meat, poultry and fish, wine and other alcoholic beverages, fruit juice, saffron and other high-value spices, spices and seasonings, vegetable oils, honey and maple syrup, grains, olive oil, coffee and tea. While the majority of fraud cases do not result in serious health consequences, the consumption of some counterfeit products can lead to food poisoning and have unexpected consequences for people with allergies.

1.2 General Characteristics of Food Analytics Related to Multi-elemental Profiles

Danezis and Georgiou (2022) characterized the application of elemental metabolomics for the determination of geographical provenance with emphasis on element transportation to food. Recently, elemental metabolomics has received considerable attention. On the basis of recent studies, it has been possible to assess many elements, with particular emphasis on ultratrace elements, mainly REEs and SIR. Analytical tools such as HR–ICP–MS appear to be significant analytical methods for the future. According to Danezis and Georgiou (2022), there is a well-known relationship between soil and plant products, but the link between soil, plants and animal products is more complex. Therefore, food-processing steps should be studied considering seasonality, climate, pollution, contact materials and organism age. An adequate number of samples is also recommended for accurate recognition of the relationship between the elemental profile in food and geographic origin (Danezis and Georgiou 2022).

Analysis of elemental food profiles is crucial for the food industry and regulatory authorities. Analytical methods are necessary tools for quality assurance and are indispensable in the evaluation of consumer intake levels for beneficial and harmful food components with respect to essential and nonessential elements, including toxic heavy metals and metalloids. Given that elevated levels of toxic elements may pose a serious health hazard, monitoring foodstuff quality is necessary.

There are numerous reports on the elemental composition of animal and plant foods (e.g. Reilly 1991, 2004, 2006, 2007, 2008; Szefer and Grembecka 2007a, b; Grembecka and Szefer 2011; De la Guardia and Garrigues 2015). The distributions of elemental pollutants and radionuclides in food have been reviewed by many authors, e.g. Poschl and Nollet (2007) and Ould-Dada (2007).

1.2.1 Validation of Analytical Measurements

Konieczka and Namieśnik (2009) defined the tools applied in QA/QC, especially the use of a statistical approach during analytical data treatment. The use of quality control systems, e.g. detailed validation parameter measurements, statistical tests, uncertainty estimation, traceability, certified reference materials (CRMs), proficiency tests and method validation, has been fundamentally requested. Food quality is considered the key factor affecting consumer acceptance and attraction to and purchase of foodstuffs. The determination of mineral elements, especially in solid food, can be a complicated process because of its complex matrix, which hinders reliable results that are consistent with the actual elemental characteristics of the analytical samples. International terms of reference are needed to improve the quality and safety of food supplies moving at the global level. Therefore, for the food control system, a basic food law supplemented by detailed regulations and administered by a capable food control organization is highly desirable. Harmonization of analysis methods at the international level is necessary and can be achieved exclusively at the international level, i.e. with contributions from the FAO/WHO Codex Alimentarius Commission, AOAC, ISO and ILAC (Szteke 2007). According to ISO standards, quality measurement systems and quality assurance, quality control has been defined as a quality measurement system for quality assurance. An essential component of laboratory practice related to the requirement of ISO 17025 for inhouse methods also includes parameters such as validation methods.

Before deciding which analytical method can be used, the method's applicability should also be taken into consideration. For monitoring the validity of tests and calibrations, the regular use of CRM and internal quality control using secondary reference materials, participation in interlaboratory comparisons or proficiency testing programs and replication of tests or calibrations via the same or different methods have been recommended. For elemental analysis in food samples, appropriate methods/techniques should be estimated by the analytical laboratory considering the main criteria according to international standards, i.e. accuracy, bias, precision (repeatability, reproducibility), true value, trueness, traceability, LOD, LOQ, selectivity/specificity, sensitivity and uncertainty (Szteke 2007; Bulska 2010; Namieśnik and Szefer 2010; Plutowska and Ramos 2012; Konieczka and Namieśnik 2009).

1.2.2 General Characteristics of Analytical Methods

After dry ashing (calcinations) or wet digestion of food samples, the content of elements released into the solution can be determined by analytical methods/techniques (Biziuk and Kuczyńska 2007).

1.2.2.1 AAS Method

AAS is one of the older techniques still being used for the determination of major and minor elemental concentrations in food, although it is much less common than simultaneous multi-element techniques. It has been reliable and simple to use and has an acceptable level of accuracy for most elements. It has been reported that 70 elements can be quantified by AAS in various sample matrices, with sensitivities ranging between the μg/g and ng/g range. Flame atomic absorption spectrometry (FAAS) is a popular tool for the analysis of essential and toxic elements in various samples, including plant and animal food samples. The main principle of this method consists of ground-state metals absorbing light at specific wavelengths.

In general, the contents of elements such as Cu, Ni, Zn, Mn and Fe have been determined by FAAS. Numerous elements can be studied by electrothermal atomization atomic absorption spectrometry (ETA–AAS), as their atoms can be evaporated in a graphite furnace. Compared with FAAS, this technique is distinguished by advantages such as higher sensitivity and lower detection limits. Moreover, it is characterized by low spectral interference and a very small sample size. However, restrictions have been applied for ETA–AAS, unlike FAAS, because of its limited dynamic range, high matrix interference and longer analysis time. However, this technique is commonly used to quantify trace elemental concentrations in environmental samples, including food samples. There are many reports in the literature on the determination of traces of Cd, Pb, Cr, Ni, Co and Al in various food products.

Metalloids such as Sb, As, Se, Te, Bi, Ge and Sn are not suitable analytes for analysis via conventional methods and can be routinely analyzed via hydride generation atomic absorption spectrometry (HG–AAS). The principle of this method involves the separation of an element from the sample matrix by the formation of volatile hydrides after reduction of the element from a higher oxidation state to its lowest state.

Another spectrometric method is cold vapour–atomic absorption spectrometry (CV–AAS). It has been used primarily to detect Hg in various types of samples, including food products (Biziuk and Kuczyńska 2007; Moreda-Piñeiro and Moreda-Piñeiro 2010; Grembecka and Szefer 2011, 2016; Capar and Szefer 2012).

López-García and Hernández-Córdoba (2015) demonstrated the advantages and limitations of AAS in the case of food analysis, including sample treatment and its potential application for speciation purposes. The authors discussed the possibility of directly using solid or undigested samples in ETAAS in an economical way and reducing the risk of contamination or analyte losses. Moreover, recent advances in instrumentation and the advantages of the combination with a solid or liquid microextraction stage have been considered. Recent studies have shown that these advances considerably increase the performance of AAS and prevent defects with regard to the sequential operation mode. AAS appears to be a useful technique for the routine determination of a large number of elements in food analysis owing to the relatively low cost of acquisition and maintenance, lower consumption of gas, versatile application and very high sensitivity in some atomization modes (López-García and Hernández-Córdoba 2015).

1.2.2.2 ICP–AES and ICP–MS Methods

Inductively coupled plasma atomic emission spectrometry (ICP–AES), also known as inductively coupled plasma–optical emission spectrometry (ICP–OES), is a type of emission spectroscopy that uses a plasma to produce excited atoms and ions that emit electromagnetic radiation at a wavelength characteristic of an appropriate element. The intensity of the emissions from various wavelengths of light is related to the concentrations of the elements in the sample studied. This technique, as a multi-element determinative technique, is used for investigating food in the presence of major and minor elements, e.g. Ca, Mg, Na, K, Mn, Cu, Sr, Al, Fe, Ba, P, S, Co, Cr, Li, Ni, Ti, V, Zn and Zr.

In contrast to ICP–AES, inductively coupled plasma–mass spectrometry (ICP–MS) is highly sensitive, ranging from two to three orders of magnitude, and is characterized by the ability to determine a wide range of metals and several nonmetals at trace and ultratrace levels, i.e. below ppt. However, similar to ICP–AES, ICP–MS allows simultaneous multi-element analysis of numerous elements in foodstuff samples. The principle of ICP–MS is based on coupling with inductively coupled plasma, which is responsible for producing ions with a mass spectrometer, making separation and detection possible. This technique can detect different isotopes of the same element; hence, it appears to be a versatile tool in isotopic labelling. A very wide range of elements in various types of food samples can be accurately determined via ICP–MS, e.g. Ce, Dy, Er, Eu, Gd, Ho, La, Lu, Nd, Pr, Sc, Sm, Tb, Th, Tm, U, Y, Yb, Ag, Al, As, Ba, Be, Bi, Ca, Cd, Co, Cr, Cs, Cu, Fe, Ga, In, K, Li, Mg, Mn, Na, Ni, Pb, Rb, Se, Sr, Tl, V and Zn (Biziuk and Kuczyńska 2007; Moreda-Piñeiro and Moreda-Piñeiro 2010; Grembecka and Szefer 2011, 2016; Capar and Szefer 2012). ICP–MS has become one of the most popular analytical techniques for elemental analysis because of its easy coupling with chromatographic techniques. In recent years, new developments in the ICP–MS technique have enabled the routine analysis of food and beverages. Moreover, ICP–MS is an effective tool for the speciation analysis of food and beverages (García Alonso et al. 2015).

1.2.2.3 Electroanalytical Methods

Among other instrumental tools, electrochemical methods, such as ion-selective electrodes (ISEs), also known as specific ion electrodes (SIEs), and anodic stripping voltammetry (ASV), which are less common in food elemental analysis, deserve attention. The principle of the ISE as a sensor is related to the conversion of the activity of a specific ion present in a solution into an electrical potential, which can be measured by a voltammetric analyzer. Four main types of ion-selective membranes can be used in ISEs: glass, solid-state, liquid-based and compound electrodes. Although the electrodes are very selective, they are not free of interference since they may be sensitive to other interfering ions having comparable physical properties, to the extent conditioned by the degree of their similarity (Biziuk and Kuczyńska 2007; Moreda-Piñeiro and Moreda-Piñeiro 2010; Capar and Szefer 2012). However, ISEs using new materials appear to be useful for the analysis of alkaline ions, alkaline earth ions and other divalent metal ions, trivalent metal ions and anions. Furthermore, sensitive voltammetric techniques with different types of electrodes and tools based on stripping analysis, as well as amperometric sensors for continuous detection, are increasingly being used in the case of these electrochemical measurements. The essential advantages of electrochemical methods are their high sensitivity, simple analytical approach, ease of use, transferability and low cost, allowing the development of efficacious procedures that are beneficial for the analysis of mineral elements in foods. However, their quantification cannot be regarded as a research priority in modern analytical electrochemistry, considering the numerous competitive traditional analytical tools in this field. According to Yáñez-Sedeño and Pingarrón (2015), automated and effective electroanalytical methods are particularly appropriate for evaluating food quality and controlling industrial technological processes in the food industry, providing new solutions for the integration of electrochemically driven devices in conventional industrial processing.

1.2.2.4 NAA Method

Neutron activation analysis (NAA), a type of nondestructive analysis, involves no or minimal preparation of a sample that is bombarded with neutrons, resulting in the conversion of a stable atomic nucleus into a radioactive nucleus. This process allows the qualitative and quantitative identification of major, minor and trace elements and REE elements. This technique enables the simultaneous determination of 50–65 elements, although some elements, such as Pb, cannot be determined. The most common neutron activation techniques used for routine food analysis include instrumental neutron activation analysis (INAA), radiochemical neutron activation analysis (RNAA), neutron capture prompt gamma-ray activation analysis (PGAA) and episodic instrumental neutron activation analysis (EINAA) (Bode 2003; Biziuk and Kuczyńska 2007; Moreda-Piñeiro and Moreda-Piñeiro 2010; Capar and Szefer 2012; Landsberger 2015).

A detailed description of specific reactions and possible nuclear and spectral interferences, including quality control measures for the NAA of food samples, was evaluated. After Landsberger (2015), advances in the analytical methodology of species analysis have been enhanced by continuing advancements in analytical devices and by progressive knowledge of the contributions of certain elemental species to human nutrition and toxicology. Owing to progress in nonchromatographic procedures, there has been continuous effort in the development of simple and effective procedures allowing the determination of element species that trigger health risks. Moreover, the priority has been using hyphenated techniques in the analysis of known species, and ICP–MS appears to be the most powerful detector owing to its very high sensitivity, effective elimination of interferences, compatibility with all dimensions of chromatographic/electrophoretic separations and unusual isotopic efficiency. The key challenge in speciation analysis is to evaluate the “native” species distribution, bearing in mind that magnetic resonance and techniques based on synchrotron radiation have been incorporated because of minimal or no sample treatment. With respect to future trends, further methodological development and production of suitable certified reference materials are expected to increase the reliability of speciation analysis of foodstuffs (Landsberger 2015).

1.2.2.5 XRF Method

X-ray fluorescence (XRF) is one of the oldest nuclear techniques. This method is based on exposing an analytical sample to electromagnetic radiation of adequate energy to remove electrons from the inner orbitals. The fluorescence radiation stops rapidly when the primary radiation is interrupted. XRF allows qualitative and quantitative determination of elements with high selectivity, although it has a lower sensitivity than NAA. In contrast to sufficient analytical technologies, which usually involve prior sample treatment to obtain appropriate solutions of mineral elements, X-ray fluorescence techniques such as NAA can be successfully used to determine mineral element contents directly from untreated food samples (Biziuk and Kuczyńska 2007; De la Guardia and Garrigues 2015). The application of the XRF technique enables fast and environmentally safe alternative methodologies for the quantification of the essential and toxic elements of foods, at least to screen the elemental profile of foodstuffs and therefore their safety in human nutrition. Special attention should be given to modern methods that enable improvements in the analytical sensitivity of measurements or favour the use of low-cost portable instrumentation. XRF spectroscopy appears to be a high-performance technique because it is a simple tool suitable for use by small laboratories in routine screening analyses. Thus, it is mandatory to have such a simple methodology for decision-makers to ensure food safety and prevent fraudulent practices. When XRF is compared with the modern use of infrared methods such as NIR and MIR, it can be concluded that the signals provided by XRF are closer to the actual mineral element distribution patterns of samples than those registered in the infrared range. Therefore, there is a need to improve the available methodologies to study the semiquantitative and quantitative capacities of inexpensive and mobile instruments (De la Guardia and Garrigues 2015). However, XRF cannot be considered a very sensitive technique, especially ED–XRF portable instrumentation. Considering all these arguments, it is necessary to examine the geometry of the instruments to achieve total reflection measurements as well as to include soft and fast sample processing. In particular, this method involves the leaching of target analytes and their pre-concentration in solid matrices, which is appropriate for improving the limit of detection (LOD) to the μg/g and ng/g levels.

1.2.2.6 IRMS Method

Multi-elemental analysis (MEA) appears to be useful for identifying geographical origin, environmental conditions of the production area (growing/breeding area) and crop management. Furthermore, it has been demonstrated that isotope-ratio mass spectrometry (IRMS) can be used as a candidate authenticity marker. A good example of a recent IRMS application, $\delta^{18}O$ and $\delta^{2}H$ stable isotopes are strongly influenced by hydrological factors; hence, they can be effective markers for discriminating food samples with respect to geographical origin. The $^{13}C/^{12}C$ isotopic ratio ($\delta^{13}C$) is dependent on climatic conditions, plant type and agricultural practices, whereas the $^{15}N/^{14}N$ isotopic ratio ($\delta^{15}N$) enables distinguishing between conventional and organic farming since the $\delta^{15}N$ content in organic fertilizers is much higher than that in synthetic fertilizers (González-Domínguez et al. 2022).

1.2.2.7 IC Method

The coupling of ion chromatography (IC) with high-sensitivity and/or high-selectivity detectors can be a promising option for improving the separation efficiency of IC, with the highest sensitivity and selectivity when MS detectors are used for all the variants. It would also be appropriate for online post-column derivatization to convert the chemical species to chemically detectable forms for determination by detectors such as spectrophotometric, fluorometric and atomic fluorescence detectors. According to Morales-Rubio and de la Guardia (2015), prospective analysis of food by IC requires interdisciplinary studies and more advanced speciation studies suitable for assessing the natural cycle of multi-element composition.

1.2.3 Element Speciation

Speciation analysis involves the various physical and chemical forms of an element and enables their identification and determination in a sample. Speciation analysis, which is based on the quantification of individual chemical species of mineral elements, is characterized by differences in bioavailability, mobility and toxicity. Recently, speciation has become a major topic for guaranteeing food safety and determining nutritional quality. Such vital and interesting speciation topics in food and environmental chemistry, with consideration of analytical techniques and methodologies, have been reported by several authors, e.g. Crews (1998), Ebdon et al. (2001), Cornelis et al. (2003), Połatajko and Szpunar (2007), de Brauvwere et al. (2010), Capar and Szefer (2012), Berton et al. (2013), Armenta and de la Guardia (2015a) and Grembecka and Szefer (2016).

Although most regulations referring to trace elements in food have been established on the basis of the total element concentration, only a few regulations concern specific forms of chemical elements; i.e. the EPA, WHO, EEC and US FDA have proposed contamination limits for various forms of elements, e.g. As, Hg and Sn. The determination of various oxidation states of elements, e.g. Cr, Fe, As, Se, or various organometallic species, e.g. TBT, CH_3Hg, AB and Me_3Pb^+, is of great interest and a serious challenge to analysts. Because speciation forms of elements are mostly unstable and occur in various food matrices at very low levels (e.g. μg/kg), sensitive and selective methods are highly desirable for their quantitative isolation and determination. Among the state-of-the-art analytical methods applied in speciation analysis, the following deserve special attention: GC and HPLC coupled with powerful detectors, i.e. AAS, MS, FPD, ICP–MS, ICP–AES and AFS. The application of ICP–MS coupled with various separation techniques, such as LC, GC, supercritical fluid chromatography (SFC) and CE, has recently received considerable attention. Additionally, electrospray ionization mass spectrometry (ESI–MS) has been successfully used in speciation analysis (e.g. ion spray MS/MS) for the determination of TBT. Importantly, hyphenated speciation techniques are distinguished by appropriate selectivity and sensitivity for specific species of elements in liquid and solid samples, including food products. These advanced techniques have been widely applied to perform speciation analyses for numerous elements, e.g. Zn, Cu, Co, Fe, Mn, Cr, Ni, V, Mo, Ag, Pt, Ca, As, Sb, Cd, Pb, Hg, Se, Sn, Tl, Te, Al, S, N, P, Cl and I (Capar and Szefer 2012; Grembecka and Szefer 2016).

Many analytical procedures have been reported for the separation and determination of As species in various samples, including food. Arsenic compounds can be classified as follows: As(III), As(V), MMA, dimethylarsinic acid (DMA) and trimethylarsine oxide (TMAO), but organic compounds such as arsenobetaine (AB) and arsenocholine (AC), which are detected mainly in seafood, are considered non-toxic. In strong contrast to AB and AC, arsenosugars cannot be categorized as non-toxic for humans and hence can pose a risk to human health. Several of these methods involve GC and LC techniques coupled with spectroscopic and electrochemical detectors, and the standard detectors are AAS, ICP–AES and ICP–MS. For example, the following analytical methods for certification studies have been applied: HPLC–UV–HG–ICP–OES, GC–CT–HG–QFAAS and HPLC–QFAAS for the quantification of DMA, whereas methods such as HPLC–ICP–MS and HPLC–ICP–OES appear to be effective in the determination of AB (Capar and Szefer 2012; Grembecka and Szefer 2016).

In the speciation analysis of Hg, its toxic form, CH_3Hg, is of great interest among analysts. Different methods have been developed mostly on the basis of the coupling of reversed-phase HPLC to CV–AAS or CV–ICP–MS. Direct GC with electron capture detection (GC–ECD) has also been a reliable tool for the analysis of CH_3Hg in fish samples. Other methods used to detect this Hg species include GC–ICP–MS, headspace analysis (HSA) GC–MIP–AES and HSA–GC–AFS. Moreover, a rapid and low-cost GC–MW-induced plasma atomic emission detector (MIP–AED) is specifically useful for the routine quantification of Hg, as well as Sn and Pb species. HPLC–CV AFS is an effective method for determining speciation forms of Hg ($HgCl_2$, CH_3ClHg, C_2H_5ClHg and C_6H_5ClHg). This method has been used to quantify the CH_3ClHg content in seafood samples (Capar and Szefer 2012; Grembecka and Szefer 2016).

With respect to the speciation analysis of Se, the range between safe and toxic doses of Se is narrow, and its bioavailability and toxicity are dependent on the Se species in foodstuffs. Higher levels of Se accumulate in the form of organic compounds; therefore, speciation analysis of this metalloid in food is extremely important. Many methods involving the use of GC coupled with AAS, AFS, AES or ICP–MS have been effective tools for the determination of Se species in various sample matrices. Pervaporation–gas chromatography–atomic fluorescence (PV–GC–AFS) has been applied for the analysis of individual Se species. This method has been applied for the determination of volatile Se species in garlic. Another method, HPLC–ICP–MS, appears to be an effective technique for the identification of five Se species, i.e. Se(IV), Se(VI), Se–M, Se–Cys and Met–Se–Cys, and other (several) species in Se-enriched garlic, onion and broccoli. Another example of successful use for the analysis of Se speciation is a three-step sample preparation method based on a proteolytic enzymatic process for the study of the Se-enriched edible mushroom *Agaricus bisporus* coupled with an HPLC–HHPN–AFS system (Capar and Szefer 2012; Grembecka and Szefer 2016).

The speciation of Sn is especially important for its organic compounds. Tributyltin (TBT), an antifouling compound that prevents the growth of microbes and fungi and prevents the attachment of barnacles and other marine organisms (e.g. algae) to ship hulls, is toxic to fish and other seafood. HPLC methods based on specific interactions, such as reversed-phase, ion-pair and ion-exchange micellar interactions, have been coupled to ICP–AES and ICP–MS detectors for organotin speciation. HS stir-bar sorptive extraction (SBSE) and thermal desorption (TD) GC–MS are the techniques of choice for speciation analysis of butyltin and methylmercury in water samples. Moreover, a rapid HPLC–HG–ICP–AES method has been used for the quantification of organotin compounds in molluscs, as has GC–FPD for butyltins, i.e. TBT, dibutyltin (DBT) and monobutyltin (MBT) in marine molluscs, fish and fish-eating birds. Another example of an effective application of organotin speciation analysis is GC coupled with a mass selective detector with multiple selected monitoring (GC/MSA–SIM), which enables the analysis of pentylated organotin species in mussels and water (Capar and Szefer 2012; Grembecka and Szefer 2016).

With respect to two speciation forms of Cr, i.e. the biophile Cr(III) and the toxic Cr(VI), it is reasonable to pay attention to effective methods based on flow injection analysis (FIA), including ion chromatography, ion pair chromatography and ion–pair reversed-phase chromatography combined with detectors such as UV VIS, GF AAS, ECD, X-ray fluorescence (XRF) and ICP–AES. Dynamic reaction cell (DRC) ICP–MS has also been used to estimate the concentration of Cr forms in lyophilized samples of milk (Capar and Szefer 2012; Grembecka and Szefer 2016).

With respect to Pb speciation analysis, its organic compounds are much more toxic than its inorganic forms. For example, HPLC–ICP–MS and HPLC–ESI–MS are effective methods for determining the contents of Cd and Pb species in peas (Grembecka and Szefer 2016).

Online determination of Sb(III) and Sb(IV) in liver tissue has been a useful approach using flow-injection HG–AAS (Capar and Szefer 2012).

SEC connected with ICP–MS has been used to study the speciation of Cu, Cd, Zn, Se, As and Ca in tissue extracts of fish and of Fe, Zn, Cu, Ag, Cd, Sn and Pb in tissue extracts of mussels. Such multi-element speciation analysis enables us to explain any interactions between multiple groups of trace elements within the various protein fractions of biological matrices, namely, fish muscle or mollusc soft tissue (Grembecka and Szefer 2016).

Considering that the analytical tools available have become more sophisticated and much more expensive, their development may be expected with reference to collaboration among researchers, scientific units and even countries (Crews 1998). Although most speciation studies are concentrated on a limited number of chemical species, their assortment, including the chemical forms of essential and toxic elements in food, is postulated to provide data to law and policymakers (Armenta and de la Guardia 2015a).

1.3 General Characteristics of Chemometrics

Chemometrics is a useful, inherently interdisciplinary method used to explain descriptive and predictive problems in experimental natural sciences, especially in chemistry. It is based on extracting important information from extensive datasets with multiple variants and has been useful in processing elemental data, e.g. those related to objects such as abiotic and biotic environmental components. Abiotic components include atmospheric deposits and marine aerosols, marine suspended matter, bottom sediments and iron–manganese concretions. Biotic compartments, categorized chemically with respect to their elemental profile, include marine seaweeds, plankton, mussels, fish and sea mammals (Szefer 2002, 2007; Szefer and Grembecka 2009).

Chemometric techniques are typically applied when data are processed via state-of-the-art analytical methods/techniques; thus, the continued improvement of chemometric methodologies, i.e. chemometric theory, methods and application development, is still desirable. Notably, multivariate techniques appear to be the key statistical and mathematical approach for studying extensive datasets. Its structure appears to be conducive to utilizing chemometric tools, e.g. principal components analysis (PCA), partial least squares (PLS) and orthogonal partial least squares (OPLS). This is why the datasets may be highly multivariate, e.g. in the case of different groups of animal and plant foods.

There is a wide range of applications of chemometrics in the assessment of food quality on the basis of chemical profiles. This development and perspective tool enables the connection of state-of-the-art analytical methods/techniques and modern chemometrics. Therefore, it is possible to assess food quality, i.e. the recognition of authenticity/adulteration of the food chain. Since food is characterized by a complex matrix related to a multivariate dataset, the use of chemometric multivariate techniques is highly recommended. Einax (1995) demonstrated the basic principles of chemometrics applied to representative issues of environmental chemistry.

It has been shown that constructive collaboration between a chemometrician and an environmental scientist is desirable for achieving accurate explanations in line with reality (Einax 1995).

Multivariate chemometrics involves understanding the different aims and backgrounds of each of the different forms of multivariate analysis. The application of multivariate chemometrics may be related to several types of univariate and multivariate analyses used to understand the relationships between variables and their suitability for solving the problem studied. Many models, e.g. multivariate analysis of variance (MANOVA), multivariate analysis of covariance (MANCOVA), multivariate regression, principal components analysis (PCA), factor analysis (FA), canonical correlation analysis (CCA), redundancy analysis (RDA), correspondence analysis (CA), canonical correspondence analysis (CCA), multidimensional scaling (MDS), discriminant analysis (DA), linear discriminant analysis (LDA), recursive partitioning (RPART), artificial neural networks (ANN), simultaneous equation models (SEM), vector autoregression (VAR) and principal response curves analysis (PRC) on RDA, have been applied. The first editions of several important books on chemometrics were also published in the 1980s, including by Malinowski and Howery (1980), Martens and Naes (1980), Sharaf et al. (1986) and Massart et al. (1988).

Multivariate data analysis has been reported extensively in books by many authors, e.g. Cooley and Lohnes (1971), Morrison (1978), Anderson (1984), Dillon and Goldstein (1984), Flury and Riedwyl (1988), Mardia et al. (1989, 2024), Deming and Morgan (1993), Beebe et al. (1998), Malinowski (2002), Wackernagel (2003), Kollo et al. (2005), Bishop et al. (2007), Izenman (2008), Varmuza and Filzmoser (2009), Tauler et al. (2009), Dugard et al. (2010), Shirali and Vasudeva (2010), Everitt and Hothorn (2011), Rencher and Christensen (2012), Walczak (2000), Härdle and Simar (2019), Brown et al. (2020), Vehkalahti and Everitt (2020) and Backhaus et al. (2021).

Some important fields of application include new domains, namely, molecular modelling and QSAR, cheminformatics, omics of genomics, proteomics, metabonomics and metabolomics, process modelling and process analytical technology (PAT).

1.3.1 Chemometrics with Respect to Food Chemistry

Multiple factors, e.g. geographical origin, variety/breed and manufacturing practices, may influence the chemical composition of foods and therefore have a considerable effect on their organoleptic, nutritional and bioactive properties. For this reason, reliable, cost-effective and powerful analytical methods are needed to guarantee the authenticity and traceability of foodstuffs. Therefore, proper data processing and analysis steps should be performed with great care. The use of advanced chemometric tools is obligatory in building multivariate models for classification, discrimination and prediction and for identifying candidate authenticity and traceability markers. The choice of the most appropriate computational tool depends on various factors associated with the type of dataset, as well as the number of samples, number of input variables, homogeneous distribution of the sample, etc. Common practice involves applying various classification techniques and evaluating their relevance to the problem under study. This situation also involves a combination of unsupervised pattern recognition methods for the first exploration of the data, with the subsequent application of supervised learning approaches. Multivariate chemometric analysis can be performed by using unsupervised or supervised methods. Unsupervised analysis involves the identification of sample interrelationships without prior information about class membership, whereas supervised methods use a training stage, allowing samples to be assigned to known groups to build mathematical models. Multivariate methods are an adequate option for troubleshooting food fraud that cannot be solved owing to the complex nature of the food matrix (González-Domínguez et al. 2022; Callao and Ruisánchez 2018; Baker and Inventado 2014; Berrueta et al. 2007; Moncayo et al. 2015). In this context, PCA and HCA are among the most commonly employed unsupervised multivariate tools, whereas LDA and PLS–DA have been shown to perform satisfactorily for discrimination purposes in the assessment of food quality. Notably, the use of state-of-the-art machine learning tools, e.g. RF, SVM and ANN, has undergone fast and significant progress in recent years. In addition to the multivariate approach, another key factor to consider is proper data processing, which involves proper preprocessing of the data matrix to increase dataset quality and remove undesirable variations before statistical analysis. Additionally, after statistical modelling, the implementation of reliable validation tools is of the highest importance in the evaluation of model performance and checking for overfitting. The multivariate methods commonly used in food authentication studies include exploratory data analysis, discrimination and classification as well as regression and prediction (González-Domínguez 2020, 2022; González-Domínguez et al. 2022). The application of advanced chemometric techniques to determine food authenticity is essential for obtaining large amounts of data (Szefer 2007). This is possible owing to the use of state-of-the-art analytical methods such as chromatographic, spectroscopic and nontargeted fingerprinting approaches (Jiménez-Carvelo et al. 2021). As follows from the literature review, the selection of the most useful computational technique is dependent on the structure of the dataset matrix. Connecting complementary tools is usually required for properly processing data complexity. State-of-the-art analytical methods usually generate extensive and complex databases that require modern chemometric techniques. Owing to the use of this approach, it is possible to extract as much valuable information as possible from the raw data. Therefore, chemometrics enables the identification of the underlying relationships between variables. This allows the construction of chemometric models that appear to be useful in interpreting the characteristics of the system on the basis of the available chemical dataset. This results in the possibility of its use for classification, discrimination and prediction purposes. Consequently, chemometrics appears to be a reliable tool for simplifying the detection of fraud and guaranteeing the authenticity and traceability of foods (Jiménez-Carvelo et al. 2021).

Exploratory data analysis (EDA) is routinely applied as a first step to visualize the main structure of a dataset. Therefore, unsupervised pattern recognition techniques are commonly applied to reduce the dimensionality of complex data matrices without losing information. This option allows the projection of samples into a low-dimensional space, e.g. scatter plots and dendrograms, enabling inspection of sample groupings, detection of outliers, etc. Exploratory data analysis (EDA) is applied as a first step to visualize the general characteristics of the database. EDA methods include PCA, FA and CA (González-Domínguez et al. 2022).

Discrimination and classification methods are employed for predicting an object belonging to a given category on the basis of its characteristics (González-Domínguez et al. 2022; Kemsley et al. 2019). Among these methods, linear discriminant analysis (LDA), partial least square discriminant analysis (PLS–DA), soft independent modelling of class analogy (SIMCA), k-nearest neighbours (kNN), artificial neural networks (ANNs), support vector machines (SVMs) and chemometric approaches based on decision trees deserve attention. The last two methods can be categorized into two and three groups, respectively (González-Domínguez et al. 2022). SVM models can be categorized into two groups according to the form of the error function: SVM type 1 and type 2, also called C-SVM and ν-SVM, respectively. The parameters C (capacity constant) and ν can be estimated via cross-validation algorithms to avoid overtraining and thereby control the model complexity. Some of the most common decision tree-based techniques are grouped into classification and regression tree (CART) analysis, chi-square automatic interaction detection (CHAID) and random forest (RF). A combination of analytical data processed by chemometrics to control different food groups has been reviewed and interpreted by several authors, e.g. Agriopoulou et al. (2023), Varrà et al. (2023), Ye et al. (2023), Smaoui et al. (2023), Szefer (2003, 2007) and Szefer and Grembecka (2022).

The abovementioned chemometric methods/techniques have been widely applied in the processing of analytical data, enabling in-depth interpretation of the observed interdependence between correlated variables in food chemistry.

1.3.2 Chemometric Data Treatments

According to Tsakovski and Simeonov (2010), chemometrics involves related scientific specialties such as mathematics, statistics and other methods that employ formal logic to assess and interpret chemical and analytical results, optimize and model chemical and analytical processes and experiments and extract a maximum amount of chemical and analytical information from a differential and complicated large data matrix.

These methods have been useful for determining the geographical provenance, type or variety of samples and for detecting adulteration in various types of foodstuffs, e.g. meat and its products, honey, milk and dairy products, olive oil and wine (Szefer 2007; Szefer and Grembecka 2022). The implementation of chemometrics in the quality evaluation of food and beverages has been comprehensively reviewed by Efenberger-Szmechtyk et al. (2018). According to Agriopoulou et al. (2023) and Tsakovski and Simeonov (2010), chemometric techniques can be categorized as unsupervised multivariate statistical methods, e.g. PCA, FA, HCA, SOMs and non-linear mapping, whereas supervised (learning) methods based on the demarcation of a certain number of classes in the classification process include kNN, SIMCA, PLS-DA, OCPLS and others. Moreover, supervised methods can be divided into two groups: classification/certification methods and calibration/quantification methods (Agriopoulou et al. 2023). The classification methods include discriminant and class modelling methods. The discriminant approach categorizes samples into one of a number of predefined classes, at least two in the case of the binary framework or more than two in the case of the multiclass framework. The discriminant classification methods include the DA, PLS–DA and SVM algorithms; local similarity-based algorithms, including kNN, N3 and BNN; tree-based algorithms (CARTs) and ANNs (CP-ANNs). The class modelling methods used include SIMCA and UNEQ, as well as potential functions and one-class SVM (SVDD). For unsupervised analysis, PCA is the most common technique used for data exploration owing to its important properties, i.e. broad implementation, simple calculation algorithm, the ability to extract sequential information and visualize the data in the form of simple diagrams. Other techniques, such as CA, MDS, MST and SOM, are alternative or complementary tools that provide information pertaining to the data structure in this way. Most of these techniques are based on the application of distance metrics to characterize the similarity interrelationship between objects. Consequently, the results can vary by controlling the type of metric used and the use of specific variable scaling functions to avoid bias owing to the use of different measurement scales for the variables (Consonni et al. 2023). Chemometrics has been demonstrated to be a necessary and powerful tool for food analysis and control. Another main topic is traceability, i.e. the possibility of verifying the geographical, animal/botanical and/or productive origin of a foodstuff; hence, the use of chemometric techniques is not only recommended but essential. Since food is characterized by a complex matrix, the strategy used to study its elemental profile should be multivariate (Marini 2013).

1.4 Conclusions

This chapter introduced the reader to the three main topics related to the scope of this book, i.e. the study object, including counterfeit foods, and the general principles of analytics and chemometrics applications. The problem of authenticity/adulteration of a rich assortment of animal and plant foods is also discussed. The use of development research tools, i.e. combined state-of-the-art analytical methods/techniques and chemometrics, including multivariate analysis, deserves special attention. Multiple factors, including geographical origin, varietal/breed characterization and manufacturing practices, can influence the elemental and stable isotopic compositions of foods and thus may influence their properties. Consequently, a powerful hybrid methodology appears to be indispensable for ensuring a reliable and valuable dataset. The use of advanced chemometric techniques is essential for building multivariate models for classification, discrimination and prediction, enabling the identification of food authenticity and the traceability of food.

References

Agriopoulou S, Tarapoulouzi M, Varzakas T (2023) Introduction. In: Agriopoulou S, Tarapoulouzi M, Varzakas T (eds) Chemometrics and authenticity of foods of plant origin. CRC Press, Taylor & Francis Group, Boca Raton, pp 1–8. https://doi.org/10.1201/9781003266167

Ahmad RS, Imran A, Hussain MB (2018) Nutritional composition of meat. In: Arshad MS (ed) Meat science and nutrition. IntechOpen, London, pp 61–77. https://doi.org/10.5772/intechopen.77045

Anderson TW (1984) An introduction to multivariate statistical analysis. Wiley, New York. ISBN 10: 0471889873

Anne K (2011) Magnesium and calcium in drinking water and heart diseases. In: Nriagu JO (ed) Encyclopedia of environmental health. Elsevier, Amsterdam, pp 535–544. https://doi.org/10.1016/B978-0-444-52272-6.00535-3

Armenta S, de la Guardia M (2015a) Speciation analysis of food. In: De la Guardia M, Garrigues S (eds) Handbook of mineral elements in food. Wiley Blackwell, Hoboken, pp 177–188. ISBN: 978-1-118-65433-0

Armenta S, De la Guardia M (2015b) Mineral content of eggs. In: De la Guardia M, Garrigues S (eds) Handbook of mineral elements in food. Willey Blackwell, Hoboken, pp 669–683. https://doi.org/10.1002/9781118654316

Backhaus K, Erichson B, Gensler S, Weiber R, Weiber T (2021) Multivariate analysis: an application-oriented introduction. Springer Gabler, Wiesbaden. ISBN 10: 3658325887

Baker RS, Inventado PS (2014) Educational data mining and learning analytics. In: Larusson JA, White B (eds) Learning analytics: from research to practice. Springer, New York, pp 61–75. https://doi.org/10.1007/s10758-015-9244-x

Beebe KR, Pell RJ, Seasholtz MB (1998) Chemometrics: a practical guide. John Wiley & Sons, New York. ISBN: 10: 0471124516

Berrueta LA, Alonso-Salces RM, Héberger K (2007) Supervised pattern recognition in food analysis. J Chromatogr A 1158:196–214. https://doi.org/10.1016/j.chroma.2007.05.024

Berton P, Martinis EM, Wuilloud RG (2013) Trace element speciation in food. In: Visakh PM, Thomas S, Iturriaga LB, Ribotta PD (eds) Advances in food science and technology. John Wiley and Sons, New York, pp 227–263. https://doi.org/10.1002/9781118659083.ch8

Bishop YM, Fienberg S, Holland PW (2007) Discrete multivariate analysis. Springer, New York. ISBN 978-0-387-72805-6

Biziuk M, Kuczyńska J (2007) Mineral components in food—analytical implications. In: Szefer P, Nriagu J (eds) Mineral components in foods. CRC Press, Taylor & Francis, Boca Raton, pp 1–31. https://doi.org/10.1201/9781420003987

Bobis O, Moise AR, Ballesteros I, Sánchez Reyes E, Sánchez Durán S, Sánchez-Sánchez J, Cruz-Quintana S, Giampieri F, Battino M, Alvarez-Suarez JM (2020) Eucalyptus honey: quality parameters, chemical composition and health-promoting properties. Food Chem 325:126870. https://doi.org/10.1016/j.foodchem.2020.126870

Bode P (2003) Nuclear analytical techniques for environmental research. In: Namieśnik J, Chrzanowski W, Żmijewska P (eds) New horizons and challenges in environmental analysis and monitoring. CEEAM, Gdańsk, pp 1–16. ISBN 8391908100

Bogdanov S, Haldimann M, Luginbühl W, Gallmann P (2007) Minerals in honey: environmental, geographical and botanical aspects. J Apic Res 46:269–275. https://doi.org/10.3896/IBRA.1.46.4.11

Brown S, Tauler R, Walczak B (eds) (2020) Comprehensive chemometrics—chemical and biochemical data analysis. Elsevier Science B.V., Amsterdam. ISBN 10: 0444641653; ISBN-13: 978-0444641656

Bulska E (2010) Quality assurance and quality control of analytical results. In: Namieśnik J, Szefer P (eds) Analytical measurements in aquatic environments. CRC Press. Taylor & Francis Group, p 389–397. https://doi.org/10.1201/9781420082692

Callao MP, Ruisánchez I (2018) An overview of multivariate qualitative methods for food fraud detection. Food Control 86:283–293. https://doi.org/10.1016/j.foodcont.2017.11.034

Capar SG, Szefer P (2012) Determination and speciation of trace elements in foods. In: Ötleş S (ed) Methods of analysis of food components and additions. CRC Press, Taylor & Francis, Boca Raton, pp 165–210. https://doi.org/10.1201/b11218

Cobos Á, Díaz O (2014) Chemical composition of meat and meat products. In: Cheung P (ed) Handbook of food chemistry. Springer, Berlin, pp 1–32. https://doi.org/10.1007/978-3-642-41609-5_6-1

Collar C (2015) Bread and bakery products. In: De la Guardia M, Garrigues S (eds) Handbook of mineral elements in food. Wiley Blackwell, Hoboken, pp 559–572. https://doi.org/10.1002/9781118654316

Consonni V, Todeschini R, Gosetti F (2023) Authenticity and chemometrics base. In: Agriopoulou S, Tarapoulouzi M, Varzakas T (eds) Chemometrics and authenticity of foods of plant origin. CRC Press, Taylor & Francis Group, Boca Raton, pp 9–35. https://doi.org/10.1201/9781003266167

Cooley WW, Lohnes PR (1971) Multivariate data analysis. John Wiley & Sons, New York. ISBN 10: 0898747813

Cornelis R, Caruso J, Crews H, Heumann K (eds) (2003) Handbook of elemental speciation: techniques and methodology. John Wiley & Sons, Ltd., Chichester. ISBN: 9780471492146

Cozzolino D (2015) Elemental composition in grapes and wine: role, analytical methods and their use. In: De la Guardia M, Garrigues S (eds) Handbook of mineral elements in food. Wiley Blackwell, Hoboken, pp 473–487. https://doi.org/10.1002/9781118654316

Crews HM (1998) Speciation of trace elements in foods, with special reference to cadmium and selenium: is it necessary? Spectrochim Acta Part B At Spectrosc 53:213–219. https://doi.org/10.1016/S0584-8547(98)00078-0

da Silva PM, Gauche C, Gonzaga LV, Oliveira CAV, Fett R (2016) Honey: chemical composition, stability and authenticity. Food Chem 196:309–323. https://doi.org/10.1016/j.foodchem.2015.09.051

Danezis GP, Georgiou CA (2022) Elemental metabolomics: food elemental assessment could reveal geographical origin. Curr Opin Food Sci 44:100812. https://doi.org/10.1016/j.cofs.2022.100812

De Brauvwere A, Gao Y, De Galan S, Baeyens W, Elskens M, Leermakers M (2010) Speciation analytics in aquatic ecosystems. In: Namieśnik J, Szefer P (eds) Analytical measurements in aquatic environments. CRC Press, Taylor & Francis Group, Boca Raton, pp 121–137. https://doi.org/10.1201/9781420082692

De la Fuente MA, Juárez M (2015) Milk and dairy products. In: De la Guardia M, Garrigues S (eds) Handbook of mineral elements in food. Willey Blackwell, New York, pp 645–668. https://doi.org/10.1002/9781118654316

De la Guardia M, Garrigues S (2015) X-ray. In: De la Guardia M, Garrigues S (eds) Handbook of mineral elements in food. Wiley Blackwell, Hoboken, pp 285–300. ISBN: 978-1-118-65433-0

Deming SN, Morgan SL (1993) Experimental design: a chemometric approach. Elsevier, Amsterdam. ISBN 10: 9780444891112

Dillon WR, Goldstein M (1984) Multivariate analysis methods and applications. Wiley & Sons, New York. ISBN 10: 0471083178

Dugard P, Todman J, Staines H (2010) Approaching multivariate analysis: a practical introduction. Routledge/Taylor & Francis Group, New York. ISBN 9780415645911

Durazzo A, Di Lena G, Gabrielli P, Santini A, Lombardi-Boccia G, Lucarini M (2022) Nutrients and bioactive compounds in seafood: quantitative literature research analysis. Fishes 7:132. https://doi.org/10.3390/fishes7030132

Dutta S, Das AK (2015) Drinking water. In: De la Guardia M, Garrigues S (eds) Handbook of mineral elements in food. Wiley Blackwell, Hoboken, pp 455–471. https://doi.org/10.1002/9781118654316

Ebdon L, Pitts L, Cornelis R, Crews H, Donard OFX, Quevauviller P (eds) (2001) Trace element speciation for environment, food and health. Royal Society of Chemistry, London. ISBN: 978-0-85404-459-7

Efenberger-Szmechtyk M, Nowak A, Kregiel D (2018) Implementation of chemometrics in quality evaluation of food and beverages. Crit Rev Food Sci Nutr 58:1747–1766. https://doi.org/10.1080/10408398.2016.1276883

Einax J (ed) (1995) Chemometrics in environmental chemistry—statistical methods. Springer, Berlin. ISBN 10: 3662148854

Everitt B, Hothorn T (2011) An introduction to applied multivariate analysis with R. Springer, New York. ISBN 978-1-4419-9649-7

Flury B, Riedwyl H (1988) Multivariate statistics—a practical approach. Chapman & Hall, London

Galani-Nikolakaki SM, Kallithrakas NG (2007) Elemental content of wines. In: Szefer P, Nriagu J (eds) Mineral components in foods. CRC Press, Taylor & Francis, Boca Raton, pp 323–338. https://doi.org/10.1201/9781420003987

García Alonso JI, Marchante-Gayón JM, Moldovan M (2015) New developments in food analysis. In: De la Guardia M, Garrigues S (eds) Handbook of mineral elements in food. Wiley Blackwell, Hoboken, pp 239–262. ISBN: 978-1-118-65433-0

Gokoglu N, Yerlikaya S (2015) Chemical composition of fish. In: Gokoglu N, Yerlikaya S (eds) Seafood chilling, refrigeration and freezing: science and technology. John Wiley & Sons, Ltd., Chichester, pp 5–37. https://doi.org/10.1002/9781118512210.ch2

González-Domínguez R (2020) Food authentication: techniques, trends and emerging approaches. Foods 9:346. https://doi.org/10.3390/foods9030346

González-Domínguez R (2022) Food authentication: techniques, trends and emerging approaches (second issue). Foods 11:926. https://doi.org/10.3390/foods11131926

González-Domínguez R, Sayago A, Fernández-Recamales Á (2022) An overview on the application of chemometrics tools in food authenticity and traceability. Foods 11:3940. https://doi.org/10.3390/foods11233940

Gonzálvez A, Armenta D, la Guardia M (2015) Mineral content of seasonings, salt and vinegar. In: De la Guardia M, Garrigues S (eds) Handbook of mineral elements in food. Wiley Blackwell, Hoboken, pp 685–698. https://doi.org/10.1002/9781118654316

Goyal MR, Ranvir S, Malik JA (eds) (2023) The chemistry of milk and milk products: physicochemical properties, therapeutic characteristics, and processing methods. Taylor & Francis Group, Boca Raton. https://doi.org/10.1201/9781003340706

Grembecka M, Szefer P (2011) Metals and metalloids in foods—essentiality, toxicity, applicability. In: Medina DA, Laine AM (eds) Food quality: control, analysis and consumer concerns. Nova Science Publishers, Hauppauge, pp 1–60. ISBN 10: 1611229170; ISBN 13: 9781611229172

Grembecka M, Szefer P (2016) Elemental trace analysis in studies of food products. In: Baranowska I (ed) Handbook of trace analysis—fundamentals and applications. Springer, Berlin, pp 203–239. https://doi.org/10.1007/978-3-319-19614-5

Härdle WK, Simar L (2019) Applied multivariate statistical analysis. Springer Nature Switzerland AG, Cham. ISBN 978-3-030-26005-7

Hellberg R, Everstine K, Steven Sklare S (eds) (2021) Food fraud—a global threat with public health and economic consequences. Academic Press, New York. ISBN: 9780128172421

Herrero-Latorre C, Barciela-García J, García-Martín S, Peña-Crecente RM (2017) The use of honeybees and honey as environmental bioindicators for metals and radionuclides: a review. Environ Rev 25:463–480. https://doi.org/10.1139/er-2017-0029

Hoo Fung LA, Antoine JMR, Grant CN, Lalor GC (2015) Vegetables and fruits. In: De la Guardia M, Garrigues S (eds) Handbook of mineral elements in food. Wiley Blackwell, Hoboken, pp 489–520. https://doi.org/10.1002/9781118654316

Izenman AJ (2008) Modern multivariate statistical techniques. Springer, New York. ISBN 978-1-4939-3832-2

Jiménez-Carvelo AM, Martin-Torres S, Cuadros-Rodríguez L, González-Casado A (2021) Nontargeted fingerprinting approaches. In: Galanakis CM (ed) Food authentication and traceability. Academic Press, New York, pp 163–193. https://doi.org/10.1016/B978-0-12-821104-5.00010-6

Kailasapathy K (2016) Chemical composition, physical, and functional properties of milk and milk ingredients. In: Chandan RC, Kilara A, Nagendra P, Shah NP (eds) Dairy processing and quality assurance. John Wiley & Sons Ltd., New York, pp 77–105. https://doi.org/10.1002/9781118810279

Kemsley EK, Defernez M, Marini F (2019) Multivariate statistics: considerations and confidences in food authenticity problems. Food Control 105:102–112. https://doi.org/10.1016/j.foodcont.2019.05.021

Kim S-K (2015) Seafood science: advances in chemistry, technology and applications. CRC Press, Taylor & Francis Group, Boca Raton. https://doi.org/10.1201/b17402

Klein C, Hurlbut CS Jr (1985) Manual of mineralogy (after Dana JD), 20th edn. John Wiley & Sons, New York

Knochen M, Morales G (2015) Edible fats and oils. In: De la Guardia M, Garrigues S (eds) Handbook of mineral elements in food. Wiley Blackwell, Hoboken, pp 573–586. https://doi.org/10.1002/9781118654316

Kollo T, Rosen D, Hazewinkel M (2005) Advanced multivariate statistics with matrices. Springer, Dordrecht. ISBN 978-1-4020-3419-0

Konieczka P, Namieśnik J (2009) Quality assurance and quality control in the analytical chemical laboratory—a practical approach. CRC Press, Taylor & Francis, Boca Raton. https://doi.org/10.1201/9781420082715

Landsberger S (2015) Neutron activation analysis of solid foods. In: De la Guardia M, Garrigues S (eds) Handbook of mineral elements in food. Wiley Blackwell, Hoboken, pp 375–453. ISBN: 978-1-118-65433-0

López-García I, Hernández-Córdoba M (2015) Atomic absorption spectrometry. In: De la Guardia M, Garrigues S (eds) Handbook of mineral elements in food. Wiley Blackwell, Hoboken, pp 189–217. ISBN: 978-1-118-65433-0

Magan JB, O'Callaghan TF, Kelly AL, McCarthy NA (2021) Compositional and functional properties of milk and dairy products derived from cows fed pasture or concentrate-based diets. Compr Rev Food Sci Food Saf 20:2769–2800. https://doi.org/10.1111/1541-4337.12751

Malinowski ER (2002) Factor analysis in chemistry, 3rd edn. Wiley & Sons, Hoboken. ISBN 10: 978-0-471-13479-4

Malinowski ER, Howery DG (1980) Factor analysis in chemistry. John Wiley & Sons Inc., Hoboken. ISBN 10: 0471058815: ISBN 13: 9780471058816

Marcovecchio JD, De Marco SG, Buzzi NS, Botté SE, Labudia AC, La Colla N, Fernández Severini MD (2015) Fish and seafood. In: De la Guardia M, Garrigues S (eds) Handbook of mineral elements in food. Willey Blackwell, Hoboken, pp 621–643. https://doi.org/10.1002/9781118654316

Mardia KV, Kent JT, Bibby JM (1989) Multivariate analysis. Academic Press, London. ISBN 10: 9780124712522

Mardia KV, Kent JT, Taylor CC (2024) Multivariate analysis. Wiley, New York. ISBN 10: 1118738020

Marini M (ed) (2013) Chemometrics in food chemistry. Elsevier B.V., Amsterdam. ISBN: 9780444595287; https://www.elsevier.com/books/chemometrics-in-food-chemistry/marini/978-0-444-59528-7

Martens H, Naes T (1980) Multivariate calibration. John Wiley and Sons, Chichester. ISBN: 978-0-471-93047-1

Massart DL, Vandeginste BGM, Deming SN, Michotte Y, Kaufman L (1988) Chemometrics: a textbook. Elsevier, Amsterdam. ISBN 0-444-42660-4

Moncayo S, Manzoor S, Navarro-Villoslada F, Caceres JO (2015) Evaluation of supervised chemometric methods for sample classification by Laser Induced Breakdown Spectroscopy. Chemom Intell Lab Syst 146:354–364. https://doi.org/10.1016/j.chemolab.2015.06.004

Morales-Rubio A, de la Guardia M (2015) Ion chromatography. In: De la Guardia M, Garrigues S (eds) Handbook of mineral elements in food. Wiley Blackwell, Hoboken, pp 313–374. ISBN: 978-1-118-65433-0

Moreda-Piñeiro J, Moreda-Piñeiro A (2010) Analytical techniques for the determination of inorganic constituents. In: Namieśnik J, Szefer P (eds) Analytical measurements in aquatic environments. CRC Press, Taylor & Francis Group, Boca Raton, pp 259–301. https://doi.org/10.1201/9781420082692

Morrison DF (1978) Multivariate statistical methods. McGraw-Hill International, Auckland. ISBN 10: 0070431868

Namieśnik J, Szefer P (eds) (2010) Analytical measurements in aquatic environments. CRC Press, Taylor & Francis Group, Boca Raton. https://doi.org/10.1201/9781420082692

Ortega-Barrales P, Fernández-de Córdova ML (2015) Meat. In: De la Guardia M, Garrigues S (eds) Handbook of mineral elements in food. Willey Blackwell, Hoboken, pp 599–619. https://doi.org/10.1002/9781118654316

Ould-Dada Z (2007) Pollutant in food—radionuclides. In: Szefer P, Nriagu J (eds) Mineral components in foods. CRC Press, Taylor & Francis, Boca Raton, pp 389–411. https://doi.org/10.1201/9781420003987

Pena RJ (2002) Wheat for bread and other foods. In: Curtis BC, Rajaram S, Macpherson HG (eds) Bread wheat—improvement and production, FAO plant production and protection series. UNFAO, Rome, p 30

Plutowska B, Ramos MD (2012) Analytical quality assurances. In: Ötleş S (ed) Methods of analysis of food components and additions. CRC Press, Taylor & Francis, Boca Raton, pp 31–52. https://doi.org/10.1201/b11218

Pohl P, Szymczycha-Madeja A, Welna M, Jedryczko D, Jamroz P (2015) Elemental composition of sugar and honey. In: De la Guardia M, Garrigues S (eds) Handbook of mineral elements in food. Wiley Blackwell, Hoboken, pp 587–597. https://doi.org/10.1002/9781118654316

Połatajko A, Szpunar J (2007) Speciation of mineral components in food—analytical implications. In: Szefer P, Nriagu JO (eds) Mineral components in foods. CRC Press, Taylor & Francis, Boca Raton, pp 33–50

Poschl M, Nollet LML (eds) (2007) Radionuclide concentrations in food and the environment. CRC Press, Taylor & Francis, Boca Raton. ISBN 13: 9780367453497

Reilly C (1991) Metal contamination of food. Elsevier Applied Science, Amsterdam. ISBN 10: 1851665404

Reilly C (2004) The nutritional trace metals. Wiley Blackwell, Hoboken. ISBN 10:1405110406

Reilly C (2006) Selenium in food and health. Springer, Berlin. ISBN 10: 038733243X

Reilly C (2007) Pollutants in food—metal and metalloids. In: Szefer P, Nriagu J (eds) Mineral components in foods. CRC Press, Taylor & Francis, Boca Raton, pp 363–388. https://doi.org/10.1201/9781420003987

Reilly C (2008) Metal contamination of food: its significance for food quality and human health. Wiley-Blackwell, Hoboken. https://doi.org/10.1002/9780470995105

Rencher AC, Christensen WF (2012) Methods of multivariate analysis. John Wiley & Sons, New York. ISBN 10: 0470178965

Ruiz-de-Cenzano M, Rochina-Marco A, Cervera ML, de la Guardia M (2015) Cereals and pulses. In: De la Guardia M, Garrigues S (eds) Handbook of mineral elements in food. Wiley Blackwell, Hoboken, pp 521–557. https://doi.org/10.1002/9781118654316

Sammut J, Gopi K, Saintilan N, Mazumder D (2021) Facing the challenges of food fraud in the global food system. In: Galanakis CM (ed) Food authentication and traceability. Elsevier, Amsterdam, pp 35–63. ISBN 9780128211045

Sharaf MA, Illman D, Kowalski BR (1986) Chemometrics. Wiley, New York. ISBN 9780471831068

Shirali S, Vasudeva HL (2010) Multivariable analysis. Springer, Berlin. ISBN 10: 0857291912

Singh M, Ranvir S, Jacob T, Sandeep Prasad GM (2023) Mineral profiles of milk and milk products: their interaction and therapeutic benefits. In: Goyal MR, Ranvir S, Malik JA (eds) The chemistry of milk and milk products: physicochemical properties, therapeutic characteristics, and processing methods. Taylor & Francis Group, Boca Raton. eBook ISBN 9781003340706

Smaoui S, Tarapoulouzi M, Agriopoulou S, D'Amore T, Varzakas T (2023) Current state of milk, dairy products, meat and meat products, eggs, fish and fishery products authentication and chemometrics. Foods 12:4254. https://doi.org/10.3390/foods12234254

Spink JW (2019) Food fraud prevention—introduction, implementation, and management. Springer, New York. https://doi.org/10.1007/978-1-4939-9621-6

Szefer P (2002) Metals, metalloids and radionuclides in the Baltic Sea ecosystem. Elsevier Science BV, Amsterdam. ISBN: 9780444539618; ISBN: 9780080536330

Szefer P (2003) Application of chemometric techniques in analytical evaluation of biological and environmental samples. In: Namieśnik J, Chrzanowski W, Żmijewska P (eds) New horizons and challenges in environmental analysis and monitoring. CEEAM, Gdańsk, pp 355–388. ISBN 8391908100

Szefer P (2007) Chemometric techniques in analytical evaluation of food quality. In: Szefer P, Nriagu J (eds) Mineral components in foods. CRC Press, Taylor & Francis, Boca Raton, pp 69–121. https://doi.org/10.1201/9781420003987

Szefer P (2013) Safety assessment of seafood with respect to chemical pollutants in European Seas. Oceanol Hydrobiol Stud 42:110–118. https://doi.org/10.2478/s13545-013-0063-1

Szefer P, Grembecka M (2007a) Mineral components in foods of animal origin and in honey. In: Szefer P, Nriagu J (eds) Mineral components in foods. CRC Press, Taylor & Francis, Boca Raton, pp 163–229. https://doi.org/10.1201/9781420003987

Szefer P, Grembecka M (2007b) Mineral components in food crops, beverages, luxury food, spices, and dietary food. In: Szefer P, Nriagu J (eds) Mineral components in foods. CRC Press, Taylor & Francis, Boca Raton, pp 231–322. https://doi.org/10.1201/9781420003987

Szefer P, Grembecka M (2009) Chemometric assessment of chemical element distribution in bottom sediments of the Southern Baltic Sea including Vistula and Szczecin Lagoons—an overview. Pol J Environ Stud 18:25–34

Szefer P, Grembecka M (2022) Bromatological, analytical and chemometric assessment of animal and plant foods based on mineral composition. Eur J Transl Clin Med 5:77–106. https://doi.org/10.31373/ejtcm/137919

Szefer P, Nriagu JO (eds) (2007) Mineral components in foods. CRC Press, Taylor & Francis, Boca Raton. https://doi.org/10.1201/9781420003987

Szteke B (2007) Criteria of evaluation of food elements analysis data. In: Szefer P, Nriagu J (eds) Mineral components in foods. CRC Press, Taylor & Francis, Boca Raton, pp 51–69. https://doi.org/10.1201/9781420003987

Tafere DA (2021) Chemical composition and uses of honey: a review. J Food Sci Nutr Res 4:194–201. https://doi.org/10.26502/jfsnr.2642-11000072

Tauler R, Walczak B, Brown S (eds) (2009) Comprehensive chemometrics—chemical and biochemical data analysis. Elsevier Science B.V., Amsterdam. ISBN-10: 0444527028; ISBN-13: 978-0444527028

The Council of the European Union (2012) Council Regulation (EC) No 1151/2012 of 21 November 2012 on quality schemes for agricultural products and foodstuff. Off J Eur Union: 1–29. https://eur-lex.europa.eu/legal-content/EN/TXT/?uri=CELEX%3A32012R1151. Accessed 3 Nov 2022

Tirkey S, Verma S, Banerjee K, Shukla KK (2023) Nutritional potential of wild culinary mushroom species. Int J Curr Microbiol Appl Sci 12:167–184. https://doi.org/10.20546/ijcmas.2023.1202.016

Tsakovski S, Simeonov V (2010) Chemometrics as a tool for treatment processing of multiparametric analytical data sets. In: Namieśnik J, Szefer P (eds) Analytical measurements in aquatic environments. CRC Press. Taylor & Francis Group, p 369–387. https://doi.org/10.1201/9781420082692

Varmuza K, Filzmoser P (2009) Introduction to multivariate statistical analysis in chemometrics. CRC Press, Taylor & Francis, Boca Raton. ISBN 9781420059472

Varrà MO, Zanardi E, Serra M, Conter M, Ianieri A, Ghidini S (2023) Isotope fingerprinting as a backup for modern safety and traceability systems in the animal-derived food chain. Molecules 28:4300. https://doi.org/10.3390/molecules28114300

Vaz JA, Barros L, Martins A, Santos-Buelga C, Vasconcelos MH, Ferreira ICFR (2011) Chemical composition of wild edible mushrooms and antioxidant properties of their water soluble polysaccharidic and ethanolic fractions. Food Chem 126:610–616. https://doi.org/10.1016/j.foodchem.2010.11.063

Vázquez M, Vélez D (2015) Other foods of plant origin. In: De la Guardia M, Garrigues S (eds) Handbook of mineral elements in food. Wiley Blackwell, Hoboken, pp 699–708. https://doi.org/10.1002/9781118654316

Vehkalahti K, Everitt BS (2020) Multivariate analysis for the behavioral sciences. CRC Press, Boca Raton. ISBN 9780367656751

Vinceti M, Crespi CM, Bonvicini F, Malagoli C, Ferrante M, Marmiroli S, Stranges S (2013) The need for a reassessment of the safe upper limit of selenium in drinking water. Sci Total Environ 443:633–642. https://doi.org/10.1016/j.scitotenv.2012.11.025

Visciano P, Schirone M (2021) Food frauds: global incidents and misleading situations. Trends Food Sci Technol 114:424–442. https://doi.org/10.1016/j.tifs.2021.06.010

Wackernagel H (2003) Multivariate geostatistic: an introduction with application. Springer, Berlin. ISBN 978-3-540-44142-7

Walczak B (ed) (2000) Wavelets in chemistry. Elsevier Science, Amsterdam. ISBN: 9780444501110; ISBN: 9780080543741

Yáñez-Sedeño P, Pingarrón JM (2015) Electroanalytical methods: application of electrochemical techniques for mineral elements analysis in food. In: De la Guardia M, Garrigues S (eds) Handbook of mineral elements in food. Wiley Blackwell, Hoboken, pp 263–283. ISBN: 978-1-118-65433-0

Ye H, Yang J, Xiao G, Zhao Y, Li Z, Bai W, Zeng X, Dong H (2023) A comprehensive overview of emerging techniques and chemometrics for authenticity and traceability of animal-derived food. Food Chem 402:134216. https://doi.org/10.1016/j.foodchem.2022.134216

Chapter 2
Meat and Its Products

Contents

2.1 Introduction

Cobos and Díaz (2015) characterized the muscle composition and the conversion of muscle into meat; the meat components, such as nitrogen compounds and enzymes, lipids, carbohydrates, vitamins, mineral elements and water; their influence on meat quality and the factors affecting meat composition.

There are numerous international studies on the chemometric evaluation of animal foods such as meat and meat products. Instrumental and/or sensory analyses have been frequently applied to correctly classify meat and meat products such as beef, pork, lamb, poultry, cured meat products and frankfurter sausages.

In recent years, significant developments in analytical and chemometric techniques have been reported owing to advanced scientific technologies. Among the applied analytical tools, the most preferred are IRMS, ICP–MS and LIBS. Other analytical methods/techniques are used, e.g. FAAS, ET–AAS, CV–AAS, HG–AAS and ICP–AES–ICP–OES (Capar and Szefer 2012; Grembecka and Szefer 2011). The reliability and correctness of analytical measurements are usually checked via

Supplementary Information The online version contains supplementary material available at https://doi.org/10.1007/978-3-031-99687-0_2.

P. Szefer, *Chemometrics in Food Control Based on Elemental Analysis*, https://doi.org/10.1007/978-3-031-99687-0_2

quality assurance tests by using CRMs with known concentrations of analytes (Bulska 2010; Szteke 2007; Grembecka and Szefer 2016). Various chemometric techniques have been applied to assess food authenticity/fraud based on the quantitative characteristics of elemental composition, e.g. ANOVA, RA, PCA, FA, CA, HCA, ANN, PLS-DA, SOM and nonlinear mapping (e.g. Szefer 2002, 2003, 2007; Tsakovski and Simeonov 2010). Beef samples have been classified according to their origin by three machine learning algorithms, i.e. MLP, RF and CRT (De Nadai Fernandes et al. 2020). Islam and Khaliduzzaman (2022) provided a brief review of the chemometric techniques that can be useful in spectroscopy-based poultry and egg research. The potential of developing predictive models for calibration and classification responses was demonstrated.

Regarding food traceability analysis, SIRA is a relatively new technology. The adequate measurement and calibration of stable isotope data depend on properly selected reference materials. However, isotope reference materials often use chemical matrices; hence, they are inappropriate, especially for food matrix origin analysis. Seasonal variation is recommended for the stable isotopic ratio authentication of beef and other animal-derived products (Bahar et al. 2008). The differentiation of stable isotopic patterns of samples from different countries on different continents shows the potential of IRMS for the verification of geographic authenticity (Horacek and Min 2010). The effectiveness of SIRA in food traceability could be improved by selecting appropriate isotope reference materials. According to Zhao and Zhao (2021), owing to cooperation with various laboratories, proper isotope reference materials can be prepared by the addition of new food matrix types to achieve more choices for analysts.

Foods are commonly investigated for trace, minor and major elements to assess whether their contents are safe or fulfil nutrient requirements. It is also important to obtain information about the contents of potentially toxic elements in meat products. According to Pereira et al. (2020), with the exception of Cd in Spain, the levels of other toxic elements, such as As, Pb and Hg, are below the maximum limits established by the EU and thus do not pose any risk for human consumption. Moreover, many elements in food products are not considered nutrients or are not toxic at the level naturally present in foods (Grembecka and Szefer 2011). It has also been emphasized that some methods are characterized by promising potential, e.g. stable isotope ratios or trace elements, in contrast to other methods, which are unsatisfactory in their discriminating abilities for assessing the geographic origin of meat. Recent analytical developments are related to the quantification of the provenience of raw meat, especially lamb, beef and poultry (Franke et al. 2005).

It has been a preferred approach for the development of applications of SIRA for meat authentication using multi-element analysis combined with multivariate statistical analysis (Monahan et al. 2012, 2018). In the opinion of the authors, "challenges lie in discriminating meat samples produced in geographically close regions, but, on the other hand, a maritime border, local geology or regional climatic differences are likely to increase the potential for discrimination and this could be exploited in certain jurisdictions". Moreover, methods based on stable isotope, α-tocopherol stereoisomer and metabolite analyses appear to be useful for authenticating grass-fed beef. An important issue is the ability of metabolically inert animal tissues to achieve archival records of changes in diet and urine. This makes it possible to noninvasively obtain information about grass-fed beef production systems (Monahan et al. 2012, 2018). According to Bandoniene et al. (2020), analytical technologies have been developed for determining the origin of meat and milk products through labelling on the basis of REE spiking of the feed for lambs and goats using SN–ICP–MS and/or LA–ICP–MS.

The reliability and correctness of analytical measurements are usually checked via quality assurance tests that use CRMs with known concentrations of analytes (Namieśnik and Szefer 2010; Grembecka and Szefer 2016). The risk of food sample contamination by heavy metals is noteworthy during their analytical processing in the laboratory. For example, LIBS was used to detect Cr and Pb in pork contaminated from this source. The traditional CC and multivariate PLS algorithms were subsequently used to validate their accuracy in predicting the contents of Cr and Pb in pork samples (Yao et al. 2017). Moreover, it is the key approach to process chemometrically exclusive validated analytical data. Progress in the preparation of stable isotope reference materials for the traceability of agricultural products, including meat and its products, has been reviewed by Zhao and Zhao (2021). Research progress has focused not only on the preparation but also on the classification and application of such materials. Furthermore, selecting reference materials would be helpful for improving the effectiveness of stable isotope ratio analysis in food traceability.

Chemometrics appears to be a useful tool for the exploration of analytical data matrices; it is helpful for the identification of similarities or dissimilarities between samples as well as between measured parameters (variables). Such multivariate approaches have been sufficient to fulfil the basic objectives of chemometrics as modern multiparametric information, making it possible to understand the knowledge hidden in a large data matrix. Multivariate analysis is considered a very efficient tool for classifying individual products. Chemometric techniques, i.e. PCA, FA, CCA and HC–PLSR, have been found to be important techniques in the discrimination of food products according to their variety and/or geographical provenance. Instrumental and/or sensory analyses have been frequently applied for authentication purposes to correctly classify meat and meat products such as beef, pork, lamb, poultry, cured meat products and frankfurter sausages (Arvanitoyannis and van Houwelingen-Koukaliaroglou 2003). Univariate, bivariate and multivariate techniques have been applied to assess food authenticity/fraud with respect to the quantitative characteristics of elemental composition, i.e. one-way and two-way ANOVA; Pearson or Spearman correlation analysis and regression analysis, e.g. PCA, FA, CA, DA, HCA, ANN and PLS–DA (e.g. Szefer 2002, 2003, 2007).

2.2 Chemometric Assessment of Meat and Its Products

2.2.1 *Geographical Differentiation*

Verifying the authenticity of the claimed geographical origin of animal food, including meat and its products, on the basis of its mineral composition has been reviewed on an international scale, e.g. by Franke et al. (2005, 2007, 2008a, b, c), Camin et al. (2016), Zhao (2016), Monahan et al. (2018) and Arvanitoyannis and van Houwelingen-Koukaliaroglou (2003).

2.2.1.1 European Countries

Sacco et al. (2005) reported that the application of multivariate analysis combined with ^{1}H HR-MAS NMR, HPIC and ICP–AES allows the discrimination of Apulian lamb meat samples in terms of their geographical origin. The first two PCs, corresponding to 67% of the total variance, indicated that the meat samples were distinguished as being from one of five Italian regions that produce Apulian lamb; this was determined largely on the basis of the fat and moisture content and the Na, Mg, Fe and Cu contents. Additionally, PCA demonstrated discriminating abilities on the basis of descriptors such as selected NMR spectra with stable isotope ratios. The application of DA revealed that the prediction ability was 60% for the dataset related to routine analysis with elemental data and 96% for the dataset including the isotope ratio and NMR data. This means that a more effective classification model was achieved with the use of spectroscopic quantification, confirming the high power of SIRA and NMR as a verifier of the geographical provenance of lamb meat.

Camin et al. (2007) used EA–IRMS to determine the SIR for the geographical origin assignment of lamb meat from 13 sampling sites in 7 European countries. The first three PCs, which explained 91.78% of the total variance loaded by descriptors such as δ^{2}H, δ^{13}C, δ^{15}N and δ^{34}S, allowed the geographical provenance assignment of lamb meat from several European regions. The data differentiated lamb meat from most production regions with a tendency towards multi-element isotope ratios to be related to parameters such as precipitation and groundwater (δ^{2}H), feeding practices and climate (δ^{13}C) and the geographical location and surface geology (δ^{34}S) of the production regions. LDA achieved 78% correct classification. Dehelean et al. (2023) combined IRMS, FAAS and ICP–MS to assess the isotopic and mineral compositions of pork meat samples (loin) originating from Romania and other European countries, i.e. Germany, Spain and Hungary. The original data obtained were processed with chemometric analysis. The three PCs corresponded to 45.7% of the total variance. By using the LDA model, better results were achieved; i.e. 91.4% was obtained in the initial classification, whereas for the CV procedure, 90.0% was recorded. δ^{2}H, K, Rb and Pd are the most representative markers responsible for the discrimination of meat samples according to geographical origin.

2.2.1.2 Asiatic Countries

SIRA followed by DF seems to be an effective approach for characterizing defatted beef, crude fat and tail hair samples from four provinces in China (Guo et al. 2010). The results demonstrate that the $\delta^{13}C$ and $\delta^{15}N$ values are powerful tools for identifying the geographical origin of Chinese cattle. DF allowed the classification of tail hair and defatted beef samples with classification success rates of 85% and 78%, respectively.

Zhao et al. (2013), on the basis of the combination of IRMS and ICP–MS, used multivariate analysis, including PCA and LDA, to trace the geographic origin of beef in China. The first three PCs, corresponding to 71.5% of the total variance, allowed them to clearly distinguish defatted beef among the four regions (Fig. 2.1). LDA perfectly differentiated interregionally samples on the basis of stable isotope ratios and elements such as $\delta^{13}C$, $\delta^{15}N$, Mg, K, Mn, Zn, Se and Zr, achieving maximum discrimination among samples, with an overall correct classification rate of 100% and a CV rate of 100%.

According to Liu et al. (2013), the combination of EA–IRMS with ANOVA and LDA allowed the classification of cattle tail hair samples according to their geographical provenance. An overall correct classification rate of 82.6% and a CV rate of 79.6% were achieved for the four beef production regions, whereas rates of 70.7% and 70.1% were achieved for the seven sampling subregions. Therefore, cattle tail may be a potentially useful marker for verifying the geographical origin of Chinese beef to ensure its authenticity and safety.

Yanagi et al. (2012) reported that EA–IRMS enabled the assessment of stable isotope profiles of beef meat and hair samples from various locations in Japan. The GLM procedure revealed that the $\delta^{13}C$ and $\delta^{15}N$ values in the samples considerably differed depending on the farm location. Both stable isotope ratios were statistically greater in the hair samples than in the meat samples, and a significant relationship between these values was observed. Therefore, the authors postulated that hair samples would be used instead of meat samples to verify the authenticity of the product.

Sun et al. (2011) applied ICP–MS to effectively determine the geographical origin of defatted mutton from Chinese pastoral and agricultural regions. The first three PCs, which explained 52.5% of the total variance, enabled interregional differentiation. This result was confirmed by CA and LDA multivariate techniques. The latter provided overall correct classification and CV rates of 93.9% and 88.9%, respectively. Moreover, mutton samples assigned to agricultural and pastoral regions were discriminated with 100% accuracy.

Zhao et al. (2016) integrated EA–IRMS and ICP–MS methods and PCA and DA techniques, resulting in improved identifiability of chicken samples among four provinces in China. The first four PCs, which explained approximately 74.9% of the total variance, allowed us to partly discriminate chicken samples with respect to the four producing regions owing to the contributions of $\delta^{13}C$, Ca, Na, Fe, Se, $\delta^{15}N$ and Cu. SDA appears to be the most effective tool for selecting analytes such as $\delta^{13}C$, $\delta^{15}N$, K, Zn and Rb. The discriminant model provided highly significant results (Fig. 2.2), with an overall correct classification rate of 100% and a CV rate of 100%.

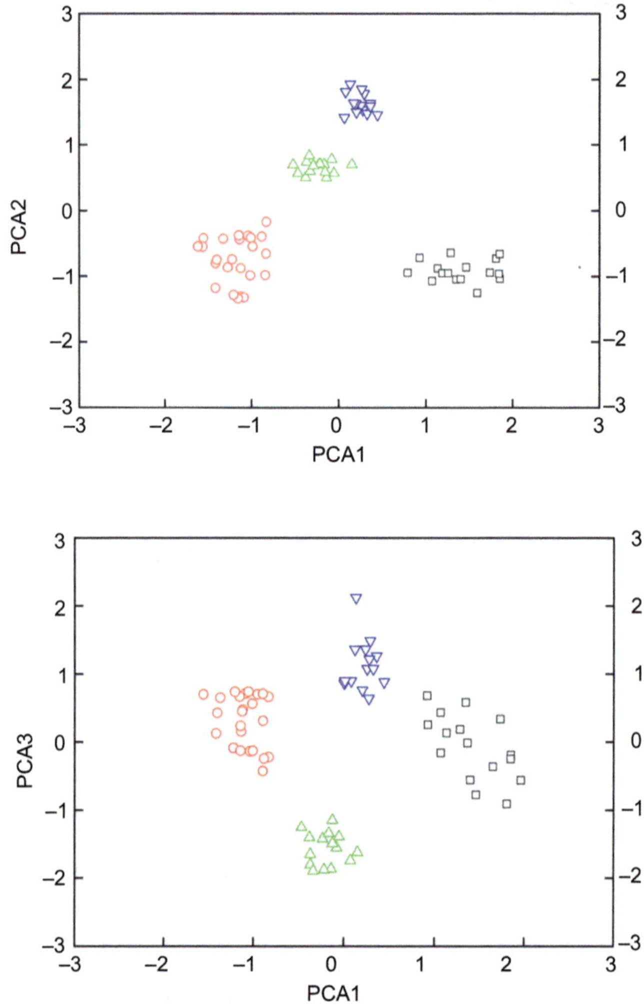

Fig. 2.1 Principal component analysis of defatted beef from four regions: (□) SD; (○) TB; (△) YN; (▽) HLJ. (After Zhao Y. et al., 2013. *J Agric Food Chem* 61:7055–7060. https://doi.org/10.1021/jf400947y. With permission)

Park et al. (2018) used two analytical methods, ICP-MS and IRMS, to verify the origin of 346 defatted Korean and non-Korean pork samples from 6 countries on the basis of trace element profiling and $\delta^{13}C$ and $\delta^{15}N$ analysis. The contents of V, Cr, As, Rb and Sr were higher in Korean pork samples, whereas Li, Co, Mn, Ni, Sr, Cd and Cs levels were higher in non-Korean samples. LDA related to trace element levels and stable isotope ratios discriminates geographical origins with the highest discrimination rate of 100%.

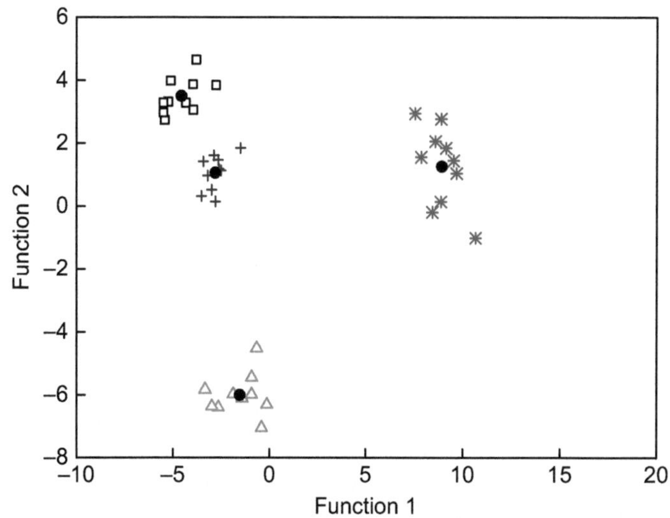

Fig. 2.2 A cross-plot showing the first two discriminate functions obtained from the stepwise canonical discriminate analysis of the stable isotope and trace element data of poultry samples. □HLJ, *SX, ΔJX, +FJ, ●Group centroid. HLJ, Heilongjiang Province; SX, Shanxi Province; JX, Jiangxi Province; FJ, Fujian Province. (After Zhao Y. et al., 2016. CyTA–J Food 14:163–168. https://doi.org/10.1080/19476337.2015.1057235. With permission)

Jiang et al. (2020) noted that a combination of IRMS, ICP–MS and ICP–OES allowed us to distinguish the geographical origin of bone samples from various provinces in China. The $\delta^{13}C$, $\delta^{15}N$ and $\delta^{2}H$ compositions and the contents of the 18-element profile in bovine bone samples were studied. A chemometric tool consisting of ANOVA, PCA and LDA revealed that all the variables analyzed were significantly different among the three regions. Total correct classification rates of 95.7% and 100% were obtained by SIRA and MEA, respectively, whereas the cross-validation rates were 91.5% and 97.9%, respectively. It was concluded that the methodology obtained could be useful for tracing the geographical origin of raw bone materials. According to the authors, it is recommended that traceability models be established via the stable isotope method in practical work on the basis of their feasibility and efficiency.

2.2.1.3 Countries Across Multiple Continents

According to Heaton et al. (2008), the results obtained by EA–IRMS and ICP–MS for SIR and MEC allowed the discrimination of raw beef samples from major cattle-producing regions in 17 countries at the level of broad geographical area, i.e. Europe, South America, North America and Australia. Owing to the use of CDA, it has been possible to differentiate a "broad" grouping of samples belonging to five European countries, three South American countries and two Australasian countries. The original and cross-validated grouped cases were correctly classified, i.e. 91.2% and 86.1%, respectively.

Nakashita et al. (2008) classified beef muscle samples from various regions in Japan, Australia and central North America using EA–IRMS ($\delta^{13}C$, $\delta^{15}N$ and $\delta^{18}O$ values). This approach is a useful potential tool for discriminating the origin of beef not only between three different countries but also among different regions within Japan.

Zhao et al. (2020a) combined EA–IRMS with multivariate tools (PCA, DA and LSD) to categorize beef samples from six countries. PCA and DA demonstrated the ability to achieve intercountry differentiation among 292 beef samples. PCA differentiated the data with respect to the isotope ratios across the four regions; i.e. the beef samples from New Zealand, Canada and China were mostly separated, but the beef samples from Argentina, Brazil and Uruguay overlapped, probably because the countries border each other and are characterized by similar crops and climates. A satisfactory classification was obtained using DA, with an accuracy of the original classification of 96.6% and an accuracy of cross-validation of 95.9%. PLS–DA also discriminates between regional beef samples, although those from Argentina and Brazil still partly overlapped. In general, the classification model was validated to satisfactorily differentiate beef samples from six countries.

Osorio et al. (2011b) successfully analyzed $\delta^{2}H$, $\delta^{13}C$, $\delta^{15}N$ and $\delta^{34}S$ via EA–IRMS in defatted beef samples from nine European and non-European countries. The CDA clearly differentiated bovine muscles in terms of regional origin using descriptors such as $\delta^{13}C$, $\delta^{15}N$, $\delta^{2}H$ and $\delta^{34}S$. After considering additional information regarding various feeding regimens used in Irish beef, it was also possible to determine the geographic origin of beef samples, i.e. between European countries, the US and Brazil, with 85.0% of the samples correctly assigned and 82.9% cross-validated. All of the Irish beef samples marketed as pasture-fed beef were correctly classified, achieving 100% cross-validation.

Nho et al. (2019) combined ICP–OES and ICP–MS methods with chemometric tools to differentiate the geographical origins of pork belly fat samples from seven countries. The PCA and LDA techniques enabled them to distinguish 98.2% of the samples. Korean pork samples were separated from other meat samples, with a discrimination rate of 98.2%. In addition, when all the meat samples except the Korean ones were used for LDA modelling, US samples were significantly separated from the remaining five regions, with a classification of 98.1%. Elements such as Li, Sr, As, Cr, V, Mn, Ni and Cd have also been demonstrated to be adequate determinants for authenticating the origin of pork belly fat. The applied methodology allowed adequate analysis of the elemental isotope ratio profiles of meat samples from South Korea and countries in the Americas and Europe. Owing to multivariate chemometric processing of the analytical results, it was possible to determine the authenticity of the geographical origins of meat samples.

The aim of the study by Shin et al. (2018) was to use ICP–MS, ICP–AES and EA–IRMS methods to quantify MEC, SIR and $^{87}Sr/^{86}Sr$ ratios in pork samples originating from seven countries. The data obtained were explored using a chemometric approach. PCA with the first three PCs accounted for 55.4% of the total variance in the classification of the Korean pork samples from the other samples. It was possible to clearly separate Korean and Mexican samples from the others (by PC1), Korean inland samples from Korean island samples (by PC2) and Canadian samples from the others (by PC3). Although discrimination of the geographic provenance of pork samples was observed, it has been suggested that more detailed investigations are required to determine whether a combination of the above approaches is valid for quantifying the geographic origin of pork.

Kim et al. (2017) reported that ICP–OES and ICP–MS appear to be useful approaches for characterizing 323 pork belly samples from 6 non-European and European countries. First, using all the element datasets or only macroelement data, PCA yielded separations of 43.38% and 65.91%, respectively. However, clear intercountry differentiation was possible through the use of exclusively trace element variables achieved by LDA, resulting in a high discrimination index of 97%. This means that all the pork samples were successfully separated according to their country of origin; the LDA model was the most useful for differentiating the origins of the analyzed samples. It has been concluded that the applied combined methodology could be useful for the authentication of the origin of pork by differentiating its intercountry elemental composition.

Kim et al. (2013) processed SIR values established by EA–IRMS and CF–IRMS via ANOVA to classify 599 pork samples from 14 countries in Asia, America and Europe. $\delta^{13}C$ and $\delta^{15}N$ in meat appear to be useful for identifying the geographical origin of lamb meat. The classification of the meat samples was useful since they were clearly assigned to three major groups, i.e. European, American and South Korean, on the basis of their $\delta^{13}C$ and $\delta^{15}N$ composition values. Thus, pork from South Korea can be distinguished in the local market from internationally supplied pork by using both stable isotopes.

According to Qie et al. (2021), in contrast to the data obtained by using each method separately, data fusion by ratio modulation of multiple techniques (EA-IRMS, ICP–MS and GC–FID) considerably improved the geographical traceability of lamb samples processed using a series of chemometric analyses, i.e. PCA, HCA and PCA–LDA. For example, the discrimination accuracy of PCA–LDA was in the order of 96.67% (stable isotope ratios) >90.00% (multi-element) >77.50% (fatty acids). This means that IRMS appears to be the most effective method for improving accuracy. Although each individual method can partially distinguish the four regions, the discrimination accuracy is still not perfect because the analytes are used separately and hence are not sufficient to achieve satisfactory geographical origin verification. Consequently, the new matrix of the multi-technique was then applied to the discriminant analysis, i.e. PCA and HCA, with the optimum ratios for stable isotope ratios relative to mineral element contents of 8:2 as well as for stable isotope ratios relative to mineral element contents and relative to fatty acid concentrations of 7:2:1. For example, Fig. 2.3 presents the PCA from data fusion by ratio modulation of multiple techniques for lamb samples from four regions. Owing to this approach, in contrast to individual techniques, the data fusion strategy can be applied as a more effective prediction model to obtain prediction results that are more reliable and accurate than those of the midlevel data fusion strategy.

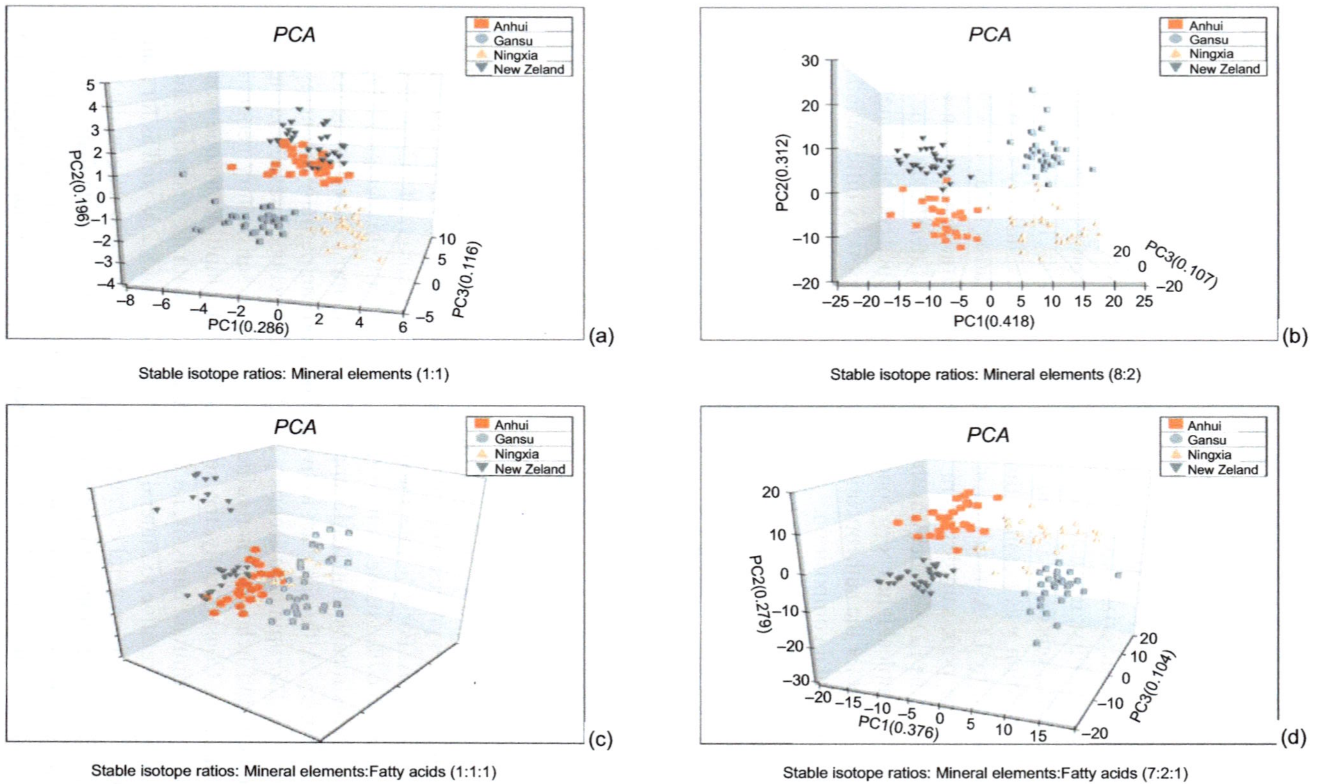

Fig. 2.3 Principal component analysis (PCA) of data fusion by ratio modulation of multiple techniques in lamb samples from four regions. These ratios are as follows: stable isotope ratios: mineral element contents (1:1) (**a**); stable isotope ratios: mineral element contents (8:2) (**b**); stable isotope ratios: mineral element contents: fatty acid concentrations (1:1:1) (**c**); and stable isotope ratios: mineral element contents: fatty acid concentrations (7:2:1) (**d**). (After Qie M. et al., 2021. Food Contr 120:107549. https://doi.org/10.1016/j.foodcont.2020:107549. With permission)

Franke et al. (2008c) reported that ICP–MS data from poultry meat and dried beef samples from 11 countries appeared to be useful for geographical authentication. For example, LDA yields a mean correct classification rate. LDA was performed on dried beef for a combination of the Ba, Ca, Cd, Dy, Fe, Ga, Li, Mg, Mn, Pd, Rb, Sr, Tl, U and V contents in a classification matrix with an average overall classification rate of 79%. The classification and validation matrix of poultry breast meat provided a worse classification rate.

Moreover, Franke et al. (2008a) used a combination of IRMS, ICP–MS and MC–ICP–MS to quantify $^{87}Sr/^{86}Sr$ and $\delta^{18}O$ variables and applied IRMS with ICP–MS methods (Franke et al. 2008b) to classify poultry breast meat and dried beef samples from nine countries. In the case of poultry, cross-validation resulted in 83% correct classification. The corresponding value for dried beef was 73%. The most descriptive elements in poultry meat samples are Se, Rb, Sr, Gd, Tl, Na, V and $\delta^{18}O$, whereas B, Ca, Cd, Cu, Fe, Li, Mn, Mo, Rb, Sr, V and $\delta^{18}O$ are descriptors for dried beef. According to the authors, owing to partial differentiation of meat samples with regard to their geographical origin, a range of methods should be developed to achieve greater discriminative effectiveness.

2.2.2 Interspecies Differentiation

Following the process performed by Bilge et al. (2016), LIBS and AAS data were used to characterize pork, beef and chicken meat to clearly discriminate these samples with chemometric support (ANOVA, PCA, PLS and LRA). PCA showed that the first two PCs explained 96.3% of the total variance for qualitatively discriminated pure beef, chicken and pork meat species (Fig. 2.4). Pork-beef and chicken-beef mixtures were also assessed with PLS. Reliable discrimination was achieved between beef, pork and chicken, with greater distinction between beef and pork

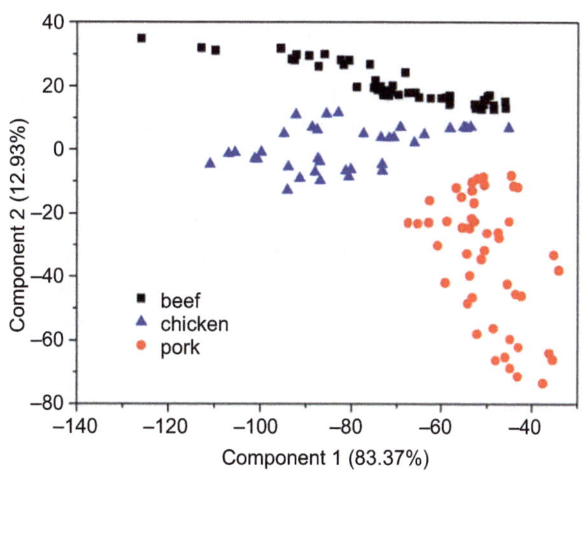

Fig. 2.4 PCA graph for qualitative discrimination of pure beef, chicken and pork meat species. (After Bilge G. et al., 2016. Meat Sci 119:118–122. https://doi.org/10.1016/j.meatsci.2016.04.035. With permission)

samples. Although the samples belonged to different carcass parts of animals originating from different regions, the same species were classified in the same area. According to the authors, this approach can be a valuable tool for quality control measurements of meat as a routine method.

Ahmad Nizar et al. (2013) used EA–IRMS, GC–MS and GC–FID to determine $\delta^{13}C$ and ten fatty acids as effective tools for the identification of various types of Malaysian animal fats, i.e. lard, chicken fat, beef fat and mutton fat. PCA data with the first two PCs explaining 94% of the total variance demonstrated that stearic, oleic and linoleic acids are the most descriptive variables for distinguishing animal fats into four subclasses. Moreover, $\delta^{13}C$ was an appropriate indicator for discriminating between the meat types examined.

Guo et al. (2016) analyzed pork samples from five provinces in China via EA–IRMS to verify their geographical origin. Classification was performed by determining the physicochemical properties of the samples through chemometric techniques, including ANOVA, PCA and DA. Among the five meat types, PCA, with the first two PCs corresponding to 57.31% of the total variance, discriminated three styles of bacon. DA differentiated traditional Chinese pork bacon, with a ratio of overall correct classification of 97.76% and that of CV amounting to 91.76%. Thus, the use of chemometric techniques combined with the physicochemical parameters of traditional Chinese pork bacon is a relatively reliable methodology for classifying and discriminating bacon varieties on the basis of their style. These findings provide a theoretical reference for the assessment of the authenticity of traditional Chinese pork bacon.

2.2.3 *Differentiation of Other Parameters*

2.2.3.1 Interregional and/or Intertype Differentiation Related to the Feeding Regime/Breeding System

Piasentier et al. (2003) reported that the most satisfactory classification was attained by using IRMS combined with chemometric analysis (*k*-means nonhierarchical clustering technique, CDA) to study lamb meat samples. CDA was performed to differentiate lamb types on the basis of six European countries of origin and after this feeding regime, i.e. Icelandic fed on pasture, British and French grazing, Karagouniko concentrate-finished, French Lacaune and Ternasco de Aragon. The fat $\delta^{13}C$ and protein $\delta^{13}C$ and $\delta^{15}N$ appear to be effective contributors to the interregional discrimination of lamb meat, enabling the corrected empirical assignment of 79.2% of the initial observations with a corrected CV of 66.7% for the individual meat samples. On the other hand, the classification of lamb types related to their feeding regime showed more satisfactory results, i.e. 91.7% of the individual meat samples were both correctly assigned and cross-validated, owing to the protein $\delta^{13}C$ and $\delta^{15}N$ of the meat fractions.

According to Sun et al. (2016), EA–IRMS data processed by chemometric analysis appear to be useful for identifying the geographical provenance of Chinese lamb meat in different regions under two feeding regimes, i.e. pastoral and agricultural. Owing to the fusion of $\delta^{13}C$, $\delta^{15}N$ and $\delta^{2}H$, the LDA achieved a total correct classification of 88.9% and 83.8%, corresponding to defatted muscle and wool samples, respectively. Moreover, a separation among regional groups corresponding to pastoral and agricultural region samples was observed. The significant relationship between the stable isotope ratio values of lamb wool and those of lamb muscle enables the use of both wool and defatted muscle for origin traceability.

Renou et al. (2004) applied multiple techniques, i.e. ^{1}H and ^{2}H NMR spectroscopy, ^{13}C NMR spectroscopy and GC–IRMS, which seem to be successful methods for evaluating the chemical profiles of French beef samples on the basis of stable isotope, α-tocopherol stereoisomer and metabolite analyses. The FDA results revealed that some parameters significantly varied with respect to the production region or feeding. Moreover, the geographic origin was verified for each diet, i.e. pasture or maize. For example, all grazing steers were correctly classified since 94% of the animals fed on maize ensilage.

Cristea et al. (2022) studied pork samples originating from Romania and other countries. EA–IRMS and ICP–MS data were processed by multivariate chemometric methods. Owing to the use of original variables in conjunction with LDA and ANNs, it was possible to classify pork samples with respect to their geographical origin on the basis of $\delta^{18}O$, Tb and Sn, whereas $\delta^{13}C$, Rb, La and Ce were effective markers for animal diet differentiation.

Perini et al. (2009) categorized Italian and Spanish lamb ham samples of various types using EA–IRMS data. Using PCA and CDA, the most effective descriptors for tracing the feeding regime were appropriate fractions of $\delta^{13}C$, $\delta^{18}O$, $\delta^{15}N$ and $\delta^{34}S$. The $\delta^{2}H$ and $\delta^{18}O$ values are correlated with the related values in meteoric waters, which are responsible for the traceability of geoclimatic factors; $\delta^{15}N$ is dependent on pedoclimatic conditions and $\delta^{34}S$ is affected by the sea spray effect and the surface geology of the production region. On the basis of CDA data, 97.7% of the individual meat samples correctly corresponded to lamb type, and 97.7% were cross-validated by a predictive model; however, $\delta^{2}H$ discriminates lamb samples with respect to their geographical provenance.

Among the numerous variables, stable isotope ratios, such as $\delta^{13}C$, $\delta^{15}N$, $\delta^{2}H$ and $\delta^{34}S$, were determined by Osorio et al. (2013) using multiple techniques, including EA–IRMS, HPLC, GC/IT–MS and reflectance spectral measurements, and were considered for chemometric selection and modelling of significant dietary biomarkers in Irish bovine muscle and adipose tissue corresponding to each dietary treatment. The analytical data processed using PCA, PLS–DA and SIMCA indicated that the PLS–DA and SIMCA achieved high levels of accuracy, suggesting their utility in cattle diet confirmation on the basis of both variable datasets (90.8–100% correct classification) and especially the reduced variable set (76.1–100% correct classification).

According to Dehelean et al. (2022), IRMS and ICP–MS combined with PLS-DA allowed the classification of chicken and pork samples according to SIR and MEC variables. Classification models were built for both types of meat samples in terms of interregional differentiation and the animal growing system (for animals from yard-rearing systems vs. industrial farms). The PLS–DA models were useful for discriminating the meat samples with respect to the animal-rearing system for pork and chicken meat with 100% and 98% accuracy, respectively. Additionally, PLS–DA allowed the discrimination of meat samples according to their provenance, achieving a predictive ability of 93.8% for chicken and 71.8% for pork. The usefulness of SIR and MEC as indicators for distinguishing meat samples in accordance with the breeding system has also been well documented.

2.2.3.2 Interregional Differentiation Related to Parent Soil and Water

The aim of the investigation by Baroni et al. (2011) was to combine QICP–MS, IRMS and TIMS to classify the chemical profiles of beef meat samples and associated soil and water samples from major Argentinean cattle-producing regions. LDA provided 100% correct classification of beef samples from three Argentinean regions on the basis of $\delta^{13}C$, $\delta^{15}N$, Rb, $^{87}Sr/^{86}Sr$ and Ca/Sr ratios (Fig. 2.5). Moreover, the dataset should be expanded by studying samples collected at different times to

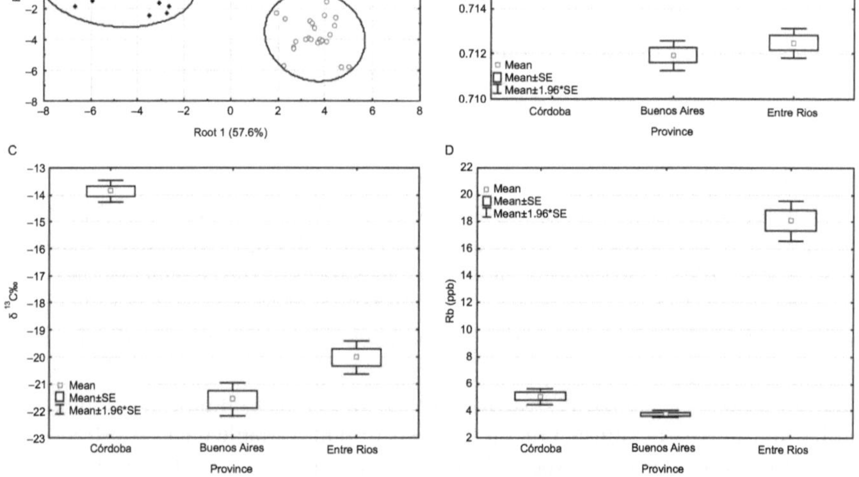

Fig. 2.5 Discriminant analysis (canonical roots) and several selected parameters for differentiating beef. (After Baroni MV et al., 2011. J Agric Food Chem 59:11117–11128. https://doi.org/10.1021/jf2023929. With permission)

consider seasonal variations, particularly with respect to C, N and S stable isotopes. Using the studied matrices, i.e. soil, water and meat, GPA and CCA indicated a consensus between them and clear differentiation between the studied areas (Fig. 2.6). This means that there is a significant relationship between the chemical-isotopic profile of meat and those corresponding to parent soil and water. A considerable connection between the meat fingerprint and that from the parent soil/water enables the effective verification of the geographic origin of beef meat.

2.2.3.3 Differentiation Between Conventional and Organic Production

According to Zhao et al. (2020b), IRMS and ICP–MS followed by PCA and OPLS–DA appeared to be effective tools for classifying pork samples originating from four Chinese locations. OPLS–DA, with contributions of $\delta^{13}C$, $\delta^{15}N$, $\delta^{2}H$ and $\delta^{18}O$ ratios combined with the contents of Na, Mg, K, Ca, Fe, Cu and Se, clearly differentiated conventional pork and organic pork from four origins (Figs. 2.7 and 2.8). This means that OPLS–DA also successfully discriminated pork samples with respect to their different origins. The CV accuracy rate was at a satisfactory level of 98.8%. According to the authors, although a limited database was used, these results can be used to verify the authenticity of organic agricultural products originating from a large area, such as a province or even a country.

Schmidt et al. (2005) processed beef cattle meat from nine countries via CF–IRMS to characterize its stable isotopic composition. MANOVA ($F_{2.65}$ = 327.3,

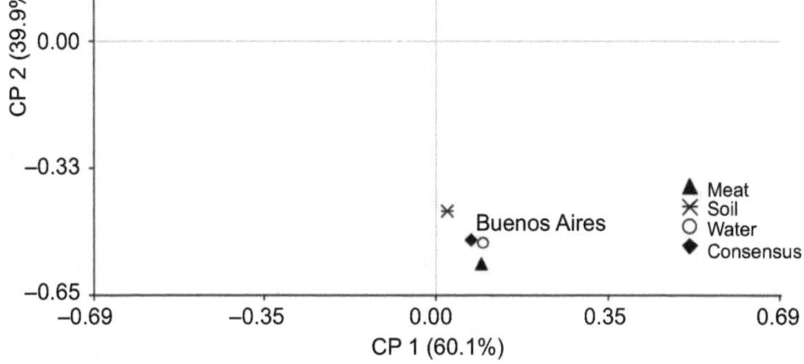

Fig. 2.6 Consensus space from generalized procrustes analysis: plot in the plane defined by the first two dimensions. (After Baroni MV et al., 2011. J Agric Food Chem 59:11117–11128. https://doi.org/10.1021/jf2023929. With permission)

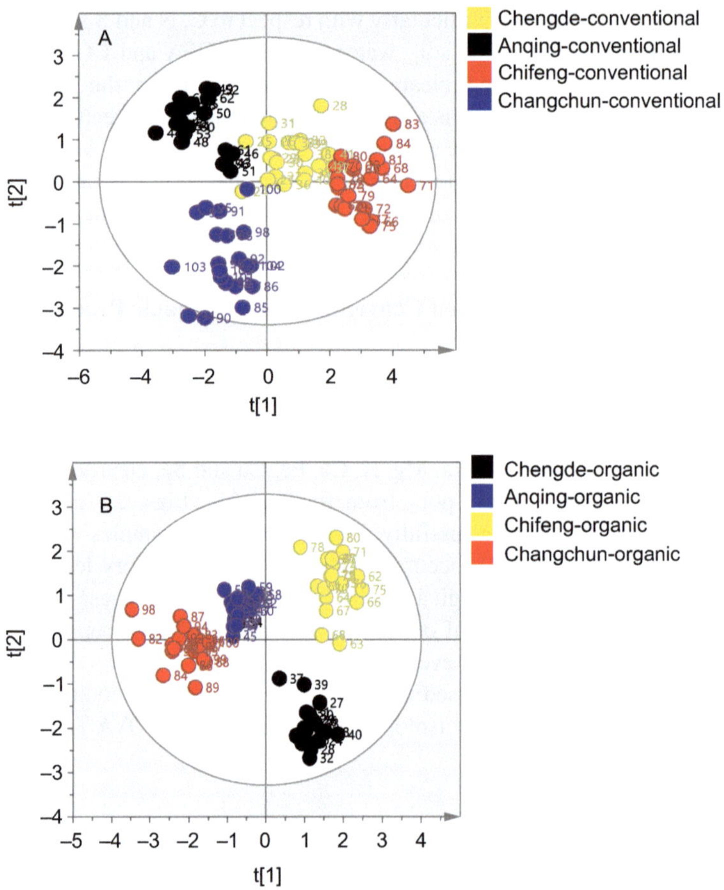

Fig. 2.7 Orthogonal partial least squares discriminant analysis (OPLS–DA) results showing the clustering of conventional pork and organic pork from four origins (Chengde, Hebei Province; Anqing, Anhui Province; Chifeng, Inner Mongolia Autonomous Region; Changchun, Jilin Province) obtained by analysis of the ratios of the stable isotopes of carbon, nitrogen, hydrogen and oxygen combined with the concentrations of seven elements (Na, Mg, K, Ca, Fe, Cu and Se). (**a**) Conventional pork; (**b**) organic pork. (After Zhao Y. et al., 2020b. Meat Sci 165:108129. https://doi.org/10.1016/j.meatsci.2020.108129. With permission)

$P < 0.001$) with contributions of $\delta^{13}C$ and $\delta^{15}N$ is an effective tool for clearly differentiating European beef, including conventional Irish, from American beef (Fig. 2.9). The unmodified conventional beef samples were also significantly different from other European beef samples (MANOVA, $F_{2,32} = 14.8$, $P < 0.001$) on the basis of $\delta^{13}C$ and $\delta^{15}N$. Owing to the application of MANOVA, the combination of stable isotopes such as $\delta^{13}C$, $\delta^{15}N$, $\delta^{13}C$ and $\delta^{34}S$ was useful in distinguishing between conventionally and organically produced Irish beef. According to the authors, the use of stable isotope profiles to authenticate the origin of beef may also be effective at relatively small geographic scales.

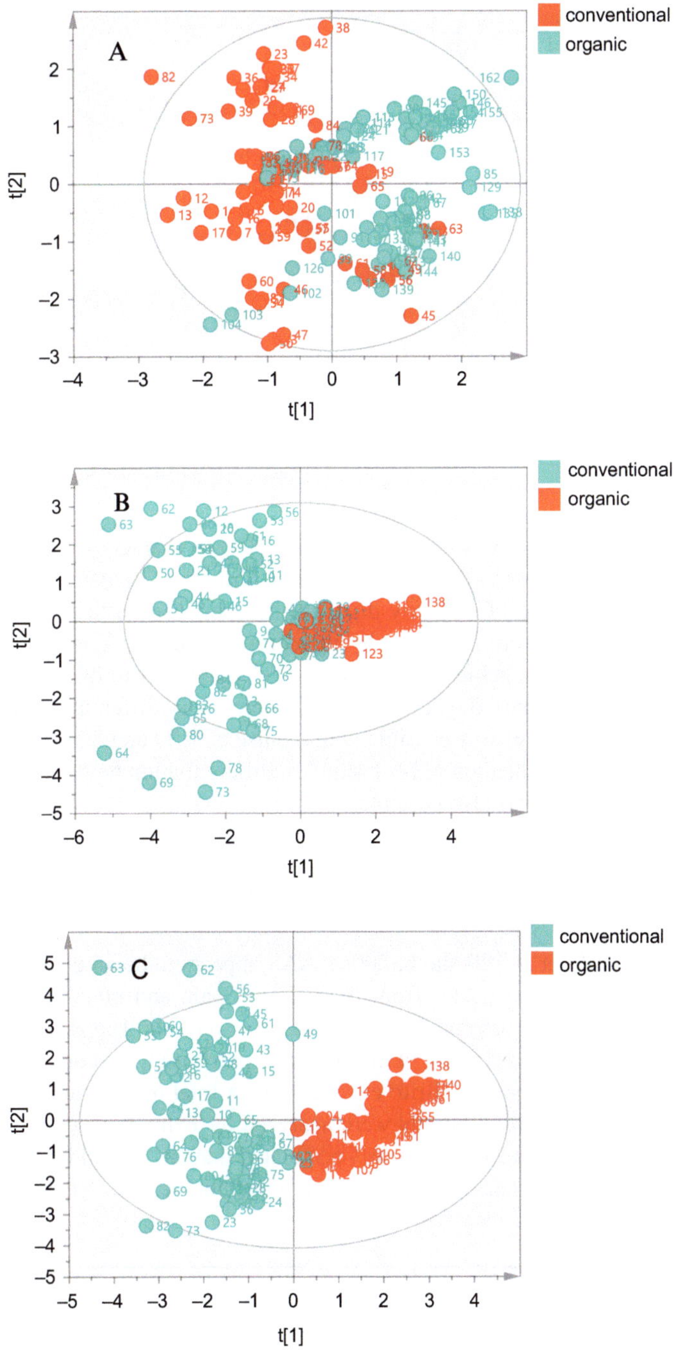

Fig. 2.8 Orthogonal partial least squares discriminant analysis (OPLS–DA) results showing the clustering of all conventional pork and organic pork samples from four areas (Chengde, Hebei Province; Anqing, Anhui Province; Chifeng, Inner Mongolia Autonomous Region; Changchun,

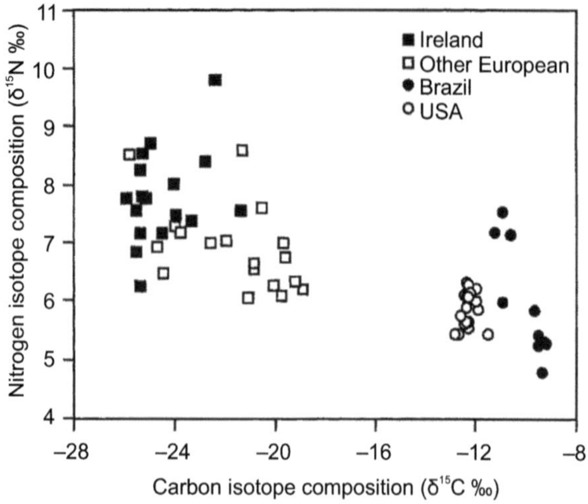

Fig. 2.9 C and N stable isotope ratios in defatted bovine muscle tissue (International study). (After Schmidt O et al., (2005). Food Chem 91:545–549. https://doi.org/10.1016/j.foodchem.2004.08.036. With permission)

EA–IRMS was used by Boner and Förstel (2004) to characterize the stable isotopic composition of beef samples from various farms in Germany and two other countries. The first two PCs clearly distinguished between northern and southern beef samples, which were improved by including the season. Local geographical differentiation was also achieved by applying $\delta^{15}N$ and $\delta^{34}S$ to distinguish among the three farms in Colonia Bay. Furthermore, intercountry differentiation (between Argentina and Germany) was possible with the use of $\delta^{18}O$ and $\delta^{2}H$. In addition to the geographical identification of beef samples, differentiation between organic and conventional farming was also possible.

2.2.3.4 Interregional Differentiation Related to Container Material

According to Brito et al. (2004a, b, 2006), AAS appears to be an effective tool for distinguishing pork liver pastes from those from Spain and other countries with respect to the container material, i.e. glass, plastic or ceramic. The separated sample groups correspond to the country of origin, brand purchased, the container material and the pork liver composites (Brito et al. 2004a). A total of 496 canned pork liver pastes from 20 commercial brands were analyzed by Brito et al. (2004a). On the basis of the multivariate analysis, it was postulated that Cr, Ni and Pb were related to the presence or absence of oxide in the can, but Fe, Cu, Zn and Co were

Fig. 2.8 (continued) Jilin Province) obtained by chemical analysis of (**a**) the ratios of the stable isotopes of carbon, nitrogen, hydrogen and oxygen; (**b**) the concentrations of seven elements (Na, Mg, K, Ca, Fe, Cu and Se); (**c**) the ratios of the stable isotopes of carbon, nitrogen, hydrogen and oxygen combined with the concentrations of seven elements (Na, Mg, K, Ca, Fe, Cu and Se). (After Zhao Y et al., 2020b. Meat Sci 165:108129. https://doi.org/10.1016/j.meatsci.2020.108129. With permission) associated with the kind of can. Moreover, chemometric techniques such as PCA, FA, LDA and PoCu have been applied to classify heat-treated liver pastes with respect to five container types (Brito et al. 2006). For example, the first three PCs explaining 60.5% of the total variance demonstrated that discriminative variables such as Fe and Zn (PC1), Cu and Mn (PC2) and Co, Cr and the factory parameter (PC3) could be used to differentiate samples according to country of origin, animal diet and the manufacturing process, respectively. LDA allowed reliable discrimination between meat pastes delivered in ferrous and nonferrous containers, achieving correct classification with 100% success. In summary, pork liver pastes were clearly categorized into four classes, namely, the origin of the sample, the manufacturer and the effect of the manufacturing process, which were taken into account to verify traceability, which is an important issue for food safety policies (Brito et al. 2006).

2.2.3.5 Identification of the Parameters Responsible for Authentication/ Adulteration

Dixit et al. (2017a, b) studied the possible adulteration of minced beef using appropriate analytical and chemometric approaches. To obtain reference values for Na and K, the authors used the AAS method. PLSR was found to be a useful chemometric technique for building prediction models. Spatial mineral element maps of the minced beef samples were generated using the predicted percentages of Na and K (Dixit et al. 2017a). LIBS was an effective tool for detecting and quantifying Rb via GF–AAS in minced beef and beef liver. PLSR was used to develop the calibration model, yielding an Rc^2 of 0.99 and an RMSEC of 0.05 ppm. Good results with LOO-CV, yielding an Rc^2 of 0.90 and an RMSEC of 0.22 ppm, were obtained. Notably, LIBS appears to be a useful potential monitoring tool for adulteration detection in the meat processing industry. Moreover, chemometrics has been successfully applied for the discrimination of pure offal and offal mixture adulterated beef samples.

According to Velioğlu et al. (2018), a combination of LIBS and ICP–OES with multivariate analysis (PCA and PLS) was used for beef meat and offal samples from Turkey. It was possible to discriminate pure beef and offal samples and to determine the adulteration of the offal mixture. For example, PCA, with the first two PCs accounting for 97.31%, demonstrated that four animal offal types were successfully discriminated from each other (Fig. 2.10). This means that beef meat was separated into beef offal; lung and heart; kidney, spleen and liver; and offal adulterated meat samples. PLS also revealed offal mixture adulteration in beef. It was concluded that verifying that pure offal and offal mixture adulterated beef samples could be achieved successfully.

As reported by Osorio et al. (2011a), clear CDA discrimination of bovine meat corresponding to dietary regimens (pasture/silage–pasture, silage–pasture/concentrate and concentrate plus straw) was achieved with the use of the muscle $\delta^{13}C$ and

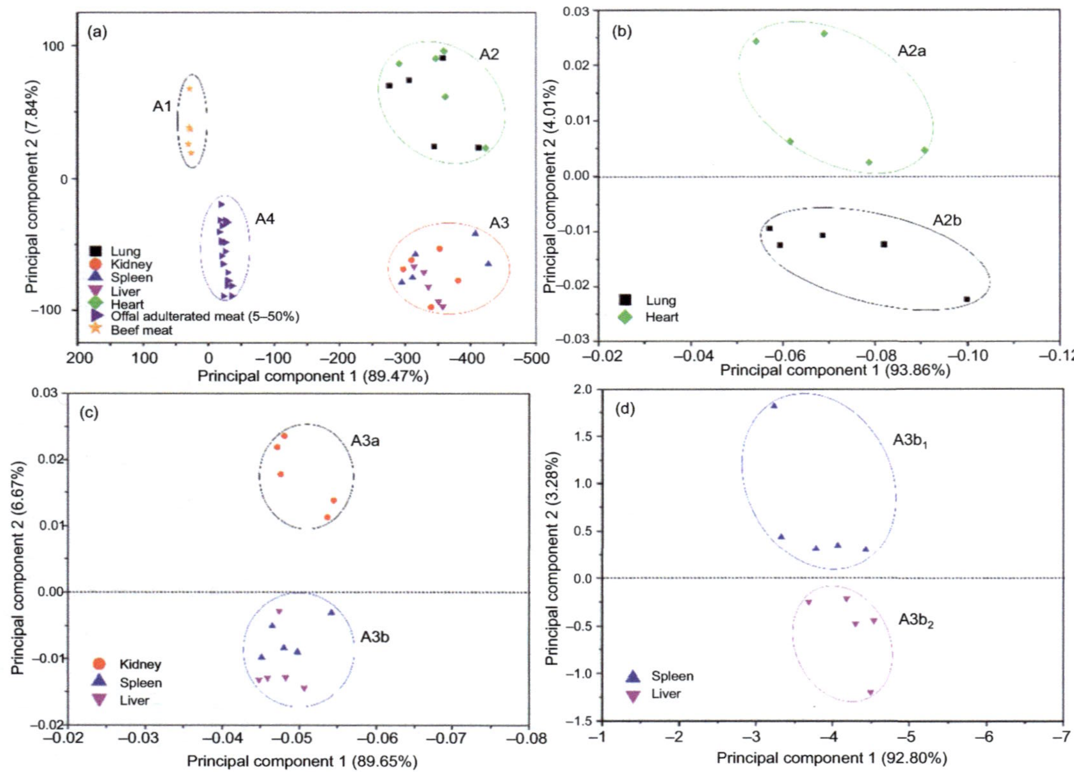

Fig. 2.10 Stages of PCA: (**a**) classification of beef meat (A1), beef offal A2 (lung and heart), A3 (kidney, spleen and liver) and A4 (offal adulterated meat sample); (**b**) separation of lung samples from heart samples; (**c**) separation of kidney samples from spleen and liver samples; (**d**) separation of spleen samples from liver samples. (After Velioğlu HM et al. (2018). Meat Sci 138:28–33. https://doi.org/10.1016/j.meatsci.2017.12.003. With permission)

$\delta^{34}S$ determined by EA–IRMS. It was found that 86.5% of the original grouped cases were correctly classified, and 86.5% of the cross-validated grouped cases were correctly categorized. Therefore, both the stable isotope ratios of meat processed by CDA appear to be useful indicators for the authentication of beef muscles. According to Osorio et al. (2011b), the $\delta^{13}C$, $\delta^{15}N$, $\delta^{2}H$ and $\delta^{34}S$ values of beef are appropriate indicators of the geographical origin of beef from nine European and non-European countries. A model was optimized using the four isotope ratios contributing to 82.2% correct classification in the cross-validation. The obtained results may be useful for developing a robust traceability model for verifying the authenticity of meat products, with eventual fraudulent mislabelling.

According to Perini et al. (2013), the use of a combination of EA–IRMS, HPLC, GC/IT–MS and reflectance spectral measurements with chemometric tools (PCA, PLS–DA and SIMCA) resulted in distinguishing Italian PDOs from Spanish hams. PCA with the first four PCs corresponding to 91.7% of the total variance revealed that PC1 discriminated the ham types according to their origin, i.e. all the Italian ham samples were separated from all the Spanish samples. Furthermore, each of the last three PCs allowed the separation of the ham samples produced following an alternative procedure in the same region. For example, along PC2, "Bellota" hams were segregated from "Campo" ones produced in Extremadura; PC3 distinguished mid-season Emilia samples from those of the end-season hams and PC4 separated Friuli hams produced in the local breed from those provided by the industrial hybrid.

LIBS, as a helpful tool, was also applied by Sezer et al. (2021) for the identification of three Turkish meat species (beef, chicken and pork) using bulk proteins and protein fractions, namely, actin and myosin. PCA and PLS analyses were performed to determine the adulteration ratio successfully. Both PCA and HCA were used as exploratory methods. Moreover, classification methods, i.e. k-NN, SIMCA, PLSDA and ANN, were used to assess the data. PCA and HCA did not efficiently discriminate the samples studied, whereas KNN and ANN achieved reliable performance for both categories, with 100% correct prediction.

Matera et al. (2014) noted that FAAS integrated with chemometric models such as PCA, HCA, PL-SDA, SIMCA, kNN and MLP-ANN appeared to be useful tools for classifying inspected and artisanal pork sausages from Brazil. SIMCA and PLSDA presented performances of 100% and 85.7% for inspected and artisanal sausages, respectively. Notably, SIMCA, PLSDA and ANN, which are based on the moisture and fat contents, may achieve the highest discriminative power. This means that the use of multivariate techniques allows accurate assessment of the quality of processed foods, including pork sausages. According to the authors, the methodology used can be useful for health agencies because they are small- and medium-sized establishments that sell inspected products as artisanal products.

2.3 Conclusions

The most applicable analytical methods/techniques, including IRMS and ICP–MS combined with multivariate techniques, are useful in food authentication as well as identification of adulteration/mislabelling and identification of food safety. The adequate application of analytical and chemometric tools for the evaluation of the nutritive and health qualities of meat and its products has a key role in their categorization with respect to geographical provenance, intertype differentiation, the influence of the environmental impact and other features. Moreover, it will be possible to detect fraud to preserve the reputation of brands that sell high-quality original products. The application of chemometrics allows a better understanding of the distributions of different assortments of minor and major elements, as well as stable isotopes in meat and its products.

References

Ahmad Nizar NN, Nazrim Marikkar JM, Hashim DM (2013) Differentiation of lard, chicken fat, beef fat and mutton fat by GCMS and EA-IRMS techniques. J Oleo Sci 62:459–464. https://doi.org/10.5650/jos.62.459

Arvanitoyannis IS, van Houwelingen-Koukaliaroglou M (2003) Implementation of chemometrics for quality control and authentication of meat and meat products. Crit Rev Food Sci Nutr 43:173–218. https://doi.org/10.1080/10408690390826482

Bahar B, Schmidt O, Moloney AP, Scrimgeour CM, Begley IS, Monahan FJ (2008) Seasonal variation in the C, N and S stable isotope composition of retail organic and conventional Irish beef. Food Chem 106:1299–1305. https://doi.org/10.1016/j.foodchem.2007.07.053

Bandoniene D, Walkner C, Ringdorfer F, Meisel T (2020) Authentication of meat and dairy products using rare earth element labeling and detection by solution based and laser ablation ICP-MS. Food Res Int 132:109106. https://doi.org/10.1016/j.foodres.2020.109106

Baroni MV, Podio NS, Badini RG, Inga M, Ostera HA, Cagnoni M, Gallegos E, Gautier E, Peral-García P, Hoogewerff J, Wunderlin DA (2011) How much do soil and water contribute to the composition of meat? A case study: meat from three areas of Argentina. J Agric Food Chem 59:11117–11128. https://doi.org/10.1021/jf2023929

Bilge G, Velioglu HM, Sezer B, Eseller KE, Boyaci IH (2016) Identification of meat species by using laser-induced breakdown spectroscopy. Meat Sci 119:118–122. https://doi.org/10.1016/j.meatsci.2016.04.035

Boner M, Förstel H (2004) Stable isotope variation as a tool to trace the authenticity of beef. Anal Bioanal Chem 378:301–310. https://doi.org/10.1007/s00216-003-2347-6

Brito G, Peña-Méndez EM, Novotná K, Díaz C, Garcíaet FJ (2004a) Differentiation of heat-treated pork liver pastes according to their metal content using multivariate data analysis. Eur Food Res Technol 218:584–588. https://doi.org/10.1007/s00217-004-0909-2

Brito G, Novotná K, Peña-Méndez EM, Díaz C, García FJ (2004b) Correlation of heavy metal concentrations with various factors in canned liver paste products using multivariate statistical strategies. J Food Protect 67:1927–1932. https://doi.org/10.4315/0362-028x-67.9.1927

Brito G, Andrade JM, Havel J, Díaz C, García FJ, Peña-Méndez EM (2006) Classification of some heat-treated liver pastes according to container type, using heavy metals content and manufacturer's data, by principal components analysis and potential curves. Meat Sci 74:296–302. https://doi.org/10.1016/j.meatsci.2006.03.024

Bulska E (2010) Quality assurance and quality control of analytical results. In: Namieśnik J, Szefer P (eds) Analytical measurements in aquatic environments. CRC Press, Taylor & Francis Group, Boca Raton, pp 389–397. https://doi.org/10.1201/9781420082692

Camin F, Bontempo L, Heinrich K, Horacek M, Kelly SD, Schlicht C, Thomas F, Monahan FJ, Hoogewerff J, Rossmann A (2007) Multi-element (H, C, N, S) stable isotope characteristics of lamb meat from different European regions. Anal Bioanal Chem 389:309–320. https://doi.org/10.1007/s00216-007-1302-3

Camin F, Bontempo L, Perini M, Piasentier E (2016) Stable isotope ratio analysis for assessing the authenticity of food of animal origin. Compr Rev Food Sci Food Saf 15:868–877. https://doi.org/10.1111/1541-4337.12219

Capar SG, Szefer P (2012) Determination and speciation of trace elements in foods. In: Ötleş S (ed) Methods of analysis of food components and additions. CRC Press, Taylor & Francis, Boca Raton, pp 111–158. https://doi.org/10.1201/b11218

Cobos Á, Díaz O (2015) Chemical composition of meat and meat products. In: Cheung PCK, Mehta BM (eds) Handbook of food chemistry. Springer, Berlin, pp 471–510. https://doi.org/10.1007/978-3-642-41609-5_6-1

Cristea G, Voica C, Feher I, Puscas R, Magdas DA (2022) Isotopic and elemental characterization of Romanian pork meat in corroboration with advanced chemometric methods: a first exploratory study. Meat Sci 189:108825. https://doi.org/10.1016/j.meatsci.2022.108825

De Nadai Fernandes EA, Sarriés GA, Bacchi MA, Mazola YT, Gonzaga CL, Sarriés SRV (2020) Trace elements and machine learning for Brazilian beef traceability. Food Chem 333:127462. https://doi.org/10.1016/j.foodchem.2020.127462

Dehelean A, Cristea G, Puscas R, Hategan AR, Magdas DA (2022) Assigning the geographical origin of meat and animal-rearing system using isotopic and elemental fingerprints. Appl Sci 12:12391. https://doi.org/10.3390/app122312391

Dehelean A, Feher I, Romulus P, Magdas DA, Covaciu FD, Kasza AM, Curean V, Cristea G (2023) Influence of geographical origin on isotopic and elemental compositions of pork meat. Foods 12:4271. https://doi.org/10.3390/foods12234271

Dixit Y, Casado-Gavalda MP, Cama-Moncunill R, Cama-Moncunill X, Markiewicz-Keszycka M, Cullen PJ, Sullivan C (2017a) Laser induced breakdown spectroscopy for quantification of sodium and potassium in minced beef: a potential technique for detecting beef kidney adulteration. Anal Methods 9:22. https://doi.org/10.1039/C7AY00757D

Dixit Y, Casado-Gavalda MP, Cama-Moncunill R, Markiewicz-Keszycka M, Cama-Moncunill X, Cullen PJ, Sulliva C (2017b) Quantification of rubidium as a trace element in beef using laser induced breakdown spectroscopy. Meat Sci 130:47–49. https://doi.org/10.1016/j.meatsci.2017.03.013

Franke BM, Gremaud G, Hadorn R, Kreuzer M (2005) Geographic origin of meat-elements of an analytical approach to its authentication. Eur Food Res Technol 221:493–503. https://doi.org/10.1007/s00217-005-1158-8

Franke BM, Haldimann M, Reimann J, Baumer B, Gremaud G, Hadorn R, Bosset J-O, Kreuzer M (2007) Indications for the applicability of element signature analysis for the determination of the geographic origin of dried beef and poultry meat. Eur Food Res Technol 225:501–509. https://doi.org/10.1007/s00217-006-0446-2

Franke BM, Koslitz S, Micaux F, Piantini U, Maury V, Pfammatter E, Wunderli S, Gremaud G, Bosset J-O, Hadorn R, Kreuzer M (2008a) Tracing the geographic origin of poultry meat and dried beef with oxygen and strontium isotope ratios. Eur Food Res Technol 226:761–769. https://doi.org/10.1007/s00217-007-0588-x

Franke BM, Hadorn R, Bosset JO, Gremaud G, Kreuzer M (2008b) Is authentication of the geographic origin of poultry meat and dried beef improved by combining multiple trace element and oxygen isotope analysis? Meat Sci 80:944–947. https://doi.org/10.1016/j.meatsci.2008.03.018

Franke BM, Haldimann M, Gremaud G, Bosset J-O, Hadorn R, Kreuzer M (2008c) Element signature analysis: its validation as a tool for geographic authentication of the origin of dried beef and poultry meat. Eur Food Res Technol 227:701–708. https://doi.org/10.1007/s00217-007-0776-8

Grembecka M, Szefer P (2011) Metals and metalloids in foods—essentiality, toxicity, applicability. In: Medina DA, Laine AM (eds) Food quality: control, analysis and consumer concerns. Nova Science Publishers, Hauppauge, pp 1–60. ISBN 10: 1611229170; ISBN 13: 9781611229172

Grembecka M, Szefer P (2016) Elemental trace analysis in studies of food products. In: Baranowska I (ed) Handbook of trace analysis—fundamentals and applications. Springer, Berlin, pp 203–239. https://doi.org/10.1007/978-3-319-19614-5

Guo BL, Wei YM, Pan JR, Li Y (2010) Stable C and N isotope ratio analysis for regional geographical traceability of cattle in China. Food Chem 118:915–920. https://doi.org/10.1016/j.foodchem.2008.09.062

Guo X, Huang F, Zhang H, Zhang C, Hu H, Chen W (2016) Classification of traditional Chinese pork bacon based on physicochemical properties and chemometric techniques. Meat Sci 117:182–186. https://doi.org/10.1016/j.meatsci.2016.02.008

Heaton K, Kelly SD, Hoogewerff J, Woolfe M (2008) Verifying the geographical origin of beef: the application of multi-element isotope and trace element analysis. Food Chem 107:506–515. https://doi.org/10.1016/j.foodchem.2007.08.010

Horacek M, Min JS (2010) Discrimination of Korean beef from beef of other origin by stable isotope measurements. Food Chem 121:517–520. https://doi.org/10.1016/j.foodchem.2009.12.018

Islam MH, Khaliduzzaman A (2022) Chemometric for spectroscopy-based egg and poultry research. In: Khaliduzzaman A (ed) Informatics in poultry production. Springer Singapore, Singapore, pp 111–126. ISBN: 978-981-19-2555-9; eBook ISBN: 978-981-19-2556-6. https://doi.org/10.1007/978-981-19-2556-6

Jiang D, Du L, Guo Y, Ma J, Li X, Han L, Xu Y, Qian Y (2020) Potential use of stable isotope and multi-element analyses for regional geographical traceability of bone raw materials for gelatin production. Food Anal Methods 13:762–769. https://doi.org/10.1007/s12161-019-01687-1

Kim KS, Kim JS, Hwang IM, Jeong IS, Khan N, Lee SI, Jeon DB, Song YH, Kim KS (2013) Application of stable isotope ratio analysis for origin authentication of pork. Korean J Food Sci Anim Resour 33:39–44. https://doi.org/10.5851/kosfa.2013.33.1.39

Kim JS, Hwang IM, Lee GH, Park YM, Choi JY, Jamila N, Khan N, Kim KS (2017) Geographical origin authentication of pork using multi-element and multivariate data analyses. Meat Sci 123:13–20. https://doi.org/10.1016/j.meatsci.2016.08.011

Liu X, Guo B, Wei Y, Shi J, Sun S (2013) Stable isotope analysis of cattle tail hair: a potential tool for verifying the geographical origin of beef. Food Chem 140:135–140. https://doi.org/10.1016/j.foodchem.2013.02.020

Matera JA, Cruz AG, Raices RSL, Silva MC, Nogueira LC, Quitério SL, Cavalcanti RN, Freiras MQ, Conte Júnior CA (2014) Discrimination of Brazilian artisanal and inspected pork sausages: application of unsupervised, linear and non-linear supervised chemometric methods. Food Res Int 64:380–386. https://doi.org/10.1016/j.foodres.2014.07.003

Monahan FJ, Moloney AP, Osorio MT, Röhrle FT, Schmidt O, Brennan L (2012) Authentication of grass-fed beef using bovine muscle, hair or urine. Trends Food Sci Technol 28:69–76. https://doi.org/10.1016/j.tifs.2012.05.005

Monahan FJ, Schmidt O, Moloney AP (2018) Meat provenance: authentication of geographical origin and dietary background of meat. Meat Sci 144:2–14. https://doi.org/10.1016/j.meatsci.2018.05.008

Nakashita R, Suzuki Y, Akamatsu F, Iizumi Y, Korenaga T, Chikaraishi Y (2008) Stable carbon, nitrogen, and oxygen isotope analysis as a potential tool for verifying geographical origin of beef. Anal Chim Acta 617:148–152. https://doi.org/10.1016/j.aca.2008.03.048

Namieśnik J, Szefer P (2010) Analytical measurements in aquatic environments. Taylor & Francis Group, LLC, Boca Raton. https://doi.org/10.1201/9781420082692

Nho EY, Choi JY, Lee CM, Dang YM, Khan N, Jamila N, Kim KS (2019) Origin authentication of pork fat via elemental composition, isotope ratios, and multivariate chemometric analyses. Anal Lett 52:1445–1461. https://doi.org/10.1080/00032719.2018.1545133

Osorio MT, Moloney AP, Schmidt O, Monahan FJ (2011a) Beef authentication and retrospective dietary verification using stable isotope ratio analysis of bovine muscle and tail hair. J Agric Food Chem 59:3295–3305. https://doi.org/10.1021/jf1040959

Osorio MT, Moloney AP, Schmidt O, Monahan FJ (2011b) Multielement isotope analysis of bovine muscle for determination of international geographical origin of meat. J Agric Food Chem 59:3285–3294. https://doi.org/10.1021/jf1040433

Osorio MT, Downey G, Moloney AP, Röhrle FT, Luciano G, Schmidt O, Monahan FJ (2013) Beef authentication using dietary markers: chemometric selection and modelling of significant beef biomarkers using concentrated data from multiple analytical methods. Food Chem 141:2795–2801. https://doi.org/10.1016/j.foodchem.2013.05.118

Park YM, Lee CM, Hong JH, Jamila N, Khan N, Jung J-H, Jung Y-C, Kim KS (2018) Origin discrimination of defatted pork via trace elements profiling, stable isotope ratios analysis, and multivariate statistical techniques. Meat Sci 143:93–103. https://doi.org/10.1016/j.meatsci.2018.04.012

Pereira V, Miranda M, Sierra J, Benedito JL, López-Alonso M (2020) Toxic and essential trace element concentrations in different tissues of extensively reared sheep in northern Spain. J Food Compos Anal 96:103709. https://doi.org/10.1016/j.jfca.2020.103709

Perini M, Camin F, Bontempo L, Rossmann A, Piasentier E (2009) Multielement (H, C, N, O, S) stable isotope characteristics of lamb meat from different Italian regions. Rapid Commun Mass Spectrom 23:2573–2585. https://doi.org/10.1002/rcm.4140

Perini M, Camin F, Sánchez del Pulgar J, Piasentier E (2013) Effect of origin, breeding and processing conditions on the isotope ratios of bioelements in dry-cured ham. Food Chem 136:1543–1550. https://doi.org/10.1016/j.foodchem.2012.06.084

Piasentier E, Valusso R, Camin F, Versini G (2003) Stable isotope ratio analysis for authentication of lamb meat. Meat Sci 64:239–247. https://doi.org/10.1016/S0309-1740(02)00183-3

Qie M, Zhang B, Li Z, Zhao S, Zhao Y (2021) Data fusion by ratio modulation of stable isotope, multi-element, and fatty acids to improve geographical traceability of lamb. Food Control 120:107549. https://doi.org/10.1016/j.foodcont.2020:107549

Renou J-P, Bielicki G, Deponge C, Gachon P, Micol D, Ritz P (2004) Characterization of animal products according to geographic origin and feeding diet using nuclear magnetic resonance and isotope ratio mass spectrometry. Part II: beef meat. Food Chem 86:251–256. https://doi.org/10.1016/j.foodchem.2003.08.021

Sacco D, Brescia MA, Buccolieri A, Caputi Jambrenghi A (2005) Geographical origin and breed discrimination of Apulian lamb meat samples by means of analytical and spectroscopic determinations. Meat Sci 71:542–548. https://doi.org/10.1016/j.meatsci.2005.04.038

Schmidt O, Quilter JM, Bahar B, Molone AP, Scrimgeour CM, Begley IS, Monahan FJ (2005) Inferring the origin and dietary history of beef from C, N and S stable isotope ratio analysis. Food Chem 91:545–549. https://doi.org/10.1016/j.foodchem.2004.08.036

Sezer B, Bjelak A, Velioglu HM, Boyaci IH (2021) Protein based evaluation of meat species by using laser induced breakdown spectroscopy. Meat Sci 172:108361. https://doi.org/10.1016/j.meatsci.2020.108361

Shin W-J, Choi S-H, Ryu J-S, Song B-Y, Song J-H, Park S, Min J-S (2018) Discrimination of the geographic origin of pork using multi-isotopes and statistical analysis. Rapid Commun Mass Spectrom 32:1843–1850. https://doi.org/10.1002/rcm.8251

Sun S, Guo B, Wei Y, Fan M (2011) Multi-element analysis for determining the geographical origin of mutton from different regions of China. Food Chem 124:1151–1156. https://doi.org/10.1016/j.foodchem.2010.07.027

Sun S, Guo B, Wei Y (2016) Origin assignment by multi-element stable isotopes of lamb tissues. Food Chem 213:675–681. https://doi.org/10.1016/j.foodchem.2016.07.013

Szefer P (2002) Metals, metalloids and radionuclides in the Baltic Sea ecosystem. Elsevier Science B.V., Amsterdam. ISBN 10: 0444503528

Szefer P (2003) Application of chemometric techniques in analytical evaluation of biological and environmental samples. In: Namieśnik J, Chrzanowski W, Żmijewska P (eds) New horizons and challenges in environmental analysis and monitoring. CEEAM, Gdańsk, pp 355–388. ISBN 8391908100

Szefer P (2007) Chemometric techniques in analytical evaluation of food quality. In: Szefer P, Nriagu J (eds) Mineral components in foods. CRC Press, Taylor & Francis, Boca Raton, pp 69–121. https://doi.org/10.1201/9781420003987

Szteke B (2007) Criteria of evaluation of food elements analysis data. In: Szefer P, Nriagu J (eds) Mineral components in foods. CRC Press, Taylor & Francis, Boca Raton, pp 51–69. https://doi.org/10.1201/9781420003987

Tsakovski S, Simeonov V (2010) Chemometrics as a tool for treatment processing of multiparametric analytical data sets. In: Namieśnik J, Szefer P (eds) Analytical measurements in aquatic environments. CRC Press, Taylor & Francis Group, Boca Raton, pp 369–387. https://doi.org/10.1201/9781420082692

Velioğlu HM, Sezer B, Bilge G, Baytur S, Boyacı IH (2018) Identification of offal adulteration in beef by laser induced breakdown spectroscopy (LIBS). Meat Sci 138:28–33. https://doi.org/10.1016/j.meatsci.2017.12.003

Yanagi Y, Hirooka H, Oishi K, Choumei Y, Hata H, Arai M, Kitagawa M, Gotoh T, Inada S, Kumagai H (2012) Stable carbon and nitrogen isotope analysis as a tool for inferring beef cattle feeding systems in Japan. Food Chem 134:502–506

Yao M, Rao G, Huang L, Liu M, Yang H, Chen J, Chen T (2017) Simultaneous analysis of Cr and Pb in contaminated pork by laser-induced breakdown spectroscopy. Appl Opt 56:8148–8153. https://doi.org/10.1364/AO.56.008148

Zhao Y (2016) Meat authenticity. In: Cummins EJ, Lyng JG (eds) Emerging technologies in meat processing: production, processing and technology. Wiley, New York, pp 391–406. https://doi.org/10.1002/9781118350676.ch15

Zhao S, Zhao Y (2021) Application and preparation progress of stable isotope reference materials in traceability of agricultural products. Crit Rev Anal Chem 51:742–753. https://doi.org/10.1080/10408347.2020.1768359

Zhao Y, Zhang B, Chen G, Chen A, Yang S, Ye Z (2013) Tracing the geographic origin of beef in China on the basis of the combination of stable isotopes and multielement analysis. J Agric Food Chem 61:7055–7060. https://doi.org/10.1021/jf400947y

Zhao Y, Zhang B, Guo B, Wang D, Yang S (2016) Combination of multi-element and stable isotope analysis improved the traceability of chicken from four provinces of China. CyTA J Food 14:163–168. https://doi.org/10.1080/19476337.2015.1057235

Zhao S, Zhang H, Zhang B, Xu Z, Chen A, Zhao Y (2020a) A rapid sample preparation method for the analysis of stable isotope ratios of beef samples from different countries. Rapid Commun Mass Spectrom 34:e8795. https://doi.org/10.1002/rcm.8795

Zhao Y, Tu T, Tang X, Zhao S, Qie M, Chen A, Yang S (2020b) Authentication of organic pork and identification of geographical origins of pork in four regions of China by combined analysis of stable isotopes and multi-elements. Meat Sci 165:108129. https://doi.org/10.1016/j.meatsci.2020.108129

Chapter 3
Fish Flesh

Contents

3.1 Introduction

Fish protein has a high nutritional value because it is rich in many bioactive peptides and essential amino acids. Lipids are responsible for the nutritional value of fish because of the presence of long-chain PUFAs that consist of omega-3 fatty acids, particularly Eicosapentaenoic Acid (EPA) and Docosahexaenoic Acid (DHA). The American Heart Association has recommended at least two servings of fish per week to reduce cardiovascular disease risk. Most micronutrients distinguished by high bioavailability occur in fish at concentrations between 0.4% and 1.5%. Fish is enriched in high-nutritional value minerals, including Ca, Fe, Zn, Se, I, P and K, in widely varying quantities. Moreover, fish is a rich source of vitamins A and D and a good source of B-group vitamins (Ali et al. 2022).

Many researchers, e.g. Julshamn and Grahl-Nielsen (1996), Molkentin et al. (2015), Ye et al. (2011), Yamashita et al. (2006), Li et al. (2015), Ahmed et al. (2019), Rahman et al. (2012) and Marpaung et al. (2020), have recognized the environmental parameters differentiating diverse assortments of worldwide populations of fish. The authenticity

---

Supplementary Information The online version contains supplementary material available at https://doi.org/10.1007/978-3-031-99687-0_3.

P. Szefer, *Chemometrics in Food Control Based on Elemental Analysis*, https://doi.org/10.1007/978-3-031-99687-0_3

of the geographical origins of animals, including fish and their products, in view of their mineral composition, has been verified on an international scale, e.g. by Yamashita et al. (2006), Iguchi et al. (2013), Chaguri et al. (2015) and Kim et al. (2015a, b).

Szefer's mini-review (2013) revealed that fish are considerably exposed to certain pollutants that enter food, e.g. MeHg, BFR and OTs. The most important toxic compounds are MeHg and dioxin-like compounds, especially for individuals who consume large amounts of seafood. The PTWI calculated for these critical compounds may be exceeded even if other sources of dietary exposure are excluded. Early life stages are more susceptible to pollutants since they are exposed to the cumulative total amount of toxic compounds in the mother's body. Women can reduce MeHg levels in their bodies by limiting their dietary intake of seafood in the months preceding and during pregnancy, in contrast to other pollutants such as PCDD/Fs and DL-PCBs. In the latter cases, their body levels can take many years to decrease considerably. Among the fish that are widely available and frequently consumed in the EU, wild tuna fish have the highest MeHg content. However, the highest levels of PCDD/Fs and DL-PCBs have been found in wild herring and salmon, which are mostly farmed (EFSA 2005, 2008). Szefer (2013) reported that a decrease in Pb concentrations in herring liver has been observed in many Baltic areas. Studies by Polak–Juszczak (2009, 2010) based on concentration data obtained for more than 1400 samples of herring, sprat, cod and flounder collected from 1994 to 2003 (and, in the case of flounder, from 1996 to 2003), supported this decreasing trend. A similar reduction was observed for hepatic Pb in Baltic cod from 1994 to 2010 (Polak-Juszczak 2013). This long-term trend was attributed to the increase in the use of lead-free fuel and the wider application of cleaner production technologies in the HELCOM region and European countries (HELCOM 2007).

The concentrations of mineral components in the edible tissues of fish appear to be affected by various factors, e.g. species, size, age, life history, sex, sexual maturity, food source, water chemistry, salinity, climate, contaminants, sampling procedures and analytical techniques (Alasalvar et al. 2002; Roy and Lall 2006; Yamashita et al. 2006).

Substantial progress in analytical chemistry has contributed to the development of innovative and efficient analytical methods and techniques (Grembecka and Szefer 2011; Capar and Szefer 2012). The reliability and correctness of analytical measurements should be checked with quality assurance tests that use CRMs (Konieczka and Namieśnik 2009; Bulska 2010; Grembecka and Szefer 2016). This has been the key approach to process chemometrically exclusively validated analytical data using techniques such as ANOVA, correlation and regression analysis, PCA, FA, CA, HCA, ANN and PLS-DA. (e.g. Forina et al. 2009; Marini 2013; Szefer 2002, 2003, 2007). Multivariate analysis has enabled a more complete interpretation of the data and exploitation of the information contained therein. It can be useful for exploratory and qualitative/quantitative modelling purposes. Since modelling is generally performed for predictive applications, a thorough model validation is always needed (Oliveri et al. 2020). Chemometric techniques can be grouped into unsupervised multivariate statistical methods and supervised (learning) methods (Tsakovski and Simeonov 2010).

According to Arechavala-Lopez et al. (2013), appropriate methods can be used to identify differences between chemical profiles, genetics, morphology and organoleptic parameters. Analytical techniques, as previously mentioned, have also been used to differentiate farmed from wild fish. The most common techniques have been based on the differentiation of the SIR, FA and MEC values related to fish. These studies generally address food science and product quality. A wide assortment of identification tools applied in the quantification of the correct provenance of capture and proper labelling of marketed fish is expected. External appearance and morphometry are required for quick evaluation and can be attained with satisfactory accuracy and at low cost. Therefore, these methods appear to be sufficient for identifying large and recent accidental escape events, are applicable in fisheries studies and can be used to ensure that the differentiation of wild and farmed fish occurs in the marketplace. Techniques based on differences in chemical or genetic composition are especially applicable for environmental monitoring owing to their relatively high accuracy and ability to detect escape events long after an escape incident. The differentiation of commercial brands and breeding and farming conditions is an important factor in the characterization of edible aquatic organism quality.

Rodushkin et al. (2007) reported that, on the basis of the MEC, it could be possible to assess chemometrically specific features of fish, including discrimination between wild and farmed fish (Li et al. 2015). Fish scales have been analyzed (Iguchi et al. 2013; Liu et al. 2020) because their MECs significantly differ between wild and farmed fish. Chemometric processing of mineral compositions related to vendace caviar has been performed by Rodushkin et al. (2007).

3.2 Chemometric Assessment of Fish Flesh and Its Products

3.2.1 Geographical Differentiation

3.2.1.1 Asiatic Countries

Ye et al. (2011) combined ICP–MS with PCA, CA and CDA and proposed that interlake variations in elements in Chinese large icefish are more substantial than intralake variations are, although the latter are sufficient. PCA clearly discriminated the fish studied with respect to geographical origin; i.e. *P. chinensis* within the four sampling areas were distinctly differentiated without overlap. Variables such as Mg, Al, Ca, Mn, Cu and Zn were highly correlated with PC1, whereas K and Se were highly correlated with PC2. In addition, CDA, which uses the same elements as PCA does, was used to correctly identify fish as originating from specific sampling areas in Taihu Lake. This was also achieved chemometrically by the EFA method. This approach has the potential to identify source populations of species and fishery products in the marketplace.

Iguchi et al. (2013) integrated ICP–MS with LDA to investigate the intermuscular bones of grilled eel fillets originating from three Asiatic countries. The country-of-origin discrimination models based on V, Ni, Sr, Ba, Pb and U were constructed with the use of grilled eel fillets. It has been possible to identify significant differences between Japanese eels and eels imported from China and Taiwan. Both the Japan–China and Japan–Taiwan discriminant models showed a 100% probability of detecting fish from Japanese waters. In the case of both models for eel samples of Chinese and Taiwanese origin, the probabilities of detection in the two models were 84.6% for samples of Chinese origin and 100% for samples of Taiwanese origin. Therefore, the discriminant models used are effective for tracing the geographical provenance of eel, which means that the fish were processed as grilled eel fillets.

3.2.1.2 Countries Across Multiple Continents

The applicability of EA–SIMS in investigations of the geographical origin of three marketable commercial fish, i.e. mackerel, yellow croaker and pollock (referring to four species), has been reported by Kim et al. (2015b). On the basis of one-way ANOVA and particularly Tukey's post hoc test, $\delta^{13}C$ and especially $\delta^{15}N$ appeared to be indicators used to authenticate the geographical provenance of fish, i.e. mackerel from the South Sea (Korea), Busan (Korea), Australia, China, Norway, Oman (western Arabian Sea), Italy (northern Adriatic Sea), Japan and China (Yellow Sea); yellow croaker (*Larimichthys polyactis*) from the South Sea, Southwest Sea, Gunsan and Mokpo (Korea) and China and pollock from Japan, Russia, the western Bering Sea (Russia) and the eastern Bering Sea, the Bering Strait and Alaska (USA). Kurle et al. (2011) noted that the EA–IRMS data corresponded to numerous species of fish, i.e. arrowtooth flounder, eulachon, Pacific cod, Pacific herring, salmon (Pink, Coho and Chinook), sand lance, walleye pollock, yellow Irish lord, two species of squid, arrowtooth flounder, Atka mackerel, Pacific cod and walleye pollock from the Gulf of Alaska and Bering Sea were processed by ANOVA and subsequent Tukey's pairwise comparison tests. More significant differences than similarities in the comparison of $\delta^{13}C$ and $\delta^{15}N$ values were obtained for geographic locations. The classification of the fish according to their size and muscle samples appeared to be useful for the geographical differentiation of the fish collected from these two geographical regions. However, owing to the small number of samples studied, the authors strongly encourage further analysis of stable isotope ratios for animals in marine ecosystems related to temporal and spatial differentiation.

Rodushkin et al. (2007) used ICP–SFMS and MC–ICP–MS methods to identify the origin of vendace and whitefish caviars from brackish and fresh water. Importantly, elemental ratios such as Sr/Ca, Sr/Mg and Sr/Ba were the most significant alternatives for caviars. Although multielemental analysis appears to be useful for evaluating the geographical provenance of vendace caviar samples, the most useful discrimination between caviars from brackish and fresh water sources (Sweden, Finland and the USA) can be achieved using the $^{87}Sr/^{86}Sr$ ratio. Therefore, Sr isotope ratio measurements have more potential for establishing geographical origin.

3.2.2 Interspecies and/or Interregional Differentiation

Mielcarek et al. (2020, 2022) reported that the elemental composition of freshwater fish, which consisted of raw fish, smoked fish and fish pickled in spirit vinegar, was determined. Among the analytical methods, the combination of FAAS and ETA–AAS with CA, DA and PCA (Mielcarek et al. 2020) and the combination of AAS–AMA, ET–AAS and ICP–MS with CA and DA were applied to quantify the concentrations of Zn, Cu and Se and As, Cd, Hg and Pb, respectively. CA enabled the differentiation of European eel (raw and smoked) with the most discriminatory variables, such as the Zn content, whereas on the basis of the Cu and Se levels, pickled common whitefish were discriminated from other species and types of fish products. Chemometric techniques such as CA and DA were helpful in discriminating and classifying the selected species of fish and fish products. The model for pickled fish classified the studied cases poorly, with 56% accuracy, whereas the results obtained for smoked fish achieved a classification accuracy of 70% for the studied cases. The classification correctness reached 100% for common whitefish and 90% for pike-perch, whereas for the remaining species, the correctness of classification was lower.

In Rampazzo et al. (2020), IRMS, GC–MS/MS and morphological and DNA analyses were used to quantify variables such as $\delta^{13}C$, $\delta^{15}N$, 12 different fatty acids and extracted DNA, allowing the differentiation of fish species (*Thunnus thynnus*, *T. alalunga*, *Auxis rochei* and *Scomber scombrus* belonging to the Scombridae family) and their geographical origins. Both stable isotope ratios allowed the confirmation of various fish diets and trophic positions, whereas FA analysis indicated that unsaturated fatty acids prevailed over saturated fatty acids with interspecies and interregional differentiation. The PCA with the first two PCs corresponding to 55.63% and 70.92% achieved good discrimination among species and their geographical catching areas, respectively. This approach could represent a promising tool for identifying commercial fish and their origin to guarantee their authenticity.

3.2.3 Differentiation of Other Parameters

3.2.3.1 Interregional and Intersex Differentiation

Yellowtail flounder (muscle, liver and gonad) from the Northwest Atlantic was studied to characterize its elemental composition. A biplot of scores was constructed on the basis of the first two PCs, which accounted for 49% of the total variance. It has been possible to differentiate fish flesh with respect to geographical provenance, i.e. inshore and offshore sampling sites. The most descriptive elements responsible for differentiating between the muscle of offshore fish and inshore fish are Al, Ca, Li and Pb, as determined by ICP–MS. Moreover, the elemental profiles of the hepatic and gonad samples differed between males and females. This means that PCA allows the identification of interregional and intersex differentiation (Hellou et al. 1996).

3.2.3.2 Interregional or Interspecies Differentiation Related to Seasonal Variations

ICP–AES was performed by Han et al. (2020) for salmonid samples collected from two sites in various aquaculture areas in China during four seasons. The concentrations of elements in fish from Yantai were stable with seasonal alternations, in contrast to those in fish from Liujiaxias, where the impacts of seasonal changes on element concentrations and compositions were considerable. In contrast to PCA, CDA revealed greater discrimination ability for the tested samples. Moreover, other multivariate statistics, including LDA, KNN and PLS–DA, were used to discriminate the fish samples from the two areas. The overall correct classification rates for the LDA, PLS–DA and KNN models were 97.56%, 97.56% and 98.78%, respectively. All the discrimination techniques could be used to differentiate samples effectively while remaining unaffected by seasonal effects.

The muscles of ten species of fish from the Bangshi River in Savar, Bangladesh, were assayed in two seasons. The ANOVA data revealed significant interspecies variation (CI = 95%), with significant positive correlations between the trace element contents in the fish muscles in both seasons. However, the seasonal variation was not significant for the ten species studied. It has been estimated that there is no possible health risk to consumers due to the intake of the studied fishes under the current consumption rate (Rahman et al. 2012).

3.2.3.3 Differentiation Between Farmed and Wild Fish and/or Related to Various Farming Systems

Busetto et al. (2008) integrated EA–IRMS and GC–FID data ($\delta^{13}C$, $\delta^{15}N$ and fatty acids) with a chemometric processing tool to characterize wild and farmed turbot from three countries. The first two PCs, which explained 55.1% of the total variance, clearly distinguished farmed from wild fish, although the latter from Denmark appeared not to be differentiated from Dutch turbot. Reliable discriminative abilities were achieved owing to the use of LDA and SIMCA. For the former, 100% of the original and cross-validated samples were correctly classified according to the production method, i.e. farmed turbot vs. wild turbot. However, the LDA technique was applied to differentiate between both types of fish in the context of their geographical provenance, i.e. Denmark versus the Netherlands versus Spain. The correctness percentages of the original and cross-validated samples were estimated to be 96.7% and 95.0%, respectively, with respect to geographical origin. The sensitivity and specificity of the SIMCA were estimated to be 100% and 87.5% for the training and testing sets, respectively, but 100% for the training and testing sets.

Notably, IRMS provides rapid results and may be a promising tool for discriminating wild fish of various origins.

According to Gopi et al. (2019), discrimination of the geographic origin of Asian seabass with the combination of CF–IRMS, SIA and XRF followed by PCA, RF and LDA is a reliable approach. PCA with the first two PCs, which explained 95.3% of the total variance, distinguished between the farmed and wild-caught samples in most cases and geographic origin (Fig. 3.1), although the samples of farmed fish from the northern territory overlapped with the farmed fish samples from Queensland. The random forest model was characterized by an accuracy of 84% when the production methods and geographic provenance were differentiated with SIA. The models using LDA were characterized by accuracies of 88% for the SIR data, 22% for the MEC data and 9% for the integrated dataset. On the basis of the elemental profile obtained via XRF using ITRAX, an accuracy of 72% was achieved, but a combination of SIA and the elemental model provided 81% accuracy in determining geographic origin.

Varrà et al. (2019) determined $\delta^{13}C$, $\delta^{15}N$ and REE data by EA–IRMS and ICP–MS in wild and farmed samples of European sea bass from 17 different European

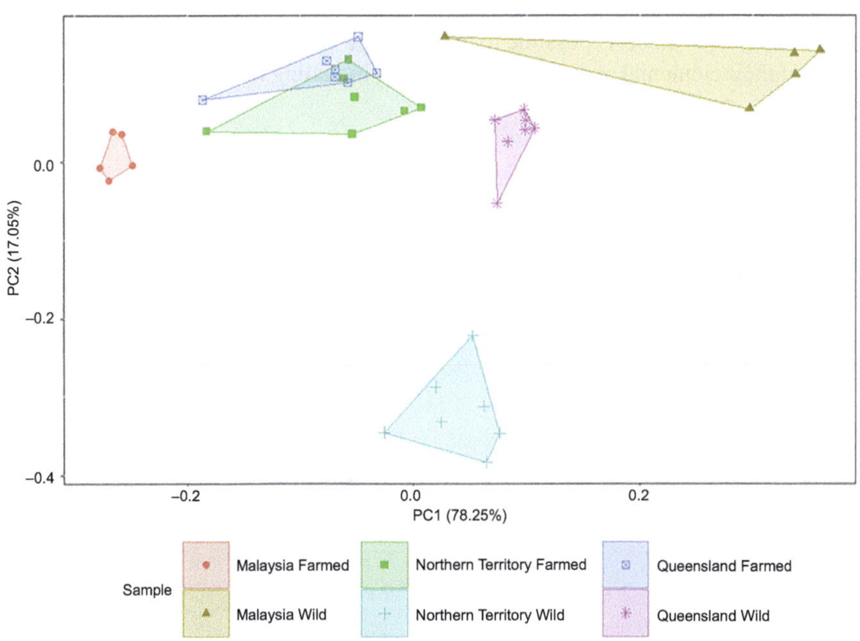

Fig. 3.1 PCA plot of the first two components showing the effectiveness of stable isotopes and relative elemental abundance for distinguishing the production method and geographic origin of *L. calcarifer*. The distance between groups indicates the differences between them. There is a clear difference between all farmed and wild-caught samples and most geographic origins, except for an overlap between the Northern Territory and Queensland farmed fish samples. (After Gopi K et al., 2019. Aquaculture 502:56–62. https://doi.org/10.1016/j.aquaculture.2018.12.012. With permission)

areas in the Mediterranean Sea Basin. The first two PCs, which explain 78% of the total variation in the original dataset, showed that the approach could be applied for subsequent elaboration. The OPLS–DA model achieved a good correct classification rate for both the production method and geographical origin. The classification performances tested on the prediction set provided an overall classification rate related to differentiating wild from farmed fish with 100% accuracy (Fig. 3.2). Additionally, fish samples originating from the western, central and eastern Mediterranean Sea were classified with 88.9% accuracy (Fig. 3.3). Both $\delta^{13}C$ and $\delta^{15}N$ were predominant discriminative stable isotopes, which enabled the distinguishing of fish via the two production methods. Furthermore, La and Ho were important contributors responsible for sample differentiation relative to their origin. It has been concluded that chemometric processing of the results obtained can be a reliable tool for accurate authentication of European sea bass.

Alasalvar et al. (2002) used 18 different FAs and MECs determined by GC–FID, FAAS and XRF in the flesh of cultured and wild sea bass. The percentages of total saturated and polyenoic fatty acids and the n-3/n-6 ratio in wild fish flesh were greater than those in cultured sea bass, whereas the related total monoenoic contribution was lower. Moreover, significant ($p < 0.05$) differences were observed between cultured and wild sea bass on the basis of the Fe, Al, Ti and V levels. This finding indicates that it is possible to discriminate cultured and wild sea bass on the basis of their elemental and total lipid contents, including fatty acid proportions, and

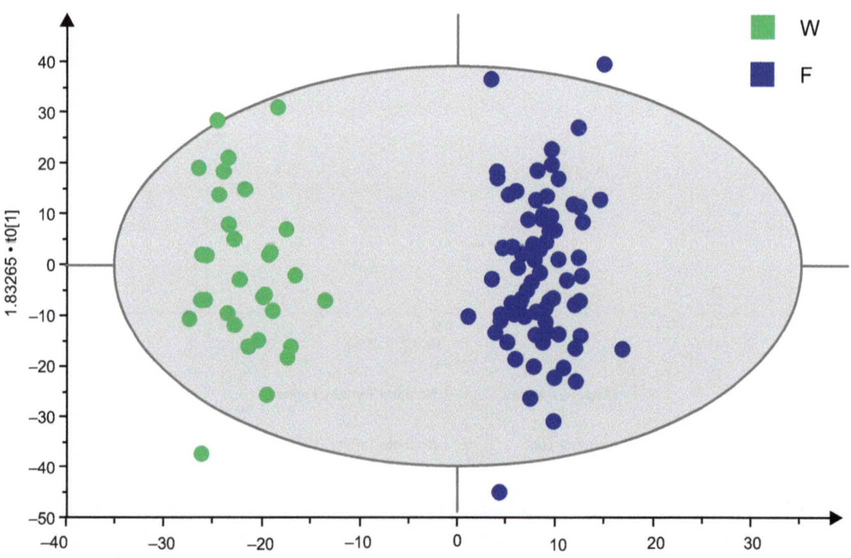

Fig. 3.2 OPLS–DA score plot discriminating wild sea bass from farmed sea bass (W, wild; F, farmed). (After Varra MO et al., 2019. Ital J Food Saf Mar 22:7872. https://doi.org/10.4081/ijfs.2019.7872. With permission)

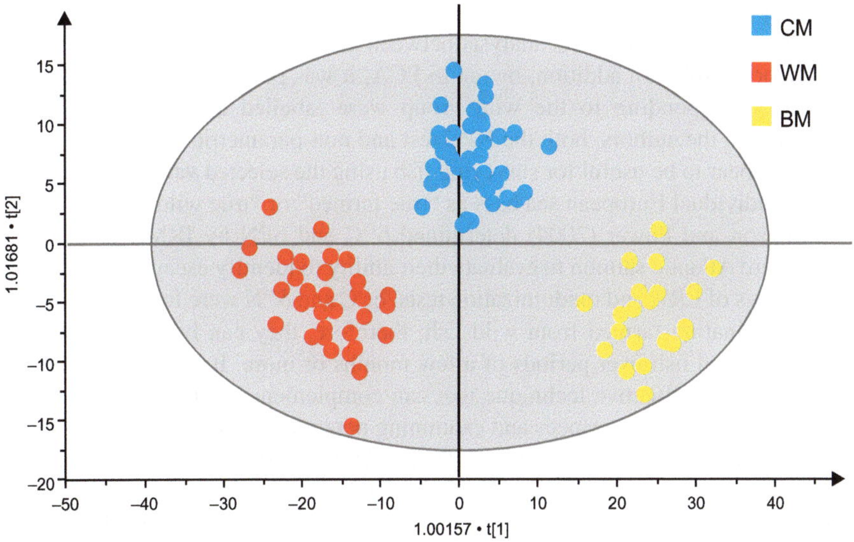

Fig. 3.3 OPLS–DA score plot discriminating CM sea bass, WM sea bass and EM sea bass (CM, central Mediterranean; WM, western Mediterranean; EM, eastern Mediterranean). (After Varra MO et al., 2019. Ital J Food Saf Mar 22:7872. https://doi.org/10.4081/ijfs.2019.7872. With permission)

such differentiation may be explained by the influence of dietary elements on the chemical composition of the fish.

Studies performed by Moreno Rojas et al. (2007) with a combination of CF–EA–IRMS and GC–MS revealed that the $\delta^{13}C$ values appeared to be muscle indicators useful in distinguishing between wild and farmed gilthead sea bream originating from the Mediterranean Sea (Italy, Greece, Croatia, Turkey and France). However, $\delta^{15}N$ was more informative with regard to the geographical origin of the fish. This finding could be related more to differences in the feed mixtures given to farmed fish than to geographical factors.

Bell et al. (2007) described the use of a combination of EA–IRMS, GC–C–IRMS and GC–FID to classify profiles of $\delta^{13}C$, $\delta^{15}N$, $\delta^{18}O$ and 13 different fatty acids of English wild and Scottish or Greek farmed European sea bass. A clear discrimination of fish from two producing methods, as well as farmed sea bass of Scottish or Greek origin, was achieved by PCA, where the first two PCs corresponded to 59.26% of the total variance.

Fasolato et al. (2010) applied IRMS and GC–FID to survey samples of wild and farmed European sea bass from the Mediterranean Sea and Atlantic (Italy and Greece). The analytical data were analyzed via ANOVA, PCA and NPC to classify the fish samples. In the case of standard fish, the farmed and wild profile types were clearly distinguished. Using rank-based cross-classification, the standard wild dataset for all the samples was estimated to be correctly classified as 100%, whereas the standard farmed dataset was achieved at the 95% level. Furthermore, the intercomparison data of standard farmed and survey samples using only significant variables obtained after NPC analysis between groups were performed, achieving satisfactory results. In addition, owing to PCA, it was possible to confirm that the samples corresponding to the wild group were labelled and considered wild. According to the authors, both the NPC test and non-parametric rank-based classification appear to be useful for surveying fish using the selected variables to categorize the individual European sea bass as "true farmed" or "true wild".

Dempson and Power (2004) determined $\delta^{13}C$ and $\delta^{15}N$ by IRMS in wild and aquaculture Atlantic salmon to evaluate their ability to identify escaped farmed fish. On the basis of LRA and randomization tests, $\delta^{13}C$ and $\delta^{15}N$ were found to be useful for discriminating farmed from wild fish; therefore, they can be used to identify escaped farmed fish over periods of a few months or more. It has been concluded that SIRA is an effective technique that can complement other methods used for accurately identifying escapees and examining potential influences on wild stocks.

Molkentin et al. (2007) combined TC/EA–IRMS and GC–MS–IRMS with an ANN model to examine $\delta^{13}C$, $\delta^{15}N$, $\delta^{18}O$ and 13 fatty acids to classify Atlantic salmon via two methods. Although distinguishing between organically farmed Atlantic salmon and conventionally farmed Atlantic salmon from wild salmon cannot be achieved using SIR or FAA profiles alone, the complete discrimination between the three fish types may be clearer when a combination of linolenic acid and $\delta^{15}N$ is used. It has been postulated that the methodology should be verified in the context of using additional samples to accommodate maximum variation in feed composition before a suitable method for organically raising fish is approved at the stage of official food control.

Anderson et al. (2010) reported that three species of wild or farm-raised salmon (king salmon, Coho salmon and Atlantic salmon) from the Pacific (USA) were usefully classified with the use of a combination of ICP–AE and EA–IRMS followed by multivariate models. As shown in Fig. 3.4, chemical profiling with PCA and CDA clearly differentiated the two fish production methods. In general, classification modelling methods frequently correctly categorize samples with a >90% success rate on average. Owing to the exclusive use of bulk C, N and their isotope ratios, LDA, QDA, NN, PNN and NNB were characterized by similar success rates, i.e. between 86% and 100%, with the most successful PNN and NNB models. Somewhat worse data were obtained with MEC.

Morrison et al. (2007) reported that integrated EA–IRMS, GC–C–IRMS and GC–FID data processed by PCA were useful for the reliable classification of gilthead sea bream from Spain and Greece. The first two PCs, accounting for 74.44% of the total variance, achieved clear discrimination between wild and farmed bream, with PC1 performing best. The authors suggested that, due to too few samples, further research should be planned on the basis of various fish species and their more numerous sampling regions.

Arechavala-Lopez et al. (2016) analyzed the concentration of MECs quantified by ICP–MS in otoliths of sea bass and sea bream from fish farms and coastal wild populations in the western Mediterranean Sea. Mineral elements such as Mg, K and Mn appeared to differ considerably between sea wild and farmed sea European

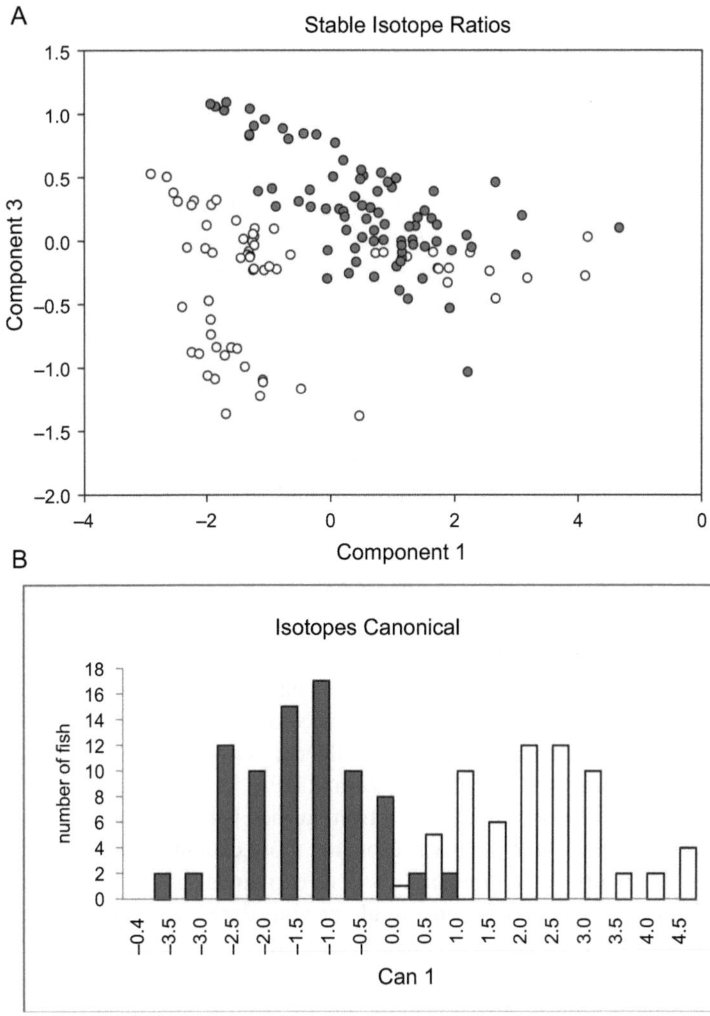

Fig. 3.4 (**a**) Principal component 1 versus principal component 3 for the stable carbon and nitrogen isotope ratios and bulk C/N ratio (n = 145). (**b**) Canonical discriminant analysis (CDA) frequency histogram using CDA1 representing the two groups of farm-raised and wild-caught salmon (n = 145) from the stable carbon and nitrogen isotopes and bulk C/N ratios. Farm-raised and wild-caught salmon are denoted by grey and white, respectively. (After Andersen KA et al., 2010. *J Agric Food Chem* 58:11768–11774. https://doi.org/10.1021/jf102046b. With permission)

bass, while the levels of Mg, K, Mn, Fe, Zn, Sr and Ba were characterized by considerably different contents between wild and farmed gilthead sea bream. Using CDA for the chemometric processing of analytical data, the researchers identified farmed fish and wild stocks with satisfactory accuracy for both fish species. A discrimination between wild stocks for European sea bass (85.7–90% of wild fish correctly classified) and for Gilthead sea bream (93.3–100% of fish correctly classified) was achieved. In contrast, it was impossible to accurately distinguish between farmed stocks of both species.

Chaguri et al. (2017) used IRMS, EDXRF, FAAS and AAS analyses to differentiate between wild and farmed meagre using many discriminative chemical parameters, such as proximate chemical composition, namely, Cl, S, Fe, Zn, Se and Br, $\delta^{13}C$ and $\delta^{15}N$ and FA contents. Significant differences in both stable isotopes between farmed and wild meagre were detected; however, no discrimination was detected between large and small farmed fish. According to the authors, complementary studies concerning the determination of species with high commercial value, seasonality and sampling site differentiation are still ongoing.

Vertebrae and muscles of wild and cultured haddock, including commercial and natural diets, were investigated by Roy and Lall (2006) to assess how natural and commercial diets influence the MEC data of the fish. The results revealed significant differences in the Al, Ba, Fe, P, K and Sr contents of muscle and vertebrae, as quantified by ICP–MS, between wild and cultured haddock (*Melanogrammus aeglefinus*). This interelemental discrimination of wild and cultured haddock may have resulted from differences in food sources affecting the mineral composition of the diet. According to the authors, life history and diet play important roles in shaping the elemental profile of fish tissues. Therefore, special attention should be given to MEC data when test diets intended for the culture of fish are being developed. In this context, the elemental distribution pattern in wild fish may be a reference point for correcting the food composition of farmed haddock.

Sant'Ana et al. (2010) employed EA–IRMS to determine $\delta^{15}N$ and $\delta^{13}C$ in wild and farmed Brazilian freshwater cachara collected in two seasons. The data obtained were processed using ANOVA, MANOVA and DFA. On the basis of DFA data, wild fish samples were clearly differentiated from farmed fish. Moreover, a clear distinction between fish caught in the dry season and those caught in the rainy season was obtained. The first two discriminant functions explained 92% of the variance in the dataset. Finally, seasonal variations should be considered when chemical and isotopic traceability of fish and fish products is applied.

Turchini et al. (2009) noted that the combination of EA–IRMS and GC–FID with DFA could be helpful in distinguishing among Murray cod from four Australian farms (indoor recirculating, outdoor floating cage and flow-through systems). FAs, tissue proximate compositions and morphological parameters appear to be useful in differentiating between farmed and wild fish and were less informative in distinguishing between fish originating from various farms. The combination of $\delta^{13}C$, $\delta^{15}N$ and $\delta^{18}O$ can discriminate between fish originating from various farms because the first two variables clearly link fish to a specific commercial diet, whereas the last one linked fish to a specific water source.

According to Li et al. (2015), the integration of ICP–AES with multivariate analysis (PCA, CDA and kNN) appears to be a useful tool for the verification of ictalurid catfish and hybrid catfish production, i.e. between fed fish and fish from fertilized ponds. The first two PCs, which explained 63.92% of the total variance, differentiated fillets from ictalurid catfish between those cultured in fertilized ponds and those cultured in ponds with feeding, with contributions from descriptors such as Ca, Cr, Cu, Fe, K, Mg, Mn, Na, P, S, Se and Zn. The *k*-NNA classification model allows the clear discrimination of fish fillets relative to the two culture methods, with 100% correct classification. It is possible to achieve this for both empirical and cross-validated grouped observations. Similarly, acceptable results were obtained by CDA. According to the authors, substantial interelemental differentiation could be applied to other fish species to verify the method of production, namely, aquaculture versus wild-caught.

Yamashita et al. (2006) investigated farm-raised and wild eel from three Asian countries with a combination of ICP–AES, CV–AAS and fluorometric methods with the FA technique. FA partially classifies fish according to their sampling sites, i.e. the Japanese prefectures of Miyazaki, Kagawa and Shizuoka, as well as two remaining countries, i.e. Taiwan and China. Factor 1 was related to the Se and Hg contents, whereas F2 was related to the Zn and Cu contents. The aquaculture methods used in Japan, Taiwan and China may also be responsible for the intercountry differentiation achieved. Most Japanese eel farms use tank-based systems, whereas in Taiwan and China, eel farms use traditional ponds for aquaculture.

Perez de Nanclares et al. (2016), via ICP–MS, tested REEs alone or in combination (As, Sm, U and Sr), i.e. La, Ce, Pr, Nd and Dy, in the scales of Atlantic salmon. This approach was evaluated to identify escaped farmed fish. The authors concluded that REEs could be added to the diet of Atlantic salmon smolts reared in closed containment systems for a short period to induce a relatively long-standing mark in the fish scales. Moreover, the elemental background contents of the scales provided valuable information that could be potentially useful for identifying farm escapees.

Farabegoli et al. (2018) reported data on biometric parameters, ICP–OES, EA–IRMS and GC–FID data, i.e. Ca, K, Mg, Na, P, S, Ag, Cd, Co, Cr, Fe, Hg, Mn, Ni, Pb, Se, Zn, $\delta^{13}C$, $\delta^{15}N$, 22 different fatty acids and other physicochemical parameters subjected to chemometric processing. PCA was used to differentiate farmed sea bass according to geographical origin (Turkey vs. other regions), and PCA was used to weakly distinguish wild sea bass according to geographical origin (Mediterranean FAO fishing subareas vs. the Atlantic Ocean). Wild and reared (intensively, semi-intensively and extensively) European sea bass from 18 southern European sources showed correct allocations according to the production method, origin and stocking density, with good approximation rates (94%, 92% and 92%, respectively) on the basis of the FA distribution profile. Less acceptable results were achieved using SIR, biometric traits and elemental profile.

Martin-Pérez et al. (2013) applied EA–IRMS and HPLC to classify juveniles of gilthead sea bream fed with six isoenergetic diets under experimental conditions. Among the obtained variables, such as $\delta^{13}C$, $\delta^{15}N$, 17 amino acids, crude protein and fat, ash and others, an indicator of fish growth and feeding balance that allowed the definition of the optimum diet for efficient growth conditions was selected. Juveniles of gilthead sea bream were fed six isoenergetic diets differing in protein-to-lipid proportions. PCA with the first two PCs accounting for 46.6% of the total variance appeared to be effective, with stable isotopic parameters having a high discrimination load. Specifically, clusters with lower protein intake and growth were clearly separated from the other, protein-related stable isotopic variables, with the largest share in the main discriminating PC. Notably, the stable isotopic composition of muscle, especially $\delta^{15}N$, can be used to evaluate the nutritional status of farmed fish.

A study performed by Wang et al. (2018) on amino acid $\delta^{13}C$ fingerprints determined by EA–IRMS and GC–C–IRMS and subjected to PCA and LDA classified wild-caught, organically and conventionally farmed salmon groups as well as salmon fed alternative diets such as insects and macroalgae. This approach is particularly well suited for tracing protein sources and the non-essentials for tracing lipid origins (terrestrial vs. aquatic). The PCA with the first two PCs, which explained 79.6% of the total variance, demonstrated that the five sample groups clustered separately (Fig. 3.5a). In the case of the laboratory-cultured experimental salmon, all the treatments were clustered into separate groups, with the most marked difference between the insect-fed salmon and the control salmon. Seven amino acids, i.e. Ser, Gly, Met, Phe, Tyr, Pro and Leu, explained 93.9% of the total variation in the first two PCs (Fig. 3.5b). On the basis of amino acid $\delta^{13}C$ fingerprints, it has been possible to discriminate among wild-caught, organically farmed and conventionally farmed salmon groups, as well as salmon fed alternative diets such as insects and macroalgae. This study provides a methodology for tracing the production of salmon and other marine fish species fed commercial diets.

3.2.3.4 Interage Differentiation and Seasonal/Temporal Variations Related to the Impact of Pollution

Trace metal analysis by FAAS, ET–AAS and AAS–Hg analysis integrated with chemometric elaboration (FA, ANOVA) is a promising approach for classifying perch samples (muscle and liver) originating from two coastal regions of the southern Baltic region (Poland). Hg in muscle and Cd, Pb and Cu in liver are descriptors for the factorial differentiation of the age groups of the fish investigated. The positive relationship between muscle Hg and age (weight–length) seems to be associated with the specific bioaffinity of CH_3Hg with a high biological half-life, which generally constitutes the dominant pool of the total Hg in the fish muscle. However, considerable seasonal variations in muscle Hg and hepatic Cd, Pb and Cu were also observed. All the data obtained were processed using FA. The first two factors explained 70.9% (for the liver) and 61.8% (for the muscle) of the total variance. FA confirmed seasonal differences in muscle and especially in hepatic samples; specifically, summer muscles were clearly differentiated from winter muscles (Figs. 3.6 and 3.7). The fish muscle corresponding to the winter season presented relatively high concentrations of Hg, Cd and Pb. On the basis of the PTWI recommended by the WHO, Hg, Cd and Pb levels in perch muscle do not constitute a threat to humans (Szefer et al. 2003).

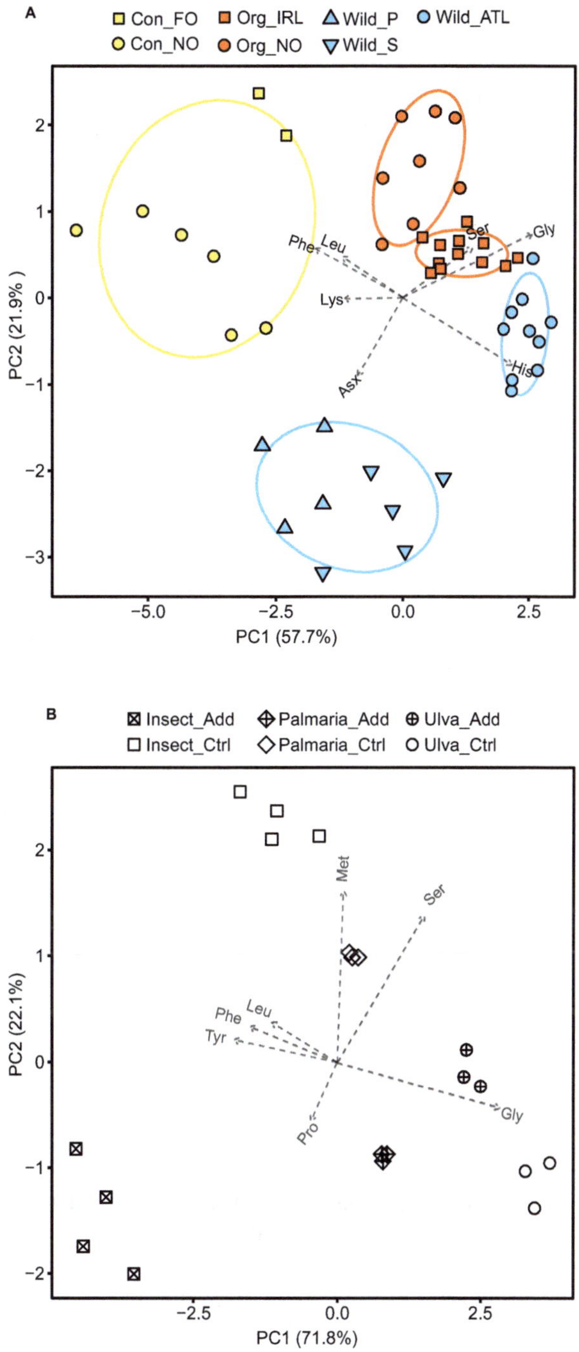

Fig. 3.5 Principal component analyses (PCAs) of the $\delta^{13}C_{AA}$ values revealed that both (**a**) wild and farmed salmon groups and (**b**) experimental salmon groups clustered separately. The values in parentheses are the percentage variations accounted for by each axis. In (**a**, **b**), the first two axes, accounting for 80% and 95% of the variations, respectively, separated salmon into distinct groups. The most important amino acids for the variations among groups are the first two coordination components. (After Wang YV et al., 2018. Food Chem 256:380–389. https://doi.org/10.1016/j.foodchem.2018.02.095. With permission)

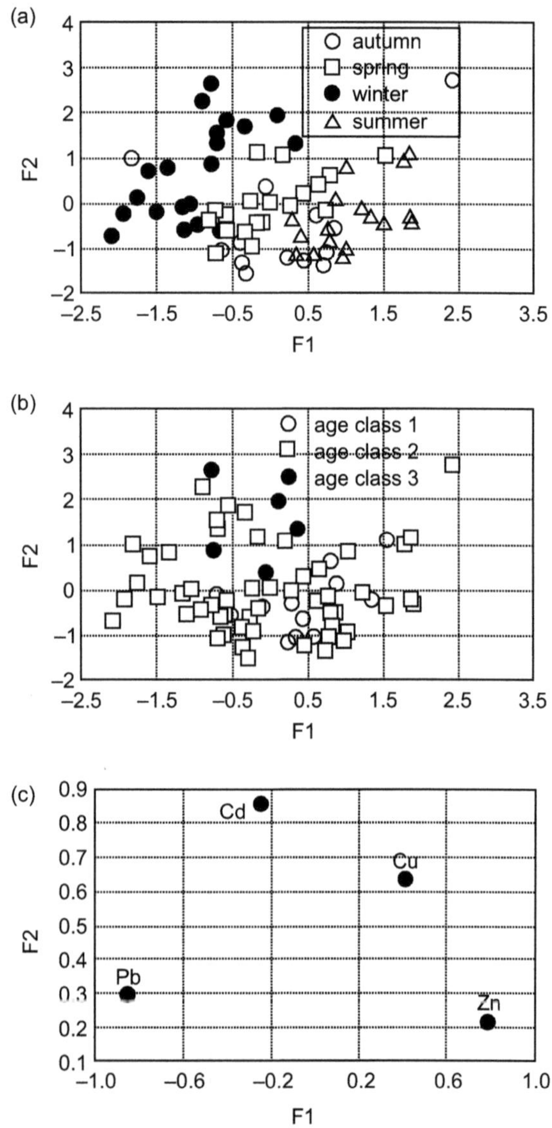

Fig. 3.6 Biplot of scores showing seasonal (**a**) and age (**b**) variations in concentrations of trace metals (**c**) in the liver of *Perca fluviatilis* from Pomeranian Bay and Szczecin Lagoon, Baltic Sea. (After Szefer P et al., 2003. Food Chem 81:73–83. https://doi.org/10.1016/S0308-8146(02)00380-1. With permission)

3.2.3.5 Interregional and/or Interspecies Differentiation Related to the Impact of Pollution

Julshamn and Grahl-Nielsen (1996) studied the distribution of trace elements quantified by a combination of FAAS, ET(GF)–AAS and CV–AAS at eight sites along Hardangerfjord (Norway) up to 100 km from industrial discharges in the fjord of saithe and flounder samples. The logarithmic concentration values were processed

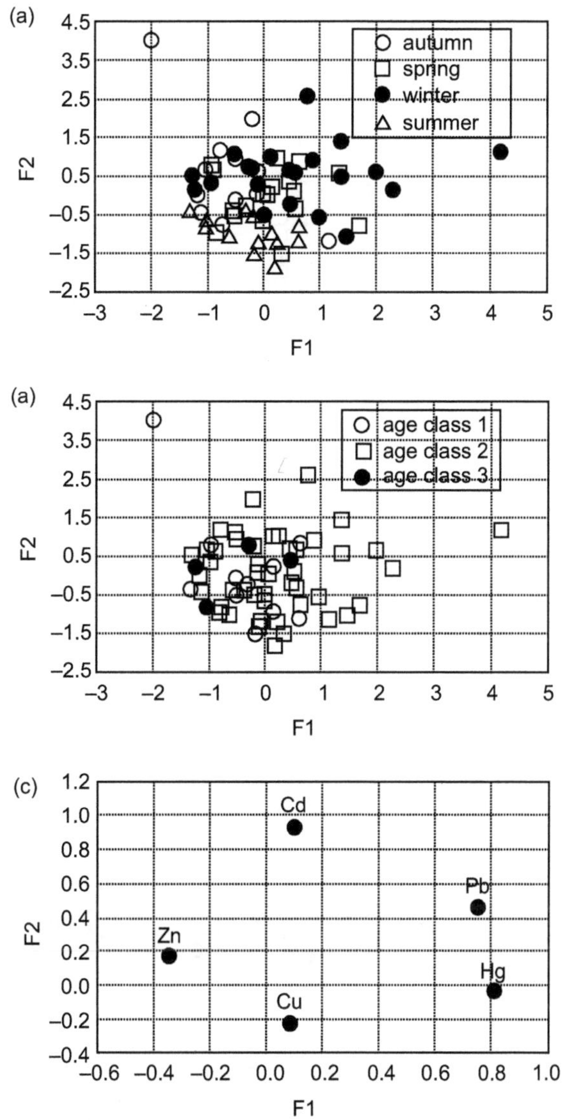

Fig. 3.7 Biplot of scores showing seasonal (**a**) and age (**b**) variations in concentrations of trace metals (**c**) in the muscle of *Perca fluviatilis* from Pomeranian Bay and Szczecin Lagoon, Baltic Sea. (After Szefer P et al., 2003. Food Chem 81:73–83. https://doi.org/10.1016/S0308-8146(02)00380-1. With permission)

by PCA. Figure 3.8 (with the first two PCs corresponding to 98.0% for muscle and 90.0% for liver) shows the distribution pattern of the object scores for the muscle and liver of saithe from various sampling locations. The samples from the different regions are similar for the two tissues of the sentinel saithe, although a clearer geographical distinction of objects was found in the case of the muscle samples. It has been concluded that a decreasing gradient in As abundance in the saithe from Hardangerfjord occurred with distance from the inner part of the fjord, although with a small increase in the outermost location. Moreover, a concentration gradient

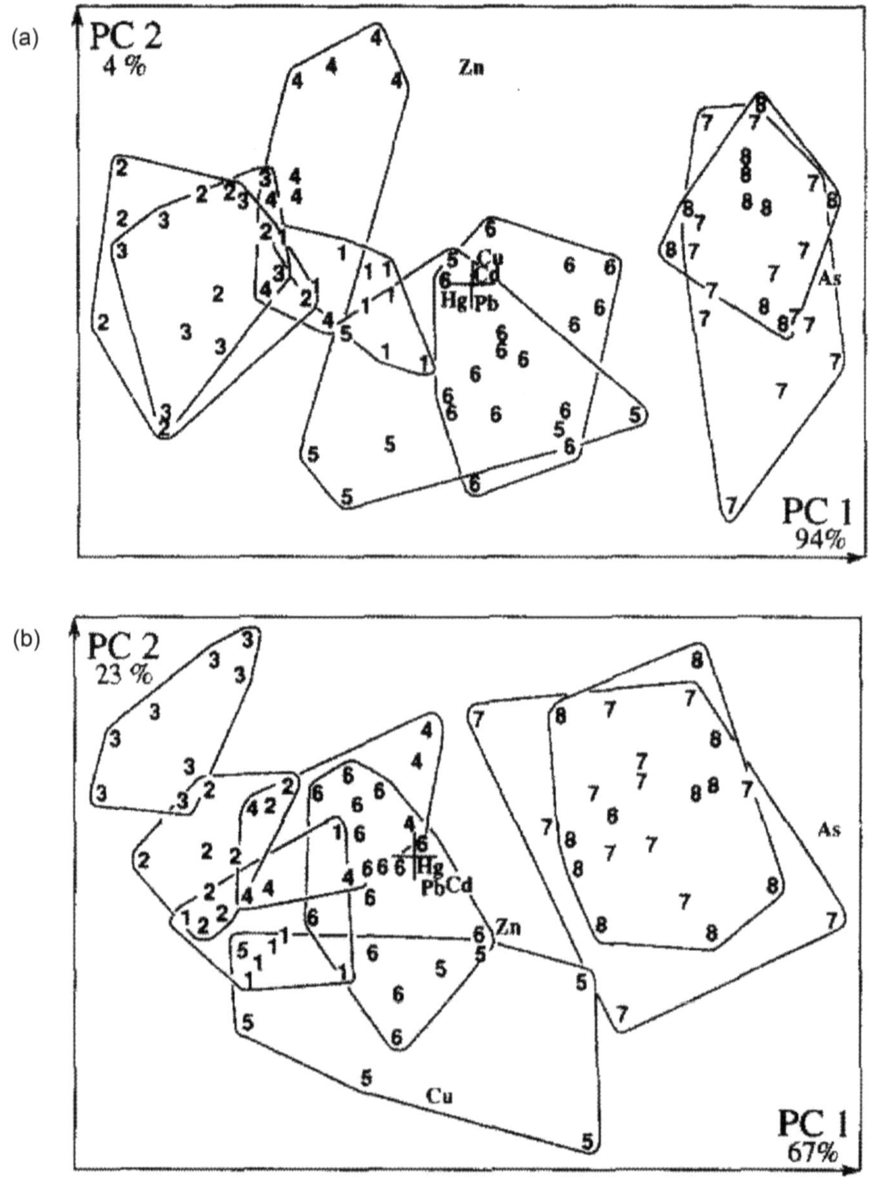

Fig. 3.8 Biplot of muscle (**a**) and liver (**b**) samples from saithe from all eight locations. Each number represents one fish. (After Julshamn K et al., 1996. Mar Pollut Bull 32:564–571. https://doi.org/10.1016/0025-326X(96)84577-2. With permission)

away from the source of pollution was detected for As in muscle and liver samples of saithe, whereas such a gradient was found for Hg in the liver samples of flounder.

According to Albuquerque et al. (2021), elemental profiles established by ICP–MS combined with chemometric processing are useful approaches for evaluating liver and muscle samples from Tucunaré and Acari species with respect to their potential use for biomonitoring purposes. PCA and HCA data demonstrated that it is possible to achieve satisfactory interspecies and interregional discrimination of liver samples from the two fish species. The Tucunaré liver is the best candidate for biomonitoring elements that accumulate through the food chain, e.g. Hg. Acari liver appears to be a promising bioindicator reflecting elements that typically accumulate in sediments, e.g. As. The very high levels of several toxic elements in some Brazilian soils (mainly Hg and As) are potentially responsible for their elevated levels in aquatic ecosystems due to anthropogenic activities, e.g. intensive agriculture, mining and deforestation and energy production in hydroelectric plants. In the context of the ability of some trace elements to serve as potential biomarkers of pollution in relation to their geographical origin, PCA clearly discriminated liver and muscle samples among the three sampling areas with various degrees of contamination, with toxic elements expected. According to the authors, since various information was obtained in terms of the toxic and trace elements in each type of sample (liver or muscle) and fish species (Acari or Tucunaré) in the various geographical areas (Amazon River, Itaituba and Porto Trombetas), it is important to properly select the most appropriate type of sample with the aim of the particular objectives of the biomonitoring study.

The quantification of OT (TBT, DBT, MBT and TPhT) content by GC–MS was performed by Filipkowska et al. (2016) using muscle and liver tissues of nine fish species (flounder (*Platichthys flesus*), round goby (*Neogobius melanostomus*), herring (*Clupea harengus*), bream (*Abramis brama*), cod (*Gadus morhua*), ziege (*Pelecus cultratus*), ruffe (*Gymnocephalus cernua*), perch (*Perca fluviatilis*) and roach (*Rutilus rutilus*)) from three Polish basins located in the southern Baltic coastal zone. PCA was used to identify interspecies variation with respect to estimated OT accumulation in different fish organs. The first two PCs, which explained 93.10% of the total variance, revealed that the round goby from the Gulf of Gdańsk and bream from the Szczecin Lagoon accumulated the highest levels of OTs in both the liver and muscles, in contrast to the other studied fish. However, the round gobies accumulated butyltins more effectively in muscle than in liver, in contrast to bream and other fish species. According to the authors, the high OT content in muscle tissues may result from the fact that compound uptake through the skin may be important.

Mielcarek et al. (2022) reported that concentrations of As, Cd, Pb and Hg determined by AAS–AMA, ET–AAS and ICP–MS were processed chemometrically by CA, DA and PCA. The DA model for pickled fish yielded a lambda value of 0.263 ($p < 0.001$), which included As ($p < 0.001$), Pb ($p < 0.01$) and Cd ($p < 0.01$). The first function corresponded to 46% of the explained variance. The highest mean levels of As were detected in smoked trout, whereas Cd was most concentrated in raw roach and pickled whitefish. The raw pike-perch was distinguished by the highest level of contamination by Hg, whereas the whitefish was characterized by the highest level of Pb. According to the authors, the consumption of the three selected species of freshwater fish studied, particularly smoked fish products, may pose non-carcinogenic and carcinogenic health risks to consumers.

3.2.3.6 Verifying Authenticity/Adulteration

According to Oliveira et al. (2011), Gadidae fish species can be sold commercially as salted fish; hence, differences in price between fish of various species may result in falsification. Therefore, the aim of this study was to discriminate between salted Atlantic cod and salted saithe based on IRMS, including processing with Tukey's post hoc test. The $\delta^{15}N$ ratio in tissues with lower turnover rates (bone and skin) and in tissues with greater turnover rates (muscle) can be useful for the authentication of salted gadoid fish species because distinguishing between Atlantic cod and saithe was achieved.

Chaguri et al. (2015) reported that the combination of IRMS, FAAS, EDXRF, CV–AAS, GC–FID and other chemical analyses was a useful tool for classifying samples of muscle croaker from two Brazilian coastal regions. The original variables, such as $\delta^{13}C$, $\delta^{15}N$, 15 macro-, micro- and trace elements, and 32 different FAs, were processed chemometrically. The first two PCs, corresponding to 69.22% of the total variance, allowed a clear differentiation between two geographical regions and seasons (winter and summer) (Fig. 3.9). The profiles of FAs, elements and SIRs processed by PCA appear to be effective approaches for the authentication

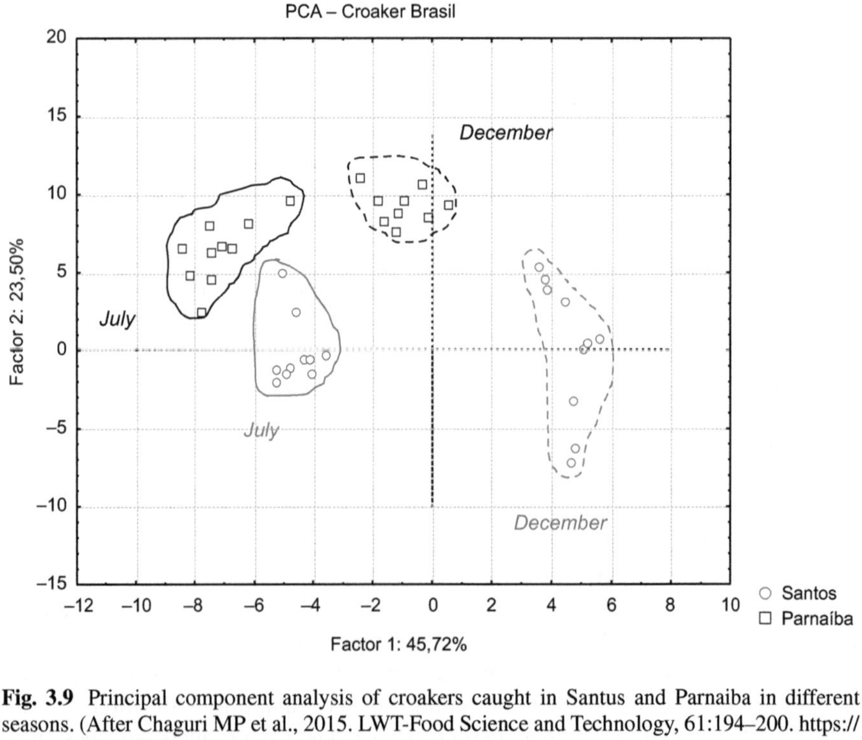

Fig. 3.9 Principal component analysis of croakers caught in Santus and Parnaiba in different seasons. (After Chaguri MP et al., 2015. LWT-Food Science and Technology, 61:194–200. https://doi.org/10.1016/j.lwt.2014.11.006. With permission)

and traceability of croakers originating from various regions and seasons, resulting in the identification of fraud as well as reducing the costs and duration of analysis.

Kim et al. (2015a) evaluated the suitability of stable isotopes and cytochrome oxidase (COI) sequencing for identifying and tracing the origins of hairtail fish (two species) and shrimp (six species). The COI method allows the identification and tracing of the origin of hairtail fish and shrimp with respect to their species characteristics. The LDA of the $\delta^{13}C$ and $\delta^{15}N$ values was useful for establishing the geographical origin of hairtail fish from Korea (South Sea), China, Japan, Thailand, Senegal and Canada. This technique categorizes individuals of the same species, considering the country of provenance. According to the authors, these two methods can be used together to identify and authenticate fishery products, and their utilization would improve the clarity and balance of trade.

Thomas et al. (2008) demonstrated that the combination of CF–EA–IRMS, CF–EA–P–IRMS and GC to determine SIR, MEC and FA contents appears to be useful for discriminating 171 samples of wild and farmed Atlantic salmon (*Salmo salar*) from 32 origins in Europe, North America and Tasmania. PCA, with the first two PCs accounting for 89% of the total variance, with a combination of $\delta^{18}O_{oil}$, $\delta^{15}N_{choline}$ and C18:2n-6%, allows successful discrimination between farmed and wild salmon. Moreover, PCA distinguished between wild salmon from various regions, especially between Ireland and other regions. However, it was not possible to differentiate fish from Scotland and Norway; hence, more samples would need to be analyzed. However, the certainty of identification of mislabelling in market-derived fish is reinforced by the contribution of the percentage of linoleic acid C18:2n-6 in the lipidic fraction. Therefore, several potential misidentifications can be identified. The method is practical and can be implemented in official laboratories as a tool for labelling verification.

In Xuan and Wang (2023), the multielemental profiles of wild and cultured fourfinger threadfins (*Eleutheronema tetradactylum*) from the southern coast of China were analyzed via ICP–MS. The LDA and RF models correctly discriminated wild and cultured fish. The classification accuracies were 100% according to LDA. The classification accuracy of identifying the source of cultured *E. tetradactylum* was 95%. Sharing the author's opinion, the established approach enables the identification of potential substitutions and eventual fraudulent mislabelling in the food market.

Vasconi et al. (2019) quantified $\delta^{13}C$, $\delta^{15}N$ and 26 FAs via a combination of EA–IRMS, CFEA–IRMS and GC–FID with PCA, sPLSDA and PLSDA processing for wild and farmed eels from Italy, Denmark and the Netherlands. PCA with the first two PCs, which explained 65% of the total variance, differentiated eel samples into four groups, i.e. the first group of wild eels from the sea, the second group containing eels from lagoons, the third group consisting of buratelli farmed in RAS in the Netherlands and Denmark and the last group corresponding to all eels farmed in pond systems and capitoni farmed in RAS systems. sPLS–DA classified the samples via a one-step procedure, effectively distinguishing potential markers. Moreover, an intercomparison assessment between sPLS–DA and PLS–DA indicates that the latter stands out for equal or better discrimination ability. The farmed eels (POND and RAS systems) were clearly differentiated from the wild eels (lagoon and sea) in X-variate 1, accounting for 33% of the total variance. Additionally, X-variate 3, which explained 10% of the total variance, enabled the differentiation of eels from lagoons and eels from the sea. The authors proposed new biomarkers for verifying the authenticity of European farmed and wild eel. This multidisciplinary approach provides practical benefits for producers and consumers, accelerating the economically sustainable growth of substantial regional aquaculture production.

Moreover, Carrera and Gallardo (2017) aimed to identify the geographical origin of 13 fish species belonging to the Merlucciidae family using IRMS analysis. On the basis of the $\delta^{13}C$ values, PCA enabled a clear determination of the geographical origin of the hake species according to latitude. HCA allowed us to obtain a clear differentiation of Polar-origin hakes (Australian and North American) from Equatorial-origin hakes (North African and South American). These results are in agreement with the PCA data. The $\delta^{13}C$ and $\delta^{15}N$ values enable the classification of fish according to fishing area, latitude and geographical origin from European, North African, South African, North American, South American and Australian hake species. According to the authors, this was the first study showing the ability of IRMS analysis to classify the geographical origin of closely related fish species. SIRA provides a good strategy for fingerprinting that might be useful for food authorities and the fish industry as an adequate control tool for ensuring food quality and safety for consumers.

3.2.3.7 Discriminatory Approach for the Selection of Fish Seeds

Martín-Pérez et al. (2011) obtained $\delta^{13}C$, $\delta^{15}N$, RNA·protein^{-1}, protein·DNA^{-1} and RNA·DNA^{-1} values from indices of growth (RNA and DNA) and energy metabolism (COX and CS activities) using IRMS. On the basis of a dataset of these variables, a discriminatory tool for the selection of fish seeds with high growth capacity could be achieved. This has been possible owing to the selection of juvenile sea bream produced in three hatcheries of the Spanish Mediterranean and Atlantic coasts. PCA with two PCs accounting for 60.19% of the total variance with bulk $\delta^{15}N$ as the highest load on F1 (0.868), provided satisfactory discrimination among fingerlings from the three hatcheries, i.e. between the Mediterranean coast, Cantabrian coast and Atlantic coast. In particular, the latter was significantly differentiated from the remaining two areas.

3.3 Conclusions

Instrumental methods coupled with chemometrics, especially multivariate techniques, are useful in the assessment of the authenticity of food and the identification of food adulteration/mislabelling. The adequate use of the combined tools to evaluate the nutritive and health qualities of fish flesh is highly important in the identification of various parameters, e.g. identification of geographical origin; differentiation between wild and farmed fish; detection of interspecies, intersex and interage variations and seasonal and temporal differentiation. It would also be possible to differentiate unprocessed and technologically processed fish products. Proper application of analytical and computer tools appears to be a helpful approach for characterizing the distribution of mineral components in fish in both key aspects, i.e. nutritional and ecotoxicological aspects.

References

Ahmed ASS, Sultana S, Habib A, Ullah H, Musa N, Hossain MB, Rahman M, Sarker SI (2019) Bioaccumulation of heavy metals in some commercially important fishes from a tropical river estuary suggests higher potential health risk in children than adults. PLoS One 14:e0219336. https://doi.org/10.1371/journal.pone.0219336

Alasalvar C, Taylor KDA, Zubcov E, Shahidi F, Alexis M (2002) Differentiation of cultured and wild sea bass (*Dicentrarchus labrax*): total lipid content, fatty acid and trace mineral composition. Food Chem 79:145–150. https://doi.org/10.1016/S0308-8146(02)00122-X

Albuquerque FEA, Herrero-Latorre C, Miranda M, Júnior Barrêto RA, Oliveira FLC, Sucupira MCA, Ortolani EL, Minervino AHH, López-Alonso M (2021) Fish tissues for biomonitoring toxic and essential trace elements in the Lower Amazon. Environ Pollut 283:117024. https://doi.org/10.1016/j.envpol.2021.117024

Ali A, Wei S, Ali A, Khan I, Sun Q, Xia Q, Wang Z, Han Z, Liu Y, Liu S (2022) Research progress on nutritional value, preservation and processing of fish—a review. Foods 11:3669. https://doi.org/10.3390/foods11223669

Anderson KA, Hobbie KA, Smith BW (2010) Chemical profiling with modeling differentiates wild and farm-raised salmon. J Agric Food Chem 58:11768–11774. https://doi.org/10.1021/jf102046b

Arechavala-Lopez P, Fernandez-Jover D, Black KD, Ladoukakis E, Bayle-Sempere JT, Sanchez-Jerez P, Dempster T (2013) Differentiating the wild or farmed origin of Mediterranean fish: a review of tools for sea bream and sea bass. Rev Aquac 5:137–157. https://doi.org/10.1111/raq.12006

Arechavala-Lopez P, Milošević-González M, Sanchez-Jerez P (2016) Using trace elements in otoliths to discriminat between wild and farmed European sea bass (*Dicentrarchus labrax* L.) and Gilthead sea bream (*Sparus aurata* L.). Int Aquat Res 8:263–273. https://doi.org/10.1007/s40071-016-0142-1

Bell JG, Preston T, Bron J, Henderson RJ, Strachan F, Bron JE, Cooper K, Morrison DJ (2007) Discrimination of wild and cultured European sea bass (*Dicentrarchus labrax*) using chemical and isotopic analyses. J Agric Food Chem 55:5934–5941. https://doi.org/10.1021/jf0704561

Bulska E (2010) Quality assurance and quality control of analytical results. In: Namieśnik J, Szefer P (eds) Analytical measurements in aquatic environments. CRC Press, Taylor & Francis Group, Boca Raton, pp 389–397. https://doi.org/10.1201/9781420082692

Busetto ML, Moretti VM, Moreno-Rojas JM, Caprino F, Giani I, Malandra R, Bellagamba F, Guillou C (2008) Authentication of farmed and wild Turbot (*Psetta maxima*) by fatty acid and isotopic analyses combined with chemometrics. J Agric Food Chem 56:2742–2750. https://doi.org/10.1021/jf0734267

Capar SG, Szefer P (2012) Determination and speciation of trace elements in foods. In: Ötleş S (ed) Methods of analysis of food components and additions. CRC Press, Taylor & Francis, Boca Raton, pp 111–158. https://doi.org/10.1201/b11218

Carrera M, Gallardo JM (2017) Determination of the geographical origin of all commercial hake species by stable isotope ratio (SIR) analysis. J Agric Food Chem 65:1070–1077. https://doi.org/10.1021/acs.jafc.6b04972

Chaguri MP, Maulvault AL, Nunes ML, Santiago DA, Denadai JC, Fogaça FH, Sant'Ana LS, Ducatti C, Bandarra N, Carvalho ML, Marques A (2015) Different tools to trace geographic origin and seasonality of croaker (*Micropogonias furnieri*). LWT-Food Sci Technol 61:194–200. https://doi.org/10.1016/j.lwt.2014.11.006

Chaguri MP, Maulvault AL, Costa S, Gonçalves A, Nunes ML, Carvalho ML, Sant'ana LS, Bandarra N, Marques A (2017) Chemometrics tools to distinguish wild and farmed meagre (*Argyrosomus regius*). J Food Process Preserv 41:e13312. https://doi.org/10.1111/jfpp.13312

Dempson JB, Power M (2004) Use of stable isotopes to distinguish farmed from wild Atlantic salmon, *Salmo salar*. Ecol Freshw Fish 13:176–184. https://doi.org/10.1111/j.1600-0633.2004.00057.x

EFSA (European Food Safety Authority) (2005) Opinion of the Scientific Panel on contaminants in the food chain [CONTAM] related to the safety assessment of wild and farmed fish Question number: EFSA-Q-2004-22. 2005. http://www.efsa.europa.eu/EFSA/efsa_locale1178620753812_1178620762697.htm

EFSA (European Food Safety Authority) (2008) Opinion of the Scientific Panel on contaminants in the food chain to a request from the European Parliament related to the Safety Assessment of Wild and Farmed Fish, Question No EFSA-Q-2004-22; adopted on 22 June 2005. EFSA J 236:1–118

Farabegoli F, Pirini M, Rotolo M, Silvi M, Testi S, Ghidini S, Zanardi E, Remondini D, Bonaldo A, Parma L, Badiani A (2018) Toward the authentication of European Sea Bass origin through a combination of biometric measurements and multiple analytical techniques. J Agric Food Chem 66:6822–6831. https://doi.org/10.1021/acs.jafc.8b00505

Fasolato L, Novelli E, Salmaso L, Corain L, Camin F, Perini M, Antonetti P, Balzan S (2010) Application of nonparametric multivariate analyses to the authentication of wild and farmed European sea bass (*Dicentrarchus labrax*). Results of a survey on fish sampled in the retail trade. J Agric Food Chem 58:10979–10988. https://doi.org/10.1021/jf1015126

Filipkowska A, Złoch I, Wawrzyniak-Wydrowska B, Kowalewska G (2016) Organotins in fish muscle and liver from the Polish coast of the Baltic Sea: Is the total ban successful? Mar Pollut Bull 111:493–499. https://doi.org/10.1016/j.marpolbul.2016.06.052

Forina M, Casale M, Oliveri P (2009) Application of chemometrics to food chemistry. In: Tauler R, Walczak B, Brown S (eds) Comprehensive chemometrics—chemical and biochemical data analysis. Elsevier, Amsterdam, pp 75–128. https://doi.org/10.1016/B978-044452701-1.00124-1

Gopi K, Mazumder D, Sammut J, Saintilan N, Crawford J, Gadd P (2019) Isotopic and elemental profiling to trace the geographic origins of farmed and wild-caught Asian seabass (*Lates calcarifer*). Aquaculture 502:56–62. https://doi.org/10.1016/j.aquaculture.2018.12.012

Grembecka M, Szefer P (2011) Metals and metalloids in foods—essentiality, toxicity, applicability. In: Medina DA, Laine AM (eds) Food quality: control, analysis and consumer concerns. Nova Science Publishers, Hauppauge, pp 1–60. ISBN 10: 1611229170. ISBN 13: 9781611229172

Grembecka M, Szefer P (2016) Elemental trace analysis in studies of food products. In: Baranowska I (ed) Handbook of trace analysis—fundamentals and applications. Springer, Berlin, pp 203–239. https://doi.org/10.1007/978-3-319-19614-5

Han C, Dong S, Li L, Wei F, Zhou Y, Gao Q (2020) The effect of the seasons on geographical traceability of salmonid based on multi-element analysis. Food Control 109:106893. https://doi.org/10.1016/j.foodcont.2019.106893

HELCOM (2007) Baltic Sea environment proceedings no. 108 heavy metal pollution to the Baltic Sea in 2004. Helsinki Commission Baltic Marine Environment Protection Commission

Hellou J, Zitko V, Friel J, Alkanani T (1996) Distribution of elements in tissues of yellowtail flounder *Pleuronectes ferruginea*. Sci Total Environ 181:137–146. https://doi.org/10.1016/0048-9697(95)04963-0

Iguchi J, Yamashita Y, Touhata K, Yabu T, Yamashita M (2013) Origin identification method by multiple trace elemental analysis of intermuscular bones in grilled eel products produced in Japan, China, and Taiwan. Fish Sci 79:531–536. https://doi.org/10.1007/s12562-013-0619-4

Julshamn K, Grahl-Nielsen O (1996) Distribution of trace elements from industrial discharges in the Hardangerfjord, Norway: a multivariate data analysis of saithe, flounder and blue mussel as sentinel organisms. Mar Pollut Bull 32:564–571. https://doi.org/10.1016/0025-326X(96)84577-2

Kim H, Kumar KS, Hwang SY, Kang BC, Moon H-B, Shin KH (2015a) Utility of stable isotope and cytochrome oxidase/gene sequencing analyses in inferring origin and authentication of Hairtail fish and shrimp. J Agric Food Chem 63:5548–5556. https://doi.org/10.1021/acs.jafc.5b01469

Kim H, Kumar KS, Shin K-H (2015b) Applicability of stable C and N isotope analysis in inferring the geographical origin and authentication of commercial fish (Mackerel, Yellow Croaker and Pollock). Food Chem 172:523–527. https://doi.org/10.1016/j.foodchem.2014.09.058

Konieczka P, Namieśnik J (2009) Quality assurance and quality control in the analytical chemical laboratory—a practical approach. CRC Press, Taylor & Francis, Boca Raton. https://doi.org/10.1201/9781420082715

Kurle CM, Sinclair EH, Edwards AE, Gudmundson CJ (2011) Temporal and spatial variation in the $\delta^{15}N$ and $\delta^{13}C$ values of fish and squid from Alaskan waters. Mar Biol 158:2389–2404. https://doi.org/10.1007/s00227-011-1741-4

Li L, Boyd CE, Dong S (2015) Chemical profiling with modeling differentiates Ictalurid cat fish produced in fertilized and feeding ponds. Food Control 50:18–22. https://doi.org/10.1016/j.foodcont.2014.08.014

Liu Z, Yuan Y, Zhao Y, Zhang Y, Nie J, Shao S, Rogers KM (2020) Differentiating wild, lake-farmed and pond-farmed carp using stable isotope and multi-element analysis of fish scales with chemometrics. Food Chem 328:127115. https://doi.org/10.1016/j.foodchem.2020.127115

Marini M (ed) (2013) Chemometrics in food chemistry. Elsevier B.V., Amsterdam. ISBN: 9780444595287

Marpaung AM, Syahrun SN, Ramli M, Idris N, Khumaeni A, Budi WS, Suyanto H, Suliyanti MM, Karnadi I, Tanra I, Pardede M, Jobiliong E, Hedwig R, Lie ZS, Kurniawan KH, Kagawa K (2020) Emission spectrochemical analysis of soft samples including raw fish by employing laser-induced breakdown spectroscopy with a subtarget at low-pressure helium gas. ACS Omega 5:16811–16818. https://doi.org/10.1021/acsomega.0c01904

Martín-Pérez M, Fernández-Borràs J, Ibarz A, Felip O, Gutiérrez J, Blasco J (2011) Stable isotope analysis combined with metabolic indices discriminates between gilthead Sea Bream (*Sparus aurata*) fingerlings produced in various hatcheries. J Agric Food Chem 59:10261–10270. https://doi.org/10.1021/jf201670t

Martin-Pérez M, Fernandez-Borras J, Ibarz A, Felip O, Fontanillas R, Gutierrez J, Blasco J (2013) Naturally occurring stable isotopes reflect changes in protein turnover and growth in Gilthead Sea Bream (*Sparus aurata*) juveniles under different dietary protein levels. J Agric Food Chem 61:8924–8933. https://doi.org/10.1021/jf402617h

Mielcarek K, Puścion-Jakubik A, Gromkowska-Kępka KJ, Soroczyńska J, Karpińska E, Markiewicz-Żukowska R, Naliwajko SK, Moskwa J, Nowakowski P, Borawska MH, Socha K (2020) Comparison of zinc, copper and selenium content in raw, smoked and pickled freshwater fish. Molecules 25:3771. https://doi.org/10.3390/molecules25173771

Mielcarek K, Nowakowski P, Puścion-Jakubik A, Gromkowska-Kępka KJ, Soroczyńska J, Markiewicz-Żukowska R, Naliwajko SK, Grabia M, Bielecka J, Żmudzińska A, Moskwa J, Karpińska E, Socha K (2022) Arsenic, cadmium, lead and mercury content and health risk assessment of consuming freshwater fish with elements of chemometric analysis. Food Chem 379:132167. https://doi.org/10.1016/j.foodchem.2022.132167

Molkentin J, Meisel H, Lehmann I, Rehbein H (2007) Identification of organically farmed Atlantic salmon by analysis of stable isotopes and fatty acids. Eur Food Res Technol 224:535–543. https://doi.org/10.1007/s00217-006-0314-0

Molkentin J, Lehmann I, Ostermeyer U, Rehbein H (2015) Traceability of organic fish—authenticating the production origin of salmonids by chemical and isotopic analysis. Food Control 53:55–66. https://doi.org/10.1016/j.foodcont.2015.01.003

Moreno Rojas JM, Serra F, Giani I, Moretti VM, Reniero F, Guillou C (2007) The use of stable isotope ratio analyses to discriminate wild and farmed gilthead sea bream (*Sparus aurata*). Rapid Commun Mass Spectrom 21:207–211. https://doi.org/10.1002/rcm.2836

Morrison DJ, Preston T, Bron JE, Hemderson RJ, Cooper K, Strachan F, Bell JG (2007) Authenticating production origin of gilthead sea bream (*Sparus aurata*) by chemical and isotopic fingerprinting. Lipids 42:537–545. https://doi.org/10.1007/s11745-007-3055-3

Oliveira EJVM, Sant'Ana LS, Ducatti C, Denadai JC, de Souza Kruliski CR (2011) The use of stable isotopes for authentication of gadoid fish species. Eur Food Res Technol 232:97–101. https://doi.org/10.1007/s00217-010-1367-7

Oliveri P, Malegori C, Casale M (2020) Chemometrics: multivariate analysis of chemical data. In: Pico Y (ed) Chemical analysis of food: techniques and applications. Academic Press, New York, pp 33–76. https://doi.org/10.1016/B978-0-12-813266-1.00002-4

Perez de Nanclares M, Dessen JE, Rorvik KA, Thomassen Y, Thomassen MS (2016) Feasibility of using rare earth elements (REEs) to mark and identify escaped farmed Atlantic salmon *Salmo salar* L. Aquac Res 47:1885–1898. https://doi.org/10.1111/are.12647

Polak-Juszczak L (2009) Temporal trends in the bioaccumulation of trace metals in herring, sprat, and cod from the southern Baltic Sea in the 1994-2003 period. Chemosphere 76:1334–1339. https://doi.org/10.1016/j.chemosphere.2009.06.030

Polak-Juszczak L (2010) Bioaccumulation and temporal trends of trace elements in flounder from the southern Baltic Sea for the 1996-2003 period. J Toxic Environ Health Part A 73:1186–1193. https://doi.org/10.1080/15287394.2010.491776

Polak-Juszczak L (2013) Trace elements in the livers of cod (*Gadus morhua* L.) from the Baltic Sea: levels and temporal trends. Environ Monit Assess 185(1):687–694. https://doi.org/10.1007/s10661-012-2584-2

Rahman MS, Molla AH, Saha N, Rahman A (2012) Study on heavy metals levels and its risk assessment in some edible fishes from Bangshi River, Savar, Dhaka, Bangladesh. Food Chem 134:1847–1854. https://doi.org/10.1016/j.foodchem.2012.03.099

Rampazzo F, Tosi F, Tedeschi P, Gion C, Arcangeli G, Brandolini V, Giovanardi O, Maietti A, Berto D (2020) Preliminary multi analytical approach to address geographic traceability at the intraspecific level in Scombridae family. Isot Environ Health Stud 56:260–279. https://doi.org/10.1080/10256016.2020.1739671

Rodushkin I, Bergman T, Douglas G, Engström E, Sörlin D, Baxter DC (2007) Authentication of Kalix (N.E. Sweden) vendace caviar using inductively coupled plasma-based analytical techniques: evaluation of different approaches. Anal Chim Acta 583:310–318. https://doi.org/10.1016/j.aca.2006.10.038

Roy PK, Lall SP (2006) Mineral nutrition of haddock *Melanogrammus aeglefinus* (L.): a comparison of wild and cultured stock. J Fish Biol 68:1460–1472. https://doi.org/10.1111/j.0022-1112.2006.001031.x

Sant'Ana LS, Ducatti C, Ramires DG (2010) Seasonal variations in chemical composition and stable isotopes of farmed and wild Brazilian freshwater fish. Food Chem 122:74–77. https://doi.org/10.1016/j.foodchem.2010.02.016

Szefer P (2002) Metals, metalloids and radionuclides in the Baltic Sea ecosystem. Elsevier Science B.V., Amsterdam. ISBN 10: 0444503528

Szefer P (2003) Application of chemometric techniques in analytical evaluation of biological and environmental samples. In: Namieśnik J, Chrzanowski W, Żmijewska P (eds) New horizons and challenges in environmental analysis and monitoring. CEEAM, Gdańsk, pp 355–388. ISBN 8391908100

Szefer P (2007) Chemometric techniques in analytical evaluation of food quality. In: Szefer P, Nriagu J (eds) Mineral components in foods. CRC Press, Taylor & Francis, Boca Raton, pp 69–121. https://doi.org/10.1201/9781420003987

Szefer P (2013) Safety assessment of seafood with respect to chemical pollutants in European Seas. Oceanol Hydrobiol Stud 42:110–118. https://doi.org/10.2478/s13545-013-0063-1

Szefer P, Domagała-Wieloszewska M, Warzocha J, Garbacik-Wesołowska A, Ciesielski T (2003) Distribution and relationships of mercury, lead, cadmium, copper and zinc in perch (*Perca fluviatilis*) from the Pomeranian Bay and Szczecin Lagoon, southern Baltic. Food Chem 81:73–83. https://doi.org/10.1016/S0308-8146(02)00380-1

Thomas F, Jamin E, Wietzerbin K, Guérin R, Lees M, Morvan E, Billault E, Derrien S, Moreno Rojas JM, Serra F, Guillou C, Aursand M, McEvoy L, Prael A, Robins RJ (2008) Determination of origin of Atlantic Salmon (*Salmo salar*): the use of multiprobe and multielement isotopic analyses in combination with fatty acid composition to assess wild or farmed origin. J Agric Food Chem 56:989–997. https://doi.org/10.1021/jf072370d

Tsakovski S, Simeonov V (2010) Chemometrics as a tool for treatment processing of multiparametric analytical data sets. In: Namieśnik J, Szefer P (eds) Analytical measurements in aquatic environments. CRC Press, Taylor & Francis, Boca Raton, pp 369–387. https://doi.org/10.1201/9781420082692

Turchini GM, Quinn GP, Jones PL, Palmeri G, Gooley G (2009) Traceability and discrimination among differently farmed fish: a case study on Australian Murray Cod. J Agric Food Chem 57:274–281. https://doi.org/10.1021/jf801962h

Varrà MO, Ghidini S, Zanardi E, Badiani A, Ianieri A (2019) Authentication of European sea bass according to production method and geographical origin by light stable isotope ratio and rare earth elements analyses combined with chemometrics. Ital J Food Saf Mar 22:7872. https://doi.org/10.4081/ijfs.2019.7872

Vasconi M, Lopez A, Galimberti C, Moreno Rojas JM, Muñoz Redondo JM, Bellagamba F, Moretti VM (2019) Authentication of farmed and wild European eel (*Anguilla Anguilla*) by fatty acid profile and carbon and nitrogen isotopic analyses. Food Control 102:112–121. https://doi.org/10.1016/j.foodcont.2019.03.004

Wang YV, Wan AHL, Lock E-J, Andersen N, Winter-Schuh C, Larsen T (2018) Know your fish: a novel compound-specific isotope approach for tracing wild and farmed salmon. Food Chem 256:380–389. https://doi.org/10.1016/j.foodchem.2018.02.095

Xuan Z, Wang W-X (2023) Chemometric analysis of fish fin ray multi-elemental signatures with differentiation of wild and farmed fourfinger threadfin fish. J Food Compos Anal 123:10559. https://doi.org/10.1016/j.jfca.2023.105597

Yamashita Y, Omura Y, Okazaki E (2006) Distinct regional profiles of trace element content in muscle of Japanese eel *Anguilla japonica* from Japan, Taiwan, and China. Fish Sci 72:1109–1113. https://doi.org/10.1111/j.1444-2906.2006.01263.x

Ye S, Yang J, Liu H, Oshima Y (2011) Use of elemental fingerprint analysis to identify localities of collection for the large icefish *Protosalanx chinensis* in Taihu Lake, China. J Fac Agr Kyushu Univ 56:41–45. https://doi.org/10.5109/19634

Chapter 4
Seafood: Molluscs and Crustaceans

Contents

4.1 Introduction

In addition to fish, seafood also includes shellfish such as shrimp, crab, clams and oysters. Seafood is a high-protein and low-calorie food that is low in total fat as well as saturated fat and Na. Seafood is also one of the few foods that contains long-chain omega-3 fatty acids and is high in vitamins and minerals. Owing to these unique nutritive properties, seafood has many beneficial health effects.

Szefer (2013) reported that the bioaccumulation of environmental pollutants in marine fish and other seafood has been introduced within the criteria of the European Union food quality standards. The health risks associated with the human consumption of seafood have been assessed mainly by the EFSA, an independent European agency funded by the EU. The potential health risks arising from the consumption of "frutti di mare" were examined from the standpoint of estimated PTWIs for organic and inorganic toxic pollutants, including BTs, MeHg, Cd, As, Pb and their organic compounds. Seafood, such as shellfish and crustaceans, which contain elevated levels of Cd, is therefore an important source of this element in consumer

Supplementary Information The online version contains supplementary material available at https://doi.org/10.1007/978-3-031-99687-0_4.

P. Szefer, *Chemometrics in Food Control Based on Elemental Analysis*, https://doi.org/10.1007/978-3-031-99687-0_4

tissues. To provide an updated assessment of exposure from foodstuffs, 140,000 data points from 2003 to 2007 on Cd occurrence in various food commodities (obtained from 20 Member States) were considered by the EFSA Panel. The highest Cd levels were detected in seaweed, fish and seafood, chocolate and foods for special dietary uses. Up to 20% of the samples had excessive MLs in celeriac, horse meat, fish and frutti di mare other than oysters and cephalopods. Locally produced food from highly polluted areas may contain relatively high Cd levels. Regular consumers of bivalve molluscs are also seriously exposed (4.6 μg Cd kg^{-1} b.w. $week^{-1}$) (EFSA 2009a, b). Interestingly, a downwards seasonal trend was noted in the concentration of Cd in four Baltic species of fish: herring, sprat, cod and flounder (Polak-Juszczak 2009, 2010). A similar downwards trend was reported by Polak-Juszczak (2013) for hepatic cadmium in Baltic cod from 1994 to 2010. Besada et al. (2009) determined the contents of Cu and Zn by AAS; Cd, Pb and total As by ETA-AAS; total Hg by the CV-AAS and inorganic As by HG-AAS in edible seaweeds commercialised for human consumption. The original data obtained were processed by PCA, CA and LDA. The investigations made it possible to determine whether seaweeds can be differentiated objectively by employing chemical measurements.

Seafood, including fish, can be a major source of As in the human diet and in animal feeds that contain products derived from fish or other marine organisms. In seafood and fish, As is present predominantly in the organic forms of arsenobetaine and arsenocholine, which, unlike inorganic forms, are virtually non-toxic. Although many regional and ethnic populations consume large quantities of fish, resulting in organoarsenic intakes of approximately 50 μg kg^{-1} b.w. day^{-1}, there are no reports of toxicity to humans or animals from the consumption of organoarsenicals in seafood. Some studies suggest that seafood is a possible contributor to inorganic As exposure. The PTWI for inorganic As is 15 μg kg^{-1} b.w. $week^{-1}$ (EFSA 2005).

According to Szefer (2013), much attention has recently been focused on the pollution of seafood by OTC. TBT, with its breakdown products, can be biomagnified along the food chain at particular trophic levels, e.g. shellfish, squid and fish, and in top predators such as whales, dolphins, seals and fish-eating waterfowl. TARLs for TBT in seafood products have been reported; they were calculated on the basis of the TDI of TBT and average seafood consumption in 24 countries (Belfroid et al. 2000). Seafood other than fish generally contains higher levels of OTC than do fish. The intakes of large-scale consumers, calculated from median and mean concentrations, were 0.037 and 0.17 μg kg^{-1} b.w. day^{-1}, respectively, which represent approximately 15% and 70% of the group TDI. The EFSA Panel noted that the consumption of fish, mussels and other marine animals from highly polluted areas, such as the vicinity of harbours and heavily used shipping routes, might lead to OTC intake exceeding the group TDI (EFSA 2004a, b). The key issue is the proper selection of the method used to quantify the geographic origin of aquaculture products, which should not be susceptible to stable elements being affected by parameters other than exclusively geographical factors. Feed should also be considered identifiable in aquaculture products since its qualitative and quantitative mineral composition is an extraordinary factor in fish nutrition. According to Gopi et al. (2019a), among the most common methods used for investigating seafood are DNA profiling, FAA and other techniques, such as ICPS and SIRA. The authentication of fishery and aquaculture products with MEA and SIRA has been characterized by Li et al. (2015).

The elemental composition of various species of seafood appears to be dependent on local environmental conditions and has been effectively employed for product authentication (Gopi et al. 2019a and others). A rich assortment of seafood (e.g. various species of mussels and shrimp), seaweed and macroalgae is consumed in Asian cultures, and these foods are becoming increasingly popular in Europe and other Western countries. However, seaweed can accumulate trace elements and therefore can be exposed to heavy metals (especially Hg, Cd and Pb), as well as As, and hence may pose a health risk (Guo et al. 2023). Moreover, methods such as blockchain, radio frequency identification and X-ray fluorescence through ITRAX (μXRF) are currently being tested for their usefulness in quantifying seafood origin (Gadd et al. 2018). However, these individually used methods have drawbacks; hence, it is preferable to apply a combination of methods, resulting in greater effectiveness in determining the origin of seafood products. The market for fishery and aquaculture products has been internationalized with an increasing number of misbranded products; hence, there is a need for the development and application of proper tools for identifying the provenance of these products. Among the analytical methods/techniques used to identify the origin of seafood products, MEA and SIRA are recommended as reliable tools for distinguishing fishery products and aquaculture products. A key issue is the validation of the analytical measurements using CRMs as well as the chemometric assessment, especially the multivariate interpretation of experimental data for the authentication of aquaculture products (Szteke 2007; Konieczka and Namieśnik 2010; Bulska 2010; Tsakovski and Simeonov 2010; Capar and Szefer 2012; Grembecka and Szefer 2011, 2016; Szefer 2002, 2003, 2007; Szefer and Grembecka 2022).

Chemometrics appears to be a reliable tool in investigations of analytical databases; it is effective in identifying the similarity or dissimilarity between objects and measured variables. Multivariate approaches have been entirely acceptable for fulfilling the principal objectives of chemometrics, as contemporary multiparametric information enables the understanding of the knowledge hidden in a large data matrix. The goal of the multivariate approach is to break down the mixed data structure into its components. It is possible to reduce relatively numerous variables to a smaller number of orthogonal factors by the use of chemometric techniques, i.e. PCA, FA, LDA, CDA, CA, DFA, PLS-DA, ANN, etc. (Szefer 2007). Crucial issues related to the use and misuse of chemometrics in dealing with classification problems have been reported by Defernez and Kemsley (1997). Notably, chemometrics is a reliable tool for the analytical assessment of fishery and aquaculture product quality. Combined techniques, in particular, may be especially useful in detecting fraud to preserve the brand reputations of original products as well as for protecting the consumer from being cheated financially.

4.2 Chemometric Assessment of Seafood

4.2.1 Geographical Differentiation

4.2.1.1 European Countries

Mollusca

Kobelja et al. (2016) investigated the content of the multi-elemental profile of Noah's Ark shell via ICP–OES. The original data obtained were processed chemometrically. The first two PCs, which explained 52.44% of the total variance, distinguished two geographical regions of the eastern Adriatic Sea, i.e. the shipping route in the sheltered Pašman channel and the marine protected area of Telašćica Bay. The former can be identified by loadings such as B, Cu, Ni, Pb, K, Mn, Cd and Na, whereas the latter is associated with discriminative variables such as Ca, Sr, Mg, Fe, Al, Zn, Ag and Co. It was concluded that shells of the Croatian mussel are useful biomarkers for classifying the samples studied with respect to their interregional origin.

Szefer et al. (2002) reported that MEC data quantified by FAAS and CV–AAS allow the registration of intersize variations in the contents of some elements; i.e. the tissue levels of Ni and Mn decreased significantly with the shell length of blue mussels from various sites within three sectors of the southern Baltic Sea, i.e. Pomeranian Bay, Słupsk Bank and the Gulf of Gdańsk. Moreover, there was also a strong relationship between the contents of Cd, Pb, Ag and Ni in the soft tissue and byssus. FA, with the first factor describing 44.0% of the total variance, allows satisfactory discrimination between the soft tissues of a mussel with respect to geographical provenance. Molluscs from the Gulf of Gdańsk were clearly differentiated from those from Pomeranian Bay and Słupsk Bank (Fig. 4.1). These findings may reflect the occurrence of different rates of deposition of clay minerals at the head of the former and the Pomeranian Bay. Elements such as Cr, Ag, Co and Fe are the most discriminative variables for identifying interspatial differentiation.

Mamede et al. (2021) evaluated ICP–MS data relative to species-specific and interspatial variability associated with the EFs of the shells of three commercially important bivalve species (*Cerastoderma edule*, *Ruditapes decussatus* and *R. philippinarum*) from five locations on the northwestern and western Iberian coasts (Spain and Portugal). A reference model based on one of these species was tested to determine whether the model could be used to reliably monitor the geographic origin of specimens from the two other species. The EFs of the shells of the three species were considerably different within each location, as well as among various locations within each species. The RF enabled the correct classification of samples, ranging from 95.3% to 99.3%. Tests for deducing whether these models could reliably identify the geographic origin of samples from another species demonstrated that the overall accuracy ranged from 50% to 82.3%. This means that the three species, especially *C. edule*, may be used as surrogates for each other. This approach

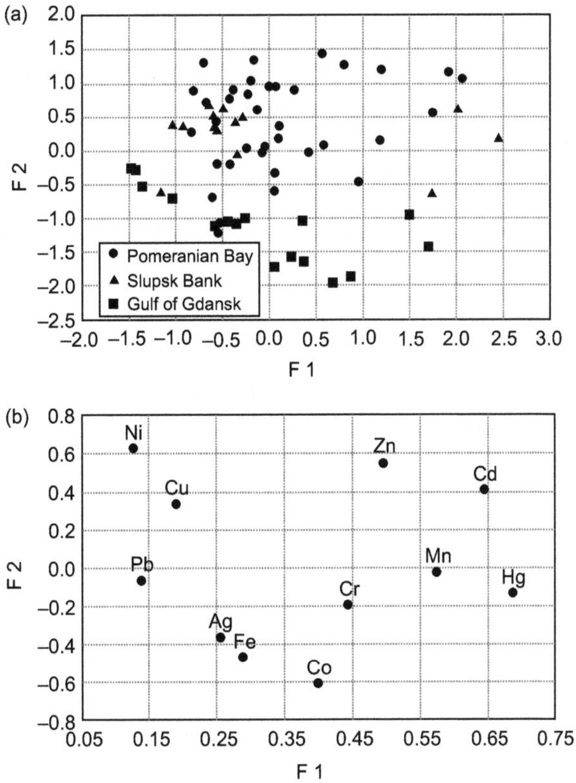

Fig. 4.1 Biplot of scores (**a**) and selected metals (**b**) corresponding to soft tissue of *Mytilus edulis trossulus* from three sectors of the southern Baltic. (After Szefer et al., 2002. Environ Pollut 120:423–444. https://doi.org/10.1016/S0269-7491(02)00111-2. With permission)

improves the cost efficiency of using the EFs of bivalve shells to trace and confirm the geographic provenance of molluscs. Przytarska et al. (2010), using FAAS, analysed the concentrations of Cu, Fe, Mn, Pb and Zn in the blue mussel *Mytilus* spp., which originated from 17 sampling sites in coastal waters of European countries, i.e. the Bay of Biscay, English Channel, Bay of Faxaflói, North Sea, Baltic Sea, Mediterranean Sea and Black Sea. Since the mussels were sampled between January 2002 and December 2004, and since the authors suspected that a given area can differ temporally, they corrected the data for the effect of season before the assessment of large-scale spatial inter-comparisons. The first two factors accounted for 64.4% of the total variance; mussels from three regions of the southern Baltic were well distinguished from the remaining samples, including those from the Black Sea (with two sampling regions) and the Bay of Faxaflói (with two sampling regions). The highest levels of Fe and Mn were observed in the North Sea and Baltic Sea, respectively, whereas elevated tissue contents of Pb were detected in the mussels from the Bay of Biscay and the Baltic Sea. However, low mussel levels of metals were identified in the Mediterranean Sea and the northern Baltic Sea. The relatively low geographic differentiation of Cu and Zn concentrations in blue mussels can be explained by the fact that mussels are able to partially regulate accumulated body concentrations, which means that Cu and Zn are, to some extent, independent of their environmental levels.

In Morrison et al. (2019), 30 minor and trace elements were determined by ICP–MS in samples of king scallops (*Pecten maximus*) collected in January 2017 from each of three bays on the west coast of Ireland, with additional individuals collected from the Ballinakill site 6 weeks later. The spatial–temporal TEF of scallops revealed the harvesting period and location (Fig. 4.2). The use of MDS in shells and soft tissues can predict geographic origin with 100% success.

Echinodermata

TXRF data processed using chemometrics were applied to the elemental fingerprints of gonads to trace the sampling locations of sea urchins originating from the North Atlantic Portuguese coast. The most promising results were achieved when the model built with CAP was used. According to the authors, the correct allocation was greater in the southern locations, i.e. Lisbon and Sines, with values of 97.5%

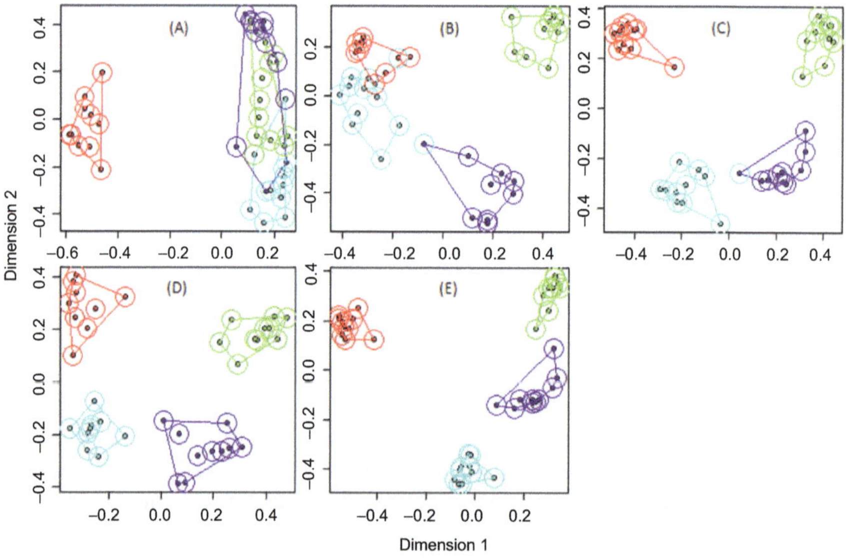

Fig. 4.2 Multidimensional scaling ordinations of proximity scores from random forest classifications based on the elemental concentrations of the scallop (**a**) adductor muscle, (**b**) clean shell, (**c**) unclean shell, (**d**) clean shell and adductor muscle and (**e**) unclean shell adductor muscle. Points are coloured according to their site of collection, and the outer ring colour indicates the site at which each individual scallop was classified from the random forest analyses. A convex hull is drawn around the a priori classes, with points coloured according to the site of collection. Red, Ballinakill Bay date 1; Green, Ballinakill Bay date 2; Blue, Castletownbere; Purple, Glengarriff. (After Morrison L et al., 2019. Sci Total Environ 697:134121. https://doi.org/10.1016/j.scitotenv.2019.134121. With permission)

and 62.5%, respectively, despite the high intra-annual variation in the gonad index of sea urchins (Mamede et al. 2022).

4.2.1.2 Asiatic Countries

Mollusca

Kang et al. (2022) used IRMS to differentiate *Mytilus edulis* samples from four major Chinese production provinces. PCA and LDA were applied to evaluate the samples with respect to classification or predictive ability. Using PCA, it has not been possible to clearly differentiate mussels in terms of their geographical origin. Fortunately, LDA achieved an overall accuracy rate of 92.8%, with a cross-validated accuracy rate of 86.7% and a predictive accuracy rate of 92.3%. It has been postulated that SIR and analysis of C, N, O and H compositions are potentially useful for determining the geographical origin of Chinese mussels.

Zhao et al. (2019) reported that IRMS integrated with LDA models enabled the identification of Chinese Yesso scallops (*Patinopecten yessoensis*) from six sampling locations. The $\delta^{13}C$ values were marginally affected by season. On the basis of ten parameters, such as the $\delta^{13}C$ values of alanine, glycine, valine, leucine, serine, phenylalanine, fucose, rhamnose, glucose and mannose, it has been possible to identify the geographical origin of scallops with an overall correct classification rate of 100.0% and a CV rate of 100.0%.

Iguchi et al. (2013) obtained MEA data (Li, V, Mn, Co, As, Rb, Mo, Ba, Pb and U) for shells of short-neck clams from various sampling sites in Japan, China and Korea. Owing to the development of four LDA models, intercountry differentiation of the samples studied was achieved. Correct classification ranging from 80.0% to 100% resulted in the identification of the country of origin.

Arthropoda

Li et al. (2017) studied shrimp from three major exporting countries, i.e. India, Vietnam and Thailand, using ICP–AES combined with chemometric tools, including PCA, SDA, the kernel method and CDA. The first two PCs, corresponding to 46.52% of the total variance, clearly differentiated the aquacultured shrimp samples from India and Vietnam owing to descriptors such as B, Cu, Zn, Mg, K, Na, P and S, with the highest weight on PC1. A group of PCA scores related to Thailand was located between two clusters formed by the Indian and Vietnamese object samples. Better results were achieved using the SDA technique. It was possible to classify shrimp effectively into three countries, with percentages of observations correctly classified as 99.4% of the empirical grouped observations and 97.2% of the CV grouped observations. SDA was also conducted for shrimp from the six sampling areas, i.e. the two sampling areas in Thailand, the three provinces in Vietnam and the single sampling area in India. The correct classification rates obtained by discriminant analysis and the CV procedure were 88.3% and 83.3%, respectively. This means that the differentiation on a smaller scale was not as reliable as that on the country scale; hence, to improve the effectiveness of this study, the use of more datasets was postulated.

Echinodermata

The combination of EA–IRMS and GC/MS SIRA with multivariate tools is a useful methodology for classifying sea cucumber samples originating from major producing areas of the northern China Sea. The $\delta^{13}C$ and $\delta^{15}N$ values, combined with FA data and then processed via PCA, resulted in the successful distinguishing of the samples according to their geographic provenance. It has been demonstrated that PCA and DA can be promising methods for determining the origins of sea cucumber samples from the northern China Sea (Zhang et al. 2017).

4.2.1.3 American Countries

Arthropoda

Li et al. (2014) reported that ICP–AES appears to be a useful tool in studies of Pacific white shrimp (*Litopenaeus vannamei*) caught in three US geographic areas, i.e. from five ponds at the Greene Prairie Aquafarm, eight ponds at the Claude Peteet Mariculture Center and five ponds at the Harlingen Shrimp Farm, as well as water from culture ponds. PCA differentiated the geographical origin of the shrimp with respect to the three regions. The CDA technique appears to be a more effective tool for distinguishing between regions. kNN was responsible for building the classification model. The shrimp from the three sampling sites were clearly differentiated, and the rate of reliability of the model estimated by a CV procedure was 100%.

4.2.1.4 Countries Across Multiple Continents

Mollusca

Iguchi et al. (2014) reported that MEA shell data of *Corbicula* clams from various sampling sites in Japan, Russia, China and Korea revealed significantly different concentrations between Japanese and Russian populations of clams with an LDA discriminant model. The correct classification of the origin was evaluated to be 89.8% for Japanese objects and 92.2% for Russian objects. Tenfold CV confirmed the effectiveness of the developed discriminant function model. It was concluded that trace element analysis of *Corbicula* clam shells could be utilized for the identification of their country of provenance.

Arthropoda

Smith and Watts (2009) presented a database of MEC data obtained by ICP–MS, including data from farm-raised shrimp (Family Penaeidae) related to more than 100 sampling sites from 8 Asian and American countries, along with the accuracy of each site relative to its known geographic origin. The intercountry discrimination was obtained using trace element profiling and multivariate chemometrics, i.e. LDA, SDA and CDA. The accuracy of all of the reference samples when one country was compared against another was demonstrated. The classification accuracy was >90%, regardless of which two countries were selected with few exceptions.

4.2.2 *Differentiation of Other Parameters*

4.2.2.1 Verifying Authenticity/Adulteration

Mollusca

Mamede et al. (2024) used ORS–ICP–MS to quantify the MEC content in *R. philippinarum* of commercial size from three estuaries along the western Portuguese coast. The potential of using the EF of Manila clam shells to monitor geographic provenance under real-case scenarios was confirmed. Two reference models, i.e. CAP and RF, were developed using the EF of the samples from the estuaries explored. The models confirmed the geographic origin of Manila clams with 100% accuracy for CAP and 97.8% accuracy for RF. The same models predominantly allocated samples from B1 and B2 to TE with 93.3% and 100% for CAP and 100% and 93.3% for RF, respectively. Mn and P appeared to be helpful in the classification of reference models. The potential of using the EF of clam shells was confirmed to be effective for identifying the geographic origin of Manila clams under real-case scenarios, aiding the prevention of false seafood labelling as well as illegal, unregulated and unreported fishing.

Costas-Rodríguez et al. (2010) focused on the PDO samples of *Mytilus. galloprovincialis* originating from five Galician Rías and samples of mussels cultivated in the Mediterranean Sea. Using MEC data, including the REE content determined via ICP–MS, correct assignments for the Galician and non-Galician samples were obtained via LDA and SIMCA. Four new CDFs, which explained 100% of the total variance, were achieved when the stepwise inclusion method was computed, and among 40 original variables, 16 were selected for categorizing the mussels related to their interregional origin. LDA with descriptive elements such as Ag, Ba, Cd, Co, Eu, Ga, Lu, Mo, Pb, Sb, Se, Th, U, V, Zn and Cr yielded a prediction ability of 95.6%. The SIMCA appears to be an inappropriate tool for classifying mussel samples according to the Rías of origin; however, in the case of non-Galician samples, none were assigned to Galician Rías. This means that SIMCA is a useful model for identifying non-Galician mussels. When ANNs were used, a classification performance of 100% was attained, improving the classification model.

Won et al. (2021) used a combination of $\delta^{13}C$, $\delta^{15}N$, $\delta^{18}O$, δD, $\delta^{34}S$, $^{87}Sr/^{86}Sr$ and $^{143}Nd/^{144}Nd$ profiles established by EA–IRMS and MC–ICP–MS to classify Manila clam samples from coastal areas of DPR Korea, Korea and China. As shown in Fig. 4.3, a robust LDA allows the clear differentiation of clam samples with respect to their geographical origin. The application of stable multi-element isotope analyses of commercially available Manila clams is useful for identifying fraudulent and mislabelled seafood products. The authors expect that this method could be useful to verify their geographical origins of other shellfish distributed around the world that are imported and exported.

del Rio-Lavín et al. (2022) examined the potential of integrated EA–IRMS and ICP–MS to verify the geographical origin of farmed *M. galloprovincialis* mussels harvested from eight different regions located in the Mediterranean Sea, the European Atlantic coast and the Chilean Pacific coast. The $\delta^{13}C$ and $\delta^{15}N$ values of mussel tissue and the contents of 16 micro- and trace elements in shells were determined. Significant differences were detected between the SIR and MEC signatures of mussels harvested at various sampling sites, especially with respect to different marine regions. The RF classification model based on combined data achieved the most accurate prediction, correctly assigning 97% of the mussel samples to their harvest locations. The most discriminative descriptors were $\delta^{13}C$, $\delta^{15}N$, Pb, Ba, Mn and Al. According to the authors, this approach allowed the development of a reliable original assignment methodology owing to the use of SIR and MEC, which will enforce food safety regulations and prevent commercial fraud.

Arthropoda

Davis et al. (2021) studied the geographic origin of shrimp from retail stores. ICP–MS was used to determine the contents of macro-, micro- and trace elements in shrimp samples collected in Thailand, India, Vietnam, Indonesia and Ecuador. A chemometric tool composed of one-way MANOVA, PCA and RF was useful in classifying the samples studied. The first two PCs explained 26.23% and 19.33% of the total variation, respectively. The shrimp samples from Ecuador were differentiated across PC1, which strongly corresponded to Ce, Co, Cu, Fe and Y, whereas the other four Asian countries did not seem to be differentiated across PC2, which was most strongly related to Al, As, K, Mg and Mn. In general, the RF of the retail samples successfully classified 71.2% of the samples with respect to their origin country via K-fold cross-validation. Elements such as Cs, Cu, Li, K and Mg were discriminative elements in the classification model. The highest rate of success was distinguished in India, where 93% of the samples were correctly assigned as being from this country by the model.

Luo et al. (2019) processed Chinese mitten crab (*Eriocheir sinensis*) from eight sampling sites using integrated ICP–MS and EA–IRMS together with LDA and SVM to distinguish the geographical origin of the crabs studied. As shown in Fig. 4.4, LDA appears to be a useful tool for differentiating crab samples according to their geographic origin. Both the LDA and SVM models showed that 99.4% of

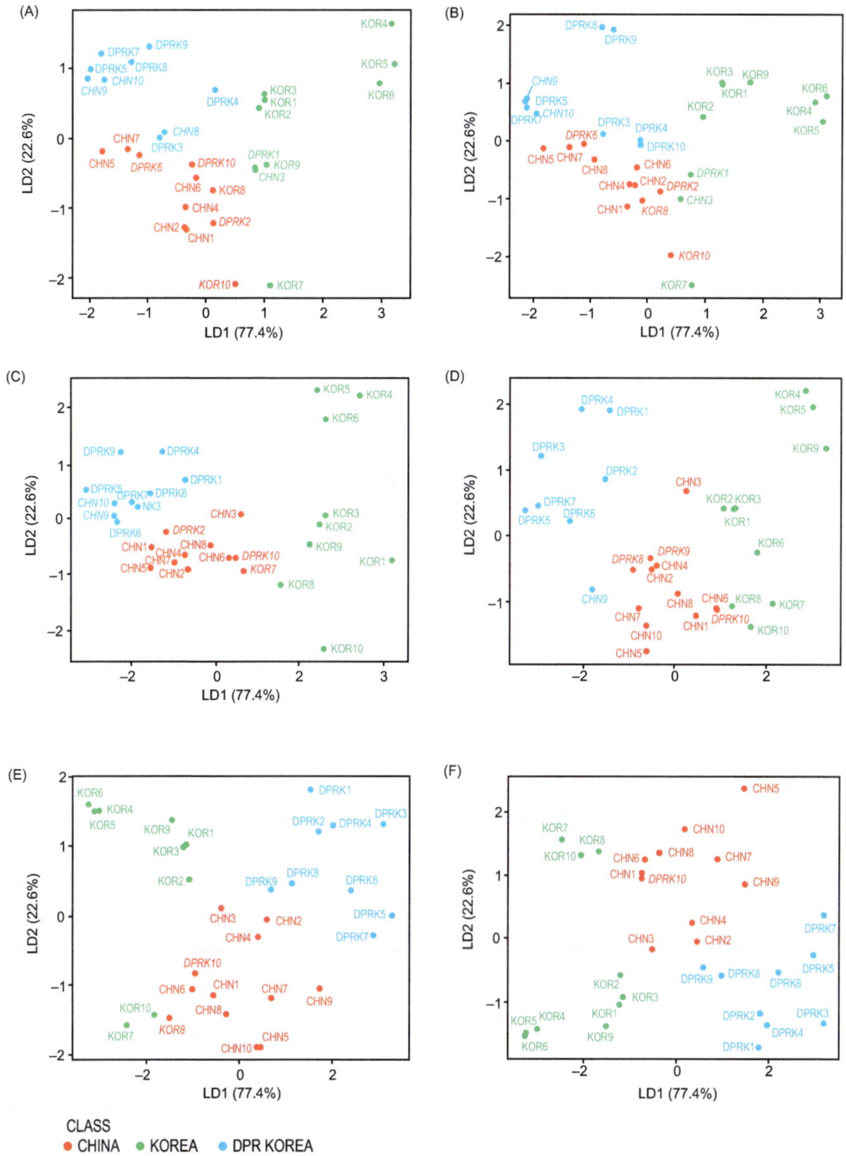

Fig. 4.3 Manila clams from three countries grouped according to LDA findings of combined basic stable isotopes. (**a**) $\delta^{13}C$ and $\delta^{15}N$; (**b**) $\delta^{13}C$, $\delta^{15}N$ and $\delta^{18}O$; (**c**) $\delta^{13}C$, $\delta^{15}N$ and δ D; (**d**) $\delta^{13}C$, $\delta^{15}N$ and $\delta^{34}S$; (**e**) $\delta^{13}C$, $\delta^{15}N$, $\delta^{34}S$ and $\delta^{18}O$ and (**f**) $\delta^{13}C$, $\delta^{15}N$, $\delta^{34}S$, $\delta^{18}O$ and Sr/Sr. Bold italics, which are shown in different colours for the same country, indicate samples that have not been classified as having the provenance that was originally claimed. (After Won E-J et al., 2021. Foods 10:646. https://doi.org/10.3390/foods10030646. With permission)

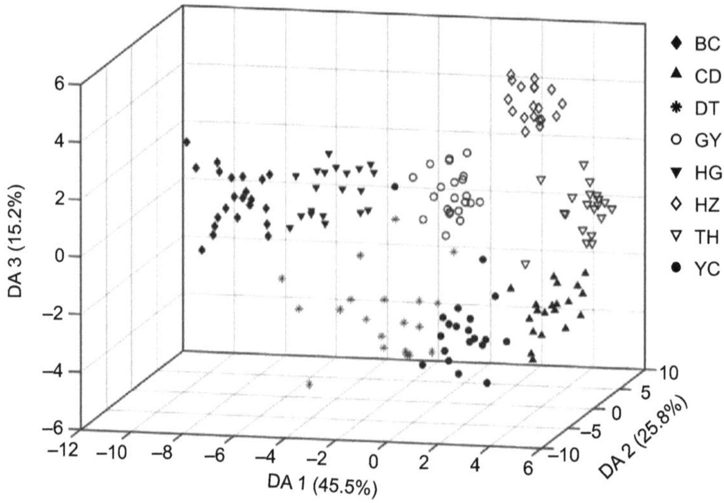

Fig. 4.4 Linear discriminant analysis plot of isotopes and elements of *Eriocheir sinensis* taken from the lower reaches of the Yangtze River, China. Bacheng aquaculture pond (BC); Changdang Lake (CD); Dongtan area of the Yangtze estuary (DT); Gaoyou Lake (GY); Honggao aquaculture pond (HG); Hongzhe Lake (HZ); Taihu Lake (TH); Yangcheng Lake (YC). (After Luo R et al., 2019. Food Chem 274:1–7. https://doi.org/10.1016/j.foodchem.2018.08.104. With permission)

the samples were correctly classified. The study revealed that MEA and SIRA could be effectively used to test the authenticity and origin of *E. sinensis*, providing a reference basis for effectively facilitating a PGI system.

Carter et al. (2015) reported that TC/EA–IRMS and ICP–MS combined with PCA and CDA may be valuable discriminating tools for Australian prawns and imported prawns from three Asian countries, i.e. China, Malaysia and Vietnam. PCA revealed that the contents of K, As, Cu and Zn enable the differentiation of Australian prawns from imported samples. In addition, CDA achieved the best classification. The SIR data with a combination of Zn and As contents made it possible to distinguish Australian and non-Australian samples with 100% correct classification. The data obtained enabled the differentiation with regard to the geographic origin of prawns labelled as Australian from those labelled as imported from Asian countries; thus, this approach is useful for verifying the authenticity of seafood products.

4.2.2.2 Interregional Differentiation Related to the Impact of Pollution

Mollusca

In Gorbi et al. (2008), caged *M. galloprovincialis* mussels, including reference organisms from an unpolluted site and those transplanted for 4 weeks in both the sampling area and the investigated platform, were investigated to evaluate the potential impact of an offshore gas platform in the central Adriatic Sea. The application of FAAS, ET–AAS, CV–AAS, GC and specific spectrophotometric methods allowed the determination of MEC content and biochemical, histological and genotoxic parameters (AOX, AChE, SOD, CAT, GST, GOxH_2O_2, GPxCHP, GSH, TOSC, GR and GSSG) as multibiomarkers for mussel tissues, with early detection of biological responses at several cellular targets. The PCA technique (with the first two PCs accounting for 55% of the total variance) enabled the differentiation of sites and sampling periods on the basis of chemical parameters and biomarker data (Fig. 4.5). Importantly, the mussels originated from the reference site before the caging period and were caged for 4 weeks at the same location, as were the mussels from the reference site, which were caged for 4 weeks to the "Giovanna" platform. The biomarker data were clearly distinguished relative to four sampling periods, i.e. between 2004 and 2005.

Favretto et al. (1997) reported that the edible tissues of *Mytilus galloprovincialis* were taken from two sampling sites, i.e. at the end of a pier near the sewage of Trieste as a polluted area of Muggia Bay and in a hatchery just off the southern end of the unpolluted bay. FAAS and CV–AAS combined with LDA better

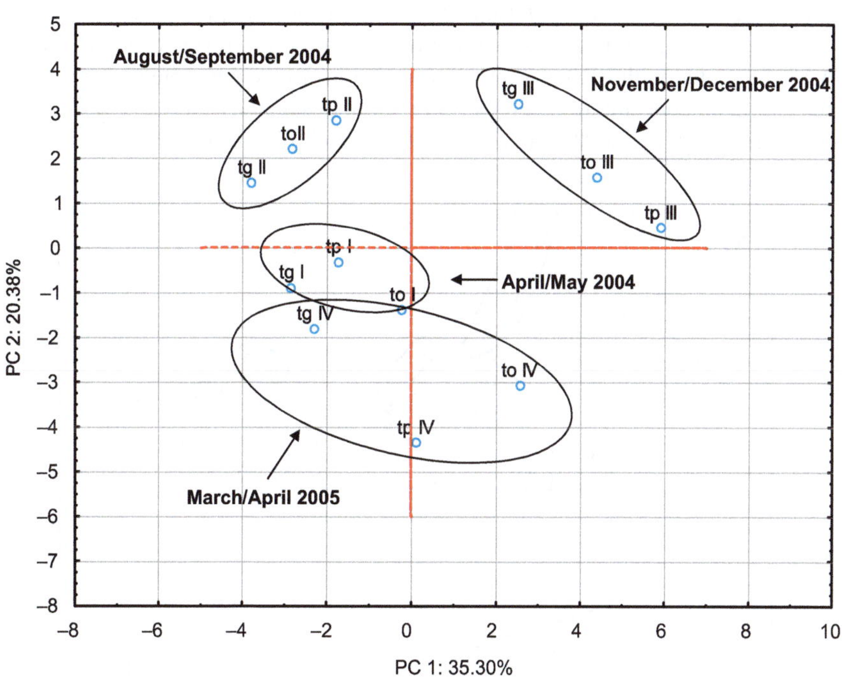

Fig. 4.5 Separation of sites and sampling periods obtained from PCA analyses of chemical parameters and biomarker data. to: mussels from the reference site before the caging period; tp: mussels from the reference site caged for 4 weeks at the same location; tg: mussels from the reference site caged for 4 weeks to the "Giovanna" platform; (I) April/May 2004, (II) August/September 2004, (III) December/January 2004, (IV) March/April 2005. (After Gorbi S et al., 2008. Mar Environ Res 65:34–49. https://doi.org/10.1016/j.marenvres.2007.07.006. With permission)

differentiated these two groups of samples, with high accuracy, 97.7%, but LPCA (with the first two eigenvectors accounting for 50% of the total variance) was not able to differentiate samples from these areas completely.

Earlier studies performed by Favretto and Favretto (1989) also differentiated contaminated molluscs from non-contaminated molluscs owing to the use of ET–AAS supported by PCA with the first three or two PCs, which explained 70–80% of the total variance.

Julshamn and Grahl-Nielsen (1996), using a combination of FAAS, ET(GF)–AAS and CV–AAS, determined heavy metals in pooled samples of blue mussels collected at eight sites and from one mussel farm. With respect to the PCA, with the first two PCs corresponding to 89.0% of the total variance, samples from the different sampling sites were clearly distinguished (Fig. 4.6). For example, the mussels from sampling site no. 3 deviated from the gradient with somewhat higher heavy metal contents than expected. This suggests that the input of some elements to the Norwegian fjord depends on the town of Norheimsund, where some small industries and a petrol station are located.

Szefer et al. (1998) determined the concentrations of ten micro- and trace elements in soft tissues and byssal threads of *Mytella strigata* and other benthic organisms, including the clam *Chione subrugosa,* from four sampling sites in a tropical mangrove lagoon, i.e. Estero de Urias Lagoon on the northwest coast of Mexico. FAAS integrated with LRA was an effective tool for revealing interregional differences in metal concentrations in *Mytella strigata*, especially Ni, Cu and Mn in the soft tissues and byssal threads. *M. strtigata* and *C. subrugosa* were effective

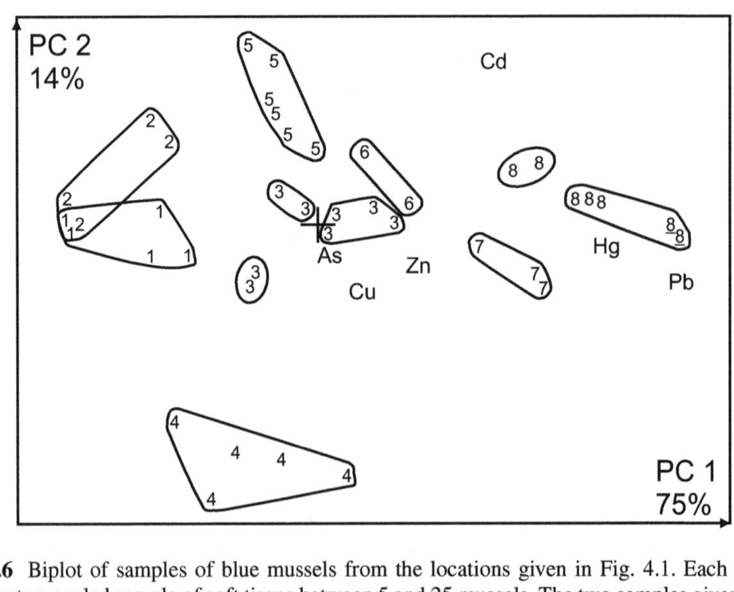

Fig. 4.6 Biplot of samples of blue mussels from the locations given in Fig. 4.1. Each number represents a pooled sample of soft tissue between 5 and 25 mussels. The two samples given in italics from location 3 represent farmed mussels. (After Julshamn K et al., 1996. Mar Pollut Bull 32:564–571. https://doi.org/10.1016/0025-326X(96)84577-2. With permission)

biomarkers for the identification of coastal waters exposed to elements such as Cd, Pb, Cu, Zn, Cr, Co, Ni, Mn and Fe in the studied region.

Szefer et al. (1997b) assayed soft tissue and byssus of blue mussel samples from the east coast of Kyushu Island via FAAS with computational analysis. Significant interregional differences in metal contents in both soft tissues and byssus, i.e. Cu, Cd and Pb, and Cu, Pb, Co, Ni, Mn and Fe, respectively, were observed. Highly significant relationships between the tissue and byssal concentrations of Pb, Cu, Zn and Mn were recorded. The high levels of Cd, especially Pb and Cu, in Saganoseki mussels and the moderately elevated concentrations of these metals in Akamizu mussels, which are related to Urashiro as a reference sampling site, may be attributed to anthropogenic emissions from a metallic refinery and an artificial fibre factory, respectively. Compared with those in soft tissue, the increase in Cu levels relative to Zn levels in mussel byssus is clearly an order of magnitude, with a slope b (Cu/Zn) of 7.5 for byssus and 0.93 for soft tissue. This finding suggests that byssus, compared with soft tissue, is a more sensitive bioindicator for Cu. According to the authors, soft tissue, especially the byssus of *M. edulis*, appears to be a good bioindicator for the identification of coastal areas exposed to metallic contaminants.

In Szefer et al. (1997a), samples of the mussel *Perna perna* (Mytilidae) originating from various sampling sites in the intertidal zone around Aden (Yemen) were characterized with respect to their elemental composition quantified by FAAS. The sewage input at Sira Island is the most likely anthropogenic source of Cu and Mn. The highest levels of tissue Cd in the Fuqum area were observed, although this area is off the harbour and located near the Bab Al-Mandab Strait, where the water from both the Red Sea and the Indian Ocean mixes and consequently changes its nature, especially during the seasonal monsoons. This specific localization is likely responsible for elevated tissue Cd concentrations in the molluscs inhabiting the Fuqum area. Furthermore, water from a fishery refrigerator and some sewage outflow may be additional anthropogenic sources of Cd in the area.

Ndiaye et al. (2020) used ICP–MS to analyse the Mn, Pb, Cd and As profiles of *Mytilus galloprovincialis* soft tissue samples from various areas along the Dakar coast to assess the pollution status of the study area and evaluate the impact of various human activities on the marine ecosystem. ANOVA allowed the identification of interregional differentiation, e.g. the highest content of As was recorded around the Port of Dakar (Senegal) in neighbouring ports' multipurpose ore loading facilities, where large quantities of arsenic ores are loaded. This type of environmental pollution is an example of the release of considerable amounts of toxic elements to the marine environment caused by spills during loading. Moreover, significant seasonal variations were detected; i.e. February samples in all the cases had higher levels of Mn, Pb, Cd and As than those collected in October.

Szefer et al. (2006) reported that approximately 5000 samples of mytilids were collected from various regions worldwide (Fig. 4.7). The individuals were sorted by size and then differentiated into soft parts and byssal threads. FAAS combined with chemometrics was used to classify the adequacy of soft tissue and byssus of mytilid species as a biomonitor for intercountry multi-elemental monitoring of metallic

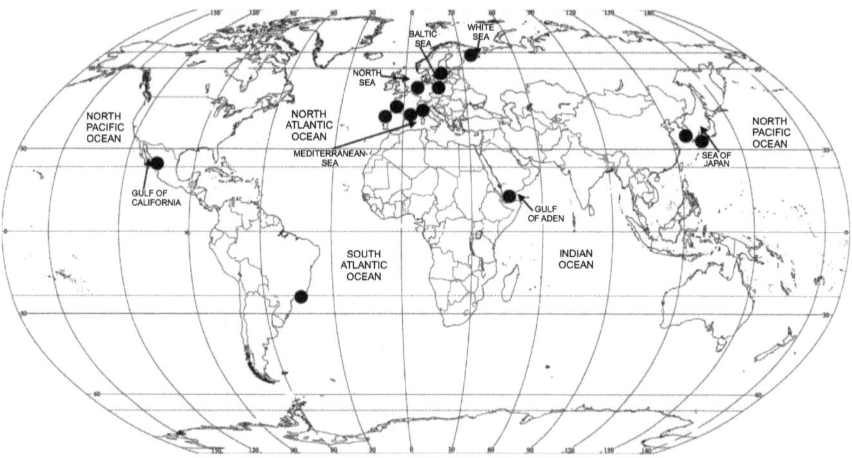

Fig. 4.7 Locations of the mussel sampling sites. (After Szefer P et al., 2006). Environ Pollut 139:70–78. https://doi.org/10.1016/j.envpol.2005.04.031. With permission)

contamination. To do this, all data previously reported by Szefer et al. worldwide were processed by FA. Therefore, most mussels in the study were sampled, and all were processed and analysed for heavy metals under the supervision of the same researcher (P.S.). This included all steps of analysis, e.g. using the same sample processing techniques, analytical methodologies and assistance of the same technical staff. For the chemometric processing of soft tissue concentration data, FA indicated that 56.7% of the total variance was explained by the first two factors, denoted F1 and F2, with eigenvalues of 2.99 and 2.11, respectively. The first two factors corresponded to 57.4% of the total variance, with eigenvalues of 3.20% and 1.97%, respectively. As shown in Fig. 4.8b, soft tissues of mussels from Mexican, French (Mediterranean coast) and partly Polish waters were isolated from the other samples, with the metal clusters Ni–Cu, Zn and Mn–Cd–Fe, respectively, responsible for their object sample discrimination (Fig. 4.8d). However, there was an overlap in the scores attributable to Polish and Korean mytilids due to similarly elevated concentrations of Fe in these samples. It was concluded that the fusion of MEA with computational tools appears to be an effective approach for clearly discriminating between mytilid species originating from various coastal regions of subarctic, temperate, subtropical and tropical marine ecosystems, including the intertidal zones of the Atlantic, Pacific and Indian Oceans. It was also postulated that the soft tissue of mytilids reflects differences in heavy metal levels in ambient waters and thus serves as a reliable biomonitor for these pollutants in a variety of coastal and estuarine areas worldwide.

In Bechmann et al. (2000), HR–ICP–MS data processed by PCA were used to evaluate spatially related trends in the elemental profiles of soft tissue mytilid samples from seven sites in Limfjord, Denmark. PCA with the first two PCs explained 90% of the total variation. The mussels were differentiated according to sampling sites. Moreover, PCA allowed observation of the differences between mussel

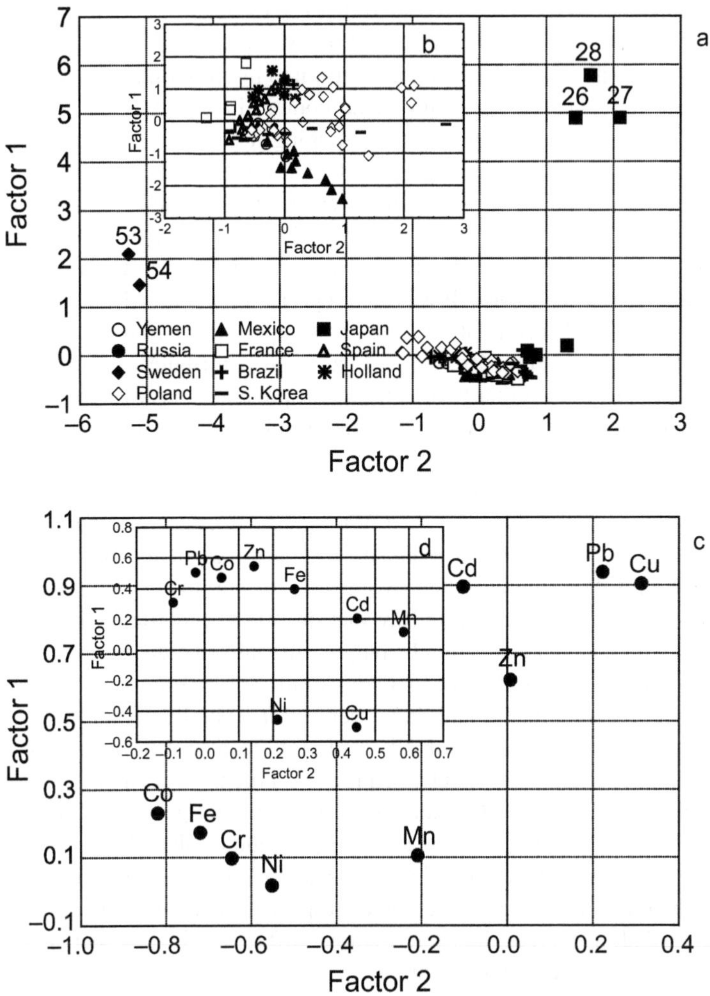

Fig. 4.8 Two-dimensional scatterplot in space determined by F1 and F2 corresponding to the metals in the soft tissue of Mytilidae. (a) Factorial distribution of all the scores (samples); (b) after excluding samples 22–28, 53 and 54, which deviate significantly from the other scores; (c) factorial distribution of all the loadings (trace elements); (d) factorial distribution of loadings after excluding samples 22–28, 53 and 54, which deviate significantly from the other scores. (After Szefer P et al., 2006. Environ Pollut 139:70–78. https://doi.org/10.1016/j.envpol.2005.04.031. With permission)

samples with respect to their size. A comparison of the content data recorded in this study with those reported for mussels analysed in 1982 revealed that trace element contamination, especially of Cd, Pb and Cr in two of the investigated areas, significantly increased. It was postulated that PCA is a powerful tool for identifying polluted regions inhabited by edible mussels, which can be simultaneously used as useful biomonitors within a small geographical area.

Szefer and Wolowicz (1993) reported that FAAS data for cockles from the Gulf of Gdańsk, the French Atlantic coast and the Mediterranean Sea allowed them to reliably distinguish seasonal factors, in addition to interregional factors, which have an important influence on the heavy metal content in cockles. The first three PCs accounted for 74% of the total variance, making it possible to clearly distinguish three geographical regions, i.e. the Gulf of Gdańsk of the Baltic Sea and Marennes–Oleron Bay in the Atlantic Ocean from other locations, although overlapping sample scores corresponding to Arcachon Bay and those from the Mediterranean Sea were observed. The samples collected in spring were different from the others and more dispersed across the plot. It was also reported that cockle Mn and Fe contents are responsible for discrimination between samples from Marennes–Oleron Bay and Arcachon Bay, whereas Zn, Cd and partly Ni contribute to distinguishing the Gulf of Gdańsk from the other two. Thus, anthropogenic sources are associated with higher levels of Zn and Cd in cockle individuals inhabiting the industrialized coastal area of the Gulf of Gdańsk.

In Szefer et al. (2004), a combination of ICP–MS and AAS MEC with chemometric processing (LRA, two-way ANOVA and FA) of concentration data was used to classify *M. galloprovincialis* from the southern part of the Korean Peninsula. Based on the FA data (with the first two factors explaining 57.1% of the total variance), mussel samples from Masan Bay were mostly differentiated from those from Ulsan Bay, which is situated near a heavily industrialized area. Significant intersize variations in the contents of trace elements in the soft tissue of mussels were also observed, e.g. for Co, Cu, Hg and Zn as well as for Fe and Sn from Masan Bay and Ulsan Bay, whereas the levels of Ni were significantly different for both bays. Additionally, seasonally dependent trends were noted. For example, mussel samples originating from Ulsan Bay and collected in winter generally had higher soft tissue levels of Cd, Cu, Mn, Co, Fe, Ti, Cr, Hg, Pb and Zn than those collected in the summer season. Thus, the dual tool applied was found to be helpful in the discrimination of Korean Peninsula mussels with respect to their intersize and seasonally dependent trends as well as their interregional differentiation, i.e. between less polluted regions and more polluted regions.

Szefer et al. (1999a) investigated the soft tissue of cockles from various regions along the Mediterranean Lagoon Etang de Thau (France) by combining FAAS and CV–AAS with LRA. Interregional differentiation was achieved on the basis of the evaluation of elemental profiles in mollusc soft tissues originating from six sampling sites. For example, elevated levels of Pb, Cr and Hg in tissue from the Etang des Eaux Blanches, an area adjacent to the harbour and a highly industrialized region, were detected, which were likely of anthropogenic origin. Moreover, seasonally dependent variations in the tissue concentrations of trace elements at Crique de l'Angle were identified. The highest levels of Cr and Pb in the tissue were identified in March 1992; the highest levels of Hg, Co and Fe were from August 1990 to January 1991; the highest level of Mn was in August 1992 and the highest level of Ag was in January 1991. Intertissue discrimination was also identified. The gills contained the maximum amounts of Co and Ni, and the mantle contained the most Fe and Hg; however, the digestive gland (hepatopancreas) contained the most Cu.

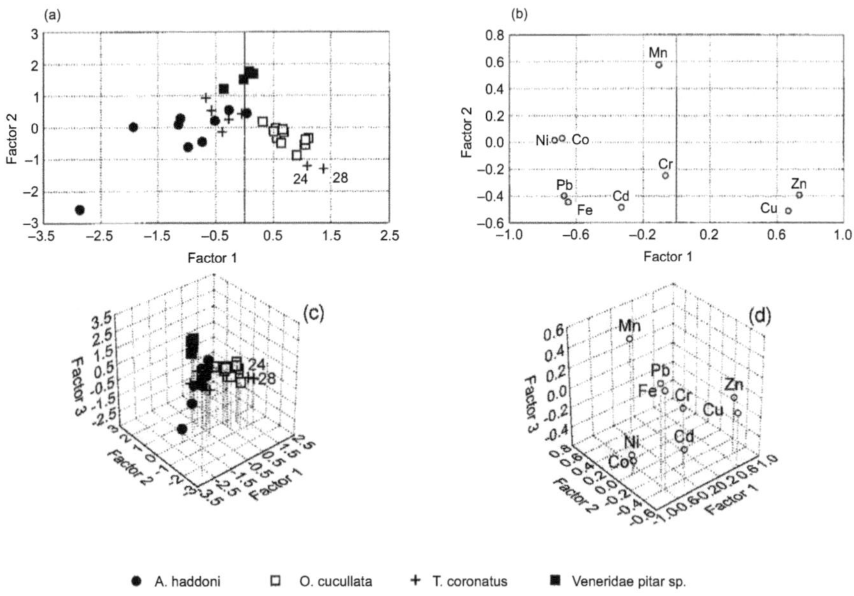

Fig. 4.9 Biplot for object scores of the first two factors of 32 mollusc samples. (**a**) Interspecies variations are illustrated by clusters of points corresponding to four species of the molluscs analysed, namely, *Acanthopleura haddoni* (Nos. 1–9) (●), *Turbo coronatus* (Nos. 22–28) (+), *Ostrea cucullata* (Nos. 10–21) (□) and *Pitar* sp. (Nos. 29–32) (■). (**b**) Location loadings reflecting statistically significant contributions of Zn and Cu as well as Mn, in distinguishing *O. cucullata* (including two samples (Nos. 24, 28) representing *T. coronatus*, unusually enriched in Zn and Cu) and *Pitar* sp., respectively, from two other species. (**c**) Three-dimensional scatterplot in the space determined by F1, F2 and F3 for the object scores. (**d**) Three-dimensional scatterplot determined by F1, F2 and F3 for loadings (metals). (After Szefer P et al., 1999b. Environ Pollut 106:299–314. https://doi.org/10.1016/s0269-7491(99)00108-6. With permission)

Szefer et al. (1999b) used FAAS for analysis of the soft tissue of four mollusc species, i.e. *Turbo coronatus*, *Acanthopleura haddoni*, *Ostrea cucullata* and *Pitar* sp., including associated surface sediments (bulk and bioavailable metal concentrations) originating from the Gulf of Aden, Yemen. Two-way ANOVA revealed significant interregional, intersize and interspecies variations in the elements in the soft tissue of the molluscs studied. Moreover, considerable spatially dependent variations in metal concentrations in both the soft tissue of the molluscs and associated sediments, as substrata, were identified. The FA data, with the first three factors accounting for 60% of the total variance, quantitatively indicated that both interspecies and ecological factors strongly affected the profiles of the selected elements in the four species of molluscs inhabiting the Gulf of Aden (Fig. 4.9). Since the molluscs investigated, especially *A. haddoni* and *T. coronatus,* have no effective ability to self-control the bioaccumulation of Pb, Cu, Cd, Cr, Fe and Ni, they appear to affect the actual pollution status of these metals in the marine ecosystem. Therefore, these zoobenthic species may be used as biomonitors of metallic pollutants.

Hendożko et al. (2010) described the application of FAAS to characterize the elemental profile of mollusc soft tissue samples with respect to extractable forms of heavy metals in adjacent sediments from the Gulf of Gdańsk and open Baltic waters. FA, with the first two factors explaining 75.66% and 93.17% of the total variance, confirms that regional variations exist between the contents of elements in *M. balthica* and their extractable forms in the associated sediments as substrata from both major regions (Figs. 4.10 and 4.11). This finding may be explained by differences in the environmental parameters between these two areas. Thus, it was concluded that

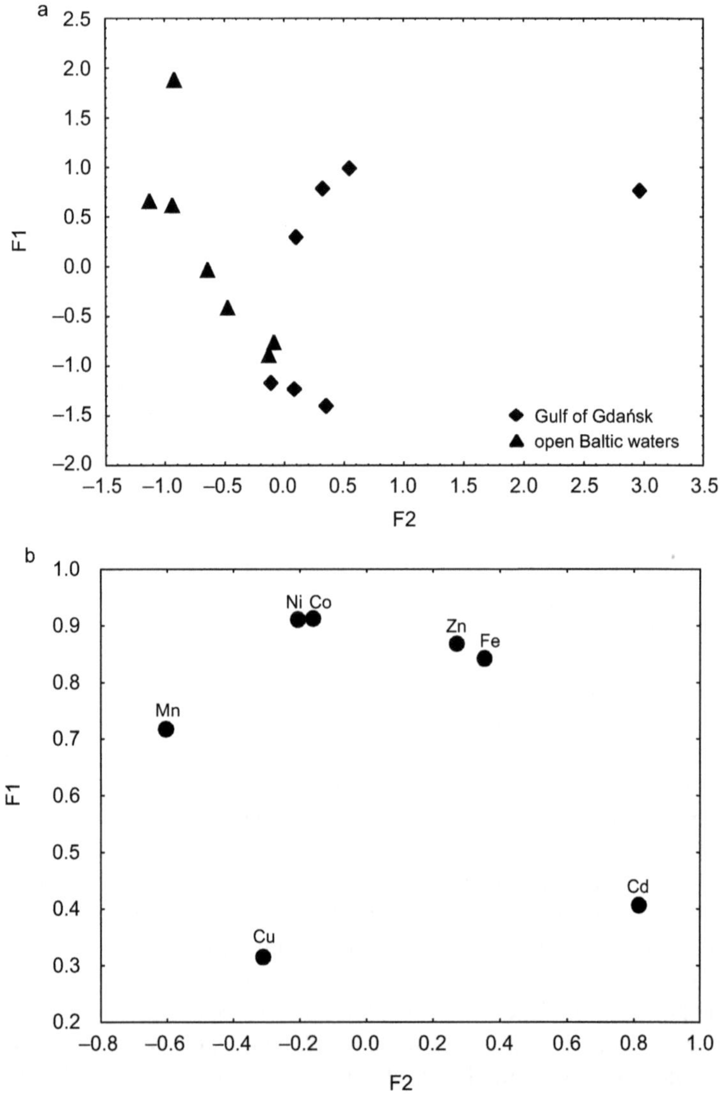

Fig. 4.10 Scatterplot of object samples (**a**) and loadings (**b**) corresponding to soft-tissue *M. balthica* from the two sectors of the southern Baltic. (After Hendożko E et al., 2010. Ecotoxic Environ Saf 73:152–163. https://doi.org/10.1016/j.ecoenv.2009.09.006. With permission)

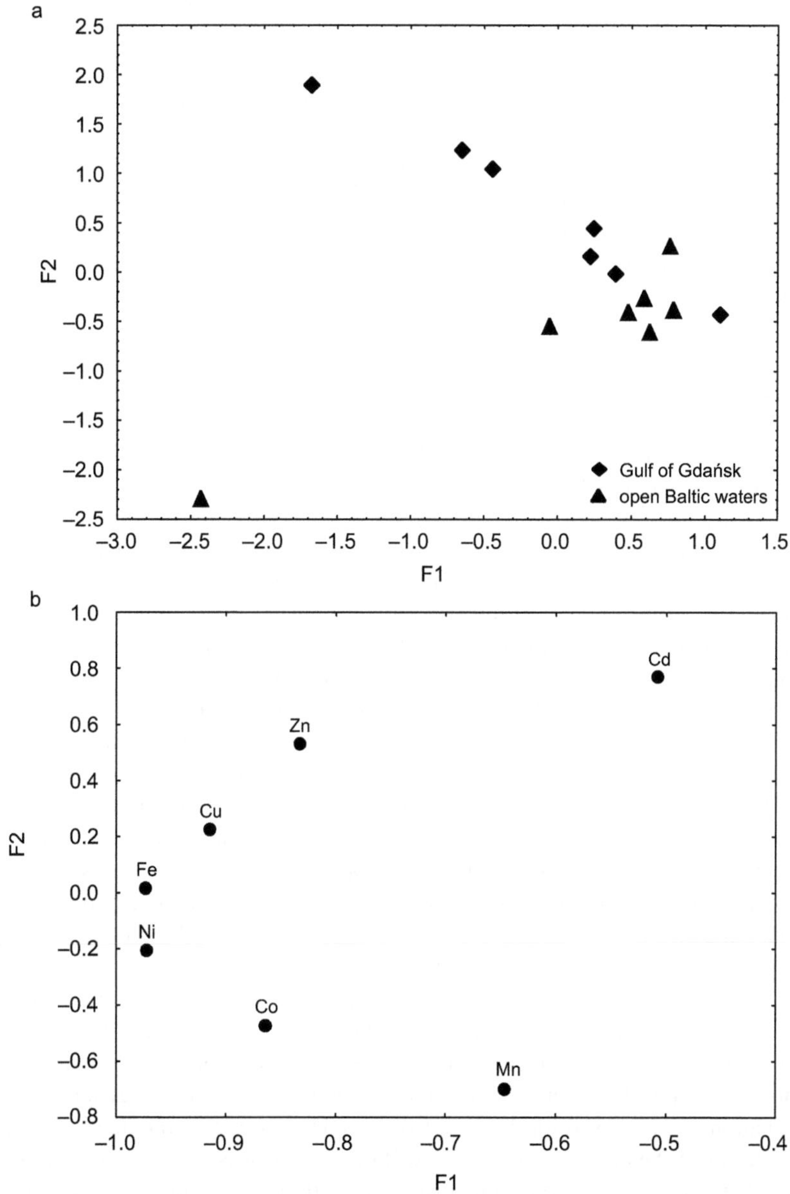

Fig. 4.11 Scatterplot of object samples (**a**) and loadings (**b**) corresponding to the extractable fraction of bottom sediments from the two sectors of the southern Baltic. (After Hendożko E et al., 2010. Ecotoxic Environ Saf 73:152–163. https://doi.org/10.1016/j.ecoenv.2009.09.006. With permission)

M. balthica may be useful as a biomarker of heavy metal bioavailability and contamination in the Baltic Sea.

In Bartolomé et al. (2010), ICP–MS data were used to examine mussels (*Mytilus galloprovincialis*) originating from ten sampling sites from Galicia to the Basque

Country at eight different sampling times. Among the statistical treatments used, PCA with the first four PCs, which explained up to 74% of the total variance, distinguished the samples with respect to spatial position (Arrigunaga, Estaca and Suances) and the sampling season. The discriminative variables related to those groups are Mn, Co, Pb and Cr. Moreover, the temporal trends of the elemental contents were studied using the Mann–Kendall test. In this context, some sampling sites presented a decreasing tendency in terms of the accumulation of elements. However, some local pollution incidents could be detected because of anthropogenic activities. Moreover, the effects of the oil spill from the Prestige could not be recognized properly.

Seaweeds

Zhao et al. (2024) developed a fast method using ICP–OES and ICP–MS with multivariate PCA and an LDA tool for tracing the geographical origins of the red seaweed *Neopyropia yezoensis* collected from five areas along the eastern and northern coast of China. *N. yezoensis* is a good source of essential dietary elements, particularly Fe, although Cd poses a high health risk regarding the food safety of seaweed. On the basis of the analysis of the elemental profiles of the samples, the PCA with the first two PCs, which explained 43.80% of the total variability, revealed that the *N. yezoensis* samples could not be differentiated between different regions; however, it was possible to achieve a clear interregional discrimination of the data obtained owing to the use of LDA with the stepwise method. The samples assigned to Liaoning and Shandong Provinces were differentiated from those originating from Jiangsu Province. Owing to the use of the LOOCV method, the predictive ability was 80.3%, indicating good performance of the elemental fingerprints in fast authentication of the origin of Chinese *N. yezoensis*. Notably, the authors simultaneously characterized the samples according to their nutritional value and safe levels for human consumption. In addition, the results obtained may be useful for producers by adding market value to products.

Kreissig et al. (2021) determined the contents of 17 elements in samples of 10 seaweed species from West, South and East Greenland via ICP–MC. The elemental profiles differed between species. Elements related to anthropogenic contamination demonstrated no clear trend with human settlement size. Broad geographic discrimination was achieved for the *Fucus* species. However, geographic identification was unclear in the cases of Sisimiut and Ilulissat near waste discharge. It was also concluded that the seaweeds studied were particularly rich sources of natural macronutrients (K, Na, Ca and Mg) and trace elements (Fe). Iodine reached high levels, particularly in the family Laminariaceae. The samples did not exceed the EU maximum levels for Cd, Hg or Pb, although some samples exceeded the stricter French regulations, especially for Cd and I. According to the authors, all ten species are promising as valuable edible seaweeds.

Miedico et al. (2017) described an ICP–MS analysis of 21 trace elements in edible seaweeds belonging to three species, i.e. *Laminaria japonica*, *Porphyra yezoensis* and *Undaria pinnatifida,* from three southeastern Asiatic countries. The original data obtained were processed by PCA. PC1 (37.0%) differentiated *Porphyra* samples from the other two species, PC2 (16.7%) differentiated two distinct subgroups of *Undaria* samples and PC3 (9.9%) was responsible for partial discrimination of *Laminaria* and *Undaria*, among others. In the case of the geographical distribution of *Porphyra*, PCA, with the first three PCs accounting for 49.6%, revealed that the three countries were differentiated sufficiently, especially the China group. Elements such as Mo, Se, Cu, Co and Cd and, less significantly, Mn, As and Fe highly discriminated this seaweed species. This means that the chemometric assessment allowed the three species studied to be differentiated. Taxonomic genus and, in part, geographic origin have a considerable effect on trace element profiles. In addition, Al and Cd can pose health risks for seaweed consumers. It has been estimated that the Cd levels of 26% of samples are inconsistent with current European regulations.

Żbikowski et al. (2006) determined micro- and macroelements by FAAS in the green alga *Enteromorpha* sp. from the coastal zone of the southern Baltic to estimate the degree of their accumulation in seaweed samples. FA and ANOVA clearly indicated interregional differentiation. *Enteromorpha* sp. from Vistula Lagoon and the southern Baltic exhibited relatively high levels of Mn and Ni and Na and K, respectively. The anthropogenic impacts of Cu, Pb and Zn, possibly originating from municipal sewage, were identified in alga samples collected in the Gulf of Gdańsk, especially near Gdynia. A comparison of the data obtained with those reported by other authors revealed that the Pb content in *Enteromorpha* sp. from the Gulf of Gdańsk decreased from 1978 to 2003, reflecting the reduced use of leaded petrol in Baltic countries during this period. It was concluded that the alga *Enteromorpha* sp. could be used for biomonitoring surveys of metal pollutants in coastal areas of the Baltic Sea.

The aim of the study by Żbikowski et al. (2007) was to obtain background data and evaluate seasonal and spatial distributions while considering relationships related to interelement correlations and CFs in *Enteromorpha* sp. and *Cladophora* sp. from the coastal zone of the southern Baltic Sea. The integration of FAAS with FA differentiated samples according to geographical origin. The algae from the southern Baltic contained relatively high levels of Na and K, whereas in the case of Cu, Pb and Zn, their anthropogenic impact was observed in *Cladophora* sp. and *Enteromorpha* sp. originating from the Gulf of Gdańsk near Gdynia. This area is exposed to the emission of heavy metals from municipal and industrial sources, with the main contributions from the shipbuilding industry and the seaport. The chemometric assessment of the data revealed that there was a correlation between the concentrations of Cu, Pb and Zn in both the green algae collected at the same time and the sampling sites in the Gulf of Gdańsk. This finding suggests that in the absence of either of these two species in the investigated area, it is still possible to continue successful biomonitoring studies with their replacement by the other species, i.e. *Cladophora* sp. by *Enteromorpha* sp. or vice versa.

Arthropoda

Fry et al. (2016) applied a combination of TC/EA–IRMS, ICP–OES and ICP–MS to investigate penaeid prawns from commercial catches of eight Asia–Pacific countries, i.e. Australia, Bangladesh, Indonesia, Myanmar, the Philippines, Pakistan, Sri Lanka and Thailand. The contents of Co, Cr, Fe, Mn, Mo and Zn with $\delta^{18}O$ values, especially $\delta^{15}N$ values, were selected from the whole dataset owing to the use of DA. Both stable isotopes, especially $\delta^{15}N$, can help distinguish between relatively polluted and unpolluted sites. When the reduced industrial-metal database (Cr, Mn, Fe, Co, Zn and Mo) was used in combination with the $\delta^{15}N$ data, the categorization considerably improved. Therefore, HCA combined with DA can help distinguish between regions that are diverse in terms of the degree of pollution.

The aim of the study by Amaral Albuquerque et al. (2020) was to evaluate the contents of 13 elements in the freshwater shrimp *M. amazonicum* from Santarem (lower Amazonia), which is affected by mining activities. The level of elements does not pose a considerable health risk. PCA with the first three PCs, which explained 66.0% of the total variance, differentiated between two clearly differentiated groups related to the Amazon and Tapajos rivers. This finding was confirmed by HCA data. This interregional differentiation can be explained by bauxite exploitation affecting the Amazon River while gold mining impacts the Tapajos River. The levels of Cr, Mn, Fe Co, Ni, Cd and Pb are also identifiers for the Amazon samples, whereas Cu, Zn, As, Se, Mo and Hg are determinants for the Tapajos samples. This methodology could be used to distinguish the origin of shrimp for commercial purposes.

In Nędzarek et al. (2019a), the contents of 12 elements in the Harris mud crab and samples of water and surface sediments from the southern Baltic Sea were measured by FAAS and GF–AAS. This zoobenthic organism is rich in Ca, Mg, Na, K, Fe, Cu and Zn and hence can be a valuable source of these elements. Low concentrations of non-essential metals such as Pb, Cd and As do not pose a threat to consumers. Parent waters and surficial sediments were also suspected to be poorly contaminated with toxic elements. Therefore, it was concluded that the Harris mud crab from this Baltic region is useful as a component of well-balanced feeds for terrestrial and aquatic animals.

Nędzarek et al. (2019b) reported that the concentrations of macro-, micro- and trace elements were different in the muscle, hepatopancreas, gonads, gills and exoskeletons of Chinese mitten crabs caught in spring and autumn in Szczecin Lagoon. Edible parts of *E. sinensis* caught in Poland can be valuable sources of essential elements important in the human diet. The levels of Pb and Cd did not exceed the established food safety level of 0.50 $\mu g\ g^{-1}$. According to the authors, commercial fishing of this species may be beneficial and has been agreed upon in the EU legal regulations related to the reduction of invasive animal populations in Europe.

Nędzarek et al. (2017) used ICP–MS to determine the group of elements in the claws, legs and carapace of Chinese mitten crabs from Lake Dąbie, Baltic Sea. Essential elements such as Fe, Cu, Zn and Mn are present at high levels. The contents of toxic metals such as Pb and Cd were 5 and 100 times lower than the threshold for foods established by Commission Regulation (EC) No 1881/2006. This means that the meat of crabs should be safe for consumers and may provide essential elements.

4.2.2.3 Interregional and/or Interspecies Differentiation Related to Various Parameters

Mollusca

Bennion et al. (2019, 2021) measured TEFs via ICP–MS and the original data obtained via MDS and RFA to identify the site of Irish blue mussel harvest, successfully revealing sampling locations related to the soft tissue and shell. Moreover, the temporal stability of TEF combined with the above computational techniques for blue mussels was tested over five harvesting dates spanning 2 years. Almost perfect temporal discrimination between harvesting dates was attained (Fig. 4.12). However, when both the clean shell and the periostracum of mussels were used, individuals could be classified into correct sampling dates of 100%, indicating that this method could identify not only the location but also the date of harvest of bivalve shellfish.

Zhang et al. (2019) provided the first $\delta^{13}C$ and $\delta^{15}N$ data quantified by EA–IRMS and processed them chemometrically to identify the geographical origin and interspecies differentiation of 725 Chinese scallop samples. LDA achieved high discrimination ability, with accuracies of 92% and 98.3% with respect to identification of the scallop provenance and species recognition. As shown in Fig. 4.13, the three scallop species identified in different seasons could be used to distinguish the origins of *P. yessoensis*, *C. farreri* and *A. irradians*. Interseasonal differences in the $\delta^{13}C$ and $\delta^{15}N$ values of scallops exist, which may be explained by variations in the composition of the food sources of scallops between different seasons. It was concluded that the obtained results could provide a theoretical basis for the traceability and authenticity of scallops.

Szefer and Wenne (1987) quantified the contents of U and Th in the soft tissue and shells of four species of molluscs, including other benthic invertebrates originating from the Gulf of Gdańsk in the southern Baltic. Some regional- and species-dependent variations in the concentrations of U, and especially Th, at the stations studied were observed. Such variations may be responsible for the feeding habits of molluscs and the structure of the associated sediment as a substratum. Additionally, some relationships between the levels of both metals and the size and age of the molluscs were obtained. As shown in Figs. 4.14 and 4.15, the data obtained by LRA revealed that the $DF_U/_{Ca}$ values, in contrast to the $DF_{Th/Ca}$ values, increased significantly with decreasing concentrations of Ca in the soft tissue of all the molluscs studied. The lack of a significant correlation between $DF_{U/Ca\ U/ca}$ or $DF_{Th/Ca}$ and the concentrations of Ca in the hard parts of molluscs could have been due to differences in the mineralogy of $CaCO_3$, i.e. the amounts of calcite and aragonite in the shells.

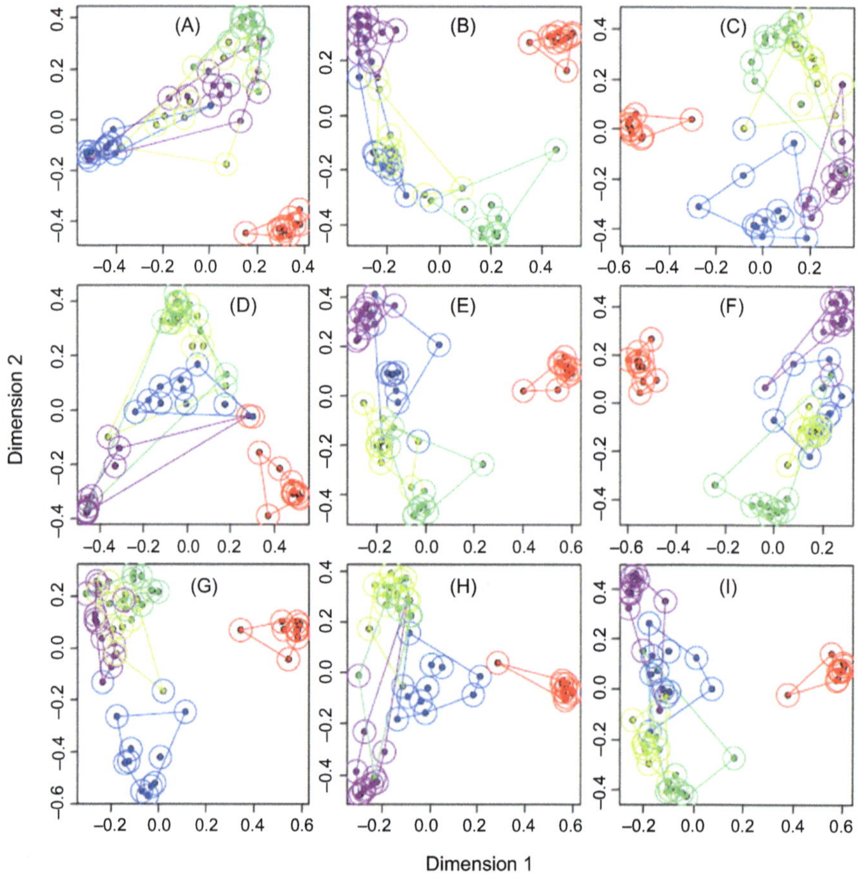

Fig. 4.12 Multidimensional scaling ordinations of proximity scores from each of the random forest classifications based on the elemental concentrations of (**a**) unclean shell; (**b**) clean shell; (**c**) foot; (**d**) periostracum; (**e**) clean shell and foot; (**f**) clean shell and periostracum; (**g**) unclean shell and foot; (**h**) periostracum and foot and (**i**) clean shell, foot and periostracum. A convex hull is drawn around the a priori classes, with points coloured according to the date of collection, and the outer ring colour refers to the date each individual was classified according to the random forest analyses. Red = January 2016; yellow = January 2017; blue = May 2017; purple = September 2017; green = January 2018. (After Bennion M et al., 2021. Food Contr 121:107515. https://doi.org/10.1016/j.foodcont.2020.107515. With permission)

Seaweeds

In Romarís-Hortas et al. (2010), ICP–OES, ICP–MS and HG–AFS were used to study edible seaweed according to type, i.e. brown, red and green seaweeds, as well as their varieties, which are commercially named Wakame, Fucus, Sea Spaghetti, Kombu, Dulse, Nori and Sea Lettuce and originated from the North Atlantic (Spanish Galicia). The data obtained were processed by a multivariate approach

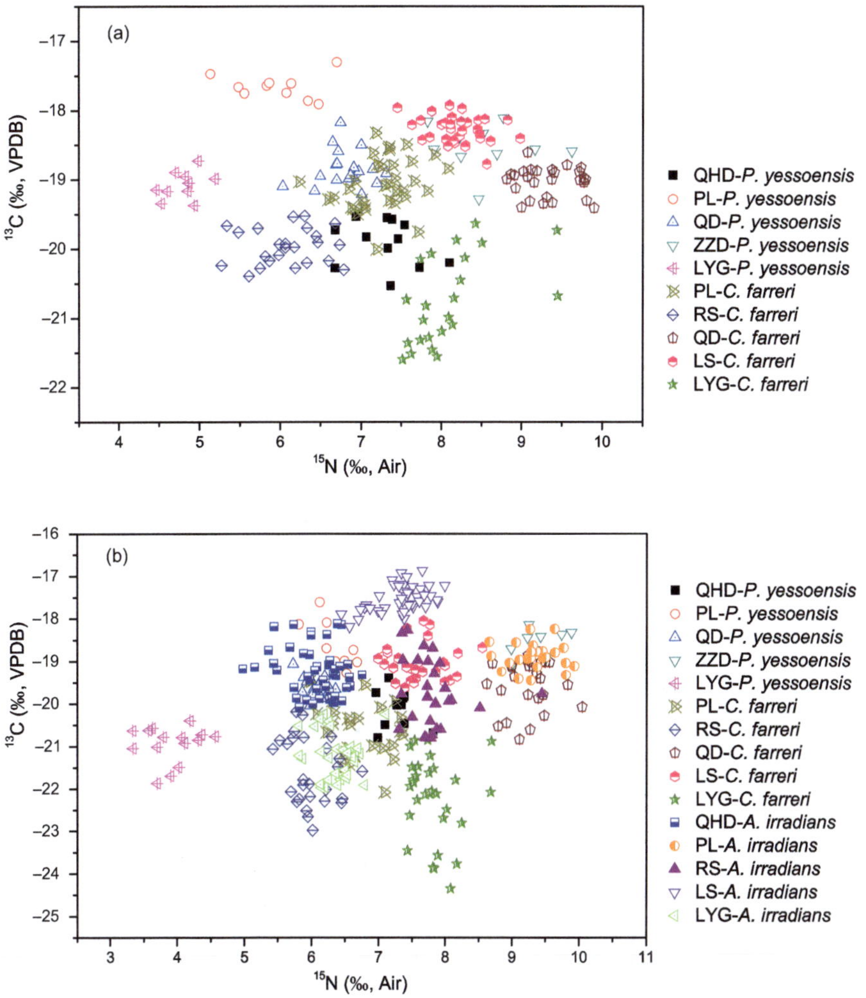

Fig. 4.13 Scatterplots of the δ ^{13}C and δ ^{15}N values of three scallop species collected in coastal areas of China. (**a**) Spring of 2016 and (**b**) autumn of 2016. (After Zhang X et al., 2019. Food Chem 299:125107. https://doi.org/10.1016/j.foodchem.2019.125107. With permission)

(PCA, CA, LDA and SIMCA). PCA was performed using CV as a validation method, and after a normalized varimax rotation, 71% of the total variance was explained by the first three PCs. PCA and CA differentiated samples with respect to their three types and seven varieties, respectively. When SIMCA was used, a highly satisfactory classification (recognition and prediction) was achieved exclusively for red seaweed, amounting to 100% of the cases correctly classified. However, in the cases of brown seaweed recognition (training set) and prediction (testing set), 89% and 80%, respectively, were attained. The application of LDA provided a correct assignment of 100% for the brown, red and green types at a significance level of 5%.

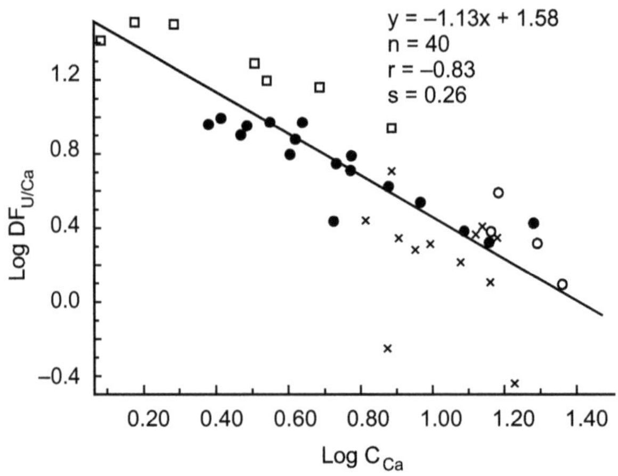

Fig. 4.14 Relationship between the log $DF_{U/ca}$. and log Ca. for the soft tissue of *Mytilus edulis* (O), *Mya arenaria* (×), *Macoma balthica* (O) and *Cardium glaucum* (O). (After Szefer P and Wenne R, 1987. Sci Total Environ 65:191–202. https://doi.org/10.1016/0048-9697(87)90172-0. With permission)

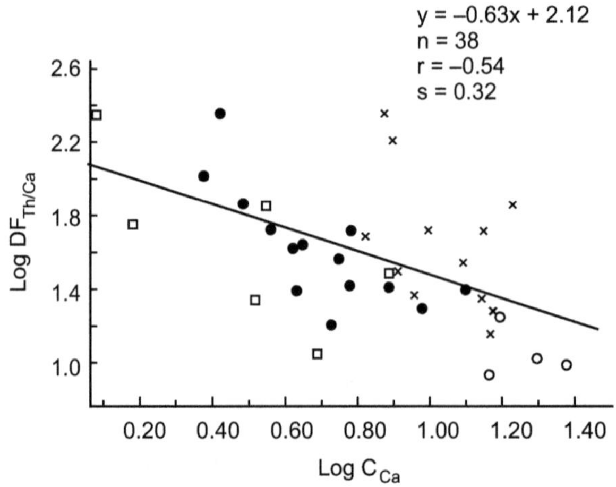

Fig. 4.15 Relationships between the log $DF_{Th/Ca}$ and log Ca for the soft tissues of *Mytilus edulis* (O), *Mya arenaria* (×), *Macoma balthica* (O) and *Cardium glaucum* (O). (After Szefer P and Wenne R, 1987. Sci Total Environ 65:191–202. https://doi.org/10.1016/0048-9697(87)90172-0. With permission)

Mollusca and Arthropoda

In Rainbow et al. (2004), samples of the blue mussel *M. trossulus* and barnacle *Balanus improvisus* from five sampling sites in the Gulf of Gdańsk, Baltic Sea, were processed for concentrations of minor and trace elements using the FAAS method.

Statistically significant relationships were found among the levels of Cu, Zn, Fe, Pb and Ni but not between the levels of Cd or Mn, in barnacles and mussels. Notably, geographical differences in trace metal bioavailability related to both barnacles and mussels were achieved. In addition to interregional variations, temporal variations were also examined. The original variables were transformed using the multivariate DA technique, which revealed that barnacles from the Baltic Sea presented greater discriminatory potential than mussels as trace metal bioindicators. Moreover, there was no significant correlation between the contents of trace elements in surficial sediments (<63 μm) and either barnacles or mussels.

Mollusca and Seaweeds

According to Struck et al. (1997), the combination of ICP–AES, ET–AAS, HG–AAS, CV–AAS and ASV methods is a useful tool for classifying the blue mussel *M. edulis* and the brown algae *F. vesiculosus* collected along the North Sea (German Bay; Island of Sylt) and the Baltic Sea (Kiel Bay, Lübeck Bay, Pomeranian Bay and Island of Rügen). The results obtained were subjected to chemometric analysis, i.e. FA, CA and DA. The multivariate techniques allowed discrimination of the Baltic Sea from the North Sea locations despite the different food habits of these two zoo- and fitobenthic organisms. Compared with the sample groups corresponding to the macroelement concentration profile associated considerably with the geographical origin as a function of water salinity, the interregional groups based on the trace element content profile revealed a less distinguishable geographical provenance. This finding indicated that the trace element content pattern exhibited a less distinctive geographical arrangement than the macroelement concentration pattern did, indicating more effective predictability of the geographical provenance according to salinity.

4.2.2.4 Assessment of Seafood Quality Related to Technological Processing

Mesa et al. (1999) combined FAAS, GF–AAS, CV–AAS and HG–AAS with the simplified algorithm for PoCu using PCA to evaluate the elemental profile, i.e. Cu, Fe, K, Mn, Na, Ni, V, Cd, Co, Pb and Hg, from soft tissue data of *M. edulis* and *Perna canaliculus* from various Spanish markets. Both the PoCu and PCA, with the first two PCs, which explained 65.17% of the total variance, were helpful tools for identifying several non-controlled factors during the periods of harvest, packing transport, etc., affecting the quality of the seafood studied.

Kwoczek et al. (2006) used a combination of FAAS, GF–AAS, HG–AAS and CV–AAS with the FA technique as a helpful tool for the analysis of a diverse assortment of seafood (oysters, mussels, prawns, surimi products, octopus, squids, octopuses, crabs and lobsters) originating from various international waters bordering eight countries, i.e. Norway, England, Spain, India, Thailand, Canada, the Philippines and New Zealand. The first two factors, which described 53% of the total variability, showed that highly processed products generally have lower values of F1, in contrast to initially processed products (just boiled), which have higher values of F1 (Fig. 4.16). Factor 2 appeared to be associated with the taxonomic groups of the raw shellfish, distinguishing, e.g. mussels, octopuses, squids and crustaceans (shrimps, lobsters and crabs). As shown in Fig. 4.17, processing technologies considerably affect the Na content in shellfish, but higher values of F2 are associated with Fe, Mn (mussels) and Cd (squids and octopus). However, lower values of F2 are related to Hg, Cu and Ca (crustaceans—shrimp, lobsters and crabs). The attained data revealed significant differences related to the degree of technological processing, i.e. preliminarily and highly processed shellfish, and the taxonomic features of seafood.

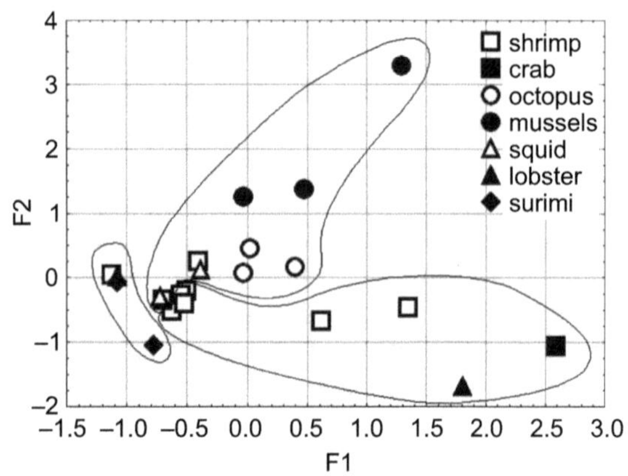

Fig. 4.16 Biplot of the distribution of object scores for seafood. (After Kwoczek M et al., 2006. *J Agric Food Chem* 54:3015–3024. https://doi.org/10.1021/jf0600511. With permission)

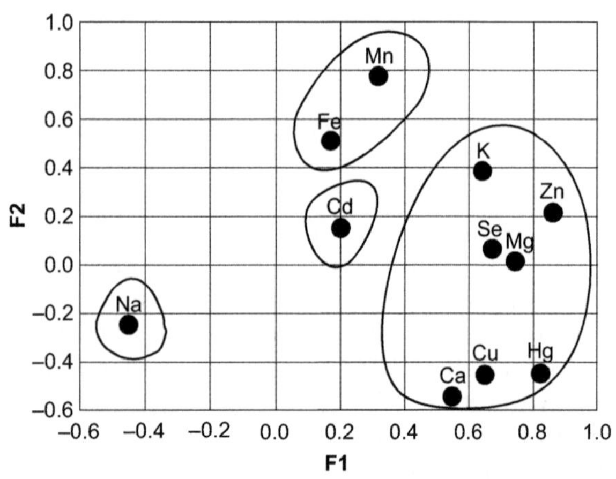

Fig. 4.17 Biplot of the distribution of loadings for seafood. (After Kwoczek M et al., 2006. *J Agric Food Chem* 54:3015–3024. https://doi.org/10.1021/jf0600511. With permission)

4.2.2.5 Interregional and/or Interspecies Differentiation Related to Farmed Versus Wild Shrimp

Arthropoda

Ortea and Gallardo (2015) used a combination of SIRA, ICP–OES and ICP–MS to trace eight species of shrimp from Argentina, North Atlantic, Mozambique, Nigeria and Senegal with respect to three factors, i.e. production method (wild or farmed), geographical origin and biological species. Multivariate chemometric tools (*k*-means and DA) were applied to evaluate the interdifferentiation of data subsets in the context of classification or predictive ability. The combination of the analytical techniques made it possible to achieve correct classification of 100% of the samples with respect to their origin (80% CV) and method of production (differentiation between wild and farmed shrimp). Moreover, DA was useful for obtaining a reliable classification of 93.5% of the various biological species relative to their origin and method of production. It was concluded that the methodology is recommended for recognizing seafood product authenticity, but more samples in each group should be used to validate these findings.

Gopi et al. (2019b) applied a combination of SIRA and XRF using ITRAX with chemometric techniques (LDA, RF and nMDS) to study black tiger prawns (*Penaeus monodon*) from the Asia–Pacific region. LDA and RF showed consistent results, and XRF effectively differentiated the production method and geographic provenance of *P. monodon*, reaching up to 100% accuracy, whereas SIRA showed 95% accuracy. According to the authors, SIRA and XRF are useful complementary approaches. The dual methodology can potentially be used for other seafood products and to profile the environmental determinants of chemical markers such as fish feed, fertilizers and water quality.

Kim et al. (2015) used a combination of EA–IRMS, molecular analysis and phylogenetic analysis ($\delta^{13}C$, $\delta^{15}N$ and COI gene sequencing) with LDA to explore six species of marketable-sized shrimp (*Pandalus borealis*, *Marsupenaeus japonicus*, *Fenneropenaeus chinensis*, *Litopenaeus vannamei*, *Penaeus monodon* and *Solenocera crassicornis*) from 11 different countries. This study validated the usefulness of the approach applied for establishing the geographical origin of specimens of the same shrimp species. Additionally, the approach can help distinguish natural and farmed shrimp from the same country on the basis of SIR data.

Gamboa-Delgado et al. (2014) used IRMS to investigate wild and farmed Pacific white shrimp (*Litopenaeus vannamei*) collected at various locations in two countries. Stable isotopes such as $\delta^{13}C$ and $\delta^{15}N$ processed by CDA showed that they effectively distinguished wild-caught shrimps from aquaculture shrimps and that open sea samples were clearly differentiated from estuarine samples, whereas there was no such evident discrimination between estuarine and farmed samples. For example, CDA clearly distinguished between semi-intensively farmed animals and shrimp caught on the open sea in Ecuador and Mexico. In the case of CDA used to classify only farmed shrimp and wild-caught shrimp (industrial fishing), 99% of the shrimp samples were correctly classified, and 100% of the shrimp originating from the open sea were correctly classified.

4.2.2.6 Interregional Differentiation of Shrimp Cultured in Freshwater and Seawater

Li et al. (2019) used SIR and MEC contents, especially REEs, measured by ICP–MS and EA–IRMS to characterize peeled and deveined Pacific white shrimp and shrimp feed from aquaculture ponds originating from freshwater and seawater. Multivariate statistics (PCA, SDA, CDA and kernel methods) were used to distinguish between shrimp cultured in freshwater and those reared in seawater. Moreover, feeds recommended for shrimp in freshwater and seawater did not differ in their elemental profiles. Among the REEs, Lu in particular appeared to be a relevant descriptive element allowing discrimination of the geographical origin of shrimp. The kernel method of SDA, which is based on six elements ($\delta^{15}N$, Sr, Ba, As, Mn and Lu), significantly differentiated between freshwater and seawater shrimp, with a classification rate of 100%. This finding is in agreement with the PCA results, which also demonstrated that REEs were more relevant for distinguishing cultured shrimp in freshwater or seawater.

4.3 Conclusions

The most useful instrumental methods, including spectroscopy combined with multivariate techniques, appear to be helpful in determining food authenticity and identifying adulteration/mislabelling. The proper use of analytical and chemometric tools for estimating the nutritive and health quality of seafood plays an important role in seafood classification in terms of geographical origin, differentiation between farmed and wild molluscs and crustaceans and identification of the intersize (age), interspecies, intertissue differences and the degree of environmental pollution. In general, such a complex approach would be useful for differentiating unprocessed and technologically processed seafood as well as detecting fraud to preserve the brand reputations of original products. The application of chemometric tools has led to a deeper understanding of the distribution and nutritive significance of mineral components in foods, which is especially important in terms of the ecotoxicological aspects of food.

References

Amaral Albuquerque FE, Hamad Minervino AH, Miranda M, Herrero-Latorre C, Barrêto Júnior RA, Costa Oliveira FL, Ribeiro Dias S, Lippi Ortolani E, López-Alonso M (2020) Toxic and essential trace element concentrations in the freshwater shrimp *Macrobrachium amazonicum* in the Lower Amazon, Brazil. J Food Compos Anal 86:103361. https://doi.org/10.1016/j.jfca.2019.103361

Bartolomé L, Navarro P, Raposo JC, Arana G, Zuloaga O, Etxebarria N, Soto M (2010) Occurrence and distribution of metals in mussels from the Cantabrian coast. Arch Environ Contam Toxicol 59:235–243. https://doi.org/10.1007/s00244-010-9476-7

Bechmann IE, Stürup S, Kristensen LV (2000) High resolution inductively coupled plasma mass spectrometry (HR-ICPMS) determination and multivariate evaluation of 10 trace elements in mussels from 7 sites in Limfjorden, Denmark. Fresenius J Anal Chem 368:708–714. https://doi.org/10.1007/s002160000576

Belfroid AC, Purperhart M, Ariese F (2000) Organotin levels in seafood. Mar Pollut Bull 40:226–232. https://doi.org/10.1016/S0025-326X(99)00241-6

Bennion M, Morrison L, Brophy D, Carlsson J, Abrahantes JC, Graham CT (2019) Trace element fingerprinting of blue mussel (*Mytilus edulis*) shells and soft tissues successfully reveals harvesting locations. Sci Total Environ 685:50–58. https://doi.org/10.1016/j.scitotenv.2019.05.233

Bennion M, Morrison L, Shelley R, Graham C (2021) Trace elemental fingerprinting of shells and soft tissues can identify the time of blue mussel (*Mytilus edulis*) harvesting. Food Control 121:107515. https://doi.org/10.1016/j.foodcont.2020.107515

Besada V, Andrade JM, Schultze F, González JJ (2009) Heavy metals in edible seaweeds commercialised for human consumption. J Mar Syst 75:305–313. https://doi.org/10.1016/j.jmarsys.2008.10.010

Bulska E (2010) Quality assurance and quality control of analytical results. In: Namieśnik J, Szefer P (eds) Analytical measurements in aquatic environments. CRC Press, Taylor & Francis Group, Boca Raton, pp 389–397. https://doi.org/10.1201/9781420082692

Capar SG, Szefer P (2012) Determination and speciation of trace elements in foods. In: Ötleş S (ed) Methods of analysis of food components and additions. CRC Press, Taylor & Francis, Boca Raton, pp 165–210. https://doi.org/10.1201/b11218

Carter JF, Tinggi U, Yang X, Fry B (2015) Stable isotope and trace metal compositions of Australian prawns as a guide to authenticity and wholesomeness. Food Chem 170:241–248. https://doi.org/10.1016/j.foodchem.2014.08.037

Costas-Rodríguez M, Lavilla I, Bendicho C (2010) Classification of cultivated mussels from Galicia (Northwest Spain) with European protected designation of origin using trace element fingerprint and chemometric analysis. Anal Chim Acta 664:121–128. https://doi.org/10.1016/j.aca.2010.03.003

Davis R, Boyd CE, Wakefield J, Shatova O, McNevin A, Harris B, Davis DA (2021) Trace element concentrations in white leg shrimp Litopenaeus vannamei from retail stores in the EU, UK, and USA and the ability to discern country of origin with classification models. Curr Res Food Sci 4:655–661. https://doi.org/10.1016/j.crfs.2021.09.004

Defernez M, Kemsley EK (1997) The use and misuse of chemometrics for treating classification problems. Trends Anal Chem 16:216–221. https://doi.org/10.1016/S0165-9936(97)00015-0

del Rio-Lavín A, Weber J, Molkentin J, Jiménez E, Artetxe-Arrate I, Pardo MA (2022) Stable isotope and trace element analysis for tracing the geographical origin of the Mediterranean mussel (*Mytilus galloprovincialis*) in food authentication. Food Control 139:109069. https://doi.org/10.1016/j.foodcont.2022.109069

EFSA (European Food Safety Authority) (2004a) Opinion of the Scientific Panel on contaminants in the food chain [CONTAM] to assess the health risks to consumers associated with exposure to organotins in foodstuffs, Question number: EFSA-Q-2003-110. 2004. http://www.efsa.europa.eu/EFSA/efsa_locale-178620753812_1178620762916.htm

EFSA (European Food Safety Authority) (2004b) Opinion of the Scientific Panel on contaminants in the food chain on a request from the commission to assess the health risks to consumers associated with exposure to organotins in foodstuffs, Question No EFSA-Q-2003-110, Adopted: 22 September 2004. EFSA J 102:1–119. https://doi.org/10.2903/j.efsa.2004.102

EFSA (European Food Safety Authority) (2005) Opinion of the Scientific Panel on contaminants in the food chain [CONTAM] related to Arsenic as undesirable substance in animal feed, Question number: EFSA-Q-2003-031, Adopted: 31 January 2005. http://www.efsa.europa.eu/EFSA/efsa_locale1178620753812_1178620762850.htm

EFSA (European Food Safety Authority) (2009a) Cadmium in food—Scientific opinion of the Panel on Contaminants in the Food Chain, Question number: EFSA-Q-2007-138. Adopted date: 30 January 2009. http://www.efsa.europa.eu/EFSA/efsa_locale1178620753812_1211902396126.htm

EFSA (European Food Safety Authority) (2009b) EFSA sets lower tolerable intake level for cadmium in food, Press Release: 20 March 2009. http://www.efsa.europa.eu/EFSA/efsa_locale1178620753812_1211902396263.htm

Favretto L, Favretto LG (1989) Principal component analysis and pollution by trace elements in a mussel survey. J Chemometrics 3:301–308. https://doi.org/10.1002/cem.1180030522

Favretto L, Favretto LG, Marletta GP, Saitta M (1989) Principal component analysis: a chemometric aid for classification of polluted and unpolluted mussels. Anal Chim Acta 220:135–144. https://doi.org/10.1016/S0003-2670(00)80257-3

Favretto L, Campisi B, Reisenhofer E, Adami G (1997) Terrigenous debris and mussel pollution—a differentiation based on trace element concentration by means of multivariate analysis. Anal Chim Acta 344:251–259. https://doi.org/10.1016/S0003-2670(97)00031-7

Fry B, Carter JF, Tinggi U, Arman A, Kamal M, Metian M, Waduge VA, Yaccup RB (2016) Prawn biomonitors of nutrient and trace metal pollution along Asia-Pacific coastlines. Isotop Environ Health Stud 52:619–632. https://doi.org/10.1080/10256016.2016.1149481

Gadd P, Gopi K, Sammut J, Neil Saintilan N, Crawford J, Mazumder D (2018) Itrax micro X-ray fluorescence (μXRF) for soft biological tissues. MethodsX 5:1267–1271. https://doi.org/10.1016/j.mex.2018.10.001

Gamboa-Delgado J, Molina-Poveda C, Godínez-Siordia DE, Villarreal-Cavazos D, Ricque-Marie D, Cruz-Suárez LE (2014) Application of stable isotope analysis to differentiate shrimp extracted by industrial fishing or produced through aquaculture practices. Can J Fish Aquat Sci 71:1520–1528. https://doi.org/10.1139/cjfas-2014-0005

Gopi K, Mazumder D, Sammut J, Saintilan N (2019a) Determining the provenance and authenticity of seafood: a review of current methodologies. Trends Food Sci Technol 91:294–304. https://doi.org/10.1016/j.tifs.2019.07.010

Gopi K, Mazumder D, Sammut J, Saintilan N, Crawford J, Gadd P (2019b) Combined use of stable isotope analysis and elemental profiling to determine provenance of black tiger prawns (*Penaeus monodon*). Food Control 95:242–248. https://doi.org/10.1016/j.foodcont.2018.08.012

Gorbi S, Lamberti CV, Notti A, Benedetti M, Fattorini D, Moltedo G, Regoli F (2008) An ecotoxicological protocol with caged mussels, *Mytilus galloprovincialis*, for monitoring the impact of an offshore platform in the Adriatic Sea. Mar Environ Res 65:34–49. https://doi.org/10.1016/j.marenvres.2007.07.006

Grembecka M, Szefer P (2011) Metals and metalloids in foods—essentiality, toxicity, applicability. In: Medina DA, Laine AM (eds) Food quality: control, analysis and consumer concerns. Nova Science Publishers, Hauppauge, pp 1–60. ISBN 10: 1611229170; ISBN 13: 9781611229172

Grembecka M, Szefer P (2016) Elemental trace analysis in studies of food products. In: Baranowska I (ed) Handbook of trace analysis—fundamentals and applications. Springer, Berlin, pp 203–239. https://doi.org/10.1007/978-3-319-19614-5

Guo Y, Lundebye A-K, Li N, Ergon Å, Pang S, Jiang Y, Zhu W, Zhao Y, Li X, Yao L, Wang L, Aakre I (2023) Comparative assessment of food safety regulations and standards for arsenic, cadmium, lead, mercury and iodine in macroalgae used as food and feed in China and Europe. Trends Food Sci Technol 141:104204. https://doi.org/10.1016/j.tifs.2023.104204

Hendożko E, Szefer P, Warzocha J (2010) Heavy metals in *Macoma balthica* and extractable metals in sediments from the southern Baltic Sea. Ecotoxic Environ Saf 73:152–163. https://doi.org/10.1016/j.ecoenv.2009.09.006

Iguchi J, Takashima Y, Namikoshi A, Yamashita Y, Yamashita M (2013) Origin identification method by multiple trace elemental analysis of short-neck clams produced in Japan, China, and the Republic of Korea. Fish Sci 79:977–982. https://doi.org/10.1007/s12562-013-0659-9

Iguchi J, Isshiki M, Takashima Y, Yamashita Y, Yamashita M (2014) Identifying the origin of *Corbicula* clams using trace element analysis. Fish Sci 80:1089–1096. https://doi.org/10.1007/s12562-014-0775-1

Julshamn K, Grahl-Nielsen O (1996) Distribution of trace elements from industrial discharges in the Hardangerfjord, Norway: a multivariate data analysis of saithe, flounder and blue mussel as sentinel organisms. Mar Pollut Bull 32:564–571. https://doi.org/10.1016/0025-326X(96)84577-2

Kang X, Zhao Y, Tan Z, Ning J, Zhai Y, Zhen G (2022) Evaluation of multivariate data analysis for marine mussels *Mytilus edulis* authentication in China: based on stable isotope ratio and compositions of C, N, O and H. J Food Compos Anal 111:104627. https://doi.org/10.1016/j.jfca.2022.104627

Kim H, Kumar KS, Hwang SY, Kang B-C, Moon HB, Shin K-H (2015) Utility of stable isotope and cytochrome oxidase/gene sequencing analyses in inferring origin and authentication of Hairtail fish and shrimp. J Agric Food Chem 63:5548–5556. https://doi.org/10.1021/acs.jafc.5b01469

Kobelja K, Nemet I, Župan I, Čulin J, Rončević S (2016) Elemental profiling of Noah's Ark shell (*Arca noae*, Linnaeus, 1758) by plasma optical spectrometry and chemometric tools. J Trace Elem Med Biol 38:157–164. https://doi.org/10.1016/j.jtemb.2016.04.011

Konieczka P, Namieśnik J (2010) Quality assurance and quality control in the analytical chemical laboratory—a practical approach. CRC Press, Taylor & Francis, Boca Raton. https://doi.org/10.1201/9781420082715

Kreissig KJ, Hansen LT, Jensen PE, Wegeberg S, Geertz-Hansen O, Sloth JJ (2021) Characterisation and chemometric evaluation of 17 elements in ten seaweed species from Greenland. PLoS One 16:e0243672. https://doi.org/10.1371/journal.pone.0243672

Kwoczek M, Szefer P, Hać E, Grembecka M (2006) Essential and toxic elements in seafood available in Poland from different geographical regions. J Agric Food Chem 54:3015–3024. https://doi.org/10.1021/jf0600511

Li L, Boyd CE, Odom J (2014) Identification of Pacific white shrimp (*Litopenaeus vannamei*) to rearing location using elemental profiling. Food Control 45:70–75. https://doi.org/10.1016/j.foodcont.2014.03.013

Li L, Boyd CE, Dong S (2015) Chemical profiling with modeling differentiates Ictalurid cat fish produced in fertilized and feeding ponds. Food Contr 50:18–22. https://doi.org/10.1016/j.foodcont.2014.08.014

Li L, Boyd CE, Racine P, McNevin AA, Somridhivej B, Minh HN, Tinh HQ, Godumala R (2017) Assessment of elemental profiling for distinguishing geographic origin of aquacultured shrimp from India, Thailand and Vietnam. Food Control 80:162–169. https://doi.org/10.1016/j.foodcont.2017.04.046

Li L, Han C, Dong S, Boyd CE (2019) Use of elemental profiling and isotopic signatures to differentiate Pacific white shrimp (*Litopenaeus vannamei*) from freshwater and seawater culture areas. Food Control 95:249–256. https://doi.org/10.1016/j.foodcont.2018.08.015

Luo R, Jiang T, Chen X, Zheng C, Liu H, Yang J (2019) Determination of geographic origin of Chinese mitten crab (*Eriocheir sinensis*) using integrated stable isotope and multi-element analyses. Food Chem 274:1–7. https://doi.org/10.1016/j.foodchem.2018.08.104

Mamede R, Ricardo F, Gonçalves D, Ferreira da Silva E, Patinha C, Calado R (2021) Assessing the use of surrogate species for a more cost-effective traceability of geographic origin using elemental fingerprints of bivalve shells. Ecol Indic 130:108065. https://doi.org/10.1016/j.ecolind.2021.108065

Mamede R, Duarte IA, Caçador I, Tanner SE, Silva M, Jacinto D, Fonseca VF, Duarte B (2022) Elemental fingerprinting of sea urchin (*Paracentrotus lividus*) gonads to assess food safety and trace its geographic origin. J Food Compos Anal 114:104764. https://doi.org/10.1016/j.jfca.2022.104764

Mamede R, Santos A, Ferreira da Silva E, Patinha C, Calado R, Ricardo F (2024) New evidence of fraudulent mislabeling and illegal harvesting of Manila clams (*Ruditapes philippinarum*) through elemental fingerprints of their shells and chemometric analyses. Food Control 163:110501. https://doi.org/10.1016/j.foodcont.2024.110501

Mesa LM, Méndez EP, Sánchez MS, Montelongo FG (1999) Interpretation of heavy metal data from mussel by use of multivariate classification techniques. Chemosphere 38:1103–1111. https://doi.org/10.1016/s0045-6535(98)00365-8

Miedico O, Pompa C, Tancredi C, Cera A, Pellegrino E, Tarallo M, Chiaravalle AE (2017) Characterisation and chemometric evaluation of 21 trace elements in three edible seaweed species imported from south-east Asia. J Food Compos Anal 64:188–197. https://doi.org/10.1016/j.jfca.2017.09.004

Morrison L, Bennion M, Gil S, Graham CT (2019) Spatio-temporal trace element fingerprinting of king scallops (*Pecten maximus*) reveals harvesting period and location. Sci Total Environ 697:134121. https://doi.org/10.1016/j.scitotenv.2019.134121

Ndiaye B, Ndiaye M, Cid BP, Diop A, Diagne I, Cissé D, Dione CT, Hanne M (2020) Trace metals in mussels *Mytilus galloprovincialis* from Dakar Coast (Senegal). Am J Anal Chem 11:137–145. https://doi.org/10.4236/ajac.2020.113011

Nędzarek A, Czerniejewski P (2021) The edible tissues of the major European population of the invasive Chinese mitten crab (Eriocheir sinensis) in the Elbe River, Germany, as a valuable and safe complement in essential elements to the human diet. J Food Compos Anal 96:103713. https://doi.org/10.1016/j.jfca.2020.103713

Nędzarek A, Czerniejewski P, Drost A, Harasimiuk F, Machula S, Tórz A, Masalski P (2017) The distribution of elements in the body of invasive Chinese mitten crabs (*Eriocheir sinensis* H. Milne-Edwards, 1853) from Lake Dąbie, Poland. J Food Compos Anal 60:1–9. https://doi.org/10.1016/j.jfca.2017.03.003

Nędzarek A, Czerniejewski P, Tórz A (2019a) Microelements and macroelements in the body of the invasive Harris mud crab (*Rhithropanopeus harrisii*, Maitland, 1874) from the central coast of the South Baltic Sea. Environ Monit Assess 191:499. https://doi.org/10.1007/s10661-019-7564-3

Nędzarek A, Czerniejewski P, Tórz A (2019b) Macro- and trace elements in Chinese mitten crabs (*Eriocheir sinensis*) from Szczecin Lagoon, Poland—implications for human health. Aquaculture 506:229–237. https://doi.org/10.1016/j.aquaculture.2019.03.042

Ortea I, Gallardo JM (2015) Investigation of production method, geographical origin and species authentication in commercially relevant shrimps using stable isotope ratio and/or multi-element analyses combined with chemometrics: an exploratory analysis. Food Chem 170:145–153. https://doi.org/10.1016/j.foodchem.2014.08.049

Polak-Juszczak L (2009) Temporal trends in the bioaccumulation of trace metals in herring, sprat, and cod from the southern Baltic Sea in the 1994-2003 period. Chemosphere 76:1334–1339. https://doi.org/10.1016/j.chemosphere.2009.06.030

Polak-Juszczak L (2010) Bioaccumulation and temporal trends of trace elements in flounder from the southern Baltic Sea for the 1996-2003 period. J Toxic Environ Health Part A 73:1186–1193. https://doi.org/10.1080/15287394.2010.491776

Polak-Juszczak L (2013) Trace elements in the livers of cod (Gadus morhua L.) from the Baltic Sea: levels and temporal trends. Environ Monit Assess 185(1):687–694. https://doi.org/10.1007/s10661-012-2584-2

Przytarska JE, Sokołowski A, Wołowicz M, Hummel H, Jansen J (2010) Comparison of trace metal bioavailabilities in European coastal waters using mussels from *Mytilus edulis* complex as biomonitors. Environ Monit Assess 166:461–476. https://doi.org/10.1007/s10661-009-1015-5

Rainbow PS, Fialkowski W, Sokolowski A, Smith BD, Wolowicz M (2004) Geographical and seasonal variation of trace metal bioavailabilities in the Gulf of Gdansk, Baltic Sea using mussels (*Mytilus trossulus*) and barnacles (*Balanus improvisus*) as biomonitors. Mar Biol 144:271–286. https://doi.org/10.1007/s00227-003-1197-2

Romarís-Hortas V, García-Sartal C, Barciela-Alonso MC, Moreda-Piñeiro A, Bermejo-Barrera P (2010) Characterization of edible seaweed harvested on the Galician coast (northwestern Spain) using pattern recognition techniques and major and trace element data. J Agric Food Chem 58:1986–1992. https://doi.org/10.1021/jf903677y

Smith RG, Watts CA (2009) Determination of the country of origin of farm-raised shrimp (family Penaeide) using trace metal profiling and multivariate statistics. J Agric Food Chem 57:8244–8249. https://doi.org/10.1021/jf901658f

Stanek M, Dąbrowski J, Różański S, Janicki B, Długosz J (2017) Heavy metals bioaccumulation in tissues of spiny-cheek crayfish (*Orconectes limosus*) from Lake Gopło: effect of age and sex. Bull Environ Contam Toxicol 98:740–746. https://doi.org/10.1007/s00128-017-2098-2

Struck BD, Pelzer R, Ostapczuk P, Emons H, Mohl C (1997) Statistical evaluation of ecosystem properties influencing the uptake of As, Cd, Co, Cu, Hg, Mn, Ni, Pb and Zn in seaweed (*Fucus vesiculosus*) and common mussel (*Mytilus edulis*). Sci Total Environ 207:29–42. https://doi.org/10.1016/S0048-9697(97)00246-5

Szefer P (2002) Metals, metalloids and radionuclides in the Baltic Sea ecosystem. Elsevier Science B.V., Amsterdam. ISBN 10: 0444503528

Szefer P (2003) Application of chemometric techniques in analytical evaluation of biological and environmental samples. In: Namieśnik J, Chrzanowski W, Żmijewska P (eds) New horizons and challenges in environmental analysis and monitoring. CEEAM, Gdańsk, pp 355–388. ISBN 8391908100

Szefer P (2007) Chemometric techniques in analytical evaluation of food quality. In: Szefer P, Nriagu J (eds) Mineral components in foods. CRC Press, Taylor & Francis, Boca Raton, pp 69–121. https://doi.org/10.1201/9781420003987

Szefer P (2013) Safety assessment of seafood with respect to chemical pollutants in European Sas. Oceanol Hydrobiol Stud 42:110–118. https://doi.org/10.2478/s13545-013-0063-1

Szefer P, Grembecka M (2022) Bromatological, analytical and chemometric assessment of animal and plant foods based on mineral composition. Eur J Transl Clin Med 5:77–106. https://doi.org/10.31373/ejtcm/137919

Szefer P, Wenne R (1987) Concentration of uranium and thorium in molluscs inhabiting Gdańsk Bay, Baltic Sea. Sci Total Environ 65:191–202. https://doi.org/10.1016/0048-9697(87)90172-0

Szefer P, Wolowicz M (1993) Occurrence of metals in the cockle *Cerastoderma glaucum* from dfferent geographical regions in view of principal component analysis. Stud Mater Ocean Mar Pollut 64:253–264

Szefer P, Geładon J, Ali AA, Bawazir A, Sad M (1997a) Distribution and association of trace metals in soft tissue and byssus of *Perna perna* from the Gulf of Aden, Yemen. Environ Int 23:53–61

Szefer P, Ikuta K, Kushiyama S, Szefer K, Frelek K, Geładon J (1997b) Distribution and association of trace metals in soft tissue and byssus of *Mytilus edulis* from the east coast of Kyushu Island, Japan. Arch Environ Contam Toxicol 32:184–190. https://doi.org/10.1007/s002449900173

Szefer P, Geładon J, Ali AA, Paez-Osuna F, Ruiz-Fernandes AC, Guerrero Galvan SR (1998) Distribution and association of trace metals in soft tissue and byssus of *Mytella strigata* and other benthal organisms from Mazatlan harbour, mangrove lagoon of the northwest coast of Mexico. Environ Int 24:359–374. https://doi.org/10.1016/S0160-4120(98)00014-2

Szefer P, Wołowicz M, Kusak A, Deslous-Paoli J-M, Czarnowski W, Frelek K, Belzunce-Segarra M-J (1999a) Distribution of mercury and other trace metals in the cockle *Cerastoderma glaucum* from the Mediterranean Lagoon, Etang de Thau. Arch Environ Contam Toxicol 36:56–63. https://doi.org/10.1007/s002449900442

Szefer P, Ali AA, Ba-Haroon AA, Rajeh AA, Geładon J, Nabrzyski M (1999b) Distribution and relationships of selected trace metals in molluscs and associated sediments from the Gulf of Aden, Yemen. Environ Pollut 106:299–314. https://doi.org/10.1016/s0269-7491(99)00108-6

Szefer P, Frelek K, Szefer K, Lee C-B, Kim B-S, Warzocha J, Zdrojewska I, Ciesielski T (2002) Distribution and relationships of trace metals in soft tissue, byssus and shells of *Mytilus edulis trossulus* from the southern Baltic. Environ Pollut 120:423–444. https://doi.org/10.1016/S0269-7491(02)00111-2

Szefer P, Kim B-S, Kim C-K, Kim E-H, Lee C-B (2004) Distribution and coassociations of trace elements in soft tissue and byssus of *Mytilus galloprovincialis* relative to the surrounding seawater and suspended matter of the southern part of the Korean Peninsula. Environ Pollut 129:209–228. https://doi.org/10.1016/j.envpol.2003.10.012

Szefer P, Fowler SW, Ikuta K, Paez Osuna F, Ali AA, Kim B-S, Fernandes HM, Belzunce M-J, Guterstam B, Kunzendorf H, Wołowicz M, Hummel H, Deslous-Paoli M (2006) A comparative assessment of heavy metal accumulation in soft parts and byssus of mussels from subarctic, temperate, subtropical and tropical marine environments. Environ Pollut 139:70–78. https://doi.org/10.1016/j.envpol.2005.04.031

Szteke B (2007) Criteria of evaluation of food elements analysis data. In: Szefer P, Nriagu J (eds) Mineral components in foods. CRC Press, Taylor & Francis, Boca Raton, pp 51–69. https://doi.org/10.1201/9781420003987

Tsakovski S, Simeonov V (2010) Chemometrics as a tool for treatment processing of multiparametric analytical data sets. In: Namieśnik J, Szefer P (eds) Analytical measurements in aquatic environments. CRC Press, Taylor & Francis, Boca Raton, pp 369–387

Won E-J, Kim SH, Go Y-S, Kumar KS, Kim MS, Yoon S-H, Bayon G, Kim J-H, Shin KH (2021) A multi-elements isotope approach to assess the geographic provenance of Manila Clams (*Ruditapes philippinarum*) via recombining appropriate elements. Foods 10:646. https://doi.org/10.3390/foods10030646

Żbikowski R, Szefer P, Latala A (2006) Distribution and relationships between selected chemical elements in green alga *Enteromorpha* sp. from the southern Baltic. Environ Pollut 143:435–448. https://doi.org/10.1016/j.envpol.2005.12.007

Żbikowski R, Szefer P, Latala A (2007) Comparison of green algae *Cladophora* sp. and *Enteromorpha* sp. as potential biomonitors of chemical elements in the southern Baltic. Sci Total Environ 387:320–332. https://doi.org/10.1016/j.scitotenv.2007.07.017

Zhang X, Liu Y, Li Y, Zhao X (2017) Identification of the geographical origins of sea cucumber (*Apostichopus japonicus*) in northern China by using stable isotope ratios and fatty acid profiles. Food Chem 218:269–276. https://doi.org/10.1016/j.foodchem.2016.08.083

Zhang X, Cheng J, Han D, Zhao X, Chen X, Liu Y (2019) Geographical origin traceability and species identification of three scallops (*Patinopecten yessoensis*, *Chlamys farreri*, and *Argopecten irradians*) using stable isotope analysis. Food Chem 299:125107. https://doi.org/10.1016/j.foodchem.2019.125107

Zhao X, Liu Y, Wang G, Tao W, Lou Y, Li N, Liu Y (2019) Tracing the geographical origins of Yesso scallop (*Patinopecten yessoensis*) by using compound-specific isotope analysis: an approach for overcoming the seasonal effect. Food Control 102:38–45. https://doi.org/10.1016/j.foodcont.2019.03.016

Zhao Y, Wu J, Kang X, Guo Y, Wang L, Sheng X, Tan Z (2024) Elemental profiling of red seaweed *Neopyropia yezoensis* used in fast authenticating the geographical origin and food safety assessment. J Food Compos Anal 125:105839. https://doi.org/10.1016/j.jfca.2023.105839

Chapter 5
Milk and Dairy Products

Contents

5.1 Introduction

Milk and dairy products are rich in nutrients that are essential for good bone health, including Ca, P, protein and vitamins D and K. Milk and dairy products also supply riboflavin and vitamin B_{12}. The assessment of the authenticity and detection of adulteration of milk and dairy products is a major challenge. Therefore, to achieve this task, the use of different analytical techniques, which are expected to involve modern and sophisticated analytical equipment and skilled operators, is necessary.

Milk and milk products have been assayed for mineral composition to assess their authenticity in view of chemometric evaluation. Published data are available related to multivariate control of milk (e.g. Favretto et al. 1989; Benincasa et al. 2008; Bilandžić et al. 2011; Di Bella et al. 2012; Król et al. 2012; Potortì et al. 2013; Miedico et al. 2016; Potočnik et al. 2016; Bandoniene et al. 2020), commercial skim milk powder sweet whey and different milk-based infant formulae (e.g. Hermansen et al. 2005; Sola-Larrañaga and Navarro-Blasco 2006; Ataro et al. 2008; Rey-Crespo

Supplementary Information The online version contains supplementary material available at https://doi.org/10.1007/978-3-031-99687-0_5.

P. Szefer, *Chemometrics in Food Control Based on Elemental Analysis*, https://doi.org/10.1007/978-3-031-99687-0_5

et al. 2013; Azcarate et al. 2015, 2017; Bilge et al. 2016; McLeod et al. 2016; Zain et al. 2016; Akele et al. 2017; Zhou et al. 2017; Chen et al. 2020), cheeses (e.g. Suhaj and Koreňovská 2007, 2008b, 2010; Moreno-Rojas et al. 2012; Osorio et al. 2015), butter, margarine and peanut butter (e.g. Marpaung et al. 2020) and eggs (e.g. de Freitas et al. 2013; Barbosa et al. 2014; Borges et al. 2015; Esposito et al. 2016). The used methodology could be applied successfully for evaluating the traceability of milk and dairy products, defending this unique product against fraud or commercial disputes.

According to Schneider et al. (2015), stable isotope analysis is a fundamental tool for food origin and authenticity testing. The use of SIRA in livestock production requires knowledge of stable isotopes for distinguishing between product and diet. The analysis of $\delta^{13}C$ discrimination between the diet and all the main milk components, i.e. milk fat, casein and lactose, will improve the use of SIRA with respect to authenticity testing and proof of origin. Capici et al. (2015) reported that IRMS could identify milk samples according to cheese manufacturing and detect the addition of milk or reconstructed powdered milk samples with significantly different isotope values. Fortunato et al. (2004) reported that MC–ICP–MS can be used for the precise determination of differences in the isotopic profile of Sr. This method was applied to verify the geographical provenance of different Emmental cheese samples. Camin et al. (2015) described the use of validated analytical methods to determine $\delta^{2}H$, $\delta^{13}C$, $\delta^{15}N$ and $\delta^{34}S$ and multi-elemental profiles of cheese samples processed in international collaborative studies in 13 different laboratories. These investigations, which were based on blind duplicates of seven hard cheeses, were performed in compliance with the IUPAC protocol and ISO Standards 5725/2004 and 13528/2005. According to the authors, the obtained data could be a valuable analytical database for standardization agencies to obtain official recognition of methods; then, the data could be used to initiate commercial disputes and legal debates.

The traceability of dairy products has been an urgent issue; therefore, assessing the authenticity and detection of adulteration of milk and dairy products is important, and the use of different analytical techniques is recommended for this purpose. These methods/techniques are expected to involve state-of-the-art and skilled analysts. Therefore, there is a need to select inexpensive and fast methods for quantifying the adulteration of these products. Therefore, spectroscopic techniques such as NIRS, MIRS, NMR and fluorescence spectroscopy in combination with multivariate analysis could be considered helpful tools (Kamal and Karoui 2015). These authors specified and discussed the advantages and disadvantages of each technique. Another spectroscopic technique, LIBS, enables rapid, high-throughput, microdestructive and multi-elemental analysis of a wide range of samples. It can be a useful tool for evaluating food authenticity. LIBS shows excellent power for at-line or portable systems; however, improvements in the sensitivity of trace element detection, sample preparation, data analysis and instrument miniaturization are still needed (Markiewicz-Keszycka et al. 2019).

The implementation of quality and safety control methods in conjunction with chemometrics, particularly in food science, revealed that this combination is a powerful tool for verifying food authenticity, reflecting geographic origin and variety. Dairy products, in particular, are subject to validation by public authorities regarding the identification of the many illnesses sometimes traced to products of specific origin. Therefore, the development of a reliable system provided by discriminatory power is highly important for ensuring public health and protecting consumers (Arvanitoyannis and Tzouros 2005).

The increased marketability and production of eggs suggest that protein production in the form of eggs is more sustainable than the production of other protein-rich foods, such as beef. However, in the case of eggs, caution should be taken due to health problems. Therefore, appropriate systems are important to ensure egg traceability and authenticity. In this context, the application of SIRA combined with alternative analytical methods for egg identification is needed for the traceability of eggs in the context of production, usage, environmental benefits, etc. (Rock 2012).

PDO labelling of cheeses has been established by the EU as a quality policy to ensure the authenticity of cheese produced in a distinct region via traditional production methods. However, currently applied technologies to differentiate and establish PDOs are limited in terms of time, cost, accuracy and ability to identify PDO adulteration. HTS methodologies have enabled, among other achievements, a more accurate identification of microbial communities that develop in fermented cheeses, contributing to the development of specific organoleptic and physiochemical properties. The applicability of these methods may provide an additional tool for identifying the essential microbial species that contribute to the unique sensorial properties of PDO cheeses, defining their typicity, which distinguishes them from various fraudulent products (Kamilari et al. 2019).

5.2 Chemometric Assessment of Milk and Dairy Products

5.2.1 Geographical Differentiation

5.2.1.1 European Countries

Milk

Favretto et al. (1989) reported that ET–AAS combined with PCA and PFA was applied to characterize Italian cow milk samples from two regions. PC1 and PC2, with EVs amounting to 1.32 and 1.01, respectively, explained 78% of the total variance. Two groups of samples related to milk originating from Modena and Karst were clearly differentiated. It was concluded that MEC processed by chemometric analysis could be successfully applied to distinguish Italian milk samples from milk with two geographical origins.

According to Renou et al. (2004), milk samples analysed by a combination of NMR, IRMS and GC–IRMS originated from two distinct INRA regions, where cows at each site fed on pasture and silage. The DA results revealed that the $\delta^{18}O$

parameter was affected by the production location and diet. For a pasture diet, the two sites, i.e. plain and mountain regions, were distinctly differentiated.

Sacco et al. (2009), on the basis of a combination of ICP–AES, HPIC–CD, IRMS and NMR with PCA and DA techniques, reported that samples of cow milk from southern Italy and six European countries were classified according to their geographical provenance. The first two PCs explained 75% of the total variance, indicating differentiation between Italian and Central–Eastern European milk samples. The classification and prediction abilities were estimated by DA to be 95% and 90%, respectively.

Potočnik et al. (2016) analysed EDXRF, k0-INAA and ICP–MS data from Slovenian raw cow milk originating from four geographical regions (Alpine, Mediterranean, Dinaric and Pannonian regions). LDA was useful only in partial regional discrimination, with a prediction ability of 66.7%. The Alpine and Pannonian samples were clearly differentiated, in contrast to the overlapping Dinaric and Mediterranean samples. It was expected that completely distinguishing between samples from all four regions could be attained by integrating MEA with SIRA. Notably, Rossmann et al. (1998) noted the possibility of performing regional origin assignment for cow milk from Alpine regions by using the SIRA results.

The application of TC/EA–IRMS, EDXRF and ICP–MS data processed by multivariate techniques (CA, OPLS–DA and DD–SIMCA) was used by Potočnik et al. (2020a) to assess the intercountry differentiation of cow milk samples, including those from four geographical regions of Slovenia collected during summer and winter. It was possible to discriminate milk samples according to the year, season and production region using DA. Furthermore, the discrimination of milk samples from three geographic regions was achieved, although the Alpine samples overlapped with the Dinaric and Pannonian samples, with the prediction ability the highest for the latter (82.1%). OPLS–DA enabled good regional predictability ($\geq$0.77), with $\delta^{13}C_{cas}$ values and Br content providing the most variance. It has been postulated that DD–SIMCA classification using a combination of SIR and MEC as parameters could be an appropriate approach for confirming the origin designation of milk, indicating authenticity or possible mislabelling.

Dairy Products

Pillonel et al. (2003) selected profiles of major and trace elements, stable isotope ratios and radioactive elements obtained by several analytical techniques, i.e. IRMS, CF–IRMS, TIMS, GF–AAS, ICP–MS and α-spectrometry. It was possible to classify Emmental cheese from five European countries (Germany, France, Finland, Switzerland and Austria) with respect to their geographical origin. As shown in Fig. 5.1, PCA revealed that PC1 and PC3, accounting for 67% of the total variance, corresponded to $\delta^{13}C$, $\delta^{15}N$, $\delta^{2}H$ and $\delta^{87}Sr$. These most discriminative variables importantly contributed to differentiating cheese samples with respect to the samples from Finland and the Bretagne and Savoie, whereas the samples from Switzerland and the Allgau and Vorarlberg regions were unsatisfactorily

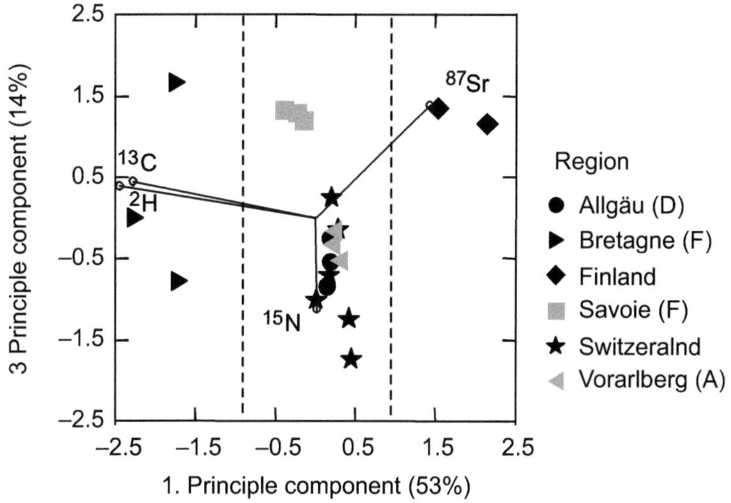

Fig. 5.1 PCA of the parameters $\delta^{13}C$, $\delta^{2}H$, $\delta^{15}N$ and $\delta^{87}Sr$. Differentiation of the groups "Finland", "Savoie" and "Bretagne". (After Pillonel L et al., 2003. Int Dairy J 15:547–556. https://doi.org/10.1016/j.idairyj.2004.07.028. With permission)

differentiated. The reason for this may be their similarity with respect to climatic, geological and agricultural parameters, which are responsible, among other factors, for variability in the feed composition.

Pillonel et al. (2005a) used FAAS, EA–IRMS, routine analyses, enzymatic test kits and GC to determine that different analytical parameters of MEA and IRMS, including other biochemical factors, also appeared to be useful for characterizing the interfive-country profile of Emmental cheese through the processing of concentration data with ANOVA and Fisher's LSD techniques.

Moreover, Pillonel et al. (2005b) used more advanced chemometric studies, i.e. initially DA and ANN, which provided comparable data when 25 analytical factors were used for all regions. With the DA and ANN techniques, 95% and 91% correctly classified all the samples in the validation procedure, respectively. In the case of ANN, the size of the database was too small to develop its overall power. By building a new DA model for each of the six pairs, i.e. Switzerland versus another region, 100% correct classification could be achieved for the Swiss samples in the jackknife validation using 15 analytical factors. It was postulated that the use of selected analytical parameters fused with DA allows the assignment of unknown Emmental cheese samples with respect to their geographical provenance with high confidence.

According to Suhaj and Koreňovská (2008a), the use of a combination of FAAS, GF–AAS and CV–AAS with multivariate processing tools (CA, PCA, FA and DA) allowed the evaluation of the elemental profiles of Emmental and Edam cow hard cheeses originating from ten European countries. Intercountry differentiation was studied using PCA, with the first three PCs accounting for 59.1% of the total variance. Moreover, CDA yielded 85.9% and 93% of the original cases and prediction ability, respectively, on the basis of elemental markers such as Ba, Ca, Cu, Cr, Hg, K, Mg, Mn, Mo, Ni and V. These classifications are represented in a scatter plot of the three most important discriminant functions. The best classification was achieved by CD and CV for cheese samples originating from Italy, Austria, Hungary, Great Britain, France, the Netherlands, the Czech Republic, Germany, Poland and Slovakia.

Comparable results for Emmental cheeses from five European countries (Austria, France, Germany, Finland and Switzerland) were reported by Pillonel et al. (2005a, b).

Manca et al. (1999, 2001) successfully used SIR and amino acid ratios quantified in cheese from different Italian regions. Manca et al. (2001) demonstrated that the combination of IRMS and HPLC with chemometric tools such as RA, PCA, LDA and CA allowed the quantification of $\delta^{13}C$, $\delta^{15}N$ and free amino acid ratios in ewe milk cheese from cheese factories, i.e. from Sardinia, Sicily and Apulia. In the latter case, a clear differentiation between the products from the three regions with 100% correct classification was achieved.

Koreňovská and Suhaj (2007) used a combination of FAAS, GF–AAS and CV–AAS with PCF and MCDA models to differentiate multiregional Slovak bryndza sheep cheeses from Polish and Romanian varieties. The classification procedure based on the DA model achieved 100% accuracy in the differentiation of all seven producing regions, whereas a less satisfactory effect was achieved for the two remaining regions, with accuracies of 75% and 60%. Intercountry differentiation of the bryndza cheese-producing regions was also attained. The Romanian cheese samples were clearly distinguished, whereas the Polish samples somewhat overlapped with the Slovak samples on the basis of the most discriminative descriptors, i.e. Cr, Hg, Mn and V. The results obtained could be explained by the close localization of the Polish samples to the Slovak samples. The origin of raw cheeses is attributable to a relatively identical geochemical Carpathian composition of subsoil bordering the producing regions of Slovakia and Poland.

5.2.1.2 Asiatic Countries

Milk

Xie et al. (2020) used a combination of EA–IRMS, ICP–MS and HPLC–PDA to trace the origin of fresh milk in small-scale districts, i.e. from 11 districts located in 5 cities of the Inner Mongolia Autonomous Region. This methodology appears to be the most effective tracing approach. The application of PCA and OPLS–DA to process multiparametric analytical data was more efficient at identifying the geographic provenance of milk samples than a single or dual combination of amino acids, stable isotopes or elemental analysis (Fig. 5.2). In the case of the triple combination, the

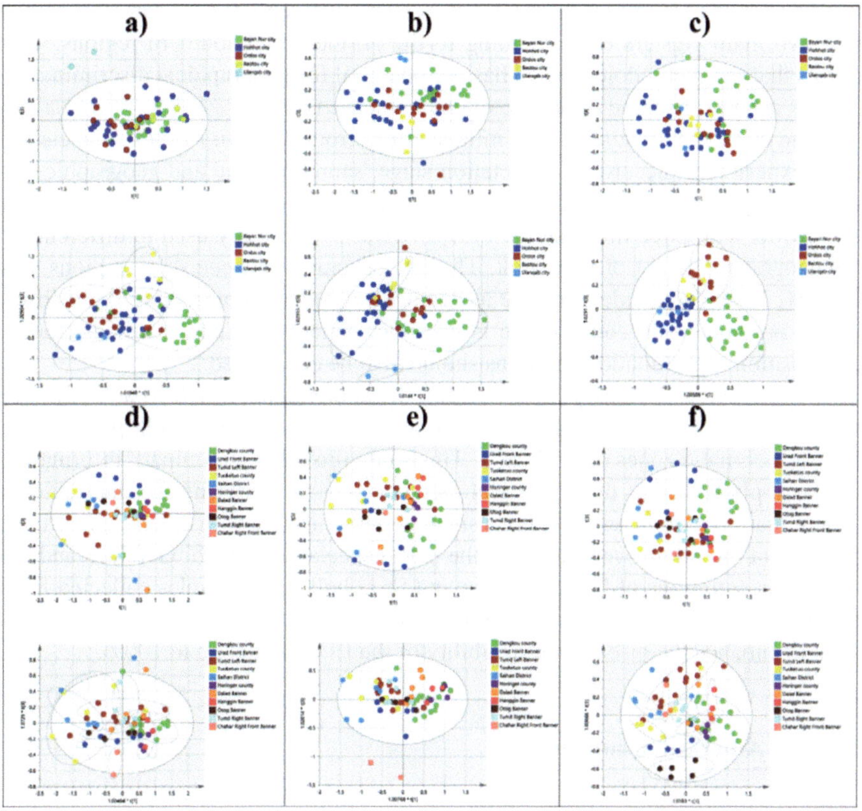

Fig. 5.2 PCA and OPLS–DA of milk from five cities with amino acid, stable isotope and element data and their combinations: (**a**) stable isotopes; (**b**) amino acids and stable isotopes; (**c**) amino acids, stable isotopes and elemental concentrations. PCA and OPLS–DA of milk from 11 countries with amino acid, stable isotope and element data and their combinations; (**d**) amino acids; (**e**) amino acids and elemental concentrations; (**f**) amino acids, stable isotopes and elemental concentrations. (After Xie L et al., 2020. Food Chem 316:126332. https://doi.org/10.1016/j.foodchem.2020.126332. With permission)

accuracies of the original classification and cross-validation were satisfactory since they reached 100% and 71.6%, respectively. Variables such as $\delta^{13}C$, $\delta^{18}O$, Rb, Sr, Lys, Tyr, Se, Mn, N, Na, Ca and Fe for 5 cities and $\delta^{13}C$, Sr, Ca, $\delta^{18}O$, Rb, Mg, Ti, K, Sc, Na, Mo and $\delta^{15}N$ for 11 districts were calculated to have high importance values (>1). This means that the combination of SIR, MEC and amino acids was the best choice, providing an effective solution to identify regions of milk in small-scale districts or cities.

Behkami et al. (2020) aimed to investigate the effectiveness of precipitation isotope information to build a milk geographical origin dataset for Peninsular Malaysia. SIR data, i.e. $\delta^{2}H$ and $\delta^{18}O$ values, were processed by FA. FA with the first two factors explained 89.5% of the total variance and differentiated the samples studied into two main clusters corresponding to the northern and southern regions. This means that the methodology used has the potential for geographical discrimination of milk samples because of the use of chemometric tools.

Zhao et al. (2020) studied cow milk samples from two dairy farms subjected to three experiments according to lactation stage, sampling time and geographic origin, i.e. four Chinese provinces, including Inner Mongolia. The combination of EA–IRMS with three-dimensional PLS–DA was successfully used to differentiate the geographic origin of cow milk. The CV accuracy between these regions was 92.11%. It was concluded that the assignment of milk samples using multi-SIRA within provinces and countrywide is possible. Notably, farming methods, forage composition and climatic conditions should also be considered.

In the opinion of Kalpage et al. (2022), the combination of IRMS and ICP–MS was an effective tool for determining the agroclimatic origin of milk produced in four agroclimatic zones of Sri Lanka. This study provided preliminary information on the applicability of this analytical methodology for verifying the interregional differentiation of milk, even within a small geographic area with diverse conditions. Significant variations were noted in the SIR, especially in the $\delta^{18}O$ values and the mean concentrations of Li, Al, Cr, Mn and Sr in bulk milk samples collected from different agroclimatic zones. Owing to such significant differentiation, it was possible to enhance the discriminating ability of the LDA model up to 100%.

5.2.1.3 American Countries

Milk

Chesson et al. (2010) reported that dairy milk, supermarket milk and restaurant milk samples from different states in the United States were studied by TC/EA–IRMS for δ^2H and $\delta^{18}O$. The RA data indicated that both stable isotope contents are useful descriptors of the geographical provenance of milk samples, whereas milk water $\delta^{18}O$ values were applied to predict potential regions of origin for restaurant samples.

Milk and Dairy Products

A pilot study using EA–IRMS and UTC–SIR–MS involved buffalo milk and two types of cheese, i.e. cream and butter cheese, from the Amazon basin of Brazil. According to the authors, this approach was used for the first time in Brazil in this study, and the approach was used especially for Amazon dairy products, although this methodology has been widely applied for several years in Northern Hemisphere countries. The applied methodology appeared to be a useful tool in the geographic characterization of dairy products to protect consumers from dairy fraud (Silva et al. 2014).

5.2.1.4 African Countries

Milk

As reported by Akele et al. (2017), raw cow milk samples from dairy farms at three study sites in Ethiopia were investigated by FAAS. Heavy metal contents processed using PCA and PLS–DA tools allowed the discrimination of milk samples according to their interregional assignment. For example, PLS–DA indicated a clear classification concerning milk sample origin on the basis of Cd and Cu, as the most effective variables reached 100% classification efficiency using external validation.

5.2.1.5 Oceania Countries

Milk

The pilot study by Crittenden et al. (2007) aimed to estimate the power of SIRA to differentiate milk samples from pasture-fed cows in seven dairy regions in Australia and New Zealand. SIRA and $^{87}Sr/^{86}Sr$ determined by EA–IRMS and TIMS, respectively, were useful tools for identifying the geographic origin of dairy products from Australasian producing regions. For example, the $\delta^{18}O$ values of milk samples were significantly related to mid-late autumn climatic conditions, which influence water turnover (rainfall/evaporation) in different areas of Australia. This association, together with $\delta^{13}C$ decreasing from north to south, enabled milk samples produced from pasture-fed cows in northern, central and southern Australasia to be distinguished. It was also reported that milk samples from Australasia tended to be considerably enriched in ^{18}O and ^{34}S compared with dairy products from northern and central Europe. Such differentiation can be useful in the assessment of the chemical profile of dairy products with respect to its interspatial variability.

According to Ehtesham et al. (2013a, b), a clear discrimination of milk samples between the South Island and North Island (New Zealand) was achieved. HCA confirmed the results obtained by PCA when the North Islands were divided into two subregions. The LDA models were applied with different combinations of variables using bulk and/or FA isotopic compositions, e.g. fatty acid δ^2H and $\delta^{13}C$ or bulk and fatty acid δ^2H and $\delta^{13}C$. It was possible to provide the best geographical discrimination, with corrected classification amounting to 100% and from 98% to 100% of the recognition and prediction abilities (estimated by CV), respectively.

Ehtesham et al. (2013b) quantified δ^2 H and 11 FA contents by composition in powder milk samples using TC/EA–IRMS, GC/MS, GC–IRMS and GC–C–IRMS. This approach was applicable to New Zealand only because it is necessary to have samples of known origin to build a model that has the ability to correlate the spatial variation in rainfall SIR composition with that of the milk produced in the region. A global "isoscape" of the d^2H values of milk (fatty acids and bulk) may considerably enhance the application of this technique for verifying the origin of dairy products.

Wijenayake et al. (2020) analysed SIRs determined from casein from cow milk by IRMS to investigate the geoclimatic variation across dairy farms from the South and North Islands of New Zealand. As uniform feeding and fertilizer practices were applied throughout the sampling period, the subtropical (North Island) and temperate (South Island) climates were reflected in the variations in $\delta^{13}C$ and $\delta^{15}N$. Notably, the $\delta^{2}H$ and $\delta^{18}O$ values did not differ in terms of climatic variation between the islands but rather in terms of separate topographical locations. It was postulated that $\delta^{15}N$ content considerably influenced climatic variability. PCA with the first two PCs accounted for 82.9% of the total variance and allowed discrimination between the North and South Islands. $\delta^{13}C$ and $\delta^{2}H$ were the main contributors to PC-1, whereas $\delta^{15}N$ and $\delta^{18}O$ were related mostly to PC-2. According to the authors, using a combination of $^{87/86}Sr$ ratios with trace elements may allow the construction of a better model.

5.2.1.6 Countries Across Multiple Continents

Milk

Xu et al. (2021) constructed a reliable discrimination model to quantify milk geographical origin, i.e. SIRs ($\delta^{13}C$, $\delta^{15}N$ and $\delta^{18}O$), 51 elements and 35 FAs in milk samples from Australia, New Zealand and Austria. The SIR levels in the milk samples from Australia were the highest, followed by those in the milk samples from New Zealand and Austria. Moreover, 14 elements and 8 FAs were significantly different in the samples from various countries ($p < 0.05$). Thus, a PLS–DA model with good robustness and predictive ability for authenticating milk origin ($R^2X = 0.693$, $Q^2 = 0.854$) was adequately constructed. The most reliable variables for milk origin discrimination were Rb, $\delta^{18}O$, Tl, Ba, Mo, Sr, $\delta^{15}N$, Cs, As, Eu, C20:4n6, Sc, C13:0, K, Ca and C16:1n7 in the multivariate model for verifying milk origin.

McLeod et al. (2016) analysed MEA and SIRA in goat whole milk powder from New Zealand, the Netherlands, Spain and France. The obtained database was integrated with MDS and CAPC. A clear distinction between milk powder samples corresponding to their origin was exhibited by a classification success rate of the LOOCV of 98.1%.

Luo et al. (2016) reported that SIRA data integrated with chemometric analysis were useful for evaluating the stable isotope profiles of pure milk samples from six European, Asiatic, North American and Australasian countries. A 3D distribution diagram of the $\delta^{13}C$ and $\delta^{15}N$ compositions of the extracted proteins, as well as the $\delta^{18}O$ composition of milk water, distinguished four sample groupings: Australia–New Zealand, Germany–France, the USA and China. It was concluded that the tools used allowed clear discrimination between milk samples corresponding to the four geographic regions, i.e. Australasia, Europe, Asia and North America.

According to Zain et al. (2016), raw and factory cow milk samples from different farms in the northern and southern regions of Malaysia and from other countries, i.e.

Iran, Turkey, Azerbaijan, Belgium, Canada, the USA, Australia and New Zealand, were investigated by ICP–MS. PCA with the first four PCs explained up to 53.1% of the total variance, especially HCA, revealed clear discrimination of Malaysian milk samples from selected regions corresponding to different countries worldwide (Figs. 5.3 and 5.4). Among the detected elements, Ca, Na, Fe, Zn, Mn, K, Ba and Mg were mainly responsible for effective differentiation of Malaysian milk samples from those originating from various intercontinental regions. However, Ca, Fe, Mg, Mn, K, Zn, Na and B were the most discriminative variables for the identification of tropical and Australasian areas, whereas the remaining variables were designated by Cu, Mo and Se.

Dong et al. (2017) used EA–IRMS to investigate Tetra Pak milk samples from four continents, including Australia with New Zealand, Europe (Germany, France), North America (US) and Asia (China). The adequacy of the $\delta^{13}C$ and $\delta^{15}N$ values for accurate interregional assessment revealed their potential as "fingerprints" of the geographical origin of milk. According to the authors, further statistical investigation based on a larger number of samples is needed to verify whether the obtained results could be generally applicable to all types of milk.

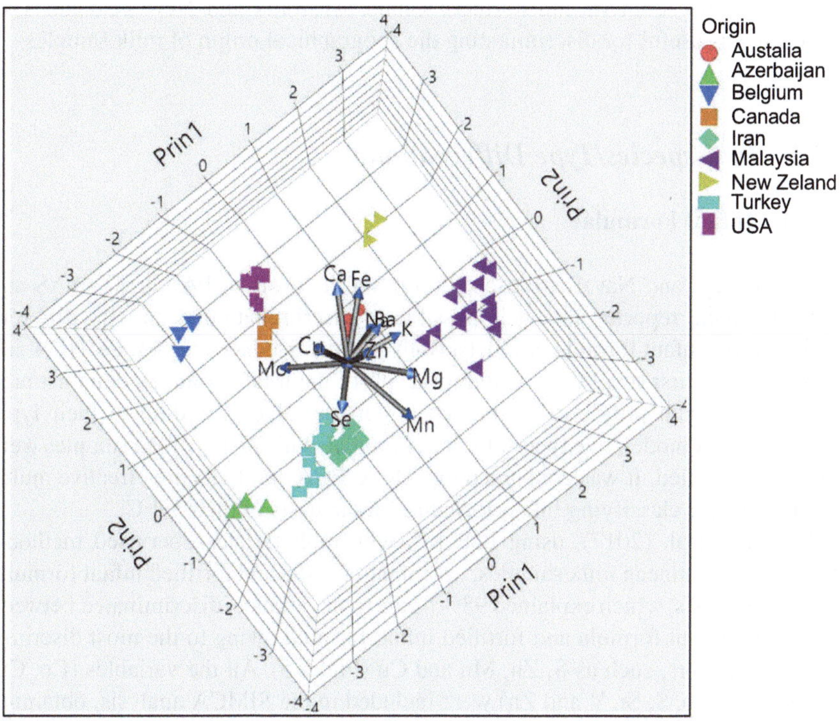

Fig. 5.3 3D varimax rotation plot of PC1 vs. PC2 vs. PC3. (After Zain SM et al., 2016. Food Contr 66:306–314. https://doi.org/10.1016/j.foodcont.2016.02.015. With permission)

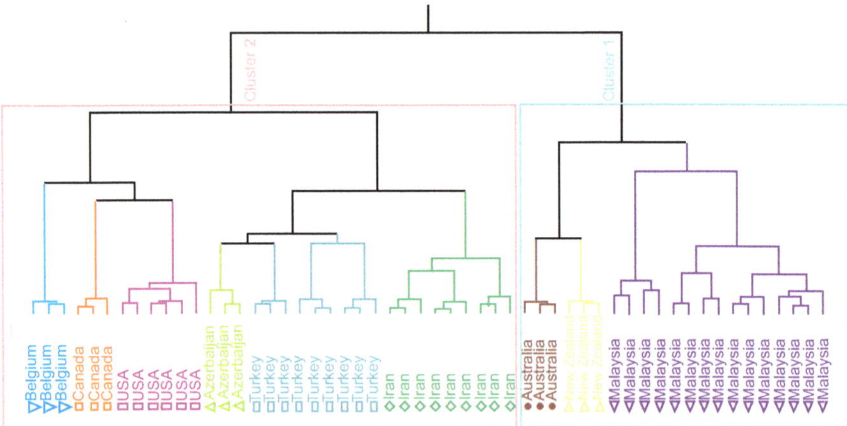

Fig. 5.4 Hierarchical graphs for all the factory milk samples. (After Zain SM et al., 2016. Food Contr 66:306–314. https://doi.org/10.1016/j.foodcont.2016.02.015. With permission)

Norgbey et al. (2017) and Huang et al. (2017) quantified δ^2H and $\delta^{18}O$ data by TC/EA–IRMS, which were processed by ANOVA for milk samples from the United States, Canada, southern and northern China, Australia and New Zealand. This approach was useful for discriminating the geographical origin of milk samples.

5.2.2 Interspecies/Type Differentiation

5.2.2.1 Infant Formulae

Sola-Larrañaga and Navarro-Blasco (2006), on the basis of FAAS, ET–AAS and ICP–AES data, reported that it is possible to discriminate and classify different groups of five infant formula products. Both unsupervised techniques, i.e. HCA and PCA (with the first two PCs explaining 42.5% of the total variance), demonstrated that the elemental profile differentiated infant formulae according to their type. Using the LDA model, five types of infant formulae with 77.1% of the samples were correctly classified. It was concluded that HCA, PCA and LDA are effective multivariate tools for classifying infant formulae on the basis of their MEC.

Azcarate et al. (2017), using ICP–MS with multivariate supervised methods, classified Argentinean milk samples, i.e. infant formula and fortified infant formula. The first two PCs, which explained 98% of the total variance, discriminated between milk-based infant formula and fortified infant formula owing to the most discriminative descriptors, such as S, Zn, Mn and Cu (Fig. 5.5). All the variables (Co, Cu, Mn, Mo, Ni, Rb, S, Sr, V and Zn) were included in the SIMCA analysis, obtaining good modelling power in the discrimination data, especially in the case of the contents of Mn and S for classes 1 and 3 differentiation and Cu and S for the discrimination of these two classes (Fig. 5.6). The classification accuracies for each supervised

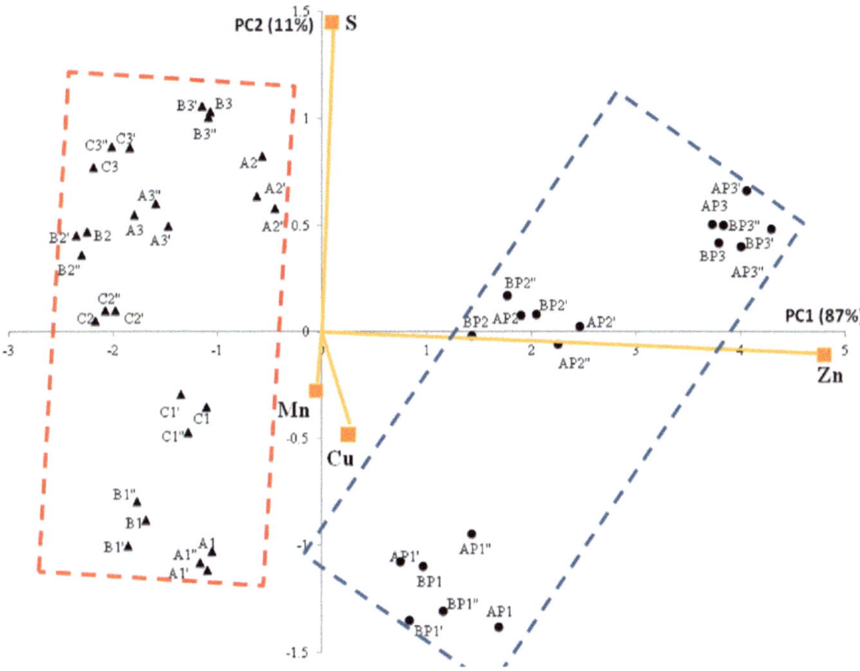

Fig. 5.5 PCA and loading plots showing the discrimination of groups through their nutritional requirements at different stages of growth, verifying that the variables that had a greater influence on each group formed for milk-based infant formula. (After Azcarate SM et al., 2017. Microchem J 130:1–6. https://doi.org/10.1016/j.microc.2016.07.016. With permission)

method model, i.e. LDA, PLS–DA and SIMCA, were estimated to be 100%, 72% and 100%, respectively. The offered approach provides a simpler, faster and cheaper classification for the study of foodomics in milk-based infant formula.

5.2.3 Interspecies and/or Interregional Differentiation

5.2.3.1 Milk

Miedico et al. (2016) applied Q–ICP–MS to investigate raw sheep and goat milk collected at four locations in southern Italy. ANOVA indicated that ten elements contributed significantly to classifying the samples with respect to interregional differentiation. PCA with the first two PCs, which explained 41.0% of the total variance, revealed that the samples from Val Basento and Manfredonia were inadequately differentiated, whereas the groups of samples corresponding to Brindisi and Tito were much better distinguished as well as for interspecies. The species and the season of production contributed marginally to the ability to differentiate the milk samples.

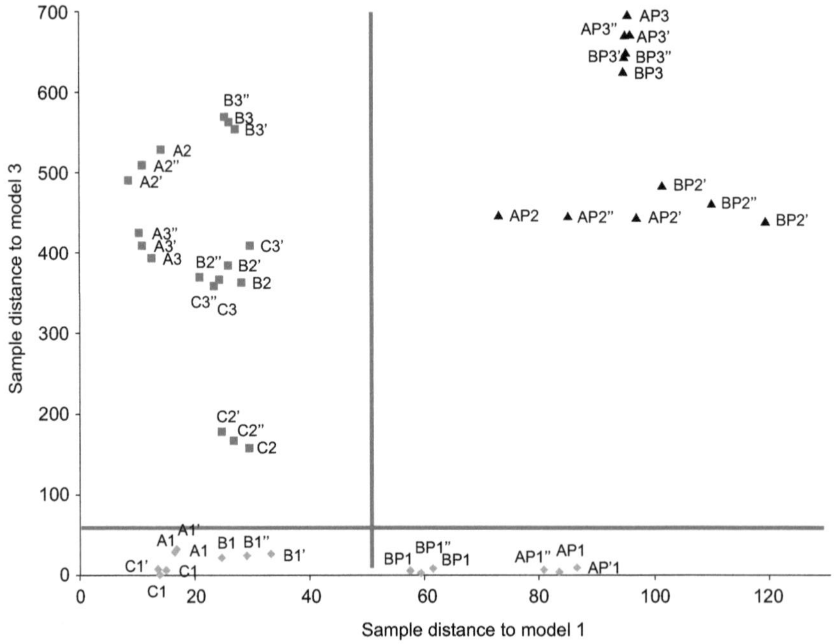

Fig. 5.6 Cooman plot depicting the sample distances of Model 3 with respect to Model 1, with a significance level of 5%. (After Azcarate SM et al., 2017. Microchem J 130:1–6. https://doi.org/10.1016/j.microc.2016.07.016. With permission)

Chen et al. (2020) reported that ICP–MS combined with chemometrics could be used to distinguish milk types, i.e. cow, goat, buffalo, yak and camel milk, from five provinces in China. PCA with the first six PCs, which explained 73% of the total variability, partially differentiated the milk samples into different clusters. Stepwise LDA was able to identify the most useful variables for a clear discrimination of five milk types, providing an overall accuracy of the original grouped cases and a CV of grouped cases of 98.6% and 96.4%, respectively.

Zhou et al. (2017) reported ICP–MS analysis results for raw, goat and buffalo milk, feed and drinking water from three Chinese provinces, i.e. Shandong, Shaanxi and Guangxi. The MEC data were transformed by PCA, with the first four PCs explaining 68.12% of the total variance. Mineral composition was dependent on interanimal species and interregional differences. It was also postulated that the contents of elements in water and feed might influence those in milk.

Magdas et al. (2016) processed chemometrically MEA and IRMS data for cow and sheep milk samples from three Transylvanian farms in Romania. By applying multivariate analysis to milk, water, grass and dried food samples, it was possible to differentiate different farms and milk species. The DA data based on Sr, Mn, $\delta^{18}O$ and $\delta^{13}C$ provided a satisfactory classification for 90.9% of the cross-validated grouped cases related to geographical origin, whereas raw milk speciation with Sr and $\delta^{2}H$ achieved a correct classification for 96.4% of the cases.

5.2.3.2 Dairy Products

Osorio et al. (2015) consolidated MEA with chemometrics to discriminate between sheep milk, goat milk and Halloumi cheese according to their geographic origins. PCA, with the first two PCs accounting for 61% and 73% of the total variance, discriminated milk and cheese samples, respectively, from three regions in Cyprus. CDA based on descriptive variables such as Ca, Sr and Zn allowed them to discriminate milk samples from the three producing regions, with 88.5% and 92.3% of the original and cross-validated grouped cases correctly classified, respectively. Better results for Halloumi cheese related to Sr, K and Mn were obtained, with 100% and 95.7% of the original and cross-validated grouped cases correctly classified, respectively. This method can provide advantageous fingerprints for the characterization of dairy products.

Suhaj and Koreňovská (2010) reported that FAAS, GF–AAS and CV–AAS data processed by CDA were useful in the discrimination of goat hard and semihard cheeses from goats and hard cheeses from cows from 13 European countries. PCA with the first three PCs accounting for 55.7% of the total variance differentiated Slovakian cow cheeses from goat cheeses. Similar results were obtained using CA and SDA. CDA, which considered the 12 elements, appeared to be the most effective model, with 99.3% of the total cheeses correctly classified. Comparison studies of international European cheeses also revealed the recognition and prediction ability of multivariate chemometrics for discriminating European cow and goat cheese species. The most descriptive elements enabled PCA differentiation, represented by Na, Ca, Mg and K. Furthermore, 97.7% of the cheese samples were correctly classified (recognition capability), and 97.5% of the cross-validated samples were correctly classified via the prediction capability procedure. Elements such as Ca, Mg, Cr and Ba were the most discriminative markers for differentiating European goat cheeses from cow cheeses. According to the authors, interspecies origin is caused by natural physiological differences in animals that produce milk, which may be intensified by the diversity of feed sources and divergent environmental farm conditions at regional and multiregional levels.

5.2.4 Differentiation of Other Parameters

5.2.4.1 Verification of Authenticity/Adulteration

Moncayo et al. (2017) used a combination of LIBS with chemometric processing to investigate milk samples for the detection of adulteration. Commercial brands of Spanish cow, goat, sheep and blended milk were examined. The first two PCs, which represented 86.02% of the variance, were unable to clearly discriminate the LIBS spectra because of overlapping (Fig. 5.7b); therefore, an NN classification model was developed. Figure 5.7c shows the output results of the self-validation step of the training process for each type of milk. Although the spectra for

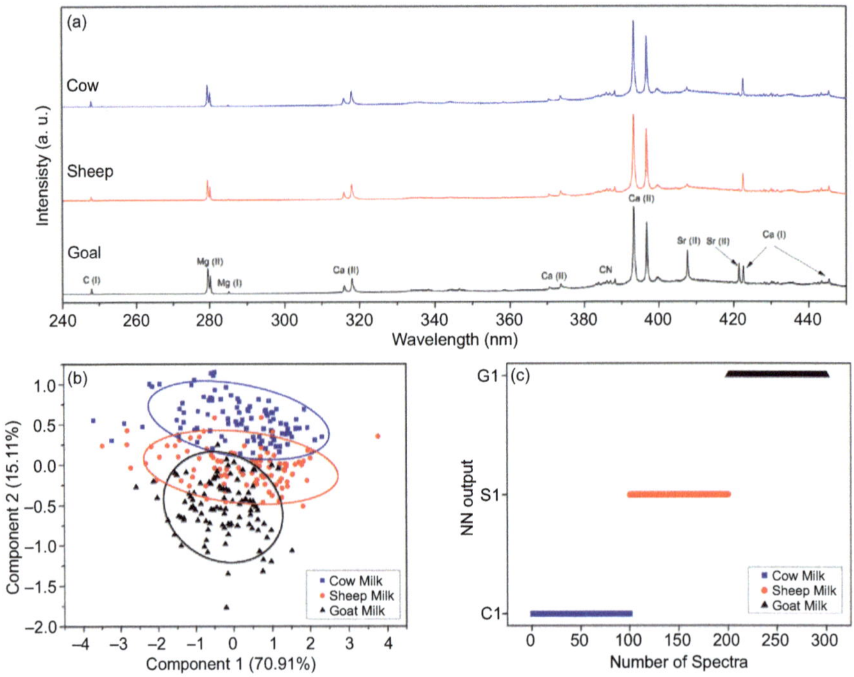

Fig. 5.7 (**a**) Typical spectra of (I) cow, (II) sheep and (III) goat samples. (**b**) PCA of cow, sheep and goat milk samples. (**c**) NN outputs of the training samples. (After Moncayo S et al., 2017. Food Chem 232:322–328. https://doi.org/10.1016/j.foodchem.2017.04.017. With permission)

calibrating the NN were identical to those used in the PCA technique, the NN model was useful for differentiating between types of milk, in contrast to the PCA data. The authors concluded that the LIBS/NN combination significantly discriminated blended milk from pure milk and distinguished pure milk with a 100% correct classification.

Brescia et al. (2005) applied ICP–AES, ICP–OES, HPIC–CD, IRMS and NMR to geographically characterize buffalo milk and mozzarella cheeses originating from two areas of southern Italy. PCA, with the first two PCs accounting for 53% of the total variance, clearly discriminated milk samples from Caserta and Foggia Provinces on the basis of highly discriminative loadings such as those of Na, Li, K and protein contents. The PCA data were confirmed by HCA. In the case of mozzarella cheese, PCA (with the first four PCs explaining 85% of the total variance) was performed using NMR and IRMS. The cheese samples were clearly distinguished with respect to their interspatial assignment. In addition, the HCA data were unambiguously consistent with the PCA and DA results. The latter applied to the first four PCs provided classification and prediction abilities of 100% and 93%, respectively. They concluded that this approach was useful in clearly distinguishing PDO samples from Campania (Caserta Province) and northern Apulia (Foggia Province).

Bilge et al. (2016) combined LIBS with ICP–MS to determine milk powder adulterated with whey powder in the food industry. PCA, with the first two PCs accounting for 85.21% and 87.8% of the total variance, allowed good discrimination between sweet whey and acid whey powders as well as between skim milk powder, acid/sweet whey powder and demineralized whey powder, respectively (Fig. 5.8). Furthermore, a prediction of the sweet and acid whey adulteration ratio in skim milk powder was obtained with PLS. According to the authors, although LIBS is a reliable and sensitive method, further studies are needed to improve its accuracy.

According to Bostic et al. (2018), the combination of EA–IRMS and ANOVA for samples of five commercially available varieties of milk from randomly selected counties across the US was useful since the $\delta^{13}C$ value varied with macronutrient content and region of purchase, highlighting its potential applications in food authentication and the detection of food adulteration. The $\delta^{13}C$ and $\delta^{15}N$ values of the bulk samples were used to estimate their variations depending on the contribution of fat content, added sugar amount and census-designated region of sampling. It was concluded that the content of the quantified stable isotopic signatures exhibited significant homogeneity, supporting the link between feeding regimens and production processes, which has been typical for the US dairy industry in recent decades.

Moreno-Rojas et al. (2010) studied 22 varieties of cheese selected from different PDOs located in northern Spain. FAAS and AES data were processed chemometrically. MEC in combination with chemometric procedures was useful for classifying cheeses produced in different zones with various manufacturing standards (PDOs). PCA and MCDA were found to be useful for discriminating among a wide variety of the studied cheeses. The former, with the first two PCs, accounted for 98.82% of the total variance, with Ca and Na as discriminative loadings. According to Moreno-Rojas et al. (2012), PCA and MCDA are suitable techniques for distinguishing numerous varieties of Spanish cheeses on the basis of Ca, P and Na levels. The PCA

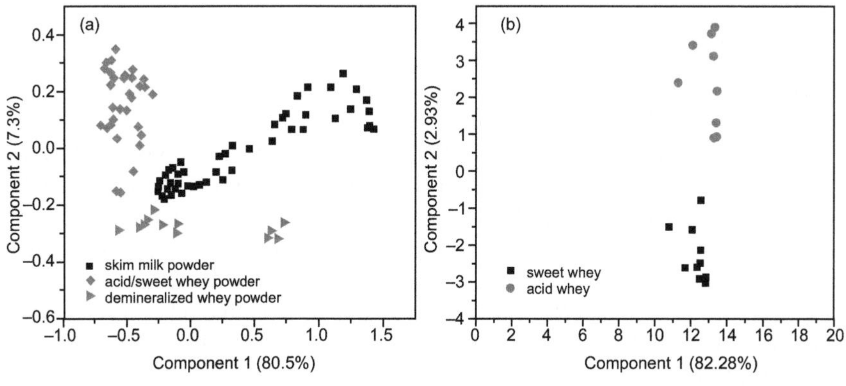

Fig. 5.8 PCA graph of skim milk, acid/sweet whey and demineralized whey powder (**a**) and PCA graph of sweet and acid whey powder (**b**). (After Bilge G et al., 2016. Food Chem 212:183–188. https://doi.org/10.1016/j.foodchem.2016.05.169. With permission)

with the first two PCs corresponded to 98.72% of the total variability. In the case of MCDA, only 3 DFs explained 74% of the total variability, indicating the satisfactory performance of the achieved discrimination model.

Sezer et al. (2018) obtained LIBS data, which were useful for the identification of adulterated milk samples upon chemometric processing. PCA with the first two PCs accounting for 97.35% of the total variance allowed the differentiation of Turkish bovine, caprine and ovine milk and adulterated samples. The combination of LIBS with PCA and PLSR achieved successful classification of the three milk species and determination of their falsification ratios. It was concluded that the developed methodology could be useful for routine analysis in quality control laboratories in the dairy industry.

Azcarate et al. (2015) reported useful ICP–MS data for classifying 11 types of Argentinean milk samples of different brands. The first two PCs used to construct the model explained more than 98% of the total variance in the dataset. As shown in Fig. 5.9a, the PCA scatter plot demonstrates the discrimination of the four groups, i.e. allowing milk classification (adult, baby or baby fortified). Among the loading plots (Fig. 5.9b), the most discriminative variables in the differentiation of the groups were S, Zn and, to a lesser degree, Sr, Rb, Mn and Cu. It was concluded that multi-element quantification in milk samples was useful for performing traceability and/or origin denomination studies and for evaluating their quality.

According to Nečemer et al. (2016), a combination of EA–IRMS and EDXRF with chemometrics was applied for the quantification of SIR and MEC of cow, sheep and goat milk as well as cheese, including milk samples with mixed origins from the Alpine, Dinaric, Pannonian and Mediterranean regions. In the case of milk and cheese samples, determinants such as P, S, Cl, K, Ca, Zn, Br, Rb, Sr, $\delta^{18}O$, $\delta^{13}C$, $\delta^{15}N$ and the same group of elements together with $\delta^{13}C$ and $\delta^{15}N$ were processed by DA. The overall classification rate was 97% for goat, sheep and cow milk, and the classification functions revealed that all three types of cheese were correctly classified (100%) into their original groups. This study emphasized the most descriptive parameters in the DA categorization of milk and cheese samples according to their interspatial and interspecies differentiation and could contribute to supporting the existence of a PDO. In summary, it will be possible to construct a traceability model that could be applied to protect PDO cheeses against fraudulent mislabelling and to verify the authenticity of commercial products.

Camin et al. (2012) used TC–EA–IRMS and ICP–MS to establish an SIR and MEC and demonstrated that one model could predict the provenance of different types of 265 European hard cheese samples, and the other could differentiate Parmigiano Reggiano cheese from European and two extra-European imitators. PCA, with the first two PCs accounting for 30.0% of the total variance, clearly differentiated the Parmigiano Reggiano cheese samples from the other hard cheeses, suggesting that PCA is useful for constructing relevant classification models (Fig. 5.10). Moreover, elements such as Sr, Cu, Mo, Re, Na, U, Bi, Ni, Fe, $\delta^{13}C$, Mn, $\delta^{34}S$, Ga, Se, $\delta^{15}N$, Er, $\delta^{2}H$, Dy, Pb and Li were identified by the RF as the most discriminative variables. This model could individually predict the geographic provenance of the seven main European hard cheese producers. Another model was

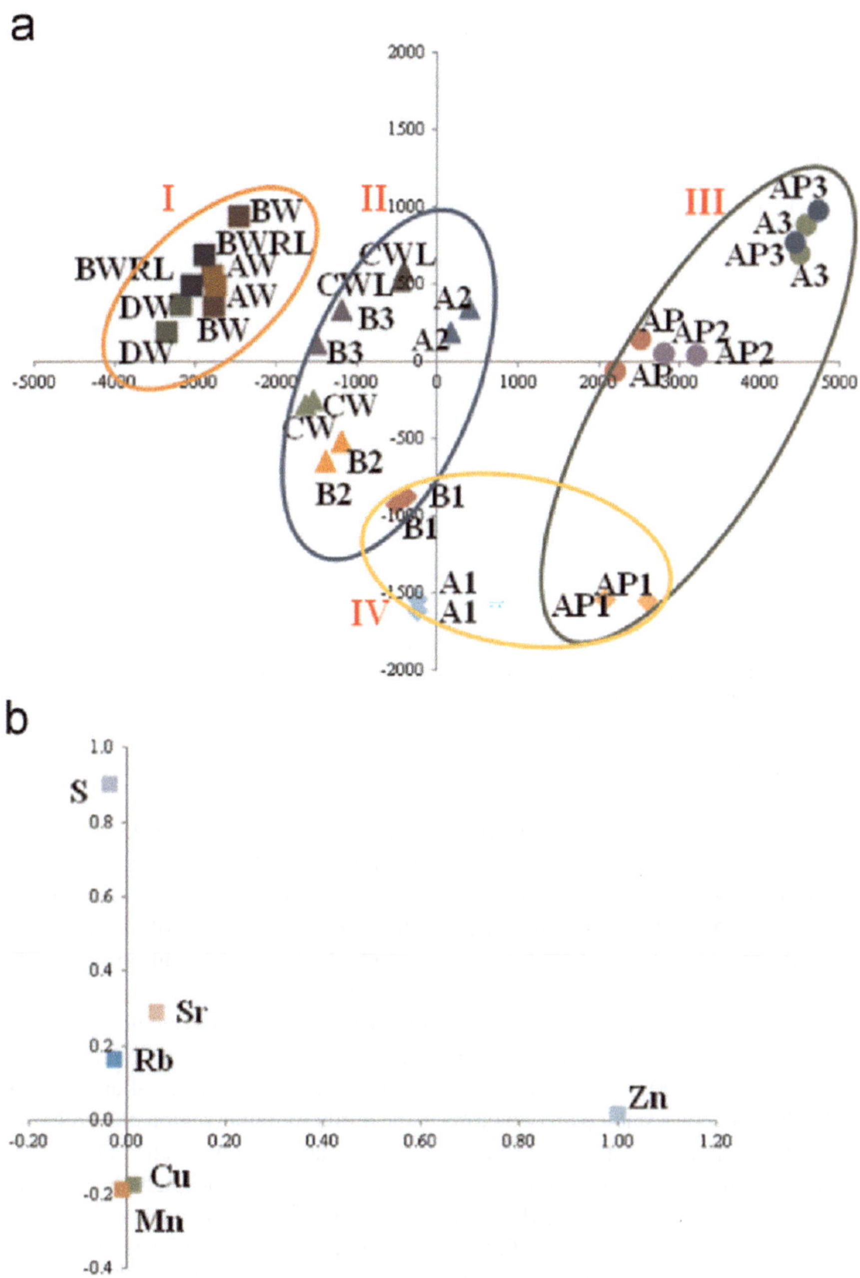

Fig. 5.9 (**a**) PCA score plot for the first two PCs. (**b**) Loadings plot for PC1 and PC2. (After Azcarate SM et al., 2015. Talanta 143:64–70. https://doi.org/10.1016/j.talanta.2015.04.088. With permission)

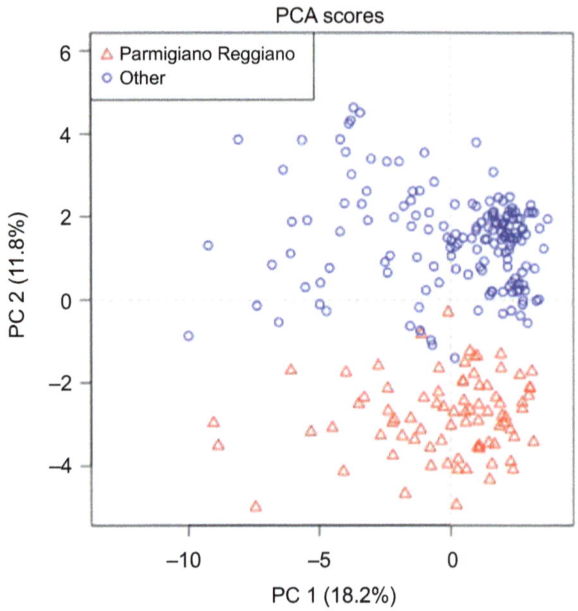

Fig. 5.10 PCA score plot, contrasting Parmigiano Reggiano cheese with other hard cheeses. (After Camin F et al., 2012. Anal Chim Acta 711:54–59. https://doi.org/10.1016/j.aca.2011.10.047. With permission)

the most important because it specifically discriminated the most famous PDO, Parmigiano Reggiano cheese, from European and two extra-European imitators on the basis of the same variables as before, except Er, Pb and Dy. For both models, very good classification success rates of 98.3% and 98.6%, respectively, were achieved.

Bontempo et al. (2011) used a combination of TC/EA-IRMS, TIMS, ICP–OES and ICP–MS to group seven kinds of bovine milk cheese (Asiago, Fontina, Montasio, Puzzone, Spressa, Toma and Vezzena) produced in alpine and pre-alpine Italian areas. A CDA with the first three CANs accounting for 91% of the variability correctly classified 94% of all the samples studied, effectively contributing to the identification of different types of cheese, supporting the high status of a PDO (Fig. 5.11). Such differentiation was possible owing to the use of the most discriminative descriptors, such as Ba, Ca, K, Mg, Rb, $\delta^{13}C_{casein}$, $\delta^{15}N_{casein}$ and $\delta^{18}O_{glycerol}$.

Camin et al. (2008) reported the $\delta^{13}C$, $\delta^{18}O$, $\delta^{2}H$ and $\delta^{18}O$ values of animal feed and milk from cows stabled on two farms in the northern Italian region of Lombardia and fed diets that consisted of various kinds of C_3 plants and were characterized by different quantitative proportions of maize. Analytical data processed chemometrically, e.g. by ANOVA, enabled the identification of mislabelled dairy products produced by pastured animals or of PDO cheeses with an attested amount of maize in the diet. Therefore, it was possible to verify the unpermitted addition of exogenous components to milk.

According to Di Bella et al. (2012), contents of Cl^-, NO_2^-, NO_3^-, PO_4^{3-} and SO_4^{2-} were determined by IEC in donkey milk samples collected from conventional Italian donkey farms, i.e. from Sicily, Calabria and Emilia-Romagna (Italy). A

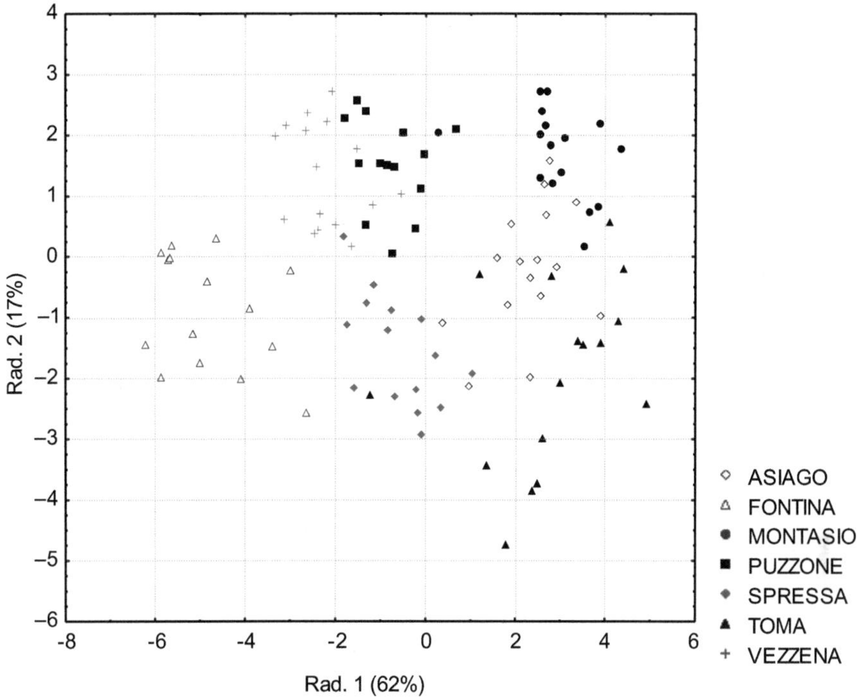

Fig. 5.11 Canonical discriminant analysis of the isotopic and elemental compositions of cheeses. Scatterplot of the first two canonical variables. (After Bontempo L et al., 2011. Inter Dairy J 21:441–446. https://doi.org/10.1016/j.idairyj.2011.01.009. With permission)

three-step DA confirmed good classification among the donkey milk samples, with 91–98% of the samples correctly classified. The methodology could be used for the identification of donkey milk, thus protecting this unique product against fraud or commercial disputes.

Potorti et al. (2013) used an approach that appeared to be useful for evaluating the traceability of donkey milk, protecting this unique product against fraud. The studies were based on the combination of ICP–MS with PCA to categorize donkey milk samples from three conventional farms. PCA clearly distinguished the inter-regional milk samples from the forage/feed samples, with the first two PCs accounting for 55.08% and 80.42% of the total variance, respectively (Fig. 5.12). Moreover, complex and different seasonal variations in various elements were detected in milk as well as in forage/feed samples. The application of both tools resulted in establishing the geographical origin of donkey milk from three Italian farms, which could be used to successfully achieve DOP or IGP trademarks.

According to Gregorčič et al. (2020), EA–IRMS data from Slovenian milk and its major constituents (water, casein and lactose) were compared with those from cow, sheep and goat milk from farms in macroregions, i.e. Alpine, Dinaric, Pannonian and Mediterranean regions. It was demonstrated that the use of

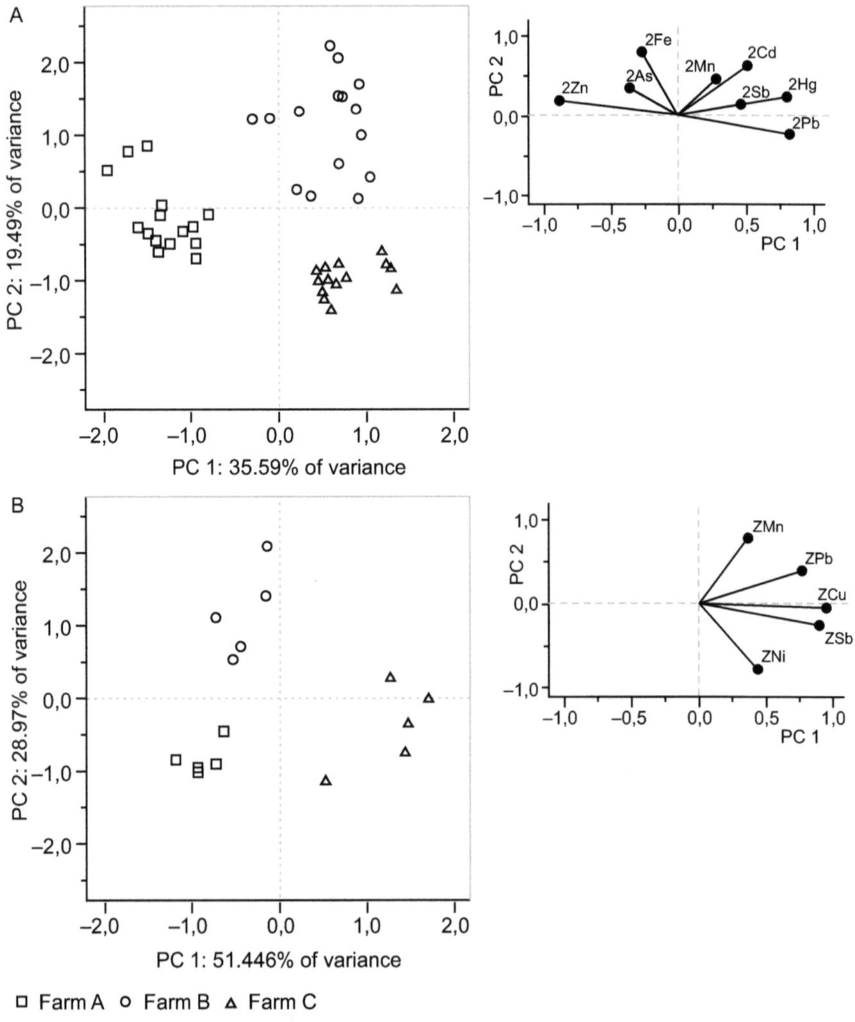

Fig. 5.12 PCA for donkey milk samples (**a**) and forage and feed samples (**b**) categorized by farm origin. Inserts: loading plots for PC1 and PC2. (After Potorti AG et al., 2013. J Food Compos Anal 31:161–172. https://doi.org/10.1016/j.jfca.2013.05.006. With permission)

$\delta^{18}O_{milk\ water}$ allowed the classification of milk samples with respect to their interseasonal and interanimal species differentiation, while interregional discrimination was limited. Diet and physiology had a strong effect on the SIR profile of animal body water and, consequently, milk. Moreover, >15% added water could be detected. Using $\delta^{18}O$ lactose values as an internal standard, the method would be even more promising for improving the detection of illegal watering of milk (>7%).

Temiz et al. (2018) reported the applicability of LIBS combined with multivariate data analysis (PCA and PLS) for determining butter adulteration with margarine. The first two PCs, which explained 97.92% of the total variance, clearly discriminated the butter and margarine samples as well as their blind samples. The intensity differences in the LIBS spectral lines of Na, K, Fe, Ca, Mg and Mn highlighted their considerable potential and contributed to effectively distinguishing between butter and margarine samples. The RMSEC and RMSE were 2.02 and 3.37 for the PLS calibration and validation models, respectively. Notably, the developed LIBS method is an environmentally friendly, safe and rapid method for the determination of butter adulteration.

Bandoniene et al. (2020) developed a method for labelling lamb meat and goat milk via the selective enrichment of Tb and Tm in the feed of animals. The use of SN–ICP–MS and LA–ICP–MS in combination with chemometric tools has shown that REEs are useful in the development of methodologies. Hence, the present method allows discrimination of labelled products from unlabelled animal products, e.g. milk, dairy products and eggs, while the REE contents in all labelled products remained low enough to avoid any health risk for the consumer. Bandoniene et al. (2018) investigated laying hens fed Tb- and Tm-supplemented feed to introduce a distinctive REE profile enabling the differentiation of labelled poultry products from unlabelled poultry products. In addition, samples of egg yolk, egg shells, meat, bones, liver, blood and faeces were also studied via SN–ICP–MS and LA–ICP–MS. Further analysis was consistent with that presented in another article by Bandoniene et al. (2020).

In the study by Barbosa et al. (2014), which was based on ICP–MS data integrated with DTs and NBs, egg classification methods were evaluated. Decision trees and NBs related to the elemental contents of eggs established a high level of accuracy, i.e. 80.8% and 90.4%, respectively, for classification between free-range and battery eggs. A significant distinction was achieved with respect to the levels of Ba, Co, Pb and Mo between free-range and battery eggs. This approach is an alternative method for adulteration evaluation.

Mi et al. (2019) reported that eggs from three chicken breeds, i.e. Deqingyuan, Taihe and crossbred black-boned silky fowl eggs, from China's retail market were successfully differentiated by comparative ICP–MS, UPLC–MS/MS, ESI–MS/MS and lipidomic analysis. The first two PCs explained 84.8% of the variance, indicating clear differentiation, as did the PLS–DA model for the three types of whole eggs. The PLS–DA, which was established on the basis of lipidomic data, enabled 100% correct classification of egg yolks. It was suggested that the distribution and abundance of chemical elements varied considerably among the eggs of the three chicken breeds. Different element marker candidates were selected for the whole egg, egg yolk and egg white, which corresponded to the three breeds of chicken. Moreover, 22 potential lipid markers were screened via chemometric analysis of the lipidomics data. A combination of the analytical method with the chemometric tool provided a convenient and powerful tool to discriminate commercially available eggs.

de Freitas et al. (2013) noted that ICP–MS combined with a statistical tool was useful for characterizing the elemental profiles of conventional and home-produced egg samples. The levels of Ba, Cd, Co, Pb, Se, Fe, Zn and Mo were significantly different in eggs with respect to both production methods. The method was effective in discriminating between conventional and home-produced eggs obtained from Brazilian cities.

According to Cristea et al. (2023), IRMS and ICP–MS were used to study egg white and yolk samples from different hen rearing systems (backyard and barn). The PLS–DA model achieved a total accuracy score of 100% for yolk samples and 96% for egg white samples according to the hen rearing system. On the basis of the human health risk assessment and estimation of non-carcinogenic risk, it could be concluded that the non-carcinogenic risk values for eating eggs studied were within the safe range for consumers, indicating no health risks from their consumption.

Bay et al. (2018) used IRMS to differentiate fresh hen eggs produced in Malaysia from eggs produced in Singapore to identify the misdeclaration of eggs in possible scenarios of fraudulent imports into Singapore during outbreaks of foodborne diseases. Monte Carlo simulations with standard spreadsheet software were used to optimize analytical strategies and to determine quantitative threshold limits for sampling uncertainty and analytical uncertainty for the investigation. Data analysis of the simulated data yielded threshold limits for a minimal number of eggs to be analysed at a given precision of isotopic analysis for an intended level of statistical significance. The chosen Monte Carlo approach permitted the optimization of analytical strategies, in general, by matching restraints in available resources, the available number of samples for analysis and target values for measurement precision in a quantitative manner.

Manuelian et al. (2017a, b) reported that fresh cheeses usually contain lower mineral levels than other cheeses, which makes mineral prediction more difficult. The prediction models using PLSRA were satisfactorily developed for Ca, P, S, Mg and Zn and for FA samples originating from different dairy types and ripening times. Spectroscopy models using PLSRA were also used to predict the major mineral content of fresh cheeses. The most accurate prediction model was offered for the Na content of mozzarella cheese using NIT spectroscopy and for P in Stracchino cheese using both NIR and NIT spectroscopy, although the mineral content was difficult to predict using these spectroscopic methods.

Chung et al. (2018) examined the monthly and seasonal variations in $\delta^{13}C$, $\delta^{15}N$, FA and vitamin E in organic and conventional milk from different regions of Korea from 2016 to 2017. The average $\delta^{13}C$ and $\delta^{15}N$ values were lower in organic milk than in conventional milk, whereas the former contained higher mean levels of α-tocopherol and FA. Furthermore, interseasonal differentiation was achieved for the $\delta^{13}C$, $\delta^{15}N$ and FA concentrations. Chemometric tools enabled the selection of reliable descriptors, especially C18:3n-3, C18:2n-6 and $\delta^{13}C_{bulk\text{-}milk}$, for clearly distinguishing organic milk in accordance with the sampling season. According to the authors, the data obtained enabled the control of fraudulent organic milk labelling in Korea, with the possible application of this approach in other countries.

According to Faberi et al. (2018), the SIR variability of PDO cheeses collected over a 3-year period from five Italian regions was studied to identify intertype differences. Five isotopic parameters were analysed, i.e. $\delta^{13}C$ and $\delta^{15}N$ in the casein fraction and cheese as a whole and $\delta^{13}C$ in the fat fraction of the cheese. The dataset, in the context of some of the parameters considered (year, season, province and altitude), was processed chemometrically. Univariate analysis revealed that single parameters were not sufficient to provide clear discrimination between the cheeses. However, PCA and PLS–DA indicated satisfactory differentiation between cheese classes, especially for the Pecorino Toscano cheese type. The potential of the applied methodology for the production of the four types of PDO cheeses for the protection of market competition and safety was suggested.

5.2.4.2 Interregional and Intertype Differentiation Related to Pasture Type

According to Bontempo et al. (2012), TC/EA–IRMS was applied to investigate milk, cheese and Alpine forage samples from two sites, i.e. Valcavera, which is located in the southwestern Italian Alps, and Malga Dosso di Sotto, which is located in southeastern Italy. A multivariate CDA with the first two CANs accounting for 98% and 99% of the total variability for milk and cheese, respectively, distinguished between milk and cheese from four pasture types. CAN1 (mainly loaded negatively by $\delta^{13}C$ and $\delta^{18}O$ in casein) clearly discriminated between the two sampling regions, whereas CAN2 (mainly loaded positively by $\delta^{13}C$ and negatively by $\delta^{15}N$) differentiated between the two pasture types in each geographical location. According to the authors, after CV, 83% of the milk samples and 80% of the cheese samples were correctly classified. With respect to an exclusively interspatial designation, it was estimated that 96% of the milk and cheese samples were appropriately assigned.

5.2.4.3 Interregional Differentiation Related to Forage Type and Parent Soil/Water or Other Parameters

Suhaj and Koreňovská (2008b) reported that bryndza sheep cheeses, soil and grass samples from nine locations in Slovakia and producers and ten European countries were classified using FAAS, GF–AAS, AMA and X–RFS with multivariate CDA. Seven producers from all nine regions studied were correctly distinguished with >90% accuracy owing to descriptors such as Ba, Cu, Cr, Hg, Mg, Mn, Ni and V. Such satisfactory geographical differentiation was attained for relatively small geographical areas despite the major geochemical similarities among Slovakian regions. In addition, the differentiation of bryndza cheese from other Central European countries was achieved.

5.2.4.4 Differentiation Between Organic and Conventional Milk, Cheese and Eggs

Molkentin and Giesemann (2007) investigated organic milk and conventional milk from Germany via SIR and FA analyses determined by EA–IRMS and GC–FID. Analysis of $\delta^{13}C$ also enabled complete differentiation of both types of milk due to the various amounts of maize in the feed. All-year variations in $\delta^{13}C$, $\delta^{34}S$ and $\delta^{15}N$ as well as seasonal variations in $\delta^{15}N$ and $\delta^{34}S$ in milk were observed. According to the authors, the study of more samples is recommended to check/verify the obtained results of pilot studies.

The study by Rey-Crespo et al. (2013) was based on the use of ICP–MS to establish the MEC content in milk from organic and conventional farms in NW Spain. The essential trace element levels in organic milk were significantly lower than those in conventional milk. Such differentiation was especially evident in the case of Zn, I and Se, which are routinely supplemented in conventional concentrated feed.

Chung et al. (2014) applied EA–IRMS to quantify $\delta^{13}C$ and $\delta^{15}N$ in Korean organic and conventional milk. Chemometric analysis with the use of GLM and LSD was an effective tool for verifying the differentiation between organic and conventional milk. However, according to the authors, this study provided preliminary data; hence, further investigations are needed. However, the methodology is relatively reliable; thus, it may be useful for limiting the risk of fraudulently labelled organic milk.

González-Martín et al. (2011a) used NIRS and ICP–OES combined with multivariate tools to discriminate between organic and non-organic Spanish sheep cheeses. Stepwise LDA allowed them to obtain a percentage of samples correctly classified via internal validation of 86.7% for organic cheeses and 90% for non-organic cheeses, with Mg, K and P selected as descriptors. Despite promising data, expanding these investigations to different geographical regions as well as to different breeds of sheep was recommended.

Rodríguez-Bermúdez et al. (2018) verified the authenticity of organic milk samples from North Spain via ICP–MS data. PCA, with the first four PCs accounting for 65.68% of the total variance, as well as HCA, revealed a natural differentiation between organic and conventional milk (Fig. 5.13). The LDA model correctly predicted 97.43% and 77.97% of the organic and conventional samples, respectively. SIMCA built a model for the organic class with a specificity of 97.34%, but the specificity of the model was only 71.19%. The ANN was capable of correctly assessing the type of milk that achieved classification hits, with predictions of 94.87% and 93.22% for the organic and conventional classes, respectively.

Chung et al. (2020) combined SIRA with a chemometric tool to identify the geographical origin of organic milk samples from four regions in Korea. The OPLS–DA models in pairwise comparisons enabled relatively satisfactory predictability for samples originating from the Boryeong, Gochang, Hoengseong and Jeju regions, with the $\delta^{13}C$ value as the most descriptive factor influencing the discrimination models developed. The results obtained may extend the global isotope ratio database for correct geographical classification of organic milk, potentially facilitating

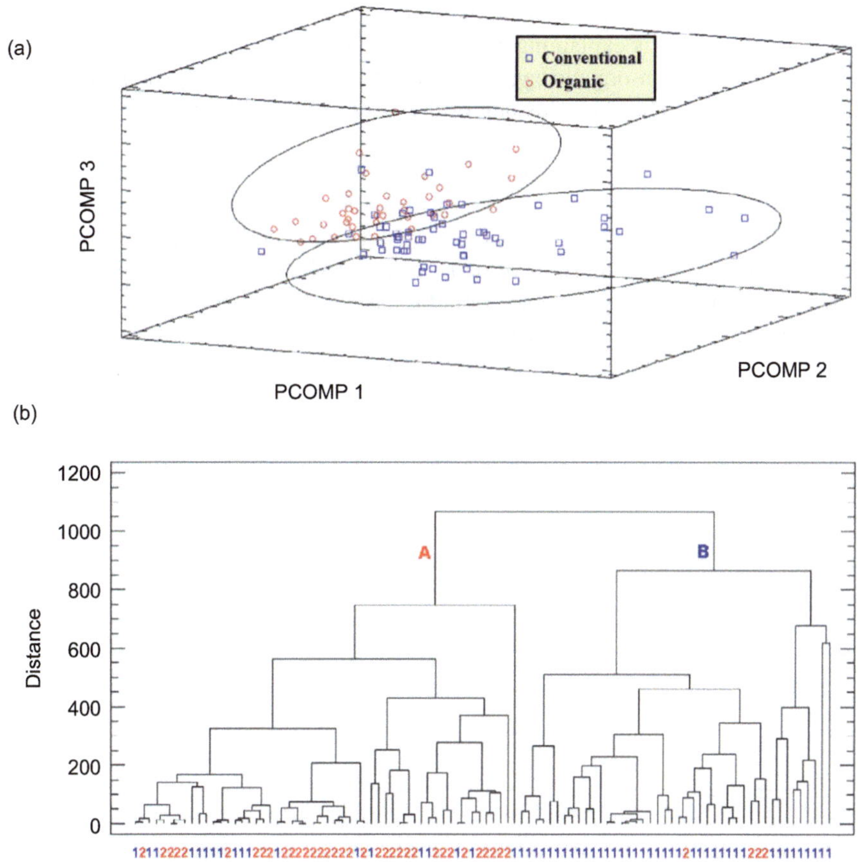

Fig. 5.13 Data display techniques. (**a**) Score plot of the milk samples in the space of the first three principal components. (**b**) HCA dendrogram of milk samples. Codes: 1, conventional; 2, organic. (After Rodríguez-Bermúdez R et al., 2018. Food Chem 240:686–693. https://doi.org/10.1016/j.foodchem.2017.08.011. With permission)

the regulation and management of improper or fraudulent labelling of the geographic assignment for organic milk traded between countries.

Rogers et al. (2015) used EA–IRMS to determine $\delta^{13}C$ and $\delta^{15}N$ in commercial poultry feeds and their corresponding egg albumen levels for Dutch and retail sales in New Zealand. The original data obtained were transformed by chemometric techniques, i.e. MANOVA and LDA. In the case of the former, to examine the effects of two or more independent variables for significant effects on two or more metric-dependent countries, farming regimes, farms, replicate analyses and year factors were included. The results obtained allowed them to discriminate between conventional and organic egg farming regimens.

Borges et al. (2015) reported ICP–MS results, which were used to examine certified organic and non-organic egg samples obtained from the Brazilian retail market.

This approach can be used as an alternative tool for the assessment of adulteration. With the first two PCs explaining 50% of the total variance, PCA clearly discriminates organic and non-organic eggs on the basis of As, Co, Cr, Fe, Mn, P, Rb, Se, Tl and V (Fig. 5.14). The Coomans plot of the SIMCA model offers 100% prediction ability and 100% specificity.

According to Orjales et al. (2018), the ICP–MS results processed by PCA (with the first three PCs explaining ca. 70% of the total variance) and HCA revealed no significant differences in trace element content in blood serum (522 samples) between organic and conventional dairy cattle (164 samples) in northern Spain. However, among the feed samples (alfalfa, pasture, hay, concentrate, grass silage and corn silage), the dietary contents of I, Cu, Se and Zn were significantly greater in the conventional system than in the organic system, which was attributable to

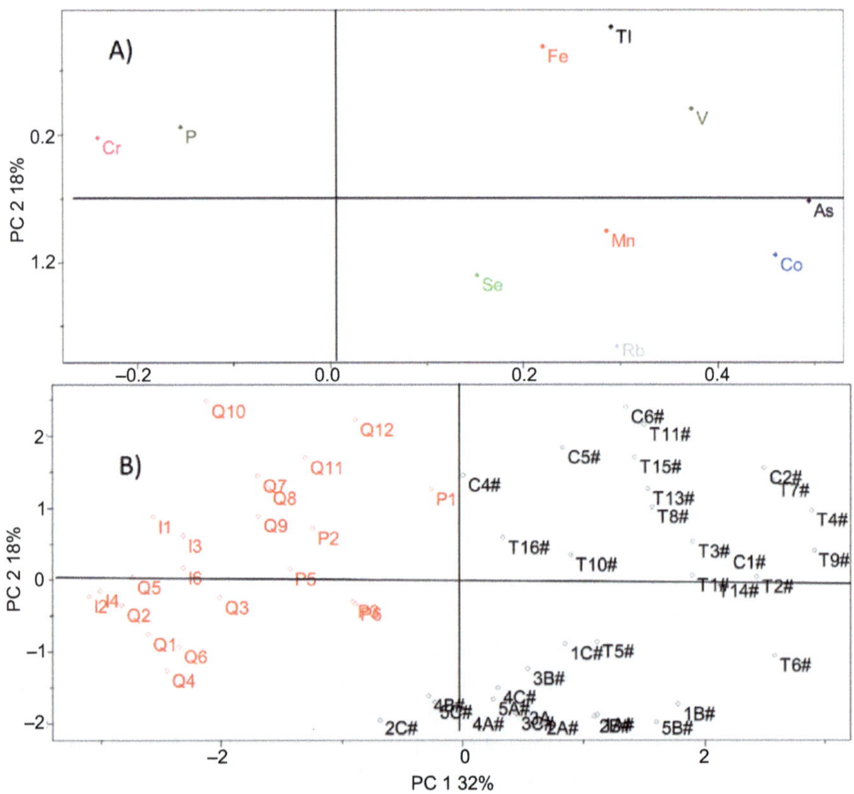

Fig. 5.14 Principal component analysis of the elemental concentrations performed on the basis of 10 element levels (As, Co, Cr, Fe, Mn, P, Rb, Se, Tl and V) and 56 egg samples (35 organic and 21 non-organic). Panels **a** and **b** show the score and loading plots, respectively. The organic samples are black and have the prefix "#"; ordinary samples are indicated in red. (After Borges EM et al., 2015. Anal Meth 7:2577–2584. https://doi.org/10.1039/C4AY02986K. With permission)

their higher concentrations in the concentrated feed. No differences in the contents of trace elements were found among the other types of feed.

Lopez-Alonso et al. (2017) applied ICP–MS to analyse the elemental profiles of 650 samples, including 522 blood samples from dairy cattle in Spain. PCA with the first three PCs representing ca. 40% of the total variance revealed a natural differentiation between samples from organic and conventional farms (Fig. 5.15a). Further examination of the PCA overlapping area allowed identification of the grazing-conventional subgroup representing a transition from organic to conventional grazing and to more intensively managed conventional farms (Fig. 5.15b). The CA results are consistent with those obtained by the PCA. It was concluded that the elemental status in conventional farming is dependent on the mineral supplementation of forage, but it cannot be directly extrapolated to organic farming, where soil particle ingestion is the main determinant affecting the metal profile.

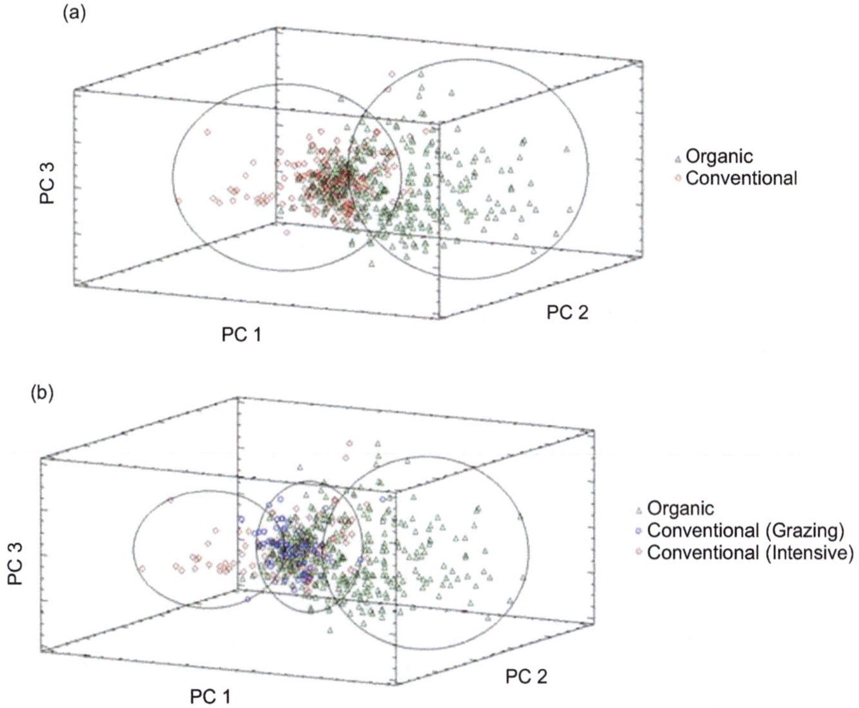

Fig. 5.15 Score plot of the samples in the space defined by the first three principal components (40% total variance) according to the type of farming system: (**a**) organic and conventional and (**b**) organic, conventional-grazing and conventional-intensive. (After Lopez-Alonso M et al., 2017. Chemosphere 185:1048–1055. https://doi.org/10.1016/j.chemosphere.2017.07.112. With permission)

5.2.4.5 Interregional and/or Intertype Differentiation Related to the Production Process

Altieri et al. (2020) investigated different brands of MBC, with and without PDO identification (milk and mozzarella cheese), and buffalo feeding from Caserta and Salerno farms (Italy). TC/EA IRMS was used to determine the sample origins. The lack of changes in the $\delta^{13}C$ and $\delta^{15}N$ values of milk and mozzarella indicated that there was no fractionation in the production process, in contrast to the cheese $\delta^{18}O$ value, which resulted in depletion with respect to the milk value. The $\delta^{13}C$ value of milk depends on the food components, whereas the milk $\delta^{15}N$ value is influenced by metabolism fractionation in ruminants. The first two PCs, which explained 78.96% of the total variance, clearly distinguished the Salerno area from Basso Lazio and Caserta. This result can be explained by the particular climatic conditions of the Salerno area, which is adjacent to the sea. Moreover, geographical differentiation is based more on $\delta^{13}C$ than on $\delta^{15}N$ and $\delta^{18}O$.

Griboff et al. (2019) reported that ICP–MS and EA–IRMS were used to differentiate the origin of milk samples from four production areas in Argentina, linking milk fingerprints with those related to soil, water and feed. GPA links water–soil–feed–milk from different areas, and CCA reveals significant correlations between milk–drinking water, milk–forage and milk–soil; hence, chemometrics is a useful tool for establishing milk provenances.

Liu et al. (2019) used EA–IRMS and HR–ICP–MS to quantify the SIR in casein, lipids in goat milk, feed and drinking water and the MEC content in milk and parent soil. Stepwise CDA was a useful tool for identifying the three geographic origins of goat milk (Fig. 5.16). A significant relationship was found between the mineral composition of milk powder and soils. Thus, $\delta^{15}N_{casein}$, $\delta^{18}O_{casein}$, Ni, Se, Rb, Sr and Ba appear to be discriminative variables for identifying the geographical origins of goat milk. Satisfactory overall correct classification and CV rates, amounting to 100% of the samples from three Chinese regions, were obtained.

According to Scampicchio et al. (2012), EA–IRMS was useful in studying raw, HTST and UHT milk originating from the mountain extensive breeding system in Italy, i.e. Mantova, Udine and Bolzano. PCA with the first two PCs accounting for 77.0% of the total data variance enabled the differentiation of samples along PC1 with respect to geographical origin on the basis of mostly discriminative $\delta^{13}C$ and $\delta^{15}N$ variables (Fig. 5.17). Moreover, considering both the stable isotope distributions along PC2, the $\delta^{15}N$ values of the whey and fat fractions appear to be good descriptive markers for distinguishing between types of milk processing. The LDA model built on the samples from Bolzano based on $\delta^{15}N_{whey}$, fat and $\delta^{13}C_{casein}$ allowed the identification of raw, HTST and UHT milk samples, with cross-validated results achieving 100% successful recognition. Finally, regression models, i.e. MLR, PCR and PLS, were applied on the basis of the full database of milk samples from three geographical regions and corresponded to the three processing types. Thus, this approach is quite helpful in discriminating between milk samples of various geographical origins and types of processing.

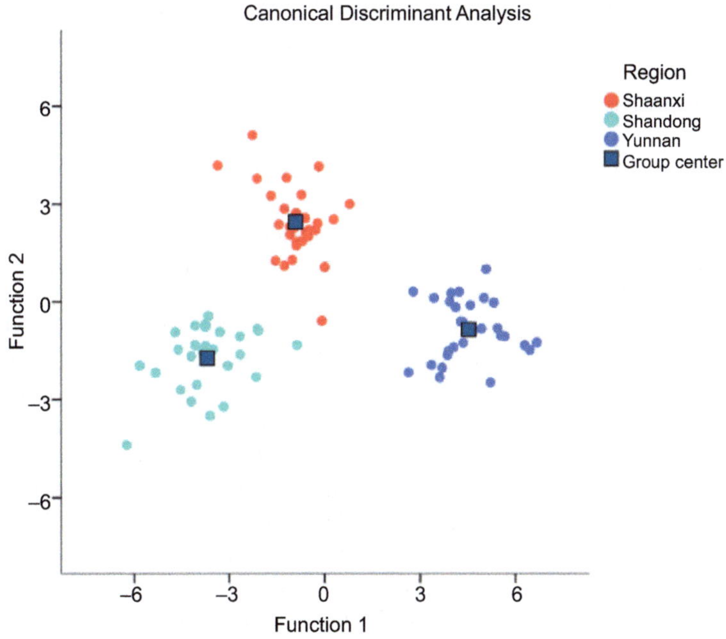

Fig. 5.16 Stepwise discriminant analysis of milk samples from three regions. (After Liu et al. 2019. Food Chem 277:448–454. https://doi.org/10.1016/j.foodchem.2018.10.144. With permission)

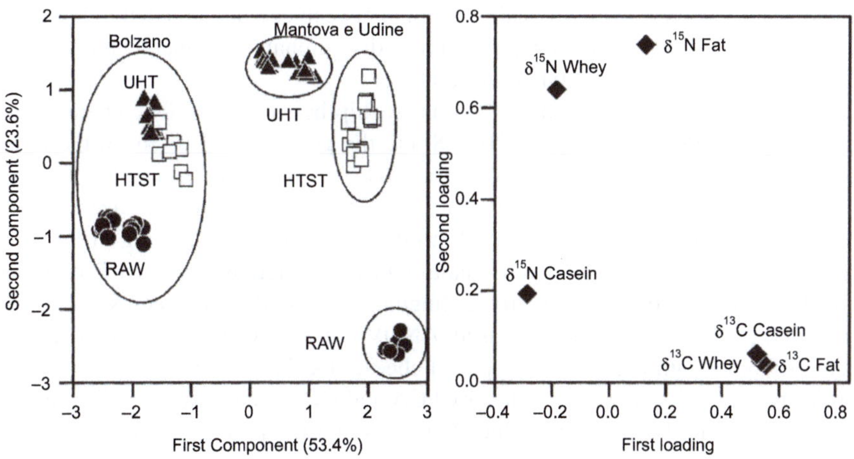

Fig. 5.17 Score (left) and loading (right) plots of the principal component analysis (a) based on a dataset containing 72 samples from Bolzano, Udine and Mantova (n = 24 raw, n = 24 HTST and n = 24 UHT). (After Scampicchio M et al., 2012. *J Agric Food Chem* 60:11268–11273. https://doi.org/10.1021/jf302846j. With permission)

Valenti et al. (2017) explored the TC/EA–IRMS data of Pecorino Siciliano cheese samples originating from three eastern Sicilian areas with green herbage present or absent in the diet of the animal. The $\delta^{34}S$ and $\delta^{15}N$ values processed by stepwise CDA enabled the best discrimination among the three areas, with 97.2% correct classification of the cheeses. The differentiation of the feeding system, resulting from the presence or absence of fresh forage in the diet of the animals, was used to correctly classify 86.1% of the cheese samples. The $\delta^{13}C$ and $\delta^{18}O$ values were the most effective parameters. It was concluded that the methodology applied could allow differentiation between cheeses produced in a narrow geographical region within a PDO area.

Erich et al. (2015) combined EA–IRMS, GC–FID, ^{13}C NMR and ^{1}H NMR with chemometric analysis to discriminate organic and conventional milk. Raw, pasteurized and UHT milk samples were collected in 2013 and 2014. The $\delta^{13}C$ values of milk protein and milk fat as well as the α-linolenic acid content were quantified. Moreover, ^{1}H and ^{13}C NMR spectroscopy was used for milk fat analysis. The chemometric analysis of the combined data resulted in more precise authentication of raw and retail milk from Germany, with a significantly increased classification rate of 95%, compared with 81% for NMR and 90% for IRMS when LDA was used.

Voica et al. (2023) used ICP–MS to obtain MEC data for eggs (egg white, yolk and eggshell) and estimated the non-carcinogenic health risk associated with the presence of heavy metals in egg samples. A chemometric assessment of the data obtained from home-produced and commercial eggs was performed. The statistical models identified several elemental discriminators that allowed them to differentiate between market and local producer samples. PLS–DA indicated the potential of fingerprinting to differentiate egg components and rearing systems. Significant efficiency was achieved owing to distinguishing between egg parts related to the yolk, albumen or shell. Moreover, it was possible to discriminate supermarket and domestic samples using 17 elemental descriptors.

Manca et al. (2006) described the relationship between the SIR content determined by EA–IRMS and TC/EA–IRMS and the geographical origin of the three types of cow milk cheese produced in local dairies. Milk was obtained from free-grazing or pasture-grazing cows in Sardinia (local dairy product), cheeses were made on an industrial scale from milk produced by intensive farming in Sardinia (factory cheese) and cheeses were made with raw materials imported from other countries (imported product). To distinguish the Sardinian cheeses from the imported product, five SIRs were used. ANOVA and Tukey's test revealed significant differences among the three types of cheese (local dairy, factory and imported products). An assessment of the data by PCA and HCA indicated that a discrimination between the local dairy-produced cheeses and the products in the other two categories was performed.

5.2.4.6 Interregional and/or Intertype Differentiation Related to the Production Process and/or Interseasonal Variations

Scampicchio et al. (2016) reported the study of milk samples from North Tyrol (raw milk), South Tyrol (raw milk and HTST) and other European regions (UHT) using a combination of EA–IRMS, MIRS, NIRS and GC–FID. PLS–DA allowed the discrimination of three groups of milk samples; i.e. the group from South Tyrol, only weakly overlapped with that of North Tyrol, whereas the class of milk originating from other EU regions was significantly differentiated from the other two groups. PLS–DA discriminated milk samples according to their geographical origin with an error below 5%. The type of processing and the season were also studied. For example, 92% of the raw, HTST and UHT milk samples were correctly classified in the fitting, and 86% were correctly classified in the prediction, although no adequate discrimination was obtained when the effects of season were considered.

Potočnik et al. (2020a) used a combination of TC/EA–IRMS, EDXRF and ICP–MS to determine the geographical origin of cow milk samples from four Slovenian regions during the summer and winter of 2012–2014 as well as from other countries, i.e. Croatia, Germany, Austria and Italy. It was possible to discriminate milk samples with respect to the year of production (2012, 2013 and 2014), sampling season (summer and winter) and production region (Alpine, Mediterranean, Pannonia and Dinaric regions) using DA, with the overall temporal prediction variabilities being 84.6% and 56.4%, respectively, for regional differences. Comparable results were obtained by OPLS–DA models. Finally, DD–SIMCA classification using a combination of SIR values and MEC as parameters was a suitable tool for verifying the declaration of the geographical milk origin and therefore was useful for determining the authenticity of this product.

Garbaras et al. (2019) reported that samples of cow milk, drinking water, precipitation and feed collected for 3 years from the same location allowed the quantification of seasonal stable isotope differentiation between cow milk, water and drinking water. It was concluded that the combination of $\delta^{18}O$, $\delta^{13}C$ and $\delta^{15}N$ analysis is a valuable tool for determining the geographic origin of milk and the dietary regime of animals, contributing to improvements in studies concerning milk product authentication. However, owing to the observed seasonal variations in the isotopic ratios, databases and comparative materials are needed to evaluate the origin of food products.

Camin et al. (2004) used EA–IRMS combined with multivariate processing analysis to study various factors, such as the animal-feeding regimen, geographical origin and climatic and seasonal conditions, to classify eight types of cow and sheep cheeses originating from various European countries. As shown in Fig. 5.18, the combination of the first two canonical variables, accounting for 82% of the variability, allowed satisfactory differentiation of the different cheese types originating from three European countries. In the case of the PDO cow and sheep cheeses, the $\delta^{18}O$ value was more dependent on the geographical origin of the cheese and on

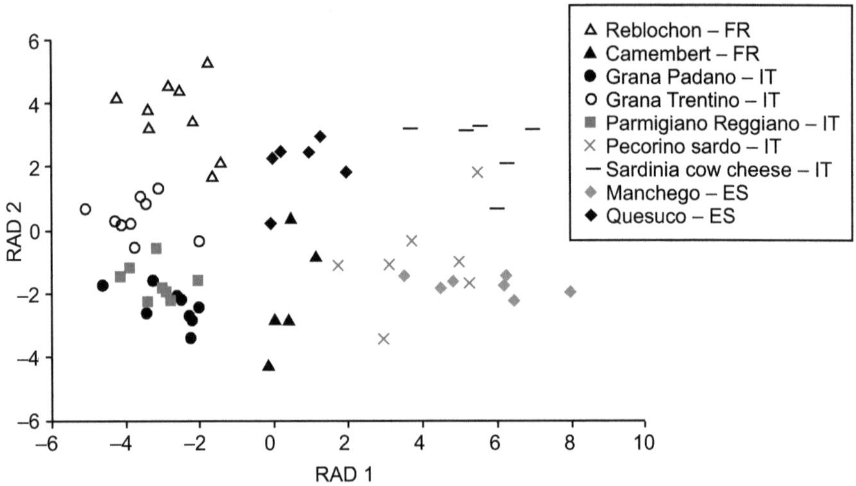

Fig. 5.18 Scatterplot of the first two canonical variables. (After Camin F et al., 2004. *J Agric Food Chem* 52:6592–6601. https://doi.org/10.1021/jf040062z. With permission)

climatic/seasonal parameters. A multivariate stepwise CDA allowed satisfactory discrimination of the different European cheeses, with a classification of ca. 95% of the samples being correctly reclassified.

Chen et al. (2017) performed an analytical procedure involving investigations of 608 milk samples from 28 farms with various feeding strategies in southern Germany with respect to the influences of ambient conditions, drinking water sources and feeding strategies on the seasonal variation in $\delta^{18}O$ in milk water. The MK model satisfactorily estimated the seasonal variation in $\delta^{8}O$ in milk. According to the authors, distinguishing between production systems may be possible only when both the season and location parameters are known.

Auerswald et al. (2015) analysed $\delta^{13}C$ and fatty acid composition by EA–IRMS and GLC for 671 milk samples from 40 farms distributed across Central Europe. The model was used to predict $\delta^{13}C$ in the bulk feed and then applied a constant feed–milk discrimination. This study considered the dietary contributions of C3 and C4 plants, the contribution of concentrates, altitude, seasonal variation in $^{12/13}CO_2$, Suess's effect and diet–milk discrimination revealed a significant correlation ($r^2 = 0.932$, $p < 0.001$) between the predicted and measured $\delta^{13}C$ values in milk. The $\delta^{13}C$ data for the milk enabled the validation of a reported C4 component with an error of <8% in 95% of all cases.

Potočnik et al. (2020b) used EA–IRMS, GC–IRMS and GC–FID data to examine raw cow milk samples taken twice per year in 2012, 2013 and 2014 in winter and summer from four distinct geographical regions in Slovenia, i.e. the Mediterranean, Alpine, Dinaric and Pannonian regions. The Mediterranean samples were differentiated from other regions on the basis of high percentages of C15:0 and C17:0 and their isomers. The CV test demonstrated 64.1% correct classification, with the highest rate of classification. The classification rate related to the year/

season of production improved when the $\delta^{13}C$ composition of individual FAs was considered, reaching 93.8% variability. However, it was postulated that $\delta^{13}C$ in FA could be more useful for tracing metabolic transformation processes in ruminants since these natural stable isotope differences in FA depend on both the diet type and the FA precursor.

Peláez Puerto et al. (2004) studied fresh and semihard cheeses for their elemental profiles established by FES and FAAS to classify the samples according to different categories, i.e. type of cheese, region of origin, season of production and goat diet. FA with the first two factors, which explained 53.57% of the total variance, allowed the differentiation of fresh cheese samples from semihard cheese. On the basis of the LDA data, the classification of the fresh and semihard cheeses was satisfactory, generally with respect to the season of production and the type of goat diet. The correct classification values of the fresh cheeses were 75.0% and 72.8% after CV, and that of the semihard cheeses was 70.7% (70.7% with CV). It was concluded that the type of goat diet might be responsible for the chemical distribution pattern of cheeses, particularly fresh cheese produced from goat milk.

Interseasonal differentiation was studied by Sola-Larrañaga and Navarro-Blasco (2009). ICP–AES, FAAS, AES and ET–AAS enabled the investigation of raw cow milk samples from different farms in three zones of the community of Navarra (northern Spain). As shown in Fig. 5.19, PCA differentiated samples relative to the seasonal criteria in contrast to the unsatisfactory geographical classification. The

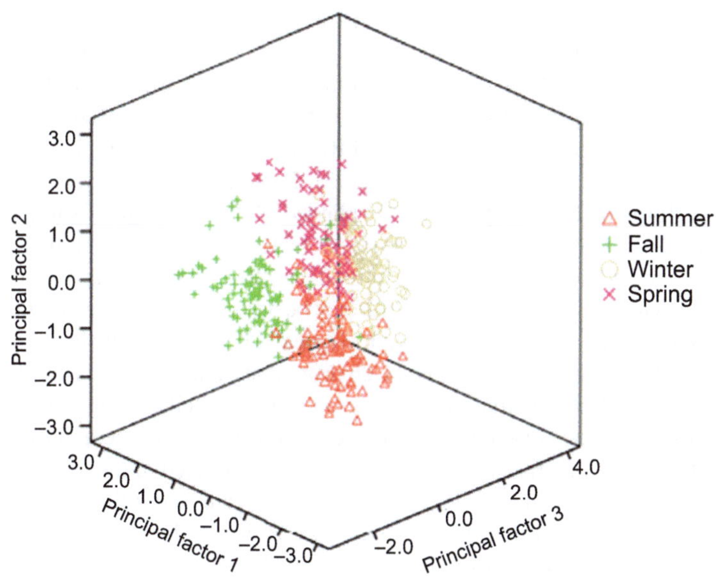

Fig. 5.19 Three-dimensional representation of the principal component scores for raw cow milk samples differentiated according to seasonal criteria. (After Sola-Larrañaga C and Navarro-Blasco I, 2009. Food Chem 112:189–196. https://doi.org/10.1016/j.foodchem.2008.05.062. With permission)

first three PCs accounted for 49.7% of the total variance. The obtained reasonable interseasonal classification was also satisfactory, amounting to 97.1% of the cases being correctly assigned with very satisfying specificity and sensitivity.

Zhao et al. (2020) examined the SIR compositions of cow milk samples from various provinces in China to evaluate the effects of lactation stage, sampling time and geographic origin on the milk samples analysed via EA–IRMS. $\delta^{13}C$, $\delta^{15}N$ and $\delta^{18}O$ data from the milk were affected by sampling time. In addition, highly significant interregional differentiation in the mean $\delta^{13}C$ and $\delta^{15}N$ values for the milk was achieved. Interregional differentiation was provided by PLSDA, with a CV accuracy between the four regions of 92.11%. In conclusion, the lactation stage had no effect on the traceability of milk, but the sampling time and geographic provenance did influence milk traceability. Interregional locations with differentiation distances exceeding 0.7 km can be differentiated with the use of SIRA.

5.2.4.7 Interregional Differentiation Related to the Impact of Pollution

Bilandžić et al. (2011) determined the concentrations of As, Cd, Cu, Hg and Pb by GF–AAS in Croatian raw milk samples from rural areas in the northern and southern regions of Croatia. Cd and Hg levels in milk samples had a significant impact on geographical origin; i.e. higher levels of these heavy metals were detected in the southern region than in the northern region. Similar interregional trends were observed for other elements, i.e. As, Cu and Pb. The mean Pb levels in both the northern and southern regions exceeded the maximum recommended level.

As reported by Diyabalanage et al. (2021), ICP–MS was used to classify raw cow milk, animal feed forage and concentrates, including water samples from 23 local farms in four agroclimatic zones. Since trace elements are used as bioindicators of agricultural pollution, among the different elements studied, toxic elements such as Cd, Pb and As were also determined. R-mode factor analysis was subsequently performed. The first three PCs explained 75.5% of the total variance, indicating that PC1 may be related to intrinsic factors such as the animal species, lactation stage and health status of the animal, together with feeding and metabolic adaptations with seasonal variations. PC2 may be associated with feed additives and dietary supplements used in animal feed, whereas PC3 may be attributed to differences caused by specific geographical environments and climatic conditions. As shown in Fig. 5.20, two distinct sampling groups were specified, suggesting general trends in the behaviour of the object studied with respect to the region of origin. It was concluded that the trace elements in cow milk are related to the feed, and therefore, their content in feed samples should be monitored. According to the authors, although there is low milk consumer exposure to toxic elements, further studies with numerous samples are strongly recommended.

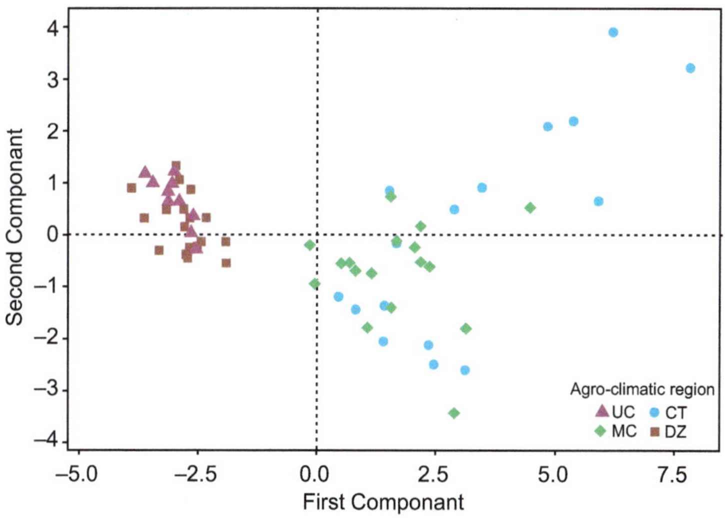

Fig. 5.20 Biplot representation of the principal component scores for raw cow milk samples differentiated according to sample location. (After Diyabalanage S et al., 2021. Biol Trace Elem Res 199:1377–1388. https://doi.org/10.1007/s12011-020-02242-4. With permission)

5.2.4.8 Interspatial and Intertype Differences in the Production Process of Dietary Feed

TC/EA–IRMS and GC–IRMS have been used to establish the δ^2H, $\delta^{13}C$ and $\delta^{18}O$ values of dairy milk, diet and drinking water from New Zealand dairy farms. The basic environmental relationships of milk fatty acids can be determined by analysis of their respective δ^2H values. This diagnostic tool appears to be useful for determining the type and origin of dietary feed utilized in milk production (Ehtesham et al. 2015).

5.2.4.9 Colostrum Elemental Profiles of Native and Non-native Goats

The ICP–OES, FRAP and DPPH methods have been used to study the colostrums of native and non-native Indian goat breeds subjected to high-altitude hypoxic conditions; the colostrums were processed chemometrically. Compared with that of non-native goats, the colostrum of native goats contained significantly greater levels of nutritive elements. On the basis of the antioxidant and elemental profiles, PCA differentiated the colostrums of native and non-native goats. This highlights the role of colostrum supplements as a natural source to increase the endurance of newborn kid goats for survival in high-altitude environments (Kumar et al. 2020).

5.2.4.10 Intertype Differentiation Related to Varying Compositions with up to 6 Months of Cheese Ripening

González-Martín et al. (2011b) studied three types of Spanish cheese, i.e. cow, ewe and goat cheeses, with respect to their elemental compositions. The MEC content was quantified by NIRS and ICP–OES. The potential for prediction was highly satisfactory for Ca, P, K, Na and Mg. This finding indicated that the models that use the MPLS regression method allow the quantification of mineral elements in samples of unknown cheese of varying compositions with up to 6 months of ripening, which can be performed using milk from different seasons.

5.2.4.11 Determination of Markers for the Characterization of Human Milk from Lactating Women

Jagodic et al. (2020) used a combination of GC–C–IRMS, ICP–MS, CVAFS and GC–FID to study milk from women inhabiting two areas of Slovenia. Chemometric analysis was performed to evaluate the discrimination between mothers living in different geographical areas (coastal and inland) and the intake of terrestrial-based food and seafood as well as different types of seafood (fresh/canned/frozen). Clustering allowed the database to be divided into two tested groupings with 90% accuracy. This approach was based on the percentage of FAs, i.e. iC17:0, C4:0, C18:2ω6t, aC17:0, CLA, C22:4ω6 and $\delta^{13}C$ of C18:1ω9c, as accurate indicators to differentiate high and low fresh seafood intake. This means that human milk samples from coastal and inland areas with different MEC, FAC and $\delta^{13}C_{FA}$ values could be used as markers of fresh seafood intake by lactating women who consume seafood in coastal areas.

In Covaciu et al. (2023), the SIR and FA profiles established by GC–FID and ICP–MS were used to characterize chemometrically human milk from lactating women. Correlation analysis, ANOVA and LDA were applied to assess the relationships among the lactation stages as a function of FAs and elemental distribution patterns to identify the most effective descriptors. LDA achieved a satisfying classification accuracy of 100% for the initial classification and 92% with CV. The latter value resulted from an overlap between two samples. The most important discriminators are Na, Ca, Zn, C24:1 and C12:0. The MEC of human milk should be used to design nutritious infant formula that has a similar nutrition profile as human milk and that does not have excess trace and toxic elements.

5.2.4.12 Interregional Differentiation of Milk and Cattle Tail Hair in the Assessment of Cow Milk Traceability

Raw and factory milk and tail hair samples from farms and supermarkets in different Malaysian localities of Peninsular Malaysia were analysed using IRMS for $\delta^{13}C$ and $\delta^{15}N$. A significant differentiation between the two types of samples from the northern and southern regions of Peninsular Malaysia was achieved. A statistically significant correlation between the $\delta^{13}C$ and $\delta^{15}N$ values in milk and cattle tail hair was obtained. This indicated that both stable isotope values of cattle tail hair are helpful descriptors in the evaluation of cow milk traceability (Behkami et al. 2017).

5.2.4.13 Prediction of Calcium Content in Ready-to-Feed Infant Formula

Cama-Moncunill et al. (2020) examined samples of ready-to-feed infant formula and ready-to-feed follow-up formula from Ireland. A combination of LIBS spectral data and AAS data with the PLSR model could predict the Ca content in ready-to-feed infant formula.

5.3 Conclusions

The most promising instrumental tools integrated with chemometric techniques appear to be useful in the evaluation of milk and milk products, authentication of milk, identification of adulterations/mislabelling and determination of milk safety. The proper use of these hyphenated tools for assessing the nutritive and health quality of this food group is important for its classification with respect to geographical origin, type of milk and its products, considering the feeding system and degree of environmental pollution. This would be useful for differentiating unprocessed and technologically processed food as well as for detecting mislabelling of products to preserve the brand reputations of original products and to protect the consumer from financial harm. The application of chemometric tools has resulted in a deeper understanding of the distribution of mineral components in milk and its products, which is especially important in the food sciences and ecotoxicological field.

References

Akele ML, Abebe DZ, Alemu AK, Assefa AG, Madhusudhan A, de Oliveira RR (2017) Analysis of trace metal concentrations in raw cow's milk from three dairy farms in North Gondar, Ethiopia: chemometric approach. Environ Monit Assess 189:499. https://doi.org/10.1007/s10661-017-6203-0

Altieri S, Saiano K, Biondi M, Ricci P, Lubritto C (2020) Traceability of 'Mozzarella di Bufala Campana' production chain by means of carbon, nitrogen and oxygen stable isotope ratios. J Sci Food Agric 100:995–1003. https://doi.org/10.1002/jsfa.10100

Arvanitoyannis IS, Tzouros NE (2005) Implementation of quality control methods in conjunction with chemometrics toward authentication of dairy products. Crit Rev Food Sci Nutr 45:231–249. https://doi.org/10.1080/10408690490478073

Ataro A, McCrindle RI, Botha BM, McCrindle CME, Ndibewu PP (2008) Quantification of trace elements in raw cow's milk by inductively coupled plasma mass spectrometry (ICP-MS). Food Chem 111:243–248. https://doi.org/10.1016/j.foodchem.2008.03.056

Auerswald K, Schaufele R, Bellof G (2015) Routing of fatty acids from fresh grass to milk restricts the validation of feeding information obtained by measuring ^{13}C in milk. J Agric Food Chem 63:10500–10507. https://doi.org/10.1021/acs.jafc.5b03646

Azcarate SM, Savio M, Smichowski P, Martinez LD, Camiña JM, Gil RA (2015) Single-step solubilization of milk samples with N,N-dimethylformamide for inductively coupled plasma-mass spectrometry analysis and classification based on their elemental composition. Talanta 143:64–70. https://doi.org/10.1016/j.talanta.2015.04.088

Azcarate SM, Gil R, Smichowski P, Savio M, Camiña JM (2017) Chemometric application in foodomics: nutritional quality parameters evaluation in milk-based infant formula. Microchem J 130:1–6. https://doi.org/10.1016/j.microc.2016.07.016

Bandoniene D, Walkner C, Zettl D, Meisel T (2018) Rare earth element labeling as a tool for assuring the origin of eggs and poultry products. J Agric Food Chem 66:11729–11738. https://doi.org/10.1021/acs.jafc.8b03828

Bandoniene D, Walkner C, Ringdorfer F, Meisel T (2020) Authentication of meat and dairy products using rare earth element labeling and detection by solution based and laser ablation ICP-MS. Food Res Int 132:109106. https://doi.org/10.1016/j.foodres.2020.109106

Barbosa RM, Nacano LR, Freitas R, Batista BL, Barbosa F Jr (2014) The use of decision trees and Naïve Bayes Algorithms and trace element patterns for controlling the authenticity of free-range-pastured hens' eggs. J Food Sci 79:C1672–C1677. https://doi.org/10.1111/1750-3841.12577

Bay LJ, Harn Chan JS, Walczyk T (2018) Optimization of analytical strategies by Monte Carlo simulation: a case study in eggs for tracing their geographical origin using stable isotope signatures. Forensic Chem 11:32–37. https://doi.org/10.1016/j.forc.2018.08.003

Behkami S, Zain SM, Gholami M, Bakirdere S (2017) Isotopic ratio analysis of cattle tail hair: a potential tool in building the database for cow milk geographical traceability. Food Chem 217:438–444. https://doi.org/10.1016/j.foodchem.2016.08.130

Behkami S, Gholami R, Gholami M, Roohparvar R (2020) Precipitation isotopic information: a tool for building the data base to verify milk geographical origin traceability. Food Control 107:106780. https://doi.org/10.1016/j.foodcont.2019.106780

Benincasa C, Lewis J, Sindona G, Tagarelli A (2008) The use of multi element profiling to differentiate between cow and buffalo milk. Food Chem 110:257–262. https://doi.org/10.1016/j.foodchem.2008.01.049

Bilandžić N, Đokić M, Sedak M, Solomun B, Varenina I, Knežević Z, Benić M (2011) Trace element levels in raw milk from northern and southern regions of Croatia. Food Chem 127:63–66. https://doi.org/10.1016/j.foodchem.2010.12.084

Bilge G, Sezer B, Eseller KE, Berberoglu H, Topcu A, Boyaci IH (2016) Determination of whey adulteration in milk powder by using laser induced breakdown spectroscopy. Food Chem 212:183–188. https://doi.org/10.1016/j.foodchem.2016.05.169

Bontempo L, Larcher R, Camin F, Hölzl S, Rossmann A, Horn P, Nicolini G (2011) Elemental and isotopic characterisation of typical Italian alpine cheeses. Int Dairy J 21:441–446. https://doi.org/10.1016/j.idairyj.2011.01.009

Bontempo L, Lombardi G, Paoletti R, Ziller L, Camin F (2012) H, C, N and O stable isotope characteristics of alpine forage, milk and cheese. Int Dairy J 23:99–104. https://doi.org/10.1016/j.idairyj.2011.10.005

Borges EM, Volmer DA, Gallimberti M, de Souza DF, de Souza EL, Barbosa F Jr (2015) Evaluation of macro- and microelement levels for verifying the authenticity of organic eggs by using chemometric techniques. Anal Methods 7:2577–2584. https://doi.org/10.1039/C4AY02986K

Bostic JN, Hagopian WM, Jahren AH (2018) Carbon and nitrogen stable isotopes in U.S. milk: insight into production process. Rapid Commun Mass Spectrom 32:561–566. https://doi.org/10.1002/rcm.8069

Brescia MA, Monfreda M, Buccolieri A, Carrino C (2005) Characterisation of the geographical origin of buffalo milk and mozzarella cheese by means of analytical and spectroscopic determinations. Food Chem 89:139–147. https://doi.org/10.1016/j.foodchem.2004.02.016

Cama-Moncunill X, Markiewicz-Kęszycka M, Cullen P, Sullivan C, Casado-Gavalda MP (2020) Direct analysis of calcium in liquid infant formula via laser-induced breakdown spectroscopy (LIBS). Food Chem 309:125754. https://doi.org/10.1016/j.foodchem.2019.125754
Camin F, Wietzerbin K, Blanch Cortes A, Haberhauer G, Lees M, Versini G (2004) Application of multielement stable isotope ratio analysis to the characterization of French, Italian, and Spanish cheeses. J Agric Food Chem 52:6592–6601. https://doi.org/10.1021/jf040062z
Camin F, Perini M, Colombari G, Bontempo L, Versini G (2008) Influence of dietary composition on the carbon, nitrogen, oxygen and hydrogen stable isotope ratios of milk. Rapid Commun Mass Spectrom 22:1690–1696. https://doi.org/10.1002/rcm.3506
Camin F, Wehrens R, Bertoldi D, Bontempo L, Ziller L, Perini M, Nicolini G, Nocetti M, Larcher R (2012) H, C, N and S stable isotopes and mineral profiles to objectively guarantee the authenticity of grated hard cheeses. Anal Chim Acta 711:54–59. https://doi.org/10.1016/j.aca.2011.10.047
Camin F, Bertoldi D, Santato A, Bontempo L, Perini M, Ziller L, Stroppa A, Larcher R (2015) Validation of methods for H, C, N and S stable isotopes and elemental analysis of cheese: results of an international collaborative study. Rapid Commun Mass Spectrom 29:415–423. https://doi.org/10.1002/rcm.7117
Capici C, Mimmo T, Kerschbaumer L, Cesco S, Scampicchio M (2015) Determination of cheese authenticity by carbon and nitrogen isotope analysis: Stelvio cheese as a case study. Food Anal Methods 8:2157–2162. https://doi.org/10.1007/s12161-015-0113-y
Chen G, Schäufele R, Auerswald K (2017) Ambient conditions and feeding strategy influence $\delta^{18}O$ of milk water in cows (*Bos taurus*). J Agric Food Chem 65(34):7387–7395. https://doi.org/10.1021/acs.jafc.7b02482
Chen L, Li X, Li Z, Deng L (2020) Analysis of 17 elements in cow, goat, buffalo, yak, and camel milk by inductively coupled plasma mass spectrometry (ICP-MS). RSC Adv 10:6736–6742. https://doi.org/10.1039/D0RA00390E
Chesson LA, Valenzuela LO, O'Grady SP, Cerling TE, Ehleringer JR (2010) Hydrogen and oxygen stable isotope ratios of milk in the United States. J Agric Food Chem 58:2358–2363. https://doi.org/10.1021/jf904151c
Chung I-M, Park I, Yoon J-Y, Yang Y-S, Kim S-H (2014) Determination of organic milk authenticity using carbon and nitrogen natural isotopes. Food Chem 160:214–218. https://doi.org/10.1016/j.foodchem.2014.01.061
Chung I-M, Kim J-K, Lee K-J, Son N-Y, An M-J, Lee J-H, An Y-J, Kim S-H (2018) Discrimination of organic milk by stable isotope ratio, vitamin E, and fatty acid profiling combined with multivariate analysis: a case study of monthly and seasonal variation in Korea for 2016-2017. Food Chem 261:112–123. https://doi.org/10.1016/j.foodchem.2018.04.017
Chung I-M, Kim J-K, Yang Y-J, An Y-J, Kim S-Y, Kwon C, Kim S-H (2020) A case study for geographical indication of organic milk in Korea using stable isotope ratios-based chemometric analysis. Food Control 107:106755. https://doi.org/10.1016/j.foodcont.2019.106755
Covaciu FD, Feher I, Molnar C, Floare-Avram V, Dehelean A (2023) Characterization of the fatty acid and elemental composition of human milk with chemometric processing to determine the nutritional value. Anal Lett 56:344–356. https://doi.org/10.1080/00032719.2022.2081857
Cristea G, Dehelean A, Puscas R, Hategan AR, Magdas DA (2023) Isotopic and elemental fingerprint of edible egg parts—the health risk assessment based on potentially toxic elements content. Molecules 28:503
Crittenden RG, Andrew AS, LeFournour M, Young MD, Middleton H, Stockmann R (2007) Determining the geographic origin of milk in Australasia using multi-element stable isotope ratio analysis. Int Dairy J 17:421–428. https://doi.org/10.1016/j.idairyj.2006.05.012
de Freitas R, Ramos Nacano L, Batista BL, Barbosa F Jr (2013) Toxic and essential elements in conventional and home-produced eggs by ICP-MS analysis. Food Addit Contam Part B Surveill 6:30–35. https://doi.org/10.1080/19393210.2012.721095
Di Bella G, Turco VL, Potortì AG, Luppino RR, Fotia V, Conte F, mo Dugo G (2012) Classification of the geographical origin of Italian donkey's milk based on differences in inorganic anions. Food Addit Contam Part A 29:1021–1029. https://doi.org/10.1080/19440049.2012.674979

Diyabalanage S, Kalpage MD, Mohotti DG, Dissanayake CKK, Fernando R, Frew RD, Chandrajith R (2021) Comprehensive assessment of essential and potentially toxic trace elements in bovine milk and their feeds in different agro-climatic zones of Sri Lanka. Biol Trace Elem Res 199:1377–1388. https://doi.org/10.1007/s12011-020-02242-4

Dong H, Xiao K, Luo D (2017) Stability of carbon and nitrogen isotopic compositions of the protein extracted from milk and their potential as "fingerprints" of geographical origin. RSC Adv 7:18946–18952. https://doi.org/10.1039/c7ra00722a

Ehtesham E, Hayman AR, McComb KA, Hale RV, Frew RD (2013a) Correlation of geographical location with stable isotope values of hydrogen and carbon of fatty acids from New Zealand milk and bulk milk powder. J Agric Food Chem 61:8914–8923. https://doi.org/10.1021/jf4024883

Ehtesham E, Baisden WT, Keller ED, Hayman AR, Van Hale R, Frew RD (2013b) Correlation between precipitation and geographical location of the δ^2H values of the fatty acids in milk and bulk milk powder. Geochim Cosmochim Acta 111:105–116. https://doi.org/10.1016/j.gca.2012.10.026

Ehtesham E, Hayman A, Van Hale R, Frew R (2015) Influence of feed and water on the stable isotopic composition of dairy milk. Int Dairy J 47:37–45. https://doi.org/10.1016/j.idairyj.2015.02.008

Erich S, Schill S, Annweiler E, Waiblinger H-U, Kuballam T, Lachenmeier DW, Monakhova YB (2015) Combined chemometric analysis of ^1H NMR, ^{13}C NMR and stable isotope data to differentiate organic and conventional milk. Food Chem 188:1–7. https://doi.org/10.1016/j.foodchem.2015.04.118

Esposito M, Cavallo S, Chiaravalle E, Miedico O, Pellicanò R, Rosato G, Sarnelli P, Baldi L (2016) Trace elements in free-range hen eggs in the Campania region (Italy) analyzed by inductively coupled plasma mass spectrometry (ICP-MS). Environ Monit Assess 188:326. https://doi.org/10.1007/s10661-016-5316-1

Faberi A, Compagnone D, Fuselli F, La Mantia A, Mascini M, Montesano C, Rocchi R, Sergi M (2018) Italian cheeses discrimination by means of $\delta^{13}C$ and $\delta^{15}N$ isotopic ratio mass spectrometry. Food Anal Methods 11:1467–1475. https://doi.org/10.1007/s12161-017-1110-0

Favretto LG, Marletta GP, Bogoni P, Favretto L (1989) Chemometric studies of some trace elements in cows' milk. Z Lebensm-Unters-Forsch 189:123–127. https://doi.org/10.1007/BF01332945

Fortunato G, Mumic K, Wunderli S, Pillonel L, Bosset JO, Gremaud G (2004) Application of strontium isotope abundance ratios measured by MC-ICP-MS for food authenticity. J Anal Atom Spectrom 19:227–234. https://doi.org/10.1039/b307068a

Garbaras A, Skipitytė R, Šapolaitė J, Ežerinskis Ż, Remeikis V (2019) Seasonal variation in stable isotope ratios of cow milk in Vilnius region, Lithuania. Animals 9:69. https://doi.org/10.3390/ani9030069

González-Martín MI, Hernández-Hierro JM, Revilla I, Vivar-Quintana A, González-Pérez C, Gómez García L, Palacios Riocerezo C, Lobos Ortega IA (2011a) Differentiation of organic and non-organic ewe's cheeses using main mineral composition or near infrared spectroscopy coupled to chemometric tools: a comparative study. Talanta 85:1915–1919. https://doi.org/10.1016/j.talanta.2011.07.021

González-Martín I, Hernández-Hierro JM, Revilla I, Vivar-Quintana A, Lobos Ortega I (2011b) The mineral composition (Ca, P, Mg, K, Na) in cheeses (cow's, ewe's and goat's) with different ripening times using near infrared spectroscopy with a fibre-optic probe. Food Chem 127:147–152. https://doi.org/10.1016/j.foodchem.2010.12.114

Gregorčič SH, Potočnik D, Camin F, Ogrinc N (2020) Milk authenticity: stable isotope composition of hydrogen and oxygen in milks and their constituents. Molecules 25:4000. https://doi.org/10.3390/molecules25174000

Griboff J, Baroni MV, Horacek M, Wunderlin DA, Monferran MV (2019) Multielemental + isotopic fingerprint enables linking soil, water, forage and milk composition, assessing the geographical origin of Argentinean milk. Food Chem 283:549–558. https://doi.org/10.1016/j.foodchem.2019.01.067

Hermansen JE, Badsberg JH, Kristensen T, Gundersen V (2005) Major and trace elements in organically or conventionally produced milk. J Dairy Res 72:362–368. https://doi.org/10.1017/S0022029905000968

Huang J, Norgbey E, Nkrumah PN, Sefah GA, Michelle R (2017) Elucidating the origin of milk products on the Chinese market using hydrogen and oxygen stable isotope technique. Integr Food Nutr Metab 4:1–3. https://doi.org/10.15761/IFNM.1000184

Jagodic M, Potočnik D, Snoj Tratnik J, Mazej D, Pavlin M, Trdin A, Eftimov T, Kononenko L, Ogrinc N, Horvat M (2020) Selected elements and fatty acid composition in human milk as indicators of seafood dietary habits. Environ Res 180:108820. https://doi.org/10.1016/j.envres.2019.108820

Kalpage M, Dissanayake C, Diyabalanage S, Chandrajith R, Frew R, Fernando R (2022) Stable isotope and element profiling for determining the agroclimatic origin of cow milk within a tropical country. Foods 11:275. https://doi.org/10.3390/foods11030275

Kamal M, Karoui R (2015) Analytical methods coupled with chemometric tools for determining the authenticity and detecting the adulteration of dairy products: a review. Trends Food Sci Technol 46:17–48. https://doi.org/10.1016/j.tifs.2015.07.007

Kamilari E, Tomazou M, Antoniades A, Tsaltas D (2019) High throughput sequencing technologies as a new toolbox for deep analysis, characterization and potentially authenticity of protection designation of origin cheeses? Int J Food Sci 2019:5837301. https://doi.org/10.1155/2019/5837301

Koreňovská M, Suhaj M (2007) Identification of Slovakian, Polish, and Romanian bryndza cheeses origin by factor analysis of some elemental data. Eur Food Res Technol 225:707–713. https://doi.org/10.1007/s00217-006-0473-z

Król J, Litwińczuk Z, Brodziak A, Kędzierska-Matysek M (2012) Content of selected essential and potentially toxic trace elements in milk of cows maintained in eastern Poland. J Elem 17:597–608. https://doi.org/10.5601/jelem.2012.17.4.04

Kumar P, Bharti VK, Mukesh M (2020) Chemometric analysis of antioxidant and mineral elements in colostrum of native and non-native goat breeds to hypoxic conditions at high altitude. Biol Trace Elem Res 196:446–453. https://doi.org/10.1007/s12011-019-01940-y

Liu H, Zhao Q, Guo X, Tan C, Yu X, Zhan T, Qin Y, Zhang J (2019) Application of isotopic and elemental fingerprints in identifying the geographical origin of goat milk in China. Food Chem 277:448–454. https://doi.org/10.1016/j.foodchem.2018.10.144

Lopez-Alonso M, Rey-Crespo F, Herrero-Latorre C, Miranda M (2017) Identifying sources of metal exposure in organic and conventional dairy farming. Chemosphere 185:1048–1055. https://doi.org/10.1016/j.chemosphere.2017.07.112

Luo D, Dong H, Luo H, Xian Y, Guo X, Wu Y (2016) Multi-element (C, N, H, O) stable isotope ratio analysis for determining the geographical origin of pure milk from different regions. Food Anal Methods 9:437–442. https://doi.org/10.1007/s12161-015-0204-9

Magdas D-A, Dehelean A, Feher I, Cristea G, Puscas R, Dan S-D, Cordea D-V (2016) Discrimination markers for the geographical and species origin of raw milk within Romania. Int Dairy J 61:135–141. https://doi.org/10.1016/j.idairyj.2016.06.003

Manca G, Franco MA, Del Caro A, Coloru GC (1999) The role of the amino acids profile in characterising "Pecorino Sardo" cheese. J Commodity Sci 38:167–179

Manca G, Camin F, Coloru GC, Del Caro A, Depentori D, Franco MA, Versini G (2001) Characterization of the geographical origin of Pecorino Sardo cheese by casein stable isotope ($^{13}C/^{12}C$ and $^{15}N/^{14}N$) ratios and free amino acid ratios. J Agric Food Chem 49:1404–1409. https://doi.org/10.1021/jf000706c

Manca G, Franco MA, Versini G, Camin F, Rossmann A, Tola A (2006) Correlation between multielement stable isotope ratio and geographical origin in Peretta cows' milk cheese. J Dairy Sci 89:831–839. https://doi.org/10.3168/jds.S0022-0302(06)72146-4

Manuelian CL, Currò S, Penasa M, Cassandro M, De Marchi M (2017a) Prediction of minerals, fatty acid composition and cholesterol content of commercial cheeses by near infrared transmittance spectroscopy. Int Dairy J 71:107–113. https://doi.org/10.1016/j.idairyj.2017.03.011

Manuelian CL, Currò S, Visentin G, Penasa M, Cassandro M, Dellea C, Bernardi M, De Marchi M (2017b) Technical note: at-line prediction of mineral composition of fresh cheeses using near-infrared technologies. J Dairy Sci 100:6084–6089. https://doi.org/10.3168/jds.2017-12634

Markiewicz-Keszycka M, Cama-Moncunill R, Casado-Gavalda MP, Sullivan C, Cullen PJ (2019) Laser-induced breakdown spectroscopy for food authenticity. Curr Opin Food Sci 28:96–103. https://doi.org/10.1016/j.cofs.2019.10.002

Marpaung AM, Abdulmadjid SN, Ramli M, Idris N, Khumaeni A, Budi WS, Suyanto H, Suliyanti MM, Karnadi I, Tanra I, Pardede M, Jobiliong E, Hedwig R, Lie ZS, Kurniawan KH, Kagawa K (2020) Emission spectrochemical analysis of soft samples including raw fish by employing laser-induced breakdown spectroscopy with a subtarget at low-pressure helium gas. ACS Omega 5:16811–16818. https://doi.org/10.1021/acsomega.0c01904

McLeod RJ, Prosser CG, Wakefield JW (2016) Identification of goat milk powder by manufacturer using multiple chemical parameters. J Dairy Sci 99:982–993. https://doi.org/10.3168/jds.2015-9627

Mi S, Shang K, Zhang C-H, Fan Y-Q (2019) Characterization and discrimination of selected chicken eggs in China's retail market based on multi-element and lipidomics analysis. Food Res Int 126:108668. https://doi.org/10.1016/j.foodres.2019.108668

Miedico O, Tarallo M, Pompa C, Chiaravalle AE (2016) Trace elements in sheep and goat milk samples from Apulia and Basilicata regions (Italy): valuation by multivariate data analysis. Small Rumin Res 135:60–65. https://doi.org/10.1016/j.smallrumres.2015.12.019

Molkentin J, Giesemann A (2007) A differentiation of organically and conventionally produced milk by stable isotope and fatty acid analysis. Anal Bioanal Chem 388:297–305. https://doi.org/10.1007/s00216-007-1222-2

Moncayo S, Manzoor S, Rosales JD, Anzano J, Caceres JO (2017) Qualitative and quantitative analysis of milk for the detection of adulteration by Laser Induced Breakdown Spectroscopy (LIBS). Food Chem 232:322–328. https://doi.org/10.1016/j.foodchem.2017.04.017

Moreno-Rojas R, Sánchez-Segarra PJ, Cámara-Martos F, Amaro-López MA (2010) Multivariate analysis techniques as tools for categorization of Southern Spanish cheeses: nutritional composition and mineral content. Eur Food Res Technol 231:841–851. https://doi.org/10.1007/s00217-010-1338-z

Moreno-Rojas R, Cámara-Martos F, Sánchez-Segarra PJ, Amaro-López MA (2012) Influence of manufacturing conditions and discrimination of Northern Spanish cheeses using multi-element analysis. Int J Dairy Technol 65:594–602. https://doi.org/10.1111/j.1471-0307.2012.00853.x

Nečemer M, Potočnik D, Ogrinc N (2016) Discrimination between Slovenian cow, goat and sheep milk and cheese according to geographical origin using a combination of elemental content and stable isotope data. J Food Compos Anal 52:16–23. https://doi.org/10.1016/j.jfca.2016.07.002

Norgbey E, Huang J, Nkrumah PN, Sefah GA, Michelle R (2017) Elucidating the origin of milk products on the Chinese market using multi-element stable isotope technique. Adv Nutr Food Sci 2:1–4

Orjales I, Herrero-Latorre C, Miranda M, Rey-Crespo F, Rodríguez-Bermúdez R, López-Alonso M (2018) Evaluation of trace element status of organic dairy cattle. Animal 12:1296–1305. https://doi.org/10.1017/S1751731117002890

Osorio MT, Koidis A, Papademas P (2015) Major and trace elements in milk and Halloumi cheese as markers for authenticity of goat feeding regimes and geographical origin. Int J Dairy Technol 68:573–581. https://doi.org/10.1111/1471-0307.12213

Peláez Puerto P, Fresno Baquero M, Rodríguez Rodríguez EM, Darias-Martín J, Díaz Romero C (2004) Chemometric studies of fresh and semi-hard goats' cheeses produced in Tenerife (Canary Islands). Food Chem 88:361–366. https://doi.org/10.1016/j.foodchem.2004.01.048

Pillonel L, Badertscher R, Froidevaux P, Haberhauer G, Hölzl S, Horn P, Jakob A, Pfammatter E, Piantini U, Rossmann A, Tabacchi R, Bosset JO (2003) Stable isotope ratios, major, trace and radioactive elements in emmental cheeses of different origins. LWT - Food Sci Technol 36:615–623. https://doi.org/10.1016/S0023-6438(03)00081-1

Pillonel L, Badertscher R, Casey M, Meyer J, Rossmann A, Schlichtherle-Cerny H, Tabacchi R, Bosset JO (2005a) Geographic origin of European Emmental cheese: characterisation and descriptive statistics. Int Dairy J 15:547–556. https://doi.org/10.1016/j.idairyj.2004.07.028

Pillonel L, Butikofer U, Schlichtherle-Cerny H, Tabacchi R, Bosset JO (2005b) Geographic origin of European emmental. Use of discriminant analysis and artificial neural network for classification purposes. Int Dairy J 15:557–562

Potočnik D, Nečemer M, Mazej D, Jaćimović R, Ogrinc N (2016) Multi-elemental composition of Slovenian milk: analytical approach and geographical origin determination. Acta Imeko 5:15–21. Identifier: IMEKO-ACTA-05 (2016)-01-05

Potočnik D, Nečemer M, Perišić I, Jagodic M, Mazej D, Camin F, Eftimov T, Strojnik L, Ogrinc N (2020a) Geographical verification of Slovenian milk using stable isotope ratio, multi-element and multivariate modelling approaches. Food Chem 326:126958. https://doi.org/10.1016/j.foodchem.2020.126958

Potočnik D, Strojnik L, Eftimov T, Levart A, Ogrinc N (2020b) Fatty acid and stable carbon isotope composition of Slovenian milk: year, season, and regional variability. Molecules 25:2892. https://doi.org/10.3390/molecules25122892

Potortì AG, Di Bella G, Turco VL, Rando R, Dugo G (2013) Non-toxic and potentially toxic elements in Italian donkey milk by ICP-MS and multivariate analysis. J Food Compos Anal 31:161–172. https://doi.org/10.1016/j.jfca.2013.05.006

Renou J-P, Deponge C, Gachon P, Bonnefoy J-C, Coulon J-B, Garel J-P, Vérité R, Ritz P (2004) Characterization of animal products according to geographic origin and feeding diet using nuclear magnetic resonance and isotope ratio mass spectrometry: cow milk. Food Chem 85:63–66. https://doi.org/10.1016/j.foodchem.2003.06.003

Rey-Crespo F, Miranda M, López-Alonso M (2013) Essential trace and toxic element concentrations in organic and conventional milk in NW Spain. Food Chem Toxic 55:513–518. https://doi.org/10.1016/j.fct.2013.01.040

Rock L (2012) The use of stable isotope techniques in egg authentication schemes: A review. Trends Food Sci Technol 28:62–68. https://doi.org/10.1016/j.tifs.2012.04.002

Rodríguez-Bermúdez R, López-Alonso M, Miranda M, Fouz R, Orjales I, Herrero-Latorre C (2018) Chemometric authenticity of the organic status of milk on the basis of trace element content. Food Chem 240:686–693. https://doi.org/10.1016/j.foodchem.2017.08.011

Rogers KM, van Ruth S, Alewijn M, Philips A, Rogers P (2015) Verification of egg farming systems from The Netherlands and New Zealand using stable isotopes. J Agric Food Chem 63:8372–8380. https://doi.org/10.1021/acs.jafc.5b01975

Rossmann A, Kornexl B, Versini G, Pichlmayer F, Lamprecht G (1998) Origin assignment of milk from alpine regions by multielement stable isotope ratio analysis (SIRA). Rivista di Scienza dell'Alimentazione 27:9–21. http://hdl.handle.net/10449/16381

Sacco D, Brescia MA, Sgaramella A, Casiello G, Buccolieri A, Ogrinc N, Sacco A (2009) Discrimination between Southern Italy and foreign milk samples using spectroscopic and analytical data. Food Chem 114:1559–1563. https://doi.org/10.1016/j.foodchem.2008.11.056

Scampicchio M, Mimmo T, Capici C, Huck C, Innocente N, Drusch S, Cesco S (2012) Identification of milk origin and process-induced changes in milk by stable isotope ratio mass spectrometry. J Agric Food Chem 60:11268–11273. https://doi.org/10.1021/jf302846j

Scampicchio M, Eisenstecken D, De Benedictis L, Capici C, Ballabio D, Mimmo T, Robatscher P, Kerschbaumer L, Oberhuber M, Kaser A, Huck CW, Cesco S (2016) Multi-method approach to trace the geographical origin of alpine milk: a case study of Tyrol Region. Food Anal Methods 9:1262–1273. https://doi.org/10.1007/s12161-015-0308-2

Schneider S, Auerswald K, Bellof G, Schnyder H (2015) ^{13}C discrimination between diet, faeces, milk and milk components. Isot Environ Health Stud 51:33–45. https://doi.org/10.1080/10256016.2014.988151

Sezer B, Durna S, Bilge G, Berkkan A, Yetisemiyen A, Boyaci IH (2018) Identification of milk fraud using laser-induced breakdown spectroscopy (LIBS). Int Dairy J 81:1–7. https://doi.org/10.1016/j.idairyj.2017.12.005

Silva AV, Hélie JF, de Andrade CF, Monarde H, Mustafa AF, Stevenson R (2014) Multi-stable isotope analysis as a tool for assessing the geographic provenance of dairy products: a case study using buffalo's milk and cheese samples from the Amazon basin. Int Dairy J 35:107–110. https://doi.org/10.1016/j.idairyj.2013.10.019

Sola-Larrañaga C, Navarro-Blasco I (2006) Preliminary chemometric study of minerals and trace elements in Spanish infant formulae. Anal Chim Acta 555:354–363. https://doi.org/10.1016/j.aca.2005.09.015

Sola-Larrañaga C, Navarro-Blasco I (2009) Chemometric analysis of minerals and trace elements in raw cow milk from the community of Navarra, Spain. Food Chem 112:189–196. https://doi.org/10.1016/j.foodchem.2008.05.062

Suhaj M, Koreňovská M (2007) Identification of cheese species origin by pattern recognition processing of elemental data. J Food Nutr Res 46:174–180

Suhaj M, Koreňovská M (2008a) Study of some European cheeses geographical traceability by pattern recognition analysis of multielemental data. Eur Food Res Technol 227:1419–1427. https://doi.org/10.1007/s00217-008-0861-7

Suhaj M, Koreňovská M (2008b) Correlation and distribution of elemental markers of origin in the production of Bryndza sheep cheese. Food Chem 107:551–557. https://doi.org/10.1016/j.foodchem.2007.08.033

Suhaj M, Koreňovská M (2010) The use of mineral and trace elements profiles for cows' and goats' cheese species prediction. J Food Nutr Res 49:178–185

Temiz HT, Sezer B, Berkkan A, Tamer U, Boyaci IH (2018) Assessment of laser-induced breakdown spectroscopy as a tool for analysis of butter adulteration. J Food Compos Anal 67:48–54. https://doi.org/10.1016/j.jfca.2017.12.032

Valenti B, Biondi L, Campidonico L, Bontempo L, Luciano G, Di Paola F, Copani V, Ziller L, Camin F (2017) Changes in stable isotope ratios in PDO cheese related to the area of production and green forage availability. The case study of Pecorino Siciliano. Rapid Commun Mass Spectrom 31:737–744. https://doi.org/10.1002/rcm.7840

Voica C, Cristea G, Iordache AM, Roba C, Curean V (2023) Elemental profile in chicken egg components and associated human health risk assessment. Toxics 11:900. https://doi.org/10.3390/toxics11110900

Wijenayake K, Frew R, McComb K, Van Hale R, Clarke D (2020) Feasibility of casein to record stable isotopic variation of cow milk in New Zealand. Molecules 25:3658. https://doi.org/10.3390/molecules25163658

Xie L, Zhao S, Rogers KM, Xia Y, Zhang B, Suo R, Zhao Y (2020) A case of milk traceability in small-scale districts-Inner Mongolia of China by nutritional and geographical parameters. Food Chem 316:126332. https://doi.org/10.1016/j.foodchem.2020.126332

Xu S, Zhao C, Deng X, Zhang R, Qu L, Wang M, Ren S, Wu H, Yue Z, Niu B (2021) Determining the geographical origin of milk by multivariate analysis based on stable isotope ratios, elements and fatty acids. Anal Methods 13:2537–2548. https://doi.org/10.1039/D1AY00339A

Zain SM, Behkami S, Bakirdere S, Koki IB (2016) Milk authenticity and discrimination via metal content clustering—a case of comparing milk from Malaysia and selected countries of the world. Food Control 66:306–314. https://doi.org/10.1016/j.foodcont.2016.02.015

Zhao S, Zhao Y, Rogers KM, Chen G, Chen A, Yang S (2020) Application of multi-element (C, N, H, O) stable isotope ratio analysis for the traceability of milk samples from China. Food Chem 310:125826. https://doi.org/10.1016/j.foodchem.2019.125826

Zhou X, Qu X, Zhao S, Wang J, Li S, Zheng N (2017) Analysis of 22 elements in milk, feed, and water of dairy cow, goat, and buffalo from different regions of China. Biol Trace Elem Res 176:120–129. https://doi.org/10.1007/s12011-016-0819-8

Chapter 6
Honey and Its Products

Contents

6.1 Introduction

Honey, a natural food, is mainly composed of sugars and water with components such as mineral elements, vitamins, carotenoids, amino and organic acids, polyphenols, flavonoids, hydroxymethylfurfural, hydrogen peroxide, methylglyoxal, proteins, enzymes and other phenolic compounds and aromatic substances. Its physical properties include colour, aroma, taste and conductivity. The physicochemical composition of honey depends on its botanical and geographical origins. Therefore, analytical methods can be used to analyze the chemical components of honeys and determine their geographical origin (Santos-Buelga and González-Paramás 2017; Boussaid et al. 2018; Mohammed 2020). A comparative study on the evolution of environmental and honey pollution with heavy metals was presented by Berinde and Michnea (2013).

Varieties of flower nectar, seasonal variations, environmental conditions and technological processing can influence the compositional profile of honey, but

Supplementary Information The online version contains supplementary material available at https://doi.org/10.1007/978-3-031-99687-0_6.

P. Szefer, *Chemometrics in Food Control Based on Elemental Analysis*, https://doi.org/10.1007/978-3-031-99687-0_6

essentially, the main nutritional components are carbohydrates, i.e. fructose and glucose. The compositional criteria for honey are regulated by the European Council Directive 2001/110/EC of 20th December 2001. There are numerous studies concerning the chemometric assessment of honey and its products.

The use of the hive as a passive sampling tool with honeybees and honey as environmental bioindicators for microelements, macroelements and radionuclides has been reported. Moreover, parameters such as sampling networks and their procedures, sample pre-treatments, analytical techniques, data analysis and other factors have also been considered (Herrero-Latorre et al. 2017). According to Dżugan et al. (2018), honeybees (*Apis mellifera*) can be considered a biological barrier for the contamination of honey by environmentally toxic metals. Aghamirlou et al. (2015) reported that the elements in honey threaten human health when in excess. Atrouse et al. (2004) reported that the Pb and Cd contents in honey samples might be related to their origin and the nature of their environmental pollution. Their results confirmed the presence of heavy metals, indicating the presence of pollution in the areas from which pollen was collected.

Fakhlaei et al. (2020) provided a critical overview of the various types of adulteration and detection methods, elucidating the impact of honey adulteration on human health. Although the number of consumers of adulterated honey has considerably increased in recent years, honey has not been evaluated or monitored regularly for safety.

The botanical and geographical characteristics of nectar are related to the chemical composition of honey. Honey originates from single or multiple plant species, and the prices of the former type of honey are much higher than those of the latter type of honey. Multiple chemical compounds can be used as markers to detect the botanical and geographical origins of honey with modern chemometric tools in combination with advanced analytical instrumentation, ensuring high potential for the in-depth evaluation of the nutritive quality of honey (Wang and Li 2011).

Specific aspects of New Zealand mānuka and kānuka honey have also been discussed, and marker substances have been proposed. Bearing in mind price differences and potential fraud, available methods and challenges have been summarized, and their suitability for distinguishing both types of honey has been assessed (Schmidt et al. 2021).

According to Arvanitoyannis et al. (2005), determining mineral components is highly important for the classification of honey of various origins. Several novel analytical methods, e.g. AAS, HPLC, GC–MS, HPAED–PAD and NMR, assisted by the application of multivariate analysis, in particular, PCA, CLA and CA, have been extremely useful for the detection of adulteration. A review by Mohammed (2020) considered the various factors affecting its physicochemical properties and chemical composition, including the mineral components of bee honey, to help beekeepers produce high-quality bee honey.

A summary of the latest methods for assessing the quality of natural bee honey was reported by Puścion-Jakubik et al. (2020). This paper briefly characterized methods recommended by the International Honey Commission (2009) and newer methods published in the last decade, assisted by chemometric techniques, e.g.

PLS, PCA and ANN. Modern analytical methods, e.g. CE–UV, ET–AAS and TXRF, enable detailed examination of the mineral composition of honey, including macroelements, microelements and toxic elements. According to the authors, prospects for expanding methods of honey quality assessment should include the development of fast, reliable, inexpensive and accurate methods as well as quick tests for detecting honey adulteration.

Honey is a popular and widely consumed natural product, mainly because of its taste qualities and nutritional value, and hence its health benefits. Owing to its unique characteristics, honey can be classified as a high-quality product. Therefore, honey has been the object of adulteration through improper or fraudulent production practices and has often been miscategorized. Authentication of honey involves several main aspects, such as production, which is the main issue, and geographical and/or botanical origin. Therefore, it is important to apply state-of-the-art and advanced molecular, chromatographic and spectroscopic methodologies for the identification of geographical and botanical origins (Soares et al. 2017).

Krakowska et al. (2015), on the basis of data obtained for honey from different countries, concluded that the application of chemometric tools supports the extraction of considerable information from analytical data.

Unifloral and multifloral honeys have been investigated on the basis of their MEC content by many authors to classify their chemical characteristics with respect to their botanical origin. There are available literature data concerning MECs and other analytes in unifloral honeys (i.e. Louppis et al. 2017; Gašić et al. 2015; Nascimento et al. 2018; Terrab et al. 2003a), multifloral honeys (i.e. Lazarević et al. 2013; Vukašinović-Pešić et al. 2020; Bilandžić et al. 2017; Conti et al. 2018; Abdulkhaliq and Swaileh 2017) and monofloral and multifloral honeys (Quinto et al. 2016; Czipa et al. 2015; Corbella and Cozzolino 2006; Conti et al. 2014; Santos et al. 2008; Bontempo et al. 2017; Terrab et al. 2003a, b; Altun et al. 2017 and others). Galician and non-Galician honeys have also been differentiated successfully by Latorre et al. (1999, 2000) and others. For example, various authors have reported different concentration data for honey samples corresponding to one or two types of honey (Czipa et al. 2019; Terrab et al. 2003a, b), three botanical origins from Slovakia (Kováčik et al. 2016) and four types of honey from Italy (Pisani et al. 2008; Conti et al. 2007, 2018), Spain (Fernandez-Torres et al. 2005), Romania (Oroian et al. 2015), Greece (Louppis et al. 2017), India (Nayik and Nanda 2016) and China (Wu et al. 2015). Five types of honey samples have been recognized to originate from Italy (Bontempo et al. 2017), Greece (Louppis et al. 2019) and Malaysia (Kek et al. 2017), as have six honeys of different botanical origins from Eastern European countries, Italy and Equatorial and other countries outside Europe (Pasquini et al. 2014). Data have also been reported for seven types of honey samples from Slovenia (Nečemer et al. 2009), Spain (Nozal Nalda et al. 2005) and Italy (Bontempo et al. 2017; Di Bella et al. 2015). Several brands of seven types of commercially available honeys and bee products from three countries and other EU countries have been analyzed by Grembecka and Szefer (2013). Moreover, eight to ten types of honey originate from Slovenia, Spain, Italy and the Western Balkan area (e.g. Golob et al. 2005; de Alda-Garcilope et al. 2012; Fermo et al. 2013).

Seventeen types of honeys from 12 European and non-European countries were investigated by Devillers et al. (2002). There are also data for 1023 samples of honey of 15 different floral types from New Zealand (Rogers et al. 2014).

Different honey products have been studied, e.g. raw propolis samples (Cantarelli et al. 2011), beeswax (Bommuraj et al. 2019), sweetener syrups and 228 fortified samples (Nespeca et al. 2020) and honeydew honey (Chudzinska and Baralkiewicz 2010; Madejczyk and Baralkiewicz 2008; Fermo et al. 2013; Nayik et al. 2018; Lachman et al. 2007; Vasić et al. 2020; Dżugan et al. 2017; Bontempo et al. 2017).

6.2 Chemometric Assessment of Honey and Its Products

6.2.1 Geographical Differentiation

6.2.1.1 European Countries

In Latorre et al. (2000), FAAS and FAES data were used to assess the effectiveness of differentiating naturally unprocessed Galician honey samples from non-Galician samples. PCA with the first three PCs (75.52%) reliably discriminated natural Galician honey samples from processed non-Galician samples. LDA, KNN and SIMCA achieved satisfactory percentages of correct classifications of two types of honey between 90% and 100%, and classifications for natural Galician honey/processed non-Galician honey of 84.6/95.0%, 83.3/90.2% and 100/76.2% recognition ability, respectively, were obtained. The percentage of prediction ability was comparable and highly acceptable (Latorre et al. 1999). Moreover, in addition to PCA, other tools have been applied, i.e. CA, ANN, the Bayes method and PLS modelling (Latorre et al. 2000). Because natural Galician honey samples were used, effective differentiation of natural Galician honey samples and non-Galician honey samples was possible. For example, on the basis of these multivariate techniques, classification of processed and unprocessed Galician honey samples ranged between 84.3% and 100% in terms of recognition success and between 80.0% and 100% in terms of prediction ability.

Terrab et al. (2004) described the combination of ICP–OES and multivariate analysis as a helpful methodology to classify the elemental profile of Thyme honeys (Lamiaceae: *Thymus* spp., mainly *T. mastichina* and *T. capitatus*) originating from coastal and mountainous regions of Spain. PCA, with the first two PCs accounting for 53.69% of the total variance, revealed a clear differentiation of honey produced in both regions (Fig. 6.1). Calcium, Cu, Li, Mg, S and V (PC1) and Al, Cr and Na (PC2) were the main discriminative variables. The classification functions of Al, As, Cr, Cu, K, Li, Mg, Na, P, S and V were achieved using SDA and were useful for correctly classifying 100% of the honey samples.

Hernández et al. (2005) classified unifloral and multifloral honey from the Canary Islands and other regions using FAAS and FAES with chemometric analysis (PCA, CA, DA, LR, MR and MD). It was possible to classify the honey samples

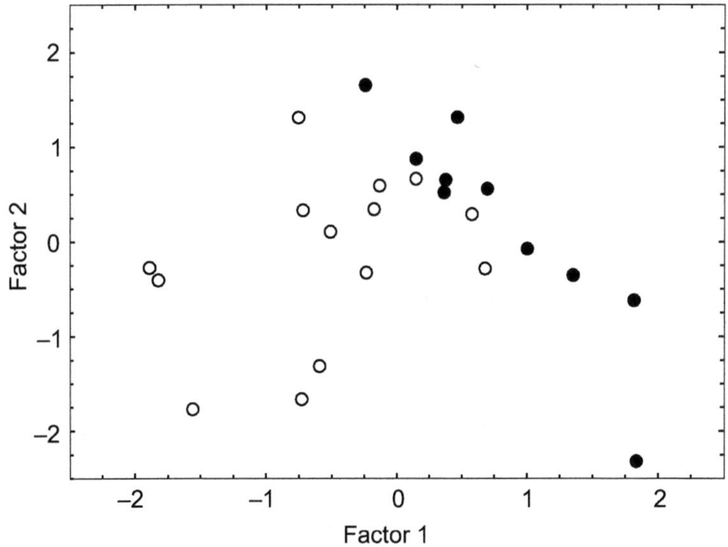

Fig. 6.1 Plot of the first factor versus the second factor for the classification of honeys according to their origin: ○, coastal; ●, mountainous. (After Terrab A et al., 2004. *J Agric Food Chem* 52:3441–3445. https://doi.org/10.1021/jf035352e. With permission)

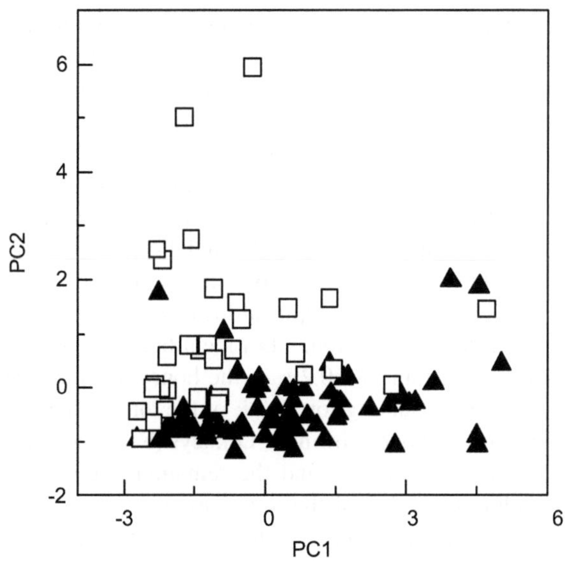

Fig. 6.2 Score plot of honey samples. (PC2 vs. PC1): ▲, Canary; □, non-Canary. (After Hernández OM et al., 2005. Food Chem 93:449–458. https://doi.org/10.1016/j.foodchem.2004.10.036. With permission)

relative to their geographic origin. As shown in Fig. 6.2, the first two PCs, accounting for 54.89% of the total variance, allowed them to discriminate between the Canary and non-Canary samples. The former samples were identified by the contents of variables such as K, Cu, Na, Mg, Ca and Sr (PC1), whereas the latter samples corresponded to Fe, Zn and Li (PC2). This means that Canary honey has relatively high levels of Na and K of marine aerosol origin and Mg and Sr of volcanic soil provenance. Moreover, DA and LR enabled clear classification of the honey according to interregional characterization, with prediction abilities of 100% and 96.3%, respectively.

Lachman et al. (2007) used ICP–OES and FAAS combined with EC to study honey samples originating from 24 regions of the Czech Republic. The CA data were used to identify two main groups of samples, i.e. nectar honey and honeydew honey. It was concluded that without including EC in the dataset, differentiating the honey samples studied in terms of their origin would not be possible.

Kováčik et al. (2016) investigated three types of Slovak honey samples via FAAS, AMA, HPLC–MRM, HPLC–DAD/UV/Vis, LC–PDA, LC–TQ–ESI–MS and UHPLC methods. The data obtained via PCA were processed on the basis of different metals and phenol contents. Different honey types originating from the three analyzed regions, i.e. the village Geča, the village Perín and the vicinity of the townlet Medzev, could be distinguished. Elements such as Ni, Mn and Cd were the main discriminative descriptors that differentiated forest honey from other honey types, whereas HBA, Ca and Cu differentiated rapeseed honey from black locust honey.

Atanassova et al. (2016) reported that ICP–AES and harmonized methods of IHC allowed the characterization of the elemental profiles of multifloral blossom honeys, including those of 64 pollen taxa from 2 Eastern Rhodope Mt. sites, i.e. serpentine and non-serpentine areas. PCA, with the first two PCs accounting for 60.5% of the total variance, distinguished between the non-serpentine and serpentine honeys. Elements such as Mg, Mn, Ni and P were used to differentiate the serpentine honey samples from the remaining samples.

Schellenberg et al. (2010) used TC/EA–IRMS ($\delta^{13}C$, $\delta^{2}H$, $\delta^{15}N$ and $\delta^{34}S$ in honey protein) to investigate 16 different floral types for 617 samples of honey from different regions of 12 European countries, i.e. Austria, Denmark, France, Germany, Greece, Ireland, Island, Italy, Poland, Portugal, Spain and the United Kingdom. The $\delta^{13}C$ values were influenced by climate, but the $\delta^{34}S$ values were considerably variable according to geographical location (sea spray effect) and the surface geology of the production region. CDA identified $\delta^{13}C$ and $\delta^{34}S$ as the most discriminative variables. A total of 60.2% of the honey samples were correctly classified. CDA, with DF1 accounting for 70.3% of the total variance, enabled the main differentiation between honeys from the Orkneys (Scotland), Ireland, Cornwall (Great Britain) and Lakonia (Greece) and the remaining authentic honey samples. DF2, corresponding to 13.1% of the total variance, allowed discrimination between regions with moderate climates, such as Franconia, Allgäu, Gäuboden (Germany), Marchfeld, Mühlviertel (Austria) and Jylland (Denmark), and regions with Mediterranean climates, such as Barcelona (Spain), Carpentras, Limousin (France), Chalkidiki (Greece) and Sicily (Italy). However, the main reason for the lack of effective prediction ability could be the comparable geoclimatic conditions resulting in overlapping sample regions. It was concluded that this approach is useful for identifying the origins of different floral types of honey.

6.2.1.2 Asiatic Countries

ICP–MS is a promising, powerful tool for differentiating Chinese vitex honey samples from different geographical origins. PCA reduced all of the variables to the first four PCs, accounting for 81.6% of the total variance. The results could be used to discriminate the vitex honey samples into three interregional groups. The overall correct classification and cross-validation rates of the built BP–ANN model were 100% and 95.4%, respectively. The prediction rate of the proposed BP–ANN model was estimated to be 75% for honey samples from each geographical provenance (Chen et al. 2015).

6.2.1.3 American Countries

Baroni et al. (2009) used FAAS, HPLC–RID and other analyses according to AOAC official methods to study honey samples collected from two geographical zones, i.e. north and south Cordoba Province (Argentina). However, a less satisfying effect indicated only 83% correct classification was attained when 15 variables were applied. The K–NNA classification procedure for the six selected SLD variables (glucose, pH, free acidity, free amino acids, Ca and Zn) achieved a highly satisfying assignment of 99% correct honey samples to their geographic provenance. This means that considering the completed database is unnecessary to achieve satisfactory geographical differentiation of the honey samples studied. For example, even one metal, such as Ca, allows good differentiation between southern and northern honey.

Cantarelli et al. (2011) demonstrated that processing of NAA data via multivariate analysis was a convenient approach for classifying the elemental profiles of propolis from three botanical sources according to their interspatial differentiation. The PC scatterplot with the first two PCs representing 65.1% of the total variance did not show clear differentiation between the geographical origins or seasonal variations of all the analyzed samples. However, it was possible to obtain a clear differentiation of honey samples corresponding to the three production zones of the central region of Argentina via S–LDA (Fig. 6.3). Eight trace elements were selected to explain the classification of propolis according to its interspatial origin (Fig. 6.4). LDA yielded satisfactory results because 97.9% and 95.8% of the original grouped cases and cross-validated grouped cases were correctly classified, respectively. A somewhat better 98% prediction ability was achieved when the kNN method was used for the reduced space formed by the previously selected eight stepwise variables.

Pellerano et al. (2012) described the use of INAA coupled with multivariate chemometrics to classify multifloral Argentinean honeys according to their regional origin. PCA, with the first two PCs explaining 58.8% of the total variance, was not sufficient to distinguish samples with respect to their geographical categorization. On the basis of the LDA results, a clear differentiation of honey samples was

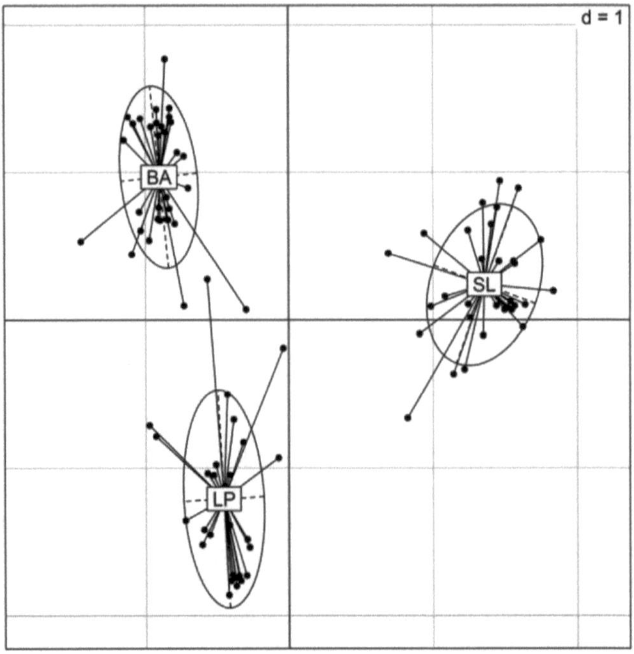

Fig. 6.3 Projections of 96 raw propolis samples analyzed according to different provenance regions in the space formed by the two discriminant functions after S–LDA. (After Cantarelli MA et al., 2011. LWT-Food Sci Technol 44:256–260. https://doi.org/10.1016/j.lwt.2010.06.031. With permission)

Fig. 6.4 Correlation circle between the selected variables (by S–LDA) and the discriminant functions. (After Cantarelli MA et al., 2011. LWT-Food Sci Technol 44:256–260. https://doi.org/10.1016/j.lwt.2010.06.031. With permission)

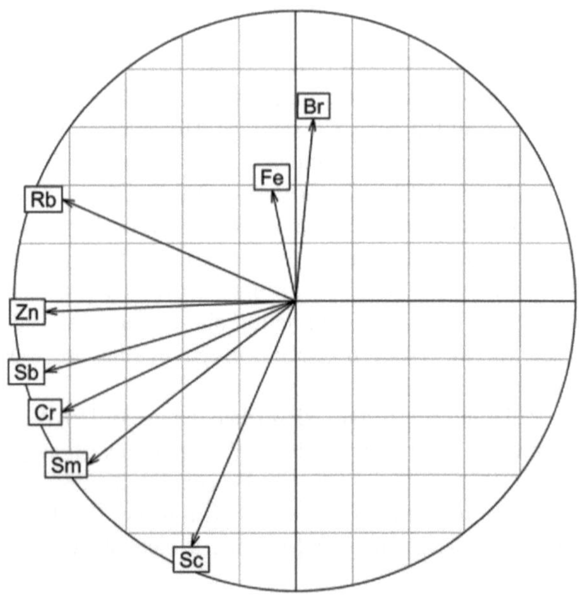

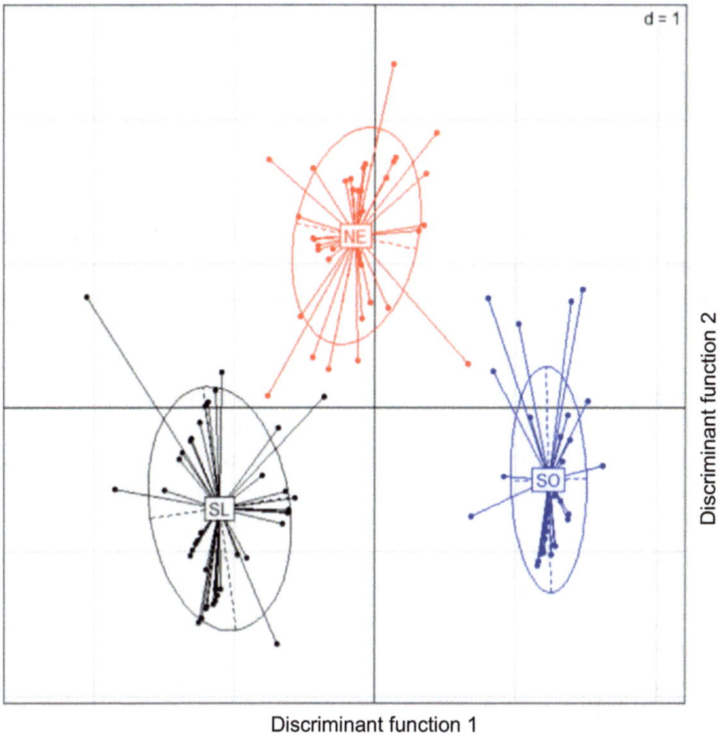

Fig. 6.5 Plot of the first two principal component scores for three geographic provenance groups. NE, northeast San Luis—SL, San Luis—SO, south San Luis and La Pampa. (After Pellerano RG et al., 2012. Food Chem 134:578–582. https://doi.org/10.1016/j.foodchem.2012.02.125. With permission)

obtained (Fig. 6.5), with 97.5% of original grouped samples and 95.9% of cross-validated grouped samples correctly classified.

Characterization of multifloral and five types of unifloral honey on the basis of ICP–MSDRC–e and other analyses was performed by Conti et al. (2014). PCA, which explained 52.03% of the total variability in the first two factors, was shown to be quite a discriminative chemometric tool. HCA also enabled the differentiation of samples belonging to three clusters related to honeys from three Argentinian macroregions, i.e. Córdoba and Tucumán Provinces (with higher levels of Zn and Fe), Entre Ríos and Santiago honeys (with higher levels of K, Mg, Mn, Cu and Ni and higher pH) and Buenos Aires Province honeys (with higher levels of Pb). DA was used for the PCA data. In summary, multivariate analysis combining CA, DA and PCA showed the appropriateness of these methods as useful tools for honey classification.

Fechner et al. (2020) used ICP–MS, GF–AAS, AES and other methods to characterize ten species of multifloral honey from four provinces. PCA, with the first two PCs explaining 49.2% of the total variance, revealed that many overlapping samples originated from different provinces. However, the LDA model, which is based on multi-element parameters, allows the three groups corresponding to the sample origin to be distinguished; however, when multi-element data were combined with physicochemical data, the interspatial differentiation was considerably improved. Owing to this dual approach, the prediction capability of the model enabled the correct classification of 94% of the samples from the three provinces in the northeast region of Argentina.

Santos et al. (2008) employed a combination of FAAS and other analyses according to AOAC official methods to study monofloral and multifloral honeys from southwestern Bahia State (Brazil). Through CA, it was possible to distinguish five groups corresponding to three regional climates. PCA, with the first three PCs achieving 71.26% of the total variance, provided comparable data to those obtained with CA, which means that both of these tools distinguish honey produced in three regional climates, i.e. semiarid, Atlantic and transitional forest zones.

Rizelio et al. (2012) investigated the use of MEA to characterize honey samples from various regions across the state of Santa Catarina (Brazil). The contents of elements such as K, Na, Ca, Mg and Mn were processed by a chemometric tool. A PCA plot (the first two PCs explained 85.15% of the total data variance) visualized the differences in honey samples according to their geographical provenance. Honey samples originating from coastal regions of the island had the highest levels of elements, probably due to the effects of salinity from marine aerosols.

Batista et al. (2012) found differences in three types of honey samples originating from various cities in Brazil using ICP–MS with three machine learning tools, i.e. SVM, MLP and RF. The chemometric classification was useful for identifying the geographical origin of the analyzed honey. The selection tools allowed a reduction from 42 element concentrations to only five, i.e. Pb, Tl, Pt, Ho and Er, to be considered in the analysis. In this case, good classification achieved accuracies of 82.8%, 66.3% and 79.3% for the MLP, SVM and RF classifiers, respectively.

6.2.1.4 Countries Across Multiple Continents

Karabagias et al. (2017a) integrated ICP–OES, HS–SPME–GC/MS and XDL–ICP–AES with analysis according to AOAC official methods to characterize *Citrus* spp. honey samples from various Mediterranean countries, i.e. Greece, Spain, Egypt and Morocco. LDA revealed that the physicochemical parameters provided a higher geographical classification rate (97.3%) than did the volatile compound data (86.5%) or mineral element data (83.8%). Karabagias et al. (2017b) reported that for thyme honey samples from four Mediterranean countries, a combination of ICP–OES with analyses according to the AOAC official method allowed classification of the samples. Geographical differentiation was achieved using LDA with respect to the cross-validation classification rate for exclusively 10 physicochemical and colour parameters (91.2%), using 12 major, minor and trace elements (88.2%) and their combination (76.5%).

According to Pasquini et al. (2014), commercial and artisanal honey samples, including six botanical species from three geographical classes, i.e. Eastern European countries, Italy and Equatorial countries (e.g. Cameroon, Kenya and Brazil), were analyzed. The contents of cationic and anionic species, as well as other compounds, were determined by IC–HPLC and spectrophotometric methods. PCA, with the first two PCs explaining 69.70% of the total variance, distinguished two clearly defined groups attributed to Italy and Eastern European countries, and Equatorial and other countries. It was also concluded that LDA provided better differentiation than did QDA, with one exception. However, CART was the most predictive tool for the geographical discrimination of the samples. This model achieved specificities of 83%, 64% and 100% and sensitivities of 56%, 75% and 100% for the Eastern European countries class, the Italian class and the Equatorial and other countries class, respectively. Moreover, the average specificity and sensitivity were 82% and 77%, respectively, and the overall specificity and sensitivity were 80% and 67%, respectively. Parameters such as Mg^{2+}, Br^- and gluconic acid were the most important for the differentiation of the various geographical classes.

6.2.2 Interspecies Differentiation

Terrab et al. (2003a) applied ICP–AES and other analyses according to AOAC official methods for the study of multifloral honey and five types of unifloral honey from northwestern Morocco. PCA, with the first PCs explaining 67.16% of the total variance, revealed that only honeydew honeys were effectively differentiated from the remaining honeys. LDA achieved better results, indicating that Eucalyptus with 92% correct classification and honeydew (100%) were clearly differentiated. The MLP result was essentially the same as that of the LDA. By applying LDA and MLP, reliable discrimination between two groups of samples, e.g. *Eucalyptus* honey and honeydew, was achieved.

According to Terrab et al. (2003c), *Eucalyptus* honey derived mainly from *E. camaldulensis* Dehnh was investigated for its elemental profile as well as other physicochemical and colour parameters by ICP–AES, GC–MS and other methods, mainly according to the AOAC methods. PCA indicated that the cumulative variance was 44.28% for the first two PCs. However, no clear distinction between the five unifloral honey types was obtained. As shown in Fig. 6.6, SDA allows satisfying differentiation of all honey types except for the *Lythrum* samples, which overlap with the Umbelliferae samples. It has been concluded that ca. 97% of the samples were correctly classified using both classifiers, i.e. SDA and MLP, and satisfactory results (100% correct) were obtained for *Eucalyptus* sp., *Citrus*, sp. Umbelliferae and honeydew honeys.

Pisani et al. (2008) used a combination of ICP–OES, ICP–MS and chemometric analysis to characterize the elemental profiles of honey samples from four botanical origins in several zones of Siena County (Italy). PCA (with the first two PCs explaining 53% of the total variance) enabled clear differentiation of the honeydew and

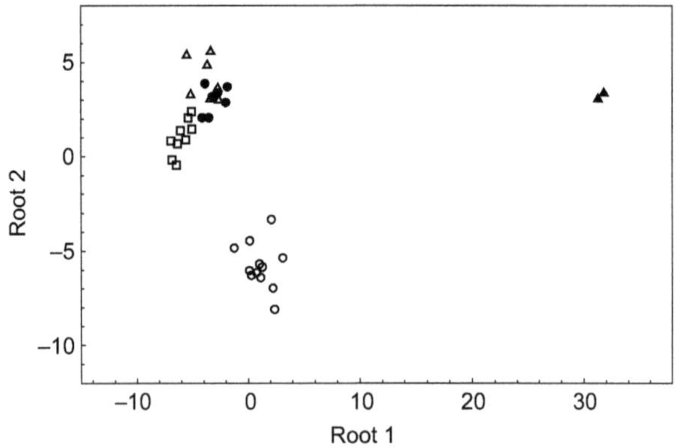

Fig. 6.6 Scatter plot for the unifloral honey samples generated via the two first discriminant functions. ○, *Eucalyptus;* □, *Citrus;* Δ, *Lythrum;* •, Umbelliferae; ▲, Honeydew. (After Terrab A et al., 2003c. Eur Food Res Techn 218:88–95. https://doi.org/10.1007/s00217-003-0797-x. With permission)

nectar samples, including multifloral, sulla and clover samples. Notably, these results confirmed the health safety of honey, especially in terms of the contents of toxic trace elements, such as As, Cd, Pb and Sb.

Chudzinska and Baralkiewicz (2010) estimated the authenticity of three types of Polish honey using ICP–MS combined with chemometrics to differentiate groups corresponding to the botanical origin of honey. The PCA data (with the first four PCs explaining 77.19% of the total variance) allowed discrimination of the honeys according to their botanical origin, i.e. for honeydew and nectar honey. The most discriminative variables were elements such as K, Al, Ni, Cd and Zn, as well as Na, Ba and Pb, which were important for the identification of honeydew and rape honey samples, respectively. Comparable results were obtained by CA. However, according to the authors, assigning characteristic mineral elements with respect to buckwheat honey was difficult.

Kędzierska-Matysek et al. (2021) reported that FS–FAAS, antioxidant activity and other parameters, including colour combined with chemometric processing analysis, were useful approaches for evaluating the chemical profiles of five honey types. On the basis of antioxidant activity and selected macro- and microelement contents, PCA, with the first three PCs accounting for 73.46% of the total variance, allowed the differentiation of buckwheat honeys from the other blossom types. It was concluded that buckwheat honey has the strongest antioxidant activity, which may be associated with the highest levels of Cu, Mn and Zn. These parameters, which are considerably influenced by the botanical origin of honey, can be useful for distinguishing buckwheat honey from other blossom types of honey.

Bilandžić et al. (2017) used ICP–MS data from different honey types from the southern Mediterranean regions of Croatia to determine the role of botanical origin in the elemental composition of monofloral honey. Honeydew honey was distinguished as having the highest content of almost all the elements analyzed, whereas mandarin orange-blossom honey had the lowest levels, except for Be, Se (sage honey) and Th (bearberry honey). Moreover, multifloral honeys presented the greatest concentrations of Ca, Cr, Mo and Se, whereas common heather and bearberry honeys presented the highest levels of Mg and Na and of Ba, Fe and Pb, respectively. Finally, the highest concentration of Ag was observed in sage. The data indicated that the botanical characteristics of honey predominantly influence the qualitative and quantitative differences in its elemental profile.

Research carried out by Rodríguez-Flores et al. (2019) involved the combination of AAS, AES, HPAEC–PAD, palynological analysis and other methods to classify Galician honey samples from Spain. The first two PCs (explaining 46.2% of the total data variability) allowed the differentiation of chestnut honeys from honeydew honeys. LDA clearly confirmed the discriminative PCA data, correctly classifying 97.6% of the honey samples.

The characterization of Moroccan spurge (*Euphorbia*) honeys via ICP–MS, UV–vis spectrophotometry, pollen analysis and other methods according to the AOAC methods was performed by Bettar et al. (2019). The first two PCs, accounting for 55.92% of the total data variance, demonstrated that the *E. officinarum* subsp. *echinus* honey samples were clearly differentiated from the *E. regis-jubae* honey samples. According to SDA, the two unifloral honey types were reliably classified with correctness in 96.29% of the samples. Among all the variables analyzed, Na, Mg, L^*, C^*_{ab} and h_{ab} were the dominant parameters. Thus, using elemental, physicochemical and colorimetric parameters was appropriate to achieve effective discrimination between the two unifloral honey classes considered.

According to González-Miret et al. (2005), the combination of ICP–OES, the spectroradiometric method and other methods allowed the characterization of the four botanical types of dark honey and the five types of light and brown honey. Owing to the application of LDA for the elemental content, it was possible to achieve a model that differentiated the honeys by their lightness. The prediction ability of the constructed model, quantified with the test set method, was 85%. The colour of dark honeys is strongly correlated with the contents of As, Cd, Fe, S, Pb and Ca, whereas light and brown honeys are highly correlated with the contents of Al and Mg. The model exclusively included Mg, Fe, Mn, Cu, Ca and As since the variables were required to explain 95% of the variance between both groups of samples, i.e. Spanish dark and light honeys.

Grembecka and Szefer (2013) employed FAAS and a spectrophotometric method combined with chemometric analysis (ANOVA, CA, FA) to characterize the elemental distributions of Polish and Italian honeys (12 honey types and honeys with natural additives and other bee products). FA (with the first two factors accounting for up to 51.5% of the total variance) was found to be useful for the differentiation of honeys with respect to their botanical origin (Fig. 6.7). Dark-coloured honeys, such as honeydew, buckwheat and heather, were clearly differentiated from light-coloured honeys, i.e. acacia, lime, rape and multifloral. Elements such as K, Cu, Mn, P and Mg were discriminative descriptors for the dark-colour honeys, whereas Zn, Fe, Ca and Na were indicators of light-colour honeys. To evaluate the data structure concerning natural and artificial honey, they carried out FA. The first two factors, which explained 49.1% of the total variance, allowed the artificial honey to be differentiated from natural honey, including syrup-feed honey and honey with natural additives (Fig. 6.8). Calcium and Na were identified as the best descriptors for

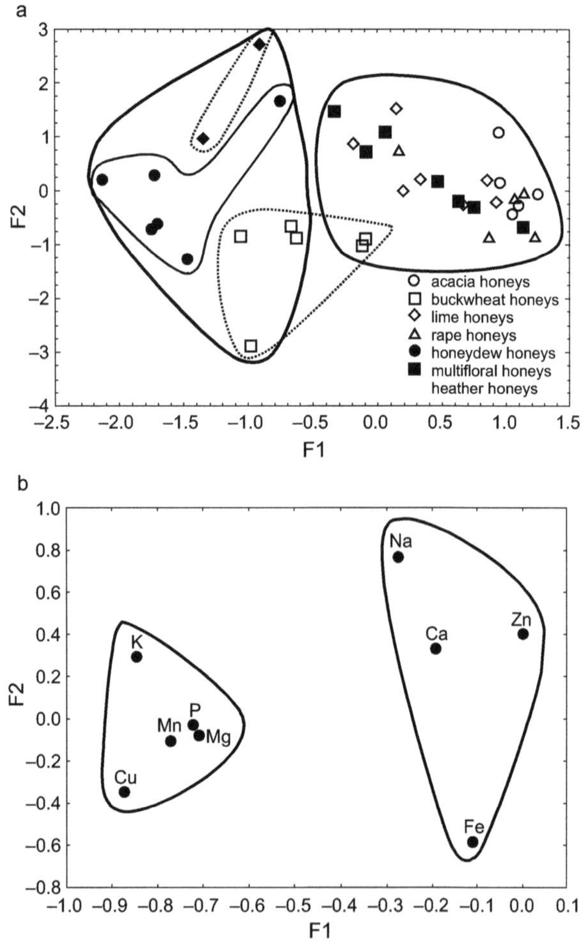

Fig. 6.7 (**a**) Scatterplot of the object scores of the two discriminant functions of exclusively natural honeys. (**b**) Scatterplot of loadings for nine elements in natural honey samples. (After Grembecka M and Szefer P, 2013. Environ Monit Assess 185:4033–4047. https://link.springer.com/article/10.1007/s10661-012-2847-y. With permission)

the determination of artificial honeys. A hierarchical dendrogram of CA confirmed the obtained PCA data.

The analytical measurements performed by Alqarni et al. (2014) with the use of AAS, spectrophotometers and other procedures according to AOAC methods enabled the characterization of 23 varieties of dark and light honeys from Saudi Arabia and other countries, i.e. Egypt, New Zealand, Germany, Yemen, Australia and Malaysia. LSD differentiated dark and lighter honeys. In particular, acacia honeys were distinguished by higher elemental contents and EC values compared with the other honeys. The heavy metal contents showed that the studied honeys were safe for human consumption.

Conti et al. (2007) investigated three types of Marche honeys collected in central Italy with respect to their elemental composition. The first two PCs explained 77.1% of the total variance, with the dominant loading contributions from Ca, Na and Mn,

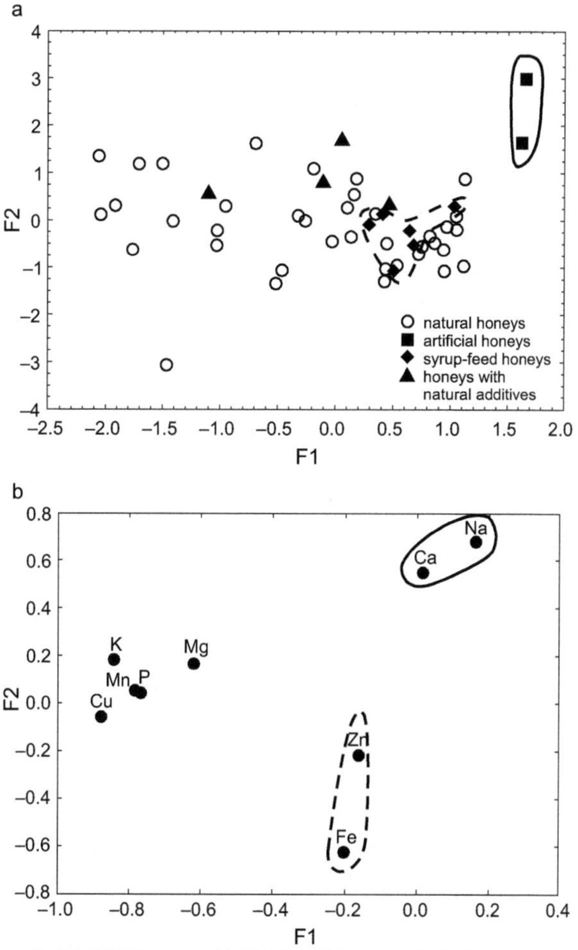

Fig. 6.8 (**a**) Scatterplot of the object scores of the two discriminant functions of natural, syrup-feed, artificial, and natural additive honeys. (**b**) Scatterplot of loadings for nine elements in the analyzed samples. (After Grembecka M and Szefer P, 2013. Environ Monit Assess 185:4033–4047. https://link.springer.com/article/10.1007/s10661-012-2847-y. With permission)

especially those from Mg and K (PC1) and pH (PC2). Three honey groups of different botanical origins were differentiated, although some multifloral and acacia honey samples overlapped. LDA revealed better differentiation for all the types of honeys since the acacia and honeydew honey samples were correctly classified according to their a priori established honey types (up to 100%), whereas the multifloral honeys showed slightly lower percentages of agreement (97.7%).

Czipa et al. (2015) explored eight types of monofloral and multifloral honeys from various geographical regions of Hungary with respect to the content of numerous elements quantified by ICP–OES and ICP–MS. In the case of LDA, both DFs were significant since they clearly categorized the three types of honey, i.e. acacia, rape and sunflower. The samples correctly classified 89.5% of the cross-validated grouped cases. Although LDA showed that three honey species groups were distinguished, no differentiation of samples was observed with respect to their geographical origin.

Nayik and Nanda (2016) used FAAS, HPLC–DAD and different spectrophotometric methods to study two types of Indian and multifloral honeys corresponding to their MEC and antioxidant properties. The first two PCs, which explained 76.13% of the total data variance, enabled differentiation of the three botanical types of honey. Furthermore, the LDA model correctly classified (100%) all four honey types. The application of both multivariate tools, with contributions from discriminative variables such as macroelements and antioxidant and sugar parameters, allowed useful differentiation of the botanical types of honey, i.e. acacia, pine honeydew and multifloral honeys. Moreover, Nayik et al. (2018) characterized three types of honey, i.e. acacia, pine honeydew and multifloral honey, from the high-altitude Kashmir valley of India with respect to macroelement profiles and antioxidant and sugar parameters. PCA, with the first two PCs accounting for 81.24% of the total variance, allowed them to clearly discriminate three types of honey relative to their botanical classification. Among the variables analyzed, most of the mineral elements and antioxidant parameters were discriminative descriptors for all the pine honeydew honey samples, whereas the sugar parameters and moisture content were linked to multifloral honey samples. The LDA model indicated that 100% of the samples were correctly classified into each honey type.

According to Vanhanen et al. (2011), ICP–OES and other routine analyses were useful for classifying ten types of locally produced monofloral honey from New Zealand. The most abundant mineral elements were K, P and Ca, the levels of which ranged from 34.8 to 3640, from 29.5 to 255 and from 7.21 to 94.3 mg/kg, respectively. Potassium constituted the largest contribution (up to 73%) to the total content of all the elements analyzed. A large range of mean total element contents were observed for honeydew honey, with the highest level (4060 mg/kg), and for viper bugloss honey, with the lowest level (126 mg/kg). Moreover, the heavy metal contents, i.e. Cd, Pb and Zn, of the monofloral honey types were very low.

Eight botanical types of honey originating from Croatia were studied by Vasić et al. (2020). Application of analytical methods such as ICP–OES, ICP–MS, IRMS and HPAEC–PAD combined with chemometrics allowed them to classify the honeydew honeys. PCA, with the first two PCs explaining 49.45% of the variance, enabled the differentiation of three botanical types of honey. As shown in Fig. 6.9, the elemental distribution pattern allows a correct PLS–DA prediction of the botanical origin of silver fir, Hungarian oak and Evergreen oak and offers an effective tool for determining the authenticity of samples. Elements such as Ba, Ca, Mg, Al, Mn, Co, Ni and Se were the most discriminative descriptors. Moreover, the $\delta^{13}C$ values for proteins could be effective indicators of the botanical origin of honey samples.

Louppis et al. (2017) differentiated 4 species of unifloral honeys from 14 regions in Greece using MEC in combination with a chemometric approach. On the basis of the 14 mineral elements and 10 physicochemical parameters (Fig. 6.10a, b), LDA was found to be an effective tool for achieving a correct prediction rate of 96.1% for determining the botanical origin (fir, pine, orange and thyme) of unifloral honeys. In

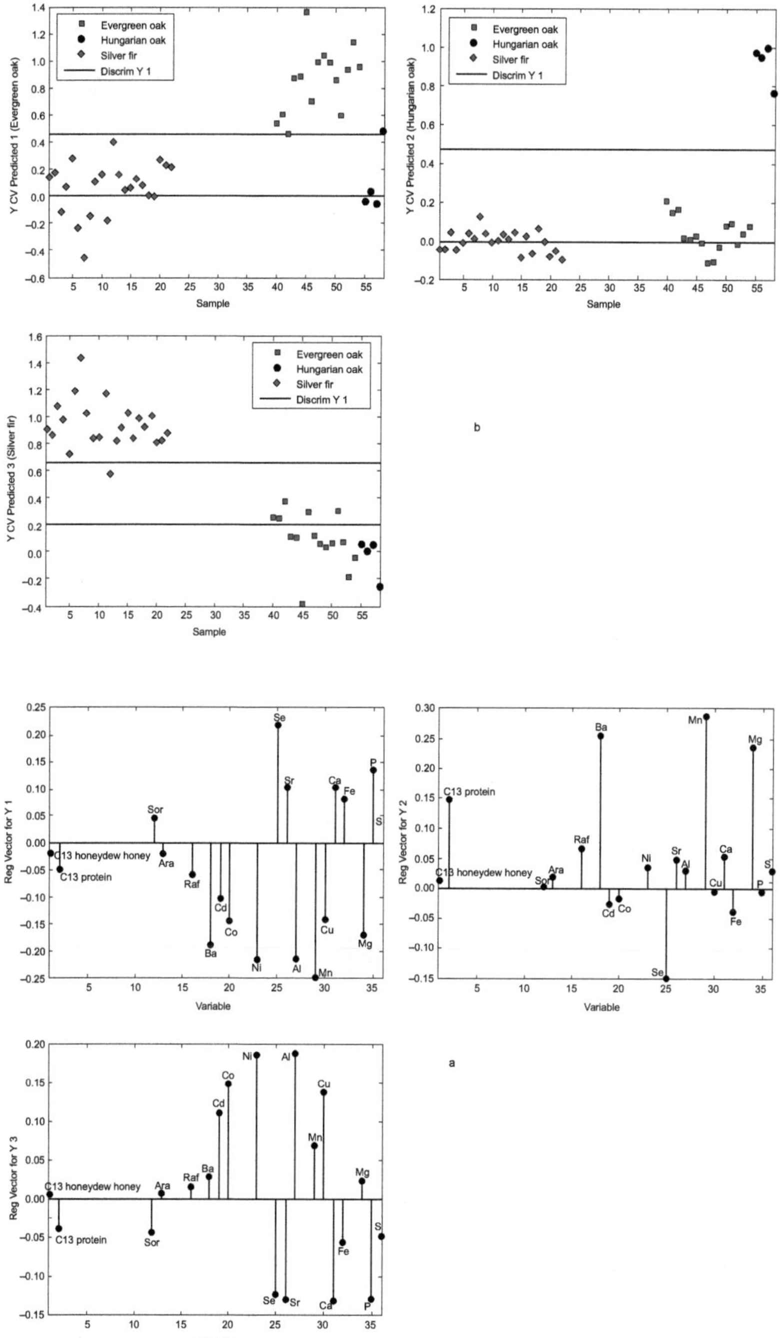

Fig. 6.9 PLS–DA model for three botanical origins of honeydew honey, (**a**) score plots of the data, (**b**) plots of the coefficients of the parameters in the models. (After Vasić V et al., 2020. Food Chem 305:125457. https://doi.org/10.1016/j.foodchem.2019.125457. With permission)

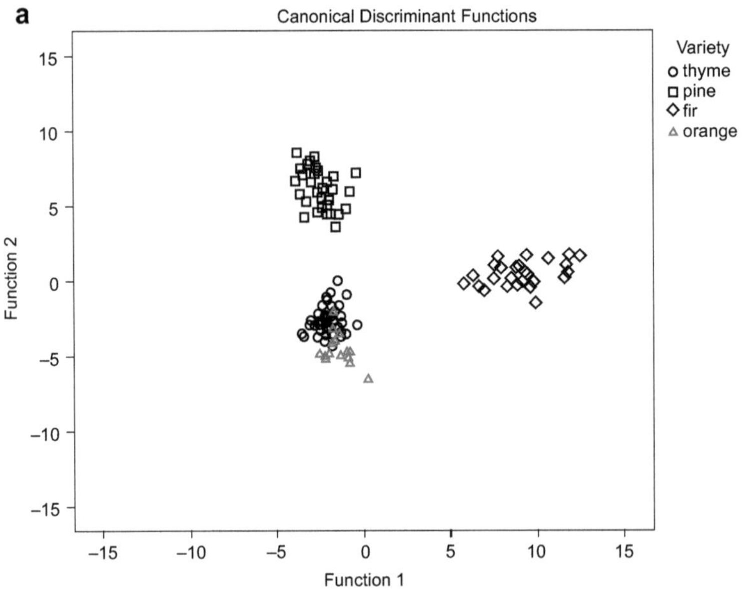

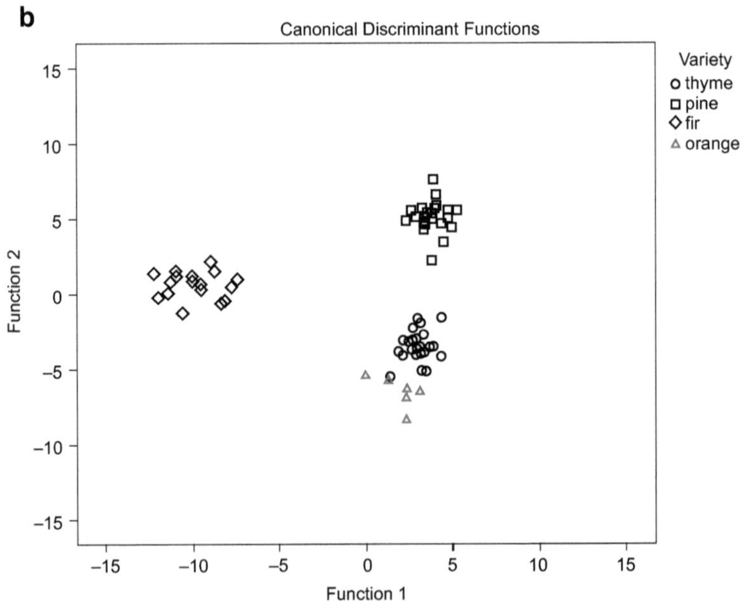

Fig. 6.10 (**a**) Botanical discrimination of Greek unifloral honeys on the basis of the combination of 14 minerals and 10 physicochemical parameters using a validated chemometric approach (application to 70% of the total honey samples); (**b**) botanical discrimination of Greek unifloral honeys on the basis of the combination of 14 minerals and 10 physicochemical parameters using a validated chemometric approach (application to 30% of the total honey samples). (After Louppis AP et al., 2017. Microchem J 135:180–189. https://doi.org/10.1016/j.microc.2017.09.004. With permission)

contrast, single-element analysis provided correct prediction rates of 92.9% (related to physicochemical parameters) and 85.7% (related to element content).

According to Fernandez-Torres et al. (2005), ICP–AES was applied for the study of four species of honey samples from different regions of Spain. PCA, with the first two PCs accounting for 71.59% of the total variance, as well as CA, allowed the discrimination of honey samples between their four botanical types. LDA data showed more reliable discrimination of four clusters corresponding to the four botanical origins of honey, i.e. eucalyptus, heather, orange-blossom and rosemary. The recognition and prediction abilities were 100% and 97%, respectively.

Gašić et al. (2015) reported that four honey species from the northeastern Adriatic region of Croatia were assayed using ICP–OES, UHPLC–DAD–MS/MS, UHPLC–MS/MS Orbitrap analysis, UHPLC–LTQ OrbiTrap MS, HPAEC–PAD and a UHPLC–DAD–TSQ Quantum Access Max triple-quadrupole spectrometer. The contents of macro-, micro- and trace elements, polyphenolics, TPC, six sugars and sugar alcohols, including RSA data, were established. PCA (with the first two PCs explaining 33.18% of the total variance) differentiated sage honey from the remaining unifloral honeys, i.e. thyme, mint and winter savoury honeys. Moreover, among all the studied unifloral honeys of the Lamiaceae species, relatively high levels of B, K, turanose and kaempferol could serve as authentication markers of unifloral common-stage honeys.

Chen et al. (2014) studied four species of Chinese honey from different producing regions using the ICP–MS method. As shown in Fig. 6.11, PCA, with the first two PCs explaining 82.38% of the total variance, clearly differentiated linden honey from the other samples. This distribution of score patterns suggested that the remaining three types of honeys did not depend on the major variables in PC1 and PC2. Therefore, to clearly differentiate the honey samples with elemental profiling, they constructed PLS–DA and BP–ANN models on the basis of the ten elements determined to predict the botanical origins of the honey samples. The recognition and prediction abilities of the PLS–DA model were 90.9% and 88.4%, respectively. For the BP–ANN model, the values were 100% and 92.6%, respectively. This means that the performance of BP–ANN was more acceptable than that of PLS–DA. With respect to the validation of the developed BP–ANN model, the overall correct prediction accuracy was 97.6% for the four groups, i.e. linden, vitex, rape and acacia honey. This confirmed the applicability of such options when the proposed BP–ANN model was used.

Nečemer et al. (2009) classified seven major types of Slovenian honey from three production years, namely, 2004, 2005 and 2006, via TXRF coupled with multivariate processing analysis. A clear differentiation of most honey types was achieved. For example, well-differentiated samples of the acacia, floral, chestnut and forest/spruce groups from 2005 were attained by PCA, with the first two PCs explaining 84.6% of the total variance. Among all the determined elements, only Cl, K, Mn and Rb were the most discriminative variables. The RDA model achieved 90.4% for

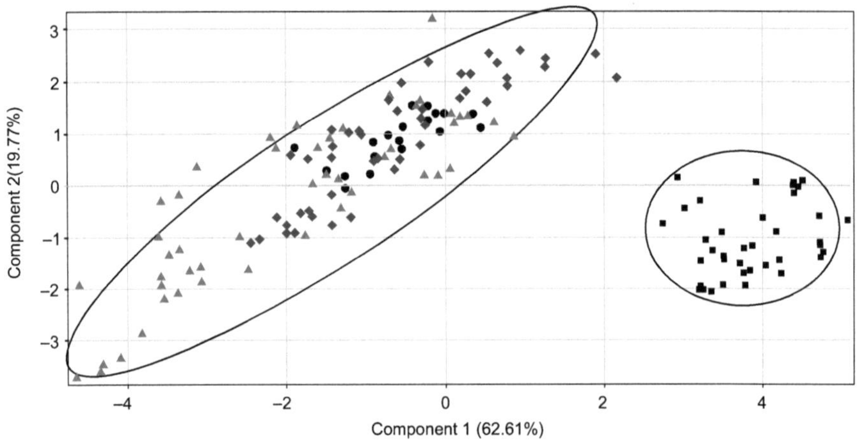

Fig. 6.11 Two first-component scores of honeys from different botanical origins: linden (■), vitex (▲), rape (●) and acacia (◆) honeys. (After Chen H et al., 2014. *J Agric Food Chem* 62:2443–2448. https://doi.org/10.1021/jf405045q. With permission)

both the classification ability (with and without CV) and prediction ability. It was concluded that TXRF, as an inexpensive, simple and fast technique integrated with PCA and RDA chemometric tools, is a fast screening method for the quantification of the botanical origin of honey.

In Oroian et al. (2015), ICP–MS was an effective tool for classifying four species of honey from the northeast region of Romania. The provided reasonable data were processed by chemometric analysis. The honey samples were differentiated into four groups by the first two PCs, which explained 74% of the total variance, with a significant contribution of the most discriminative elements, such as K, Mg and Ca. The SDA model with the cross-validation method allowed them to estimate that 92.7% of the honey samples were correctly classified. The combined multi-elemental and multivariate methodology used is a suitable approach for predicting the botanical origin of acacia, sunflower, tilia and honeydew honeys.

Wu et al. (2015) integrated TC/EA–IRMS with ICP–MS, which provided a more effective and accurate analytical approach for classifying four species of Chinese honeys relative to their botanical origins. A two-dimensional plot of δ^2H and δ^{18}O isotope ratios allowed the classification of four botanical origins of Chinese honeys, i.e. rape, acacia, vitex and jujube (Fig. 6.12). Furthermore, a tri-dimensional scatter plot for the first three PCs, which explained 54.7% of the total variance, explained the characteristics of the four botanical origins of the Chinese honeys. Using a combination of both suitable stable isotopic and elemental compositions, the LDA model resulted in the most satisfactory rates for all cases (100%) and for cross-validation (96.5%).

Voica et al. (2020) combined ICP–Q–MS with chemometric analysis to investigate four varieties of honey (linden, acacia, rape and sunflower) originating from

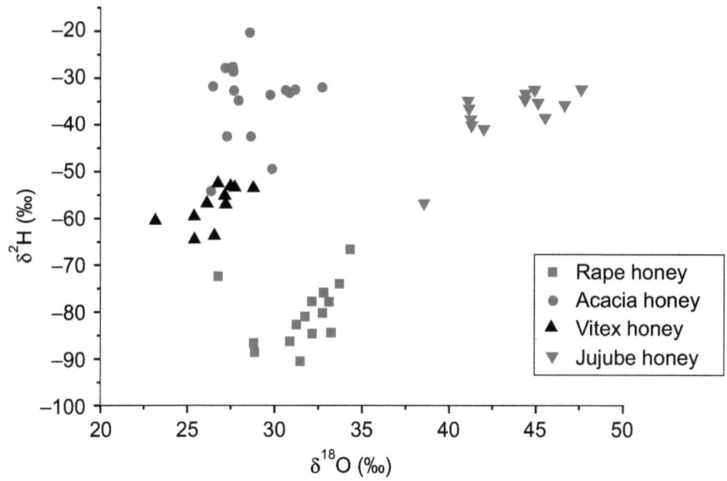

Fig. 6.12 Plot of the hydrogen isotope ratio (δ^2H) versus the oxygen isotope ratio ($\delta^{18}O$) for Chinese honey samples of four floral origins. (After Wu Z et al., 2015. *J Agric Food Chem* 63:5388–5394. https://doi.org/10.1021/acs.jafc.5b01576. With permission)

Romania. The first two PCs (48.06%) allowed the honey samples to be differentiated into four distinct botanical groups according to their elemental pattern profiles. Among all the variables, K, As and Mn are among the most discriminative descriptors for predicting the botanical origin of honey.

Louppis et al. (2019) evaluated the usefulness of ICP–OES for characterizing the elemental profile of beekeepers' honey samples from various producing regions in Greece. On the basis of the first two PCs (71.98%) with EV >1, it was possible to differentiate honey samples according to their botanical origin. Furthermore, LDA and SDA allowed the discrimination of asfaka, cotton, fir, flower, forest flower and orange blossom honeys, with correct classification rates of 75% and 100%, respectively.

Nozal Nalda et al. (2005) used ICP–AES, FAES and spectrophotometric methods to characterize seven species of honey from Spain. PCA, with the first two PCs (46.5%), made it possible to differentiate ling and heather honeys from all other honeys. By employing a DA scheme of eight steps, they were able to correctly classify up to 90% of all honeys following the cross-validation procedure.

6.2.3 Interspecies and Interregional Differentiation

Kropf et al. (2010a) reported that a combination of TXRF with IRMS was used for the investigation of three honey types originating from four Slovenian locations (Alpine, Dinaric, Pannonnian and Mediterranean regions). On the basis of the LDA data, all the lime honey samples were correctly classified (100%) according to their three-region origins, whereas black locust and chestnut honeys were correctly assigned in 98.2% and 94.6% of the samples, respectively. According to Kropf et al. (2010b), $\delta^{13}C_{honey}$, $\delta^{13}C_{protein}$ and $\delta^{15}N$ were determined in 271 honey samples of 7 types (black locust, multifloral, lime, chestnut, forest, spruce and fir honeys) originating from 4 natural geographical regions of Slovenia. A combination of IRMS (SCIRA and ISCIRA) with chemometric processing was used. Owing to the LDA classification without considering the origin of the honey botanical origin, 55.0% of the samples were correctly classified. When the honey type was considered, the classification rate achieved more satisfying results, e.g. 84.6% and 88.9% for black locust and lime honeys, respectively. This means that the approach used is useful for identifying the geographical origin of a honey, although botanical origin generally did not have any major influence on the honey isotope profiles. This was possible because honey in Slovenia is mostly a mixture of nectar and/or honeydew from different C3 plants.

Chudzinska and Baralkiewicz (2011) investigated three types of honey samples originating from different areas in Poland via ICP–MS with multivariate analysis. In contrast to the clear distinction between various types of honey obtained via the CA and PCA techniques, unsatisfactory differentiation was achieved between honey samples with respect to their geographic provenance. Moreover, owing to the use of LDA, a clear differentiation of the three honey types was achieved. This model achieved an efficiency of 100%. Importantly, the C&RT model discriminated honey samples from southern, northern, western and northwestern Poland, although none of the samples from two major regions, i.e. Dolnośląskie and Opolskie, as well as Mazowieckie, Łódzkie and Lubelskie, respectively, were identified correctly. However, the LDA model was much more effective at determining the geographic origin of honey, especially its botanical origin.

According to Madejczyk and Baralkiewicz (2008), ICP–MS, FAAS and AES data integrated with CA processing obtained for two types of Polish honey were helpful in the classification of samples into two types, i.e. honeydew and rape honeys. The latter contained lower levels of Mn than in honeydew honeys, which contained much greater concentrations of Al, Cu, K, Fe and Ni. It was also possible to distinguish the particular subgroups of variables associated with the geographical origin of the studied honey samples.

Yücel and Sultanoğlu (2013) reported that ICP–OES data transformed using chemometrics were useful for investigating eight honey types from different provinces in Turkey and determining the impact of their geographical and botanical origins on the contents of mineral components. The distribution pattern of scores along the first two PCs (40.7%) showed that the samples were clustered into four groups and assigned to different geographic origins of honey (Fig. 6.13). However, eight clusters of CA corresponded to each botanical origin, i.e. citrus, blossom, gossypium, calluna-capparis-blossom, citrus-calluna-blossom, gossypium-blossom, pinus-citrus and petroselinum-pinus honeys. CA and PCA were found to be useful techniques for the discrimination of honey sample authenticity using the profile of element contents, highlighting the relationship between the elemental distribution pattern and honey type.

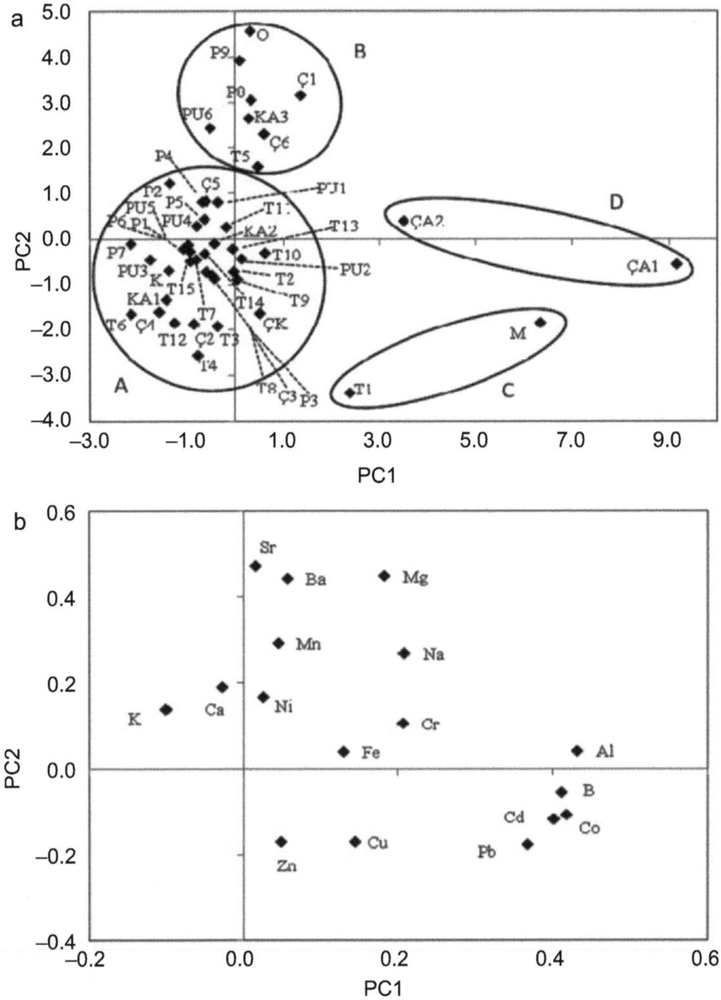

Fig. 6.13 Principal component analysis (PC2 vs. PC1). (**a**) Distribution of samples on the score plot, (**b**) distribution of variables on the loading plot. (After Yücel Y and Sultanoğlu P, 2013. Food Chem 140:231–237. https://doi.org/10.1016/j.foodchem.2013.02.046. With permission)

Di Bella et al. (2015) classified geographical differentiation of Italian honey using ICP–OES and ICP–MS with a chemometric approach. PCA (PC1 and PC2 cumulatively explained 74.65% of the total variance) differentiated Sicilian and Calabrian honey. CDA also allowed the differentiation of these two types of honeys of the same botanical origin, i.e. chestnut and lemon blossom. CDA showed that 100% of the total samples were correctly classified, which was also supported by a CV procedure.

Bontempo et al. (2017) studied Italian honey samples from 7 botanical sources using a combination of EA–IRMS, SNIF–NMR and ICP–OES. PC1 and PC2,

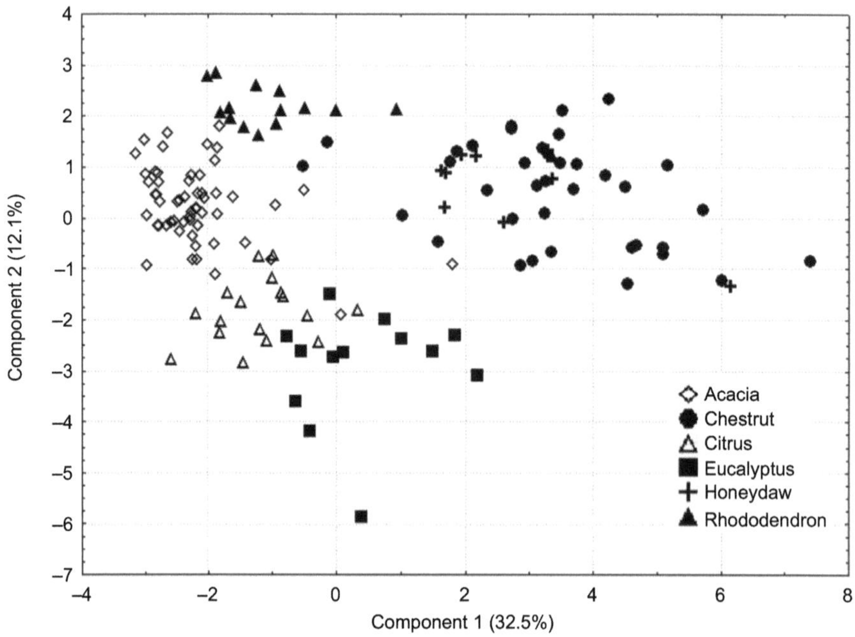

Fig. 6.14 Plot of the first two principal components obtained from the principal component analysis of Italian honey stable isotopes and elemental data according to the botanical species. (After Bontempo L et al., 2017. Measurement 98:283–289. https://doi.org/10.1016/j.measurement.2015.11.022. With permission)

which explained 44.6% of the total variance, demonstrated that different botanical types of honey could be discriminated, although not completely (Fig. 6.14). In particular, the rhododendron honey samples were differentiated from other varieties of honey, with an evident tendency for citrus, acacia and eucalyptus in the discrimination. Moreover, honeydew and chestnut honey tended to form different groups. Moreover, the plot of the first two PCs (39.8%) revealed the total differentiation between chestnut honey from northern macroareas and southern macroareas, whereas honey from the central macroarea was not separable from the abovementioned two sample groupings (Fig. 6.15). This means that PCA enabled the differentiation of the studied honey samples according to their geographical and botanical origin, although not completely. In addition, ANOVA revealed that the species significantly influenced all the stable isotopic ratios as well as most of the analyzed elements, except Pb, Ni and Cr. The latter heavy metals are closely associated with honey pollution originating from contaminated environments and can be attributed to the use of non-food-grade beekeeping materials.

According to Sajtos et al. (2019), 11 different types of Hungarian honey and special bee products from 7 regions were analyzed by MP–AES and ICP–OES methods. CDA allowed discrimination between five honey types, i.e. acacia, forest, chestnut, linden and sunflower. The classification results were satisfactory for 100%

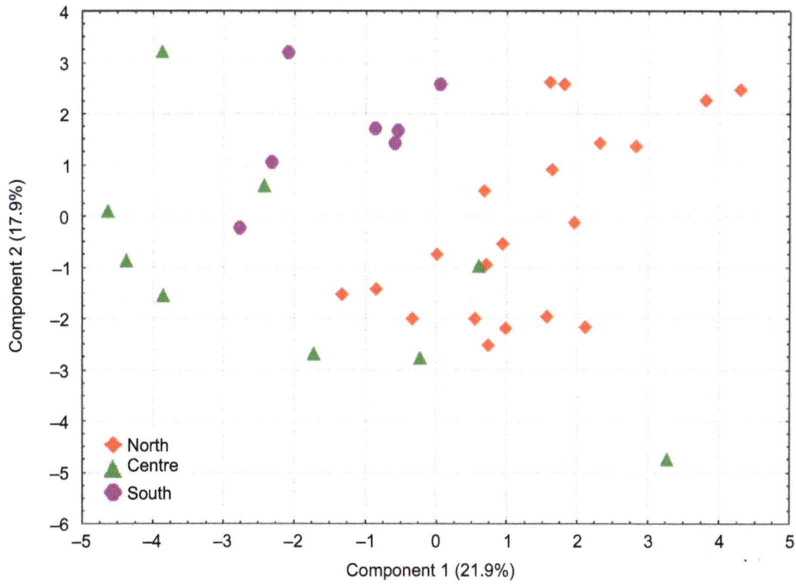

Fig. 6.15 Plot of the first two principal components obtained from the principal component analysis of Italian chestnut honey stable isotopes and elemental data according to geographical origin. (After Bontempo L et al., 2017. Measurement 98:283–289. https://doi.org/10.1016/j.measurement.2015.11.022. With permission)

of the samples, with DF1 and DF2 accounting for 67% and 16%, respectively, of the discriminating ability. The canonical correlation values were 0.973 and 0.900, respectively. For the assessment of interregional differentiation, CDA showed partial differentiation of honeys from Central Hungary and Western Transdanubia from the other regions, with discriminating abilities of DF1 and DF2 of 75% and 14%, respectively, and canonical correlation values of 0.876 and 0.611, respectively. Greater diversity was found among the honey types than among the regions, suggesting that botanical origin dominated the impact of geological factors on the elemental composition of the studied honeys.

Drivelos et al. (2021) analyzed ICP–MS data, including REEs, for eight botanical types of honey from five European and Asiatic countries. The classification results achieved by CA correctly differentiated most of the honey samples, with minor exceptions, with respect to their botanical type. Using a PNN, 85.3% of the samples were correctly classified by the network according to their geographical origin, and 73.3% were classified according to their floral characterization. A PLS model was constructed, providing a prediction accuracy of >95%.

Chua et al. (2012) combined ICP–MS and ICP–OES with multivariate analysis to study 18 honey species originating from different regions of 9 countries, i.e. Italy, Poland, Slovenia, Spain (Tenerife), Spain (non-Canary), India, Malaysia, Turkey, Syria and Egypt. Considering that honey is produced by bees under relatively uncontrolled conditions and that its composition has a highly complicated structure, the first PC, accounting for 93.1% of the total variance, is generally an appropriate tool for identifying the geographical origin of honey samples from around the world in relation to their botanical origin. It has been postulated that K and Na can be markers to differentiate honey origin. Egyptian clover and orange honeys, as well as Turkish natural honey, have higher Na levels than the remaining honey samples. The elemental profiles of tualang and gelam honeys from Malaysia are comparable to those of sesame honeys from Egypt and multifloral honeys from India. The PCA distribution of forest honey from Malaysia was similar to that of avocado honeys from Spain and multifloral honey from India. Syrup-feed honeys usually contain higher concentrations of Co, Fe, K, Mg, Mn, Pb and Na, which originate as contaminants from the industrial processing of syrups.

Kalaycıoğlu et al. (2017) reported that CE–DAD, ICP–MS and spectrophotometric methods, including the Folin–Ciocalteu method, were used to explore honey of different botanical origins (honeybee pollens: Anzer, Chestnut, Buckwheat, Oak, Abana, Bayburt and Balıkesir) collected from seven regions of Turkey. The variables obtained, i.e. MEC, sugars, organic acids, TPC and antioxidant activities, were processed chemometrically. Score distribution pattern of the studied samples on PC1/PC3, which explained 52.91% of the total variance, classified pollen samples from different botanical origins and regions.

Adgaba et al. (2017) integrated FAAS, GF–AAS and HPLC–RID methods and analyses according to AOAC official methods with PCA, HCA, LDA and SDA models to investigate Ethiopian honey (pollen grains of honey from four groups). Significant discrimination among different groups of honey samples corresponding to four geographical regions (north, central, southern and southwestern) was obtained. This finding was achieved via HCA and PCA (with the first two PCs explaining 57.6% of the total variance). Furthermore, the SDA model, which included 11 variables (TDS, EC, acidity, total ash, colour, Na, Mg, K, Ca, Mn and Zn), also confirmed the grouping of the honey samples into 4 distinct cluster groups on the basis of their botanical and geographical origins. LDA correctly classified 100% of the samples related to each group. The jackknife procedure achieved an identical classification rate.

6.2.4 *Differentiation of Other Parameters*

6.2.4.1 Assessment of Authenticity/Adulteration

Elflein and Raezke (2008) studied seven types of 451 honey samples tested for adulteration, including pre-trade honeys made for analysis and packed honeys collected from supermarkets. The products originated from the following geographical regions of commercial honey production: Argentina, Austria, Bulgaria, Canada, China, the Czech Republic, El Salvador, France, Germany, Great Britain, Greece, Hungary, India, Italy, Japan, Malaysia, Mexico, Romania, Slovakia, Slovenia, Spain, Turkey, Ukraine, Uruguay and Vietnam. Importantly, the newly developed

EA/LC-IRMS method was a considerably improved tool compared with existing methods. The detection of honey adulteration with inverted sugar syrups from various C3 and C4 plant sources was performed owing to the use of modern methodology. On the basis of the obtained results and considering a confidence level of 99.7%, limits for the $\delta^{13}C$ values of authentic honey were proposed.

In de Alda-Garcilope et al. (2012), chemometric techniques such as CA and ANOVA were used to differentiate PDO "Miel de Granada" honeys relative to their geographical origin on the basis of MEC data determined by FAAS and AES. CA allowed specific discrimination of honey samples of seven botanical origins and confirmed the determination of only five elements. ANOVA clearly revealed differences between rosemary honeys from different geographical regions because the K and Ca contents are descriptive variables. This methodology was useful for assessing the quality of honey samples, confirming the authenticity of PDO labelling as a Granada product.

Five species of raw honey samples were classified by Kek et al. (2017) on the basis of the combination of MEA, including predominant sugars and HMF content, as well as proximate composition and diastase activity. The variable data were quantified via ICP–MS, AAS, HPLC–UV, HPLC–ELSD and other analyses. PCA (with the first two PCs accounting for 77.45% of the total variance) and HCA enabled the distinguishing of Malaysian honey by its bee species, i.e. honey bees and stingless bees. Potassium and Na were the most discriminative variables for Tualang, Gelam and Pineapple honeys, whereas Kelulut honey was associated with high protein and sucrose contents and low fructose and glucose levels. The elemental content was greater for honeys collected from forests or jungles than for those collected from farm plantations. According to the authors, the chemometric evaluation of raw honeys produced by honey bees and stingless bees facilitated honey identification for product labelling. Czipa et al. (2019) reported that the MEC content, diastase activity, HMF content, TPC, EC, three sugar contents and colour were measured by ICP–OES, ICP–MS, HPLC–RID and other standardized methods in Hungarian acacia honeys. The effects of collection year and location, storage, centrifugation and filtration, heating and the addition of sugar syrup were studied. LDA and LSD tests revealed that the method and region of honey collection, and the duration of storage and processing, including adulteration with sugar syrups, had major effects on the quality of acacia honey. Although sugar syrup degrades honey quality, it is difficult to identify in the finished product.

Peng et al. (2020) used LIBS integrated with chemometric techniques to quickly quantify adulterant content. Two common types of adulteration, i.e. mixing acacia honey with high HFCS and rape honey, were determined with PLSR. Emissions from Mg, Ca, Na and K are strongly related to adulterant content. Many acceptable models for identifying the adulteration ratios of HFCS 55, HFCS 90 and rape honey were obtained by SR–PLSR, VIP–PLSR and VIP–PLSR, with RMSEs of 8.9%, 8.2% and 4.8%, respectively.

According to Liu et al. (2021), it is possible to estimate the chemical profile of honeys of six botanical origins, syrups and honeys adulterated with syrup via ICP–OES. Twelve mineral elements combined with sPLS–DA may be used for floral origin differentiation. The accuracies of three types of honey, i.e. jujube, linden and *T. cochinchinensis* honey samples, were 100%, 92.9% and 100%, respectively. Five monofloral chemometric models based on sPLS–DA with a discrimination accuracy greater than 93% indicated that the methodology applied might be a powerful tool for differentiating adulterated honey from pure honey.

Geană et al. (2020) performed analytical measurements of general physicochemical properties, major sugar compositions and SIR via EA–IRMS, U–HPLC–DAD, HPLC–ELSD and other standardized methods to assess the authenticity of honeys from beekeepers, commercial honeys of different botanical origins (acacia, polyfloral, honeydew, sunflower, rape and linden) and various industrial syrup adulterants. Syrups with different botanical origins and brands originated from plants such as corn and cane, beet with some quantities of sugars (corn), grapes, fruits and wheat. PCA with the first two PCs explaining 46.01% of the total variance revealed that differentiation among the acacia, rape and honeydew honey samples was achieved. Moreover, both HCA and PCA were found to be reliable tools for estimating the degree of adulteration of honey. It was concluded that 9.7% of all the honey samples were adulterated with C4 sugar syrups.

According to Dong et al. (2016a, b), the adulteration of 8 commercial types of 800 samples of honey with C-4 sugar content could be identified on the basis of SIR values measured by means of EA–IRMS and LC–RID assisted by PRP analysis. The developed procedure first and successfully provided favourable evidence for the authenticity and identification of honeys with a C-4 sugar content of <0%. According to the authors, "honey samples with −7 <C-4 content (%) <0 could be identified as free of adulteration of C-4 sugars, while samples with C-4 content (%) >7 and C-4 content (%) <−7 were considered to be adulterated with C-4 sugars or C-3 sugars". According to the AOAC 998.12 method, honey contains significant amounts of C-4 sugars with a C-4 sugar content of >7%, which are naturally identified as adulterations.

Stocker et al. (2006) reported that CF–IRMS is a useful method for the detection of royal jelly adulteration with high-fructose corn syrup as a sugar source or yeast powder as a protein source. Stable isotope ratios of $\delta^{13}C$ and $\delta^{15}N$ are acceptable for distinguishing authentic and adulterated royal jelly. Moreover, it has been suggested that royal jelly protein, as an internal standard, and pollen analysis may be useful indicators in the context of identifying food adulteration within a range of uncertainty. The authors recommended continuing studies using isotope ratios of Sr, S, O and H since such an approach has been successfully used in other food science fields.

Stefas et al. (2021) applied LIBS to study the possibility of identifying adulteration of four types of honey with glucose syrup. Three machine learning methods, namely, MDS, LDA and ERT, are useful tools for differentiating honey from honey adulterated with glucose. Some inorganic constituents of honey, i.e. Ca, Na and K, differed significantly in terms of the degree of adulteration. LDA enabled the classification of accuracy, amounting to 90.2%, and the cross-validation procedure of the training data. The external validation accuracy was 90.5%. The built predictive models exhibited satisfactory classification with 90% correctness. Notably, the methodology used appears to have considerable potential for verifying honey adulteration.

Nespeca et al. (2020) assayed Brazilian pure honey, sweetener syrups and 228 fortified samples via LIBS. The identification of adulterants was achieved with 100% accuracy. The PLS model (with the variables selected by iPLS) was highly useful for the quantification of adulterants. The calibration model had good accuracy, linearity, sensitivity and selectivity with satisfactory precision for samples with adulterant contents exceeding 20% (w/w). This methodology highlights the power of the analytical tool for honey authenticity certification, offering relevant benefits, i.e. fast, uncomplicated and clean analytical quantification without sample pre-treatment. According to the authors, LIBS can be used for exploring other food assortments, i.e. condensed milk, mayonnaise, ketchup, etc.

Rogers et al. (2014) studied IRMS data ($\delta^{13}C$ honey and $\delta^{13}C$ protein) by EA–IRMS and the apparent C-4 sugar contents of 1023 New Zealand honey samples from 15 different floral types. The aim of this investigation was to determine which honey is prone to sugar failure via the AOAC 998.12 C-4 sugar method. On the basis of the results of a very extensive experimental database, recommendations for future interpretation of this method were proposed by the authors. Moreover, the following change to the AOAC 998.12 method was suggested: "pure honey (free of corn or cane sugars) yields a value of 'apparent' C-4 sugars of $\leq$7%. New Zealand's Manuka honey may exceed this value but will have honey $\delta^{13}C$ values that are in the normal range (more negative than −24.7‰) and an apparent C-4 sugar content no greater than 13.0% (this study and White et al. (1991))".

6.2.4.2 Interregional and Interseasonal Differentiation

MP–AES and AMS data for Hungarian honey samples collected from 1994 to 2018, almost annually, from the same nectar-producing area and the same species (acacia) were processed by CDA. A clear differentiation of honey samples according to time-dependent variations, i.e. into the three last decades of origin, was achieved. The first CDA function corresponds to 70% of the discriminating ability of the discriminative variables, whereas the second function corresponds to 30%. The cumulative percentages were estimated to be 89.2 (CDA1) and 97.9 (CDA2). The age-specific differentiation of the CDA data suggests that honey samples are useful for tracing long-term changes since environmental factors are preserved in samples often collected by apiarian specialists and honey museums (Varga et al. 2020).

6.2.4.3 Interregional Differentiation Related to the Influence of Soil and Water

The relationships among the soil, water and honey compositions were studied to evaluate the geographical provenance of Argentinean honey by QICP–MS, IRMS and TIMS analyses. LDA allowed a clear distinction between honey samples originating from Buenos Aires, Córdoba and Entre Ríos. On the basis of the eight variables (Rb, K/Rb, B, U, $^{87}Sr/^{86}Sr$, Na, La and Zn), 98.7% of the samples were classified correctly with respect to their origin, but in the case of C&RT, with the use of Rb and K/Rb loadings, 100% and 95% prediction powers were attained, respectively. The GPA data provided evidence of the correspondence between the three studied matrices, i.e. the soil, water and honey samples. The chemical data for the honey showed a significant consensus (91.5%) with those obtained from soil and water since the three datasets predicted the different geographical origins in the same way onto the plane defined by its CP1 and PC2 axes (Baroni et al. 2015).

6.2.4.4 Interregional and Interspecies Differentiation Related to the Impact of Pollution

According to Quinto et al. (2016), ICP–MS data were processed by chemometric techniques, including PCA and CA, which partially distinguished honey samples (multifloral, chestnut, sunflower, sulla, cilantro, acacia and ivy honeys) originating from eight different Italian regions. The first two PCs, representing 47.4% of the total variance, enabled discrimination between heavy metals originating from the most polluted zones and the remaining heavy metals, confirming that honey can be a valuable bioindicator of environmental pollution. However, for honey samples from regions with different pollution levels, an insignificant influence of the botanical origin of honey on the element level may be expected.

Rodríguez García et al. (2006) investigated the classification of honey samples from Galicia with different environmental origins, i.e. rural, urban and industrial areas, owing to the use of ETAAS and ICP–AES methods. The dataset of the honey elemental profiles was studied to perform a preliminary evaluation of honey as an environmental indicator in Galicia for monitoring pollution. Both the CA model and PCA with the first three PCs distinguished urban and industrial samples from rural samples. The next step of the study involved the prediction procedures for modelling the environment surrounding the hive. In this context, different classification rules for linking the metal concentration of honey with that of its surrounding environment via chemometric pattern recognition procedures were required. The classification methods developed by the NN were able to provide more satisfactory results than traditional pattern recognition procedures. The elemental composition of honey can provide useful information by allowing categorization criteria to classify samples relative to their surrounding environment. Therefore, honey may be a suitable pollution indicator for the Galician area.

Devillers et al. (2002) analyzed 17 botanical types of honey from 11 different European and non-European countries by combining ICP–AES with chemometric techniques such as PCA, CFA and HCA. It was impossible to discriminate between the contaminant content in the honey samples and their geographical provenance. On the basis of the distribution of object samples along the first two PCs (46.24%), an approximate relationship was determined between the profile of chemical elements in the honeys and their floral origin. However, observing the influence of the elements studied on the distribution pattern of particular object scores in the PC1/PC2 rectangular coordinate system was possible. Honeys sold in France exhibited significantly high variability related to their macro-, micro- and trace element contents, depending on their botanical origin. Some were strongly contaminated with heavy metals and other industrial xenobiotics at concerning concentrations for health. For example, honey from caducous trees was less contaminated than honey from other botanical origins; aromatic plants are more inclined to concentrate pollutants than herbaceous plants.

Yayinie and Atlabachew (2022) applied a combination of ICP–OES with multivariate processing analysis to characterize the elemental profiles of honey samples originating from apiarists in different administrative zones of the Amhara Region (Ethiopia). PCA with the first four PCs explaining 80.16% of the total variance partially discriminated samples with respect to their interregional assignment. Using LDA, they could clearly classify the honey samples into 5 major regions with an overall average discrimination power of 89.91%, while the HCA model was unsatisfactory. The sorting model was verified by the cross-validation method, which achieved 92.11% recognition power and 93.33% prediction ability. The contents of toxic elements such as Cd, Pb and As can be considered to cause disturbing pollution levels in the environment.

Raeymaekers (2006) established a monitoring method to trace long-term changes in anthropogenic contamination in a district of Bavaria (Germany). Expected anomalies from the industry or residential areas in the sampled district could not be detected. PCA with the first four PCs accounting for 76% of the total variance indicated that biogenic, anthropogenic and unknown impacts may be responsible for the environmental status of the district of Bavaria. According to the author, the methodology may be useful for classifying honey samples originating from more spacious sampling areas, e.g. several European countries.

The multi-elemental composition integrated with chemometrics for the study of citrus honey samples from different regions of Hungary has been used for classification with respect to the distance between the sampling site and the locations of factories. PCA with the first two PCs explained 43.52% of the total variance. On the basis of the PCA and DA data, it was possible to distinguish between industrialized areas and two non-industrialized areas. This means that attention should be given to preventing industrial pollution effects in honey production (Yücel and Sultanoğlu 2012).

Nascimento et al. (2018) used the ICP–OES data of Brazilian honey from stingless bees (classified as multifloral with an expressive representation of pollen types of the families Fabaceae and *Mimosa caesalpiniifolia*) from different areas (urban, semirural, rural and non-urban) that were processed chemometrically. There was a clear differentiation of polluted areas from the remaining areas, i.e. rural and non-urban regions. The authors reported that the honey of *M. scutellaris* was helpful for evaluating the presence of environmental contaminants and can therefore be considered an effective indicator of environmental contamination.

de Andrade et al. (2014) noted that the contents of Cd, Cr and Pb in honey samples from different regions of Paraná (Brazil) were mainly responsible for significant discrimination of samples with respect to their geographic origin at the macroregional scale. The most discriminative variables, i.e. Cd and Pb, significantly corresponded to PC1, which explained 43.59% of the total variance, whereas PC2 (33.35%) differentiated the honey samples according to their Cr levels. According to the authors, honey can be used as a bioindicator of environmental contamination.

According to Grainger et al. (2021), a combination of ICP–MS with multivariate analysis was used to investigate New Zealand monofloral and multifloral honey collected from six regions of the North Island to determine interregional differentiation (e.g. agricultural, rural and urban). PCA, with the first two PCs providing 75.2% of the total variance, discriminated samples between the 6 regions owing to the use of elements influenced by the soil, i.e. Ca, K, Mg, Mn and Na. Heavy metals of anthropogenic origin, such as Cd and Pb, were detected in specific areas because of their activity, e.g. agriculture or high-traffic areas, within the region. Finally, honey was categorized according to region of origin using LDA, with 73.5% accuracy when all the elements were included in the analysis.

Bogdanov et al. (2007) analyzed element concentration data for honey from Swiss production areas to identify possible contamination by common air pollution or other pathways. The levels of Pb and Cd did not indicate their origins related to the impact of contamination. Chemometric analyses, e.g. ANOVA, non-parametric Kruskal–Wallis ANOVA, PCA and LDA, enabled interregional differentiation with respect to the Fe and Cr contents of honey. According to the authors, all of the trace elements analyzed could have originated from natural sources (soil and plants) or anthropogenic sources. Notably, compared with geographical and environmental features, botanical features seem to be more significant for the discrimination of trace element levels in various honey types.

de Oliveira Resende Ribeiro et al. (2014) reported that among honey samples collected from four regions of Rio de Janeiro State, only those from Teresópolis generally presented higher levels of elements than did those from other regions. Moreover, a low level of contamination in all the regions studied was observed. According to de Oliveira Resende Ribeiro et al. (2015), Brazilian honey collected during spring, summer, autumn and winter for 2 years was investigated. Owing to the use of CA, it was possible to sort samples according to season. Significant seasonal patterns in the profile of essential elements in honey were observed, with relatively high levels, especially K and Ca, in samples collected in spring and summer. No seasonal differences in the contents of Cr, Ni, Se and Ti were observed, which could be explained by a low degree of contamination in the investigated regions in any season.

Bosancic et al. (2020) analyzed heavy metals, i.e. Zn, Pb, Cd and Ni, in honey samples from three environmental production systems. Pollen analysis was conducted using PCA to determine if it was possible to obtain grouping patterns corresponding to foraging plant species. PCA, with the first two PCs accounting for

76.27% and 72.4% of the total variance, discriminated samples into three clusters related to different sampling regions and various types of pollen studied. The honey samples were collected from the apiaries near the thermal power plant and the apiary from certified organic production and conventional production facilities, which are semi-controlled. Significantly higher levels of Pb, Cd and Zn were detected in the studied honeys from the area near the thermal power plant than in those from the other locations. The sources of these heavy metals in honey included contaminated forage plants and foraging honeybees. Honey from certified organic production differed significantly from those from the other two types of production. Consequently, certified organic honey has a clear advantage over other bee products contaminated with heavy metals. Consequently, honey can be used as an effective indicator of environmental pollution.

Bilandžić et al. (2011) determined the concentrations of As, Cd, Cu, Hg and Pb via GF–AAS and AMA in multifloral honey samples collected from five regions of Croatia. Cu and Pb appear to be the most abundant heavy metals in the central region, ranging from 0.108 to 41.271 μg/g and from 0.022 to 0.440 μg/g, respectively. The levels of arsenic, Cd and Hg were highest in the southern, southwest and northeast regions, respectively. Importantly, a higher content of toxic Pb was found in Croatian honey than in honey from other European countries. According to the authors, attention should be given to determining the location of honey production hives in regions far from highways and railways.

Stankowska et al. (2008) performed MEA analysis via FAAS and ET–AAS of honey samples from different regions of the Republic of Macedonia. On the basis of a comparison of chemometric parameters, the geographical differentiation of particular elements in Macedonian honey and the results of the RMFA (with the first three Fs corresponding to 61% of the total variance) allowed the identification of two natural associations and one anthropogenic geochemical association. The natural geochemical associations [Mg, Mn, Ca, K and Fe, Zn, Ca, K (–F2), Na (–F2)] are impacted mainly by lithology. The anthropogenic associations [Cd and Cu (–F3)] predominantly resulted from metallurgical activities, namely, Pb production in the town of Veles.

6.2.4.5 LIBS Combined with PLSRA as a Strategy for Predicting Essential Element Contents in Honey

Se et al. (2019) used ICP–OES and LIBS combined with PLSR for the assessment of Ca, Mg and Na levels in stingless bee honeys. Cross-validation assessments of the three individual PLSR models, based on Ca, Mg and Na and the coefficient of determination for the calibration, were performed. Owing to the use of the constructed models, it was possible to predict the contents of Ca, Mg and Na with RSDs of 4.56%, 6.14% and 5.47%, respectively. The proposed strategy appears to be an efficient approach for elemental composition analyses of stingless bee honeys.

6.3 Conclusions

Analytical methods, including spectroscopy combined with chemometric techniques, are very powerful for the identification of authenticity/adulterations or mislabelling, as well as the quantification of honey and its products. The appropriate use of a combination of analytical and chemometric devices for assessing the nutritive quality of honey plays a key role in quality control, its categorization with respect to geographical and botanical origin and the degree of environmental pollution. Both of these tools would be effective for detecting fraud to preserve the brand reputation of original products and to protect the consumer from being deceived financially. The application of chemometric tools has led to a deeper understanding of the distribution of mineral components in foods, which is especially important in food science and ecotoxicology.

References

Abdulkhaliq A, Swaileh KM (2017) Physico-chemical properties of multi-floral honey from the West Bank, Palestine. Int J Food Prop 20:447–454. https://doi.org/10.1080/10942912.2016.1166128

Adgaba N, Al-Ghamdi AA, Getachew A, Tadesse Y, Belay A, Ansari MJ, Radloff SE, Sharma D (2017) Characterization of honeys by their botanical and geographical origins based on physico-chemical properties and chemometrics analysis. Food Measure 11:1106–1117. https://doi.org/10.1007/s11694-017-9487-4

Aghamirlou HM, Khadem M, Rahmani A, Sadeghian M, Mahvi AH, Akbarzadeh A, Nazmara S (2015) Heavy metals determination in honey samples using inductively coupled plasma-optical emission spectrometry. J Environ Health Sci Eng 13:39. https://doi.org/10.1186/s40201-015-0189-8

Alqarni AS, Owayss AA, Mahmoud AA, Hannan MA (2014) Mineral content and physical properties of local and imported honeys in Saudi Arabia. J Saudi Chem Soc 18:618–625. https://doi.org/10.1016/j.jscs.2012.11.009

Altun SK, Dinç H, Paksoy N, Temamoğullari FK, Savrunlu M (2017) Analyses of mineral content and heavy metal of honey samples from south and east region of Turkey by using ICP-MS. Int J Anal Chem 2017:6391454

Arvanitoyannis IS, Chalhoub C, Gotsiou P, Lydakis-Simantiris N, Kefalas P (2005) Novel quality control methods in conjunction with chemometrics (multivariate analysis) for detecting honey authenticity. Crit Rev Food Sci Nutr 45:193–203. https://doi.org/10.1080/10408690590956369

Atanassova J, Pavlova D, Lazarova M, Yurukova L (2016) Characteristics of honey from serpentine area in the Eastern Rhodopes Mt., Bulgaria. Biol Trace Elem Res 173:247–258. https://doi.org/10.1007/s12011-015-0616-9

Atrouse OM, Oran SA, Al-Abbadi SY (2004) Chemical analysis and identification of pollen grains from different Jordanian honey samples. Int J Food Sci Technol 39:413–417. https://doi.org/10.1111/j.1365-2621.2004.00798.x

Baroni MV, Arrua C, Nores ML, Fayé P, del Pilar Díaz M, Chiabrando GA, Wunderlin DA (2009) Composition of honey from Córdoba (Argentina): assessment of north/south provenance by chemometrics. Food Chem 114:727–733. https://doi.org/10.1016/j.foodchem.2008.10.018

Baroni MV, Podio NS, Badini RG, Inga M, Ostera HA, Cagnoni M, Gautier EA, Peral García P, Hoogewerff J, Wunderlin DA (2015) Linking soil, water, and honey composition to assess the geographical origin of Argentinean honey by multielemental and isotopic analyses. J Agric Food Chem 63:4638–4645. https://doi.org/10.1021/jf5060112

Batista BL, da Silva LRS, Rocha BA, Rodrigues JL, Berretta-Silva AA, Bonates TO, Gomes VSD, Barbosa RM, Barbosa F (2012) Multi-element determination in Brazilian honey samples by inductively coupled plasma mass spectrometry and estimation of geographic origin with data mining techniques. Food Res Int 49:209–215. https://doi.org/10.1016/j.foodres.2012.07.015

Berinde ZM, Michnea AM (2013) A comparative study on the evolution of environmental and honey pollution with heavy metals. J Sci Arts 23:173–180

Bettar I, González-Miret ML, Hernanz D, Marcon A, Heredia FJ, Terrab A (2019) Characterisation of Moroccan Spurge (Euphorbia) honeys by their physicochemical characteristics, mineral contents and colour. Arab J Chem 12:2052–2060. https://doi.org/10.1016/j.arabjc.2015.01.003

Bilandžić N, Đokić M, Sedak M, Solomun Kolanović B, Varenina I, Končurat A, Rudan N (2011) Determination of trace elements in Croatian floral honey originating from different regions. Food Chem 128:1160–1164. https://doi.org/10.1016/j.foodchem.2011.04.023

Bilandžić N, Gajger IT, Kosanović M, Čalopek B, Sedak M, Kolanović BS, Varenina I, Luburić ĐB, Varga I, Đokić M (2017) Essential and toxic element concentrations in monofloral honeys from southern Croatia. Food Chem 234:245–253. https://doi.org/10.1016/j.foodchem.2017.04.180

Bogdanov S, Haldimann M, Luginbühl W, Gallmann P (2007) Minerals in honey: environmental, geographical and botanical aspects. J Apic Res 46:269–275. https://doi.org/10.3896/IBRA.1.46.4.11

Bommuraj V, Chen Y, Klein H, Sperling R, Barel S, Shimshoni JA (2019) Pesticide and trace element residues in honey and beeswax combs from Israel in association with human risk assessment and honey adulteration. Food Chem 299:125123. https://doi.org/10.1016/j.foodchem.2019.125123

Bontempo L, Camin F, Ziller L, Perini M, Nicolini G, Larcher R (2017) Isotopic and elemental composition of selected types of Italian honey. Measurement 98:283–289. https://doi.org/10.1016/j.measurement.2015.11.022

Bosancic B, Zabic M, Mihajlovic D, Samardzic J, Mirjanic G (2020) Comparative study of toxic heavy metal residues and other properties of honey from different environmental production systems. Environ Sci Pollut Res 27:38200–38211. https://doi.org/10.1007/s11356-020-09882-y

Boussaid A, Chouaibia M, Rezigb L, Hellalc R, Donsìa F, Ferraria G, Hamdib S (2018) Physicochemical and bioactive properties of six honey samples from various floral origins from Tunisia. Arab J Chem 11:265–274. https://doi.org/10.1016/j.arabjc.2014.08.011

Cantarelli MÁ, Camiña JM, Pettenati EM, Marchevsky EJ, Pellerano RG (2011) Trace mineral content of Argentinean raw propolis by neutron activation analysis (NAA): assessment of geographical provenance by chemometrics. LWT - Food Sci Technol 44:256–260. https://doi.org/10.1016/j.lwt.2010.06.031

Chen H, Fa C, Chang Q, Pang G, Hu X, Lu M, Wang W (2014) Chemometric determination of the botanical origin for Chinese honeys on the basis of mineral elements determined by ICP-MS. J Agric Food Chem 62:2443–2448. https://doi.org/10.1021/jf405045q

Chen H, Fan CL, Chang QY, Pang GF, Cao YF, Jin LH, Hu XY (2015) Application of ICP-MS method in the determination of mineral elements in vitex honey for the classification of their geographical origins with chemometric approach. Spectrosc Spectr Anal 35:212–216. https://doi.org/10.3964/j.issn.1000-0593(2015)01-0212-05

Chua LS, Abdul-Rahaman N-L, Sarmidi MR, Aziz R (2012) Multi-elemental composition and physical properties of honey samples from Malaysia. Food Chem 135:880–887. https://doi.org/10.1016/j.foodchem.2012.05.106

Chudzinska M, Baralkiewicz D (2010) Estimation of honey authenticity by multielements characteristics using inductively coupled plasma-mass spectrometry (ICP-MS) combined with chemometrics. Food Chem Toxicol 48:284–290. https://doi.org/10.1016/j.fct.2009.10.011

Chudzinska M, Baralkiewicz D (2011) Application of ICP-MS method of determination of 15 elements in honey with chemometric approach for the verification of their authenticity. Food Chem Toxicol 49:2741–2749. https://doi.org/10.1016/j.fct.2011.08.014

Conti ME, Stripeikis J, Campanella L, Cucina D, Tudino MB (2007) Characterization of Italian honeys (Marche Region) on the basis of their mineral content and some typical quality parameters. Chem Cent J 1:14. https://doi.org/10.1186/1752-153X-1-14

Conti ME, Finoia MG, Fontana L, Mele G, Botrè F, Iavicoli I (2014) Characterization of Argentine honeys on the basis of their mineral content and some typical quality parameters. Chem Cent J 8:44. https://doi.org/10.1186/1752-153X-8-44

Conti ME, Canepari S, Finoia MG, Mele G, Astolfi ML (2018) Characterization of Italian multifloral honeys on the basis of their mineral content and some typical quality parameters. J Food Compos Anal 74:102–113. https://doi.org/10.1016/j.jfca.2018.09.002

Corbella E, Cozzolino D (2006) Classification of the floral origin of Uruguayan honeys by chemical and physical characteristics combined with chemometics. LWT - Food Sci Technol 39:534–539. https://doi.org/10.1016/j.lwt.2005.03.011

Czipa N, Andrási D, Kovács B (2015) Determination of essential and toxic elements in Hungarian honeys. Food Chem 175:536–542. https://doi.org/10.1016/j.foodchem.2014.12.018

Czipa N, Phillips CJC, Kovács B (2019) Composition of acacia honeys following processing, storage and adulteration. J Food Sci Technol 56:1245–1255. https://doi.org/10.1007/s13197-019-03587-y

de Alda-Garcilope C, Gallego-Picó A, Bravo-Yagüe JC, Garcinuño-Martínez RM, Fernández-Hernando P (2012) Characterization of Spanish honeys with protected designation of origin “Miel de Granada” according to their mineral content. Food Chem 135:1785–1788. https://doi.org/10.1016/j.foodchem.2012.06.057

de Andrade CK, dos Anjos VE, Felsner ML, Torres YR, Quináia SP (2014) Relationship between geographical origin and contents of Pb, Cd, and Cr in honey samples from the state of Paraná (Brazil) with chemometric approach. Environ Sci Pollut Res 21:12372–12381. https://doi.org/10.1007/s11356-014-3175-2

de Oliveira Resende Ribeiro R, Mársico ET, Oliveira de Jesus EF, da Silva Carneiro C, Conte Junior CA, Almeida E, do Nasciemnto Filho VF (2014) Determination of trace elements in honey from different regions in Rio de Janeiro State (Brazil) by total reflection X-ray fluorescence. J Food Sci 79:T738–T742. https://doi.org/10.1111/1750-3841.12363

de Oliveira Resende Ribeiro R, Mársico ET, da Silva Carneiro C, Siqueira Simoes J, da Silva Ferreira M, Oliveira de Jesus EF, Almeida E, Conte Junior CA (2015) Seasonal variation in trace and minor elements in Brazilian honey by total reflection X-ray fluorescence. Environ Monit Assess 187:96. https://doi.org/10.1007/s10661-015-4284-1

Devillers J, Doré JC, Marenco M, Poirier-Duchêne F, Galand N, Viel C (2002) Chemometrical analysis of 18 metallic and nonmetallic elements found in honeys sold in France. J Agric Food Chem 50:5998–6007. https://doi.org/10.1021/jf020497r

Di Bella G, Turco VL, Potortì AG, Bua GD, Fede MR, Dugo G (2015) Geographical discrimination of Italian honey by multi-element analysis with a chemometric approach. J Food Compos Anal 44:25–35. https://doi.org/10.1016/j.jfca.2015.05.003

Dong H, Luo D, Xian Y, Luo H, Guo X, Li C, Zhao M (2016a) Adulteration identification of commercial honey with the C-4 sugar content of negative values by an elemental analyzer and liquid chromatography coupled to isotope ratio mass spectroscopy. J Agric Food Chem 64:3258–3265. https://doi.org/10.1021/acs.jafc.6b00691

Dong H, Xiao K, Luo D, Xian Y, Luo H, Guo X, Li C, Zhao M (2016b) Correction to adulteration identification of commercial honey with the C-4 sugar content of negative values by an elemental analyzer and liquid chromatography coupled to isotope ratio mass spectroscopy. J Agric Food Chem 64:8071. https://doi.org/10.1021/acs.jafc.6b04050

Drivelos SA, Danezis GP, Halagarda M, Popek S, Georgiou CA (2021) Geographical origin and botanical type honey authentication through elemental metabolomics via chemometrics. Food Chem 338:127936. https://doi.org/10.1016/j.foodchem.2020.127936

Dżugan M, Zaguła G, Wesołowska M, Sowa P, Puchalski C (2017) Levels of toxic and essential metals in varietal honeys from Podkarpacie. J Elem 22:1039–1048. https://doi.org/10.5601/jelem.2016.21.4.1298

Dżugan M, Wesołowska M, Zaguła G, Sowa P, Kaczmarski M, Czernicka M, Puchalski C (2018) Honeybees (*Apis mellifera*) as a biological barrier for contamination of honey by environmental toxic metals. Environ Monit Assess 190:101. https://doi.org/10.1007/s10661-018-6474-0

Elflein L, Raezke K-P (2008) Improved detection of honey adulteration by measuring differences between $^{13}C/^{12}C$ stable carbon isotope ratios of protein and sugar compounds with a combination of elemental analyzer—isotope ratio mass spectrometry and liquid chromatography—isotope ratio mass spectrometry ($\delta^{13}C$-EA/LC-IRMS). Apidologie 39:574–587. https://doi.org/10.1051/apido:2008042

Fakhlaei R, Selamat J, Khatib A, Razis AFA, Sukor R, Ahmad S, Babadi AA (2020) The toxic impact of honey adulteration: a review. Foods 9:1538. https://doi.org/10.3390/foods9111538

Fechner DC, Hidalgo MJ, Ruiz Díaz JD, Gil RA, Pellerano RG (2020) Geographical origin authentication of honey produced in Argentina. Food Biosci 33:100483. https://doi.org/10.1016/j.fbio.2019.100483

Fermo P, Beretta G, Facino RM, Gelmini F, Piazzalunga A (2013) Ionic profile of honey as a potential indicator of botanical origin and global environmental pollution. Environ Pollut 178:173–181. https://doi.org/10.1016/j.envpol.2013.03.029

Fernandez-Torres R, Perez-Bernal JL, Bello-Lopez MA, Callejon-Mochon M, Jimenez-Sanchez JC, Guiraum-Perez A (2005) Mineral content and botanical origin of Spanish honeys. Talanta 65:686–691. https://doi.org/10.1016/j.talanta.2004.07.030

Gašić UM, Natić MM, Mišić DM, Lušić DV, Milojković-Opsenica DM, Tešić ŽL, Lušić D (2015) Chemical markers for the authentication of unifloral *Salvia officinalis* L. honey. J Food Compos Anal 44:128–138. https://doi.org/10.1016/j.jfca.2015.08.008

Geană E-I, Ciucure CT, Costinel D, Ionete RE (2020) Evaluation of honey in terms of quality and authenticity based on the general physicochemical pattern, major sugar composition and $\delta^{13}C$ signature. Food Control 109:106919. https://doi.org/10.1016/j.foodcont.2019.106919

Golob T, Doberšek U, Kump P, Necemer M (2005) Determination of trace and minor elements in Slovenian honey by total reflection X-ray fluorescence spectroscopy. Food Chem 91:593–600. https://doi.org/10.1016/j.foodchem.2004.04.043

González Paramás AM, Gómez Bárez JA, Garcia-Villanova RJ, Rivas Pala T, Ardanuy Albajar R, Sánchez Sánchez J (2000) Geographical discrimination of honeys by using mineral composition and common chemical quality parameters. J Sci Food Agric 80:157–165. https://produccioncientifica.usal.es/documentos/5e68afbf29995232672a1b1a

González-Miret ML, Terrab A, Hernanz D, Fernández-Recamales MA, Heredia FJ (2005) Multivariate correlation between color and mineral composition of honeys and by their botanical origin. J Agric Food Chem 53:2574–2580. https://doi.org/10.1021/jf048207p

Grainger MNC, Klaus H, Hewitt N, French AD (2021) Investigation of inorganic elemental content of honey from regions of North Island, New Zealand. Food Chem 361:130110. https://doi.org/10.1016/j.foodchem.2021.130110

Grembecka M, Szefer P (2013) Evaluation of honeys and bee products quality based on their mineral composition using multivariate techniques. Environ Monit Assess 185:4033–4047. https://link.springer.com/article/10.1007/s10661-012-2847-y

Hernández OM, Fraga JMG, Jiménez AI, Jiménez F, Arias JJ (2005) Characterization of honey from the Canary Islands: determination of the mineral content by atomic absorption spectrophotometry. Food Chem 93:449–458. https://doi.org/10.1016/j.foodchem.2004.10.036

Herrero-Latorre C, Barciela-García J, García-Martín S, Peña-Crecente RM (2017) The use of honeybees and honey as environmental bioindicators for metals and radionuclides: a review. Environ Rev 25:463–480. https://doi.org/10.1139/er-2017-0029

International Honey Commision (2009) World Network of Honey Science. Harmonised methods of the International Honey Commission. 2009. http://ihc-platform.net/ihcmethods2009.pdf. Accessed 19 Aug 2019

Kalaycıoğlu Z, Kaygusuz H, Döker S, Kolaylı S, Bedia Erim F (2017) Characterization of Turkish honeybee pollens by principal component analysis based on their individual organic acids, sugars, minerals, and antioxidant activities. LWT 84:402–408. https://doi.org/10.1016/j.lwt.2017.06.003

Karabagias IK, Louppis AP, Karabournioti S, Kontakos S, Papastephanou C, Kontominas MG (2017a) Characterization and geographical discrimination of commercial *Citrus* spp. honeys produced in different Mediterranean countries based on minerals, volatile compounds and physicochemical parameters, using chemometrics. Food Chem 217:445–455. https://doi.org/10.1016/j.foodchem.2016.08.124

Karabagias IK, Louppis AP, Karabournioti S, Kontakos S, Papastephanou C, Kontominas MG (2017b) Characterization and classification of commercial thyme honeys produced in specific Mediterranean countries according to geographical origin, using physicochemical parameter values and mineral content in combination with chemometrics. Eur Food Res Technol 243:889–900

Kędzierska-Matysek M, Teter A, Stryjecka M, Skałecki P, Domaradzki P, Rudaś M, Florek M (2021) Relationships linking the colour and elemental concentrations of blossom honeys with their antioxidant activity: a chemometric approach. Agriculture 11:702. https://doi.org/10.3390/agriculture11080702

Kek SP, Chin NL, Tan SW, Yusof YA, Chua LS (2017) Classification of honey from its bee origin via chemical profiles and mineral content. Food Anal Methods 10:19–30. https://doi.org/10.1007/s12161-016-0544-0

Kováčik J, Grúz J, Biba O, Hedbavny J (2016) Content of metals and metabolites in honey originated from the vicinity of industrial town Košice (eastern Slovakia). Environ Sci Pollut Res 23:4531–4540. https://doi.org/10.1007/s11356-015-5627-8

Krakowska A, Muszyńska B, Reczyński W, Opoka W, Turski W (2015) Trace metal analyses in honey samples from selected countries. A potential use in bio-monitoring. Int J Environ Anal Chem 95:855–866. https://doi.org/10.1080/03067319.2015.1055475

Kropf U, Korošec M, Bertoncelj J, Ogrinc N, Nečemer M, Kump P, Golob T (2010a) Determination of the geographical origin of Slovenian black locust, lime and chestnut honey. Food Chem 121:839–846. https://doi.org/10.1016/j.foodchem.2009.12.094

Kropf U, Golob T, Nečemer M, Kump P, Korošec M, Bertoncelj J, Ogrinc N (2010b) Carbon and nitrogen natural stable isotopes in Slovene honey: adulteration and botanical and geographical aspects. J Agric Food Chem 58:12794–12803. https://doi.org/10.1021/jf102940s

Lachman J, Kolihová D, Miholová D, Košata J, Titěra D, Kult K (2007) Analysis of minority honey components: possible use for the evaluation of honey quality. Food Chem 101:973–979. https://doi.org/10.1016/j.foodchem.2006.02.049

Latorre MJ, Peña R, Pita C, Botana A, García S, Herrero C (1999) Chemometric classification of honeys according to their type. II. Metal content data. Food Chem 66:263–268. https://doi.org/10.1016/S0308-8146(98)00217-9

Latorre MJ, Peña R, García S, Herrero C (2000) Authentication of Galician (N.W. Spain) honeys by multivariate techniques based on metal content data. Analyst 125:307–312. https://doi.org/10.1039/a905978d

Lazarević KB, Trifković JD, Andrić F, Tešić Ž, Anđelković IB, Radović DI, Nedić NM, Milojković-Opsenica DM (2013) Quality parameters and pattern recognition methods as a tool in tracing the regional origin of multifloral honey. J Serb Chem Soc 78:1875–1892. https://doi.org/10.2298/JSC130701099L

Liu T, Ming K, Wang W, Qiao N, Qiu S, Yi S, Huang X, Luo L (2021) Discrimination of honey and syrup-based adulteration by mineral element chemometrics profiling. Food Chem 343:128455. https://doi.org/10.1016/j.foodchem.2020.128455

Louppis AP, Karabagias IK, Kontakos S, Kontominas MG, Papastephanou C (2017) Botanical discrimination of Greek unifloral honeys based on mineral content in combination with physicochemical parameter analysis, using a validated chemometric approach. Microchem J 135:180–189. https://doi.org/10.1016/j.microc.2017.09.004

Louppis AP, Karabagias IK, Papastephanou C, Badeka A (2019) Two-way characterization of beekeepers' honey according to botanical origin on the basis of mineral content analysis using ICP-OES implemented with multiple chemometric tools. Foods 8:210. https://doi.org/10.3390/foods8060210

Madejczyk M, Baralkiewicz D (2008) Characterization of Polish rape and honeydew honey according to their mineral contents using ICP-MS and F-AAS/AES. Anal Chim Acta 617:11–17. https://doi.org/10.1016/j.aca.2008.01.038

Mohammed MEA (2020) Factors affecting the physicochemical properties and chemical composition of bee's honey. Food Rev Int 38:1330–1341. https://doi.org/10.1080/87559129.2020.1810701

Nascimento AS, Chambó ED, Oliveira DJ, Andrade BR, Bonsucesso JS, Carvalho CAL (2018) Honey from stingless bee as indicator of contamination with metals. Sociobiology 65:727–736. https://doi.org/10.13102/sociobiology.v65i4.3394

Nayik GA, Nanda V (2016) A chemometric approach to evaluate the phenolic compounds, antioxidant activity and mineral content of different unifloral honey types from Kashmir, India. LWT - Food Sci Technol 74:504–513. https://doi.org/10.1016/j.lwt.2016.08.016

Nayik GA, Suhag Y, Majid I, Nanda V (2018) Discrimination of high altitude Indian honey by chemometric approach according to their antioxidant properties and macro minerals. J Saudi Soc Agric Sci 17:200–207. www.sciencedirect.com. https://doi.org/10.1016/j.jssas.2016.04.004

Nečemer M, Kosir IJ, Kump P, Kropf U, Jamnik M, Bertoncelj J, Ogrinc N, Golob T (2009) Application of total reflection X-ray spectrometry in combination with chemometric methods for determination of the botanical origin of Slovenian honey. J Agric Food Chem 57:4409–4414. https://doi.org/10.1021/jf900930b

Nespeca MG, Vieira AL, Júnior DS, Gomes Neto JA, Ferreira EC (2020) Detection and quantification of adulterants in honey by LIBS. Food Chem 311:125886. https://doi.org/10.1016/j.foodchem.2019.125886

Nozal Nalda MJN, Bernal Yagüe JLB, Diego Calva JCD, Martín Gómez MTM (2005) Classifying honeys from the Soria Province of Spain via multivariate analysis. Anal Bioanal Chem 382:311–319. https://doi.org/10.1007/s00216-005-3161-0

Oroian M, Sonia S, Amariei A, Leahu A, Gutt G (2015) Multi-element composition of honey as a suitable tool for its authenticity analysis. Pol J Food Nutr Sci 65:93–100. https://doi.org/10.1515/pjfns-2015-0018

Pasquini B, Goodarzi M, Orlandini S, Beretta G, Furlanetto S, Dejaegher B (2014) Geographical characterisation of honeys according to their mineral content and antioxidant activity using a chemometric approach. Int J Food Sci Technol 49:1351–1359. https://doi.org/10.1111/ijfs.12436

Pellerano RG, Uñates MA, Cantarelli MA, Camiña JM, Marchevsky EJ (2012) Analysis of trace elements in multifloral Argentine honeys and their classification according to provenance. Food Chem 134:578–582. https://doi.org/10.1016/j.foodchem.2012.02.125

Peng J, Xie W, Jiang J, Zhao Z, Zhou F, Liu F (2020) Fast quantification of honey adulteration with laser-induced breakdown spectroscopy and chemometric methods. Foods 9:341. https://doi.org/10.3390/foods9030341

Pisani A, Protano G, Riccobono F (2008) Minor and trace elements in different honey types produced in Siena County (Italy). Food Chem 107:1553–1560. https://doi.org/10.1016/j.foodchem.2007.09.029

Puścion-Jakubik A, Borawska MH, Socha K (2020) Modern methods for assessing the quality of bee honey and botanical origin identification. Foods 9:1028. https://doi.org/10.3390/foods9081028

Quinto M, Miedico O, Spadaccino G, Paglia G, Mangiacotti M, Li D, Centonze D, Chiaravalle AE (2016) Characterization, chemometric evaluation, and human health-related aspects of essential and toxic elements in Italian honey samples by inductively coupled plasma mass spectrometry. Environ Sci Pollut Res Int 23:25374–25384. https://doi.org/10.1007/s11356-016-7662-5

Raeymaekers B (2006) A prospective biomonitoring campaign with honey bees in a district of Upper Bavaria (Germany). Environ Monit Assess 116:233–243. https://doi.org/10.1007/s10661-006-7389-8

Rizelio VM, Gonzaga LV, da Silva Campelo Borges G, Maltez HF, Oliveira Costa AC, Fett R (2012) Fast determination of cations in honey by capillary electrophoresis: a possible method for geographic origin discrimination. Talanta 99:450–456. https://doi.org/10.1016/j.talanta.2012.06.009

Rodríguez García JC, Iglesias Rodríguez R, Peña Crecente RM, Barciela García J, García Martín S, Herrero Latorre C (2006) Preliminary chemometric study on the use of honey as an environmental marker in Galicia (Northwestern Spain). J Agric Food Chem 54:7206–7212. https://doi.org/10.1021/jf060823t

Rodríguez-Flores MS, Escuredo O, Míguez M, Seijo MC (2019) Differentiation of oak honeydew and chestnut honeys from the same geographical origin using chemometric methods. Food Chem 297:124979. https://doi.org/10.1016/j.foodchem.2019.124979

Rogers KM, Sim M, Stewart S, Phillips A, Cooper J, Douance C, Pyne R, Rogers P (2014) Investigating C-4 sugar contamination of Manuka Honey and other New Zealand honey varieties using carbon isotopes. J Agric Food Chem 62:2605–2614. https://doi.org/10.1021/jf404766f

Sajtos Z, Herman P, Harangi S, Baranyai E (2019) Elemental analysis of Hungarian honey samples and bee products by MP-AES method. Microchem J 149:103968. https://doi.org/10.1016/j.microc.2019.103968

Santos JS, Santos NS, Santos ML, Santos S, Lacerda JJ (2008) Honey classification from semi-arid, Atlantic and transitional forest zones in Bahia, Brazil. J Braz Chem Soc 19:502–508. https://doi.org/10.1590/S0103-50532008000300018

Santos-Buelga C, González-Paramás AM (2017) Chemical composition of honey. In: Alvarez-Suarez JM (ed) Bee products—chemical and biological properties. Springer, Cham, pp 43–82. https://doi.org/10.1007/978-3-319-59689-1

Schellenberg A, Chmielus S, Schlicht C, Camin F, Perini M, Bontempo L, Heinrich K, Kelly SD, Rossmann A, Thomas F, Jamin E, Horacek M (2010) Multielement stable isotope ratios (H, C, N, S) of honey from different European regions. Food Chem 121:770–777. https://doi.org/10.1016/j.foodchem.2009.12.082

Schmidt C, Eichelberger K, Rohm H (2021) New Zealand mānuka honey—a review on specific properties and possibilities to distinguish mānuka from kānuka honey. LWT 136:1110311

Se KW, Ghoshal SK, Wahab RA (2019) Laser-induced breakdown spectroscopy unified partial least squares regression: an easy and speedy strategy for predicting Ca, Mg and Na content in honey. Measurement 136:1–10. https://doi.org/10.1016/j.measurement.2018.12.052

Soares S, Amaral JS, Oliveira MB, Mafra I (2017) A comprehensive review on the main honey authentication issues: production and origin. Compr Rev Food Sci Food Saf 16:1072–1100. https://doi.org/10.1111/1541-4337.12278

Stankowska E, Stafilov T, Šajn R (2008) Monitoring of trace elements in honey from the Republic of Macedonia by atomic absorption spectrometry. Environ Monit Assess 142:117–126. https://doi.org/10.1007/s10661-007-9913-x

Stefas D, Gyftokostas N, Kourelias P, Nanou E, Kokkinos V, Bouras C, Couris S (2021) A laser-based method for the detection of honey adulteration. Appl Sci 11:6435. https://doi.org/10.3390/app11146435

Stocker A, Rossmann A, Kettrup A, Bengsch E (2006) Detection of royal jelly adulteration using carbon and nitrogen stable isotope ratio analysis. Rapid Commun Mass Spectrom 20:181–184. https://doi.org/10.1002/rcm.2287

Terrab A, González AG, Díez MJ, Heredia FJ (2003a) Mineral content and electrical conductivity of the honeys produced in Northwest Morocco and their contribution to the characterisation of unifloral honeys. J Sci Food Agric 83:637–643. https://doi.org/10.1002/jsfa.1341

Terrab A, Díez MJ, Heredia FJ (2003b) Palynological, physico-chemical and colour characterization of Moroccan honeys: I. River red gum (*Eucalyptus camaldulensis* Dehnh) honey. Int J Food Sci Technol 38:379–386. https://doi.org/10.1046/j.1365-2621.2003.00715.x

Terrab A, González AG, Diez MJ, Heredia FJ (2003c) Characterisation of Moroccan unifloral honeys using multivariate analysis. Eur Food Res Technnol 218:88–95. https://doi.org/10.1007/s00217-003-0797-x

Terrab A, Hernanz D, Heredia FJ (2004) Inductively coupled plasma optical emission spectrometric determination of minerals in thyme honeys and their contribution to geographical discrimination. J Agric Food Chem 52:3441–3445. https://doi.org/10.1021/jf035352e

Vanhanen LP, Emmertz A, Savage GP (2011) Mineral analysis of mono-floral New Zealand honey. Food Chem 128:236–240. https://doi.org/10.1016/j.foodchem.2011.02.064

Varga T, Sajtos Z, Gajdos Z, Timothy Jull AJ, Molnár M, Baranyai E (2020) Honey as an indicator of long-term environmental changes: MP-AES analysis coupled with ^{14}C-based age determination of Hungarian honey samples. Sci Total Environ 736:139686. https://doi.org/10.1016/j.scitotenv.2020.139686

Vasić V, Đurđić S, Tosti T, Radoičić A, Lušić D, Milojkovi-Opsenica D, Tešić Ž, Trifković J (2020) Two aspects of honeydew honey authenticity: application of advance analytical methods and chemometrics. Food Chem 305:125457. https://doi.org/10.1016/j.foodchem.2019.125457

Voica C, Iordache AM, Ionete RE (2020) Multielemental characterization of honey using inductively coupled plasma mass spectrometry fused with chemometrics. J Mass Spectrom 55:e4512. https://doi.org/10.1002/jms.4512

Vukašinović-Pešić V, Blagojević N, Brašanac-Vukanović S, Savićand A, Pešić V (2020) Using chemometric analyses for tracing the regional origin of multifloral honeys of Montenegro. Foods 9:210. https://doi.org/10.3390/foods9020210

Wang J, Li QX (2011) Chemical composition, characterization, and differentiation of honey botanical and geographical origins. In: Taylor S (ed) Advances in food and nutrition research. Academic Press, New York, pp 89–137. ISBN 10: 0123859891

White JW, Bryant WM, Jones JG (1991) Adulteration testing of southwestern desert honeys. Am Bee J 131:123–126

Wu Z, Chen L, Wu L, Xue X, Zhao J, Li Y, Ye Z, Lin G (2015) Classification of Chinese honeys according to their floral origins using elemental and stable isotopic compositions. J Agric Food Chem 63:5388–5394. https://doi.org/10.1021/acs.jafc.5b01576

Yayinie M, Atlabachew M (2022) Multi-element analysis of honey from Amhara region-Ethiopia for quality, bioindicator of environmental pollution, and geographical origin discrimination. Biol Trace Elem Res 200:5283–5297. https://doi.org/10.1007/s12011-021-03088-0

Yücel Y, Sultanoğlu P (2012) Determination of industrial pollution effects on citrus honeys with chemometric approach. Food Chem 135:170–178. https://doi.org/10.1016/j.foodchem.2012.04.061

Yücel Y, Sultanoğlu P (2013) Characterization of Hatay honeys according to their multi-element analysis using ICP-OES combined with chemometrics. Food Chem 140:231–237. https://doi.org/10.1016/j.foodchem.2013.02.046

Chapter 7
Cereals and Their Products

Contents

7.1 Introduction

Cereals are nutritionally important sources of dietary protein, Fe, B vitamins (niacin, riboflavin and thiamine), vitamin E, carbohydrates and fibre, and they contain traces of mineral elements that are important for humans.

As progress has been made concerning the cereal grain market and the fast transportation of food worldwide, the possibility for the creation of mislabelled products has increased, which highlights the need for consumers to look for high-quality cereal grains with a known and approved geographical origin. Adequate analytical methods/techniques are necessary to protect consumers from fraud. Stable isotope ratio analysis and multi-element analysis have become common tools for cereal grain traceability. Recently, numerous research papers have been published on the assessment of cereal grains by SIRA and MEA combined with other instrumental analytical methods (Li et al. 2016a, b).

Supplementary Information The online version contains supplementary material available at https://doi.org/10.1007/978-3-031-99687-0_7.

P. Szefer, *Chemometrics in Food Control Based on Elemental Analysis*,
https://doi.org/10.1007/978-3-031-99687-0_7

The major contribution (over 90%) to the world's rice production is Asia, where contamination of agricultural lands by heavy metals is often detected. In addition to such anthropogenic sources, other environmental factors, such as natural sources, agrochemicals, wastewater irrigation, sewage sludge application, livestock manures, mining and fly ash, may be identified. The accumulation of heavy metals and metalloids in different parts of rice plants, i.e. roots, straw, hulls and grains, has been reported to differ; rice grains are generally distinguished by the accumulation of considerably lower levels of toxic metals than other plant parts. Numerous studies have reported that metal levels in rice are within the allowable limits of the Codex recommendations of the Joint Food and Agriculture Organization/World Health Organization Food Standards Programme and/or food regulations imposed by the respective governments. However, given that rice is a main route of human exposure to toxic metals, appropriate protective and restorative programmes should be undertaken in regions at potential risk of metal pollution (Arunakumara et al. 2013).

It is highly important to quantify trace, micro- and macroelements in cereal products of different origins for classification purposes, i.e. geographical and varietal parameters. Numerous analytical methods/techniques are frequently used for the detection of adulteration of food, such as FAAS, ET–AAS, HG–AAS, CV–AAS, ICP–MS, ICP–AES, ICP–OES, LIBS and NAA, which are assisted by the application of multivariate analysis (Szefer 2003, 2007; Grembecka 2007; Grembecka and Szefer 2016). Issues related to the authentication of organically grown plants while considering the advantages, developments and limitations of atomic spectroscopy for multi-element and stable isotope analysis have been discussed by Laursen et al. (2014).

To perform adequate analytical and chemometric assessments of original data, it is important to apply novel integrated tools that strictly follow the rules and research criteria. The dynamics of their development have been observed owing to new and advanced technologies that can be used to obtain useful information from fundamental experimental data (Szefer 2007). To achieve reliability and accuracy, the use of exclusively validated analytical data for statistical processing via chemometric analysis is recommended. According to Maione and Melgaço Barbosa (2018), reliable results have been obtained by multivariate data analysis and data mining techniques in combination with specific parameters for ascertaining authenticity and many other useful properties of rice, such as quality and yield. The data obtained from image processing, molecular and atomic spectroscopy, elemental fingerprinting, genetic markers, molecular content and other methods are promising sources of information regarding geographical provenance, variety and other aspects of rice, which have been widely processed via multivariate chemometric analysis. PCA and LDA are the preferred methods, although several other data classification techniques, such as SVM and ANN, are also frequently applied, resulting in high performance for the discrimination of rice samples.

Marini et al. (2004) successfully used CP–ANN to classify 1779 Italian rice samples according to their 11 varieties using physical measurements routinely determined for the commercial classification of the product. In contrast to PCA, mapping depending on the Kohonen network has considerably greater potential for class differentiation which is impossible when PCA is used. It was concluded that the optimal CP–ANN was useful for correctly predicting >90% of the test set samples.

Senesi et al. (2019) used LIBS for the investigation of crop plants and their products in combination with recent methodological and statistical approaches that have been widely applied. Specifically, such a combined approach is an efficient tool for evaluating the differentiation of various grains and seeds, including wheat, maize, rice and some of their derivatives, rice and wheat flours. Moreover, the potential of recent advanced LIBS approaches, including DP–LIBS, fs–LIBS, micro-LIBS, NE–LIBS, and 3D elemental imaging, for agricultural products has been discussed.

The basic goal of a biofortified wheat project is to develop nutritionally enhanced cultivars of common wheat to increase the intake of Zn and Fe, which are essential elements for human health. Zn and Fe deficiencies in humans are observed in developing countries, where there is high child and adult mortality (Joshi et al. 2010; Velu et al. 2012; Xu et al. 2012). Therefore, a global biofortification initiative was launched to breed and disseminate crops for better nutrition (Pfeiffer and McClafferty 2007). The goal of this global fortification initiative was to reduce micronutrient malnutrition among poor populations in developing countries, including Asia, by improving food security and increasing quality of life. Enriching wheat grains with greater Fe and Zn levels has been beneficial for reducing malnutrition in regions such as South Asia. Multiplicative (linear–bilinear) models, including SREG, are suitable for grouping sites and cultivars without cultivar rank changes. This approach includes the effects of the genotype plus G × E interaction. Consequently, SREG biplots could be used for identifying subsets of sites and for identifying genotypes with non-crossover G × E interactions. Chemometric models incorporating numerous multienvironmental and genotypic variables into the analysis of multiple environmental trials are useful for exploring and explaining the G × E interaction. Among these models, two are the FR and PLS regression methods (Joshi et al. 2010).

7.2 Chemometric Assessment of Cereals and Their Products

7.2.1 Geographical Differentiation

7.2.1.1 Asiatic Countries

Liu et al. (2016) used a combination of $^{87}Sr/^{86}Sr$ and $\delta^{13}C$, $\delta^{15}N$ and $\delta^{2}H$ data determined by TIMS and EA–IRMS to identify the geographical origins of three genotypes of winter wheat and soil in China. A clear discrimination between wheat samples originating from the three regions was achieved. It was concluded that CDA enables satisfactory interregional differentiation, providing a correct classification rate of 98.1%.

Liu et al. (2017) reported that HR–ICP–MS was used to characterize the elemental distribution pattern of ten Chinese wheat varieties with respect to their interregional origins. The content of Ba was influenced mostly by the genotype, accounting for 27.3%, whereas the other elements were affected by the harvest year. Mn, Sr, Mo and Cd are useful elements for establishing a robust discriminant model with an overall correct classification rate of 98.5% to identify the geographical origin of wheat. Owing to the use of LDA, it was possible to correctly classify wheat samples with respect to their interregional differentiation.

Wadood et al. (2019) utilized NIR and EA–IRMS to assess stable isotope profiles for the classification of wheat whole kernels from China and their products (break, reduction, noodles and cooked noodles) with regard to interregional differentiation. Functions 1 and 2 could verify the differentiation of wheat, flour, noodles and cooked noodles according to geographical origin. By applying LDA to the whole database, the generated model correctly classified over 91% of the samples. Therefore, $\delta^{13}C$, $\delta^{15}N$ and $\delta^{18}O$ isotopes could be effective markers for tracing the geographic provenance of wheat kernels and their products.

According to Li et al. (2012), ICP–OES and ICP–MS integrated with chemometric techniques (PCA, DFA and FIA) were applied to characterize the elemental distribution patterns of paddy rice samples from nine prefectures of Chinese Fujian Province. Compared with the traditional approaches of PCA and DFA, significant improvements in rice classification can be achieved owing to the use of FIA. Calcium, Ni, Fe and Cd were the elements with the highest discriminatory power. This new technique is a potentially effective tool for differentiating rice samples according to their geographical origin at previously unattainable resolutions.

Shao et al. (2015) investigated rice samples from three Chinese provinces via a combination of IRMS and ICP–MS tools. PCA was unable to achieve a clear assessment of interregional distinctions in the rice samples. However, PCA combined with LDA yielded a correct interspatial assignment of 91% for all the samples across the three provinces, i.e. Heilongjiang, Liaoning and Jiangsu.

According to Liang et al. (2018), the combination of ICP–MS and proximate composition is an effective tool for characterizing foxtail millet originating from various provinces in China. Considerable differences in chemical composition were noted for Hebei samples with relatively high protein levels; however, Henan and Shandong were distinguished by relatively high fat and ash contents and dietary fibre and amylose contents, respectively. On the basis of the integration of both methods, DA allowed optimal interregional differentiation of millet samples with a 95.2% classification rate.

Bui et al. (2022) integrated ICP–MS with a chemometric method to study Sengcu rice and other types of rice origins. Elements such as As, Ba Sr, Pb, Ca and Se were identified as the key variables for geographical traceability in the case of Sengcu rice among the three fields of Lao Cai (Vietnam). Al, Ca, Fe, Mg, Ag and As were the major discriminative variables used to distinguish between Sengcu from other types of rice. Owing to the use of the LDA and PLS–DA models, the overall correct identification rates discriminating between Sengcu and other types of rice were approximately 100% in both the training and validation tests. Furthermore, to differentiate the geographical origins of the Sengcu rice samples, these rates ranged from 80% to 99%.

Yasui and Shindoh (2000) reported that the combination of ICP–AES and ICP–HRMS with multivariate chemometric processing is a reliable approach for achieving interspatial differentiation of the rice samples studied. PCA and CA categorized the objects into particulate groups. Barium, Cu, Fe, Mn, Mo, Ni, Rb, Sr and Zn were the most descriptive variables distinguishing geographical rice samples cultivated in the studied producing regions of Japan. For example, rice samples cultivated in the Tohoku/Kanto area were differentiated from those in Hokuriku, but rice grown in Tohoku was distinguished from that originating from Kanto.

Yang et al. (2020) investigated using combinations of FAAS, HG–AFS, amino acid analyzer, GC and other chemical tools to characterize the chemical profiles of common buckwheat samples from the eastern, central and western regions of the Loess Plateau. LDA allowed the discrimination of buckwheat samples originating from the three Chinese regions. The correct classification rate and cross-validation rate of the LDA model were 97.4% on the basis of nine variables (Zn, Mn, Cu, Se and the relative contents of Asp, Glu, Val, Leu and Lys). The two functions, which are based on Wilks' lambda values, explained 100% of the variance. This means that the established methodology using multivariate analysis appeared to be highly effective for identifying the geographical origin of common buckwheat.

Pandey et al. (2016) used ICP–AES data and grain protein content data to quantify bread wheat genotypes of two-country origin. PCA (with the first three PCs accounting for 60.8% of the total variance) and the HCA heatmap divided all the genotypes into two main clusters. Figure 7.1 shows the two main groups of Indian and Turkish genotypes, although overlapping Indian and Turkish genotypes could result from the direct or indirect mixing of these varieties in global wheat exchange

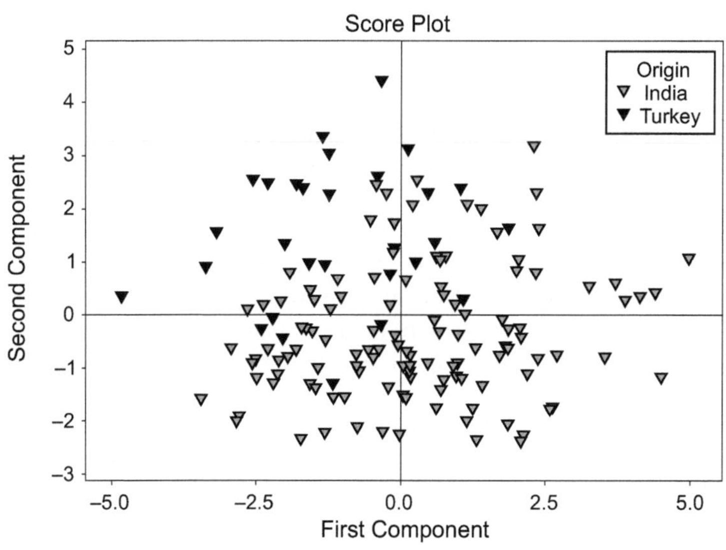

Fig. 7.1 Scatterplot of first versus second principal components showing the grouping of 121 Indian and 29 Turkish bread wheat genotypes. (After Pandey A et al. 2016. SpringerPlus 5:1912. https://doi.org/10.1186/s40064-016-3586-2. With permission)

programs. Therefore, such a biofortification strategy is also used for wild and cultivated Turkish wheat, which is widespread abroad.

Wang et al. (2020) reported $\delta^{13}C$, $\delta^{15}N$, $\delta^{18}O$, $^{207}Pb/^{206}Pb$ and $^{208}Pb/^{207}Pb$ data quantified by EA–IRMS and ICP–MS of rice samples from six rice-producing provinces in China and four other Asian rice-producing countries. Significant differences in the $\delta^{15}N$ values for the two varieties of rice, i.e. Indica and Japonica rice, were obtained. The PCA (with PC1 and PC2 explaining 66.28% of the total variance) showed relatively clear discriminatory power for the geographical origins of rice from different Asian countries (Fig. 7.2). For samples originating from various provinces in China, the first two PCs contributing 60.95% of the total variance allowed the differentiation of rice samples from South China from those of the northeastern region. A correct discrimination rate of 78.43% for rice cultivated in different countries was achieved by using DA with a combination of $\delta^{13}C$, $\delta^{15}N$ and $\delta^{18}O$, whereas in the case of different Chinese provinces, it reached over 85%. However, according to the authors, further verification is needed to improve the accuracy, reliability and practicability of the constructed discriminant model.

7.2.1.2 American Countries

A combination of ICP–MS and chemometric analysis was used to classify Brazilian white rice samples from two important producing states. The differences in the Cd and Rb levels in the samples from the two regions were obtained. The data were standardized, and three classification models, i.e. SVM, RF and MLP, achieved accuracy rates of 93.66%, 93.83% and 90.00%, respectively (Maione et al. 2016).

7.2.1.3 Countries Across Multiple Continents

Suzuki et al. (2008) determined the geographical origin of polished rice from 12 different regions of Japan and from two other countries using combined TC/EA–IRMS with a pentagonal radar plot, which clearly differentiated polished rice samples from Australia and the USA from Japanese samples. Additionally, the data obtained for Japanese rice possibly reflect even small interspatial differences within this country. However, Uonuma rice, which is the most premium rice in Japan, was clearly distinguished from the remaining rice samples. The data obtained could be effective for quick and routine analyses of the geographical provenance of rice.

Jeong et al. (2023) integrated LIBS and NIR spectroscopy with multivariate processing (LRA, PCA and LDA) to differentiate soybean paste samples according to geographical provenance (Korea, China, USA and Canada). When the areas of the C and Ca peaks in the LIBS spectra were used as the inputs for LDA, the accuracy reached 95.4%. An accuracy of 92.1% was achieved when the PC scores were obtained by PCA (with the first six PCs describing 99.1% of the total spectral variation) of the NIR spectra (Fig. 7.3). Furthermore, 2T2D correlation analysis was adopted to enhance NIR discrimination, enabling the recognition of minute spectral

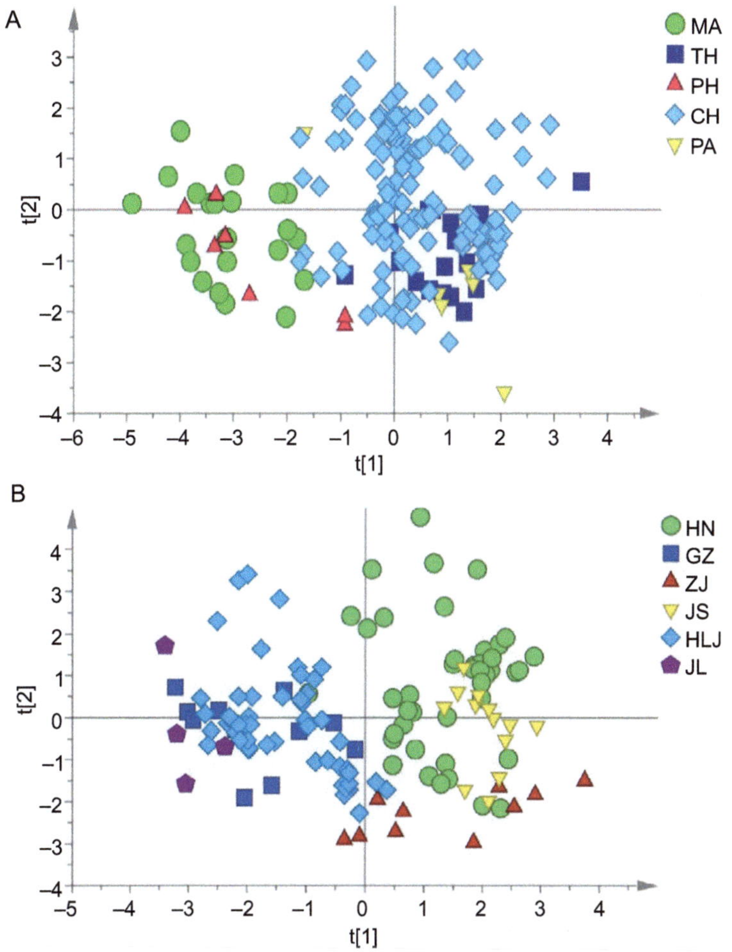

Fig. 7.2 PCA classification results for rice samples: (**a**) Scatter plot of the first two principal score components (PC1 and PC2) for polished rice imported from various countries; note: TH, Thailand; MA, Malaysia; PH, Philippines; PA, Pakistan (PA) and CH, China. Some data tags were covered with each other due to the sample volume. (**b**) Scatter plot of the first two principal score components (PC1 and PC2) for polished rice from different production regions in China. *Note*: HLJ, Heilongjiang; JL, Jilin; JS, Jiangsu; ZJ, Zhejiang; HN, Hunan; GZ, Guizhou. Some data tags were covered with each other due to the sample volume. (After Wang J et al., 2020. Food Chem 313:126093. https://doi.org/10.1016/j.foodchem.2019.126093. With permission)

differences. Owing to the use of the PC1/PC2 scores of 2T2D slice spectra, it was possible to achieve an improved accuracy of 95.0%. A combination of the ratios of A_{Ca}/A_C and the PC2 scores of the samples resulted in the accuracy increasing to 99.6%.

Otaka et al. (2014) noted that XRF and ICP–MS methods combined with multivariate tools (PCA, LDA, CA and SIMCA) and related to soybean samples from four various countries were applied to classify the objects studied according to their

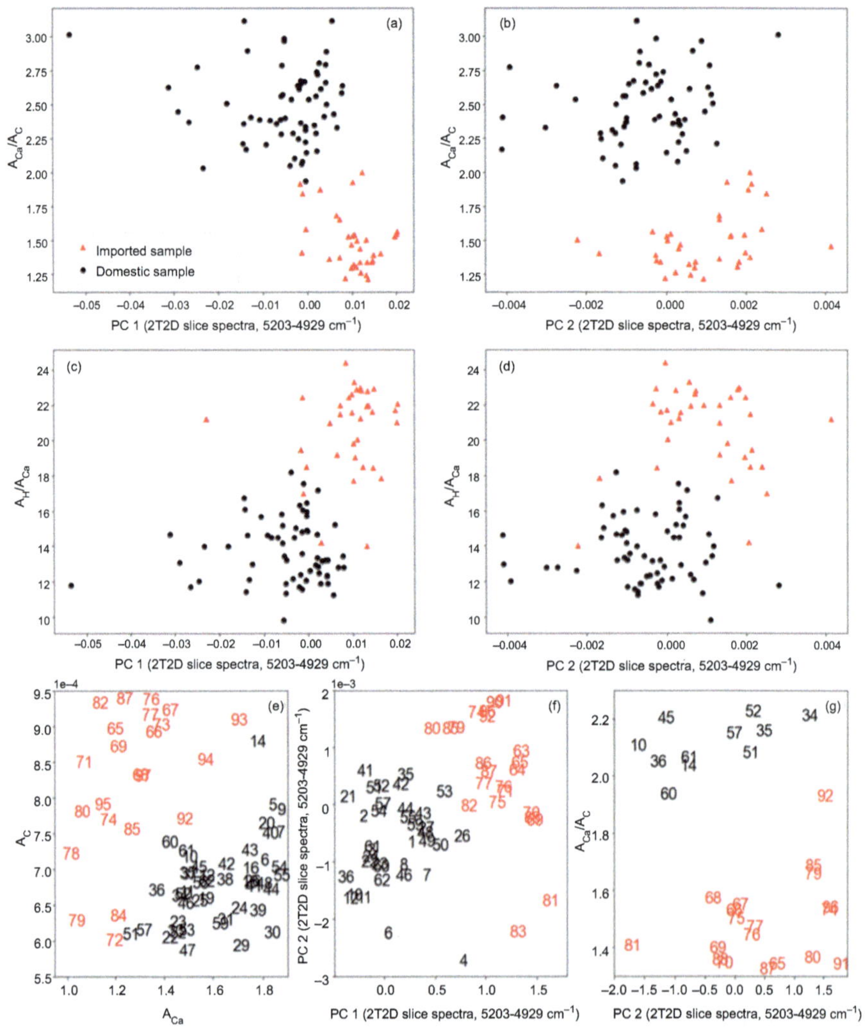

Fig. 7.3 Distributions of the A_{Ca}/A_C versus first PC score (**a**), A_{Ca}/A_C versus second PC score (**b**), A_{Ca}/A_H versus first PC score (**c**) and A_{Ca}/A_H versus second PC score (**d**). The PC scores were obtained from the 2T2D NIR slice spectra in the 5203–4929 cm^{-1} range. Additionally, only samples along the boundary in the domains of A_C and A_{Ca} in the LIBS spectra (**e**), first and second PC scores of 2T2D slice spectra in the 5203–4929 cm^{-1} range (f) and A_{Ca}/A_C and second PC scores (**g**) are shown. The sample numbers with red and black colours correspond to the imported and domestic samples, respectively (For interpretation of the references to colour in this figure legend, the reader is referred to the web version of this article). (After Jeong S et al. 2023. Food Chem 399:133956. https://doi.org/10.1016/j.foodchem.2022.133956. With permission)

origin. PCA (with the first two PCs accounting for 58.8% of the total variance) discriminated between soybeans from Japan, China, the USA and Canada. The LDA prediction accuracies for all samples were 91.3% correct for Japanese samples and 91.3% correct for samples from the remaining countries and mixed soybean.

Elements such as Mg, P, Cl, K, Mn, Cu, Br and Ba were the discriminative parameters used as origin identifiers. This means that the approach is sufficiently accurate to be considered a screening method for reliably discriminating Japanese samples from imported soybeans.

According to Luo et al. (2015), EA–SIRMS results processed by chemometrics enabled the partial differentiation of wheat samples from Australia, the United States, Canada and even two Chinese provinces, i.e. Jiangsu and Shandong. According to the authors, it is questionable, however, whether the achieved differentiation of $\delta^{13}C$ and $\delta^{15}N$ values is generally applicable to all cereal grains.

As reported by Kelly et al. (2002), by combining ICP–MS and Py–CF–IRMS with CDA and RDA, it was possible to differentiate significantly between European, Indian/Pakistani and USA rice samples, achieving 100% correct classification of the three groups of object samples. Among the processed variables, nine were selected as more descriptive parameters, i.e. $\delta^{13}C$, $\delta^{18}O$, B, Mg, Se, Rb, Gd, Ho and W. The elemental compositions of rice samples due to topographic/soil parameters and/or possible input from production processes are factors responsible for the discrimination of samples in terms of their geographical origin. Specifically, in the Basmati region, irrigation with Himalayan waters plays an important role in the differentiation of stable isotopes in rice grown in this area.

Branch et al. (2003) categorized wheat samples from Europe (France and Germany), the USA and Canada using HR–ICP–MS, IRMS and TIMS to determine Cd, Pb, Se, Sr, $^{208}Pb/^{206}Pb$, $^{207}Pb/^{206}Pb$, $^{87}Sr/^{86}Sr$, $\delta^{13}C$ and $\delta^{15}N$ variables. The integrated data obtained were processed using CDA. This methodology allows partial differentiation of wheat samples according to their geographical origin. The best result was obtained by the DA model based on Se and Cd, although it incorrectly classified 50% of the USA and Canadian wheat objects. According to the authors, owing to the limited size of the dataset, it was not possible to validate the model with a test set. However, this pilot study could be geographically classified because a single analyte, $\delta^{13}C$, was used. Moreover, the methodology presents some possibilities for identifying various cultivars and blends within a given sampling location.

Brescia et al. (2002b) studied samples of durum wheat semolina from different regions of Italy, Turkey, Canada and Australia by EA–IRMS. On the basis of ANOVA, differentiation of samples was achieved for all four countries considered. The data obtained confirmed the usefulness of chemometric processing of stable isotope element data for the geographical identification of durum wheat semolina. It was postulated that the limited size of the database excludes the use of multivariate chemometric analysis to achieve clear interregional discrimination of samples.

Cheajesadagul et al. (2013) classified polished rice grains from Asiatic and European countries using HR–ICP–MS fingerprinting combined with chemometric analysis. PCA (with the first two PCs explaining 40.71% of the total variance) provided a discrimination pattern between Thai rice samples and samples originating from France, India, Italy, Japan and Pakistan. However, only partial differentiation of jasmine rice from the three regions of Thailand was achieved. This result was generally confirmed by radar plots and DA. The latter was much more effective than PCA, providing 100% correct classification of objects originating from Asia, Europe and Thailand, and 90.3% correct classification of Thai jasmine rice samples from the northern, northeastern and central regions.

Oda et al. (2001) investigated brown rice samples from different locations in five countries using the $^{11}B/^{10}B$ and $^{87}Sr/^{86}Sr$ values quantified by ICP–QMS and MC–ICP–MS. The data indicated that ratios of the isotopes of B and Sr could distinguish rice samples from Japan, Australia, China–Vietnam, as well as from the USA (California). Furthermore, it was possible to clearly differentiate rice samples from northwestern and southwestern Japan. The $^{87}Sr/^{86}Sr$ values of Japanese volcanic rocks could be divided into northwest and southwest Japan by the boundary of the Itoigava–Shizuoka tectonic line. Additionally, discrimination between rice originating from northeast and southwest Japan and the abovementioned four countries was reported by Kawasaki et al. (2002), who used MC–ICP–MS to establish $^{87}Sr/^{86}Sr$ values.

Ariyama et al. (2011) analyzed samples of barley, rice and wheat from six countries (China, Japan, Thailand, the USA, Canada and Australia) via HR–ICPMS to determine the ratios of $^{87}Sr/^{86}Sr$, $^{204}Pb/^{206}Pb$, $^{207}Pb/^{206}Pb$ and $^{208}Pb/^{206}Pb$. LDA and SIMCA were helpful in discriminating the production countries of samples, although they could not completely differentiate between all the countries, especially samples from the USA and Canada, because of the neighbourhood of the production areas in these two countries. According to the authors, a combination of other variables, such as MEC and SIR values, is recommended.

Ariyama et al. (2012) applied HR–ICPMS to classify rice samples originating from four countries, i.e. Japan, China, Thailand and the United States. Multivariate techniques were constructed to evaluate the effectiveness of quantifying the country of origin. PCA with the first two PCs explaining 55% of the total variability allowed them to achieve intercountry differentiation with respect to the rice samples. Moreover, other techniques, such as LDA, SIMCA and KNN, were applied since their combination was expected to be effective. The methodology was effective, achieving approximately 97% accuracy according to tenfold cross-validation, although the performances of the techniques differed by country. According to the authors, this approach could be applied to other agricultural products, although selecting more adequate element profiles in addition to Sr and Pb isotope ratios was recommended.

Korenaga et al. (2010) used EA–IRMS to analyze δ^2H, $\delta^{13}C$, $\delta^{15}N$ and $\delta^{18}O$ and the elemental contents of H, C, N and O in rice samples from four Asiatic and two non-Asiatic countries. PCA indicated that Japanese rice samples were clearly distinguished from the Australian and American rice samples because many discriminative markers, such as $\delta^{13}C$, $\delta^{15}N$ and $\delta^{18}O$, were used. This differentiation may be responsible for the interregional discrimination of isotopic signatures related to the climate, utilized nutrition and/or quality of irrigation water specified for farming countries. The combined methodology is useful for distinguishing rice samples grown in Japan from those cultivated in other countries.

7.2.2 Intervarietal/Type Differentiation

A study by Ribeiro et al. (2020) based on LIBS and FTIR results analyzed by chemometric methods (PCA, SVM, LDA, QDA, DT and KNN) was used to classify the chemical distribution patterns of four types of rice from Brazil, Uruguay and Colombia. The PCA data suggested that protein, fatty acids and Mg contributed the most to sample differentiation. The first three PCs (74.21% of the total variance) corresponded to the DV1–SNV–LIBS dataset, in contrast to the SNV–FTIR dataset. A clear differentiation of the white rice group from the coloured rice groups was achieved, whereas the black and red rice samples bordered the brown group and were far from the white group. A supervised ML method applied to LIBS and ATR–FTIR data achieved 100% accuracy, sensitivity and specificity for both datasets, although discrimination was more evident owing to the use of LIBS data. It was concluded that LIBS and FTIR combined with multivariate analysis are promising alternatives to traditional analytical techniques for both composition analysis and differentiation when subtle distinctions between chemical variations exist.

Ranulfi et al. (2018) reported that soybean leaves from healthy and sick plants of two varieties affected by GSFR disease from two regions of Brazil were studied by LIBS and AAS. The 2-D score plots based on the first two PCs indicated that most of the information corresponded to 55.8% for the Xingu variety and almost 82.5% for the Parnaíba variety. The data distribution demonstrated the tendency of healthy and sick leaf samples to differentiate into two clusters for both variety sets. Moreover, the classification via the CVR method associated with PLSR achieved a success rate for cross-validation, reaching 94.2% for the Xingu set and 87.3% for the Parnaíba set. It was demonstrated that LIBS could be used as an effective analytical tool to differentiate between healthy and GSFR-infected plants through the analysis of Ca, K and Mg, providing an early GSFR diagnostic procedure.

7.2.3 Interspecies and Interregional Differentiation

Brescia et al. (2002a) reported that EA–IRMS, ^{1}H HR–MAS NMR, GC–FID and HPLC–UV/VIS are effective methods for the quantification of $\delta^{13}C$, $\delta^{15}N$, $\delta^{18}O$, FAs, α- and β-carotenes and α- and γ-tocopherols in two Italian durum wheat semolina cultivars. The first three PCs, which explained 78% of the total data variance, allowed the discrimination of samples according to their geographical origin and two cultivars. Better results were obtained with the use of DA, which effectively differentiated samples originating from northern, central and southern Italy. The isotopic data mostly discriminated the samples according to geographical origin. The prediction ability of the model reached 100% for the training set. Additionally, this model, which was applied to the validation set, exhibited 100% success. A correct classification was also possible owing to the application of DA for the differentiation of the samples relative to the cultivar.

Brandolini et al. (2006) used genetic analysis, AAS and GC–MS to analyze FA, MEC and total antioxidant capacity to estimate the genetic variety among four rice varieties by means of PCR–RAPD markers. PCA was used to classify the rice samples according to their variety and district of production on the basis of genetic and chemical analyses. The classification (with the first three PCs corresponding to 83.2% of the total variance) of the rice samples with respect to their producing region was achieved by using RAPD data. Moreover, genomic data provide the most reliable information on intervariety differentiation. PCA (with the first three PCs accounting for 96.1% of the total variance) demonstrated that analysis of the eighty-two RAPD bands was helpful in distinguishing between closely related species. This means that the RAPD data confirmed the presence of a genetic distance among the Italian rice varieties investigated, whereas the compositional analyses were not influenced by genetics exclusively but were attributed to the origin area. It was concluded that compositional analyses depend not only on genetic characteristics but also on the production area.

Brescia et al. (2007) used EA–IRMS and NMR to investigate loaves of bread together with the durum wheat flours used in their production from four Italian regions. A combination of SIR, ^{1}H HR–MAS NMR 400 MHz spectra and other parameters with multivariate analysis was found to be a reliable approach. PCA enabled partial interregional differentiation between the Sardinian region and the other regions. Based on LDA, clear differentiation between dough samples, especially flours, was achieved for four regions of origin, with 87% classification ability and 76% predictive capacity. The interspatial variation was less significant when the bread samples were considered, resulting in a decrease in the predictive ability to 60%. This deterioration of the obtained data could be explained by the undesirable impact of chemical modifications during processing, particularly after cooking. However, owing to the relatively high δ^{15}N values of the Bonorva samples, the bread samples were clearly differentiated from the other samples.

Li et al. (2022) combined EA–IRMS and ICP–MS to classify 900 Japonica and Indica rice samples from 17 provinces in four regions: the Middle–Lower Yangtze Plain (Y–R), Northeast China (N–E), Southwest China (S–W) and Southeast China (S–E). This study was the largest ever performed in China, covering 96% of the main production regions. Owing to the use of BP–NN models with a combined stable isotope and elemental approach, it was possible to classify Japonica and Indica rice from different regions with high accuracies of 97.2% and 77.9%, respectively. Similar results were also obtained with MFA. According to the authors, their study provides the first extensive SIR and MEC geographical database for Chinese rice and offers promising multivariate chemometrics for the effective differentiation of two key production regions in China.

Wu et al. (2015) applied EA–SIRMS to investigate samples of five kinds of cereal grains (early polished rice, late polished rice, glutinous rice, late long-shaped rice and early long-shaped rice) from three provinces of China. The wheat samples originated from the USA, Australia, Canada and Jiangsu and Shandong Provinces of China. The δ^{13}C values for Australian, USA and Canadian samples and Chinese provinces (Jiangsu, Shandong) were significantly different; hence, the $\delta^{13}C$ value is a useful descriptor of the geographical origin of cereal grains. However, chemometric analysis of samples of five Chinese cultivars of cereal grains revealed that their differentiation was not significant; i.e. they were not sufficiently detectable by measurements with the use of $\delta^{13}C$.

Li (2011) analyzed 12 varieties of rice grown in various 15 intercontinental countries via IRMS and ICP–MS to assess the capacity for classifying their geographical origins. CDA provided a clear differentiation between the samples of various rice types as well as between broad geographical areas (Asia, Australia, Europe, India, Pakistan, North America and Southeast Asia), enabling 90.7% correct classification for the model built.

The aim of the study by Phuong et al. (1999) was to quantify 14 elements in rice samples originating from Australian and Vietnamese sampling sites. These data were processed chemometrically by Kokot and Phuong (1999) to estimate the impact of environmental and crop cultivation variables on the elemental profile of rice. ICP–AES, ICP–MS and FAAS, combined with multivariate analysis (PCA, MDM, PROMETHEE and GAIA procedures), were reported to classify white/polished and brown/unpolished rice samples from two countries. A PC1 versus PC2 biplot accounting for 72.6% of the total variance revealed the differences between brown and white rice samples. Moreover, PCA enabled clear differentiation of the Australian white rice samples from the Vietnamese samples, mostly by Mn and Mo. Based on PROMETHEE and GAIA, it was concluded that some Australian rice varieties have somewhat better elemental nutritional contents than Vietnamese varieties. Furthermore, Vietnamese special and sticky rice has a better mineral nutritional distribution pattern than does common and non-sticky rice. In general, a wet season seems to be more favourable for achieving a better mineral nutritional profile.

7.2.4 Differentiation of Other Parameters

7.2.4.1 Interregional Differentiation Related to Varieties, Harvest Years and Agricultural Practices

ICP–MS was integrated with multivariate processing of 38 species of wheat samples from four major wheat-producing regions in China. A clear differentiation between these regions was achieved, although LDA provided classification of samples attaining only 73.8% and 71.1% correct classification for original and cross-validation, respectively. There were significant influences of varieties, harvest years and agricultural practices on the wheat element profiles across the region. According to the authors, the use of this tool could be an optimal choice for identifying the geographical provenance of the samples studied despite the impacts of the abovementioned parameters at the elemental level in wheat (Zhao et al. 2011).

7.2.4.2 Verifying Authenticity/Adulteration

Goiter Asfaha et al. (2011) developed methods using geochemical markers for verifying the authenticity of the geographical origin of cereal samples originating from 11 different European countries. Extensive analytical measurements were carried out as part of an international cooperation. Investigations of 527 samples collected over 2 years from 17 sampling sites with 36 descriptive variables were performed. The Sr isotope ratios and SIR (n (^{87}Sr)/n (^{86}Sr) and $\delta^{13}C$, $\delta^{15}N$, $\delta^{18}O$ and $\delta^{34}S$ values, alone or with key element concentrations of Na, K, Ca, Cu and Rb) were determined by numerous instrumental tools, such as MC–TIMS, MC–ICPMS, TC/EA–IRMS and Q–ICP–MS, combined with chemometric methods. Multiple-step multivariate statistics for numerous samples originating from many representative locations across Europe, representing an extensive range of geographical and environmental features, were used. PCA provided a rich dataset; however, no clear trends or relationships with respect to the interregional classification of cereal samples were obtained. Consequently, PLS–DA was applied on the basis of supervised classification and to compare two models, using the five isotopic variables alone and in combination with the Sr isotope ratios and mineral element results. Importantly, the chemometric models tested made it possible to individualize 15 of the sampling sites. The classification of sampling sites according to the latitude (north and south classes) and proximity to a marine environment (Atlantic, Mediterranean and inland classes) was particularly efficient when ten variables were chosen carefully.

Since the SIR composition of rice could be strictly dependent on the influence of water sources on its geographical origin, Zhang et al. (2024) analyzed SIRs in Indica and Japonica rice via pot experiments. Under the three water isotope treatments, the $\delta^{2}H$ values of polished Indica rice were significantly different from the $\delta^{13}C$ and $\delta^{15}N$ values. Moreover, the $\delta^{2}H$ and $\delta^{18}O$ values varied during subsequent phases of Indica rice leaf growth; i.e. they decreased at the heading stage of leaves, increased at the filling stage and then decreased at the harvest stage. A similar trend was observed in Japonica rice. The results provide useful information to improve the authentication of the geographical origin of rice via SIRA, which is important for food safety and consumer confidence.

Paolini et al. (2015) used GC–C–IRMS to determine stable isotopes and amino acid contents in grain samples of winter wheat and five cultivars of durum wheat from Denmark and Italy. It was possible to assess differentiation between organically and conventionally grown plants treated with synthetic nitrogen fertilizers, animal manures or green manures from nitrogen-fixing legumes. As shown in Fig. 7.4, the LDA score plot of the $\delta^{15}N$ and $\delta^{13}C$ analyses of amino acids clearly differentiated conventional and organic agricultural systems, explaining 91.2% of the total variance. This approach is useful for verifying the authenticity of organic crops. It was concluded that SIRA is a novel analytical methodology for improving control procedures in the organic sector.

Gonzálvez et al. (2011) studied rice grain samples originating from Spain and three other countries on the basis of the combination of ICP–OES with a multivariate chemometric tool. The majority of the variability between groups was explained

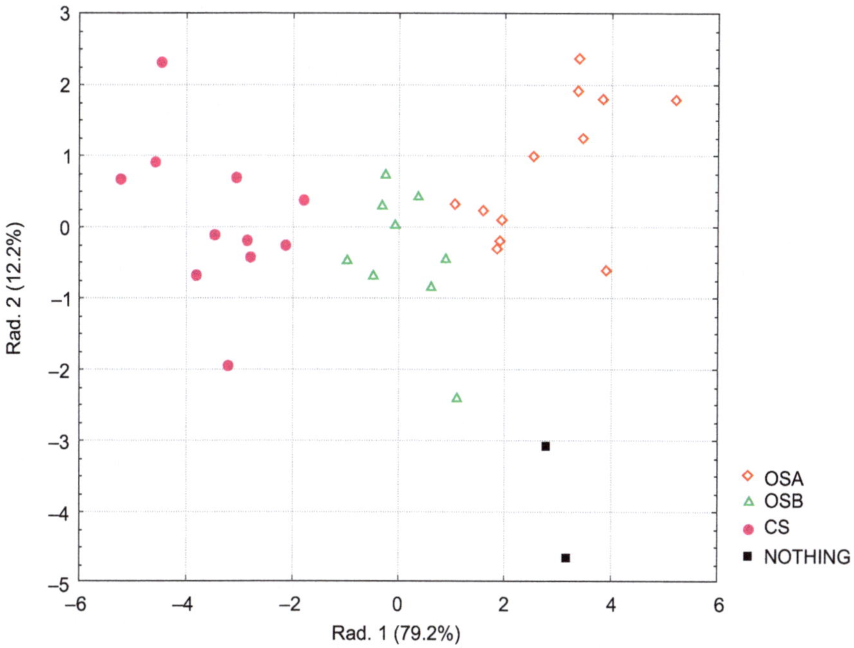

Fig. 7.4 LDA score plot of nitrogen and carbon isotope analysis of amino acids for organic and conventional wheat samples. (After Paolini M et al., 2015. J. Agric. Food Chem 63:5841–5850. https://doi.org/10.1021/acs.jafc.5b00662. With permission)

cumulatively by the first three CFs, amounting to 73.7%. On the basis of the distribution pattern in 3D space, a complete differentiation of the Spanish samples with respect to the Brazilian, Indian and Japanese samples was achieved, indicating that 91.59% of the samples were correctly classified. Moreover, LDA differentiated "Arròs de València" PDO rice samples from Spanish rice, such as "Arroz del Delta del Ebro", "Arroz de Calasparra", and rice samples from Extremadura and India, Japan and Brazil, allowing a 91.30% correct classification for an independent validation set. Lanthanides, such as Cd and Co, were the most appropriate indicators of the interregional differentiation of rice samples owing to their various contents in parent soil as well as their effective bioaccumulation by plants. It was concluded that the trace element composition could be used for the verification of rice authenticity.

Lin et al. (2023) determined the geographical origins of crops using deep learning-assisted LIBS to quickly realize food traceability and guarantee the interests of consumers. However, this technique was not useful for practical applications; therefore, a transfer learning-assisted LIBS method was developed to identify the geographical origins of crops. Consequently, a maximum accuracy of 93.81% between six transfer combinations was achieved. It was possible to introduce further methodological innovations for improving identification accuracy owing to the use of DaNN and DAN to classify crops for the first time.

Quinn et al. (2022) used ICP–MS–MS to determine the MRC content in rice samples from different regions of three different Asian countries. The original data obtained were processed by PCA, PLS–DA, fivefold CV, unpaired t tests, ROC curves, DD–SIMCA and K–NN. PLS–DA showed satisfactory discrimination ability. Moreover, the DD–SIMCA and K–NN models were distinguished by acceptable sensitivity (98%) and specificity (100%). According to the authors, such assessment of the geographical origins of rice for its authenticity is especially important in terms of protecting businesses and consumers since the COVID-19 pandemic has influenced the food industry and consumers, resulting in production gaps, shipping delays and changes in supply and demand with an increased risk of food fraud.

Laursen et al. (2009) used ICP–MS in combination with PCA to classify unpolished brown rice samples from an experimental field station (Laos) with respect to the discrimination of three rice genotypes. Based on 30 elements, PCA allowed a clear distinction of all three genotypes. The semi-quantitative approach may therefore efficiently contribute additional elements to the chemometric data analysis. Owing to the use of 29 variables, PCA with PC1 and PC2 (explaining 31% and 18% of the total variance, respectively) also enabled the full discrimination of the three rice genotypes. Therefore, semi-quantitative ICP–MS combined with chemometrics is a promising analytical approach that is powerful for use as a routine procedure for authenticity and adulteration testing of food samples.

Arif et al. (2021) demonstrated that the combination of ICP–MS with DD–SIMCA has high potential for distinguishing different varieties of Pakistani rice grown inside and outside the recognized Basmati-growing region. Owing to the use of such methodology, it was possible to verify the authenticity of Basmati rice. Satisfactory sensitivity and specificity values of 100% and 98%, respectively, were achieved with the DD–SIMCA model.

Laursen et al. (2011) reported that ICP–OES and Q–ICP–MS combined with multivariate processing allowed the evaluation of the elemental distribution patterns of wheat and barley crops grown in three Danish cropping systems. A conventional system and two organic systems, i.e. organic with animal manure and organic with cover crops, were used. PCA of multi-elemental fingerprints or pairwise comparisons and their confidence intervals, which were adjusted for multiple comparisons, was applied. PCA of full- or semi-quantitative ICP elemental contents in barley grains allowed clear discrimination of Jyndevad samples from exclusively two other regions (Fig. 7.5). It was found that multi-elemental fingerprinting with semi-quantitative ICP–MS and chemometrics can be used to verify the authenticity of organic crops. However, further improvements in the authentication of crops should be performed.

Barbosa et al. (2016) achieved recognition of Brazilian organic and conventional rice samples via the combination of q–ICP–MS MEC with SVM. It was possible to predict the authenticity of organic rice samples with an accuracy of 98% using 19 elements, although an accuracy of 96% was achieved when only the Ca and Cd contents were used.

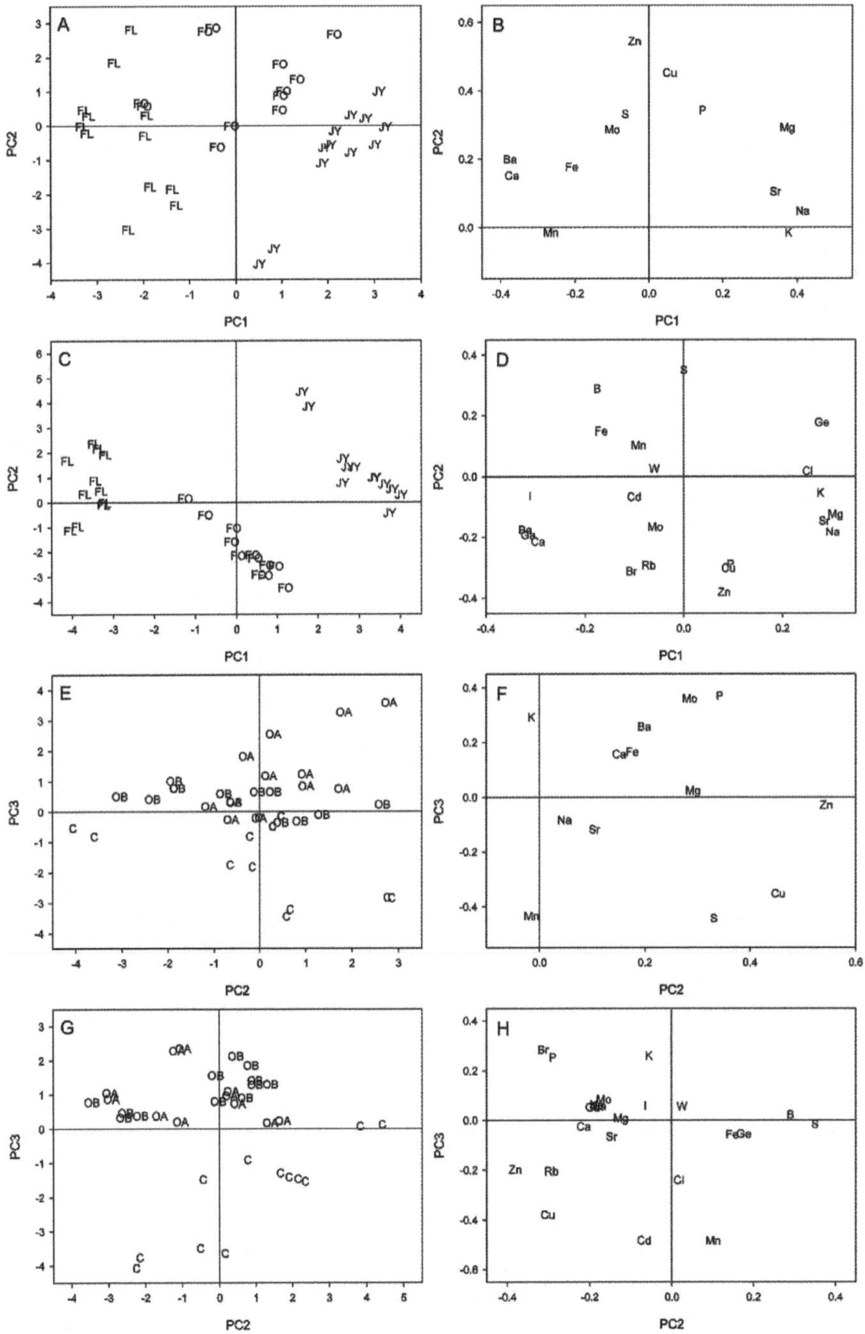

Fig. 7.5 Principal component analysis of full- or semi-quantitative ICP elemental concentrations in barley grains in year 1. Panels A and B show the score and loading plots from fully quantitative ICP–OES data (13 elements). Grouping according to geographical location is shown by PC1 and

Borges et al. (2015) coupled ICP–MS with multivariate techniques, i.e. PCA, SIMCA, HCA and KNN, to discriminate Brazilian organic samples from ordinary rice samples as alternative methods for adulteration evaluation. PCA was used separately and in the SIMCA model, with the first two PCs explaining 60% of the total variance. It was then possible to differentiate ordinary samples through PC2. Consequently, SIMCA correctly classified all samples with 100% prediction ability for ordinary and organic rice samples. A Coomans plot confirmed the high predictability and specificity of SIMCA by achieving a specificity of 100% in all cases. Moreover, KNN classified two classes, i.e. organic and ordinary rice samples, with 100% prediction ability and 100% specificity in all cases.

Zhang and Xu (2017) used a combination of FAAS, HAFS, spectrophotometric methods, amino acid analysis, GC and photocolorimetric methods for the quantification of micro- and macroelements, vitamins and amino acids in common buckwheat samples from Inner Mongolia, Shanxi and Shaanxi Provinces in China. The data obtained were transformed by PCA, CA and LDA. The latter enabled clear interregional differentiation of common buckwheat samples, with a high correct classification rate of 95.8% and a satisfactory cross-validation rate of 91.7%. Variables such as Mn, VPP, Se, Gly, Cu, Asp, Fe and Ala were effective discriminators. In the authors' opinion, the proposed method is a powerful tool for controlling the geographical origin of common buckwheat by governmental administration and protecting consumers from incorrect domestic labelling. Moreover, the authors stipulated that the discriminant method requires further validation on the basis of more reliable data.

Chung et al. (2015) integrated ICP–AES with chemometrics to determine the food authenticity of brown rice from three Asiatic countries. PCA (with the first two PCs revealing 43.2% of the total variance) allowed them to differentiate only samples belonging to the Philippines from those from other regions, although it was not possible to distinguish between rice samples originating from Korea and those from China. However, PLS-DA, a supervised method, achieved the maximum geographical differentiation between rice groups originating from Korea, China and the Philippines. They found that Cu, Ag, Zn, Cr, Ca, Ba, Cd, Bi, K, Pb and In are highly descriptive elements for building a discrimination model to identify the geographical provenance of rice grains. According to the authors, the approach applied can be

Fig. 7.5 (continued) PC2, which explained 35% and 22% of the variance, respectively. Panels C and D show the score and loading plots of the semi-quantitative ICP–MS data (22 elements). Grouping according to geographical location is shown by PC1 and PC2, which explained 35% and 17% of the variance, respectively. Panels E and F illustrate score and loading plots of the same fully quantitative data as in Panels A and B, but with grouping according to cropping system shown by PC2 and PC3, which explained 22% and 19% of the variance, respectively. Panels G and H illustrate score and loading plots from the same semi-quantitative data as in Panels C and D, but with grouping according to cropping system shown by PC2 and PC3, which explained 17% and 14% of the variance, respectively. FL, Flakkebjerg; FO, Foulum; JY, Jyndevad; C, conventional; OA, organic with animal manure; OB, organic with cover crops. (After Laursen KH et al., 2011. J Agric Food Chem 59:4385–4396. https://pubs.acs.org/doi/10.1021/jf104928r. With permission) effective for controlling the geographical origin of rice by governmental administration and protecting consumers from incorrect domestic labelling.

Chen et al. (2016) assayed the variation in SIR by IRMS in the superior and inferior grains of rice in two regions in China, i.e. Fujin and Wuchang. $\delta^{18}O$ and δD could discriminate between rice from these two rice cultivation areas. Since Daohuaxiang, an excellent-quality rice, is also cultivated in other areas, the adulteration and mislabelling of Wuchang Daohuaxiang has caused major problems due to unintentional or intentional cases. According to the authors, further studies are needed, e.g. concerning systematic consideration and verification, including issues of mechanisms regulating the fractionation of stable isotopes for different grains.

Yang et al. (2018b) used several chemometric algorithms, i.e. PCA, DT, RF, PLS–DA, LDA and SVM, supported by LIBS technology, and reported that these tools may be useful and promising for studying the classification performance for determining geographic origins of rice. A different assortment of samples, including twenty kinds of rice samples originating from China and Thailand, was detected using LIBS without any pretreatment processes. The fivefold cross-validation classification accuracies with 30 input variables for DT, RF, PLS–DA, LDA and SVM were 83.75, 97.95, 94.75, 98.35 and 99.25%, respectively. It was concluded that LDA was the most efficient tool for geographic rice provenance classification. This means that all these methods indicate a classification accuracy greater than 80%, achieving the best accuracy of 99.25%, whereas SVM had the highest operation time. Combining this multivariate technique with LIBS as an analytical method, which offers high accuracy and analytical speed, was also recommended. According to the authors, this elaborate dual methodology is highly powerful for the rapid identification of adulterated products in agriculture because it uses an inexpensive and non-destructive LIBS method.

According to Chung et al. (2016), EA–IRMS integrated with PCA (the first three PCs amounting to 83.5% of the total variance) enabled the discrimination of the geographical origin of brown rice grown in three Asiatic countries. The first PC differentiated between rice cultivated in the Philippines and that originating from China and Korea. A combination of $\delta^{15}N$, $\delta^{18}O$ and $\delta^{34}S$, in contrast to $\delta^{13}C$ in rice, demonstrated the relatively strong ability of these stable isotope elements to classify samples with respect to the intercountry discrimination of samples. The lack of distinction among the three rice cultivars could be explained by the specific genetic properties of Japonica-type and Tong-il-type rice.

ICP–MS and IRMS coupled with multivariate processing analysis were used to construct elementomic models to identify the geographical indications of 14 varieties of rice from six Asian countries (Chung et al. 2018). The unsupervised PCA model, with the first two PCs accounting for 43.2% of the total variance, clearly differentiated the Chinese, Indian and Vietnamese rice samples, although the partial overlap in the Chinese and Vietnamese rice samples could be explained by the fact that China and Vietnam are adjacent countries with similarities in geology, climate and/or farming practices. Orthogonal projection to OPLS–DA provided effective discrimination between rice cultivated in Korea and other countries, and $\delta^{34}S$, Mn and Mg could be biomarkers for the geographic authentication of Asiatic rice.

According to Chung et al. (2016, 2018), this methodology has the potential to control improper or fraudulent labelling with respect to the geographical origin of rice worldwide.

Liu et al. (2022) analytically processed two high-quality special rice products developed for Shanghai urban agriculture to characterize their elemental profile. The first two PCs, which represent 98.8% of the variance, allowed the differentiation of Songjiang rice from Chongming rice and other rice regions. After NSD treatment, the discrimination between the two main types of rice and rice from other regions improved. EA–IRMS and NIR combined with the optimized PLSDA and LVQNN models were useful for distinguishing Songjiang rice and Chongming rice from other regions. Complete differentiation was observed with respect to rice cultivation methods. The Songjiang CPB and NCPB rice samples could be clearly distinguished (with a validation set error rate of 0%) using a PLSDA model of the raw NIR spectra. Moreover, somewhat less effective discrimination was possible owing to the use of $\delta^{15}N$ and $\delta^{18}O$ values because they are affected by fertilizer type, irrigation water use efficiency and rice paddy respiration. The optimal identification accuracy rate was 87.5% for the NCFP Chongming rice method versus the CFP agricultural method. The authors reasonably concluded that protecting these products from geographical origin mislabelling or cultivation method fraud is a major challenge.

7.2.4.3 Differentiation Between Organic and Conventional Cereal Products

Laursen et al. (2011, 2013) reported that ICP–OES, Q–ICP–MS, HR MC–ICP–MS and IRMS enabled the determination of numerous mineral elements and $\delta^{13}C$, $\delta^{15}N$, $\delta^{18}O$, $\delta^{34}S$, $\delta^{2}H$, $\delta^{25}Mg$ and $\delta^{26}Mg$ to characterize the chemical profiles of winter wheat and spring barley samples produced in rigorously controlled long-term field trials. The $\delta^{2}H$ analysis enabled the complete distinguishing of organic and conventional wheat and barley grains originating from Denmark. To the best of the authors' knowledge, this study was the first time that compound-specific isotope ratios of nitrate-derived N and O were successfully used for food authentication, and it was the first attempt to use MC–ICP–MS for differentiating organic and conventional plant products. It was concluded that multi-isotopic analysis has the potential to reveal fraudulent substitutions of organic plants with conventionally cultivated plants (Laursen et al. (2011, 2013)).

Liu et al. (2020) characterized Chinese rice related to parent soil samples using ICP–MS/MS (Li, Be, Na, Mg, Al, K, Ca, V, Cr, Mn, Fe, Co, Ni, Cu, Zn, Ga, Rb, Sr, Mo, Ag, Cd, In, Cs, Tl, Pb and Bi) combined with EA–IRMS ($\delta^{13}C$ and $\delta^{15}N$). PLSDA, which integrates SIR and MEC, was found to be an effective approach for correctly distinguishing organic rice from green and conventional rice samples, with an accuracy of up to 100% over the 4-year study. However, it was difficult to classify green and conventional rice completely. In the case of a two-class discrimination model of rice, organic rice could be clearly discriminated from either green or conventional rice (with 100% accuracy, although some minor overlap between green and conventional rice was still observed, with an accuracy of 85.33%). Therefore, this strategy offers a more rigorous alternative tool for fighting fraudulent mislabelling of organic rice.

Bontempo et al. (2016) quantified $\delta^{13}C$, $\delta^{15}N$, $\delta^{2}H$, $\delta^{18}O$ and $\delta^{34}S$ contents using the EA–IRMS method. The assessment of multi-isotopic data using CDA was useful for determining the geographical origin of Italian organic and conventional pasta and its raw materials, as 98% of the samples were correctly classified relative to interregional differentiation. When both farming systems and geographical regions were considered simultaneously, 80% of the samples were successfully classified.

Yuan et al. (2018) used EA–IRMS and ICP–MS to differentiate organically grown rice from green and conventionally grown rice in experimental field trials in China using chemometric methods. Analytes such as $\delta^{13}C$, $\delta^{15}N$, $\delta^{2}H$, $\delta^{18}O$, $^{208/206}Pb$, $^{206/207}Pb$, ^{86}Sr and other elements were determined. Organic rice was much lower in ^{13}C than green or conventionally grown rice. PCA and LDA were used to discriminate the three farming methods and are useful tools for the authentication of farming methods, providing intercultivar distinctions between rice fertilized with animal manures, green composts and synthetic fertilizers originating from China or somewhere else.

7.2.4.4 Interregional and/or Intergenotype Differentiation

Brandolini et al. (2006) used genetic, AAS and GC–MS analyses to categorize four rice cultivars according to their variety and district of production using PCA of genomic and compositional data. Chemometric classification of the samples with respect to their Italian-producing region was possible using elemental profile data. Owing to the use of PCR–RAPD markers, genomic data provided the most reliable information on intervarietal differentiation. PCA (with the first three PCs accounting for 96.1% of the total variance) demonstrated that analysis of the 82 RAPD bands was helpful in distinguishing among closely related species, indicating that the RAPD data confirmed the presence of genetic distance among the Italian rice varieties investigated, whereas the compositional analyses were not influenced exclusively by genetics but were attributed to the origin area.

According to Zhao et al. (2012), a combination of HR–ICP–MS and XRF with chemometric analysis was used for the classification of ten wheat varieties related to parent soils from three regions of China. Origin and genotype, including their interaction, had considerable impacts at the element level. As shown in Fig. 7.6, LDA enabled them to clearly distinguish wheat samples originating from three regions. Na, Ca, Fe, Zn and Mo were associated with the origin, whereas Mg, Mn, Cu and Ba were associated with the genotype. Therefore, on the basis of the LDA model, elements such as Na, Ca, Fe, Zn and Mo were shown to be reliable indicators for identifying wheat interspatial discrimination, allowing the correct classification of 91.1% of samples from different regions by cross-validation.

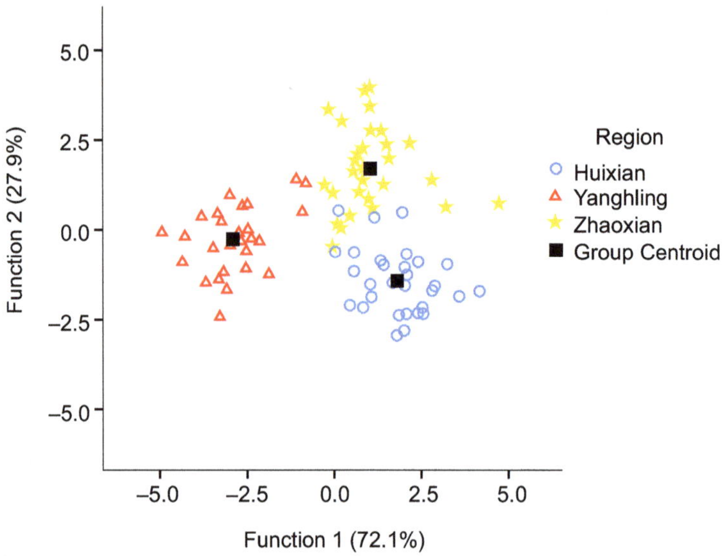

Fig. 7.6 Scatter plot of discriminant functions 1 and 2 of wheat samples. (After Zhao H et al., 2012. J Agric Food Chem 60:10957–10962. https://doi.org/10.1021/jf3021283. With permission)

Qian et al. (2019) studied three genotypes of Wuchang rice and parent soil samples from China to characterize their elemental distribution patterns via EA–IRMS and NIR. With respect to element differentiation between the two seasons, with the exception of V, Cu, Cd and Pb, the other elements significantly differed in this respect, whereas the Cu levels exhibited significant variability among the different genotypes. The complete differentiation of Songjiang and Chongming rice from other geographic origins and the discrimination of their cultivation methods were achieved. The overall correct identification rates of the rice samples from the three regions obtained in 2013 and 2014 via LDA were 96.6% and 89.6%, respectively, and the overall correct identification rate for Wuchang rice was 100%.

Research carried out by Husted et al. (2004) enabled the characterization of three spring barley genotypes, i.e. Orthega, Barke and Bartok, grown in three typical Danish agricultural soils. A combination of ICP–MS, IRMS and NIR and multivariate chemometric analysis (ANOVA, PCA and PLS) was used to study the stable isotopes and elemental characteristics of the barley genotypes at different times during the full growing season. Several PCA approaches (with the first two PCs explaining 56 to 81% of the total variance) were performed to study the critical content and composition of the elements required to achieve a clear differentiation relative to the soil type. PCA differentiated multi-elemental data with respect to soil type when plants of similar physiological age were compared, in contrast to such options when plants of all ages were compared simultaneously. Analytes such as $\delta^{15}N$ were also useful discriminants for the classification of samples with respect to soil type. However, parameters such as soil chemistry and possibly climate and agricultural methods had too much significance to allow a clear distinction between the three genotypes. Therefore, it was successfully substantiated by analyzing the MEC of grains from two additional genotypes, i.e. Otira and Barthos, grown at the northern and eastern locations, respectively, which were comparable with PCA related to parent soil based on data corresponding to the genotypes Barke, Orthega and Bartok. Comparable results could be obtained when $\delta^{15}N$ was used.

7.2.4.5 Intertype Differentiation Related to Culinary Processing

The combination of ICP–OES and multivariate chemometric processing analysis was applied to classify Brazil rice samples with respect to their intertype differentiation. Inspection of the scatter plot of PC1 (68.1%) versus PC2 (19.8%) allowed the differentiation of score groups corresponding to brown rice, parboiled rice and white rice samples. A slight differentiation between raw and cooked samples could also be achieved. HCA generally confirmed the results obtained by PCA. Moreover, the application of PCA and HCA, which were used separately for the subgroups of the rice samples, clearly distinguished between raw and cooked samples with reduced contents of Ca, K, Mg, Na, P and Fe during cooking. This means the use of PCA and HCA for the subgroups of rice samples revealed a clear differentiation of rice brands, i.e. white, brown and parboiled rice, as well as between raw and cooked samples (da Silva et al. 2013).

7.2.4.6 Differentiation Between Transgenic and Non-Transgenic Cereal Grains

Jiao et al. (2010) processed rice samples from China with respect to three kinds of transgenic (T) rice and their non-transgenic (C) counterparts. Combined methods, i.e. ICP–AES, NIR, GC–MS, HPLC and the AOAC official method, were applied to determine specific key nutrients, i.e. mineral elements, fatty acids, protein, amino acids, vitamins and antinutrient phytic acid, as recommended by the OECD. Using PCA (with the first two PCs explaining 68.33% of the total variance), it was possible to clearly differentiate T rice from C rice. The most discriminative variables that correlated significantly with PC1 and PC2 were Mg, Mo, Co, Ni, Fe, V, Cu and K, and Se, Zn and Ca. This means that these descriptors corresponding to PC1 and PC2 clearly differentiated between T and C rice samples (Fig. 7.7). According to the authors, some unintended compositional changes in T rice concerning the analyzed nutrients, including several elements, varied to different extents, in contrast to the antinutrient phytic acid, whose content in T rice did not change significantly. For example, in the case of T2 and T3 rice seeds, the Mg and Fe levels decreased to half of those in C2 and C3 rice seeds, although the values were still within the range of normative values reported for rice.

Liu et al. (2019) reported that LIBS coupled with a chemometric tool was applied to identify Chinese transgenic maize. All the models (PCA, PLS–DA, SIMCA and ELM) attained very good identification rates, especially the PCA loading-ELM

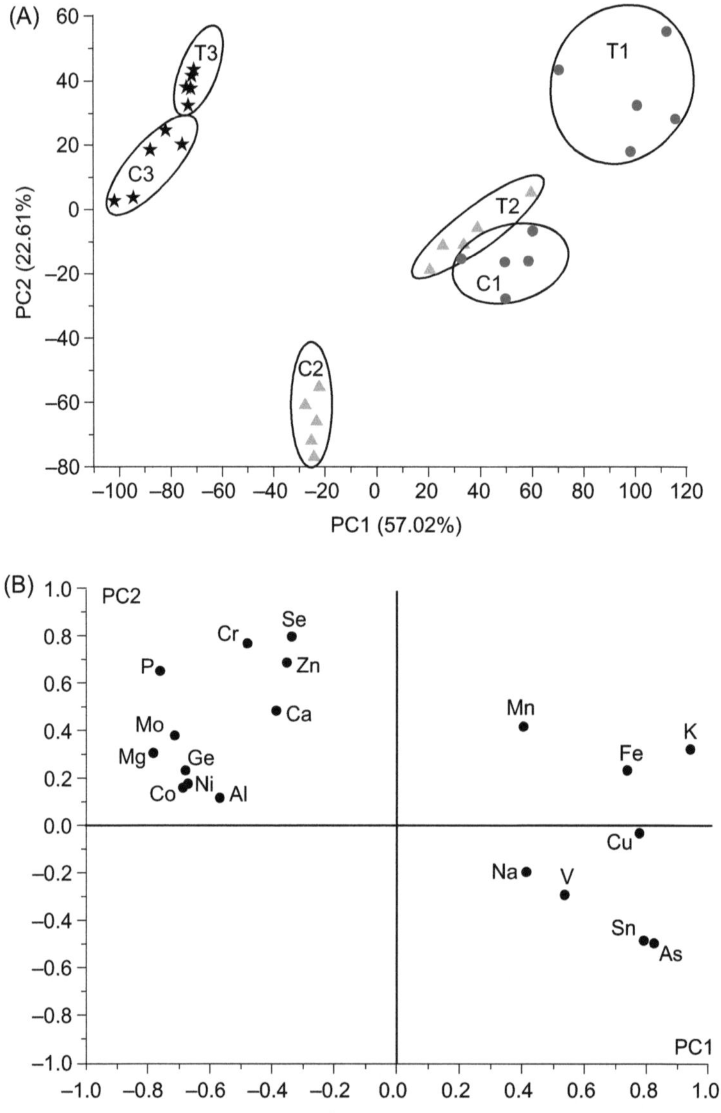

Fig. 7.7 Score plot of PC1 and PC2 (**a**) and loading plot (**b**) for elemental composition from three pairs of transgenic rice and non-transgenic rice by PCA. (After Jiao Z et al., 2010. J Agric Food Chem 58:1746–1754. https://doi.org/10.1021/jf902676y. With permission)

model and PLS–DA, which showed the best performance for the detection of GM maize, with 100% identification accuracy for both the calibration and prediction sets. Compared with the PLS–DA and ELM models, SIMCA achieved 97.5% identification accuracy for the calibration set and 92.3% identification accuracy for the prediction set. Moreover, PCA, with the first three PCs representing 86% of the total variance, indicated that PC1 could clearly differentiate the GM and non-GM maize.

These findings indicate that LIBS coupled with computational tools is an encouraging approach for the identification of transgenic maize.

7.2.4.7 Interregional Differentiation Related to Parent Soil and/or Water

Zhao et al. (2013) used HR–ICP–MS and XRF data processed by multivariate chemometric processing analysis to classify wheat grain and parent soils from the main wheat-growing areas in Hebei and Henan Provinces, China. PCA with PC1 (32.8%) and PC3 (15.3%) revealed a clear differentiation between the two producing regions. The LDA model demonstrated that the correct classifications of 88.9% and 100% of test samples from both provinces were obtained owing to the presence of metals such as Cr, Mn, Ga, Rb, Sr, Zr and Cd. The content of these elements in wheat was closely related to soil origin; hence, these elements were found to be reliable indicators of the geographical provenance of wheat, accounting for 95% of the total correct classification of test samples.

Lagad et al. (2017) integrated Q–ICP–MS and MC–ICP–MS to quantify $^{87}Sr/^{86}Sr$ ratios and REE contents of Indian Basmati rice, soil and water samples. Basmati rice samples from the southern Uttar Pradesh region had higher levels of radiogenic Sr than those from Uttarakhand, Delhi, Haryana and Punjab. On the basis of datasets available in the literature on the $^{87}Sr/^{86}Sr$ ratios of rice samples originating from six other countries, the data were compiled and statistically assessed. The Sr isotope data for worldwide rice samples, i.e. Indian, Chinese, Vietnamese, Japanese and Californian (USA) samples, are convenient as geographical tracers, except for Australian and Thai rice. It was concluded that chemometric analysis revealed a strong relationship between the $^{87}Sr/^{86}Sr$ ratios of Indian rice and its substrata, i.e. the silicate and carbonate fractions of soil and water, indicating their uptake in rice from water.

7.2.4.8 Differentiation of Some Interactive Flour Components

Li and Zhu (2017) studied MEC, including diverse physicochemical characteristics, via ICP–MS and other methods of whole grain flour, as a new product from quinoa samples in terms of flour properties correlated with flour composition as well as the properties of isolated quinoa starches through chemometrics. Li et al. (2016a, b) collected commercial quinoa seeds from different countries and suppliers. In the case of mineral components, the first two PCs explained 86.6% of the total variance, whereas the cumulative percentages of the total variance for PC1 and PC2 accounted for 70.8% and 64.3% with respect to other chemical components and physicochemical properties, respectively. It was postulated that the thermal properties and enzyme susceptibility of quinoa flour are highly affected by starch interactions with non-starch components, e.g. lipids, protein, dietary fibre, phenolics and MEC, which strongly affect flour properties.

Akin et al. (2020) analyzed element concentrations to evaluate the potential of employing LIBS to discriminate different flour types and quantify the addition of white wheat flour to rye and oat flour and breads. PCA and PLS techniques were used for analyzing LIBS spectra, whereas ICP–OES and AAS were used as reference methods for elemental analysis. The first two PCs accounting for 99.18% of the total variance clearly demonstrated the clustering pattern of the identified flour types. Five clearly differentiated sample groups, i.e. bran, rye, oat, whole wheat and refined wheat flour samples, were distinguished. It was also demonstrated that PC1 explains the variance since high loading values (Na, Ca, Mg and K) correspond to peaks at 270–300 nm, 390–400 nm, 580–590 nm and 760–770 nm in the LIBS spectrum. PC2 explains the variance with high loading values (Mg, Ca and K) related to peaks in the 270–300, 390–400 and 760–770 nm spectra. LIBS combined with PCA provided successful classification of the five flour types, and PLS provided quantification of unacceptable amounts of refined wheat flour added to rye and oat flour and bread samples. Moreover, according to Akın et al. (2021), the addition of various levels of sorghum flour to mixtures of corn-sorghum flour was used for PLS evaluation. PLS–DA determined the samples to be pure corn, pure sorghum and corn-sorghum mixtures. Various amounts of sorghum flour were added to mixtures of corn and sorghum flour for PLS analysis. The coefficients of determination of the calibration and validation PLS models were 0.979 and 0.965, respectively, with a limit of detection of 4.36%.

7.2.4.9 Assessment of the Degree of Allele Size Affinity of Microsatellites Related to Different Parameters

Zeng et al. (2009b) used ICP–AES to study phylogenetic relationships for classification traits for 789 accessions of rice landraces with significantly lower contents of K, Ca, Mg, Fe and Cu than improved cultivars. Owing to the eight mineral elements in brown rice that clustered Yunnan rice into rice landraces and improved cultivars within one rice landrace group, five subgroups could be distinguished, and purple rice did not tend to be associated with either group. The relationship between the allele sizes of microsatellites and phenotypic variations in rice landraces was studied by Zeng et al. (2009a), whereas Zeng et al. (2010) characterized the genotypic variation of eight element concentrations in brown rice from Yunnan Province in 11 different grades of brown rice owing to the number and distance coefficients of 282 alleles for 20 SSR markers. A total of 628 Yunnan landraces formed the core collection. The allele size of microsatellites related to the phenotypic traits of rice landraces in Yunnan (China) was also studied on the basis of 20 SSR markers. Moreover, 23 phenotypic traits of rice landraces in Yunnan were investigated by using 20 SSR markers and 23 phenotypic traits, as well as eight mineral elements within the core collection of 629 accessions. A significant correlation for 182 of 620 pairs among these markers and traits, as well as elements, was obtained. The allele size of microsatellites was more related to grain or panicle traits than to plant traits or element levels in brown rice (Zeng et al. (2009a, b, 2010)).

7.2.4.10 Assessment of Industrial and/or Agricultural Pollution Impacts

Jamali et al. (2008) reported that FAAS/ETAAS coupled with multivariate analysis was a useful tool for investigating five Indian sorghum varieties grown in soil changed from untreated sewage sludge of industrial origin. HCA and PCA were used to study variations in the uptake of trace and toxic elements by various varieties in the same experimental plots. CA demonstrated the differentiation in the uptake of essential and toxic elements from plants grown in soil transformed with untreated industrial wastewater sewage sludge (SUIS), although they corresponded to the same experimental plot. This means that CA indicates that the elements have adequate explanatory potential to identify a suitable variety for growing on SUIS. The high uptake of As by sorghum varieties is attributed to the anthropogenic impact of industrial sewage sludge, whereas Cu, Mn and Zn present in SUIS are essential for plants as constituents of coenzymes. The applied chemometric tool is a powerful tool for studying potential natural and/or anthropogenic sources that are reliable for determining elemental levels in different sorghum varieties.

According to Yang et al. (2019a), ICP–MS combined with chemometric analysis enabled the investigation of Chinese wheat grain and adjacent soil as substrata for the assessment of the pollution status in major wheat cultivation areas in China. PCA was used, with the first two PCs corresponding to 51.68% of the total data variance. PC1 corresponded to agricultural activities (including fertilization, wastewater irrigation and pesticide spraying), with the main discriminative variables such as Cd, Cr, Cu, Zn and Pb. PC2 was closely associated with industrial activity, i.e. related to Ni and Ti manufacturing, and was responsible for the emission of Ni to the surrounding environment. This means that the source of Ni mainly originated from industry, whereas the other heavy metals were mostly derived from agricultural activities. Furthermore, it was possible to evaluate the dietary health risks caused by consuming wheat grains to prevent food pollution. On the basis of the metal pollution index analysis, wheat grain consumption poses no direct threat to human health. The health risk assessment revealed that there was a non-carcinogenic risk to adults and children from wheat consumed in the study area.

Xu et al. (2023) analyzed heavy metals (Cu, Fe, Cd, Zn, Cr, Pb, Ni and Mn) in soil–wheat systems in two watersheds in Baiyin City, Gansu Province, China. The combination of XRD, STEM–EDS and XPS indicated the presence of smelting by-products originating from the Pb/Zn smelters. The toxic metals in agricultural soils considerably exceeded the typical values for Cd, Pb, Cu and Zn. In addition, Cd was the dominant pollutant in the soil–wheat linkage, which inhibited the transport of Cu, Mn and Pb from the soil to the wheat roots. According to the authors, –OH and C–H stretching, N–H amide and C–O bending are responsible for detoxifying heavy metals in wheat. The mean values of non-carcinogenic and carcinogenic risks from consuming wheat grain were 9.1 and 1.4E-02 for adults and 11.3 and 3.3E-03 for children, respectively, indicating a noteworthy health risk. Notably, the methodology allowed them to observe the microscopic mechanisms of heavy metal tolerance in wheat.

Kim et al. (2012) applied LIBS and ICP–OES methods to determine Mg, Ca, Na and K in spinach and rice from a local Korean market. The analytical data were transformed by multivariate chemometrics. The LIBS technique with PLS–DA could be a useful tool for quickly discriminating pesticide-contaminated samples from pesticide-free samples even though they have similar elemental profiles.

Peng et al. (2014) combined ICP–OES and chemometric analyses to classify 16 varieties of buckwheat samples, including seeds from China. PCA accounted for 67.0% of the total variance. PCA and HCA provided comparable results. The detection of toxic heavy metals in commercial Tartary buckwheat tea suggested that safety issues associated with such buckwheat products should be seriously considered. Among the toxic elements, the Pb concentrations in several samples were relatively high (up to 18.38 μg/g). The Cd and Hg concentrations were generally low and did not pose a health risk.

7.2.4.11 Elaboration of Improved Methodology

Sezer et al. (2017) processed Turkish wheat milling products (wheat flour, whole wheat meal and semolina) with different ash contents using LIBS, and PLS was used to evaluate the spectra. The results were highly correlated with the data of the standard ash analysis. It was postulated that this methodology is a useful tool for routine ash analysis of wheat flour fractions. This approach is reliable for quality control laboratories of flour mills, pasta manufacturers and other related food processing companies.

Yang et al. (2019b) described a simple and inexpensive sample preparation method (SLST) for analysis by LIBS of toxic heavy metals such as Cd and Pb in rice. Compared with that of the conventional pelletizing method, the spectral intensity of both trace elements was significantly greater when LIBS was used. The SLST method had obvious advantages over the pellet method for improving the analytical sensitivity.

Martelli et al. (2010) differentiated wheat outer tissues during grain ablation using a combination of LIBS and univariate/multivariate analysis. A chemometric tool consisting of MW2D and PLS–DA was applied to provide more in-depth information. The dual methodology is a promising and useful technique for rapid structural analysis of vegetal materials.

Yang et al. (2018a) performed an analysis of four sample preparation methods to compare the results of the classification of 20 different kinds of rice origins via LIBS coupled with PCA and SVM. The classification accuracies of the PCA and SVM combination for the four sample preparation methods were 92.70%, 95.70%, 98.80% and 99.20%, respectively. It was concluded that LIBS combined with SVM is a promising tool for effectively discriminating rice samples from China and Thailand, even though they have similar distribution patterns.

7.2.4.12 Interregional Differentiation Related to Genotype and Seasonal Variability

According to Liu et al. (2015), EA–IRMS coupled with chemometric analysis was a useful tool for studying wheat kernel samples, including ten genotypes collected from three regions of China from 2011 to 2013 in three experimental fields. The influences of region of origin, genotype, harvest year and their interactions on the $\delta^{13}C$ and $\delta^{15}N$ profiles were observed. In the case of $\delta^{2}H$, both region and harvest year significantly influenced its content in wheat, but no significant variations were noted between different genotypes within each region. Moreover, the $\delta^{2}H$ values were also significantly influenced by the interaction of region and year. Generally, the region constituted the largest proportion of the total variability, explaining 47.57%, 58.02% and 27.96% of the total variability in $\delta^{13}C$, $\delta^{15}N$ and $\delta^{2}H$, respectively. According to the authors, the results obtained could provide a potential theoretical basis for the geographical recognition of wheat and other plant foodstuffs, even if they were combined with various genotypes and originated from different years.

7.2.4.13 Prospects of Utilizing the Elemental Profile of Wheat Plants Related to Different Parameters

Velu et al. (2012) applied a methodology allowing the evaluation of high yields across locations in Mexico and South Asia for the expression, stability and magnitude of the G × E interaction with respect to grain Zn and Fe contents, grain yield and other agronomic traits to determine the feasibility of developing biofortified wheat varieties. Two SREG biplots, with the first two PCs corresponding to 61.28% and 62.44% of the total variance for nine locations and forty lines, enabled the characterization of the grain yield potential of the first HPYT as well as the grain Zn content, respectively. It was suggested that although the G × E interaction was significant, the ranking of entries remained the same across locations, indicating that it is possible to select the best-adapted entries with high Zn and Fe concentrations. Pooled data across locations showed increases of 28% and 25% over the checks for Zn and Fe, respectively. A considerable number of entries exceeded intermediate to full-breeding target Zn contents, showing that it is possible to develop Zn-biofortified varieties with competitive yields and that other farmers prefer agronomic traits. According to the authors, the positive and moderately high correlation between Zn and Fe concentrations suggests good prospects for simultaneous improvement in both micronutrients.

Joshi et al. (2010) reported that agronomic (yield and days to heading) data and ANOVA, SREG, FRA and PLS chemometric techniques were examined to evaluate the magnitude of nature and genotype, location and G × L interaction effects for agronomic traits and grain Zn and Fe contents in the EGP of South Asia. Moreover, the contributions of soil and meteorological factors to the agronomic and nutrient levels of the genotypes were studied, and an assessment of the Zn and Fe contents in the grains of the newly developed genotypes at CIMMYT, along with their agronomic performance compared with those of the established varieties from South Asia, was performed. The GE parameter was significant for grain Fe and Zn, and established varieties of EGP were not inferior to the CIMMYT germplasm tested.

Khoshgoftarmanesh et al. (2012) tested Iranian spring bread wheat genotypes with almost all commercial wheat genotypes cultivated at two field sites during two consecutive growing seasons. Significant variation in the grain Zn and Fe concentrations was detected among the genotypes, both with and without added Zn fertilization. The location had a significant effect on the classification of the wheat genotypes. CA was used to classify wheat genotypes into different groups on the basis of the STI and grain Zn and Fe concentrations as a basis for the biofortification of wheat through breeding. For example, in the case of the dataset corresponding to Karaj, five genotype groups were identified (Fig. 7.8). The groups differed in terms of grain Fe concentration and STI but not in terms of grain Zn. Group I (eight genotypes) had a medium Fe concentration, Groups II (three genotypes) and III (six genotypes) had high STI values and Groups IV (11 genotypes) and V (two genotypes) had the highest STI values and Fe levels. It was concluded that the CA technique is an efficient approach for classifying wheat genotypes as a target for biofortification breeding because of the interactions of traits with location.

Krishnappa et al. (2017), using a combination of ED–XRF, ICP–MS, PCR and QTL methods, performed investigations with 286 RILs developed by single-seed descent from a cross between a cultivated bread wheat variety and a synthetic derivative. The means and ranges of GFeC, GZnC PC and TKW in the RIL population

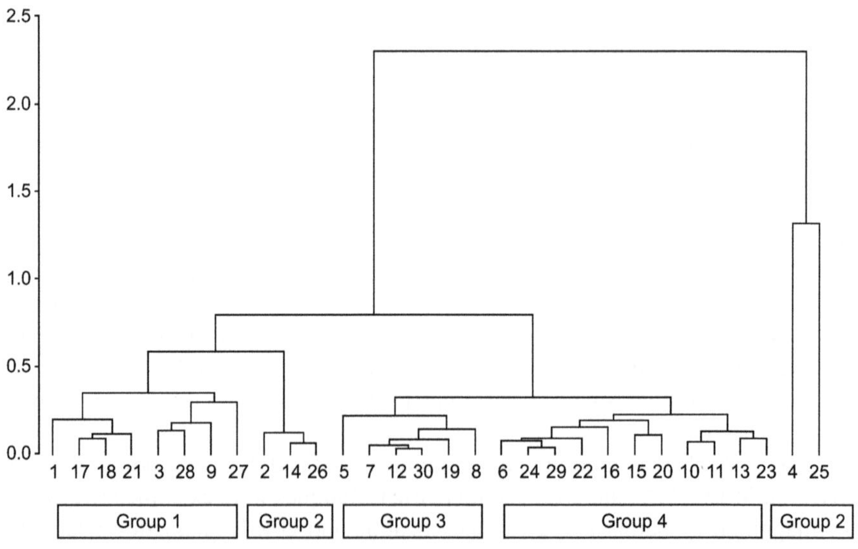

Fig. 7.8 Cluster analysis of genotypes based on the 2-year averages of the STI and the grain Zn and Fe concentrations in the treatment with Zn fertilization at Karaj. (After Khoshgoftarmanesh AH et al., 2012. J Geochem Explor 121:49–54. https://doi.org/10.1016/j.gexplo.2012.06.002. With permission)

were determined, and these parameters were found to be quantitatively inherited traits whose expression is strongly influenced by the environment. The identified QTLs represent novel genomic regions associated with Fe, Zn, PC and TKW. At least three chromosomes of the A genome are important in determining the accumulation of Fe, Zn, PC and TKW; therefore, fine mapping of the regions involved in these chromosomes may be satisfactory.

Gopalareddy et al. (2015) examined wheat samples in three environments over two years using EDXRF. Considerable genetic variation in the micronutrient contents of both grain types was present. The environmental effect was the most important source of variation in the grain Fe and Zn concentrations, explaining 37.42% and 57.78%, respectively, of the total sum of squares. The magnitude of the G × E interaction was relatively high for Fe and Zn, accounting for 29.46% and 23.24% of the total sum of squares, respectively. High heritabilities were obtained for Fe and Zn levels, indicating non-crossover interactions. The moderate relationship between the Fe and Zn contents allowed the simultaneous improvement of these micronutrients.

Krishnappa et al. (2021) used combined methods such as ED–XRF, ICP–MS, PCR and QTL analysis to explore the genomic regions governing GFeC, GZnC, GPC and TKW in a set of 163 RILs derived from a cross between cultivated wheat variety WH542 and a synthetic derivative (*Triticum dicoccon* PI94624/*Aegilops tauschii* [409]//BCN). The RIL population was genotyped using 100 SSR and 736 SNP markers and phenotyped in six environments. The constructed genetic map had a total genetic length of 7057 cM. Twenty-one novel QTLs were recognized on 13 chromosomes representing all three genomes of wheat. The identification of the best QTL combinations was achieved by evaluating the additive effects of the stable QTLs for GFeC, GZnC and GPC. Eight RILs for GZnC and three for GPC with favourable QTL combinations could be useful approaches as potential donors to develop bread wheat varieties with increased micronutrients and protein.

Using ICP–OES, Gomez-Becerra et al. (2010) studied each mineral nutrient for genotype-by-environment (G × E) interactions, genotype stability and correlations between mineral elements and their stability. PCA revealed relationships between P, Mg, K and S and between Zn and Fe. Copper, Mn and Ca were unrelated to the other elements. The coassociation between the Zn and Fe contents was more significant in the case of the PCA (with the first two PCs explaining 64.48% of the total variance) than in the correlation analysis. The G × E interaction explained between 20% and 40% of the total variation for all the mineral elements, except for S and Zn, whose contributions accounted for less than 16%. However, the genotype effect was also significant for Mg, Mn, Zn and S, with the latter two having the largest heritability values of 72% and 77%, respectively. Iron exhibited low heritability and high ratios between the G × E interaction and genotype variance components ($\delta^2_{GE}/\delta^2_{G}$). These findings suggest that specific adaptations to Fe could be positively exploited. The wild emmer germplasm revealed several outstanding accessions (MM 5/4 and 24–39) with respect to grain Zn and Fe levels and environmental stability that can be utilized as possible donors to improve grain micronutrient concentrations in wheat.

Zeng et al. (2010) identified genotypic variation in element concentrations in brown rice grains from 628 landraces from Yunnan landraces in China by means of combined methods (ICP–AES, AAS and microsatellite analysis, including PCR amplification procedures, morphological traits and different routine analyses). On the basis of the number and distance coefficients of alleles for 20 SSR markers (DNA markers of choice for genetic analysis in rice, *Oryza sativa* L.) for the landraces, such interelement differentiation was significant ($P < 0.05$). Increasing the concentrations of Ca, Fe and Zn in rice grains may be helpful for alleviating chronic Ca, Zn and Fe deficiencies in many areas of the world. According to the authors, further study of the relationships between mineral elements and gene diversity is needed.

Choi et al. (2013) reported that ICP–AES combined with HCA was efficient at classifying 21 Korean wheat cultivars, resulting in six clusters corresponding to similar mineral compositions. Among them, the fourth cluster included six wheat cultivars high in Cu and K, whereas the fifth cluster contained four cultivars with the highest Zn, Mg and P contents, and the sixth cluster included only one cultivar with the highest Fe and Mn levels. The differentiation of mineral compositions can be used in wheat breeding programs to develop new wheat cultivars with high essential element contents, consequently improving the nutritional profile of wheat grains.

Tripathi et al. (2012) determined the quantification/deposition patterns of Si and other elements in different morphological parts of wheat (*Triticum aestivum*) plants using LIBS and phytolith analysis. The LIBS spectra provided the spectral signatures of the elements analyzed. Phytolith analysis revealed intermorphological differences in the Si content of wheat plants. The results of LIBS analysis and phytolith analysis were in good agreement. PCA explaining 100% of the variance in the dataset (with PC1 and PC2 accounting for 99% and 1%, respectively) was applied to the LIBS spectral dataset of the different wheat plant parts. It was possible to discriminate between the vegetative and fertile parts of the wheat plants. In addition, the presence of Si in each morphological part of the wheat plants also indicates that plant residues could be used in eco-friendly biocement production.

7.3 Conclusions

The quality of different cereal products, e.g. wheat, barley, rye, triticale, oat and corn samples, has been assessed chemometrically relative to their preferred mineral composition and stable isotope ratios. Most studies have focused on rise analyses. Appropriate instrumental analytical tools combined with multivariate processing are highly reliable for assessing cereal product authentication as well as adulteration/mislabelling with determination of their safety. The adequate use of both of these approaches to evaluate the health quality of this food is of considerable importance for their categorization according to geographical origin, intervarietal susceptibility to industrial impact, differentiation in culinary/technological processing, the relationship between cereal cultivars and their nutritional profile as well as between genotypes and their geographic origin, etc. This approach will also be useful for differentiating conventional and organic crops and farming systems as well as for detecting the mislabelling of products to preserve the brand reputations of original products.

References

Akin PA, Sezer B, Sanal T, Apaydin H, Koksel H, Boyaci İH (2020) Multi-elemental analysis of flour types and breads by using laser-induced breakdown spectroscopy. J Cereal Sci 92:102920. doi:https://doi.org/10.1016/j.jcs.2020.102920

Akın PA, Sezer B, Bean SR, Peiris K, Tilley M, Apaydın H, Boyacı İH (2021) Analysis of corn and sorghum flour mixtures using laser-induced breakdown spectroscopy. J Sci Food Agric 101:1076–1084. https://doi.org/10.1002/jsfa.10717

Arif M, Chilvers G, Day S, Naveed SA, Woolfe M, Rodionova OY, Pomerantsev AL, Brodie C, Mihailova A, Abrahim A, Cannavan A, Kelly SD (2021) Differentiating Pakistani long-grain rice grown inside and outside the accepted Basmati Himalayan geographical region using a 'one-class' multi-element chemometric model. Food Contr 123:107827. https://doi.org/10.1016/j.foodcont.2020.107827

Ariyama K, Shinozaki M, Kawasaki A, Ishida Y (2011) Strontium and lead isotope analyses for determining the geographic origins of grains. Anal Sci 27:709–713. https://doi.org/10.2116/analsci.27.709

Ariyama K, Shinozaki M, Kawasaki A (2012) Determination of the geographic origin of rice by chemometrics with strontium and lead isotope ratios and multielement concentrations. J Agric Food Chem 60:1628–1634. https://doi.org/10.1021/jf204296p

Armanino C, Festa MR (1996) Characterization of wheat by four analytical parameters. A chemometric study. Anal Chim Acta 331:43–51. https://doi.org/10.1016/0003-2670(96)00164-X

Arunakumara KKIU, Walpola BC, Yoon MH (2013) Current status of heavy metal contamination in Asia's rice lands. Rev Environ Sci Biotechnol 12:355–377. https://doi.org/10.1007/s11157-013-9323-1

Barbosa RM, de Paula ES, Pauleili AC, Moore AF, Souza Oliveira JM, Lemos Batista B, Campiglia AD, Barbosa F Jr (2016) Recognition of organic rice samples based on trace elements and support vector machines. J Food Compos Anal 45:95–100. https://doi.org/10.1016/j.jfca.2015.09.010

Bontempo L, Camin F, Paolini M, Micheloni C, Laursen KH (2016) Multi-isotopic signatures of organic and conventional Italian pasta along the production chain. J Mass Spectrom 51:675–683. https://doi.org/10.1002/jms.3816

Borges EM, Lafayette Neves Gelinski JM, de Oliveira Souza VC, Barbosa F Jr, Batista BL (2015) Monitoring the authenticity of organic rice via chemometric analysis of elemental data. Food Res Int 77:299–309. https://doi.org/10.1016/j.foodres.2015.06.008

Branch S, Burke S, Evans P, Ben Fairman B, Wolff Briche CSJ (2003) A preliminary study in determining the geographical origin of wheat using isotope ratio inductively coupled plasma mass spectrometry with ^{13}C, ^{15}N mass spectrometry. J Anal At Spectrom 18:17–22. https://doi.org/10.1039/b207055n

Brandolini V, Coïsson JD, Tedeschi P, Barile D, Cereti E, Maietti A, Vecchiati G, Martelli A, Arlorio M (2006) Chemometrical characterization of four Italian rice varieties based on genetic and chemical analyses. J Agric Food Chem 54:9985–9991. https://doi.org/10.1021/jf061799m

Brescia MA, Di Martino G, Fares C, Di Fonzo N, Platani C, Ghelli S, Reniero F, Sacco A (2002a) Characterisation of Italian durum wheat semolina by means of chemical analytical and spectroscopic determinations. Cereal Chem 79:238–242. https://doi.org/10.1094/CCHEM.2002.79.2.238

Brescia MA, Di Martino G, Guillou C, Reniero F, Sacco A, Serra F (2002b) Differentiation of the geographical origin of durum wheat semolina samples on the basis of isotopic composition. Rapid Commun Mass Spectrom 16:2286–2290. https://doi.org/10.1002/rcm.860

Brescia MA, Sacco D, Sgaramella A, Pasqualone A, Simeone R, Peri G, Sacco A (2007) Characterisation of different typical Italian breads by means of traditional, spectroscopic and image analyses. Food Chem 104:429–438. https://doi.org/10.1016/j.foodchem.2006.09.043

Bui MQ, Quan TC, Nguyen QT, Tran-Lam TT, Dao YH (2022) Geographical origin traceability of Sengcu rice using elemental markers and multivariate analysis. Food Addit Contam B 15:177–190. https://doi.org/10.1080/19393210.2022.2070932

Cheajesadagul P, Arnaudguilhem C, Shiowatana J, Siripinyanond A, Szpunar J (2013) Discrimination of geographical origin of rice based on multi-element fingerprinting by high resolution inductively coupled plasma mass spectrometry. Food Chem 141:3504–3509. https://doi.org/10.1016/j.foodchem.2013.06.060

Chen T, Zhao Y, Zhang W, Yang S, Ye Z, Zhang G (2016) Variation of the light stable isotopes in the superior and inferior grains of rice (*Oryza sativa* L.) with different geographical origins. Food Chem 209:95–98. https://doi.org/10.1016/j.foodchem.2016.04.029

Choi I, Kang C-S, Hyun J-N, Lee C-K, Park K-G (2013) Mineral compositions of Korean wheat cultivars. Prev Nutr Food Sci 18:214–217. https://doi.org/10.3746/pnf.2013.18.3.214

Chung I-M, Kim J-K, Le J-K, Kim S-H (2015) Discrimination of geographical origin of rice (*Oryza sativa* L.) by multielement analysis using inductively coupled plasma atomic emission spectroscopy and multivariate analysis. J Cereal Sci 65:252–259. https://doi.org/10.1016/j.jcs.2015.08.001

Chung I-M, Kim J-K, Prabakaran M, Yang J-H, Kim S-H (2016) Authenticity of rice (*Oryza sativa* L.) geographical origin based on analysis of C, N, O and S stable isotope ratios: a preliminary case report in Korea, China and Philippine. J Sci Food Agric 96:2433–2439. https://doi.org/10.1002/jsfa.7363

Chung I, Kim J, Lee K, Park S, Lee J, Son N, Jin Y, Kim S (2018) Geographic authentication of Asian rice (*Oryza sativa* L.) using multi-elemental and stable isotopic data combined with multivariate analysis. Food Chem 240:840–849. https://doi.org/10.1016/j.foodchem.2017.08.023

da Silva DG, Scarminio IS, Anunciação DS, Souza AS, da Silva EGP, Ferreira SLC (2013) Determination of the mineral composition of Brazilian rice and evaluation using chemometric techniques. Anal Meth 5:998–1003. https://doi.org/10.1039/c2ay26158h

Goiter Asfaha D, Quétel CR, Thomas F, Horacek M, Wimmer B, Heiss G, Dekant C, Deters-Itzelsberger P, Hoelzl S, Rummel S, Brach-Papan C, Van Bocxstaele M, Jamin E, Baxter M, Heinrich K, Kelly S, Bertoldi D, Bontempo L, Camin F, Larcher R, Perini M, Rossmann A, Schellenberg A, Schlicht C, Froeschl H, Hoogewerff J, Ueckermann H (2011) Combining isotopic signatures of n(^{87}Sr)/n (^{86}Sr) and light stable elements (C, N, O, S) with multi-elemental profiling for the authentication of provenance of European cereal samples. J Cereal Sci 53:170–177. https://doi.org/10.1016/j.jcs.2010.11.004

Gomez-Becerra HF, Yazici A, Ozturk L, Budak H, Peleg Z, Morgounov A, Fahima T, Saranga Y, Cakmak I (2010) Genetic variation and environmental stability of grain mineral nutrient concentrations in *Triticum dicoccoides* under five environments. Euphytica 171:39–52. https://doi.org/10.1007/s10681-009-9987-3

Gonzálvez A, Armenta S, de la Guardia M (2011) Geographical traceability of “Arròs de Valencia” rice grain based on mineral element composition. Food Chem 126:1254–1260. https://doi.org/10.1016/j.foodchem.2010.11.032

Gopalareddy K, Singh AM, Ahlawat AK, Singh GP, Jaiswal JP (2015) Genotype-environment interaction for grain iron and zinc concentration in recombinant inbred lines of a bread wheat (*Triticum aestivum* L.) cross. Indian J Genet 75:307–313. https://doi.org/10.5958/0975-6906.2015.00048.6

Grembecka M (2007) Bromatological and chemometric assessment of plants food based on its mineral composition. Dissertation, Gdańsk Medical Academy

Grembecka M, Szefer P (2016) Elemental trace analysis in studies of food products. In: Baranowska I (ed) Handbook of trace analysis—fundamentals and applications. Springer, pp 203–239. https://doi.org/10.1007/978-3-319-19614-5

Husted S, Mikkelsen BF, Jensen J, Nielsen NE (2004) Elemental fingerprint analysis of barley (*Hordeum vulgare*) using inductively coupled plasma mass spectrometry, isotope-ratio mass spectrometry, and multivariate statistics. Anal Bioanal Chem 378:171–182. https://doi.org/10.1007/s00216-003-2219-0

Jamali MK, Kazi TG, Arain MB, Afridi HI, Jalbani N, Sarfraz RA, Baig JA (2008) A multivariate study: variation in uptake of trace and toxic elements by various varieties of *Sorghum bicolor* L. J Hazard Mater 158:644–651. https://doi.org/10.1016/j.jhazmat.2008.02.007

Jeong S, Seol D, Kim H, Lee Y, Nam SH, An JM, Chung H (2023) Cooperative combination of LIBS-based elemental analysis and near-infrared molecular fingerprinting for enhanced discrimination of geographical origin of soybean paste. Food Chem 399:133956. https://doi.org/10.1016/j.foodchem.2022.133956

Jiao Z, Si X-X, Li G-K, Zhang Z-M, Xu X-P (2010) Unintended compositional changes in transgenic rice seeds (*Oryza sativa* L.) studied by spectral and chromatographic analysis coupled with chemometrics methods. J Agric Food Chem 58:1746–1754. https://doi.org/10.1021/jf902676y

Joshi AK, Crossa J, Arun B, Chand R, Trethowan R, Vargas M, Ortiz-Monasterio I (2010) Genotype×environment interaction for zinc and iron concentration of wheat grain in eastern Gangetic plains of India. Field Crop Res 116:268–277. https://doi.org/10.1016/j.fcr.2010.01.004

Kawasaki A, Oda H, Hirata T (2002) Determination of strontium isotope ratio of brown rice for estimating its provenance. Soil Sci Plant Nutr 48:635–640. https://doi.org/10.1080/00380768.2002.10409251

Kelly S, Baxter M, Chapman S, Rhodes C, Dennis J, Brereton P (2002) The application of isotopic and elemental analysis to determine the geographical origin of premium long grain rice. Eur Food Res Technol 214:72–78. https://doi.org/10.1007/s002170100400

Khoshgoftarmanesh AH, Sharifi HR, Afiuni D, Schulin R (2012) Classification of wheat genotypes by yield and densities of grain zinc and iron using cluster analysis. J Geochem Explor 121:49–54. https://doi.org/10.1016/j.gexplo.2012.06.002

Kim G, Kwak J, Choi J, Park K (2012) Detection of nutrient elements and contamination by pesticides in spinach and rice samples using Laser-Induced Breakdown Spectroscopy (LIBS). J Agric Food Chem 60:718–724. https://doi.org/10.1021/jf203518f

Kokot S, Phuong TD (1999) Elemental content of Vietnamese rice Part 2. Multivariate data analysis. Analyst 124:561–569. https://doi.org/10.1039/A808799G

Korenaga T, Musashi M, Nakashita R, Suzuki Y (2010) Statistical analysis of rice samples for compositions of multiple light elements (H, C, N, and O) and their stable isotopes. Anal Sci 26:873–878. https://doi.org/10.2116/analsci.26.873

Krishnappa G, Singh AM, Chaudhary S, Ahlawat AK, Singh SK, Shukla RB, Jaiswal JP, Singh GP, Solanki IS (2017) Molecular mapping of the grain iron and zinc concentration, protein content and thousand kernel weight in wheat (*Triticum aestivum* L.). PLoS One 12:e0174972. https://doi.org/10.1371/journal.pone.0174972

Krishnappa G, Rathan ND, Sehgal D, Ahlawat AK, Singh SK, Singh SK, Shukla RB, Jaiswal JP, Solanki IS, Singh GP, Singh AM (2021) Identification of novel genomic regions for biofortification traits using an SNP marker-enriched linkage map in wheat (*Triticum aestivum* L.). Front Nutr 8:669444. https://doi.org/10.3389/fnut.2021.669444

Lagad RA, Singh SK, Rai VK (2017) Rare earth elements and $^{87}Sr/^{86}Sr$ isotopic characterization of Indian Basmati rice as potential tool for its geographical authenticity. Food Chem 217:254–265. https://doi.org/10.1016/j.foodchem.2016.08.094

Laursen KH, Hansen TH, Persson DP, Schjoerring JK, Søren Husted S (2009) Multi-elemental fingerprinting of plant tissue by semi-quantitative ICP-MS and chemometrics. J Anal At Spectrom 24:1198–1207. https://doi.org/10.1039/B901960J

Laursen KH, Schjoerring JK, Olesen JE, Askegaard M, Helekoh U, Husted S (2011) Multielemental fingerprinting as a tool for authentication of organic wheat, barley, faba bean, and potato. J Agric Food Chem 59:4385–4396. https://doi.org/10.1021/jf104928r

Laursen KH, Mihailova A, Kelly SD, Epov VN, Bérail S, Schjoerring JK, Donard OFX, Larsen EH, Pedentchouk N, Marca-Bell AD, Halekoh U, Olesen JE, Husted S (2013) Is it really organic?—Multi-isotopic analysis as a tool to discriminate between organic and conventional plants. Food Chem 141:2812–2820. https://doi.org/10.1016/j.foodchem.2013.05.068

Laursen KH, Schjoerring JK, Kelly SD, Husted S (2014) Authentication of organically grown plants—advantages and limitations of atomic spectroscopy for multi-element and stable isotope analysis. Trac-Trend Anal Chem 59:73–82. https://doi.org/10.1016/j.trac.2014.04.008

Li A (2011) Determination of the geographical origin of rice using stable isotope ratio and multi-element analyses. MSc Thesis, The University of York

Li G, Zhu F (2017) Physicochemical properties of quinoa flour as affected by starch interactions. Food Chem 221:1560–1568. https://doi.org/10.1016/j.foodchem.2016.10.137

Li G, Nunes L, Wang Y, Williams PN, Zhen M, Zhang Q, Zhu Y (2012) Profiling the ionome of rice and its use in discriminating geographical origins at the regional scale, China. J Environ Sci 25:144–154. https://doi.org/10.1016/S1001-0742(12)60007-2

Li C, Dong H, Luo D, Xian Y, Fu X (2016a) Recent developments in application of stable isotope and multi-element analysis on geographical origin traceability of cereal grains. Food Anal Methods 9:1512–1519. https://doi.org/10.1007/s12161-015-0328-y

Li G, Wang S, Zhu F (2016b) Physicochemical properties of quinoa starch. Carbohydr Polym 137:328–338. https://doi.org/10.1016/j.carbpol.2015.10.064

Li C, Nie J, Zhang Y, Shao S, Liu Z, Rogers KM, Zhang W, Yuan Y (2022) Geographical origin modeling of Chinese rice using stable isotopes and trace elements. Food Contr 138:108997. https://doi.org/10.1016/j.foodcont.2022.108997

Liang K, Liang S, Lu L, Zhu D, Cheng L (2018) Geographical origin traceability of foxtail millet based on the combination of multi-element and chemical composition analysis. Inter J Food Properties 21:1769–1777. https://doi.org/10.1080/10942912.2018.1506479

Lin P, Wen X, Ma S, Liu X, Xiao R, Gu Y, Chen G, Han Y, Dong D (2023) Rapid identification of the geographical origins of crops using laser-induced breakdown spectroscopy combined with transfer learning. Spectrochim Acta Part B: Atom Spectrosc 206:106729. https://doi.org/10.1016/j.sab.2023.106729

Liu H, Guo B, Wei Y, Wei S, Ma Y, Zhang W (2015) Effects of region, genotype, harvest year and their interactions on $\delta^{13}C$, $\delta^{15}N$ and δD in wheat kernels. Food Chem 171:56–61. https://doi.org/10.1016/j.foodchem.2014.08.111

Liu H, Wei Y, Lu H, Wei S, Jiang T, Zhang T, Guo B (2016) Combination of the $^{87}Sr/^{86}Sr$ ratio and light stable isotopic values ($\delta^{13}C$, $\delta^{15}N$ and δD) for identifying the geographical origin of winter wheat in China. Food Chem 212:367–373. https://doi.org/10.1016/j.foodchem.2016.06.002

Liu H, Wei Y, Zhang Y, Wei S, Zhang S, Guo B (2017) The effectiveness of multi-element fingerprints for identifying the geographical origin of wheat. Inter J Food Sci Technol 52:1018–1025. https://doi.org/10.1111/ijfs.13366

Liu X, Feng X, Liu F, Peng J, He Y (2019) Rapid identification of genetically modified maize using laser-induced breakdown spectroscopy. Food Bioprocess Technol 2:347–357. https://doi.org/10.1007/s11947-018-2216-0

Liu Z, Yuan Y, Xie T, Zhang Y, Shao S, Nie J, Xia W, Rogers KM, Zhang W (2020) Long-term agricultural effects on the authentication accuracy of organic, green, and conventional rice using isotopic and elemental chemometric analyses. J Agric Food Chem 68:1213–1225. https://doi.org/10.1021/acs.jafc.9b06847

Liu X, Bai B, Rogers KM, Wu D, Qian Q, Qi F, Zhou J, Yao C, Song W (2022) Determining the geographical origin and cultivation methods of Shanghai special rice using NIR and IRMS. Food Chem 394:133425. https://doi.org/10.1016/j.foodchem.2022.133425

Luo D, Dong H, Luo H, Xian Y, Wan J, Guo X, Wu Y (2015) The application of stable isotope ratio analysis to determine the geographical origin of wheat. Food Chem 174:197–201. https://doi.org/10.1016/j.foodchem.2014.11.006

Maione C, Melgaço Barbosa R (2018) Recent applications of multivariate data analysis methods in the authentication of rice and the most analyzed parameters: a review. Crit Rev Food Sci Nutr 59:1868–1879. https://doi.org/10.1080/10408398.2018.1431763

Maione C, Batista BL, Campiglia AD, Barbosa F Jr, Melgaço Barbosa R (2016) Classification of geographic origin of rice by data mining and inductively coupled plasma mass spectrometry. Comput Electron Agric 121:101–107. https://doi.org/10.1016/j.compag.2015.11.009

Marini F, Zupan J, Magrí AL (2004) On the use of counterpropagation artificial neural networks to characterize Italian rice varieties. Anal Chim Acta 510:231–240. https://doi.org/10.1016/j.aca.2004.01.009

Martelli MR, Brygo F, Sadoudi A, Delaporte P, Barron C (2010) Laser-induced breakdown spectroscopy and chemometrics: a novel potential method to analyze wheat grains. J Agric Food Chem 58:7126–7134. https://doi.org/10.1021/jf100665u

Oda H, Kawasaki A, Hirata T (2001) Determination of the geographic origin of brown-rice with isotope ratios of $^{11}B/^{10}B$ and $^{87}Sr/^{86}Sr$. Anal Sci 17:1627–1630

Otaka A, Hokura A, Nakai I (2014) Determination of trace elements in soybean by X-ray fluorescence analysis and its application to identification of their production areas. Food Chem 147:318–326. https://doi.org/10.1016/j.foodchem.2013.09.142

Pandey A, Khan MK, Hakki EE, Thomas G, Hamurcu M, Gezgin S, Gizlenci O, Akkaya MS (2016) Assessment of genetic variability for grain nutrients from diverse regions: potential for wheat improvement. Springer Plus 5:1912. https://doi.org/10.1186/s40064-016-3586-2

Paolini M, Ziller L, Laursen KH, Husted S, Camin F (2015) Compound-specific $\delta^{15}N$ and $\delta^{13}C$ analyses of amino acids for potential discrimination between organically and conventionally grown wheat. J Agric Food Chem 63:5841–5850. https://doi.org/10.1021/acs.jafc.5b00662

Peng L-X, Huang Y-F, Liu Y, Zhang Z-F, Lu L-Y, Zhao G (2014) Evaluation of essential and toxic element concentrations in buckwheat by experimental and chemometric approaches. J Integr Agric 13:1691–1698. https://doi.org/10.1016/S2095-3119(13)60724-8

Pfeiffer WH, McClafferty B (2007) HarvestPlus: breeding crops for better nutrition. Crop Sci 47(S3):S88–S105. https://doi.org/10.2135/cropsci2007.09.0020IPBS

Phuong TD, Chuong PV, Khiemc DT, Kokot S (1999) Elemental content of Vietnamese rice Part 1. Sampling, analysis and comparison with previous studies. Analyst 124:553–560. https://doi.org/10.1039/A808796B

Qian L, Zuo F, Liu H, Zhang C, Chi X, Zhang D (2019) Determination of geographical origin of Wuchang rice with the geographical indicator by multielement analysis. J Food Qual 8396865. https://doi.org/10.1155/2019/8396865

Quinn B, McCarron P, Hong Y, Birse N, Wu D, Elliott CT, Ratnasekhar C (2022) Elementomics combined with DD-SIMCA and K-NN to identify the geographical origin of rice samples from China, India, and Vietnam. Food Chem 386:132738. https://doi.org/10.1016/j.foodchem.2022.132738

Ranulfi AC, Senesi GS, Caetano JB, Meyer MC, Magalhaes AB, Villas-Boas PR, Milori DMBP (2018) Nutritional characterization of healthy and *Aphelenchoides besseyi* infected soybean leaves by laser-induced breakdown spectroscopy (LIBS). Microchem J 141:118–126. https://doi.org/10.1016/j.microc.2018.05.008

Ribeiro MCS, Senesi GS, Cabral JS, Cena C, Marangoni BS, Kiefer C, Nicolodelli G (2020) Evaluation of rice varieties using LIBS and FTIR techniques associated with PCA and machine learning algorithms. Appl Opt 59:10043–10048. https://doi.org/10.1364/AO.409029

Senesi GS, Cabral J, Menegatti CR, Marangoni B, Nicolodelli G (2019) Recent advances and future trends in LIBS applications to agricultural materials and their food derivatives: An overview of developments in the last decade (2010–2019). Part II. Crop plants and their food derivatives. Trac-Trend Anal Chem 118:453–469. https://doi.org/10.1016/j.trac.2019.05.052

Sezer B, Bilge G, Sanal T, Koksel H, Boyaci IH (2017) A novel method for ash analysis in wheat milling fractions by using laser-induced breakdown spectroscopy. J Cereal Sci 78:33–38. https://doi.org/10.1016/j.jcs.2017.04.002

Shao S, Chen Y, Zhang Y, Hu G, Zhu J, Yuan Y (2015) Determination of the geographic origin of rice by PCA-LDA based on the stable isotopes and multi-elements concentrations. J Nucl Agric Sci 29:119–127. https://doi.org/10.11869/j.issn.100-8551.2015.01.0119

Suzuki Y, Chikaraishi Y, Ogawa NO, Ohkouchi N, Korenaga T (2008) Geographical origin of polished rice based on multiple element and stable isotope analyses. Food Chem 109:470–475. https://doi.org/10.1016/j.foodchem.2007.12.063

Szefer P (2003) Application of chemometric techniques in analytical evaluation of biological and environmental samples. In: Namieśnik J, Chrzanowski W, Żmijewska P (eds) New horizons and challenges in environmental analysis and monitoring. CEEAM, Gdańsk, pp 355–388

Szefer P (2007) Chemometric techniques in analytical evaluation of food quality. In: Szefer P, Nriagu J (eds) Mineral components in foods. CRC Press/Taylor & Francis, Boca Raton, pp 69–121

Tripathi DK, Kumar R, Pathak AK, Chauhan DK, Rai AK (2012) Laser-induced breakdown spectroscopy and phytolith analysis: an approach to study the deposition and distribution pattern of silicon in different parts of wheat (*Triticum aestivum* L.) plant. Agric Res 1:352–361. https://doi.org/10.1007/s40003-012-0042-6

Velu G, Singh RP, Huerta-Espino J, Peña RJ, Arun B, Mahendru-Singh A, Mujahid MY, Sohu VS, Mavi GS, Crossa J, Alvarado G, Joshi AK, Pfeiffer WH (2012) Performance of biofortified spring wheat genotypes in target environments for grain zinc and iron concentrations. Field Crop Res 137:261–267. https://doi.org/10.1016/j.fcr.2012.07.018

Wadood SA, Guo B, Wei Y (2019) Geographical traceability of wheat and its products using multielement light stable isotopes coupled with chemometrics. J Mass Spectrom 54:178–188. https://doi.org/10.1002/jms.4312

Wang J, Chen T, Zhang W, Zhao Y, Yang S, Chen A (2020) Tracing the geographical origin of rice by stable isotopic analyses combined with chemometrics. Food Chem 313:126093. https://doi.org/10.1016/j.foodchem.2019.126093

Wu Y, Luo D, Dong H, Wan J, Luo H, Xian Y, Guo X, Qin F, Han W, Wang L, Wang B (2015) Geographical origin of cereal grains based on element analyser-stable isotope ratio mass spectrometry (EA-SIRMS). Food Chem 174:553–557. https://doi.org/10.1016/j.foodchem.2014.11.096

Xu Y, An D, Liu D, Zhang A, Xu H, Bin Li B (2012) Molecular mapping of QTLs for grain zinc, iron and protein concentration of wheat across two environments. Field Crop Res 138:57–62. https://doi.org/10.1016/j.fcr.2012.09.017

Xu J, Li Y, Wang S, Long S, Wu Y, Chen Z (2023) Sources, transfers and the fate of heavy metals in soil-wheat systems: the case of lead (Pb)/zinc (Zn) smelting region. J Hazard Mater 441:129863. https://doi.org/10.1016/j.jhazmat.2022.129863

Yang P, Zhu Y, Yang X, Li J, Tang S, Hao Z, Guo L, Li X, Zeng X, Lu Y (2018a) Evaluation of sample preparation methods for rice geographic origins classification using laser-induced breakdown spectroscopy. J Cereal Sci 80:111–118. https://doi.org/10.1016/j.jcs.2018.01.007

Yang P, Zhou R, Zhang W, Tang S, Hao Z, Li X, Lu Y, Zeng X (2018b) Laser-induced breakdown spectroscopy-assisted chemometric methods for rice geographic origin classification. Appl Opt 57:8297–8302. doi:https://doi.org/10.1364/AO.57.008297

Yang W, Wang D, Wang M, Zhou F, Huang J, Xue M, Dinh QT, Liang D (2019a) Heavy metals and associated health risk of wheat grain in a traditional cultivation area of Baoji, Shaanxi, China. Environ Monit Assess 191:428. https://doi.org/10.1007/s10661-019-7534-9

Yang P, Zhou R, Zhang W, Yi R, Tang S, Guo L, Hao Z, Li X, Lu Y, Zeng X (2019b) High-sensitivity determination of cadmium and lead in rice using laser-induced breakdown spectroscopy. Food Chem 272:323–328. https://doi.org/10.1016/j.foodchem.2018.07.214

Yang J, Wang DQ, Zhang Q, Xu JG (2020) Origin identification of buckwheat from different areas in loess plateau by chemical analysis and data mining techniques. AJMPCP 3:40–49. https://journalajmpcp.com/index.php/AJMPCP/article/view/46

Yasui A, Shindoh K (2000) Determination of the geographic origin of brown-rice with trace-element composition. Bunseki Kagaku 49:405–410. https://www.jstage.jst.go.jp/article/bunsekikagaku/49/6/49_6_405/_article/-char/en

Yuan Y, Zhang W, Zhang Y, Liu Z, Shao S, Zhou L, Rogers KM (2018) Differentiating organically farmed rice from conventional and green rice harvested from an experimental field trial using stable isotopes and multi-element chemometrics. J Agric Food Chem 66:2607–2615. https://doi.org/10.1021/acs.jafc.7b05422

Zeng Y, Zhang H, Yang S, Du J, Pu X, Wang L, Liu J, Xiao F, Li Z (2009a) Correlation between allele sizes of microsatellites and phenotypic variations in rice landraces. Front Agric China 3:130–139. https://doi.org/10.1007/s11703-009-0033-y

Zeng Y, Wang L, Du J, Liu J, Yang S, Pu X, Xiao F (2009b) Elemental content in brown rice by inductively coupled plasma atomic emission spectroscopy reveals the evolution of Asian cultivated rice. JIPB 51:466–475. https://doi.org/10.1111/j.1744-7909.2009.00820.x

Zeng Y, Zhang H, Wang L, Pu X, Du J, Yang S, Liu J (2010) Genotypic variation in element concentrations in brown rice from Yunnan landraces in China. Environ Geochem Health 32:165–177. https://doi.org/10.1007/s10653-009-9272-3

Zhang Q, Xu J-G (2017) Determining the geographical origin of common buckwheat from China by multivariate analysis based on mineral elements, amino acids and vitamins. Sci Rep 7:9696. https://doi.org/10.1038/s41598-017-08808-y

Zhang M, Li C, Liu Y, Zhang Y, Nie J, Shao S, Mei H, Rogers KM, Zhang W, Yuan Y (2024) Effects of water isotope composition on stable isotope distribution and fractionation of rice and plant tissues. J Agric Food Chem 72:8955–8962. https://doi.org/10.1021/acs.jafc.3c08451

Zhao H, Guo B, Wei Y, Zhang B, Sun S, Zhang L, Yan J (2011) Determining the geographic origin of wheat using multielement analysis and multivariate statistics. J Agric Food Chem 59:4397–4402. dx.doi.org. https://doi.org/10.1021/jf200108d

Zhao H, Guo B, Wei Y, Zhang B (2012) Effects of wheat origin, genotype, and their interaction on multielement fingerprints for geographical traceability. J Agric Food Chem 60:10957–10962. https://doi.org/10.1021/jf3021283

Zhao H, Guo B, Wei Y, Zhang B (2013) Multi-element composition of wheat grain and provenance soil and their potentialities as fingerprints of geographical origin. J Cereal Sci 57:391–397. https://doi.org/10.1016/j.jcs.2013.01.008

Chapter 8
Olive Oils and Oilseeds

Contents

8.1 Introduction

Olive oil, which is rich in oleic and linoleic acids, has been a major component of the Mediterranean diet for thousands of years. It is known to exert a protective effect against various types of cancer and cardiovascular diseases. Therefore, its consumption has been constantly increasing worldwide. The chemical composition of olive oil depends on many factors, e.g. olive oil type and climatic, agronomic, cultivation, production and storage conditions (Karabagias et al. 2013). According to Cabrera-Vique et al. (2012b), virgin argan oil is rich in both oleic acid and linoleic acid, providing essential omega-6 polyunsaturated fatty acids. Moreover, γ-tocopherol, a potent antioxidant with anti-inflammatory properties, is present in much greater amounts in argan oil than in any other oil. This composition plays a major role in explaining the recommended protective effects of argan oil against various civilization diseases, e.g. cancer, diabetes and coronary heart disease, and thus has potential as a functional food.

Supplementary Information The online version contains supplementary material available at https://doi.org/10.1007/978-3-031-99687-0_8.

P. Szefer, *Chemometrics in Food Control Based on Elemental Analysis*, https://doi.org/10.1007/978-3-031-99687-0_8

To guarantee authenticity and promote the marketing of virgin olive oil, the European Union has adopted a series of regulations to certify that virgin olive oils conform to EU schemes known as PDO, PGI and TSG. Such designations also guarantee that the quality of the product is closely linked to its territorial and botanical origin and consequently increases its market value (EC 2006; Karabagias et al. 2013).

The physical, chemical and organoleptic characteristics of olive oil (OO) are regulated by the EU by Reg. EEC 2568/91, as an amendment, including established methods for analysis. Although the OO sector is highly regulated, it is accepted that there is still a problem concerning fats and oils, including OOs, which are ranked third after meat with meat products and fish with fish products according to the 2016 EU Food Fraud report. Therefore, EU legislation, which continuously pursues emerging frauds and the process of offering new methods or examining the current methods, has been constantly in progress. According to the author, the replacements of specific steps within the present EU methods with the use of more advanced and effective analytical modifications allow time and/or solvent consumption to be reduced. Existing regulatory methods and standards are criticized as well as weaknesses, and possible solutions to safeguard the consumer and protect the OO market have been highlighted (Conte et al. 2020).

In the last decade, the global market for imported, expensive foods has grown enormously. Multiple countries request the origin label of food to protect consumer interests. Regulating authorities are searching for an extensive and updated list of analytical techniques for the certification of authentic oils and to support law implementation. Various analytical devices have been combined with multivariate chemometric analysis to trace the geographical provenance of oils. Popular analytical methods/techniques have been considered, and scientometric evaluation underlining research trends in interregional authentication and preferred journals used for dissemination has been demonstrated (Tahir et al. 2022).

Geographical evaluation of EVOO is needed worldwide to protect brand reputations. Such assessment is also related to mandatory legislation and/or the obligations of true labelling in some countries. However, the development of official methods for controlling the specific authentication of EVOOs is still needed. NMR and MS combined with chemometric analysis have been widely used as effective tools for verifying geographical origin. According to Calò et al. (2022), the use of these analytical methods, supported by multivariate analysis, has been well documented. In the case of EVOO origin assessment, some specific countries, which are generally the main producers worldwide, are especially interested in provenance evaluation and certification compared with other countries. Moreover, some specific producers, such as Italian EVOO producers and consumers, may have been interested in this context. Since there are no official methods for such specific EVOO authentication, both NMR- and MS-based techniques, as general validation methods, should be considered to establish mandatory legislation and specific labelling rules for the characterization of recognized worldwide EVOOs (Calò et al. (2022)).

The presence of heavy metals as environmental pollutants in olives may affect oil quality since their trace levels may catalyze oxidation reactions, resulting in deleterious effects on oil flavour and storage. For example, Pb is the only heavy metal that has a fixed MRL for olive oil (0.10 mg/L according to Commission Regulation (EC) No 1881/2006). Higher levels of Pb are correlated with the use of organophosphorus pesticides (Angioni 2010).

Trace element analysis has key significance for the assessment of olive oil and adulteration detection. Important sources of major and minor elements in oil may include the parent soil, manufacturing or packaging of the product. Therefore, characterization of the trace element patterns in edible oil products with respect to their provenance is recommended. The application of suitable statistical processing of trace element data to obtain the geographical characterization of olive oils is recommended. With respect to analytical devices helpful in this context, stable isotope ratio analysis is a powerful tool for evaluating the authenticity of edible oil products (Karabagias et al. 2013).

Currently, different official analytical methods are used for the detection and quantification of common adulterants in EVOOs. However, these procedures are laborious and time-consuming and require sample preparation before analysis. Hence, the need arose for the development of rapid, precise and green analytical methods to determine adulteration in EVOOs (Meenu et al. 2019).

According to Valdés et al. (2018), targeted procedures, such as chromatography and DNA-based methods coupled with chemometric analysis, have high accuracy, sensitivity and selectivity, enabling the simultaneous evaluation of multiple analytes. Moreover, non-targeted methods based on spectroscopic techniques have been applied to determine the geographic provenance of food products with quick responses, low costs, non-destructive characteristics and the possibility of miniaturization.

Recent progress in analytical chemistry has led to the development of innovative and efficient modern analytical approaches. Recently, significant development of analytical methods/techniques owing to advanced scientific technologies has been achieved. Among them, methods/techniques such as FAAS, ET–AAS, CV–AAS, HG–AAS, ICP–MS, ICP–AES, ICP–OES and ETs are among the most applied tools for tracing the levels of elements in foods (Rodríguez-Méndez et al. 2010; Capar and Szefer 2012; Grembecka and Szefer 2011, 2016). ETs are defined as ranks of non-specific chemical sensors with partial sensitivity to various constituents, allowing the analysis of complex liquids. Sensors applied in electronic tongues can measure several parameters by means of mass detection, optical transduction, or electrochemical sensors. In particular, the latter, i.e. voltammetric, impedimetric or potentiometric sensors, are the most widely used (Rodríguez-Méndez et al. 2010).

The quality and correctness of analytical measurements should be strictly tested via a quality assurance procedure involving the use of certified reference materials (CRMs) (Konieczka and Namieśnik 2009; Namieśnik and Szefer 2010; Grembecka and Szefer 2016). It is the key undertaking to process chemometrically exclusively validated analytical data. Multivariate analyses, e.g. PCA, FA, CA, CDA, ANN, SIMCA and LDA, have been mostly recommended for assessing various foods (Szefer 2002, 2003, 2007; La Pera et al. 2010; Marini et al. 2010). Various chemometric techniques are widely used in the tracing and authentication of edible oils.

An overview of the chemometric methods for the authentication of the geographical and varietal origin of olive oils has been reported by Marini et al. (2010). The authors focused on several chemometric approaches for the prediction and verification of the geographical and botanical provenance of extra virgin oil samples.

According to Tahir et al. (2022), PLS–DA and LDA provide better results than traditional chemometrics for oil authentication. Olive oil and extra virgin olive oil are generally the most authentic oils.

8.2 Chemometric Assessment of Olive Oils and Oilseeds

8.2.1 Geographical Differentiation

8.2.1.1 European Countries

Olive Oil

Portarena et al. (2017) analyzed $\delta^{13}C$, $\delta^{18}O$ and RR spectra by IRMS and RRS obtained from EVOO samples from seven coastal regions of Italy. PCA with the first two PCs corresponding to 86.0% of the total variance partially differentiated the Latium and Tuscany oil samples from the remaining samples. LDA achieved clearer discrimination of samples according to their regional provenance, correctly classifying 82% of EVOOs, especially in the case of different regions, i.e. Sicily, Sardinia and Liguria. Moreover, Latium, Tuscany and Molise were grouped satisfactorily, except for the Apulia samples, which were not sufficiently grouped. The low-prediction rates for Latium and Apulia could be explained by the morphological and climatic inhomogeneity of the sampling locations.

Benincasa et al. (2007) used ICP–MS to study two cultivars of VOO from different production regions of four Italian macroregions (Calabria, Apulia, Umbria and Abruzzo). DA allowed effective differentiation between five clusters corresponding to the five production regions of Rossano, Andria, Lamezia, Spoleto and Pescara. Such discrimination was achieved owing to Fe, Sr, As, Ca and Mg, with *F* removal values of 123.5337, 37.1311, 30.2325, 27.6391 and 25.4295, respectively.

Faberi et al. (2014) combined GC/FID, EA–IRMS and GC/C/IRMS methods to determine $\delta^{13}C$ and FA contents in EVOO samples from different regions of three Italian zones. PCA with the first three PCs explaining 65% and 69% of the total variance revealed different distribution patterns depending on the number of variables used. Generally, only partial differentiation was achieved. PLS–DA allowed clear geographical differentiation related to the three zones for only the five isotopic variables to be attained. It was estimated that 96% of the samples were correctly classified using only $\delta^{13}C$.

According to Paolini et al. (2017), the integration of advanced multi-instrumental methodologies such as EA–IRMS, TC/EA–IRMS, GC–C/Py–IRMS, GC–MS and NMR has enabled the determination of $\delta^{13}C$, $\delta^{2}H$ and FAMEs in Italian EVOOs. The methods applied allowed satisfactory analytical data to be processed by

LRA. This approach is useful for distinguishing products according to their geographical origins. Moreover, stable isotopic analysis of FAs could be proposed as a procedure for improving this ability. To the best of the authors' knowledge, an analytical method was developed to quantify the $\delta^{13}C$ values and, for the first time, the $\delta^{2}H$ values of FAMEs.

Jiménez et al. (2004) effectively used the ICP–MS and FAAS methods to study four types of VOOs originating from five Spanish regions. PCA with the first two PCs explained 74% of the total variance. Elements such as Al, Mn and Ba were the descriptors differentiating samples from Bajo Aragón and Catalonia from those corresponding to Baena, whereas Na, Ca and Mg were loadings that were used to differentiate between oil samples from southern Spain and those from northern Spain.

Cabrera–Vique et al. (2012a) investigated EVOOs from two PDOs of Granada and Jaén provinces in Andalusia (Spain) via ETA–AAS to determine several microelements. The concentration data obtained via DA were transformed such that 95.55% of all the samples analyzed were correctly classified. It was concluded that Cu, Cr, Fe and Ni were responsible for significantly distinguishing olive oils with respect to their geographical provenance.

According to Zeiner et al. (2005), ICP–AES and ET–AAS were applied to study edible oils from different sampling sites in three Croatian regions, i.e. industrial regions, national parks and rural areas. Although no interspatial differentiation on the basis of the complete set of elements was achieved, selected metals such as Al, Ca, Co, Cu, K, Mn and Ni, in contrast to Fe, Mg, Na and Zn, significantly differed in geographical origin. Therefore, according to the authors, further studies in this field should be performed, including more samples from each region, to enable assessment by multivariate analysis. Moreover, comprehensive information concerning the cultivar, growing conditions, year of olive harvest and method of production and storage should be considered and applied for data processing.

Farmaki et al. (2012) applied a combination of ICP–MS-based REE quantification with a chemometric multivariate approach to investigate olive oil samples from four different Greek regions, including ANN models (MLPs and RBF) with DA and CTs. As a guide in the learning sample dataset, the two LDFs were useful for providing the models with the most acceptable information for their training. Compared with other supervised techniques, the ANN model (MLPs with one hidden layer) was the best model for achieving much more satisfactory results. It provided 73.5% and 72.9% correct percentages for the learning and test sample datasets, respectively, whereas the CTs (LCMs) yielded 81.6% and 60.4% correct percentages for the learning and test datasets, respectively.

Karabagias et al. (2013) used ICP–AES, EA–IRMS, GC–FID and traditional instrumental methods to process six cultivars of VOO samples from four Greek islands. Figure 8.1 shows the differentiation of olive oil samples according to their geographical provenance. On the basis of LDA, the OCC rates were 93.5% and 82.6% for the original and cross-validated methods, respectively. Correct classification of 100% was achieved for Kerkyra, Lefkada and Zakynthos, but for Kefalonia, the correct classification rate was estimated to be 76.9%. This means that a

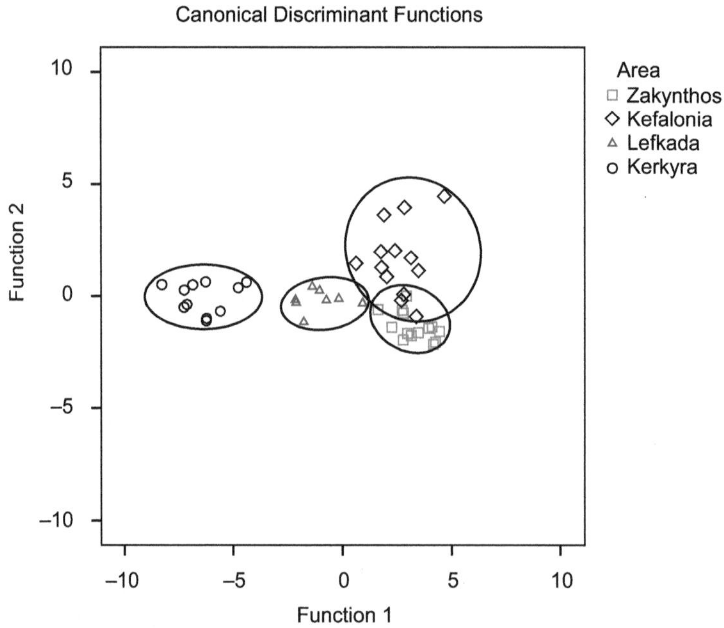

Fig. 8.1 Classification of olive oil samples from western Greece according to the geographical origin using 14 variables. (After Karabagias I et al., 2013. https://doi.org/10.1016/j.foodres.2013.09.023. With permission)

classification of oil olive samples according to geographic origin was obtained using 14 variables.

Gazeli et al. (2020) used LIBS assisted by MLA to recover the information contained in LIBS spectra to achieve a stable, uncomplicated and rapid methodology for the classification of olive oils regarding the differentiation of acidity and geographical origin. The samples originated from northern, southern, eastern, western and central Rethymno in Crete, Greece. The different statistical models, i.e. PCA, LDA, SVM and RFC, were effective classificatory tools, resulting in accuracies ranging between 90% and 99.2%. The integration of the LIBS technique with MLA provides a very effective approach for the reliable control of olive oil.

Nuts

The SIR established by EA–IRMS is a useful parameter for characterizing walnut samples from different geographical regions of Germany, distinguishing walnuts from northern regions from those from southern regions. This study emphasized the importance of harvest time for the comparison of walnuts from different provenances. Therefore, it was postulated that samples originating from the same sites should be provided for more than two consecutive years to obtain a more explicit view of the influence of annual variations in weather conditions. Moreover, German walnuts could be differentiated from French walnuts, resulting in significantly higher δ^2H values. Moreover, organic walnuts could be clearly distinguished from conventional walnuts owing to the use of $\delta^{15}N$ values, which could correspond to a cultivation practice common in organic farming. Considering the topographic characteristics between the sampling locations and precipitation data, it was possible to differentiate walnuts from two bordering countries, such as Germany and France (Krauß et al. 2020).

8.2.1.2 Asiatic Countries

Olive Oil

According to Gumus et al. (2017), the combination of ICP–MS and EA–IRMS was applied to study VOOs (four varieties and mixtures of these varieties) from different locations, including the Marmara, North Aegean, Central Aegean and South Aegean regions (Turkey). Based on the scatterplot quantified by the first two PCs, which explained 65.76% of the total variance, Fe, Zn and Cu more effectively corresponded to Manisa and İzmir than to the other locations, whereas the $\delta^{13}C$ ratios were mostly related to Edremit Bay. Moreover, Ca strongly influenced the discrimination of VOOs from other locations. Based on PCA and HCA, it was possible to classify olive oil samples into four well-differentiated clusters according to their geographical origin.

Oilseeds

Jamwal et al. (2023) investigated mustard samples from ten different provinces of India via ICP–MS and subjected them to chemometric analysis. PCA and HCA were used to group the samples according to elemental composition. PCA (with PC1 and PC2 accounting for 43.5% and 22% of the variance, respectively) clearly differentiated samples from different provinces. HCA provided comparable results to those achieved by PCA. Finally, LDA classified 100% of the mustard oil samples accurately into their particular category or class, and through a cross-validation step, it was possible to correctly classify all the classes with 100% efficiency. Among all the elements studied, Mg, P, Ca, Fe, Cr, Cu and Se were the descriptive variables that enabled identification of the geographical origin of the mustard samples.

8.2.1.3 American Countries

Olive Oil

PCA was used to distinguish three cultivars of EVOOs into three groups, i.e. central–western Argentina, central Argentina and Uruguay, on the basis of nine stable isotope ratios and the oleic–linoleic acid concentration ratio quantified by the modern instrumental triad of GC–MS, EA–IRMS and GC/C/IRMS. LDA was a useful tool for differentiating these olive samples owing to the use of the bulk δ^2H and $\delta^{18}O$ values as well as the $\delta^{13}C$ values of palmitoleic and linoleic acids. The kind of soil water accessible to the olive tree affected the δ^2H_{bulk} and $\delta^{18}O_{bulk}$ values of the oils. The analytical data processed by two-way ANOVA using the GLM procedure demonstrated a lack of significant interactions between geographical provenance and cultivar or all variables. This means that the cultivar effects on the determined bulk and molecular isotopes and fatty acid distribution patterns seemed to be the same for olive oils from northwestern Argentina and Uruguay. Combining chemometric data with stable isotope ratios of bulk and fatty acids may be a useful approach for geographic identification of olive oils from Argentina and Uruguay (Spangenberg and Lantos 2024).

8.2.1.4 Countries Across Multiple Continents

Olive Oil

Isotopes of ^{84}Sr, ^{86}Sr, ^{87}Sr and ^{88}Sr were determined by TIMS in PDO olive oils to evaluate their origin. Encouraging results were achieved for the geographical discrimination and identification of PDO olive oils. Discrimination between the studied Nîmes PDO olive oils from France and olive oils from Morocco was evaluated. The $^{87}Sr/^{86}Sr$ ratios of Moroccan olive oils are much greater than those of other oils and could thus be distinguished from the values attained for Nîmes olive oil (Medini et al. 2015).

According to Nasr et al. (2022), ICP–MS coupled with multivariate processing analysis is a useful tool for classifying mixtures of different varieties of olive oils with respect to their geographical provenance. The score plot of PC1 versus PC2, which explained 55.7% of the total variance, allowed them to obtain a differentiation between Tunisian and European olive oils. However, the scatter plot of PC3 (15.9%) and PC1 (34.6%) clearly distinguished olive oil samples according to their origin from Tunisia, Spain and France. Copper, Cr, Fe, Mn, Sr, V and Zn were the highly discriminative elements. Furthermore, Mn, Sr, Mg and Ni were significantly related to the bioavailable soil composition. The data obtained are helpful and suggest a rapid and reliable approach for the geographical recognition of olive oils on the basis of trace element composition.

Angerosa et al. (1999) performed SIRA for olive oil, and some of its fractions originated from six countries (Greece, Morocco, Spain, Italy, Tunisia and Turkey). The statistical data show the trend of grouping the samples relative to the different climatic areas in which the olive fruits were growing. Some confusion was observed for samples from neighbouring countries that have similar climates.

According to Bontempo et al. (2019), the combination of TC/EA–IRMS and GC–C\Py–IRMS was used to characterize the distribution patterns of chemical variables in different cultivars of authentic EVOOs originating from 13 countries. Variables such as $^{2}H/^{1}H$, $^{13}C/^{12}C$ and $^{18}O/^{16}O$ from the bulk were combined with $^{13}C/^{12}C$ and $^{2}H/^{1}H$ of fatty acids. The first two PCs, which explained 61.1% of the total variance, could partially differentiate between EU and non-EU sample clusters, although it was impossible to clearly distinguish between them (Fig. 8.2). The combination of stable isotopic data from bulk samples and those from FAs allowed clear interregional discrimination. Based on the PCA and RF models, the characteristics of the production region, e.g. water availability and temperature range between

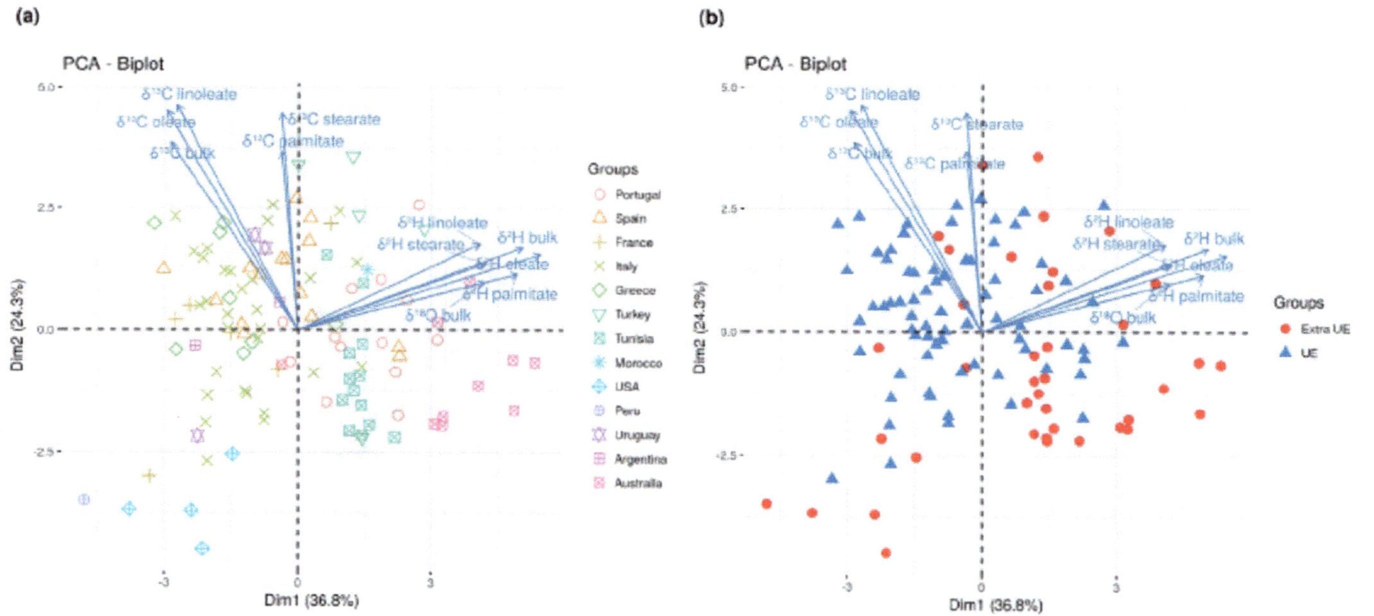

Fig. 8.2 PCA of the $\delta^{13}C_{bulk}$, $\delta^{18}O_{bulk}$, $\delta^{2}H_{bulk}$, $\delta^{13}C_{FAs}$ and $\delta^{2}H_{FAs}$ values of extra virgin olive oil samples. In **a**, the samples are labelled according to the country of origin, whereas in **b**, they are labelled according to the EU and non-EU provenances. The plot also represents the variable correlation plot with the contribution of each variable. (After Bontempo L et al., 2019. Food Chem 276:782–789. https://doi.org/10.1016/j.foodchem.2018.10.077. With permission)

day and night, may also influence the biosynthetic pathway of the most influential variables, i.e. fatty acids and especially the stable isotopic ratios of linoleic and oleic acids, which are biosynthesized by the plant.

Oilseeds

Joebstl et al. (2010) found that the REE composition determined by ICP–MS and processed by LDA is promising for differentiating pumpkin seed oil samples with respect to their origin. A classification was performed successfully between Lower Austria, Styria and three samples from China (Fig. 8.3). The data obtained demonstrate that it is possible to use REEs as fingerprints for identifying the geographical origin of pumpkin seed oil from even smaller regions, such as Lower Austria and Styria. Although Chinese oils were clearly differentiated from Austrian oils, the dataset was too limited for a more accurate statistical assessment.

Anderson and Smith (2005) used a mixture of ICP–AES, CE and IRMS (which provided SIR values together with inorganic anion and organic acid contents) to study five varieties of pistachios. The obtained data were integrated with multivariate processing by CDA, DFA and PCA. The first three PCs accounted for approximately 73% of the total variance; however, owing to the significant contribution of PC3, it was possible to achieve some visual differentiation of samples from California and a combination of samples from Turkey and Iran with the most discriminative elements, such as Mg, Mn and P. The first two canonical variables could also clearly discriminate trace elements in pistachios from different growing regions. The number of correctly classified cross-validations was >88% for all three regions,

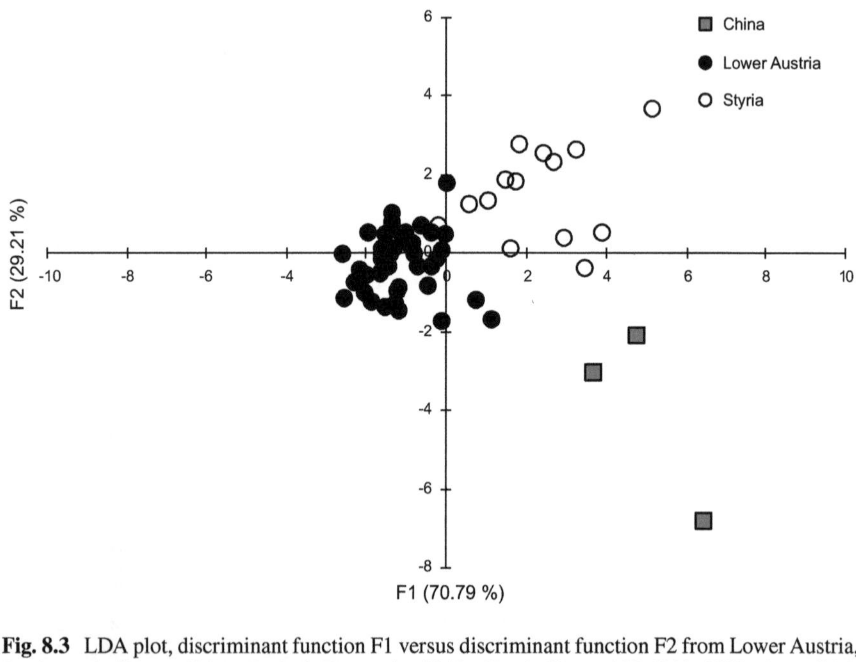

Fig. 8.3 LDA plot, discriminant function F1 versus discriminant function F2 from Lower Austria, Styria and China. (After Joebstl D et al., 2010. Food Chem 123:1303–1309. https://doi.org/10.1016/j.foodchem.2010.06.009. With permission)

whereas Californian pistachios were classified correctly, with a success rate of approximately 98%. Moreover, several LD function models classified the dataset with 95% or higher accuracy.

Anderson and Smith (2006) investigated four varieties of pistachios from different regions via IRMS combined with chemometric processing analysis. This method was helpful for investigating PC1 and PC2, which account for 65% and 31% of the total variance, respectively. It was possible to provide a clear distinction between samples originating from the three countries. Moreover, regional pistachios were subdivided into subregional units and then further subdivided into sublocations. Three subregional growing areas were identified for Iranian pistachios, i.e. northern, central and southern, in terms of $\delta^{15}N$ and $\delta^{13}C$ ratios. In the case of the Turkish samples, four subregions and sublocations were geographically designated. Moreover, seasonal differences in US pistachio samples were detected.

According to Bandoniene et al. (2013), the ICP–MS data of pumpkin seeds and pumpkin seed oils from three countries were processed chemometrically. PCA (with the first two PCs explaining 91% of the total variance) achieved reliable discrimination between Austrian and Chinese samples (Fig. 8.4), although oil samples from Austria and Russia were not clearly differentiated (Fig. 8.5). In accordance with further considerations by the authors, a larger dataset could be established for accurate traceability of the origin of pumpkin seeds and pumpkin seed oil.

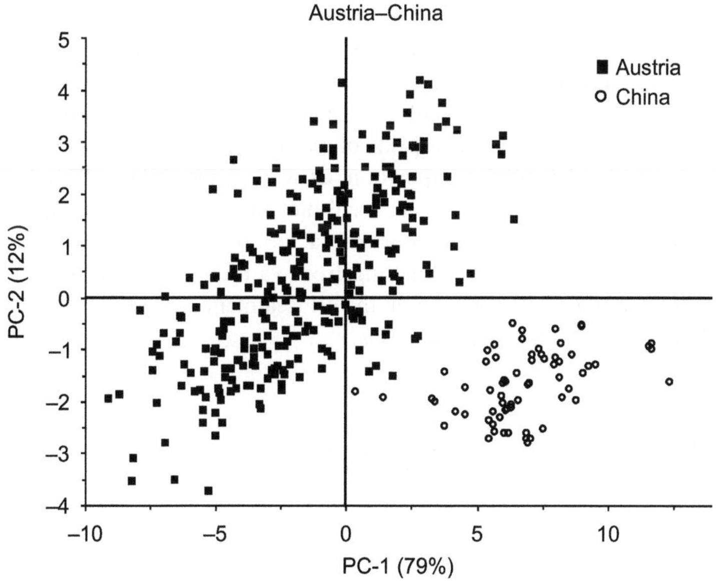

Fig. 8.4 The principal component analysis (PCA) model of Austria/China clearly distinguishes geographical origins. (After Bandoniene D et al., 2013. Food Chem 136:1533–1542. https://doi.org/10.1016/j.foodchem.2012.06.040. With permission)

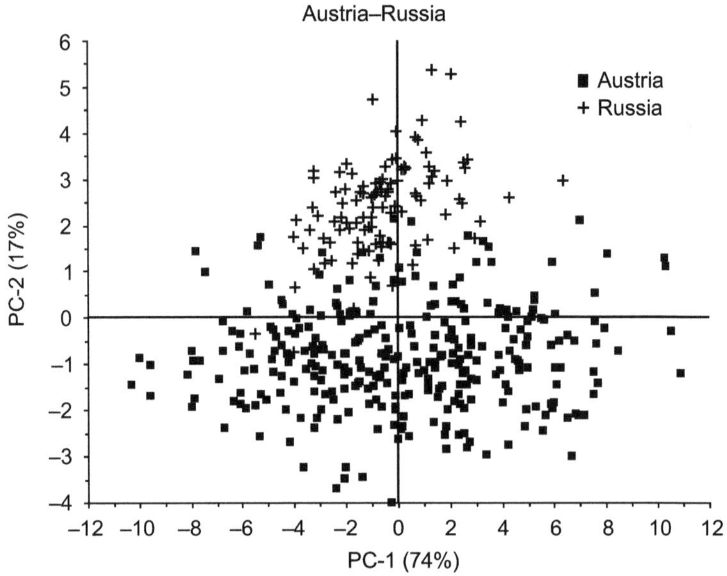

Fig. 8.5 The PCA model for Austria/Russia is less accurate than that for Austria/China. (After Bandoniene D et al., 2013. Food Chem 136:1533–1542. http://doi.org/10.1016/j.foodchem.2012.06.040. With permission)

Olive Oil and Oilseeds

Banerjee et al. (2015) used a combination of instrumental methods such as ICP–OES, ICP–MS and TC/EA–SIRA to study several sample groups, such as edible oils, including olive oils (extra virgin oils and olive oils), sesame oils, sunflower oils, palm oils, walnut oils, coconut oils, pumpkin seed oils, mustard oils, soybean oils, canola oils, hemp seed oils and vegetable butters, originating from 14 countries. The authors studied various aspects of extraction procedures of olive oils and their effects on the elemental composition, as well as the $\delta^{18}O$ and $\delta^{2}H$ values of edible oils and their relationships with local meteoric water. For example, the net oxygen and hydrogen isotope fractionation factors ($\alpha^{18}O$ and $\alpha^{2}H$) between olive oils and local meteoric waters originating from various regions of Italy indicated that the isotope ratios do not depend on the oil treatment procedure and hence may be applied to tracing geographic locations.

8.2.2 *Intervarietal/Type Differentiation*

Olives

ICP–OES combined with chemometrics was applied to study eight cultivars of Spanish table olives according to cultivar. PC1 (22.75%) was attributable to Cd, Co and S, but PC2 (14.89%) corresponded to Li, Sr, Al, B or Zr. On the basis of the distribution pattern of scores, it was impossible to obtain a clear differentiation according to styles or cultivars. Therefore, the database was subsequently analyzed via CDA. This procedure achieved an overall correct classification of 86% with 80% accuracy. After cross-validation. PCA revealed relationships among several elements, whereas LDA clearly distinguished green olives from directly brined olives or ripe olives; however, directly brined olives and ripe olives were clearly distinguished from one another (Lopez-Lopez et al. 2008).

Oilseeds

Grembecka and Szefer (2022) characterized elemental distribution profiles determined by FAAS and the spectrophotometric method of legumes and seven varieties of oilseeds from various geographical origins (European, Asiatic and North American products) available on the Polish market. In general, oilseeds had higher mean levels of Ca, Mg, Na, P, Zn, Cu, Fe, Mn and Cr than legumes. The Cd level in every sample, except for soybean and hulled wheat, exceeded the permissible European standards. The results of the Kruskal–Wallis test revealed that the Mg, Na, K, P, Zn, Cu, Mn, Cr and Cd contents are related to the type of product studied, whereas for the botanical provenance, such relationships were detected for most of the analyzed variables, except for Fe, Cr and Co. CA, like FA, classified the samples with respect to their botanical species and type on the basis of their mineral composition (Fig. 8.6).

Aguilar et al. (2011) reported that ICP–OES combined with PCA, CA, LDA and PLS–DA models was used to classify amaranth seed samples according to their three species. For all the multivariate methods, the elements Ba, Cr, Li, Mn, Ni, S

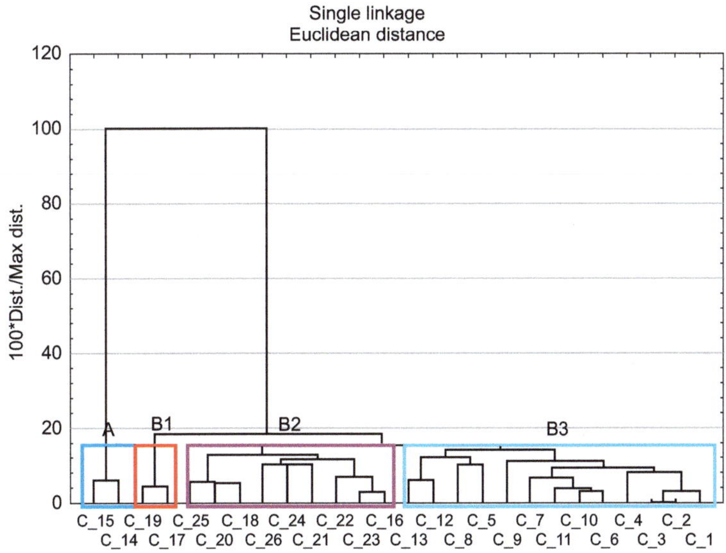

Fig. 8.6 Hierarchical dendrogram for legume and oleic acid seed samples based on 13 elements, including P, K, Cd, Fe, Mn, Ni, Co, Cr, Cu, Zn, Na, Mg and Ca. (After Grembecka M and Szefer P 2022. Appl Sci 12:1577. https://doi.org/10.3390/app12031577. With permission)

and Sr were the most discriminative variables. Although non-supervised methods enabled distinguishing among the three Argentinean botanical amaranth species, PLS–DA, a supervised method, demonstrated the best prediction ability (Fig. 8.7); i.e. 100% correct classification was obtained in the calibration (training) and validation (prediction) test sets.

Bolaños et al. (2016) processed seed samples, i.e. amaranth, chia, sesame, linen and quinoa, from different Argentinian sampling sites via MIP–OES assisted by PCA. The levels of toxic elements such as As, Cd and Pb were below the allowable limits set by the WHO. All three multivariate processing techniques enabled the correct differentiation of each type of seed. A PC1/PC2 score plot (explaining 99.6% of the variance) clearly discriminated all the seed samples, i.e. amaranth, chia, linen, quinoa and sesame. Elements such as Ca, Mg, K, P and Na are effective descriptors for this chemometric technique. The CA results confirmed the PCA data. Finally, PLS–DA, which is also based on the contents of the five major elements, clearly differentiated the five seed types and classified all the samples with 100% certainty. It was concluded that this approach could be useful for the analysis of raw materials for quality control in food factories and government control laboratories.

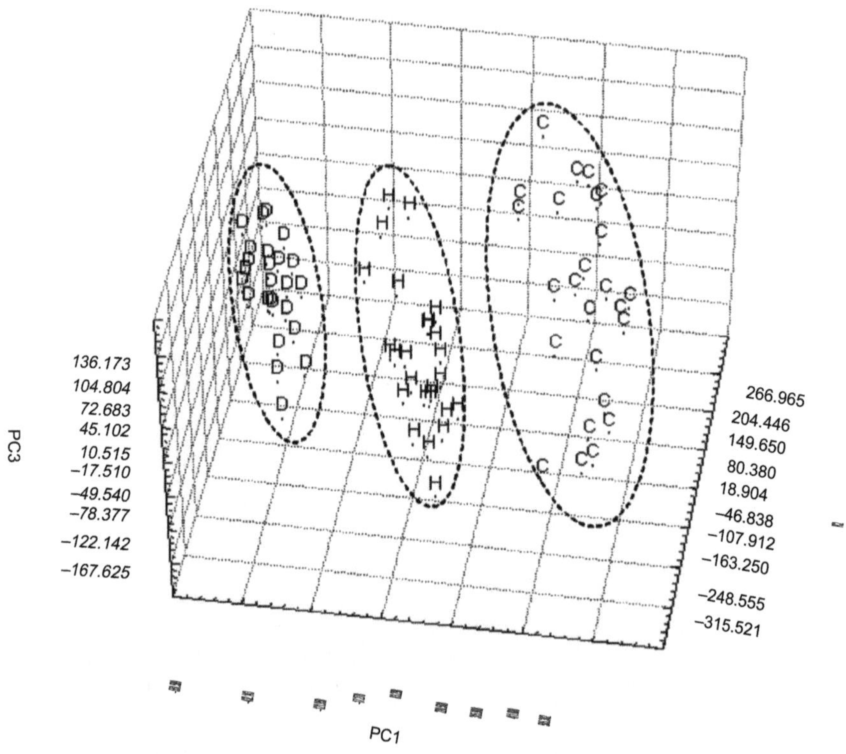

Fig. 8.7 PLS–DA 3D score plot for 75 samples of amaranth species: *A. cruentus* (C), *Amaranthus hypochondriacus* (H) and *A. dubius* (D). (After Aguilar EG et al., 2011. J Agric Food Chem 59:9059–9064. https://doi.org/10.1021/jf202610t. With permission)

Bolaños et al. (2019) reported that MIP–OES was used to characterize the elemental distribution patterns of sunflower, poppy and pumpkin seeds. With PCA, the first two PCs explained 93% of the variance. It was possible to differentiate three score groups corresponding to each type of seed according to botanical origin. Elements P, Mg and Zn were responsible for pumpkin sample selection, whereas Al, B, Ba, Ca, Cu, Na, Si, Sr and Ti were associated with poppy grouping. The sunflower seed cluster was differentiated from the remaining cluster owing to the low level of mineral elements compared with the poppy and pumpkin seeds.

Singh et al. (2017) investigated the use of LIBS and FAAS to differentiate the seed kernels of cucurbits (pumpkin, ash gourd, watermelon and muskmelon) using multivariate analysis. PCA with the first two PCs explaining 64% of the total variance revealed four types of cucurbit seeds. Notably, not only were the compositional differences between these seed species achieved, but differences in the matrix were also observed.

Kafaoğlu et al. (2014) integrated ICP–MS and ICP–OES with computational tools (PCA, CA, LDA) as a reasonable approach for the study of ten varieties, mostly raw and dry roasted nuts and seeds. They were purchased from a local Turkish shop or imported from five other countries. The nuts and seeds were classified into five groups by PCA and CA. For example, the first three PCs, which explained 61.4% of the total variance, classified nuts and seed samples into five groups, i.e. Brazil nuts (I), white sunflower seeds, black sunflower seeds and cashew nuts (II). These included pumpkin seeds and roasted pumpkin seeds (III), almonds and roasted almonds (IV), hazelnuts, roasted hazelnuts, peanuts, roasted peanuts, walnuts, pistachio nuts, chickpeas and roasted chickpeas (V). LDA also revealed that such clustering was correctly classified as 100.0%. The various types of nuts and seeds could be grouped according to their elemental composition independently of their country of origin.

Richter et al. (2010) studied seven cultivars of rapeseed oils and five cultivars of flax oils via modern analytical methods such as EA/IRMS, TC–EA/IRMS, GC/C/IRMS, GC/FID and GC/MS. Analytes such as $\delta^{13}C$, $\delta^{2}H$ and $\delta^{18}O$ and 15 different FAs were detected in oils differing in cultivar, site of growth and harvest year. Five types of oils, i.e. flax and false flax oils as well as poppy, sunflower and safflower oils, were selected and investigated. The $\delta^{13}C$ values of palmitic acid and α-linolenic acid significantly differed among rape, flax and poppy oils. Significant differences in the $\delta^{13}C$ values of FAs were also observed. The $\delta^{18}O$ and $\delta^{2}H$ levels of rape oil significantly differed between cultivars. Unlike PCA with PC2, PCA with PC1, which explained 45.1% of the total variance, indicated that samples from various harvest locations were not completely discriminated and sometimes overlapped. This unclear differentiation of the six clusters could be explained by their close geographical distribution patterns and similar climate conditions.

Mi et al. (2022b) investigated 53 chemical elements, 32 VOCs, 40 FAs and 283 metabolites in black and white sesame seeds. A combination of ICP–MS, HS–GC–IMS, GC–MS and UHPLC–Q–TOF/MS analytical methods with chemometric tools (one-way ANOVA, PCA and PLS–DA) was used. Univariate and multivariate statistics revealed a distinct difference between the two groups of sesame seeds. The most discriminative descriptors belong to 16 chemical elements, three VOCs, eight FAs and 54 metabolites with p values <0.05 and variable importance in projection scores >1. Generally, the data obtained revealed the impact of genotype on the chemical profile of sesame seeds. It was been demonstrated that the hybrid model of the methodology allowed black and white sesame seeds to be discriminated.

Nuts

According to Markiewicz–Zukowska et al. (2022), FAAS, GF–AAS and FAES tools were used to characterize the elemental profiles of nut samples. Chemometric analysis was found to be a useful tool for classifying various types of nuts. PCA with the first two PCs explaining 64% of the total variance allowed them to partially differentiate nut samples according to their types. Better results were obtained by DA (Fig. 8.8). The intertype discrimination was highly significant (Wilks' lambda: 0.004, $F = 29.995$, $p < 0.0001$). It was concluded that the studied nuts are valuable products that can be used to supplement the diet with certain essential elements and can therefore be labelled functional products that are fundamental for disease prevention.

Olives and Oilseeds

Llorent–Martínez et al. (2011) investigated various types of edible oils (virgin olive, olive, pomace-olive, sunflower, soybean and corn oils) from Spain using ICP–MS. PCA differentiated oil samples with various provenances (virgin olive, olive, pomace-olive, sunflower, soybean and corn oils) from the remaining samples; the former always presented negative PC1 values (46.37%), whereas the remaining samples presented positive values, including samples of corn and soybean oils were differentiated from the samples of sunflower oils. In contrast, PC2 (28.93%) differentiated the olive pomace samples from the other samples. A slight differentiation between olive and virgin oils was also obtained across PC2, but a larger number of samples are needed to confirm this observation. Owing to the use of the element levels, it was possible to differentiate the oil types by means of PCA.

8.2.3 *Interregional and Intervarietal Differentiation*

Beltrán et al. (2015) combined ICP–MS with chemometrics (PCA and SLDA) to investigate VOOs and olive pomaces (three cultivars) from four Spanish regions. Three cultivars in each zone could be differentiated using K, Cu, Fe and Mg. The discrimination of the four geographical regions revealed that 93% of the studied samples were correctly classified, but not influenced by the cultivars of the orchards. However, the results were not as satisfactory as those obtained for orchard and olive pomace soils.

Popescu et al. (2015) reported that ^{1}H NMR and ^{13}C NMR spectroscopy combined with PCA of data was used to study samples of nine types of pure and combined binary blends of vegetable oils with different botanical and geographical origins. Virgin olive oils clustered closely together despite their different

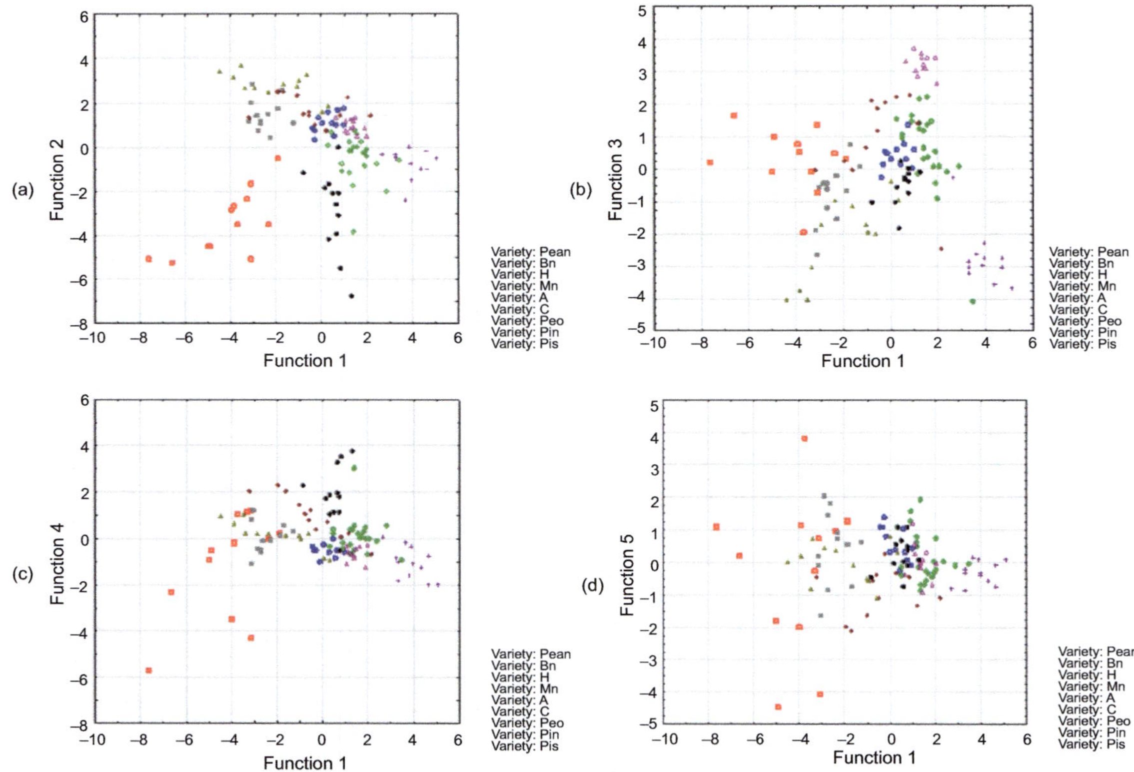

Fig. 8.8 Discriminant analysis scatter plots for Factors 1 and 2 (**a**), 1 and 3 (**b**), 1 and 4 (**c**) and 1 and 5 (**d**). Abbreviations: *A* almonds, *Bn* Brazil nuts, *C* cashews, *H* hazelnuts, *Mn* macadamia nuts, *Pean* peanuts, *Pec* pecans, *Pin* pine nuts, *Pis* pistachio nuts, *W* walnuts. (After Markiewicz-Zukowska R et al., 2022. Foods 11:3152. https://doi.org/10.3390/foods11203152. With permission)

geographical origins, whereas walnut and linseed oils were clearly distinguished from the other oil types, corresponding to their country of origin. Thus, the combination of NMR and chemometrics is useful for assessing the quality of vegetable oils in terms of their botanical origin.

8.2.4 *Differentiation of Other Parameters*

8.2.4.1 Interregional and/or Intercultivar Differentiation Related to Interannual Variability

In studies by Jiménez-Morillo et al. (2020a), EA–IRMS was an effective method for characterizing georeferenced VOO samples from three countries during two cultivation years. It was possible to obtain valuable data via chemometric analysis. One-way ANOVA for $\delta^{13}C$, $\delta^{2}H$ and $\delta^{18}O$ values revealed significant differences between harvest years and geoclimatic conditions. PCA enabled sufficient grouping of the samples into three clusters depending on the country. LDA, such as PCA, seemed to be a helpful tool for clearly distinguishing French and Portuguese olive oils. This distribution pattern is visible in Fig. 8.9, which shows LDA data for olive oils from the three countries in each year. Moreover, VOO samples may be differentiated not only between geoclimatic regions but also among cultivation years.

Alonso-Salces et al. (2010) analyzed ^{1}H NMR spectra combined with SIR data of 963 VOO samples from seven countries, i.e. from the Mediterranean basin, i.e. Greek and non-Greek VOOs (three harvests). PCA, LDA and PLS–DA were used to differentiate among geographical provenances at the national, regional and PDO levels. Greek and non-Greek VOOs were reliably classified by PLS–DA, with 96.8% and 91.1% of the samples, respectively, correctly predicted via CV and external validation. The PCA plot shows that the three clusters (corresponding to PC2, PC3 and PC4, explaining 31% of the total variance) partially overlap. This can be explained by the fact that 70% of the samples corresponded to Italy, and the remaining countries in the Mediterranean region were seasonally affected. This interaction apparently affected all the samples in the same way, independent of their geographical origin. Moreover, the PLS–DA model allowed the classification of VOOs from Italy, Spain and Greece, with 79.2%, 67.6% and 90.6% correctly predicting the CV, respectively, and such classification based on external validation achieved 73.8%, 78.6% and 83.9%, respectively. Therefore, in modelling for the authentication of agricultural food products, it is important to consider the seasonal variability.

Portarena et al. (2014) used EA–IRMS combined with two-way ANOVA to characterize the geographic origin of Italian EVOO samples from north to south that were produced in 2009, 2010 and 2011. Samples from Sicily and Sardinia (southern regions) were distinguished by higher $\delta^{13}C$ and $\delta^{18}O$ ratios than those from the northern regions. Moreover, the annual effect was significant for both the $\delta^{13}C$ and $\delta^{18}O$ variables. It was also concluded that SIR fractionations of both stable isotopes allowed the characterization of geographic provenance, year of harvest and meteorological conditions.

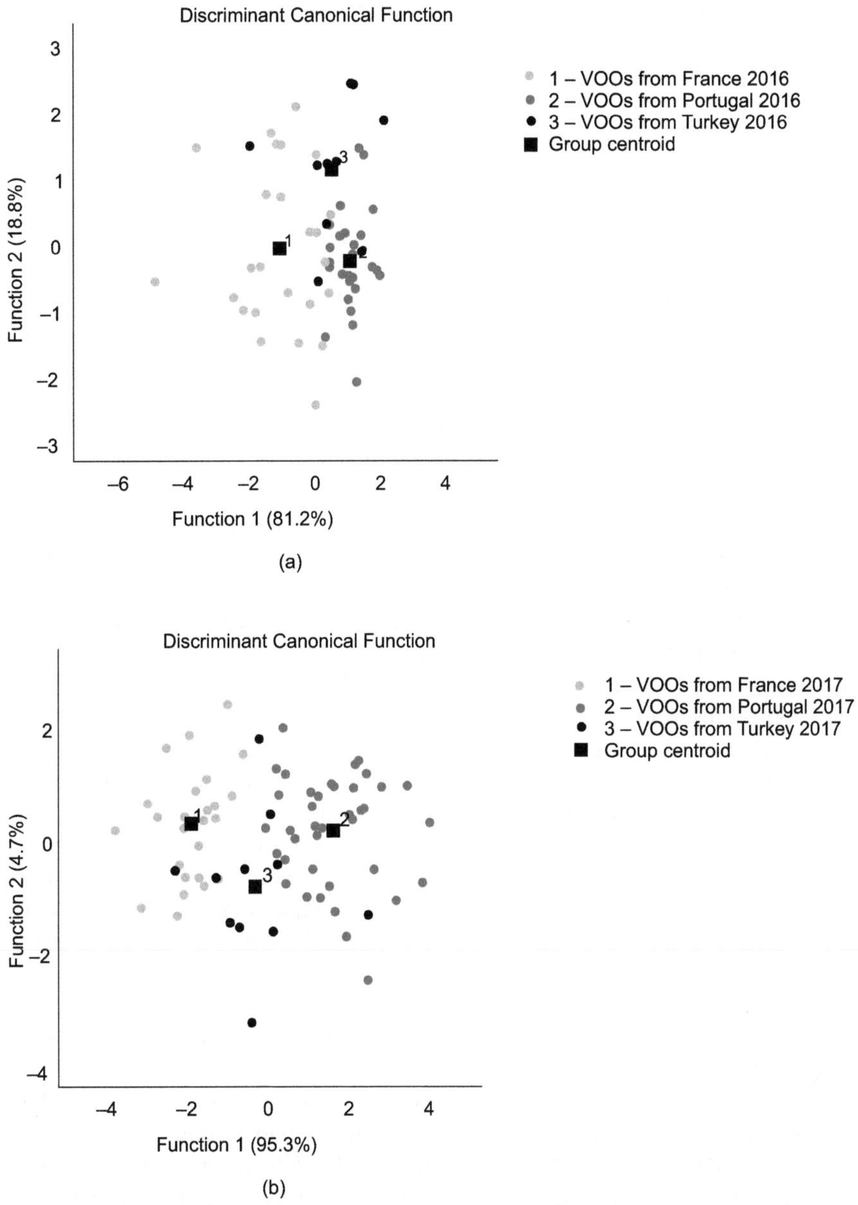

Fig. 8.9 Linear discriminant analysis of the isotopic compositions of EVOO samples ($\delta^{13}C$, $\delta^{18}O$ and $\delta^{2}H$) from 2016 (**a**) and 2017 (**b**). (After Jiménez-Morillo NT et al. (2020a). Foods 9:1855. https://doi.org/10.3390/foods9121855. With permission)

Iacumin et al. (2009) used the TC/EA–IRMS method to differentiate different olive varieties from eight Italian regions that were collected during three consecutive years of production. Statistical analyses revealed that the geographical provenance of the $\delta^{18}O$ data was much better than that of the $\delta^{13}C$ data. Moreover, a correlation between the $\delta^{18}O$ values and the distance from the sea to the production site was also observed. However, the $\delta^{13}C$ data showed a significant relationship exclusively with latitude but was helpful for discriminating between various cultivars, i.e. in the case of Frantoio and Moraiolo from the same area. The data obtained reflect a reliable relationship of the $\delta^{18}O$ of oil with latitude, mean annual temperature and mean relative humidity at the sampling site. It was concluded that climatic factors influence the isotopic composition of Italian olive oils and geographic characteristics.

Moreover, Camin et al. (2010a) used TC/EA–IRMS and ICP–MS to process 539 EVOO samples officially collected from 2000 to 2005 from different regions of North China, Central China, South-1 China and South-2 China. Non-parametric tests (Kruskal–Wallis tests and multiple bilateral comparisons) were used because of the low and unequal numbers of samples per group and because the data were not always normally distributed. The $\delta^{13}C$, $\delta^{18}O$ and δD values of the olive oil samples tended to increase from Trentino to Sicily, distinguishing northern Italy from Sicily and Calabria each year. Moreover, significant differences were observed among the years of harvest and, in some cases, between PDOs from the same region. For the first time, the δD values quantified in 2005 enabled reliable effectiveness in the geographical discrimination of Italian olive oils.

Popescu et al. (2019) applied the analytical combination of ^{1}H-NMR, EA–IRMS and TCD to investigate 34 cultivars of walnut kernel oil and nine advanced selections to study Romanian cultivars. The obtained data were processed chemometrically (DA and HCA) to assess the differentiation of cultivars according to composition, geographical region of provenance and year of harvest, including a comparative study at an international scale. For example, based exclusively on the $\delta^{13}C$ values, walnuts could be generally discriminated relative to their geographical provenance, whereas the year of harvest discrimination provided only 73.33% correct classification of the samples. In the case of DA, the first two most important factors, with 83.59% of the variance, achieved 100% correct classification of the walnut samples.

Aramendía et al. (2010) distinguished between olives grown in irrigated and dry land by investigating samples originating from the previous series and others collected during the 2005/06, 2006/07, 2007/08 and 2008/09 seasons. Analytical methods, including ^{1}H NMR, ^{13}C NMR and TC/EA–IRMS, were used to determine numerous analytes, such as $\delta^{13}C$, $\delta^{2}H$, $\delta^{18}O$ and FA, i.e. linolenic, linoleic, monounsaturated and saturated fatty acids, MUFA/PUFA and (MUFA + PUFA)/SAFA. Compared with those from irrigation-grown olives, oil samples from dryland-grown olives presented increased $\delta^{13}C$ values. The $\delta^{18}O$ and $\delta^{2}H$ values were dependent on the olive variety; however, in the case of the $\delta^{13}C$ levels, a weaker tendency was observed. Moreover, oil samples from olives harvested in the 2005/06 season in Italy were differentiated from those from Spain in terms of their isotopic values, especially δ^2H.

8.2.4.2 Verification of Authentication/Adulteration

Alonso-Salces et al. (2015) classified VOOs and their unsaponifiable fractions from six countries in the Mediterranean Basin using ^{1}H-NMR spectra and TC/EA–IRMS. PCA with the first two PCs accounting for 58.0% of the total variance did not differentiate between the samples with respect to their geographical provenance. However, δ^2H and/or δ^{13}C data provided complementary geographical characteristics of ^{1}H-NMR data in the PLS–DA binary classification models for VOOs from Greece, Spain, Italy, Izmir (Turkey), Crete (Greece) and the PDOs Riviera Ligure (Italy) as well as Huile d'olive d'Aix-en-Provence (France). The PLS–DA models allowed the authentication of VOOs from Greece and the detection of non-Greek VOOs, which yielded more than 93% correct predictions. It was also possible to detect the mislabelling of the VOO origin or the mislabelling of non-PDO oils as PDO VOOs.

Portarena et al. (2014) investigated 387 Italian EVOOs using EA–IRMS. Chemometric analysis revealed relationships between oil isotopic compositions and latitude, mean annual temperature, mean annual precipitation and the xerothermic index. Moreover, the year of harvest effect was estimated to be significant for both δ^{18}O and δ^{13}C. IRMS data were processed chemometrically by Chiocchini et al. (2016), who applied model variance estimation via spatial interpolation of δ^{13}C and δ^{18}O values and a combination of kriging and regression methods. A geospatial model of stable isotopes was developed for the authentication and verification of the geographical origin of extra virgin olive oil. The model differentiated samples from four distinct areas, i.e. north, south–central Tyrrhenian, central Adriatic and islands (Sicily and Sardinia). This approach provides valuable information necessary for the certification of the origin of EVOOs and prevents food fraud.

Spangenberg et al. (1998) characterized olive oils and pomace oils, maize, groundnut, sunflower, hazelnut, walnut, rape and vegetable oils from different countries via the modern instrumental triad of GC–MS, EA–IRMS and GC/C/IRMS, establishing δ^{13}C values of individual fatty acids. The oils labelled "CP" were compared with refined olive and pomace oils as well as other types of vegetable oils from Turkey, Morocco, Tunisia and the USA. PCA with the first two PCs accounting for 56.9% of the total variance clearly differentiated the refined oils from the CP oils, revealing a trend for the origins of the olive oils. However, the isotopic compositions of the major FAs from oils of different origins, particularly mazes, some olive, groundnut, sunflower and rape oils, were similar within the analytical uncertainty. The blending of olive oil with edible oils with slightly different FA profiles (olive pomace, sunflower and hazelnut) could be identified via a combination of GC/MS and the δ^{13}C composition of the bulk oil.

Spangenberg and Ogrinc (2001) studied olive oil, six various oilseeds and blends with refined oil or other vegetable oils using GC–MS, EA–IRMS and GC/C/IRMS tools. PCA with the first two PCs explaining 67.2% of the total variance did not allow for a clear differentiation between oil samples with respect to their geographical origin. However, PCA clearly differentiated the samples from the same olive cultivars and different seasons. Furthermore, olive oils were clearly differentiated from oil seeds, i.e. pumpkin, sunflower, soybean maize, rape and sesame. It was stated that the $\delta^{13}C_{16:0}$ versus $\delta^{13}C_{18:1}$ covariation is useful for assessing the impurity or admixture of different oils as a consequence of thermal or oxidative degradation during processing, storage or admixture of a CP oil with refined oils. Moreover, the FAs of Slovenian and Croatian olive oils were compared with those from Mediterranean producers from four other European countries.

Spangenberg (2016) coupled EA/IRMS, TC–EA/IRMS, GC/MS and GC/FID to classify six types of oil samples (maize, olive, sunflower, rice, groundnut and soybean oils) from nine countries in the southern and northern hemispheres. PC1 (41.9%) and PC2 (20.8%) partially differentiated the southern and northern hemispheres in terms of maize and sunflower oils. It was concluded that the use of $\delta^{13}C$, $\delta^{2}H$ and $\delta^{18}O$ profiles of vegetable oils for their geographic origin might be expanded to a worldwide interhemispheric scale. The interspecies differentiation can be explained by variations in the lipid biosynthetic pathways and blends of vegetable oils with different FA compositions and $\delta^{13}C_{FA}$ values. The methodology applied could be used to detect and source impurities in valuable vegetable oils commercialized worldwide.

Esteki et al. (2017) used a combination of ICP–OES with chemometric techniques to investigate Iranian almond and peanut samples in mixtures. The first two PCs, which explained 97.96% of the total variation, revealed that the multi-elemental profile distinguished almond and peanut samples. Consequently, PCA–LDA was used to distinguish the almond, peanut and adulterated samples. The adulteration of almonds was detected using the method, with 100% correct classification for both calibration and CV and 98% accuracy for the validation set samples. The PLS and LS–SVM models were developed to determine the adulteration ratios of almond, demonstrating the advantages of the LS–SVM model.

According to Kim et al. (2015), the $\delta^{13}C$, ^{1}H NMR spectra and FA profiles (12 FAs, including SFA, UFA and TFA) were determined by ^{1}H NMR spectroscopy, EA–IRMS and GLC–FID. The obtained data were used to characterize sesame oil and adulterated sesame oil with corn, soybean, sunflower, perilla, cottonseed and canola oils. This study found a reliable approach for verifying the authenticity of sesame oils. OPLS–DA was applied to select variables that most effectively verified the authenticity of the sesame oil; i.e. a clear differentiation between the authentic and adulterated sesame oil samples was achieved. The authenticity of 65 of 70 blind samples was correctly verified by applying the range of the eight variables identified in the authentic sesame oil samples, suggesting that triple analysis is a useful approach to verify the authenticity of sesame oil. It was concluded that this approach allows the successful use of these eight variables as new specifications and standards for sesame oils distributed in Korea.

Remaud et al. (1997) processed mustard oils containing natural and synthetic allyl isothiocyanate, i.e. those of natural origin extracted from natural mustard seeds from two counties and of synthetic origin produced chemically. SNIF–NMR spectra and IRMS data (δ^2H, δ^{13}C, δ^{15}N and δ^{34}S) combined with PCA were used. This approach enabled the differentiation of natural and synthetic allyl isothiocyanate as well as the identification of the geographical origin of natural mustard oils. Therefore, distinguishing natural mustard oil from adulterated natural mustard oils with the addition of synthetic allyl isothiocyanate is important.

In studies by Potočnik et al. (2016), ten FAMEs and δ^{13}C values determined by GC–FID and GC/C/IRMS were useful for the identification of pumpkin seed oil samples originating from 14 countries. To determine adulteration, rapeseed, sunflower and soybean oils were added to pumpkin seed oil in varying proportions (1–10%). A 100% correct classification of both geographical and botanical origin was achieved on the basis of the FA composition and δ^{13}C values. PCA with the first two PCs accounting for 89.6% of the total variance demonstrated that, on the basis of five variables (δ^{13}C values of C16:0, C18:0, C18:1 and C18:2 fatty acids and bulk oil), it was possible to differentiate pure individual oils (rapeseed, sunflower and soybean) from adulterated samples (e.g. soybean oil). Moreover, PCA and the scores used as the input data for RDA clearly discriminated samples originating from Slovenia, Austria and Croatia.

Jiménez-Morillo et al. (2020c) explored EVOO samples from 11 olive varieties from two Portuguese regions (Alentejo and Trás-os-Montes) using the combination of EA–IRMS and computational techniques. The scatterplot of the first two PCs, accounting for 75.93% of the total variance, demonstrated that the δ^2H values were significantly positively correlated with climatic conditions, mainly temperature (0°) and the geographic longitude of the production area. Multivariate analysis, including MLR, indicated that bulk δ^{13}C, δ^2H and δ^{18}O values allowed the prediction of altitude, latitude, longitude, temperature, rainfall and sea distance. This work revealed that the assessment of the isotopic composition of EVOO provides information on its geographic origin and environmental conditions. This means that significant differentiation between two Portuguese regions provides valuable and precise information for evaluating the geographic origin of EVOOs and therefore could be helpful in fighting against food fraud in the Portuguese olive oil sector.

Jiménez-Morillo et al. (2020b) evaluated the suitability of δ^2H Py–CSIA for the categorization of EVOO samples produced in the Mediterranean Basin according to their geographic origin. The scatterplot of the loadings of PC1 (51.10%) versus PC2 (12.73%) indicated that the δ^2H values of the organic compounds were significantly associated with the geographic longitude of the production area and its mean annual temperature. Moreover, MLR revealed that the δ^2H values of pyrolysis compounds enabled significant predictions of not only longitude and mean annual temperature but also distance to the sea. It was concluded that the applied procedure offers an innovative, fast, reproducible and reliable authentication approach that includes EVOO from many other locations in the Mediterranean Basin. According to the authors, this research was the first to evaluate the δ^2H composition directly from the pyrolysis products of olive oil using Py–CSIA (Py–GC–TC/IRMS).

According to Mi et al. (2022a), multiple elements and volatiles quantified by ICP–MS and HS–GC were useful variables for classifying sesame seeds cultivated in five countries. Owing to this approach, it was possible to discriminate and authenticate fingerprinting analysis combined with chemometrics. Clear differentiation was achieved, allowing the classification of seed samples with respect to their country of origin. The first two PCs explained 38.73% and 16.72% of the total variance, respectively (Fig. 8.10a). As shown in Fig. 8.10b, the score plot of PLS–DA, which was based on the chemical elements in the sesame seeds, allowed partial discrimination of the seed samples. Chemical profiles with p values <0.05 and VIP scores >1 were recommended for discriminating sesame seeds from China, Togo, Sudan, Mozambique and Ethiopia. Discrimination models were established with a 100% correction rate for the authentication of the investigated sesame seeds. It was concluded that an analytical database including proximate composition and multi-element and volatile profiling combined with chemometrics is highly applicable for authenticating the geographical origins of sesame seeds.

Zettl et al. (2017), owing to the use of a combination of ICP–MS and chemometric models such as PCA, PLS, SIMCA and SVM, enabled the establishment of 482 pumpkin seed and pumpkin seed oil samples from different regions of Austria, China and Russia. The calibrated and validated PLS model of the pumpkin seed oil database was useful for predicting test Set 1, and all 100% pure oils of Set 1 were predicted correctly. PLS and SVM classification provided satisfying predictions for samples that were not necessarily in the calibration area but were marginally sited. However, SIMCA classified such samples as outliers. For the pumpkin seed dataset, discrimination methods such as PLS and SVM were more powerful. Therefore, the PLS, SIMCA and SVM techniques were able to establish the origin of pumpkin seeds and pumpkin seed oil, whereas the PLS and SVM methods were in accordance with the SIMCA models. However, the classification task could be solved by PLS only. Notably, this study was performed to establish an analytical method supported by appropriate chemometric tools to protect authentic Austrian products from fraud and mislabelling.

The aim of the study of Joergensen et al. (2019) was to use IRMS for controlling fraud related to inorganic N fertilization in organic farming systems. Sunflower samples (*Helianthus annuus* L.) were obtained from trade or from farms. The research focused on determining the effects of urea fertilization, legumes as a component of crop rotation and soil types rich in soil organic matter on the $\delta^{15}N$ values of sunflower samples. According to the authors, measurements of $\delta^{15}N$ values in sunflower samples were useful but not ideal approaches for detecting fraudulent inorganic N fertilization. At the organically controlled level, 97.5% of European samples were > 2.8% and > 3.2% at the Bio + level, and it was postulated that 3.0% could be accepted as a standard for further action. It was concluded that their use of this elaborated methodology was the first, and its improvement was suggested by using an increased amount of data exclusively related to farms at organically controlled and Bio + levels.

According to Tarapoulouzi et al. (2021), IRMS was helpful in categorizing three olive oil cultivars originating from Central Greece and Peloponnese. Using this dual

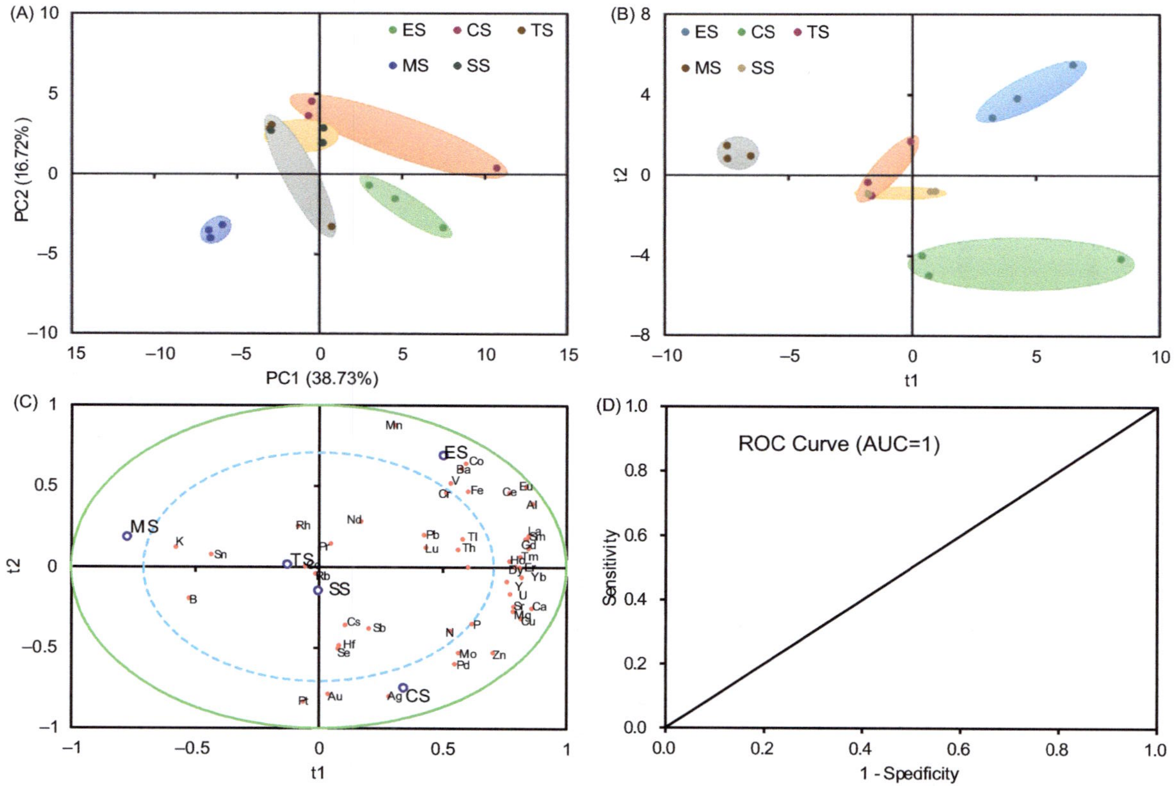

Fig. 8.10 (**a**) Projection of the first two principal components (PC1 and PC2) on the basis of the chemical elements. (**b**) Score plot from the PLS–DA model established on the basis of the chemical element data (R2 X = 0.99, R2 Y = 0.99 and Q2 = 0.67). (**c**) Correlation circle between chemical elements and sesame seeds. (**d**) ROC curve of the chemical elements. CS, sesame seeds from China; TS, sesame seeds from Togo; SS, sesame seeds from Sudan; MS, sesame seeds from Mozambique; ES, sesame seeds from Ethiopia; ROC, receiver operating characteristic curve. (After Mi S et al., 2022a. Food Chem 397:133779. https://doi.org/10.1016/j.foodchem.2022.133779. With permission)

approach was a reliable option for assessing the effects of the botanical and geographical origins of samples on their discrimination through $\delta^{13}C$, $\delta^{18}O$ and $\delta^{2}H$ data. OPLS–DA modelling produced clusters distinguished mainly by olive variety and, partially, by geographical provenance. On the basis of the model, correct recognition and correct prediction of 93.75% and 100%, respectively, were achieved on the test dataset. The overall correct classification of the model was 91%. It was concluded that the authenticity of olive oil on the basis of the studied olive cultivars could be established by means of this approach.

8.2.4.3 Interregional Differentiation Related to Associated Soil Geochemistry

Camin et al. (2010b) processed olive oils (267 samples) and surface waters (314 samples) from eight European sites in five countries using ICP–MS and TC/EA–IRMS methods integrated with a chemometric tool. Stable isotopes, including $\delta^{13}C$, $\delta^{18}O$ and $\delta^{2}H$ of olive oils, were quantified, and the ratios of $\delta^{2}H$ and $\delta^{18}O$ of the relevant surface waters were related to the climate (mainly temperature), whereas the contents of 14 elements, including Mg, K, Ca, V, Mn, Zn, Rb, Sr, Cs, La, Ce, Sm, Eu and U, were helpful for differentiating the geological origin of the olive oils. Overall, 76% of the olive oil samples were correctly classified into three geological typologies; i.e. 83% of the limestone samples, 63% of the acid magmatic samples and 72% of the clay/shale samples were classified correctly (Fig. 8.11). As shown

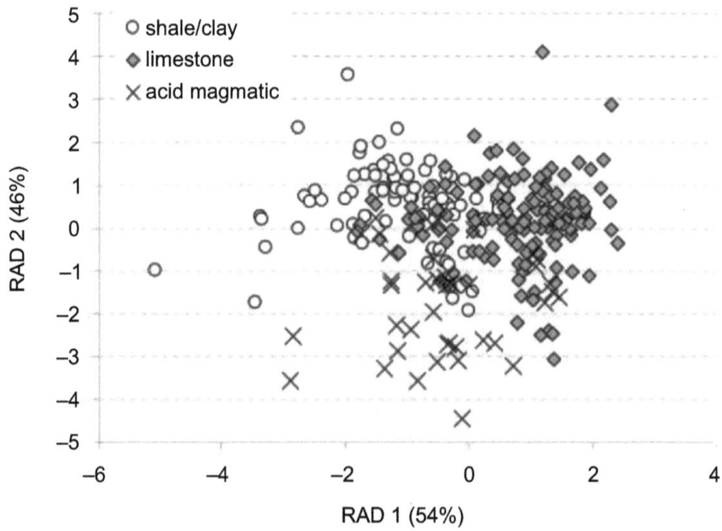

Fig. 8.11 Canonical discriminant analysis of the elemental content of olive oils from three geological origins: scatterplot of the first two canonical variables. (After Camin F et al., 2010b. J Agric Food Chem 58:570–577. https://doi.org/10.1021/jf902814s. With permission)

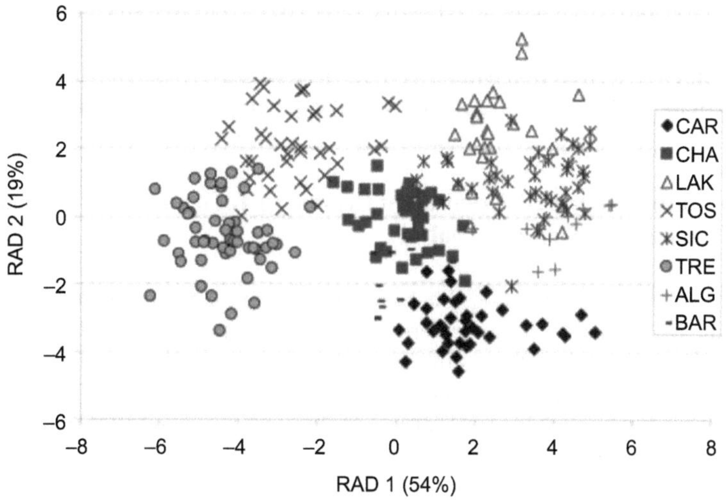

Fig. 8.12 Canonical discriminant analysis of the isotopic and elemental compositions of olive oils from eight European sites: scatterplot of the first two canonical variables. (After Camin F et al., 2010b. J Agric Food Chem 58:570–577. https://doi.org/10.1021/jf902814s. With permission)

in Fig. 8.12, CDA enabled good discrimination between olive oils from eight European sites, with 95.5% of the samples correctly classified into the production region.

Lučić et al. (2023) analyzed the contents of 29 elements and $\delta^{13}C$ via HR-ICP–MS and EA–IRMS in EVOO samples and associated soils from three regions along the Croatian Adriatic coast. The LD for the soils and olive oils explained 100% of the total variance. The geographical provenance of the olive fruits studied was differentiated by connecting the multi-element profile to the regional location with the soil geochemical composition, which may affect the elemental profile of EVOO. S, Mo, Rb, Mg, Pb, Mn, Sn, K, V and $\delta^{13}C$ were found to be the most discriminative variables for distinguishing EVOOs from all three production regions. It was concluded that LDA and k-NN are potential tools allowing the correct classification of EVOO, providing a powerful background for the authentication of its geographical origin. LDA showed that the methodology applied has great potential for the authentication of the interspatial differentiation of EVOOs.

Attempts have also been made by Damak et al. (2019) to differentiate six types of olive oils and their paired soils originating from four regions, as supported by LA–ICP–MS and XRF. It was found that it was possible to discriminate between elements such as Fe, Rb, Mg and P in olive oils and associated soils. PCA with the first four PCs explaining 75% of the total variance in the dataset suggested that the elemental profile of olive oils is substantially dependent on geochemical processes. The first two canonical LD functions accounted for 98% of the total variance. The classification of olive oils with respect to their origin was successful. In this context, LDA reached classification and prediction rates of 92.1% and 87.3%, respectively.

This relationship was not hindered by the effect of the cultivar since the geographical provenance reflects the advantage over cultivar influence on the variation in the multi-elemental distribution pattern of olive oils. The results obtained confirmed the validity of multiple elements as markers for the geographic authenticity of olive oils.

8.2.4.4 Interregional Differentiation Related to Total and Fractionated Elemental Content Analysis

Gómez–Ariza et al. (2006, 2007) used a combination of SEC–UV and ICP–MS to determine MECs and their molecular-size distribution patterns for the study of pine nut samples from different areas of Portugal and Spain. Different composition data of the variables were processed chemometrically, and five PC1/PC2 scatter plots were obtained, which yielded PC1 (from 41.34 to 80.73%) and PC2 (from 15.20 to 36.72%) of the total variance. A clear distinction between samples according to their geographic origins was possible. Metal size fractionation was also found to be a more helpful approach for identifying pine nut provenances than the traditional total element composition. The combination of these options seemed to be the most convenient approach for segregating the objects studied. It was concluded that SEC–UV–ICP–MS coupled with PCA is a useful approach to differentiate the geographic provenance of the pine nut samples studied on the basis of both multi-element fractionation and total element analysis (Gómez-Ariza et al. 2006).

8.2.4.5 Interregional Differentiation Related to the Impact of Pollution

Kötschau et al. (2014) used ICP–OES and ICP–MS combined with chemometric techniques to study sunflower (roots, stems, leaves, heads) and parent oils from different regions of Germany. HCA allowed them to obtain two clusters, differentiating the samples from a contaminated site from those from a non-contaminated site. LDA based on Mn, Ca, Fe, Ni, U and Zn also enabled discrimination between tissue types from the contaminated and non-contaminated sites, with a correct classification of 91.7%. In summary, it was possible to differentiate plant parts with respect to their origin on the basis of the element profiles of sunflowers via both multivariate techniques.

Yücel and Kılıçoğlu (2020) assayed the elemental composition of olive samples collected from different regions (industrialized and non-industrialized) of Hatay Province in Turkey via HR–CS AAS. Figure 8.13 shows the three-way PCA scatter plot using the first three PCs, which explained 70.83% of the total variation. Moreover, dendrogram results demonstrated that olive samples generally create three clusters that are similar to the PCA results. Thus, the PCA and CA tools allowed the classification of the olive samples into three groups: unpolluted, industrially polluted and non-industrially polluted. It was concluded that the model could be a useful tool for determining the authenticity of olive products in the olive industry.

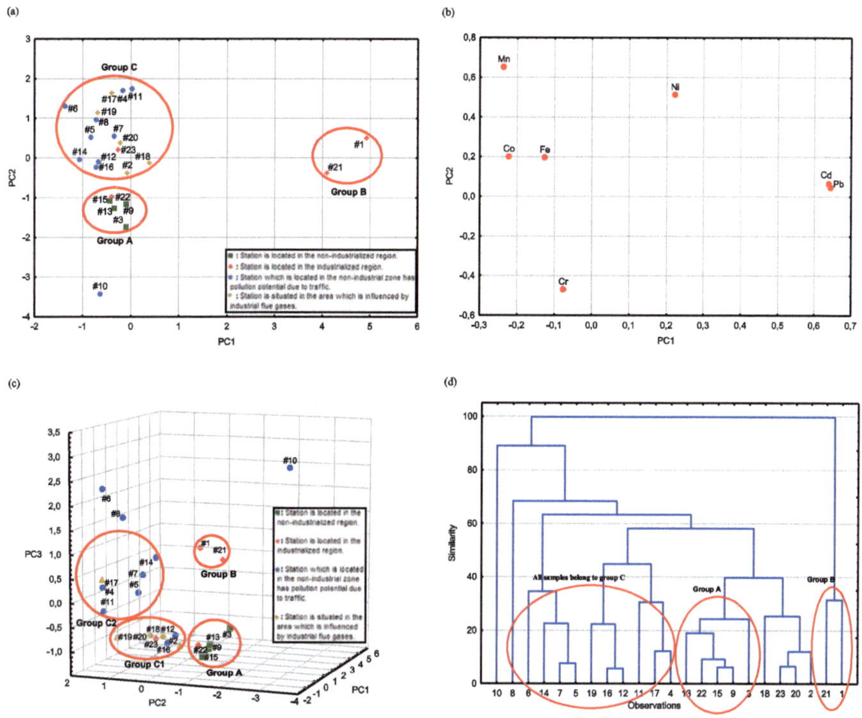

Fig. 8.13 Chemometric analysis: (**a**) PCA score plot generated with the dataset of the olive samples; (**b**) PCA loading plot generated with the dataset of the olive samples; (**c**) three-way PCA score plot generated with the dataset of the olive samples; (**d**) dendrogram resulting from complete linkage cluster analysis. (After Yücel Y and Kılıçoğlu AL, 2020. Inter J Environ Anal Chem 100:922–934. https://doi.org/10.1080/03067319.2019.1645841. With permission)

8.2.4.6 Differentiation of Walnut Cultivars Relative to Their Nutritional/ Antioxidant Qualities

The contents of polyphenols, flavonoids and vitamin E and macro- and microelements in ten main walnut cultivars from Shanxi Province were investigated via ICP–MS, ICP–OES, HPLC GC–FID and spectrophotometric methods. The latter was the major factor influencing the extensive qualities of the walnuts. Clusters of cultivars were formed on the basis of specific nutritive characteristics via the MCA method. It was also shown that microelements are the major factors influencing the comprehensive qualities of walnuts (Zheng et al. 2020).

Walnut shells constitute by-products of walnut processing and provide valuable bioactive compounds. For this reason, the aim of the study by Simsek and Süfer (2021) was to investigate the effects of infusion conditions on selected physicochemical parameters of walnut (*Juglans regia* L.) shell tea. The biological activity of walnut shell tea, similar to that of phenolic acids and tannins, is associated with the antioxidant potential of beverages. Therefore, it may be a functional, valuable and/or alternative drink to black tea infusion.

8.2.4.7 Evaluation of Element Bioaccessibility/Bioavailability

Kafaoglu et al. (2016) determined the bioaccessible amounts of trace elements by ICP–MS, ICP–AES and the modified PBET method in the gastric and intestinal phase extractions of ten varieties of nuts and seeds using an *in vitro* gastrointestinal method employing enzymes and dilute hydrochloric acid. Based on an ingestion rate of 10 g per day, the amounts of B, Ca, Co, Cu, Fe, Mn, Mg, Ni and Zn from the nuts and seeds accessible to the body were estimated to be lower than the TUIL. The data obtained were processed by multivariate *chemometric* techniques such as PCA and LDA. The latter demonstrated that groupings made using PCA in gastric and intestinal digestions were 100% correctly classified. The purpose of this approach was to classify nuts and seeds in terms of the bioaccessibility of these elements and to determine which elements are distinguished by greater bioaccessibility in gastric and intestinal ingestions. The bioaccessible ratio (%) was greater for Zn than for the other elements, whereas its value was the lowest for B. The bioaccessible Mg and Ca contents were greater than those of the other elements for most of the nuts and seeds.

Moreda–Piñeiro et al. (2016) assessed the bioavailability of essential and toxic elements in edible nuts and seeds by using an *in vitro* dialyzability approach and a microwave-assisted acid digestion pre-treatment. The samples studied included walnuts, Brazil nuts, Macadamia nuts, pecans, hazelnuts, chestnuts, cashews, peanuts, pistachios and seeds (almond, pine, pumpkin and sunflower). The quantification of analytes was performed via MAD/ICP–MS and ICP–MS supported by PCA and a quadratic surface model. Elemental dialyzability is negatively affected by fat content but positively influenced by carbohydrate content. No interaction of protein or dietary fibre content with metal dialyzability was observed. The predicted dialyzability for some elements related to fat and protein content could also be established.

8.3 Conclusions

Appropriate instrumental methods coupled with multivariate tools seem to be helpful in the classification of olive oils and oilseeds with respect to their authenticity. The identification of food adulteration/mislabelling enables the determination of food safety. The proper use of analytical and computational tools for evaluating the nutritive and health qualities of edible oil products allows their classification in view of geographical origin, cultivar identification, differentiation between natural and adulterated products, the relationship between olive oils and adjacent soils, the influence of climatic factors on the geographic characterization of olive oils, etc. The results obtained allowed the identification of mislabelling of products to preserve the brand reputation of original products, protecting the consumer from financial harm. Using chemometric techniques, it has been possible to determine the mineral elemental profile of oils and oilseeds, which is an especially important feature in the fields of food control and ecotoxicology.

References

Aguilar EG, Cantarelli MA, Marchevsky EJ, Escudero NL, Camiña JM (2011) Multielemental analysis and classification of amaranth seeds according to their botanical origin. J Agric Food Chem 59:9059–9064. https://doi.org/10.1021/jf202610t

Alonso-Salces RM, Moreno-Rojas JM, Holland MV, Reniero F, Guillou C, Héberger K (2010) Virgin olive oil authentication by multivariate analyses of ^{1}H NMR fingerprints and δ^{13}C and δ^2H data. J Agric Food Chem 58:5586–5596. https://doi.org/10.1021/jf903989b

Alonso-Salces RM, Segebarth N, Garmon-Lobato S, Holland MV, Moreno-Rojas JM, Fernandez-Pierna JA, Baeten V, Fuselli SR, Gallo B, Berrueta LA, Reniero F, Guillou C Heberger K (2015) ^{1}H-NMR and isotopic fingerprinting of olive oil and its unsaponifiable fraction: geographical origin of virgin olive oils by pattern recognition. Eur J Lipid Sci Technol 117:1991–2006. https://doi.org/10.1002/ejlt.201400243

Anderson KA, Smith BW (2005) Use of chemical profiling to differentiate geographic growing origin of raw pistachios. J Agric Food Chem 53:410–418. https://doi.org/10.1021/jf048907u

Anderson KA, Smith BW (2006) Effect of season and variety on the differentiation of geographic growing origin of pistachios by stable isotope profiling. *J Agric Food Chem* 54:1747–1752. https://doi.org/10.1021/jf052928m

Angerosa F, Bréas O, Contento S, Guillou C, Reniero F, Sada E (1999) Application of stable isotope ratio analysis to the characterization of the geographical origin of olive oils. J Agric Food Chem 47:1013–1017. https://doi.org/10.1021/jf9809129

Angioni A (2010) Olive cultivar, period of harvest, and environmental pollution on the contents of Cu, Cd, Pb, and Zn: Italian perspectives, in: Olives and Olive Oil in Health and Disease Prevention, Eds Preedy VR and Watson RR, Chap. 33, pp. 307–311. https://doi.org/10.1016/B978-0-12-374420-3.00033-4

Aramendía MA, Marinas A, Marinas JM, Sánchez E, Urbano FJ, Guillou C, Moreno Rojas JM, Moalem M, Rallo L (2010) A nuclear magnetic resonance (^{1}H and ^{13}C) and isotope ratio mass spectrometry (δ^{13}C, δ^2H and δ^{18}O) study of Andalusian olive oils. Rapid Commun Mass Spectrom 24:1457–1466. https://doi.org/10.1002/rcm.4538

Bandoniene D, Zettl D, Meisel T, Maneiko M (2013) Suitability of elemental fingerprinting for assessing the geographic origin of pumpkin (*Cucurbita pepo var. styriaca*) seed oil. Food Chem 136:1533–1542. https://doi.org/10.1016/j.foodchem.2012.06.040

Banerjee S, Kyser TK, Vuletich A, Leduc E (2015) Elemental and stable isotopic study of sweeteners and edible oils: constraints on food authentication. J Food Compos Anal 42:98–116. https://doi.org/10.1016/j.jfca.2015.03.011

Beltrán M, Sánchez-Astudillo M, Aparicio R, García-González DL (2015) Geographical traceability of virgin olive oils from south-western Spain by their multi-elemental composition. Food Chem 169:350–357. https://doi.org/10.1016/j.foodchem.2014.07.104

Benincasa C, Lewis J, Perri E, Sindona G, Tagarelli A (2007) Determination of trace element in Italian virgin olive oils and their characterization according to geographical origin by statistical analysis. Anal Chim Acta 585:366–370. https://doi.org/10.1016/j.aca.2006.12.040

Bolaños D, Heredia JZ, Moldes CA, Marchevsky EJ, Camiña JM (2019) Assessment of mineral profile in edible seeds by Microwave-Induced Plasma Optical Emission Spectrometry (MIP OES): Ionomic approach. J Nutri Food Sci Forecast 2:1011

Bolaños D, Marchevsky EJ, Camiña JM (2016) Elemental analysis of amaranth, chia, sesame, linen, and quinoa seeds by ICP-OES: assessment of classification by chemometrics. Food Anal Methods 9:477–484. https://doi.org/10.1007/s12161-015-0217-4

Bontempo L, Paolini M, Franceschi P, Ziller L, García-González DL, Camin F (2019) Characterisation and attempted differentiation of European and extra- European olive oils using stable isotope ratio analysis. Food Chem 276:782–789. https://doi.org/10.1016/j.foodchem.2018.10.077

Cabrera-Vique C, Bouzas PR, Oliveras-López MJ (2012a) Determination of trace elements in extra virgin olive oils: a pilot study on the geographical characterisation. Food Chem 134:434–439. https://doi.org/10.1016/j.foodchem.2012.02.088

Cabrera-Vique C, Marfil R, Giménez R, Martínez-Augustin O (2012b) Bioactive compounds and nutritional significance of virgin argan oil—an edible oil with potential as a functional food. Nutr Rev 70:266–279. https://doi.org/10.1111/j.1753-4887.2012.00478.x

Calò F, Girelli CR, Wang SC, Fanizzi FP (2022) Geographical origin assessment of extra virgin olive oil via NMR and MS combined with chemometrics as analytical approaches. Foods 11:113. https://doi.org/10.3390/foods11010113

Camin F, Larcher R, Nicolini G, Bontempo L, Bertoldi D, Perini M, Schlicht C, Schellenberg A, Thomas F, Heinrich K, Voerkelius S, Horacek M, Ueckermann H, Froeschl H, Wimmer B, Heiss G, Baxter M, Rossmann A, Hoogewerff J (2010b) Isotopic and elemental data for tracing the origin of European olive oils. J Agric Food Chem 58:570–577. https://doi.org/10.1021/jf902814s

Camin F, Larcher R, Perini M, Bontempo L, Bertoldi D, Gagliano G, Nicolini G, Versini G (2010a) Characterisation of authentic Italian extra-virgin olive oils by stable isotope ratios of C, O and H and mineral composition. Food Chem 118:901–909. https://doi.org/10.1016/j.foodchem.2008.04.059

Capar SG, Szefer P (2012) Determination and speciation of trace elements in foods. In: Ötleş S (ed) Methods of analysis of food components and additions. CRC Press/Taylor & Francis, Boca Raton, pp 165–210. https://doi.org/10.1201/b11218

Chiocchini F, Portarena S, Ciolfi M, Brugnoli E, Lauteri M (2016) Isoscapes of carbon and oxygen stable isotope compositions in tracing authenticity and geographical origin of Italian extra-virgin olive oils. Food Chem 202:291–301. https://doi.org/10.1016/j.foodchem.2016.01.146

Conte L, Bendini A, Valli E, Lucci P, Moret S, Maquet A, Lacoste F, Paul Brereton P, García-González DL, Moreda W, Gallina Toschi TG (2020) Olive oil quality and authenticity: a review of current EU legislation, standards, relevant methods of analyses, their drawbacks and recommendations for the future. Trends Food Sci Technol 105:483–493. https://doi.org/10.1016/j.tifs.2019.02.025

Damak F, Asano M, Baba K, Suda A, Araoka D, Wali A, Isoda H, Nakajima M, Ksibi M, Tamura K (2019) Interregional traceability of Tunisian olive oils to the provenance soil by multielemental fingerprinting and chemometrics. Food Chem 283:656–664. https://doi.org/10.1016/j.foodchem.2019.01.082

Esteki M, Vander Heyden Y, Farajmand B, Kolahderazi Y (2017) Qualitative and quantitative analysis of peanut adulteration in almond powder samples using multi-elemental fingerprinting combined with multivariate data analysis methods. Food Contr 82:31–41. https://doi.org/10.1016/j.foodcont.2017.06.014

Faberi A, Marianella RM, Fuselli F, La Mantia A, Ciardiello F, Montesano C, Mascini M, Sergi M, Compagnone D (2014) Fatty acid composition and $\delta^{13}C$ of bulk and individual fatty acids as marker for authenticating Italian PDO/PGI extra virgin olive oils by means of isotopic ratio mass spectrometry. J Mass Spectrom 49:840–849. https://doi.org/10.1002/jms.3399

Farmaki EG, Thomaidis NS, Minioti KS, Ioannou E, Georgiou CA, Efstathiou CE (2012) Geographical characterization of Greek olive oils using rare earth elements content and supervised chemometric techniques. Anal Lett 45:920–932. https://doi.org/10.1080/00032719.2012.655656

Gazeli O, Bellou E, Bellou E, Couris S (2020) Laser-based classification of olive oils assisted by machine learning. Food Chem 302:125329. https://doi.org/10.1016/j.foodchem.2019.125329

Gomez-Ariza JL, Arias-Borrego A, García-Barrer T (2007) Combined use of total metal content and size fractionation of metal biomolecules to determine the provenance of pine nuts (*Pinus pinea*). Anal Bioanal Chem 388:1295–1302. https://doi.org/10.1007/s00216-007-1331-y

Gómez-Ariza JL, Arias-Borrego A, García-Barrera T (2006) Multielemental fractionation in pine nuts (*Pinus pinea*) from different geographic origins by size-exclusion chromatography with UV and inductively coupled plasma mass spectrometry detection. J Chromatograph A 1121:191–199. https://doi.org/10.1016/j.chroma.2006.04.025

Grembecka M (2007) Bromatological and chemometric assessment of plants food based on its mineral composition. Dissertation, Gdańsk Medical Academy

Grembecka M, Szefer P (2011) Metals and metalloids in foods—essentiality, toxicity, applicability. In: Medina DA, Laine AM (eds) Food quality: control, analysis and consumer concerns. Nova Science Publishers, Inc, Hauppauge, pp 1–60. ISBN 10: 1611229170; ISBN 13: 9781611229172

Grembecka M, Szefer P (2016) Elemental trace analysis in studies of food products. In: Baranowska I (ed) Handbook of trace analysis—fundamentals and applications. Springer, pp 203–239. https://doi.org/10.1007/978-3-319-19614-5

Grembecka M, Szefer P (2022) Elemental profiles of legumes and seeds in view of chemometric approach. Appl Sci 12:1577. https://doi.org/10.3390/app12031577

Grembecka M, Gurzyńska A, Dybek K, Szefer P (2009) The applicability of chemometric techniques in assessing the differences in quality of vegetable products. In: Cubadda F, Aureli F, Ciardullo S, Patriarca M (eds) 3 International IUPAC Symposium on Trace Elements in Food, Rome, Italy, 1–3 April 2009, Abstract Book ISSN 0393-5620; TRN: IT0900518090505

Gumus ZP, Celenk VU, Tekin S, Yurdakul O, Ertas H (2017) Determination of trace elements and stable carbon isotope ratios in virgin olive oils from Western Turkey to authenticate geographical origin with a chemometric approach. Eur Food Res Technol 243:1719–1727. https://doi.org/10.1007/s00217-017-2876-4

Iacumin P, Bernini L, Boschetti I (2009) Climatic factors influencing the isotope composition of Italian olive oils and geographic characterisation. Rapid Commun Mass Spectrom 23:448–454. https://doi.org/10.1002/rcm.3896

Jamwal R, Amit SM, Jamwal R (2023) Determination of geographical origin of mustard oil based on multi-elemental fingerprinting using inductively coupled plasma mass spectrometry (ICP-MS) and chemometric analysis. Food Chem Adv 2:100233. https://doi.org/10.1016/j.focha.2023.100233

Jiménez MS, Velarte R, Gomez MT, Castillo JR (2004) Multielement determination using on-line emulsion formation and ICP-MS/FAAS for the characterization of virgin olive oils by principal component analysis. At Spectrosc 25:1–12

Jiménez-Morillo NT, Cabrita MJ, Barrocas-Dias C, González-Vila FJ, González-Pérez JA (2020b) Pyrolysis-compound-specific hydrogen isotope analysis (δ^2H Py-CSIA) of Mediterranean olive oils. Food Contr 110:107023. https://doi.org/10.1016/j.foodcont.2019.107023

Jiménez-Morillo NT, Palma V, Garcia R, Barrocas-Dias C, Cabrita MJ (2020a) Combination of stable isotope analysis and chemometrics to discriminate geoclimatically and temporally the virgin olive oils from three Mediterranean countries. Foods 9:1855. https://doi.org/10.3390/foods9121855

Jiménez-Morillo NT, Palma V, Garcia R, Pereira JA, Barrocas-Dias C, Cabrita MJ (2020c) Multivariate geostatistical analysis of stable isotopes in Portuguese varietal extra virgin olive oils. Microchem J 157:105044. https://doi.org/10.1016/j.microc.2020.105044

Joebstl D, Bandoniene D, Meisel T, Chatzistathis S (2010) Identification of the geographical origin of pumpkin seed oil by the use of rare earth elements and discriminant analysis. Food Chem 123:1303–1309. https://doi.org/10.1016/j.foodchem.2010.06.009

Joergensen RG, Toncea I, Boner M, Heß J (2019) Evaluation of organic sunflower fertilization using δ^{15}N values. Org Agr 9:365–372. https://doi.org/10.1007/s13165-018-00241-3

Kafaoğlu B, Fisher A, Hill S, Kara D (2014) Chemometric evaluation of trace metal concentrations in some nuts and seeds. Food Addit Contam A 31:1529–1538. https://doi.org/10.1080/19440049.2014.947331

Kafaoglu B, Fisher A, Hill S, Kara D (2016) Determination and evaluation of element bioaccessibility in some nuts and seeds by in vitro gastro-intestinal method. J Food Compos Anal 45:58–65. https://doi.org/10.1016/j.jfca.2015.09.011

Karabagias I, Michos C, Badeka A, Kontako S, Stratis I, Kontominas MG (2013) Classification of Western Greek virgin olive oils according to geographical origin based on chromatographic, spectroscopic, conventional and chemometric analyses. Food Res Int 54:1950–1958. https://doi.org/10.1016/j.foodres.2013.09.023

Kim J, Jin G, Lee Y, Chun HS, Ahn S, Kim BH (2015) Combined analysis of stable isotope, ^{1}H NMR, and fatty acid to verify sesame oil authenticity. J Agric Food Chem 63:8955–8965. https://doi.org/10.1021/acs.jafc.5b04082

Konieczka P, Namieśnik J (2009) Quality assurance and quality control in the analytical chemical laboratory—a practical approach. CRC Press/Taylor & Francis. https://doi.org/10.1201/9781420082715

Kötschau A, Büchel G, Einax JW, Meißner R, von Tümpling W, Merten D (2014) Element pattern recognition and classification in sunflowers (*Helianthus annuus*) grown on contaminated and non-contaminated soil. Microchem J 114:164–174. https://doi.org/10.1016/j.microc.2013.12.006

Krauß S, Vieweg A, Vetter W (2020) Stable isotope signatures (δ^{2}H-, δ^{13}C-, δ^{15}N-values) of walnuts (*Juglans regia* L.) from different regions in Germany. J Sci Food Agric 100:1625–1634. https://doi.org/10.1002/jsfa.10174

La Pera L, Dugo G, Lo V, Rando TR, Di Bella G (2010) Classification of Sicilian olive oils according to heavy metal and selenium levels using canonical discriminant analysis (CDA). In: Preedy VR, Watson RR (eds) Olives and olive oil in health and disease prevention, Elsevier, p 155–163. ISBN 978-0-12-374420-3

Llorent-Martínez EJ, Ortega-Barrales P, Fernández-de Córdova ML, Domínguez-Vidal A, Ruiz-Medina A (2011) Investigation by ICP-MS of trace element levels in vegetable edible oils produced in Spain. Food Chem 127:1257–1262. https://doi.org/10.1016/j.foodchem.2011.01.064

Lopez-Lopez A, Lopez R, Madrid F, Garrido-Fernandez A (2008) Heavy metals and mineral elements not included on the nutritional labels in table olives. J Agric Food Chem 56:9475–9483

Lučić M, Špika MJ, Mikac N, Pošćić F, Rengel Z, Romić M, Begić HB, Fiket Ż, Turk MF, Bačić N, Leder R, Petric IV, Urlić B, Žanetić M, Runjić M, Selak GV, Vitanović E, Klepo T, Rošin J, Perica S (2023) Traceability of Croatian extra virgin olive oils to the provenance soils by multielement and carbon isotope composition and chemometrics. Food Chem 424:136401. https://doi.org/10.1016/j.foodchem.2023.136401

Marini M, Bucci R, Magrì AL, Magrì AD (2010) An overview of the chemometric methods for the authentication of the geographical and varietal origin of olive oils. In: Preedy VR, Watson RR (eds) Olives and olive oil in health and disease prevention. Elsevier, pp 569–579. ISBN 978-0-12-374420-3

Markiewicz-Zukowska R, Puścion-Jakubik A, Grabia M, Perkowski J, Nowakowski P, Bielecka J, Soroczyńska J, Kańgowski G, Bołtryk JM, Socha K (2022) Nuts as a dietary enrichment with selected minerals—content assessment supported by chemometric analysis. Foods 11:3152. https://doi.org/10.3390/foods11203152

Medini S, Janin M, Verdoux P, Techer I (2015) Methodological development for ^{87}Sr/^{86}Sr measurement in olive oil and preliminary discussion of its use for geographical traceability of PDO Nîmes (France). Food Chem 171:78–83. https://doi.org/10.1016/j.foodchem.2014.08.121

Meenu M, Cai Q, Xu B (2019) A critical review on analytical techniques to detect adulteration of extra virgin olive oil. Trends Food Sci Technol 91:391–408. https://doi.org/10.1016/j.tifs.2019.07.045

Mi S, Wang Y, Zhang X, Sang Y, Wang X (2022a) Authentication of the geographical origin of sesame seeds based on proximate composition, multi-element and volatile fingerprinting combined with chemometrics. Food Chem 397:133779. https://doi.org/10.1016/j.foodchem.2022.133779

Mi S, Wang Y, Zhang X, Sang Y, Wang X (2022b) Discrimination of black and white sesame seeds based on targeted and non-targeted platforms with chemometrics: from profiling towards identification of chemical markers. Foods 11:2042. https://doi.org/10.3390/foods11142042

Moreda-Piñeiro J, Herbello-Hermelo P, Domínguez R, Bermejo-Barrera P, Moreda-Piñeiro A (2016) Bioavailability assessment of essential and toxic metals in edible nuts and seeds. Food Chem 205:146–154. https://doi.org/10.1016/j.foodchem.2016.03.006

Namieśnik J, Szefer P (eds) (2010) Analytical measurements in aquatic environments. CRC Press/ Taylor & Francis Group. https://doi.org/10.1201/9781420082692

Nasr EG, Epova EN, de Diego A, Souissi R, Hammami M, Abderrazak H, Donard OFX (2022) Trace elements analysis of Tunisian and European extra virgin olive oils by ICP-MS and chemometrics for geographical discrimination. Foods 11:82. https://doi.org/10.3390/foods11010082

Paolini M, Bontempo L, Camin F (2017) Compound-specific $\delta^{13}C$ and $\delta^{2}H$ analysis of olive oil fatty acids. Talanta 174:38–43. https://doi.org/10.1016/j.talanta.2017.05.080

Popescu R, Costinel D, Dinca OR, Marinescu A, Stefanescu I, Ionete RE (2015) Discrimination of vegetable oils using NMR spectroscopy and chemometrics. Food Contr 48:84–90. https://doi.org/10.1016/j.foodcont.2014.04.046

Popescu R, Ionete RE, Botoran OR, Costinel D, Bucura F, Geana EI, Alabedallat YFJ, Botu M (2019) ^{1}H-NMR profiling and carbon isotope discrimination as tools for the comparative assessment of walnut (*Juglans regia* L.) cultivars with various geographical and genetic origins—a preliminary study Molecules 24:1378. https://doi.org/10.3390/molecules24071378

Portarena S, Baldacchini C, Brugnoli E (2017) Geographical discrimination of extra-virgin olive oils from the Italian coasts by combining stable isotope data and carotenoid content within a multivariate analysis. Food Chem 215:1–6. https://doi.org/10.1016/j.foodchem.2016.07.135

Portarena S, Gavrichkova O, Lauteri M, Brugnoli E (2014) Authentication and traceability of Italian extra-virgin olive oils by means of stable isotopes techniques. Food Chem 164:12–16. https://doi.org/10.1016/j.foodchem.2014.04.115

Potočnik T, Ogrinc N, Potočnik D, Košir IJ (2016) Fatty acid composition and $\delta^{13}C$ isotopic ratio characterisation of pumpkin seed oil. J Food Compos Anal 53:85–90. https://doi.org/10.1016/j.jfca.2016.09.005

Remaud GS, Martin Y-L, Martin GG, Naulet N, Martin GJ (1997) Authentication of mustard oils by combined stable isotope analysis (SNIF-NMR and IRMS). J Agric Food Chem 45:1844–1848. https://doi.org/10.1021/jf960448c

Richter EK, Spangenberg JE, Kreuzer M, Leiber F (2010) Characterization of rapeseed (*Brassica napus*) oils by bulk C, O, H, and fatty acid C stable isotope analyses. J Agric Food Chem 58:8048–8055. https://doi.org/10.1021/jf101128f

Rodríguez-Méndez ML, Apetrei C, De Saja JA (2010) Electronic tongues purposely designed for the organoleptic characterization of olive oils. In: Preedy VR, Watson RR (eds) Olives and olive oil in health and disease prevention. Elsevier, pp 525–532. ISBN 978-0-12-374420-3

Simsek M, Süfer Ö (2021) Infusion of walnut (*Juglans regia* L.) shell tea: multi response optimization and antioxidant potential. J Appl Res Med Aromat Plants 20:100278. https://doi.org/10.1016/j.jarmap.2020.100278

Singh J, Kumar R, Awasthi S, Singh V, Rai AK (2017) Laser-induced breakdown spectroscopy: A rapid tool for the identification and quantification of minerals in cucurbit seeds. Food Chem 221:1778–1783. doi:https://doi.org/10.1016/j.foodchem.2016.10.104

Spangenberg JE (2016) Bulk C, H, O, and fatty acid C stable isotope analyses for purity assessment of vegetable oils from the southern and northern hemispheres. Rapid Commun Mass Spectrom 30:2447–2461. https://doi.org/10.1002/rcm.7734

Spangenberg JE, Lantos I (2024) Fingerprinting of monovarietal olive oils from Argentina and Uruguay via stable isotope, fatty acid profile, and chemometric analyses. Food Chem 139194. https://doi.org/10.1016/j.foodchem.2024.139194

Spangenberg JE, Macko SA, Hunziker J (1998) Characterization of olive oil by carbon isotope analysis of individual fatty acids: implications for authentication. J Agric Food Chem 46:4179–4184. https://doi.org/10.1021/jf980183x

Spangenberg JE, Ogrinc N (2001) Authentication of vegetable oils by bulk and molecular carbon isotope analyses with emphasis on olive oil and pumpkin seed oil. J Agric Food Chem 49:1534–1540. https://doi.org/10.1021/jf001291y

Szefer P (2002) Metals, metalloids and radionuclides in the Baltic Sea ecosystem. Elsevier. ISBN 10: 0444503528

Szefer P (2003) Application of chemometric techniques in analytical evaluation of biological and environmental samples. In: Namieśnik J, Chrzanowski W, Żmijewska P (eds) New horizons and challenges in environmental analysis and monitoring. CEEAM, Gdańsk, p 355–388. ISBN 8391908100

Szefer P (2007) Chemometric techniques in analytical evaluation of food quality. In: Szefer P, Nriagu J (eds) Mineral components in foods. CRC Press/Taylor & Francis, Hauppauge, pp 69–121. https://doi.org/10.1201/9781420003987

Tahir HE, Arslan M, Mahunu GK, Mariod AA, Hashim SBH, Xiaobo Z, Jiyong S, El-Seedi HR, Musa TH (2022) The use of analytical techniques coupled with chemometrics for tracing the geographical origin of oils: a systematic review (2013–2020). Food Chem 366:130633. https://doi.org/10.1016/j.foodchem.2021.130633

Tarapoulouzi M, Skiada V, Agriopoulou S, Psomiadis D, Rébufa C, Roussos S, Theocharis CR, Katsaris P, Varzakas T (2021) Chemometric discrimination of the geographical origin of three Greek cultivars of olive oils by stable isotope ratio analysis. Foods 10:336. https://doi.org/10.3390/foods10020336

Valdés A, Beltrán A, Mellinas C, Jiménez A, Garrigós MC (2018) Analytical methods combined with multivariate analysis for authentication of animal and vegetable food products with high fat content. Trends Food Sci Technol 77:120–130. https://doi.org/10.1016/j.tifs.2018.05.014

Yücel Y, Kılıçoğlu AL (2020) Determination of heavy metals in olive fruits as an indicator of environmental pollution. Inter J Environ Anal Chem 100:922–934. https://doi.org/10.1080/03067319.2019.1645841

Zeiner M, Steffan I, Juranovic Cindric I (2005) Determination of trace elements in olive oil by ICP-AES and ETA-AAS: a pilot study on the geographical characterization. Microchem J 81:171–176. https://doi.org/10.1016/j.microc.2004.12.002

Zettl D, Bandoniene D, Meisel T, Wegscheider W, Rantitsch G (2017) Chemometric techniques to protect the traditional Austrian pumpkin seed oil. Eur J Lipid Sci Technol 119:1600468. https://doi.org/10.1002/ejlt.201600468

Zheng Y, Wu S, Wang R, Wu Y, Zhang W, Han Y, Tang F, Shen D, Liu Y (2020) Analysis and correlationship of chemical components of various walnut (*Juglans regia* L.) cultivars. Food Meas Charact 14:3605–3614. https://doi.org/10.1007/s11694-020-00603-0

Chapter 9
Vegetables, Legume Seeds and Spices

Contents

9.1 Introduction

A dramatic increase in the production and consumption of vegetables was observed at the beginning of 2010. At this time, global growth in production was much greater than that for the production of other edible plant commodities. Vegetables are integral components of a well-balanced and healthy diet and provide substantial amounts of essential nutrients that prevent diseases, contributing to improved health. Increases in the production of vegetable crops is expected to continue. Moreover, researchers anticipate that improved varieties of edible plants may be important for increasing the productivity and quality of vegetable crops (Prohens-Tomás and Nuez 2007a, b).

Edible plants are considered healthy, safe and environmentally friendly, and consequently, adulteration and fraudulent activities for economic profit have been increasing. This has contributed to the dynamic progress of analytical procedures

---

Supplementary Information The online version contains supplementary material available at https://doi.org/10.1007/978-3-031-99687-0_9.

P. Szefer, *Chemometrics in Food Control Based on Elemental Analysis*,
https://doi.org/10.1007/978-3-031-99687-0_9

for testing authenticity. The use of multielement and stable isotope analysis based on atomic spectroscopy for distinguishing between organically and conventionally cultivated edible plants has been preferable. Methods such as ICP–MS and IRMS are promising tools for tracing the fertilization history of organic plants. However, these techniques are limited by the potential diversity of fertilization practices used for organic and conventional plant production. Therefore, analytical techniques integrated with chemometric tools are recommended for the development of statistical models that allow accurate classification of agricultural resources, including vegetables, with respect to geographic origin (Laursen et al. 2014). According to such trends and recommendations, the combination of using different instrumental techniques has emerged as an approach for progressing the accuracy of classification and prediction of specifications of food, including edible plants, in contrast to the use of a single analytical technique. Among the three levels of data combination, low- and mid-level combination approaches are among the most common. It is expected that data combination could not only increase the global classification/prediction ability but also minimize the uncertainty of each individual result and provide more effective prediction (Borràs et al. 2015).

Recently, published papers have focused on newer analytical methods, such as ICP–MS, ICP–OES, CE–UV, TXRF, LIBS and ^{1}H NMR spectroscopy, coupled with chemometric techniques, e.g., CDA, PLS–DA, PCA and ANN. For example, ^{1}H NMR, which is based on metabolomics and integrated with multivariate chemometrics, can be applied to identify the geographical origin of cabbage (Kim et al. 2013). The currently used combination approaches include systems such as LIBS–SIRA or ICP—MS/ICP–OES–SIRA, which are highly reliable in the determination of the authenticity/adulteration of vegetables and spices. According to Drivelos and Georgiou (2012), an increasing number of scientific papers published in the beginning of 2000 involved investigations of the elemental content and stable isotope ratios combined with strontium and lead isotope ratios as well as multielemental content. This classification has been used to evaluate the interspatial and intervariety/cultivar differentiation of edible plants. The determination of trace, micro- and macroelement distributions in food products, including vegetables with different, sometimes unknown origins, is important. Szefer and Grembecka (2007) presented the average concentration and range of macro, micro- and trace elements, including toxic elements, in various vegetables with different origins worldwide. Numerous analytical methods/techniques combined with multivariate analysis have been successfully used for the detection of adulteration in different food products (Szefer 2007; Grembecka and Szefer 2011, 2016).

9.2 Chemometric Assessment of Vegetables, Legume Seeds and Spices

9.2.1 Geographical Differentiation

9.2.1.1 European Countries

Longobardi et al. (2011) analysed SIRA and volatile compounds determined by IRMS and HS–SPME/GC–MS, which were helpful for determining the characteristics of three potato varieties from three locations in southern and northern Italy. PCA, with the first two PCs accounting for 53.4% of the total variance, revealed that Sicilian potato samples differentiated from the remaining samples and that Tuscany samples were not differentiated from Apulia samples. On the basis of DFA data from both the volatile dataset and the complete dataset, a reliable differentiation with classification results of 100% between Italian potato samples cultivated in Sicily, Tuscany and Apulia was achieved.

Opatić et al. (2018a) reported that EA–IRMS combined with EDXRF and ICP–MS were applied to process potato samples from four macroregions, i.e., the Alpine, Dinaric, Mediterranean and Pannonian regions. The DA model revealed that 100% of the potato samples were correctly classified. Isotopic and elemental fingerprinting, including all 29 variables or exclusively REEs, achieved excellent discrimination ability for organically grown potatoes according to their macroregions of origin. This means that all the samples studied were properly labelled a priori with respect to their geographical provenance. Variables such as δ^{13}C and δ^{18}O correlated with F1 more significantly in the Mediterranean region than in the other regions. However, the highest levels of Mg were found in the Alpine region, whereas greater contents of Ca and Cl were found mainly in the Mediterranean region.

Opatić et al. (2017b) integrated EA–IRMS with EDXRF to investigate garlic samples from four Slovenian macroregions, i.e., the Alpine, Dinaric, Mediterranean and Pannonian regions. DA demonstrated clear discrimination between these four regions, with an overall success rate of correctly reclassified samples of 77% (on average). The most common interspatial descriptive parameters were δ^{13}C, δ^{34}S, Zn, P and Cl. However, some overlap was observed among garlic samples originating from both the Alpine macroregion and the Dinaric macroregion, whereas samples from both the Pannonian and Mediterranean macroregions were well differentiated. The overlapping samples may be explained by the comparable geological and climate characteristics in both the Alpine and Dinaric regions. It was concluded that this model and this methodology could be promising approaches for rapid, inexpensive and robust screening to trace the origin of garlic samples.

Pepi et al. (2019) coupled ICP–MS and XRF to assess the elemental distribution patterns of hop plants growing in two locations in the Trentino region. The relationships between the chemical elements in the plant parts and the soil substratum were observed. PCA (with the first two PCs explaining 55.26% of the total variance) enabled the discrimination of strobili samples between two producing locations, i.e., Cavalese and Imeř. PLS–DA of the strobili samples also allowed differentiation between the two locations. Owing to the use of PCA and PLS–DA, the most discriminative elements, Ca, Cr, Fe, Mg, N, Ni, Ti and Zr, were selected as possible interregional geochemical markers.

Jandric et al. (2021) employed three advanced methods, namely, ICP–QMS, MC ICP–MS and LC–TOF–MS for characterizing carrots from five specific production regions in Austria ("Genussregionen"). Various multivariate tools (PCA–X, PLS–DA, OPLS–DA, SIMCA and DD–SIMCA) demonstrated reliable discrimination of the carrot samples according to their geographical origin. The first three PCs explaining 72.1% of the total variance were differentiated into five clusters corresponding to carrots from five production regions. A predictive ability of 97% or better (depending on the classification technique) was attained on the basis of integrated $^{86}Sr/^{87}Sr$ ratios and multielemental data. However, the application of data combination strategies resulted in 99–100% highly efficient and correct classification. Thus, it was possible to distinguish carrot samples originating from five Austrian regions effectively. Importantly, when class modelling techniques were used, satisfying performance characteristics of carrot classification models were achieved in the case of DD–SIMCA (99–100% efficiency), which outperformed the SIMCA models (93–100% efficiency) for all datasets. It was concluded that the DD–SIMCA model of combined Sr isotopic ratios and elemental and metabolomic data could provide a reliable approach for distinguishing carrot samples of different geographical provenances, thereby enabling the control of geographical origin requirements.

Using ICP–OES combined with chemometric analysis, D'Archivio et al. (2019) classified red garlic samples originating from four Italian production territories on the basis of MEC. All the calibration and prediction samples were correctly categorized by means of LDA, enabling the differentiation of garlic according to its origin. The most discriminative elements, Ba, Ca, Fe, Mg, Mn, Na and Sr, were also significant in class modelling via SIMCA. The sensitivity for the external data samples (80–83%) was somewhat lower than the sensitivity for the training data (91–100%). The specificity of the various class models was relatively satisfactory; i.e., 96–100% of extraneous garlic samples were rejected by the various SIMCA class models.

Pianezze et al. (2019) analysed SIR data obtained by TC/EA–IRMS from red garlic cultivated in different Italian territories (Abruzzo, Lazio and Sicily). PCA with the first four PCs explained 95.57% of the total variance. A visualization of the variable loadings allowed the differentiation of garlic categories with respect to variations in the stable isotope ratios to be explained. A model with satisfactory predictive ability for geographical classification via the LDA and k-NN methods was constructed. However, preliminary class modelling based on SIMCA confirmed the better usefulness of the data for geographical traceability of garlic because it showed acceptable sensitivity and good specificity.

Swoboda et al. (2008) tested and applied Sr isotopic signatures for the fast and reliable authentication and traceability of *Asparagus officinalis* originating from Marchfeld (Austria) using MC–ICP–MS after optimized Rb/Sr differentiation. The asparagus samples originated from six counties. Austrian asparagus was clearly (100%) distinguished from the neighbouring countries of Hungary and Slovakia, with a probability of more than 80%.

9.2.1.2 Asiatic Countries

Ariyama et al. (2004a) reported that methodical triad methods, such as FAAS, ICP–AES and ICP–MS, which quantify 19 inorganic elements, were applied to study 244 Welsh onion samples. Chemometric techniques were useful for distinguishing between two countries, i.e., Japan and China. Among the two models, LDA classified 101 samples studied for modelling 97% correctly and predicted another 119 samples, excluding 24 nonauthentic samples that classified 93% correctly. SIMCA classified 101 samples; 220 samples from known production areas, including samples used for modelling and excluding 24 nonauthentic samples, were predicted 92% correctly. In another study, Ariyama et al. (2004b) processed Welsh onion samples originating from Japan and China and used 22 elements. LDA and SIMCA were applied on the basis of 20 elements, and onion samples originating from Wales were differentiated from those originating from Japan and China.

Ariyama et al. (2017) performed intercomparison analyses of $^{87}Sr/^{86}Sr$ ratios and Sr concentrations for five vegetable species grown in Japan and China. In addition to the relationships between the $^{87}Sr/^{86}Sr$ ratios in the vegetables, the soil-exchangeable pool, irrigation water and fertilizer from 20 Japanese agricultural areas were also examined. Consequently, a prefectural map of the strontium isotopic ratio of Japanese vegetables was generated. This method could be useful for identifying the growing area of vegetables. The vegetable isotope ratios in Japan were similar for all species within a given agricultural area, although they tended to be low in northeast Japan and high in southwest Japan. The median $^{87}Sr/^{86}Sr$ ratio in Japanese vegetables was similar to that in fertilizer, suggesting that Japanese vegetables contain some Sr derived from fertilizer and precipitation. It was concluded that significant discrimination between Japanese and Chinese samples could be improved by suing the vegetable Sr isotope ratio.

Kobayashi et al. (2011) analysed taro samples by INAA and IC to evaluate this vegetable produced in Japan and China. The first three PCs, accounting for 53.51% of the total variance, discriminated the samples into three clusters: Japanese taro grown under excess fertilization conditions, popular Japanese taro and popular Chinese taro. The most discriminative variables, $H_2PO_4^-$ and Co, were identified. DA was performed to identify the most efficient combination of elements and compounds to discriminate taro geographic origin, with the highest percentage of correct classification being 100% for Japanese samples and 93.75% for Chinese samples. The highest contribution of correct classification was achieved with a two-variable model including the abovementioned $H_2PO_4^-$ and Co for Japanese and Chinese taro, respectively.

According to Nakamura et al. (2012), three instrumental methods, namely, ICP–AES, ICP–MS and INAA, were tested to classify fresh and frozen taro samples from Japan and China. LDA was used to distinguish between Chinese and Japanese taro with elements such as Mg, Ni, Zn, Rb and Cd. The sensitivities for Chinese fresh and frozen taro were 79.4% and 73.2%, respectively, with false-positive rates estimated to be 1.7% and 2.5% for Japanese fresh and frozen taro, respectively.

Liu et al. (2012) studied sea cucumber samples from three water environments, i.e., the Bohai Sea, Yellow Sea and East China Sea (China), via ICP–MS coupled with chemometric processing. PCA, CA and LDA distinguished between samples belonging to three water environments. The first two PCs, which explained 84.2% of the total variability, clearly distinguished the three score groupings without any overlap (Fig. 9.1). The CA results were generally in agreement with the classification results achieved by PCA. A LDA also provided an excellent correct classification rate and a CV rate, with both at 100.0%. This classification model could reliably differentiate samples from three water environments.

Bong et al. (2012b) integrated ICP–AES and ICP–MS to determine analytes, including the $^{87}Sr/^{86}Sr$ ratio, in cabbage heads from five Korean provinces and five Chinese districts. The first two DFs, DF1 and DF2, classified 88.47% of the samples and clearly differentiated all the Korean cabbage samples from the Chinese samples. The most discriminative elements responsible for distinguishing between the two countries were detected as Al for Korea and Na, Sr, Ba, Fe and Li for China. LDA, CA and AHC differentiated samples of cabbage according to their geographic origin; however, according to the LDA diagram, the samples from two Chinese regions slightly overlapped with those from some of the Korean provinces.

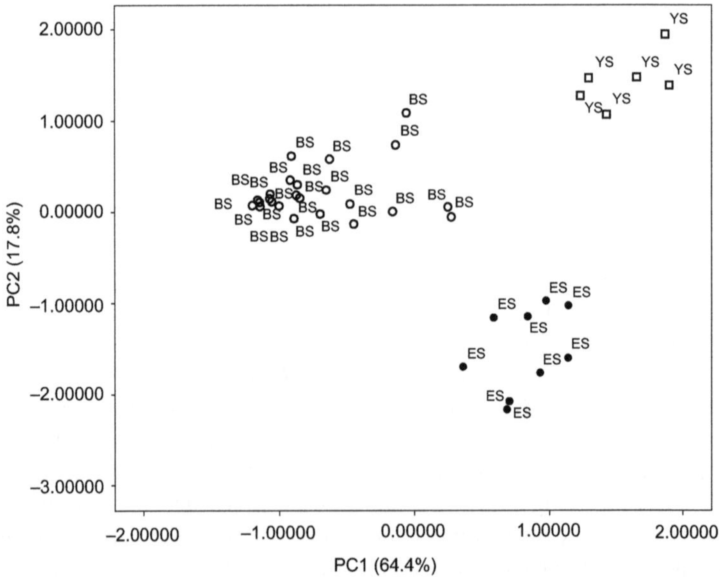

Fig. 9.1 2-D score plots of the PCA results for sea cucumber samples according to the first two principal components. BS, Bohai Sea samples, “o”; YS, Yellow Sea samples, “□”; ES, East China Sea samples, “✳”. (After Liu X et al., 2012. Food Contr 23:522–527. https://doi.org/10.1016/j.foodcont.2011.08.025. With permission)

Nevertheless, the data highlight the potential for assessing the geographical origin of Chinese cabbage.

Bong et al. (2013) investigated 363 cabbage samples from seven Korean provinces and seven Chinese areas using a combination of ICP–AES and MC–ICP–MS to determine the MEC with $^{87}Sr/^{86}Sr$ ratio. The analytical data were subjected to chemometric multivariate analysis. Cabbage samples produced in Korea were grouped relatively well by local production area, although some samples overlapped. It was also shown that the geographic origin of cabbage produced in China could be successfully distinguished within local production areas. Additionally, DA clearly discriminated cabbage originating from two countries. The lithology of cabbage-producing areas well reflects the chemical composition of cabbage.

According to Bong et al. (2012a), the MEC contents and $^{87}Sr/^{86}Sr$ values for Kimchi samples were analysed using ICP–AES, ICP–MS and MC–ICP–MS coupled with PCA and LDA. PCA with the first two PCs accounting for 51.50% of the total variance differentiated Korean and Chinese kimchi samples. On the basis of LDA, it was estimated that 96.55% of the samples were correctly reclassified between the two countries of origin, with a maximum 100% possible accuracy achieved for Korean samples relative to the value of 92.86% obtained for Chinese samples. Elements K, Na, Ca, Ba, Sr, Zn, Li and Sc were mainly responsible for obtaining such reliable intercountry differentiation. Moreover, the effective differentiation of unwashed and washed Korean kimchi from unwashed and washed Chinese kimchi was observed.

Zhang et al. (2017) processed sea cucumber samples originating from major producing areas of the northern China Sea via EA–IRMS and GC/MS methods. The $\delta^{13}C$ and $\delta^{15}N$ values did not sufficiently differ between samples from two sampling locations because of their partial overlap. As shown in Fig. 9.2, FA compositions combined with PCA successfully distinguished the samples according to their geographic provenance. PCA and DA are thus promising methods for discriminating between different origins of sea cucumber samples from the northern China Sea.

Choi et al. (2020) classified garlic samples from different Korean cultivation regions using three instrumental methods, including advanced tools such as TC/EA–IRMS, ICP–AES and ICP–MS. Based exclusively on stable isotopic compositions or the elemental concentrations, DA correctly classified 63.6% and 69.1% of the samples, respectively. However, a considerable improvement in classification was possible owing to the use of a combination of stable isotopic and element profiles. A total of 89.1% of the garlic samples were correctly classified with respect to the region of cultivation, which had different lithological and climatic conditions.

Park et al. (2019) applied the TC/EA–IRMS method to classify onion samples from different Korean regions. PCA with the first two PCs accounting for 44.3% of the total variance did not allow them to clearly differentiate the samples; therefore, LDA was conducted to maximize the covariance among the groups with elemental compositions and isotope ratios of C, H, O, N, C/N, H/C and O/C. Owing to this approach, it was possible to correctly classify 74.6% of the onion samples (63.6% after cross-validation). Satisfactory discrimination of the geographical provenance of samples was obtained by combining groups located at similar latitudes. The

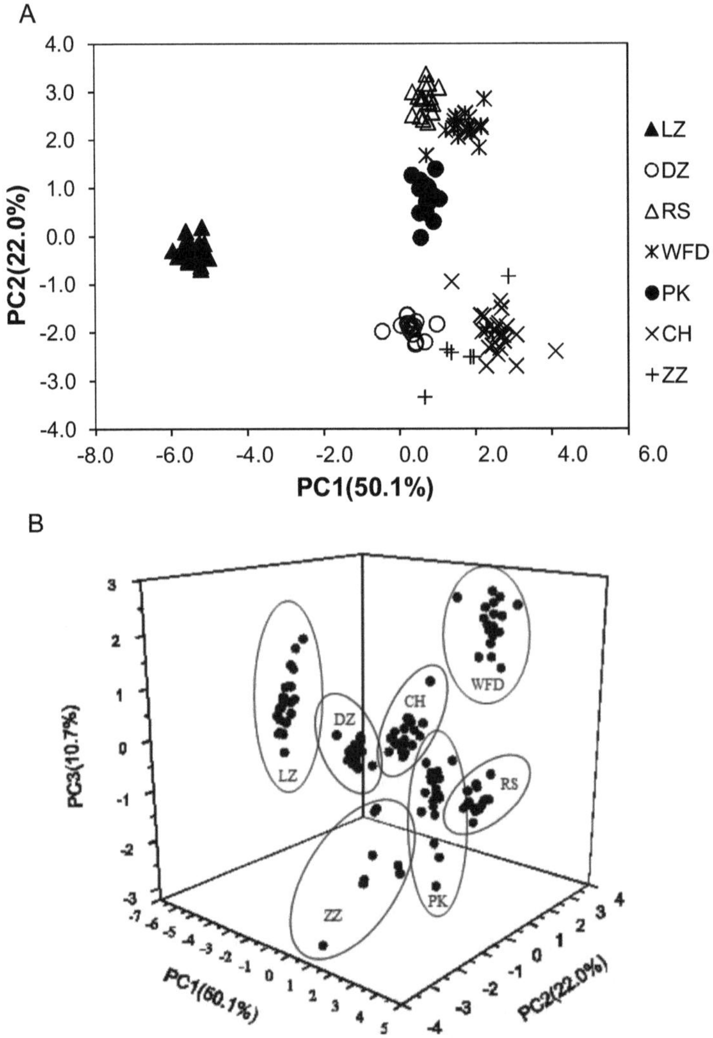

Fig. 9.2 Principal component analysis of the fatty acid composition of *A. japonicus* with various geographic origins in the northern China Sea. (**a**) Two-dimensional plot and (**b**) three-dimensional plot of the fatty acid profiles of *A. japonicas* investigated via PCA. LZ, DZ, RS, WFD, PK, CH and ZZ represent Laizhou, Danzi Island, Rushan, Wafangdian, Pikou, Changhai Island and Zhangzi Island, respectively. (After Zhang X et al., 2017. Food Chem 218:269–276. https://doi.org/10.1016/j.foodchem.2016.08.083. With permission)

values of $\delta^2H/\delta^{13}C$ and $\delta^{13}C/\delta^{15}N$ as variable ratios improved the discriminative potential for identifying the geographical origin of onion samples.

The study by Ariyama and Yasui (2006) involved the determination of MEC by ICP–OES and ICP–MS in Welsh onion. LDA discriminated samples originating from Japan and China (Shandong, Shanghai, Fujian), allowing the classification of an appropriate number of samples used for modelling with between 95 and 97%

correction and in the case of predicted samples with between 89 and 94% correction (Ariyama et al. 2004a, b). SIMCA achieved comparable effectiveness in terms of discrimination ability, including samples used for modelling and excluding nonauthentic samples, with 92% accurate predictions (Ariyama et al. 2004a). Moreover, in the case of LDA and SIMCA, it was possible to correctly distinguish samples originating from both countries.

Hondrogiannis et al. (2012) processed cumin spice samples from four countries by means of WDXRF and ICP–AES to differentiate them chemometrically by ANOVA and LSD. The LSD model correctly predicted the geographical origin of 100% of the Chinese, Syrian and Turkish samples and 87.50% of the six Indian samples with slight overlaps in elemental content in cumin from Turkey and India on the basis of the content of the most discriminative elements, i.e., Ca, K, Mg, Fe, Na, Zn, Sr and Mn.

9.2.1.3 American Countries

Nassar et al. (2012) transformed chemometrically ICP–OES data to characterize samples of 16 Canadian-grown potato cultivars collected from five growing sites. The data treated by CDA were classified relative to geographical locations using the CANDISC procedure of SAS. The canonical scores from the analysis were used for HCA for further unsupervised classification of cultivars within locations. It was concluded that variations in the elemental content of potato were detected between cultivars in each farming region among specific cultivars at different sites and collectively between sites. The "Russet Burbank" or "Yukon Gold" cultivars are beneficial for consumers because of their improved nutritional content due to essential bioelements.

Anderson et al. (1999) used ICP–AES to study 608 potato samples collected from the United States and Canada to develop a method for confirming the geographical authenticity of Idaho-labelled potatoes as Idaho-grown potatoes. The first three PCs explained only 49% of the total variability, although modest differentiation was achieved when the first two PCs were considered. More satisfactory results were obtained using CDA. However, among the PCA, CDA, DFA, *k*-NN and ANN techniques, the latter was effective for the classification of the samples into two geographic regions, i.e., Idaho and non-Idaho, with 98 to 99% degrees of accuracy when a bagging strategy was used.

Oliveira et al. (2014) analysed caigua (achoch) samples from five farms located in southwestern Bahia (Brazil) to determine the MEC content via ICP–OES. The first two PCs, which explained 66.92% of the total variance, enabled the differentiation of samples from one farm from the four others. It was also demonstrated that P, Na, Cu, Fe, Zn and Mn were the variables contributing to the differentiation between samples. This result was confirmed by HCA data. According to the authors, caigua can be strongly recommended as a nutritional supplement because high levels of K, Ca, P and Mg have mineral profiles similar to those of watermelon, cucumber, chayote, pumpkin and melon, which also belong to the Cucurbitaceae family.

dos Santos et al. (2011) analysed Chinese cabbage samples from 13 Brazilian cities via ICP–OES using MEC data transformed by chemometrics. PCA, with the first three PCs corresponding to 67.2% of the total variance, demonstrated that P, K, Cu, Fe and Mn contributed to the majority of the variability among the Chinese cabbage samples. However, PCA and HCA revealed that no systematic discrimination samples originating from 13 different Brazilian cities were obtained. Notably, the contents of macro- and micronutrients determined in the Chinese cabbage samples indicated that the vegetable could be used as a nutritional supplement.

Using the ICP–OES method, Hidalgo et al. (2016) processed *Sechium edule* (Chayote) fruits from local growers in four Argentinean cities. The score distribution pattern of the first two PCs explained 44.8% of the variability of the data matrix. A systematic differentiation of the *S. edule* fruits originating from the four regions was achieved. The most discriminative elements were Ba, Sr, Fe, K, Mn, Na, Ca and Mg. Furthermore, LDA, k-NN, PLS–DA and SVM also enabled satisfactory differentiation between samples originating from all these geographical locations. However, among these models, LDA was the most effective, with a success rate of 89.1%, whereas the SVM model was slightly less reliable, with a success rate of 87.0%. These findings suggest the great potential of the elemental composition for geographical traceability of *S. edule* fruits, with good nutritional contributions confirmed by RDA.

9.2.1.4 Countries Across Multiple Continents

Trincherini et al. (2014) used TIMS to investigate six types of potatoes from various sites in Italy and China irrespective of their type, e.g., fresh or processed. The analytical technique significantly differentiated the $^{87}Sr/^{86}Sr$ ratios and $\delta‰$ values between Chinese and Italian tomato samples. Therefore, geographical assessment of the region of origin for unknown samples is important. No significant distinctions within the interspatial variation in potato $^{87}Sr/^{86}Sr$ ratios for southern and northern Italy were obtained. For the Italian samples, the method allowed the differentiation of tomato products from soils of different geological ages. It was concluded that the methodology could be used as a commercial approach for legislative purposes.

Keppler and Hamilton (2008) reported that δ^2H determined by GC–P–IRMS and TC/EA–IRMS was used to investigate tubers of early potatoes from ten countries across Europe and regions close to the Mediterranean Sea. This vegetable originated from Middle Europe can be reliably differentiated from those of Mediterranean regions and from those of Northern Europe. The δ^2H values of the methoxyl groups could be used to identify the geographical origin of potato tubers. For example, the obtained δ^2H values revealed that potato tubers could not have been cultivated in Egypt but rather must have their geographical origin somewhere in Middle Europe. However, such a conclusion would not have been reached if the δ^2H values of the bulk biomass of potato tubers instead of their methoxyl groups had been quantified.

Fragni et al. (2015) differentiated samples from three countries, i.e., Spain, Italy and China, via ICP–oa–TOF–MS. LDA clearly differentiated samples originating from three countries, with 99.9% of the samples correctly classified. Elements Li, Co, Cu and Sr were the most discriminative variables. It was expected that the preliminary results could be used as a fast, precise and accurate tool for multielement quantification for food fingerprinting.

According to Opatić et al. (2018b), a combination of EA–IRMS, EDXRF, CFA and HPLC–DAD instrumental methods was applied to classify commercial tomato samples from four countries. DA based on LOOCV revealed that the best overall success rate was achieved when either the combination of all three sets of parameters (SIR, MEA and chemical markers including different antioxidants) or only the elemental content data with correct classification was obtained for 80% of the samples. Significant discrimination was obtained between the Slovenian and Italian samples, with overlap between the Spanish and Moroccan groups, although these samples were distinguished from the other samples.

Using the DRC–ICP–MS method, Lo Feudo et al. (2010) characterized tomatoes and paste from four Italian regions and three concentrated tomato pastes from three other countries. Satisfactory data were attained for LDA, SIMCA and KNN, especially for the latter, which provided a correct classification of all samples for both groups, i.e., Italian and non-Italian samples. In the case of tomato samples, even if KNN produced a model enabling a good percentage for prediction of 88.2%, LDA and SIMCA were preferred since these techniques correctly classified all samples for each category. For triple-concentrated tomato paste samples, the most reliable results were obtained by all the models; in particular, KNN correctly classified all the samples for both categories, i.e., Italian and non-Italy tomato pastes. Moreover, the classification was successfully unchanged with respect to the production year for the concentrated paste and the harvesting years for fresh tomatoes.

He et al. (2020) reported that the integration of EA–IRMS and ICP–MS was a good approach for studying maca from different large and small interregional scales in two countries. The 3D scatter plot of the $\delta^{13}C$, $\delta^{15}N$ and $\delta^{18}O$ values revealed a strong ability to discriminate maca grown in Peru and China (Fig. 9.3a). As shown in Fig. 9.3c, PCA (with the first three PCs corresponding to 75.45% of the total variance) clearly differentiated Peruvian maca from Chinese maca. In the case of differentiation between Peruvian maca and Chinese maca, overall discriminatory accuracies of 100 and 96.2% were obtained when SIR and MEC were used, respectively. With respect to maca from different Chinese regions, the order of discriminatory accuracy was as follows: MEC (80.2%) = SIR combined with MEC (80.2%) > SIR (71.9%). K, B, Mn, Fe, Mo, Cd and As were the main descriptors for identifying maca from different Chinese regions. Moreover, PCA with the first three PCs explaining 72.78% of the total variance revealed significant differences among the Chinese maca samples (Fig. 9.3D).

According to Ariyama et al. (2007), the combination of FAAS, ICP–OES and ICP–MS was used to estimate the geographical provenance of onion samples originating from primary onion-growing areas in Japan and those from six other countries. The models established on the basis of LDA were useful for differentiating onions originating from Hokkaido, Hyogo and Saga from those originating from other countries. The classification of these samples was estimated to be 87%

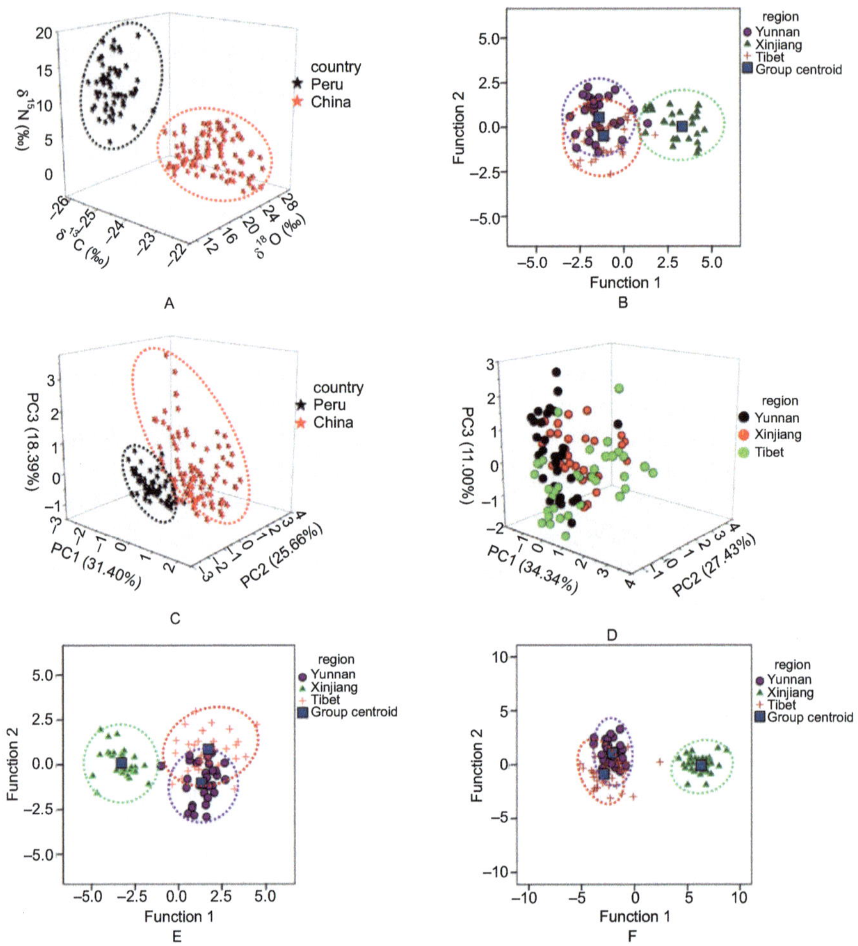

Fig. 9.3 Scatter plot of maca samples between Peru and China and among different Chinese regions. (**a**) Scatter plot of the $\delta^{13}C$, $\delta^{18}O$ and $\delta^{15}N$ values of maca samples from Peru and China. (**b**) Scatter plot of the stable isotope ratios (SIRs) of samples from different Chinese regions based on the two discriminant functions. (**c**) Scatter plot of mineral element concentrations in samples from Peru and China based on the PCA scores. (**d**) Scatter plot of mineral element concentrations in samples from different Chinese regions based on the PCA scores. (**e**) Scatter plot of mineral element concentrations in samples from different Chinese regions based on the two discriminant functions. (**f**) Scatter plot of the combination of mineral elements and SIR fingerprints of samples from different Chinese regions based on the two discriminant functions. (After He Y et al., 2020. Food Chem 311:126058. https://doi.org/10.1016/j.foodchem.2019.126058. With permission)

accurate in total, with the prediction by CV being 85% accurate in total. This model is based on Na, Mg, P, Co, Cu, Zn, Rb, Sr, Mo, Cd, Cs and Ba. Moreover, there was no significant change in the element profile from a production area due to variations in fertilization, crop year, variety, soil type or production year. This means that the approach is effective for use as a screening technique.

Longobardi et al. (2015) integrated EA–IRMS with TC/EA–IRMS methods to study dried lentil seed samples from different geographic regions in two countries. A clear PCA discrimination was observed when PC1 was used (64.0%) vs. PC3 (7.1%). It was then possible to classify the lentil samples with respect to their origin from Canada and Italy. The most reliable chemometric techniques were *k*-NN and PLS–DA. Both models demonstrated satisfactory performance with external prediction abilities of 93.1 and 93.1%, respectively, and with prediction abilities (CV 10) of 95.2 and 93.6%, respectively, indicating the suitability of the developed methods. This means that samples originating from Italy were effectively differentiated from those originating from Canada.

Kokkinofta et al. (2020) investigated the use of ICP–OES, HPLC-RID and other analytical methods with multivariate analysis to monitor changes in the fruit and seeds of carobs from seven countries in the Mediterranean region. Among the multivariate chemometric tools, PCA, PLS–DA and OPLS–DA were used to process concentration data. PCA with the first two PCs accounting for 57.1% of the total variation achieved a clear differentiation of carob seed samples. Many descriptive variables were proteins, dietary fibres, fat, Ca, Zn, ash, Cu, Mn, P, Fe and Mg had the highest values and moisture, K, Na, sucrose, glucose and fructose had the lowest values. Samples of different geographical origins were discriminated by PLS, with 78% success in total. Moreover, 100% correct classification was achieved for samples from Cyprus, Greece, Italy and Spain, but Jordanian with Palestinian carobs were presented together in one group, and Turkish carobs were classified as Greek. The lack of clear differentiation of samples from the latter three countries may have resulted from the small number of samples of this origin.

According to Opatić et al. (2017a), EA–IRMS and EDXRF were effective for categorizing lettuce, sweet pepper and tomato samples from six countries. Discriminant analysis models *per* kind using LOOCV allowed the correct classification of these three types of vegetable samples, i.e., 86.2, 71.1 and 74.4%, respectively. Parameters such as $\delta^{13}C$, $\delta^{15}N$, $\delta^{18}O$, $\delta^{34}S$, P, S, Cl, K, Ca, Mn, Fe, Zn, Br, Rb and Sr were useful for distinguishing lettuce, sweet pepper and tomato in terms of their intercountry origin.

Wakefield et al. (2019) reported that an analytical workshop constructed from the instrumental triad of ICP–MS, EA–IRMS and TC/EA IRMS was useful for classifying saffron samples. PCA with the first three PCs representing 71.9% of the variance enabled the exploration of sample origins. LDA showed clear discrimination between saffron originating from two countries (Fig. 9.4). The prediction abilities following these validation procedures were 80 and 81% for LOOCV and k-fold cross-validation, respectively.

Potortì et al. (2020) used ICP–MS data processed by chemometrics to classify ten types of spices and aromatic herbs originating from two countries. The specific elemental profile significantly differed with respect to geographical provenance for the samples studied. It was possible to discriminate between Italian and Tunisian samples for each spice and aromatic herb (Figs. 9.5 and 9.6). The intake of essential elements through the samples was small, but the content of Pb in some samples

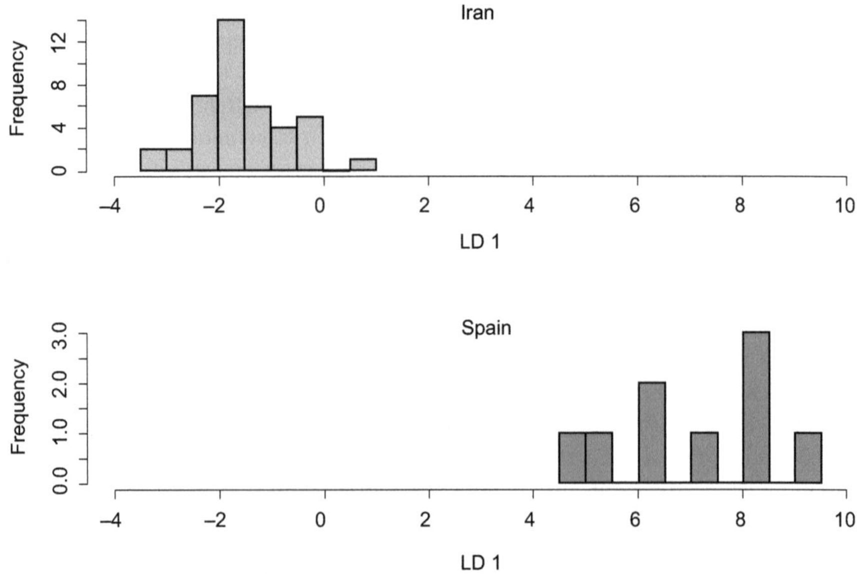

Fig. 9.4 The spread of scores, shown as histograms on Linear Discriminant Function 1 (LD 1), for saffron samples originating from each country. (After Wakefield J et al., 2019. Food Contr 106:106699. https://doi.org/10.1016/j.foodcont.2019.06.025. With permission)

exceeded the maximum allowable level, and any safety risk for consumers was excluded.

Ocvirk et al. (2019) applied EDXRF assisted by multivariate and univariate chemometrics to determine the MEC of hop plants from nine countries. Multielemental fingerprinting was used to identify the geographical origin of the hops. It was possible to distinguish between European samples and between non-European samples. For example, within the rest of the world, i.e., Australia, New Zealand, the United States and RSA, the overall correct classification for this dataset was 93% according to the LOO–CV results.

9.2.2 *Intercultivar/Varietal Differentiation*

Romarís–Hortas et al. (2010) investigated the use of ICP–OES, ICP–MS and HG–AFS for categorizing edible seaweeds from the North Atlantic (Spanish Galicia) according to type, such as brown, red and green seaweeds, as well as varieties, which are commercially named Wakame, Fucus, Sea Spaghetti, Kombu, Dulse, Nori and Sea Lettuce. Concentration data were processed using a multivariate approach (PCA, CA, LDA and SIMCA). PCA was performed using CV as a validation method, and after a normalized Varimax rotation, 71% of the total variance was explained by the first three PCs. PCA and CA differentiated samples with respect to their three types and seven varieties, respectively. When the SIMCA was used,

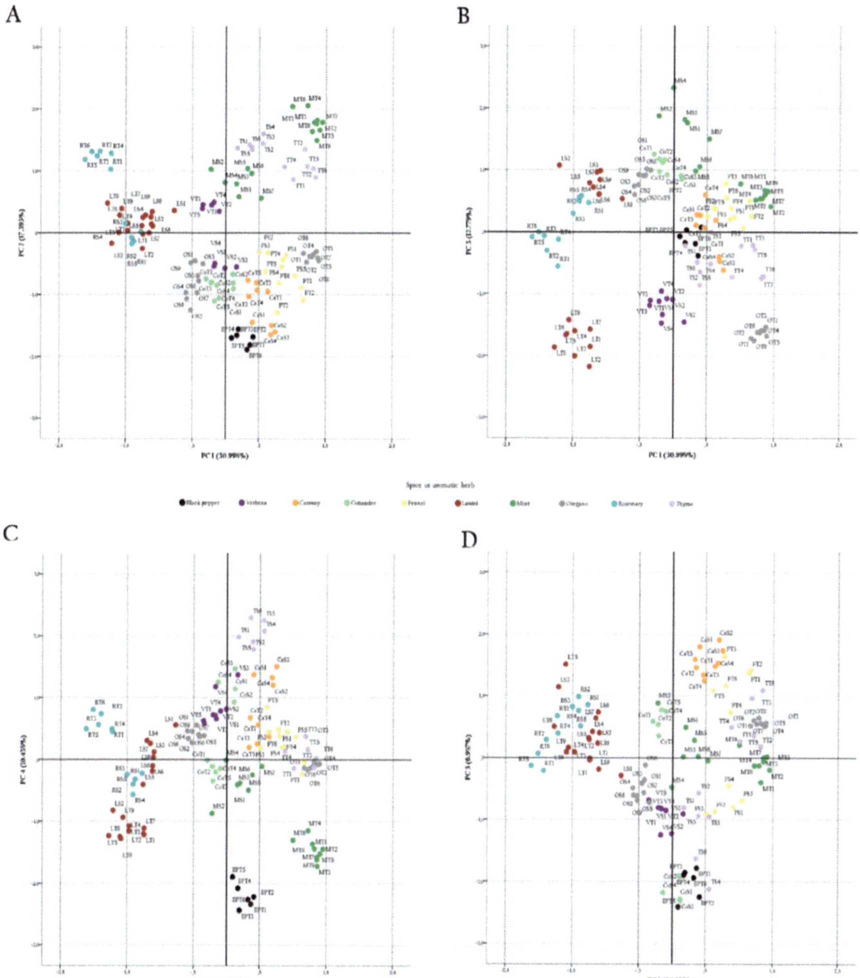

Fig. 9.5 2D score plots for the 118 spice and aromatic herb samples categorized as botanical species. (**a**, PC2 against PC1 plot; **b**, PC3 against PC1 plot; **c**, PC4 against PC1 plot; **d**, PC5 against PC1 plot). (After Potorti AG et al., 2020. Food Chem 313:126094. https://doi.org/10.1016/j.foodchem.2019.126094. With permission)

satisfactory classification (recognition and prediction) was achieved exclusively for red seaweed, amounting to 100% of the cases correctly classified, whereas in the cases of brown seaweed recognition (training set) and prediction (testing set), 89 and 80%, respectively, were attained. The application of LDA provided a correct assignment of 100% for the brown, red and green types at a significance level of 5%.

Using FAAS and a spectrophotometric method, Grembecka and Szefer (2022) characterized the elemental profiles of ten species of legumes and oilseeds from different geographical origins (European, Asiatic, North American and Oceania products) available on the Polish market. In general, legumes had lower mean

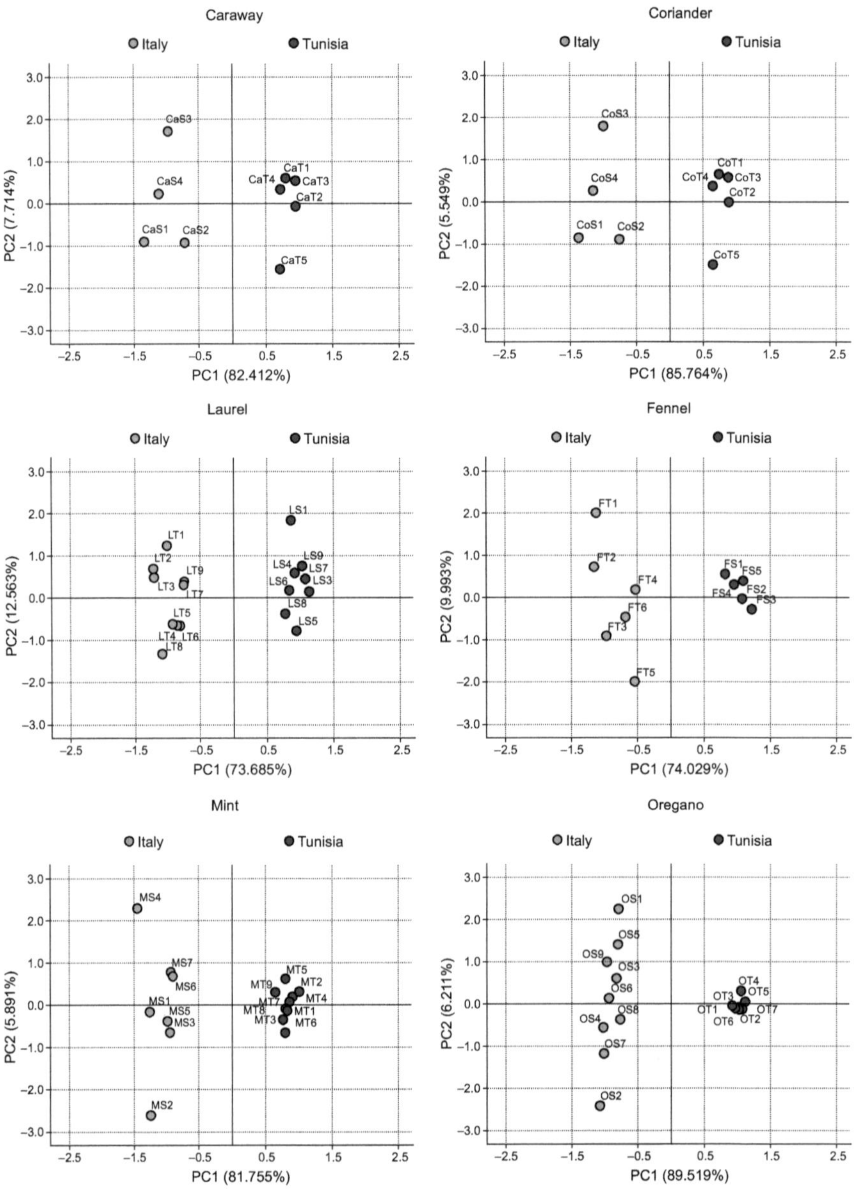

Fig. 9.6 Score plots for PC1 and PC2 showing the differentiation between Italian and Tunisian spices and aromatic herbs. (After Potorti AG et al., 2020. Food Chem 313:126094. https://doi.org/10.1016/j.foodchem.2019.126094. With permission)

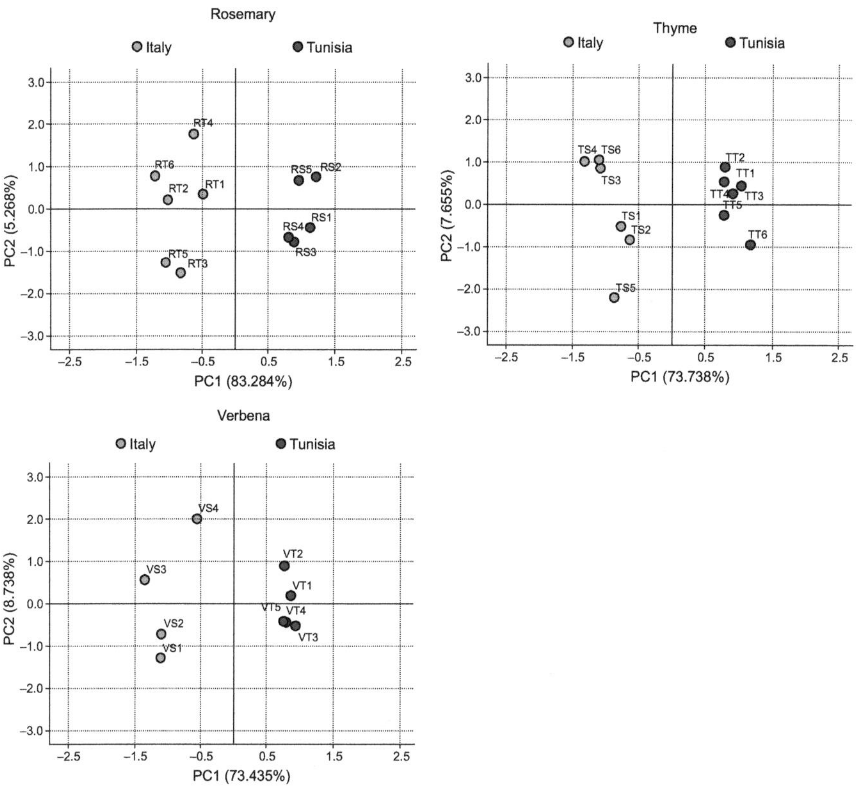

Fig. 9.6 (continued)

concentrations of Ca, Mg, Na, P, Zn, Cu, Fe, Mn and Cr than oilseeds. The Cd level in every sample, except for soybean and hulled wheat, exceeded the permissible European standards. According to the Kruskal–Wallis test results, the Mg, Na, K, P, Zn, Cu, Mn, Cr and Cd contents depended on the type of the analysed product, whereas in the case of the botanical provenance, such a relationship was registered for most of the analysed variables, except for Fe, Cr and Co. FA and CA allowed the classification of the studied samples with respect to botanical species and type on the basis of their elemental composition (Fig. 8.2 in Chap. 8).

Niu et al. (2022) used a histological approach to investigate the multielement profiles determined by ICP–MS and UHPLC–Q–TOF/MS and the metabolite profiles of two eggplant cultivars from China. Multivariate analysis revealed a clear differentiation of the two eggplant species. Figure 9.7a shows the PCA biplot, with the first two PCs accounting for 76.49% of the total variance. A clear differentiation between the peel and flesh of the two eggplants was attained. Figure 9.7b shows the PLS–DA biplot for the multielement dataset, indicating effective differentiation of the peel and flesh of eggplant species into four groups. The combination of MEA, metabolomic analysis and chemometric models was found to be a useful approach for the identification of candidate chemical markers of MVSP and MVEF.

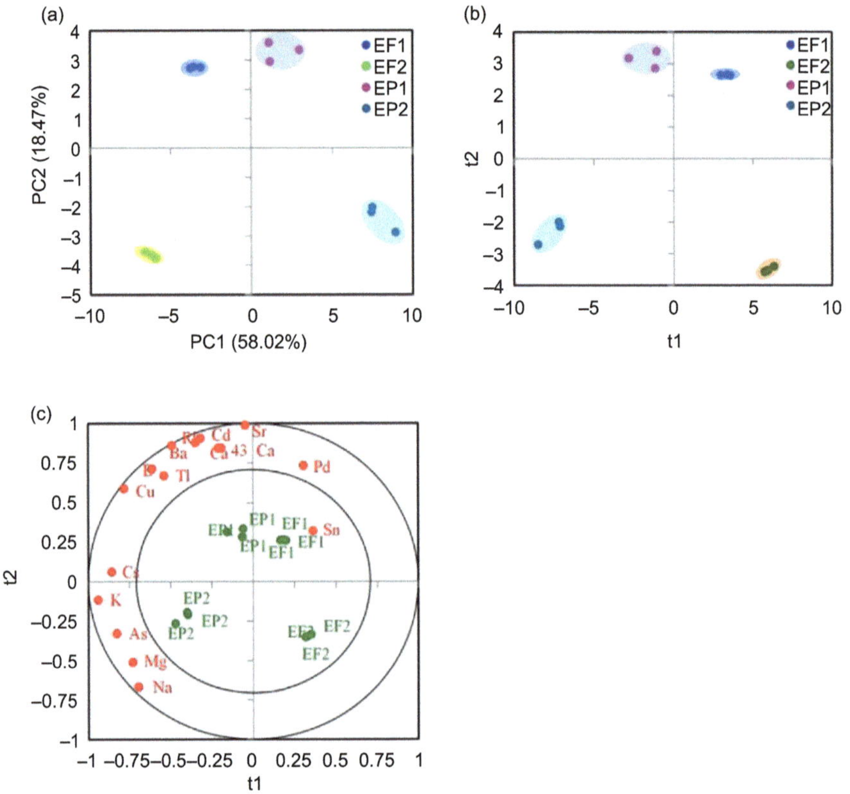

Fig. 9.7 Multivariate statistical analysis of multifactorial data: (**a**) PCA score plots for EP1, EP2, EF1 and EF2 comparisons; (**b**) PLS–DA score plots for EF1, EF2, EP1 and EP2 comparisons (two principal components, $R2X = 0.943$, $R2Y = 0.999$, $Q2 = 0.964$); (**c**) chemical element and eggplant peel and flesh correlations between samples; EP1, MVS peel; EP2, MVE peel; EF1, MVS fresh; EF2, MVE fresh. (After Niu X et al., 2022. Food Res Inter Part A 162:111976. https://doi.org/10.1016/j.foodres.2022.111976. With permission)

Grembecka (2007) and Grembecka et al. (2009) reported that seeds for sprouts were obtained from different companies that are available to consumers in Poland. After being seeded and grown for 7 days, the MEC contents of the edible sprout samples were investigated via FAAS combined with the spectrophotometric method, and content data were processed with chemometric analysis. Inspection of the scatter plot of the PCs, with the first two PCs explaining 75.5% of the variance, revealed two isolated object sample groupings enclosed mainly from the Fabaceae and Brassicaceae families. The species from other families did not create clusters, which could be explained by the small number of samples representing the Amaryllidaceae and Poaceae families, although the long distance between the points representing the species of these families on the plot suggests fewer similarities between them. Data for most of the studied sprouts, e.g., rucola, leek, onion and beetroot, were reported for the first time (e.g., rucola, leek, onion and beetroot).

Pérez-Rodríguez et al. (2019) reported that ICP–MS combined with chemometrics allowed the construction of classification models that effectively predicted the varieties among five varieties of Argentinean cowpea bean on the basis of their element profiles. Among the five supervised learning techniques, i.e., LDA, PLS–DA, k-NN, RF and SVM, the latter was the best classifier, with a 93% accuracy rate. Using the model developed by the SVM technique, it was possible to correctly differentiate the samples into the five cowpea varieties, where 100% sensitivity was achieved for most of the predicted classes. The methodology applied was shown to be a promising and noteworthy tool for testing the authenticity of cowpea beans on a global scale for differentiating various cowpea genotypes.

According to Hwang et al. (2017), the combination of ICP–MS, ICP–OES and DMA was used to investigate greenhouse vegetables, including 600 samples of two varieties of tomatoes, eight varieties of peppers and three varieties of cucumbers from South Korea supermarkets. LDA showed partial discrimination of samples according to varieties of bell peppers and corresponded to trace essential and toxic elements clearly differentiated from each other. For the analysis of intervarietal differentiation, both tomato varieties were not sufficiently differentiated. However, in the case of cucumbers, the Chwi chung variety was clearly differentiated from the Baekdadagi and yellowish Overripe varieties, which clustered together. The pepper samples were divided into two groups, i.e., green bell peppers, red bell peppers and yellow bell peppers together and Shishitho, green Serrano, green, and Chungyang red and red peppers together. Furthermore, the levels of Al, As, Cd, Pb and Hg did not pose any particular health risks resulting from the THQ values. However, the percentages of As in cucumber, Al in tomato and Cd with Hg in pepper were greater than those of other elements considering the total health HI.

Moyib et al. (2015) used RP–HPLC, FAAS and FP tools to characterize and assess several popular Nigerian common beans (three varieties). *Z*-statistics tests revealed that Nigerian beans are nutritionally analogous to French beans and soybeans. ANOVA indicated that seed coat colour varied with proximate nutrients, Ca, Fe and vitamin C. PCA, with the first two PCs accounting for 56.8% of the total variation, indicated intervarietal differentiation of the studied samples. A clear discrimination between small pecked white beans and large white beans was achieved, whereas the small brown bean was an outlier at lower PC2 and PC1 values. This means that chemometric analysis is useful for grouping beans for determining seed coat colour. In this study, a moderate genetic distance (GD) facilitating the transfer of useful genes and intercrossing among the beans was estimated. The methodology has the potential to integrate French beans and soybeans into genetic improvement programs for Nigerian common beans.

Chope and Terry (2009) aimed to quantify the effects of field application of additional Ca and/or S treatments on the elemental content (quantified by ICP–AES) of onion bulbs grown from seeds and to characterize genotypic differences in the elemental concentrations of onion bulbs of short-, intermediate- and long-storing onion cultivars, viz. SS1, Ailsa Craig and Renate, respectively. The data were processed by CVA to distinguish cultivars with minimal overlap between the 95% confidence limits for each cultivar. In the case of onions of different cultivars growing at the same site (with the same soil and under the same climatic conditions), there were still significant variations in the elemental composition of the onion bulbs. It was therefore expected that the source of such differentiation is genotypic.

Anunciação et al. (2011) combined ICP–OES with multivariate analysis to classify white and red species of cabbage (*Brassica oleracea*) samples from 24 Brazilian cities. Both the PCA and CA techniques indicated that no significant difference between the two cabbage types was achieved. These findings indicated that the elemental composition was not useful for distinguishing between red and white cabbage. It has been recommended that cabbage may be useful as a nutritional supplement.

Dobrowolska-Iwanek et al. (2022) aimed to quantify the elemental profiles of 25 sprouts from various botanical families to identify highly promising species with respect to their importance in human diets. A combination of FAAS with a multivariate chemometric tool was used to classify the studied samples. PCA with the first two PCs accounting for 75.5% of the variance allowed them to differentiate the scores into two differentiate clusters that generally corresponded to the Fabaceae and Brassicaceae families, with the exceptions of alfalfa and soy from the former and bittercress from the latter. Ca, Mg and Zn were discriminators of the Brassicaceae samples, whereas Cu and Fe were effective identifiers of the Fabaceae cluster. According to the authors, this methodology may be useful for preselecting the most prospective sprouts for further processing, which could be included in so-called functional foods.

9.2.3 Interregional and Intervarietal Differentiation

Camargo et al. (2010) combined NAA and PCA to classify garlic samples (ten cultivars) from three Argentinean geographic regions, i.e., the La Consulta region (Mendoza), the Esquel region (Chubut) and the Ushuaia region (Tierra del Fuego). PCA based on Cs, Fe and Br, with PC1 and PC3 accounting for 56.2% of the total variance, enabled the differentiation of cultivars according to their physiological groups. PCA based on Br, Co, Cs, Fe and Rb, with the first three PCs explaining 80.9% of the total variability, enabled the differentiation of samples among the three geographic regions, i.e., La Consulta, Esquel and Ushuaia. This means that a four-element PCA allowed the classification of cultivars with respect to their physiological groups, whereas a five-element PCA was appropriate for recognizing geographic origin.

Bontempo et al. (2011) reported that various instrumental tools, such as EA–IRMS, TC/EA–IRMS, ICP–MS and IC, allowed the effective processing of samples of tomato, tomato juice, passata and paste from eight farms in three Italian regions. To evaluate the discrimination efficiency relative to sample provenance, they used LDA using LOOCV to achieve prediction abilities of 88.46%, 91.03%, 95.15% and 92.98% for tomato, juice, passata and paste, respectively. Cr, U, Gd, Cs, Tm, Eu, La, Ni and δ^{34}S were identified to affect all four sets, suggesting their usefulness as geographical origin identifiers. Moreover, LDA of all the data was performed, and they used LOOCV to obtain an over 95% success rate. This means that the model was sufficiently robust, importantly distinguishing the three Italian regions.

Rodríguez Galdón et al. (2012) studied ten cultivars of potato from two Tenerife regions (Spain) by means of FAAS with physicochemical measurements according to the AOAC. LDA contributed to correct classification of the samples originating from both locations, i.e., La Cañada and El Castillo. A complete 100% classification (100% after cross-validation) of the potato samples according to geography was achieved by using proteins, Na, K, Fe, Cu and Mn or all the variables. This chemometric technique also was effective for the differentiation of tomatoes with respect to their cultivar for both locations and achieved correct classification and after CV, i.e., 100% and 86.7% for La Cañada and 100 and 100% for El Castillo.

Rodríguez Galdon et al. (2008) used FAAS and HG–AAS with a spectrophotometric method to process five traditional onion cultivars from Tenerife (Guayonje, San Juan de la Rambla, Carrizal Alto, Carrizal Bajo and Masca) and a commercial cultivar grown under the same agronomic, soil and climatic conditions. FA, with the first three factors accounting for 66.4% of the total variance, demonstrated that the onion samples belonging to the Texas cultivar were differentiated from the remaining cultivars, whereas the samples from San Juan de la Rambla had inadequate differentiation between elemental profiles; i.e., the elemental profiles were insufficiently related to seed origin. It was suspected that environmental characteristics and agronomic practices could influence the genetics of seeds, modifying their elemental composition. The onion samples related to the Texas, Masca and Carrizal Alto cultivars were properly categorized, although the samples of the remaining cultivars overlapped. When all the variables were used, a correct classification of 90.5% was achieved according to cultivar and seed origin. In addition, LDA was a useful tool for differentiating onion samples according to the studied accessions. For all the variables used, 94% (83.3% after CV) of the onion samples were classified.

Di Giacomo et al. (2007) studied Italian potato samples via ICP–MS regarding their geographic origin and variety. The elemental data were processed chemometrically. LDA allowed them to classify and distinguish Fucino potatoes from those of other areas of Italy and from those of the four provinces of Abruzzo. In addition, LDA correctly assigned cases with 96.7% differentiation of the three potato varieties cultivated in the Fucino Basin. It was concluded that, on the basis of MEC, interregional and intervarietal discrimination of Fucino potatoes was possible.

Lombardo et al. (2013) combined FAAS and HPLC–UV to study early potato cultivation in Sicily (southern Italy), which originated from three regions by using MEC and antioxidant contents. Soil samples were also investigated. PCA was useful for cultivar differentiation; the Antea cultivar contained the highest levels of Mn and Ca, and Cu mostly accumulated in Romanze, whereas Fe mostly accumulated in Valeria. Erika and Everest accumulated Mg to the greatest extent; however, Everest had the most amount of Na, in contrast to the levels of Na in Labadia and Safrane. PCA, with the first two PCs accounting for 72.7% of the total variance, was an effective tool for clearly discriminating potatoes from the Torregrotta, Giarre and Cassibile regions relative to the essential element and antioxidant contents of nine early crop cultivars. The most discriminative variables were K, Ca, Mg and phenolics. This means that PCA clearly differentiated cultivars according to location.

Potortì et al. (2021) investigated spices and aromatic herbs from two districts of Algeria according to their ten botanical species and regional origin. PCA with the first two PCs explaining 25.0% and 22.4% of the total variance, respectively, differentiated the samples studied with respect to their elemental distribution patterns. Noticeable differences were observed among the spices and aromatic herbs analysed, depending on the plant type. However, for each spice or aromatic herb, the trend of differentiation was much clearer, showing the possibility of using PCA to differentiate samples according to their geographical origin.

Studies by Krauß and Vetter (2019) used the integration of EA–IRMS with GC–C–IRMS to process red bell pepper samples from three European countries. Red bell pepper samples collected in September were distinguished from each other in terms of their bulk $\delta^{15}N$ values. Both light and dark red bell peppers could also be significantly distinguished from each other. The seasonal variations in the $\delta^{13}C$ values of the German samples were tentatively attributed to varying CO_2 enrichment during the annual production cycle. Moreover, the $\delta^{13}C$ values of the Dutch samples were also related to greenhouse production, whereas those of the Spanish samples most likely originated from open field cultivation.

Adamo et al. (2012) used advanced multi-instrumental methodologies such as MC–ICP–MS, HPLC–UV–DAD, LC/QTOF/MS/MS and DNA fingerprinting to test analytical data for elemental concentrations of Mn, Cu, Zn, Rb, Sr and Cd and $^{87}Sr/^{86}Sr$ ratios integrated with biological data of 11 secondary metabolites and DNA as markers to authenticate the origin of early potatoes at small geographical scale levels in Italy. DNA fingerprints through 12 simple sequence repeat (SSR) primer pairs enabled the identification of cultivars, confirming the discrimination ability of the molecular markers. Multivariate chemometrics, i.e., PLS–DA, allowed them to assign all the potato samples to the proper sampling sites. The models developed were satisfactorily validated.

Ocvirk et al. (2018) reported that EA–IRMS was an effective tool for classifying hop samples according to their geographical origin. The $\delta^{15}N$ and $\delta^{34}S$ values were the most discriminating parameters for ANOVA, allowing for distinct grouping of 8 out of the 10 hop-growing regions. The results also revealed a limited capacity for $\delta^{13}C$, $\delta^{15}N$ and $\delta^{34}S$ values for discriminating between different varieties of hop.

9.2.4 *Differentiation of Other Parameters*

9.2.4.1 Verification of Authentication/Adulteration

According to Peña et al. (2001), FAAS and FAES data combined with chemometric treatment were used to determine the characteristics of potato samples originating from Galicia, with guaranteed origin and indications of the CBOQ, and those from other Spanish areas outside Galicia. PCA with (PC1 and PC2 explaining 53.75% of the total variance) and CA clearly distinguished between Galician potatoes and non-Galician potatoes corresponding to CBOQ and non-CBOQ samples, respectively. The dendrogram of CA was used to detect fraud to preserve the quality name of the CBOQ product. Moreover, the combination of the two types of neural networks, i.e., MLF–ANN and SOAN–MLF–ANN, also was highly satisfactory in achieving classification (Padín et al. 2001). According to Peña et al. (2001), in the case of the analysis of Galician and non-Galician potatoes, the data obtained for the VQBCP approach were better (recognition and prediction abilities of >97%) than those obtained by KNN and were comparable in advantages with those achieved via LDA, SIMCA and MLF–ANN.

Furia et al. (2011) reported that the use of ICP–MS enabled the differentiation of tropea red onion samples with PGI samples from Calabria and non-Calabrian onion samples from regions not cultivated following production regulations. The data obtained were processed by LDA, SLDA, SIMCA and BP–ANN. Clear discrimination between tropea red onions and onions from a field not related to the Calabrian production area could be achieved. The cross-validation procedure for all four chemometric techniques enabled acceptable results because all models provided satisfactory prediction abilities, with accuracies of 94.0% (LDA), 94.5% (SLDA), 95.5% (SIMCA) and 91.5% (BP–ANN).

Bontempo et al. (2014) performed an extensive study in which EA–IRMS was used to assay 1179 different Italian tomato product samples, i.e., tap water, juice, passata and paste samples collected in different years (2004–2012). The aim of this study was to establish the appropriate range of $\delta^{18}O_{water}$ values for authentic passata using a very extensive database, including many harvest years and wide interregional diversity. In accordance with the authors' intention, the approach was intended to "establish a threshold limit that could eventually become official and be adopted to implement the Ministerial Decree". It was possible to predict geographical differences on the basis of vegetal water $\delta^{18}O$ values. Differentiation of the geographical origin of the products studied was clearly achieved for juices and passata with Brix values below 7.5, whereas this trend was less visible in the case of passata with Brix values above 7.5 and for pastes due to water loss resulting from evaporation during the production cycle. The $\delta^{18}O$ value of −1.4‰ in water could be proposed as a limit for passata that has never been diluted, whereas −2.9‰ is appropriate for passata that has been concentrated up to 11.9 Brix. Such values "could be proposed as legal limits and officially used to determine whether passata has been obtained by squeezing fresh tomatoes and partial evaporation of the juice obtained or by diluting tomato paste with tap water".

Potortì et al. (2022) applied ICP–MS to characterize the elemental profiles of PGI and non-PGI cherry tomato samples as well as parent soil samples from Italy. Both multivariate techniques, i.e., PCA with the first two PCs explaining 71.41% of the total variance, as well as CDA, allowed the differentiation of the PGI group and not the PGI group. Moreover, the combination of elemental profiles with a chemometric approach was used to determine clear differences between PGI and non-PGI tomato cherry samples as well as between non-PGI tomato cherry samples originating from two soils, i.e., limestone and volcanic soils. CDA based on Zn, Cd, Mn and Ca as effective markers could correctly classify 100% of the samples.

Studies by Brunner et al. (2010) focused on the combination of HR–ICPSFMS (MEC and $^{87}Sr/^{86}Sr$ data) as a useful tool to study authentic paprika as a declared PDO (EU PDO/PGI) originating from the Szeged regions (Hungary) and seven other countries. PCA partially distinguished samples according to their geographic origin. Samples from China, Senegal, Romania, Italy, Germany, Spain and France were differentiated from authentic Szegedi paprika samples, although it was impossible to differentiate between paprika samples from France and Spain. When CDA was used, the CV grouped cases classified 76.6% of the dataset correctly, but for the authentic samples, the classification score was 100%.

Richter et al. (2019) reported that the ICP–MS data of 319 asparagus samples harvested between 2014 and 2017 from seven countries were processed using a chemometric tool. Using a SVM combined with NCV, a prediction accuracy of 91.2% was obtained when the country of origin was classified, but the accuracy increased to 98% for subsets of samples with high SVM prediction scores. In addition, the multielemental method provided specific fingerprints of asparagus cultivation sites, such as those that grew German asparagus samples, with an accuracy of 82.6% without incorporating any significant influence of the year of harvest or the asparagus variety in the classification. Notably, the study focused not only on German asparagus, including those with a PGI attribute, but also on samples originating from other countries.

According to Foschi et al. 2020, ICP–OES was applied to the geographical classification and authentication of lentils produced in three relatively close areas of the Central Apennines (Italy) and samples imported from Canada. Discriminant and modelling pattern-recognition methods were undertaken by using LDA and PLS–DA, which was useful for discriminating lentil samples relative to origin with good accuracy. For prediction, 100.0% (LDA) and 92.6% (PLS–DA) of the external samples were correctly assigned to the target classes. The class models generated by SIMCA exhibited high sensitivity (100%) and slightly worse specificity (91.8–100%) because most of the nonacceptable samples were removed from each of the four modelled classes. It was concluded that the approach applied is a highly effective tool for the geographical traceability of legumes as well as for the authentication of certified lentil varieties.

Fragni et al. (2018) integrated ICP–OES and ICP–oa–TOF–MS with chemometric processing analysis to study 183 samples of five tomato-based products, i.e., tomato juices, passata and semiconcentrated, double- and triple-concentrated tomato pastes collected in 2013, 2015 and 2017 from four countries. A LDA allowed satisfactory differentiation between Italian and non-Italian tomato samples. The Li/Cu, Co/Rb and Sr/Cd content ratios were highly discriminative in distinguishing tomatoes with Italian provenances from tomatoes with non-Italian provenances, with 100% prediction ability of the model and 98.8% cross-validation. This means that the ratios between elements were more important than single elements in the assessment of interregional differentiation. It was emphasized that a novel mineral element database was proposed for the authentication of Italian-processed tomatoes to differentiate Italian from non-Italian tomato-based products.

Drivelos et al. (2014) aimed to discriminate "Fava Santorinis" from other yellow split peas originating from six countries by utilizing REEs measured through ICP–MS. Four classification chemometric methods, OPA, MD, PLS–DA and KNN, were used. The results of the four classification processes based on the combination of these processes were presented. The OPA model was the best approach, with up to 100% accuracy, on the basis of the combination of the REE data with the trace element data. The PDO "Fava Santorinis" was distinguished from other yellow split peas with 100% accuracy by using multivariate chemometrics as a suitable tool for assessing food authenticity.

de Rijke et al. (2016) used GC–qMS and GC–C–IRMS to classify Dutch pepper (from two greenhouses) and non-Dutch pepper (originating from three other countries). The $\delta^{18}O$ values of the fruit water and corresponding source water samples and the regional GNIP values exhibited good relationships with the pepper data. Specific $\delta^{2}H$ data were useful for classification by modelling the probability densities of both groups using KDE. The probability densities of each group (Dutch and non-Dutch) were first modelled using a Gaussian kernel. The combined values of $\delta^{2}H$ and complimentary information regarding the relative abundance of the four main alkanes in bell pepper fruit water reliably distinguished the geographic provenances of the pepper samples. Misclassification rates were estimated using a Monte Carlo five-fold cross-validation procedure. It was concluded that LDA provided a satisfactory overall classification performance (93% accuracy) despite simplified class distribution assumptions. A density-based method was used for the same classification task. Similar results were obtained concerning the overall classification accuracy, although some pepper samples were mistyped.

Naccarato et al. (2016) used ICP–MS to quantify the contents of macro- and trace elements in authentic Calabrian chili pepper samples from six Italian farms and non-Calabrian chili peppers from eight other countries. PCA (PC1 and PC2 explained 50.94% of the total variance) discriminated Calabrian and non-Calabrian samples owing to the most discriminative elements, such as Sc, Cr, V, Ga, Fe, Al, Y and lanthanides. S–LDA as well as class-modelling techniques, i.e., SIMCA, UNEQ, PF and MRM, were also applied. MRM was a more effective approach than SIMCA with respect to the forced model efficiency of 96.5% after the removal of worthless variables by S–LDA. It was possible to distinguish Calabrian from non-Calabrian chili pepper models with generally higher efficiency.

Vadalà et al. (2016) reported that ICP–MS data were processed via a multivariate tool to differentiate "Nubia" garlic samples (labelled NRG samples) and "Non-Nubia" garlic samples of various six cultivars originating from different areas of Sicily, Spain and Tunisia. As shown in Fig. 9.8, PCA (PC1 and PC2 corresponding to 70% of the total variance) enabled the selection of certain elements as indicators responsible for the discrimination of samples with respect to their geographical provenance.

Mottese et al. (2018) coupled ICP–MS to GC–FID MEA to investigate capers and caper berries from different Sicilian localities, which were marked by PGI and PDO along with corresponding non-Sicilian commercial samples. PCA and CDA were used to differentiate and classify the samples according to their geographical

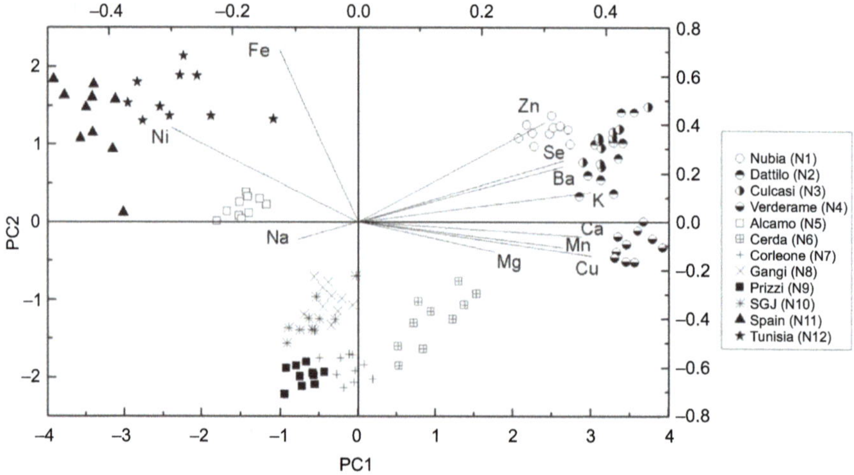

Fig. 9.8 Score (left and bottom axes) and loading (right and top axes) plot showing the results of the PCA performed on all the analysed garlic samples. (After Vadalà R et al., 2016. Foods 5:20. https://doi.org/10.3390/foods5010020. With permission)

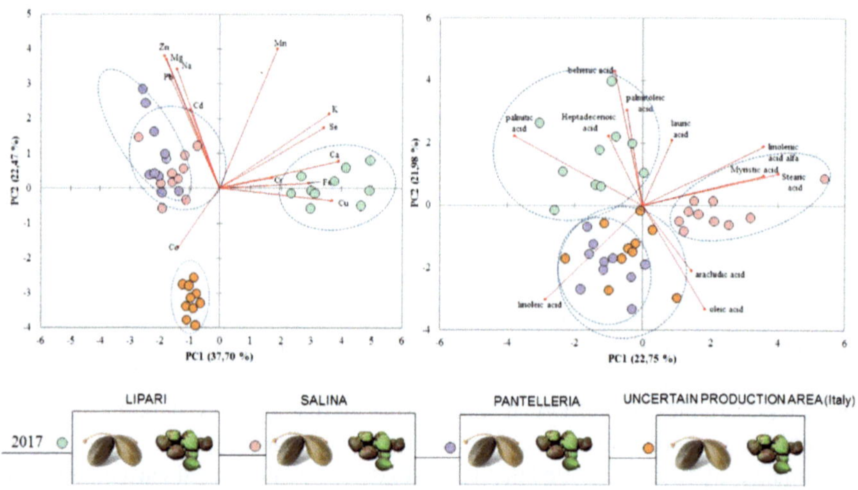

Fig. 9.9 Two-dimensional PCA biplots of inorganic element ratios (left) and FA ratios (right) over four production areas of capers and caper berries. The drawn ellipses suggest the natural grouping of samples (10 aliquots per site) according to the production area. (After Mottese AF et al., 2018. J Food Compos Anal 72:66–74. https://doi.org/10.1016/j.jfca.2018.05.009. With permission)

provenance. With respect to elements and FAs as variables, caper berries and capers were not completely distinguished relative to the geographical origin within the component space determined by the first two PCs, explaining 60.1% (related to elements) and 44.73% (related to FAs) of the total variance. It was concluded that the PCA-contributing elements appear to be more suitable than FA descriptors for assessment of the geographical provenance of capers and caper berries (Fig. 9.9).

However, clear CDA discrimination and prediction over the four locations revealing specific variables of classification were achieved by using K, Ca and Mg. In contrast to the FA variables, these data confirmed that 100% of the cross-validation cases were correctly classified according to geographical origin, allowing protection of the authenticity of Sicilian PGI and PDO capers and caper berries.

According to Chung et al. (2016), IRMS combined with chemometric techniques was a useful methodology for the characterization of potato samples collected from five Korean macroregions. PCA, with the first two PCs explaining 56.4% of the total variance, demonstrated more significant discrimination among cultivar regions than between potato cultivars. Additionally, OPLS–DA enabled the classification of potatoes on the basis of their geographical origins. It was concluded that the approach could appropriately authenticate the geographical origins of potatoes in Korea to prevent incorrect labelling, with possible application in other countries following validation.

Mir-Marqués et al. (2016a, b) combined FT–NIR and ED–XRF with ICP–OES or ICP–OES and FT–NIR with ED–XRF to process artichoke samples from Spain with the protected designation of origin (PDO) "Alcachofa de Benicarló" for the discrimination of PDO samples from those produced in other towns surrounding Valencia and Murcia outside the PDO area. According to Mir-Marqués et al. (2016b), PCA and HCA were applied to the spectral dataset as well as to the elemental profile of the samples, which revealed no clear differences between the origin groups of the samples considered. PLSDA, which is based on XRF spectra, was used to differentiate the samples into three classes with respect to Benicarlo, Murcia and Valencia, providing classification rates of 100%, 89% and 93% for calibration, cross-validation and external validation, respectively. This approach enabled the authentication of the PDO "Alcachofa de Benicarlo" at the 100% correct level. It was concluded that PLS models constructed from NIR and XRF spectra are more environmentally friendly than conventional ICP–OES.

Wilde et al. (2020) reported that SPME–GC–C–IRMS was a useful method for the differentiation of commercial food samples, including vanilla sugar, dairy and soy products, with labels describing vanilla pods/extract/powder and/or natural flavours. The sample preparation procedure was tested on selected products spiked with vanillin with a known $\delta^{13}C$ value. The ability to simulate the $\delta^{13}C$ range of vanillin from vanilla pods by combining it with vanillin originating from inexpensive sources has resulted in an increased risk of fraud being undiscovered. This study demonstrated the possibilities and limitations of the authentication of vanilla flavour by measuring the $\delta^{13}C$ value of vanillin.

Comparable results were obtained by Schipilliti et al. (2017), who also applied SPME-GC–C–IRMS data ($\delta^{13}C$) to distinguish between synthetic and biosynthetic vanillin. Such differentiation has not always been possible, but natural differentiation can be easily discriminated, although it is rarely used in different food preparations. There are numerous articles concerning the determination of the authenticity/adulteration of vanilla (e.g., Krueger and Krueger 1983, 1985; Lamprecht et al. 1994; Kaunzinger et al. 1997; Bensaid et al. 2002; Scharrer and Mosandl 2002; John and Jamin 2004; Tenailleau et al. 2004; Cicchetti et al. 2010; Greule et al.

2010, 2015; Gassenmeier et al. 2013; Geißler et al. 2017; Perini et al. 2019; Guyader et al. 2019; Wilde et al. 2019).

Hansen et al. (2014) verified the authenticity and traceability of vanilla flavours originating from ten countries by using GC–C–IRMS and GC–P–IRMS methods. Vanilla flavours produced by chemical synthesis and fermentation and extracted from two species of vanilla orchid were studied. The obtained data allowed them to discriminate between vanillin of natural origin and vanillin produced by biotechnology or chemical synthesis. On the basis of $\delta^{13}C$ values vs. V–PDB, it was possible to differentiate two groups, i.e., a first group for geographical origins growing only *V. planifolia* (Madagascar, Mexico, India, Africa, Indonesia, New Caledonia) and a second group growing only *V. tahitensis* (French Polynesia, New Guinea). ANOVA followed by Tukey's post hoc test revealed that the δ^2H values for vanillin originating from Indonesia could be differentiated from those for vanillin from mainland Africa and Madagascar. Moreover, the δ^2H values vs. V-SMOW values for vanilla pods originating from French Polynesia and New Guinea were significantly different. The relationship between $\delta^{13}C$ and δ^2H revealed that vanillin extracted from pods of adjacent geographic origins was grouped together; therefore, clearly recognizing its interspecies and interregional origin differentiation is difficult. A comment on this study entitled "Authenticity and traceability of vanilla flavours by analysis of stable isotopes of carbon and hydrogen" was made by Greule et al. (2015).

Jurado et al. (2005) indicated that the elemental profile established by ICP–AES was useful for determining the authenticity of the three brands of origin of aniseed drinks. On the basis of the scores of the samples in the three-dimensional space (with the first three PCs representing 88.7% of the total variance), three sample groupings corresponding to the three certified brands of origin were obtained. Loadings of Mg and Si were selected as discriminative variables for sample classification. Other chemometric methods, such as LDA and SIMCA were found to be very useful tools. LDA enabled a complete differentiation of the three considered classes, and the recognition ability relative to a posteriori probabilities was 100% for each class. The LOOCV procedure was used to evaluate the classification performance and the prediction ability was 100% for all classes. As shown in Cooman's plot (Fig. 9.10), the efficiency of the constructed SIMCA model was achieved for the classification of aniseed drinks of the three brands of origin.

Varliklioz Er et al. (2017) quantified saffron adulteration with some plant adulterants (safflower, marigold and turmeric) using three spectroscopic techniques, i.e., ATR–FTIR, RS and LIBS. PCA with the first two PCs accounting for 77.47% of the total variance allowed complete discrimination between saffron and the three adulterants on the basis of the LIBS dataset (Fig. 9.11c). PLS also enabled effective differentiation between saffron and plant adulterants. Three latent variables were selected in the PLS model, accounting for 96.10% and 99.26% of the cumulative variance. It was concluded that LIBS combined with the PLS model allowed sensitive determination of plant adulterants in saffron under 10%, which was difficult to detect based exclusively on the reference UV–Vis spectroscopic method.

Papaefstathiou et al. (2018) coupled ICP–OES with HPLC–PDA and characterized four groups of carob-based products as functional food products (powder/

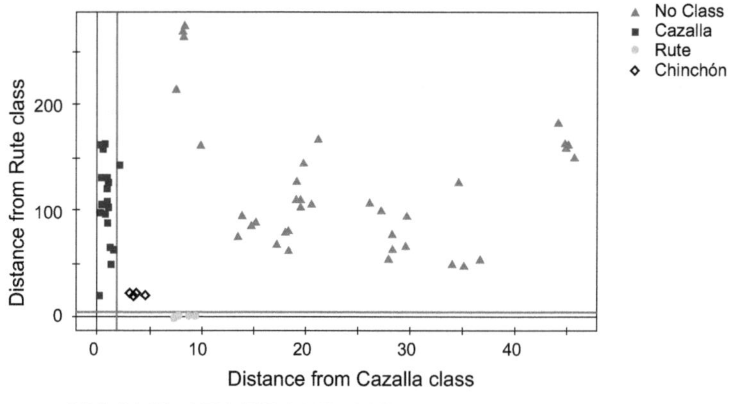

Fig. 9.10 Cooman's plot. (After Jurado JM et al., 2005. Talanta 66:1350–1354. https://doi.org/10.1016/j.talanta.2005.01.052. With permission)

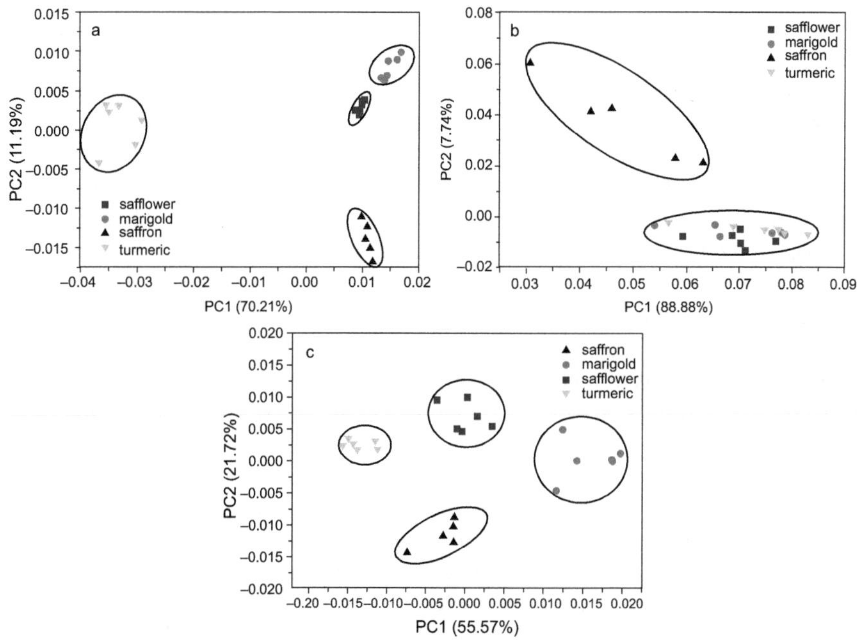

Fig. 9.11 PCA score plots of saffron, safflower, marigold and turmeric obtained from different spectroscopic methods: (**a**) ATR–FTIR, (**b**) Raman and (**c**) LIBS. (After Varliklioz Er S et al., 2017. Food Anal Methods 10:1547–1555. https://doi.org/10.1007/s12161-016-0710-4. With permission)

coffee, creamy products, carob syrup, pasteli/soutzoukkos) to make intercomparison assessments with those of domestic carob fruits, which are widely found on the island of Cyprus. With respect to PCA, the first two PCs, which explained 54.9% of the data variance, clearly differentiated the four score groups corresponding to powder/coffee, creamy product, carob syrup and pasteli/soutzoukkos, although wide variation existed within the group of the creamy carob products. Such differentiation could be dependent on the raw material that is affected by the specific geoclimatic conditions present in the production area of the carobs. Many reliable results were also obtained by PLS–DA. It was concluded that PCA and PLS–DA clearly discriminated all four types of products, although the group of creamy carob products showed wide variation.

9.2.4.2 Effects of Fertilization on Elemental Composition Related to Agricultural Practices

Bibak et al. (1998a) used the HR–ICP–MS method to characterize onion samples originating from 11 Danish sites uncontaminated by human activities in contrast to other normal agricultural practices. These results provide important nutritional information on the trace elements present in onion. According to Bibak et al. (1999b), a multielement composition was characterized for Danish cabbage and sprout from ten background areas for each crop cultivated with intense use of fertilizers and pesticides. The methodology presented could be efficiently used to study elements in agricultural crops. The results provide nutritional information on the background contents of major and trace elements in cabbage and sprouts. For example, the Pb levels were greater in cabbage than in sprouts, while sprouts contained greater content of Cd than did cabbage. It was concluded that the data are useful for nutritionist assessment, of which weight cabbage and sprouts should be a constituent part of the Danish diet. Bibak et al. (1998b) described a methodology based on the quantification of multielement composition, which was useful for providing nutritional information on the background levels of trace elements in peas collected from ten background areas intensively cultivated using fertilizer and pesticides. The variation in the levels of some elements in peas reflects the variation in parent soil properties. The results provide useful biological, nutritional and background information on peas.

Bibak et al. (1999a) applied HR–ICP–MS and INAA methods to classify potatoes grown under two fertilization practices (with pig slurry and calcium ammonium nitrate) at three levels of N fertilization. PLS enabled the potatoes to be split into three groups corresponding to three levels of N fertilizer depending on the fertilization practice. PLS component 1 (PC1) explained 24%, and PLS component 2 (PC2) explained 15% of the variation. PC1 and PC2 explained 24% and 15% of the variation, respectively, but in the other approach, the two PLS components explained 35% and 7% of the variation, respectively. This means that mean-normalized and nonnormalized data provide a similar splitting in the score plots. According to Bibak et al. (1999c), the levels of Cd, Co, Fe, Mn, Pb, V and Zn in cabbage, onion, pea and sprout negatively correlated with pH, whereas the correlation for Mo was positive. It was also suggested that any kind of acidification process due to agricultural practices leads to a decrease in soil pH. Additionally, the concentration of Mo tends to decrease in edible crops, whereas for the other elements, an overall increase would be observed.

9.2.4.3 Assessment of Interelemental Differentiation Related to the Farming System

According to Gundersen et al. (2001), all Danish greenhouse tomato samples, including parent soils, were studied for their multielemental profiles and concentrations via ICP–AES and HR–ICP–MS. One fruit of medium size per plant from each experimental unit (i.e., 12 fruits) was harvested. In the PCA score plot for harvest 3 (Fig. 9.12), a clear differentiation between soil and rockwool existed, and a grouping between the two rockwool treatments was achieved. In the individual score plots for harvests 1 and 2, a clear distinction between the soil and rockwool treatments was obtained, but no clear grouping between the two rockwool treatments was clearly achieved. The PCA data were found to be consistent with the results of the comparative statistical tests. It was possible to differentiate the two groups attributable to soil-grown and rockwool-grown fruits.

Suárez et al. (2007) reported that FAAS and HPLC combined with chemometric analysis were effective approaches for characterizing tomato samples belonging to five cultivars that originated from different farms on the island of Tenerife. On the basis of stepwise DA, data for the Boludo and Dorothy cultivars distinguished samples relative to their geographical provenance, i.e., southern and western cultivation

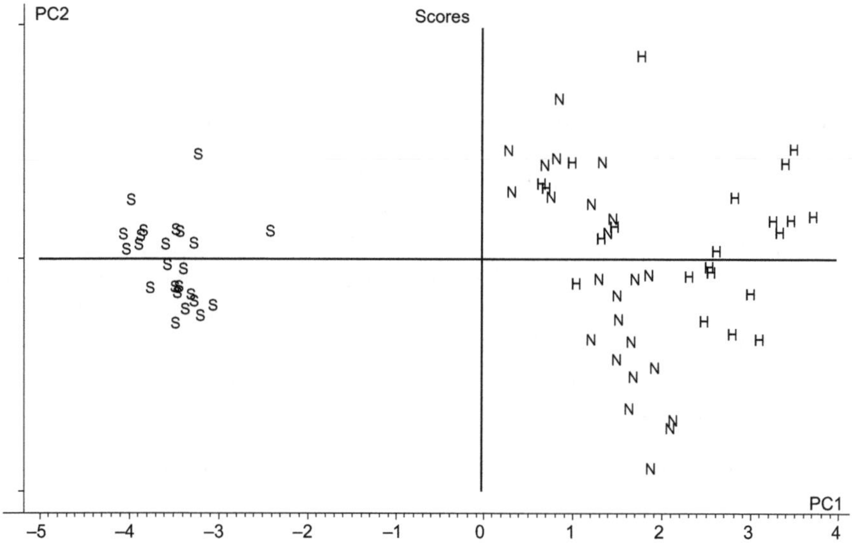

Fig. 9.12 PCA score plot, trace elements, harvest 3. The letters in the plot refer to the treatment (S, soil; N, rockwool normal EC; H, rockwool high EC). (After Gundersen V et al., 2001. J Agric Food Chem 49:3808–3815. https://doi.org/10.1021/jf0103774. With permission)

regions. On the basis of Na and Mg as selected descriptors, the samples were well classified, which means that 82.1% of the samples were classified (82.1% after cross-validation) and 88% of the samples were classified (88.0 after cross-validation) relative to the two zones and the same for the Boludo and Dorothy cultivars, respectively. Moreover, a correct classification of tomatoes was obtained with respect to cultivation method (intensive, organic, hydroponic) and sampling period.

Suárez et al. (2011) used HPLC, FAAS and other traditional instrumental techniques to demonstrate the importance of the chemical variables corresponding to metabolic relationships. PCA combined with ANOVA was a suitable tool for investigating the complex interactions between variables connected to the production–agronomy and chemical composition of tomatoes. Suárez et al. (2016) coupled FAAS with HPLC to study two tomato cultivars grown in Gran Canaria (Spain) either conventionally or hydroponically. HJ-biplot multivariate analysis, as a new approach for identifying the links among bioactive compounds of tomatoes, was applied to the chemical compositional data. Because HJ-biplot provides a better quality of representation than Gabriel biplots do (GH and JK biplots), the compositional HJ-biplot provides the most successful representation of the compositional data, which is more appropriate than the "form biplot" and "covariance biplot" offered by Aitchison and Grenacre.

Ariyama et al. (2006) processed onion samples grown under various conditions, including specific factors (fertilization, crop year, variety and provenance), from 52 fields in 18 farms in Hokkaido (Japan) to evaluate the differentiation of elements (determined by ICP–AES and ICP–MS). Multivariate chemometric techniques, such as CA, were used. The sample fertilization system and crop year influenced the variations in P, Ni, Cu, Rb, Sr, Mo, Cs and Tl contents in the onions. Moreover, intervariety differentiation was also observed owing to the high element levels. However, such variations were smaller than those achieved between production origins, with a few exceptions.

Rivero et al. (2003) reported that AAS and AES combined with multivariate analysis were used to characterize potatoes originating from eight cultivars in Tenerife (Spain). FA and CA clearly distinguished potatoes related to the farming system as well as to interregional and interspecies differentiation. The potatoes belonging to the two species could be differentiated. Similarly, FA, with the first three factors explaining 70.4% of the total variance, enabled reliable discrimination between potato samples originating from farms located in North and South Tenerife. Moreover, it was possible to differentiate traditional and recently imported potatoes as well as vegetable farms with and without irrigation systems.

9.2.4.4 Differentiation Between Conventionally and Organically Grown Vegetables

Novak et al. (2019) combined ICP–OES, IC and EA–IRMS of plant-derived sulphate with those of bulk plant tissue and plant-derived nitrate for the evaluation of organic and conventional potato, carrot and cabbage from thoroughly controlled long-term field trials and retail potatoes. The $\delta^{18}O$ values of sulphate from organic vegetables were significantly lower than those of their conventional counterparts, which is directly linked to the fertilization strategy. This means that $\delta^{18}O$ analysis of plant-derived sulphate represents a promising new approach for the determination of authenticated Danish organic vegetables.

According to Krejčová et al. (2016), ICP–OES, ICP–MS and single-purpose thermo-oxidative mercury analyses were used to quantify conventionally, organically and self-grown carrots throughout the Czech market. Significant differences in the levels of Na, K, S, Al, Mn, Ni, As and Cd were detected between conventional and organic carrots, although PCA with PC1 and PC2, which explained 55.49% of the total variance, revealed no significant differences between conventionally and organically grown carrots on the basis of the complete elemental composition.

Magdas et al. (2018) integrated EA–IRMS with ICP–MS, including REEs, which was an appropriate tool for investigating carrot samples grown in four areas of Transylvania and 12 countries other than Romania. LDA revealed that owing to $\delta^{15}N$ along with La and Ga contents, the initial classification of organically versus conventionally grown carrots was 83.3%, whereas the CV was 81%. For distinguishing between the Transylvanian samples and the samples from other EU countries, the achieved discriminant variables were Mn and Tb, with the initial and CV percentage of 75.7%, whereas for differentiating carrot samples from different areas in Transylvania, the $\delta^{2}H$ value was the most powerful marker. It was postulated that the Mn level is also correlated with the increased content of REEs such as Sc, La, Ce, Pr, Nd, Lu and Th, which are found in organically grown carrots.

Sinkovič et al. (2020) used a combination of EDXRF, EA–IRMS and CFA to differentiate chicory samples (five cultivars) from Slovenia. Six fertility management practices (control, two organic, two mineral and a combination of organic and mineral fertilizers) were used to produce five chicory cultivars in a glasshouse pot experiment. DA enabled the differentiation of organic and conventionally grown chicory according to six fertility management practices. At the minimum, parameters such as TPC, $\delta^{15}N$, $\delta^{34}S$, Zn, Br, Mo, P, S and K were required to adequately differentiate organic chicory from conventionally grown chicory. It was concluded that such a method should be included in EU regulations for organic farming systems.

Studies by Bontempo et al. (2020) were based on EA–IRMS and GC–C–IRMS data analysis of soil, tomato and tomato passata samples collected from two Italian regions that were produced organically or conventionally over two years. They were studied along all steps of the production chain of passata. PCA with the first two PCs corresponding to 86.0% of the total variance revealed differences between conventional and organic agricultural practices corresponding to two regions, i.e., Basilicata and Emilia Romagna. The highly effective discrimination was possible because of the use of amino acid $\delta^{15}N$ and $\delta^{13}C$ values compared with the bulk tomato data. Moreover, no statistically significant discrimination was achieved relative to the variety or the production year for all the isotope ratios.

Gundersen et al. (2000) analysed MEC content determined by HR–ICP–MS to classify onion and pea samples from 21 (onion) and 19 (pea) Danish cultivated sites.

The scatter plot of PC1 vs. PC2, which explained only 29% of the total variance, allowed them to differentiate onion samples according to cultivation type, i.e., between conventionally and organically cultivated sites (Figs. 9.13 and 9.14). For the pea samples, PCA, with PC1 and PC2 accounting for 19%, demonstrated only a weak ability to differentiate organically grown peas, i.e., conventional and organic farms (Fig. 9.15). This coincidence may be because the former used cattle slurry and is therefore more similar to organic farms than to other conventional farms that use other fertilizers or nothing. Compared with the PCA model for onions, the differentiation between farming methods for peas was less clear when the model based on individual sample data was used. A comparative chemometric tool that combines the element profiles of organically and conventionally cultivated onion and pea showed that the cultivation method influences the concentrations of some elements in crops.

Laursen et al. (2011, 2013) used advanced multi-instrumental methodologies such as HR MC–ICP–MS, MC–ICP–MS, ICP–OES and EA–IRMS for the investigation of Danish faba bean and potato samples produced in rigorously controlled long-term field trials comprising 144 experimental plots. Chemometric analysis data, such as PCA of multielemental fingerprints or pairwise comparisons and their confidence intervals adjusted for multiple comparisons, were interpreted. For example, the relationship between $\delta^{18}O$ of NO_3^- and $\delta^{15}N$ of NO_3^- or bulk $\delta^{15}N$ data for potato tubers as well as multielemental fingerprinting allowed for the discrimination of organic and conventional potato when combined with chemometrics. However, for faba beans, the differentiation was incomplete, which was probably due to a low

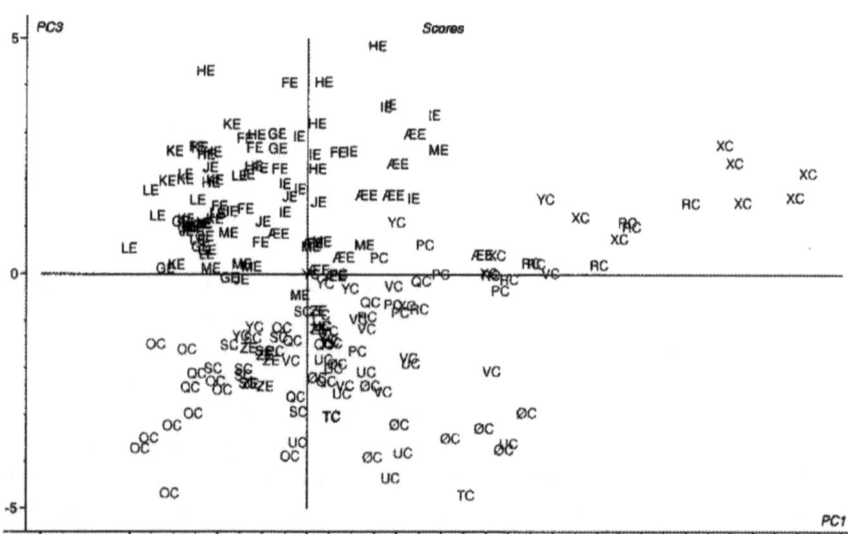

Fig. 9.13 Score plots for the first and third principal components of the PCA model for individual onion samples. Sites with E as the second letter are organic, and sites with C as the second letter are conventional. (After Gundersen V et al., 2000. J Agric Food Chem 48:6094–6102. https://doi.org/10.1021/jf0009652. With permission)

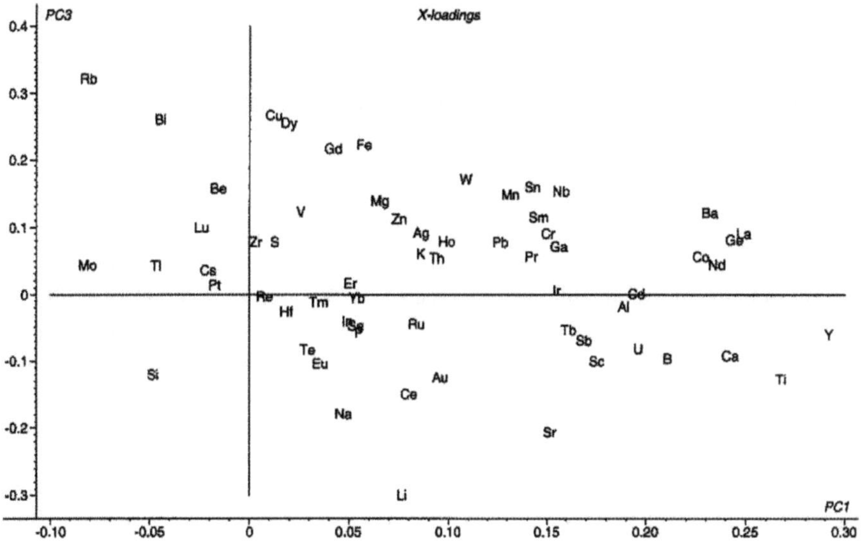

Fig. 9.14 Scatter plot of loadings for the first and third principal components in the PCA model for onions. (After Gundersen V et al., 2000. J Agric Food Chem 48:6094–6102. https://doi.org/10.1021/jf0009652. With permission)

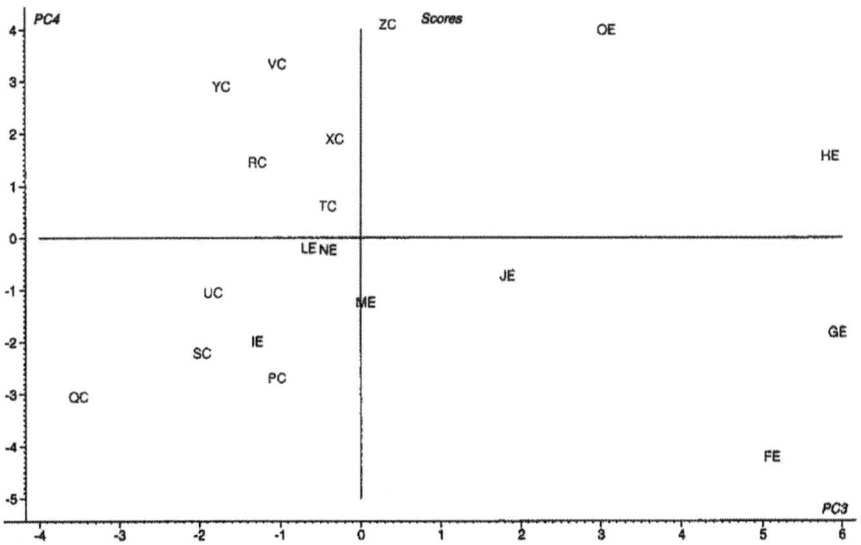

Fig. 9.15 Score plot for the third and fourth principal components of the PCA model for peas. The model is based on the mean values from each site. Sites with E as the second letter are organic, and sites with second letter C as the second letter are conventional. (After Gundersen V et al., 2000. J Agric Food Chem 48:6094–6102. https://doi.org/10.1021/jf0009652. With permission)

fertilizer input. It was concluded that the use of multielemental profiles with semi-quantitative ICP–MS combined with chemometric tools is a useful methodology for verifying the authenticity of organic crops.

Yuan et al. (2016) combined EA–IRMS and ICP–MS methods as a chemometric approach to discriminate between organic and conventional *Brassica chinensis* planted in experimental test pots. The first three PCs, representing 69.1% of the total variance, discriminated samples corresponding to chemical and nonchemical treatments, including organic and organic–chemical samples (Fig. 9.16). However, no significant differences between *Brassica* vegetables subjected to organic and organic–chemical treatments were detected. LDA correctly classified 79.0%, 94.2% and 81.3% of the organic, chemical and organic–chemical samples, respectively. Although, they correctly discriminated organic and organic–chemically treated samples at a lower confidence level than the chemically treated samples were (94.2%), LDA improved the confidence level beyond the limit of individual elements or isotopes; therefore, conventionally grown produce would not be confused with organic or organic–chemical produce. It was concluded that LDA based on PCA is a useful methodology for the discrimination of *B. chinensis* plants according to their fertilizer treatments, i.e., organic, chemical or organic–chemical methods.

Flores et al. (2013) coupled advanced multi-instrumental methodologies such as EA–IRMS, ICP–OES, IC and HPLC–RID with HPLC–MS/MS to characterize

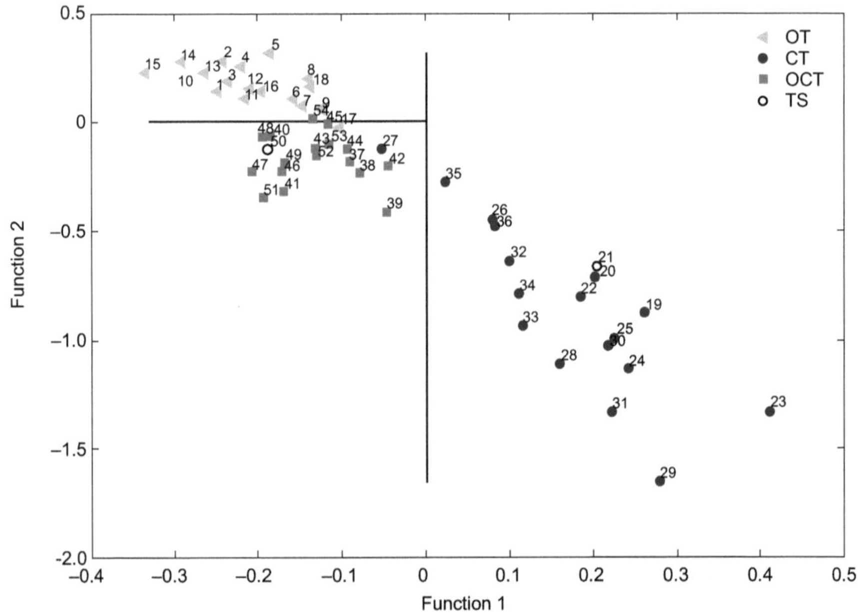

Fig. 9.16 Results of LDA analysis. OT, CT and OCT represent samples from organic, chemical and organic–chemical treatments, respectively. TS represents samples from the testing set. (After Yuan Y et al., 2016. J Agr Food Chem 64:5633–5643. https://doi.org/10.1021/acs.jafc.6b00453. With permission)

pepper and lettuce samples as a reliable approach using the CDA model. The application of preharvest parameters (plant and soil $\delta^{15}N$ and plant weight) correctly classified 78.8% and 65.4% of the pepper and lettuce samples, respectively. In pepper, this percentage was reduced (69.2%) when only postharvest sampling parameters were used to build the classification model. In the case of lettuce, the use of biomarkers considerably improved the model, allowing 90.4% of cases to be correctly classified. This methodology has considerable potential for use as an authenticity control tool for organic cultivation.

Mihailova et al. (2014) used a combination of EA–IRMS, IC and GC–IRMS of greenhouse-grown lettuce as well as retail potatoes and tomatoes coupled with chemometric analysis (ANOVA and CDA). It has been reported that conventional crops grown with low fertilizer concentrations are comparable to organic crops grown with relatively high fertilizer concentrations. The difference in the $\delta^{15}N$ and $\delta^{18}O$ values of nitrate or bulk $\delta^{15}N$ values from these crops may be less significant than that when similar rates of synthetic and organic fertilizers are applied. Discrimination between retail organic and conventional potato samples, especially tomato samples, was achieved in their study. Moreover, differences in the SIR values among lettuce leaves, roots and substrates were obtained. It was suggested that $\delta^{18}O$ analysis of edible plant-derived nitrate allowed discrimination between plants grown with synthetic nitrate fertilizers or manure, although it may be less effective when samples originating from geographical regions with significant climatic differentiation are used.

According to Georgi et al. (2005), IRMS of lettuce, Chinese cabbage and onion grown in the experimental field was applied to estimate the degree of differentiation between samples. The data indicate that there was discrimination in the $\delta^{15}N$ and $\delta^{13}C$ signatures between vegetables grown under organic and integrated production systems. Fertilization and cultivation practices had no effect on the $\delta^{34}S$ and $\delta^{18}O$ values in plant tissue.

Kelly and Bateman (2010) distinguished between conventional and organic cultivation tomatoes and lettuces by using EA–IRMS, IC and ICP–MS, which were subjected to multivariate analysis. It was possible to improve the correct classification of tomato samples, although a weaker differentiation occurred for lettuce. It was also suggested that regular differences in the contents of certain elements, e.g., Mn, Ca, Cu and Zn, might occur between crops cultivated under organic and conventional regimes, presumably because of the presence of an increased number of AMF in soils cultivated organically. It was concluded that this finding may be useful as an indicator of authenticity, although the methodology could have some limitations for horticultural crops that have significant differences in agricultural practices, e.g., conventional–hydroponic versus soil-grown organic tomatoes.

According to Bhatt et al. (2017), the LIBS technique was used to identify and compare the profiles of major nutrient elements in organic and conventional vegetables, i.e., different parts of cauliflowers and broccoli samples. The obtained data (LIBS spectra related to Ca, Mg, K, Na and Fe) were transformed by univariate and multivariate chemometric techniques for the classification of organic and conventional vegetables. The intensities of selected atomic lines of different elements and their intensity ratios with some reference lines of organic cauliflower and broccoli samples were compared with those of conventional lines. Moreover, an intercomparison assessment of different parts of cauliflower and broccoli was performed with respect to the intensity and intensity ratio of the elemental lines.

Cristea et al. (2017) determined SIR and MEC contents via EA–IRMS and ICP–MS in parsley, celery and parsnip root, cucumber, vegetable marrow, onion and pepper samples to discriminate between Romanian vegetables and imported varieties. Other parameters have also been differentiated, such as organic and conventional agriculture practices, eggplants and peppers grown in greenhouses and those grown in the field. Some interregional differentiation was achieved; e.g., $\delta^{15}N$ was the most powerful marker between organic and conventional onion samples. Red and white onions were effectively distinguished by the use of $\delta^{13}C$, V and Sr, whereas Na was designated the optimum marker for geographical areas of cultivation.

LDA was used to assess agricultural practices and the geographic provenance of the vegetables.

9.2.4.5 Interelemental Differences Related to Their Distribution in Potato Tubers

A combination of ICP–MS with ANOVA and LRA was used to determine essential elements in the skin and flesh of potato tubers. The concentrations of most bioelements were greater in the skin than in the flesh of tubers. After the orientation of the harvested tuber in the soil was determined, its variation in three dimensions was observed. It was concluded that the three-dimensional distributions of biophilic elements in potato tubers enable recognition of their transport and deposition mechanisms, which is highly important because many of these mineral elements are essential to a healthy human diet. Characterizing their distribution within the potato tuber will be helpful in the appropriate utilization of this essential vegetable (Subramanian et al. 2011).

9.2.4.6 Interelemental Differentiation Related to After-Cooking Changes

LeRiche et al. (2009) used ICAP and VP–SEM/EDS to determine the spatial distribution of elements in potato tubers, resulting in the prediction of after-cooking darkening, one of the most important unwanted potato features with respect to tuber quality. The predictor models for this severity had R^2 values ranging from 0.398 to 0.480 with respect to the datasets used. These models could be helpful for the potato processing industry to predict the existence of after-cooking darkening and for advancing agronomic treatments to reduce this phenomenon. It was concluded that the integration of the obtained data with the contents of organic molecules involved in this phenomenon, such as chlorogenic acid, might provide more promising predictive models.

dos Santos et al. (2013a) used ICP–OES to study raw and cooked okra marketed in the state of Bahia (Brazil) for differentiation using multivariate analysis. No tendency towards differentiation of organic and conventional samples was achieved by PCA, with PC1 or PC2 accounting for 79.71% of the total variance. However, PCA discriminated raw okra from cooked okra. Comparable data were obtained by HCA. A clear loss of mineral elements after baking was observed; therefore, the consumption of raw or sautéed food is the best option to maintain the nutritional value of this vegetable.

dos Santos et al. (2013b) investigated broccoli samples from four Brazilian cities with respect to their mineral composition via ICP–OES. PCA (with the first two PCs explaining 73.63% of the total variance) indicated that the broccoli samples could not be distinguished with respect to their city of origin, sampling station or type of culture. This could be because the metal content in the parent soil of origin is mutually comparable, resulting in a lack of clear differentiation in broccoli samples. However, a tendency towards differentiation between summer and winter samples as well as between raw and cooked samples was observed, with the first two PCs accounting for 73.63% of the total variance. The PCA data were in agreement with the results obtained by HCA. The raw samples had higher levels of all the mineral elements as a result of their partial leaching from the broccoli to the water used for cooking.

9.2.4.7 Interelemental Differentiation Related to Parent Soil Geochemistry

Pinto et al. (2014) coupled FAAS, ICP–MS and traditional spectrophotometric techniques with chlorophyll a + b and carotenoid analysis in a study of five stages of lettuce plants in three greenhouse experimental fields. Parent soil samples were also assayed to determine their effects on lettuce composition and, consequently, on variations in nutritional value during plant growth. The obtained MEC data were processed with chemometric techniques. PCA, with the first two PCs accounting for 70.7% of the total variance, revealed differentiation of all the lettuce samples corresponding to their three experimental fields. For example, PC1 was closely related to Ca, Mg, K, Mn, Co, Zn, As, Se, Cd, Sb, chlorophyll and carotenoids, and these variables were at relatively high or low levels in lettuce, independent of the experimental field. The concentrations of elements (selected by PC2), i.e., Fe, Al, V, Cr and Pb, increased during the growth period of lettuce, indicating increased soil-to-plant transfer. MLRA and SVS showed that soil composition plays an important role in element transfer to lettuce, including elements with low TFs corresponding to their low bioavailability in soil.

Zampella et al. (2011) combined Q–ICP–MS and (MC)–ICPMS with chemometric analysis to classify four varieties of early potatoes from three neighbouring administrative regions in southern Italy related to the soil substrate (alluvial sediments, volcanic substrates and carbonate rocks). The PLS–DA model was a useful approach with optimal recognition and prediction abilities for potatoes grown in soils developed on volcanic substrates. A clear discrimination between samples classified relative to geological substrates, i.e., alluvial sediment and volcanic substrate, was achieved. However, the prediction abilities for potatoes corresponding to soils covering carbonate rocks were not satisfactory, probably because of the processes responsible for soil formation.

According to Pepi et al. (2018), XRF and ICP–MS were used to establish a relationship between the geological and geochemical profile of the soil and the chemical composition of the caper buds growing at three locations in two Aeolian Islands (Italy). For the PCA applied to soil and caper, the first two PCs explained 83.39 and 56.65% of the total variance, respectively, allowing them to clearly discriminate the soil and caper samples from the three regions of origin. Finally, PCA was used to evaluate the soil and caper data to establish suitable geochemical markers of geographical provenance. Al, Ba, Ca, Cd, Co, Cr, Fe, Li, Mg, Mo, Mn, Na, Ni, P, Pb, Sr, Th, Ti, V, Y and Zr significantly corresponded to F1 (77.0%), whereas K, Rb and Zn were strongly related to F2 (15.9%). This means that PCA enabled the establishment of a reliable relationship between geolithological features of the soil and the chemical composition of capers from Lami, Leni and Pianogreca.

Mohamed et al. (2003) investigated the GFAAS and FP of 12 different species of vegetables (cucumber, vegetable marrow, tomato, potato, green pepper, eggplant, carrot, parsley, lettuce, spinach, salq, onion, leek, watercress and cabbage) from Saudi Arabia (the Al-Taif district). CA, PCA and transfer factor analysis of the elements from the soil to the plant were applied to the element database, which produced good interpretations of the data. The element contents depended on the type of vegetable and were related to differences in element selectivity and accumulation in the soil solution. Moreover, interregional differentiation of elements for the same vegetable was achieved with respect to the elemental concentrations in the soils at each site.

Song et al. (2014) measured Sr isotope ratios ($^{87}Sr/^{86}Sr$) via MC–ICP–MS to study soils, water (stream water and groundwater) and plants such as hot pepper and rice collected from all regions of South Korea. Although the bedrock–soil–water–plant system is closely linked, chemometric analysis has indicated that the $^{87}Sr/^{86}Sr$ ratios of plants are closely related to those of water and the exchangeable fraction of soil. This study was the first report of $^{87}Sr/^{86}Sr$ isoscapes in South Korea and the first to indicate that agricultural products are strongly linked with the exchangeable fractions of soil and water.

9.2.4.8 Interregional and/or Intercultivar Differentiation Related to Seasonal Variability

According to Drivelos et al. (2016), the ICP–MS data of MEA, including REE analysis, were used to study "Fava Santorinis" samples as PDOs originating from Greece. Data processed chemometrically demonstrated the superiority of the DA and PLS methods over the k-NN and PNN methods, respectively, when trace element data were used. However, the trace elemental profile can change from one harvesting year to another within each production region. In contrast, the REE fingerprint was found to be minimally influenced by harvesting year. Generally, the OPA was a useful method, achieving up to 100% accuracy when the combined data were used. This means that the effect of seasonal variation should be taken into account when the trace elemental composition is considered instead of the REE composition.

Fadigas et al. (2010) reported that using ICP–OES was a helpful approach to process kale samples grown in four cities in Bahia (Brazil). PCA and CA revealed a powerful tendency for differentiating between the elemental compositions of the samples collected in the summer and those collected in the winter. Interspatial discrimination was also achieved; kale samples from Conceição de Jacuipe were differentiated from those originating from the other three cities.

Suárez et al. (2007) processed FAAS and AES Spanish tomato samples belonging to five cultivars (Dorothy, Boludo, Dunkan, Dominique and Thomas) produced on the island of Tenerife. The contribution to the intake of macro- and trace elements was estimated to be generally low, except for K and Mg. The major elements seemed to be more influenced by the cultivar than the major elements, and the cultivation method affected the major element contents more than the trace element contents. Using DA may be useful for classifying tomato samples with respect to the cultivation method, period of sampling and region of production. In the case of considering tomato cultivars with respect to each cultivation method, DA allowed the complete classification (100%) of the tomato samples, although the percentage of this classification in the CV considerably decreased because of the small amount of data within each set.

9.2.4.9 Intervariety Differentiation Related to Packaging Type

Ozturkoglu-Budak and Aksahin (2016) processed analytical ICP–MS data related to unpacked fresh tomatoes and six tomato-based products, i.e., tomato paste, tomato juice, minced tomato, tomato puree, grated tomato and dried tomato. The products of the 11 brands were packed into 5 types of packaging, i.e., can, cardboard, glass, vacuumed plastic, doy-packs and nylon bags. For example, the first four PCs, explaining 75% of the total variance, allowed the differentiation of four groups of elements corresponding to their possible sources in samples, i.e., those originating from atmospheric deposition, soil substratum, mainly from fertilizer or food materials. The application of a combination of PCA, CA and ANOVA resulted in the categorization of samples according to product variety, establishing nutritional relationships among products and identifying healthy nutritional sources. This means that the elemental profile has an effect on the quality of the product; hence, tomato juice and minced tomato were found to be the best substitutes for fresh tomato.

9.2.4.10 Interregional Differentiation Related to the Impact of Pollution

Yang et al. (2021) reported that the combination of LIBS and AAS supported the investigation of uniform mulberry leaves from the "Dashi" mulberry variety from a mulberry fruit garden in China, which was coupled with the construction of a stable and accurate model. A scree plot of the first two PCs for Cu and Cr as well as a sample plot between Hotelling's T2 and Q residuals for the two metals accounted for 95.35 and 85.39% of the total variance, respectively. PCA accurately differentiated uncontaminated samples from contaminated samples, although samples with different levels of contamination could not be completely differentiated. The developed methodology improved model accuracy and robustness and provided a highly accurate and rapid alternative to the traditional method for tracing trace elements in mulberry leaves, resulting in appropriate recognition of their quality.

Gergen and Harmanescu (2012) used FAAS-processed samples of different vegetables (parsley roots, carrot roots, onion bulbs, parsley leaves, carrot leaves, cabbage, lettuce, cucumber and green beans) and the adjacent soils collected from various Romanian sampling sites. PCA differentiated between contaminated and uncontaminated areas, and more specifically, between soil samples originating from two contaminated areas corresponding to Pb and Cu contents. It was concluded that this chemometric tool helped optimize the number and type of the most differentiating data in terms of the performance of HM contamination in soil and vegetables. Additionally, it was useful for selecting the vegetable species that presents the highest/minimum risk of a negative impact on the food chain and human health.

Carlosena et al. (1999) processed FAAS samples of lettuce and cabbage from three regions of Spain. The levels of Cd, Cu and Pb were transformed by chemometric techniques (PCA, CA, CV and SIMCA) for classifying vegetable samples according to their geographical provenance. PCA with the first two PCs accounting for 84% of the total variance allows them to clearly discriminate four classes of samples. This result was strictly confirmed by CA and CV, which discriminated commercial samples with negligible motor vehicle influence, domestic samples with low–negligible traffic intensity, domestic allotments and high traffic flow, and combined groups with medium/medium–high traffic intensity. Various groups of vegetables were interpreted as a function of both the agricultural conditions (private versus commercial) and the different traffic intensities supported by the cultivated areas.

Mitic et al. (2015a) used ICP–OES for the quantification of numerous elements in parsley roots from 12 locations in five districts of Slovenia. The first two PCs, which explained 68.85% of the total variance, revealed two potential sources of pollutants, i.e., anthropogenic sources, including human activities and vehicular emissions, and natural sources related to lithogenic origins. Both multivariate chemometric analyses, PCA and CA, can be used effectively to identify sources of elements in plants. ICP–OES has been reported to be a convenient tool for characterizing carrots originating from 13 locations in five districts in Serbia. The first two PCs, which described 58.95% of the total variance, allowed the selection of the most discriminative variables, such as Cu, Sb, Pb and Tl, to differentiate among contaminated areas. CA and PCA enabled the classification of samples according to their geographical provenance. Mitic et al. (2015b) reported data concerning chemometric approach to evaluate heavy metals' concentration in Daucus carota from different localities in Serbia.

9.2.4.11 Genetic Differentiation of Mineral Element Contents

According to Bozokalfa et al. (2011a), ICP, AAS and FP were appropriate tools for providing information on the genetic variability in the mineral concentration of Swiss chard samples from 12 countries. Both HCA and PCA, with the first three PCs accounting for 49.86% of the total variance, enabled the genetic differentiation of Swiss chard accessions for some nutritional elements. The genetic resources suggest that possibly important accessions could be used as a gene source owing to their high K, Ca, Cu and Zn levels in breeding programmes.

Bozokalfa et al. (2011b) used AAS and FP to study rocket accessions from 12 countries via multivariate analysis. PCA with the first three PCs explaining 64.24% of the total variance allowed the classification of samples according to their elemental profile. The PCA classification was confirmed by the HAC distribution pattern. Significant differences in element contents among the studied rocket accessions made selection of those genotypes possible owing to certain elements.

9.2.4.12 Intertype Differentiation of Supplements

Brzezińska-Rojek et al. (2022) reported that MP–AES coupled with chemometrics was an effective tool for classifying samples of beetroot and its supplements. Beetroots are a significant source of elements, especially K, Na, Mg, Ca and P, in contrast to supplements, which contain insignificant amounts of these elements, except for iron-enriched products, which provide significant levels of Fe. Both FA and CA differentiated beetroot and its supplements with respect to their type (vegetable, supplement or iron-enriched supplement), origin, type of cultivation (conventional or organic) and form (capsule or tablet) on the basis of their elemental composition (Figs. 9.17 and 9.18). It was concluded that more stringent supplement control is needed since they could pose a considerable health risk to consumers.

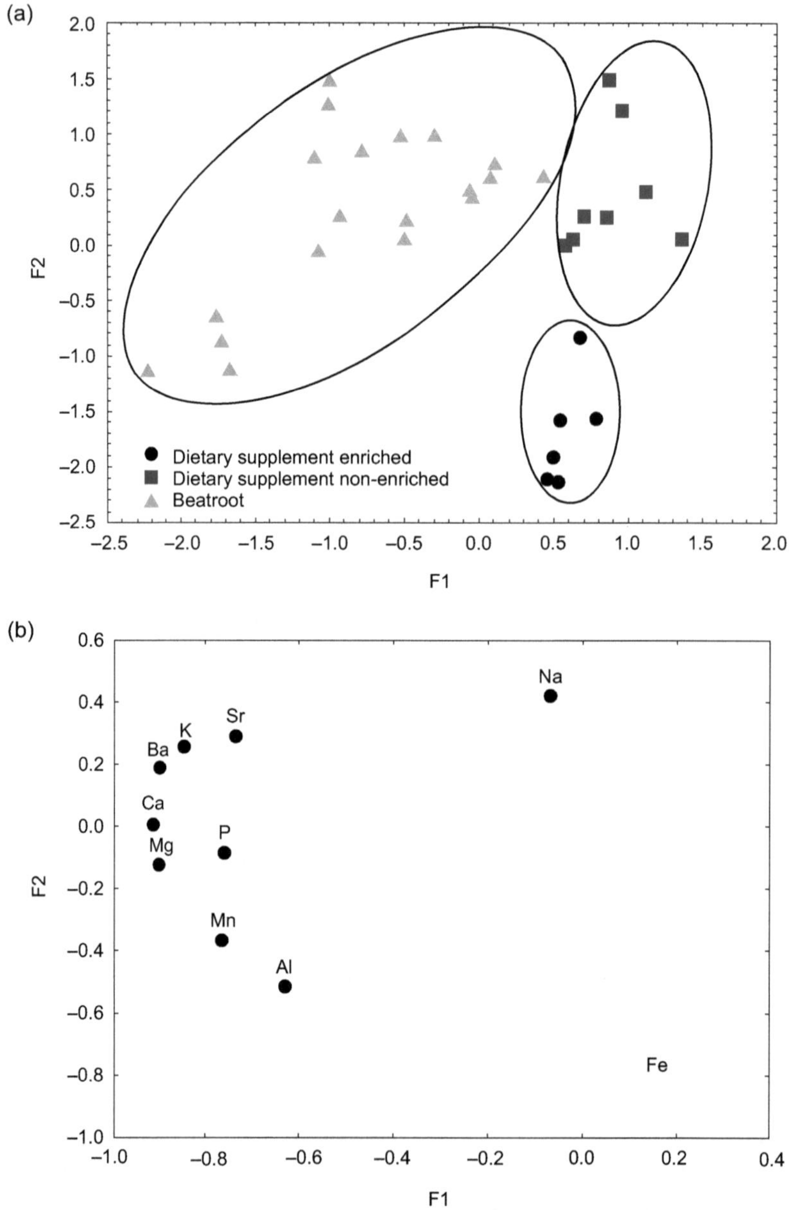

Fig. 9.17 (**a**) Scatter plot of object samples of two factors for the dataset of beetroots and all dietary supplements. (**b**) Scatter plot of loading for elements in all the analysed samples. (After Brzezińska-Rojek J et al., 2022. Nutrients 14:106. https://doi.org/10.3390/nu14010106. With permission)

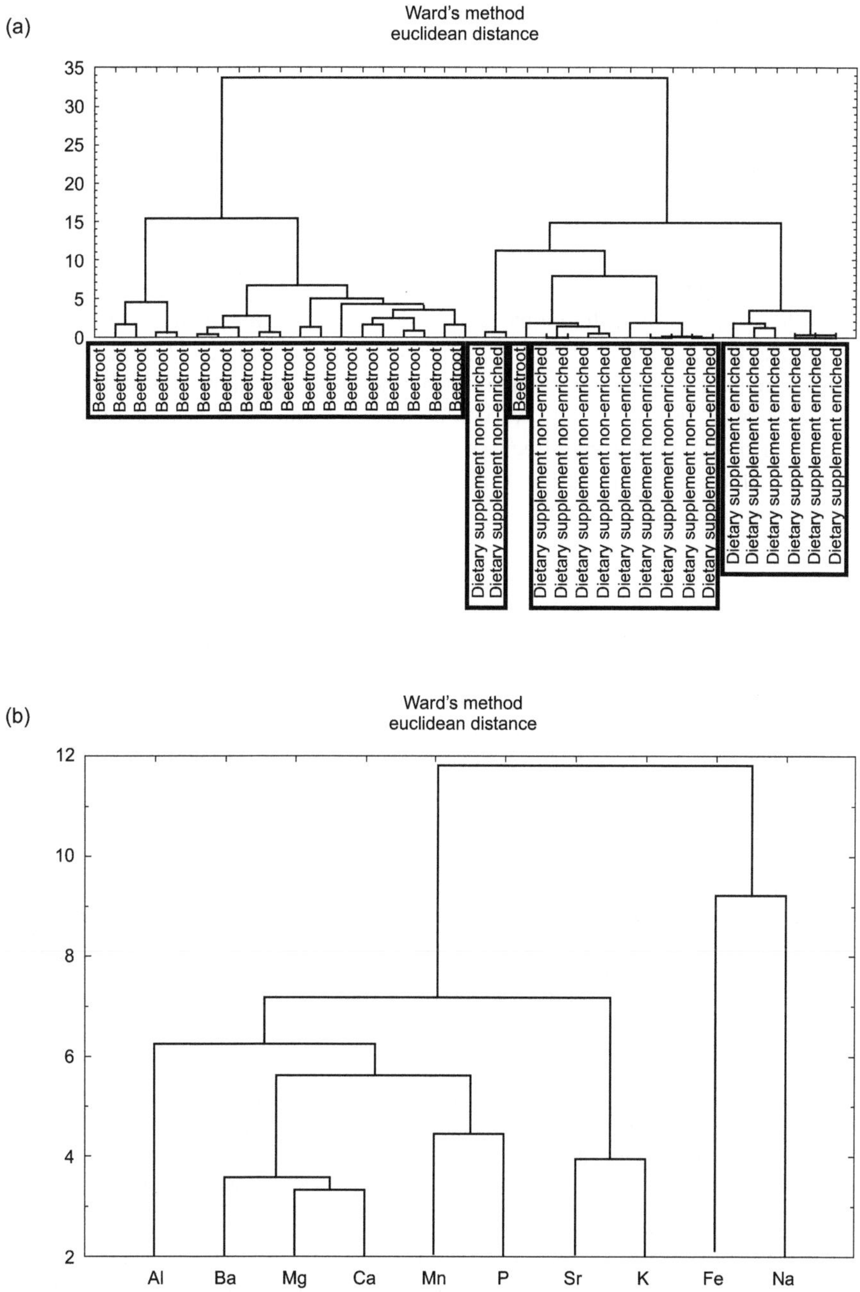

Fig. 9.18 (**a**, **b**). Hierarchical dendrogram for all products (beetroot and Fe-enriched/nonenriched dietary supplements). (After Brzezińska-Rojek J et al., 2022. Nutrients 14:106. https://doi.org/10.3390/nu14010106. With permission)

9.3 Conclusions

Recognition of food authenticity contributes to the assessment of food safety. To accomplish this task, appropriate analytical methods or techniques coupled with chemometric analysis are effective approaches for evaluating agricultural product adulteration/mislabelling. The thoughtful choice of these combined methods is a strategic task for investigating a wide variety of vegetables. Among the routinely used reliable methods, the combination of LIBS with SIRA has recently been increasingly applied to determine plant food authenticity. Therefore, it has been possible to effectively differentiate conventionally grown vegetables from those that are organically cultivated. In addition, the integration of analytical tools and chemometrics enables fingerprints of the influence of different parameters on the quality of edible plant resources, e.g., variations found at annual or seasonal scales, the degree of pollution and the effects of culinary/technological processing. Accurate results enable the identification of falsely labelled products and hence preserve the reputations of brands.

References

Adamo P, Zampella M, Quétel CR, Aversano R, Dal Piaz F, De Tommasi N, Frusciante L, Iorizzo M, Lepore L, Carputo D (2012) Biological and geochemical markers of the geographical origin and genetic identity of potatoes. J Geochem Explor 121:62–68. https://doi.org/10.1016/j.gexplo.2012.07.006

Anderson KA, Magnuson BA, Tschirgi ML, Smith B (1999) Determining the geographic origin of potatoes with trace metal analysis using statistical and neural network classifiers. J Agric Food Chem 47:1568–1575. https://doi.org/10.1021/jf980677u

Anunciação DS, Leão DJ, de Jesus RM, Ferreira SLC (2011) Use of multivariate analysis techniques for evaluation of analytical data—determination of the mineral composition of cabbage *Brassica oleracea*. Food Anal Methods 4:286–292. https://doi.org/10.1007/s12161-010-9172-2

Ariyama K, Aoyama Y, Mochizuki A, Homura Y, Kadokura M, Yasui A (2007) Determination of the geographic origin of onions between three main production areas in Japan and other countries by mineral composition. J Agric Food Chem 55:347–354. https://doi.org/10.1021/jf062613m

Ariyama K, Horita H, Yasui A (2004a) Application of inorganic element ratios to chemometrics for determination of the geographic origin of welsh onions. J Agric Food Chem 52:5803–5809. https://doi.org/10.1021/jf049333w

Ariyama K, Horita H, Yasui A (2004b) Chemometric techniques on inorganic elements composition for the determination of the geographic origin of welsh onions. Anal Sci 20:871–877. https://doi.org/10.2116/analsci.20.871

Ariyama K, Nakano T, Shin K-C, Izawa A, Morita S (2017) Variation of strontium stable isotope ratios and origins of strontium in Japanese vegetables and comparison with Chinese vegetables. Food Chem 237:1186–1195. https://doi.org/10.1016/j.foodchem.2017.06.027

Ariyama K, Nishida T, Noda T, Kadokura M, Yasui A (2006) Effects of fertilization, crop year, variety, and provenance factors on mineral concentrations in onions. J Agric Food Chem 54:3341–3350. https://doi.org/10.1021/jf0525481

Ariyama K, Yasui A (2006) The determination technique of the geographic origin of Welsh onions by mineral composition and perspectives for the future. Jpn Agric Res Quart 40:333–339. https://doi.org/10.6090/jarq.40.333

Bensaid FF, Wietzerbin K, Martin GJ (2002) Authentication of natural vanilla flavorings: isotopic characterization using degradation of vanillin into guaiacol. J Agric Food Chem 50:6271–6275. https://doi.org/10.1021/jf020316l

Bhatt CR, Alfarraj B, Ghany CT, Yueh FY, Singh JP (2017) Comparative study of elemental nutrients in organic and conventional vegetables using lase-rinduced breakdown spectroscopy (LIBS). Appl Spectrosc 71:686–698. https://doi.org/10.1177/0003702817692810

Bibak A, Behrens A, Stürup S, Knudsen L, Gundersen V (1998b) Concentrations of 55 major and trace elements in Danish agricultural crops measured by inductively coupled plasma mass spectrometry. 2. Pea (*Pisum sativum* Ping Pong) J Agric Food Chem 46:3146–3149. doi:https://doi.org/10.1021/jf980191d

Bibak A, Behrens A, Stürup S, Knudsen L, Gundersen V (1998a) Concentrations of 63 major and trace elements in Danish agricultural crops measured by inductively coupled plasma mass spectrometry. 1. Onion (*Allium cepa* Hysam). J Agric Food Chem 46:3139–3145. https://doi.org/10.1021/jf971103c

Bibak A, Stürup S, Gundersen P, Gundersen V (1999c) Effect of N-fertiliser levels and soil pH on uptake of trace elements by plants. In: Gissel-Nielsen J (ed) Plant nutrition—molecular biology and genetics. Kluwer Academic Publishers, pp 21–24. https://doi.org/10.1007/978-94-017-2685-6_3

Bibak A, Stürup S, Haahr V, Gundersen P, Gundersen V (1999a) Concentrations of 50 major and trace elements in Danish agricultural crops measured by inductively coupled plasma mass spectrometry. 3. Potato (*Solanum tuberosum* Folva). J Agric Food Chem 47:2678–2684. https://doi.org/10.1021/jf980606v

Bibak A, Stümp S, Knudsen L, Gundersen V (1999b) Concentrations of 63 elements in cabbage and sprouts in Denmark. Commun Soil Sci Plant Analysis 30:2409–2418. https://doi.org/10.1080/00103629909370382

Bong Y-S, Gautam MK, La M-R, Lee K-S (2012a) Geographic origins of Korean and Chinese kimchi determined by multiple elements. Biosci Biotechnol Biochem 76:2096–2100. https://doi.org/10.1271/bbb.120471

Bong Y-S, Shin W-J, Gautam MK, Jeong Y-J, Lee A-R, Jang C-S, Lim Y-P, Chung G-S, Lee K-S (2012b) Determining the geographical origin of Chinese cabbages using multielement composition and strontium isotope ratio analyses. Food Chem 135:2666–2674. https://doi.org/10.1016/j.foodchem.2012.07.045

Bong Y-S, Song B-Y, Gautam MK, Jang CS, An HJ, Lee K-S (2013) Discrimination of the geographic origin of cabbages. Food Contr 30:626–630. https://doi.org/10.1016/j.foodcont.2012.09.008

Bontempo L, Camin F, Manzocco L, Nicolini G, Wehrens R, Ziller L, Larcher R (2011) Traceability along the production chain of Italian tomato products on the basis of stable isotopes and mineral composition. Rapid Commun Mass Spectrom 25:899–909. https://doi.org/10.1002/rcm.4935

Bontempo L, Ceppa FA, Perini M, Tonon A, Gagliano G, Marianella RM, Marega M, Trifirò A, Camin F (2014) Use of $\delta^{18}O$ authenticity thresholds to differentiate tomato passata from diluted tomato paste. Food Contr 35:35:413–35:418. https://doi.org/10.1016/j.foodcont.2013.07.018

Bontempo L, van Leeuwen KA, Paolini M, Laursen KH, Micheloni C, Prenzler PD, Ryan D, Camin F (2020) Bulk and compound-specific stable isotope ratio analysis for authenticity testing of organically grown tomatoes. Food Chem 318:126426. https://doi.org/10.1016/j.foodchem.2020.126426

Borràs E, Ferré J, Boqué R, Mestres M, Aceña L, Busto O (2015) Data fusion methodologies for food and beverage authentication and quality assessment—a review. Anal Chim Acta 891:1–14. https://doi.org/10.1016/j.aca.2015.04.042

Bozokalfa KM, Esiyok D, Yagnur B (2011b) Use of multivariate analysis in mineral accumulation of rocket (*Eruca sativa*) accessions. Genetika 43:437–448. https://doi.org/10.2298/GENSR1103437B

Bozokalfa MK, Yağmur B, Aşçıoğul TK, Eşiyok D (2011a) Diversity in nutritional composition of Swiss chard (*Beta vulgaris* subsp. L. var.*cicla*) accessions revealed by multivariate analysis. Plant Genet Resour 9:557–566. https://doi.org/10.1017/S1479262111000876

Brunner M, Katona R, Stefánka Z, Prohaska T (2010) Determination of the geographical origin of processed spice using multielement and isotopic pattern on the example of Szegedi paprika. Eur Food Res Technol 231:623–634. https://doi.org/10.1007/s00217-010-1314-7

Brzezińska-Rojek J, Rutkowska M, Brzezicha J, Konieczka P, Prokopowicz M, Grembecka M (2022) Mineral composition of dietary supplements-analytical and chemometric approach. Nutrients 14:106. https://doi.org/10.3390/nu14010106

Camargo AB, Resnizky S, Marchevsky EJ, Luco JM (2010) Use of the Argentinean garlic (*Allium sativum* L.) germplasm mineral profile for determining geographic origin. J Food Compos Anal 23:586–591. https://doi.org/10.1016/j.jfca.2010.01.002

Carlosena A, Andrade JM, Tomas X, Fernandez E, Prada D (1999) Classification of edible vegetables affected by different traffic intensities using potential curves. Talanta 48:795–802. https://doi.org/10.1016/S0039-9140(98)00096-4

Choi S-H, Bong Y-S, Park J-H, Lee K-S (2020) Geographical origin identification of garlic cultivated in Korea using isotopic and multi-elemental analyses. Food Contr 111:107064. https://doi.org/10.1016/j.foodcont.2019.107064

Chope GA, Terry LA (2009) Use of canonical variate analysis to differentiate onion cultivars by mineral content as measured by ICP-AES. Food Chem 115:1108–1113. https://doi.org/10.1016/j.foodchem.2008.12.090

Chung III-M, Kim J-K, Jin Y-Ik, Oh Y-T, Prabakaran M, Youn K-J, Kim S-H (2016) Discriminative study of a potato (*Solanum tuberosum* L.) cultivation region by measuring the stable isotope ratios of bio-elements. Food Chem 212:48–57. doi:https://doi.org/10.1016/j.foodchem.2016.05.161

Cicchetti E, Silvestre V, Fieber W, Sommer H, Remaud G, Serge S, Chaintreau A (2010) Procedure for the isolation of vanillin from vanilla extracts prior to isotopic authentication by quantitative ^{13}C-NMR. Flavour Fragr J 25:463–467. https://doi.org/10.1002/ffj.2006

Cristea G, Ioana Feher ID, Alina Magdas DA, Cezara Voica C, Romulus Puscas R (2017) Characterization of vegetables by stable isotopic and elemental signatures. Anal Lett 50:2677–2690. https://doi.org/10.1080/00032719.2016.1263312

D'Archivio AA, Foschi M, Aloia R, Maggi MA, Rossi L, Ruggieri F (2019) Geographical discrimination of red garlic (*Allium sativum* L.) produced in Italy by means of multivariate statistical analysis of ICP-OES data. Food Chem 275:333–338. https://doi.org/10.1016/j.foodchem.2018.09.088

de Rijke E, Schoorl JC, Cerli C, Vonhof HB, Verdegaal SJA, Vivó-Truyols G, Lopatka M, Dekter R, Bakker D, Sjerps MJ, Ebskamp M, de Koster CG (2016) The use of δ^2H and $\delta^{18}O$ isotopic analyses combined with chemometrics as a traceability tool for the geographical origin of bell peppers. Food Chem 204:122–128. https://doi.org/10.1016/j.foodchem.2016.01.134

Di Giacomo F, Del Signore A, Giaccio M (2007) Determining the geographic origin of potatoes using mineral and trace element content. J Agric Food Chem 55:860–866. https://doi.org/10.1021/jf062690h

Dobrowolska-Iwanek J, Zagrodzki P, Galanty A, Fołta M, Kryczyk-Kozioł J, Szlósarczyk M, Rubio PS, Saraiva de Carvalho I, Paśko P (2022) Determination of essential minerals and trace elements in edible sprouts from different botanical families—application of chemometric analysis. Foods 11:371. https://doi.org/10.3390/foods11030371

dos Santos AMP, Oliveira AC, Souza AS, de Jesus RM, Sergio L, Ferreira C (2011) Determination and evaluation of the mineral composition of Chinese cabbage (*Beta vulgaris*). Food Anal Methods 4:567–573. doi:https://doi.org/10.1007/s12161-011-9205-5

dos Santos IF, dos Santos AMP, Barbosa UA, Lima JS, dos Santos DC, Matos GD (2013a) Multivariate analysis of the mineral content of raw and cooked okra (*Abelmoschus esculentus* L.). Microchem J 110:439–443. doi:https://doi.org/10.1016/j.microc.2013.05.008

dos Santos AMP, Lima JS, Anunciação DS, Souza AS, dos Santos DCMB, Matos GD (2013b) Determination and evaluation employing multivariate analysis of the mineral composition of Broccoli (*Brassica oleracea L.* var. *Italica*). Food Anal Methods 6:745–752. https://doi.org/10.1007/s12161-012-9475-6

Drivelos A, Georgiou CA (2012) Multi-element and multi-isotope-ratio analysis to determine the geographical origin of foods in the European Union. Trac-Trend Anal Chem 40:38–51. https://doi.org/10.1016/j.trac.2012.08.003

Drivelos SA, Danezis GP, Haroutounian SA, Georgiou CA (2016) Rare earth elements minimal harvest year variation facilitates robust geographical origin discrimination: the case of PDO "Fava Santorinis". Food Chem 213:238–245. https://doi.org/10.1016/j.foodchem.2016.06.088

Drivelos SA, Higgins K, Kalivas JH, Haroutounian SA, Georgiou CA (2014) Data fusion for food authentication. Combining rare earth elements and trace metals to discriminate "Fava Santorinis" from other yellow split peas using chemometric tools. Food Chem 165:316–322. https://doi.org/10.1016/j.foodchem.2014.03.083

Fadigas JC, Santos AMP, de Jesus RM, Lima DC, Fragoso WD, David JM, Ferreira SLC (2010) Use of multivariate analysis techniques for the characterization of analytical results for the determination of the mineral composition of kale. Microchem J 96:352–356. https://doi.org/10.1016/j.microc.2010.06.006

Flores P, López A, Fenoll J, Hellín P, Kelly S (2013) Classification of organic and conventional sweet peppers and lettuce using a combination of isotopic and bio-markers with multivariate analysis. J Food Compos Anal 31:31:217–31:225. https://doi.org/10.1016/j.jfca.2013.05.015

Foschi M, D'Archivio AA, Rossi L (2020) Geographical discrimination and authentication of lentils (*Lens culinaris* Medik.) by ICP-OES elemental analysis and chemometrics. Food Contr 118:107438. https://doi.org/10.1016/j.foodcont.2020.107438

Fragni R, Trifirò A, Nucci A (2015) Towards the development of a multi-element analysis by ICP-oa-TOF-MS for tracing the geographical origin of processed tomato products. Food Contr 48:96–101. https://doi.org/10.1016/j.foodcont.2014.04.027

Fragni R, Trifirò A, Nucci A, Seno A, Allodi A, Di Rocco M (2018) Italian tomato-based products authentication by multi-element approach: a mineral elements database to distinguish the domestic provenance. Food Contr 93:211–218. https://doi.org/10.1016/j.foodcont.2018.06.002

Furia E, Naccarato A, Sindona G, Stabile G, Tagarelli A (2011) Multielement fingerprinting as a tool in origin authentication of PGI food products: Tropea red onion. J Agric Food Chem 59:8450–8457. https://doi.org/10.1021/jf201556e

Gassenmeier K, Binggeli E, Kirsch T, Otiv S (2013) Modulation of the $^{13}C/^{12}C$ ratio of vanillin from vanilla beans during curing. Flavour Fragr J 28:25–29. https://doi.org/10.1002/ffj.3122

Geißler K, Greule M, Schäfer U, Hans J, Geißler T, Meier L, Keppler F, Krammer G (2017) Vanilla authenticity control by DNA barcoding and isotope data aggregation. Flavour Fragr J 32:228–237. https://doi.org/10.1002/ffj.3379

Georgi M, Voerkelius S, Rossmann A, Graßmann J, Schnitzler WH (2005) Multielement isotope ratios of vegetables from integrated and organic production. Plant Soil 275:93–100. https://doi.org/10.1007/s11104-005-0258-3

Gergen I, Harmanescu M (2012) Application of principal component analysis in the pollution assessment with heavy metals of vegetable food chain in the old mining areas. Chem Cent J 6:156. https://doi.org/10.1186/1752-153X-6-156

Grembecka M (2007) Bromatological and chemometric assessment of plants food based on its mineral composition. Dissertation, Gdańsk Medical Academy

Grembecka M, Szefer P (2011) Metals and metalloids in foods—essentiality, toxicity, applicability. In: Medina DA, Laine AM (eds) Food quality: control, analysis and consumer concerns. Nova Science Publishers, Inc., Hauppauge, pp 1–60. ISBN 10: 1611229170; ISBN 13: 9781611229172

Grembecka M, Szefer P (2016) Elemental trace analysis in studies of food products. In: Baranowska I (ed) Handbook of trace analysis—fundamentals and applications. Springer, pp 203–239. https://doi.org/10.1007/978-3-319-19614-5

Grembecka M, Szefer P (2022) Elemental profiles of legumes and seeds in view of chemometric approach. Appl Sci 12:1577. https://doi.org/10.3390/app12031577

Grembecka M, Gurzyńska A, Dybek K, Szefer P (2009) The applicability of chemometric techniques in assessing the differences in quality of vegetable products. In: Cubadda F, Aureli F, Ciardullo S, Patriarca M (eds) 3 International IUPAC Symposium on Trace Elements in Food, Rome, Italy, 1–3 April 2009, Abstract Book ISSN 0393-5620; TRN: IT0900518090505

Greule M, Mosandl A, Hamilton JTG, Keppler F (2015) Comment on authenticity and traceability of vanilla flavors by analysis of stable isotopes of carbon and hydrogen. J Agric Food Chem 63:5305–5306. https://doi.org/10.1021/jf506172q

Greule M, Tumino LD, Kronewald T, Hener U, Schleucher J, Mosand A, Keppler F (2010) Improved rapid authentication of vanillin using $\delta^{13}C$ and $\delta^{2}H$ values. Eur Food Res Technol 231:933–941. https://doi.org/10.1007/s00217-010-1346-z

Gundersen V, McCall D, Bechmann IE (2001) Comparison of major and trace element concentrations in Danish greenhouse tomatoes (*Lycopersicon esculentum* Cv. Aromata F1) cultivated in different substrates. J Agric Food Chem 49:3808–3815. https://doi.org/10.1021/jf0103774

Gundersen V, Bechmann IE, Behrens A, Stürup S (2000) Comparative investigation of concentrations of major and trace elements in organic and conventional Danish agricultural crops. 1. Onions (*Allium cepa* Hysam) and peas (*Pisum sativum* Ping Pong). J Agric Food Chem 48:6094–6102. https://doi.org/10.1021/jf0009652

Guyader S, Thomas F, Jamin E, Grand M, Akoka S, Silvestre V, Remaud GS (2019) Combination of ^{13}C and ^{2}H SNIF-NMR isotopic fingerprints of vanillin to control its precursors. Flavour Fragr J 34:133–144. https://doi.org/10.1002/ffj.3486

Hansen A-MS, Fromberg A, Frandsen HL (2014) Authenticity and traceability of vanilla flavors by analysis of stable isotopes of carbon and hydrogen. J Agric Food Chem 62:10326–10331. https://doi.org/10.1021/jf503055k

He Y, Sun Q, Zhang X, Bao X, Wang Y, Rasheed M, Guo B (2020) Authentication of the geographical origin of Maca (*Lepidium meyenii* Walp.) at different regional scales using the stable isotope ratio and mineral elemental fingerprints. Food Chem 311:126058. https://doi.org/10.1016/j.foodchem.2019.126058

Hidalgo MJ, Fechner DC, Marchevsky EJ, Pellerano RG (2016) Determining the geographical origin of *Sechium edule* fruits by multielement analysis and advanced chemometric techniques. Food Chem 210:228–234. https://doi.org/10.1016/j.foodchem.2016.04.120

Hondrogiannis E, Peterson K, Zapf CM, Roy W, Blackney B, Dailey K (2012) The use of wavelength dispersive X-ray fluorescence and discriminant analysis in the identification of the elemental composition of cumin samples and the determination of the country of origin. Food Chem 135:2825–2831. https://doi.org/10.1016/j.foodchem.2012.07.003

Hwang IM, Choi JY, Nho EY, Dang YM, Jamila N, Khan N, Seo H-Y, Kim KS (2017) Determination of essential and toxic elements in vegetables from South Korea. Anal Lett 50:663–681. https://doi.org/10.1080/00032719.2016.1194426

Jandric Z, Tchaikovsky A, Zitek A, Causon T, Stursa V, Prohaska T, Hann S (2021) Multivariate modelling techniques applied to metabolomic, elemental and isotopic fingerprints for the verification of regional geographical origin of Austrian carrots. Food Chem 338:127924. https://doi.org/10.1016/j.foodchem.2020.127924

John TV, Jamin E (2004) Chemical investigation and authenticity of Indian vanilla beans. J Agric Food Chem 52:7644–7650. https://doi.org/10.1021/jf049461k

Jurado JM, Alcázar A, Pablos F, Martín MJ, González AG (2005) Classification of aniseed drinks by means of cluster, linear discriminant analysis and soft independent modelling of class analogy based on their Zn, B, Fe, Mg, Ca, Na and Si content. Talanta 66:1350–1354. https://doi.org/10.1016/j.talanta.2005.01.052

Kaunzinger A, Juchelka D, Mosandl A (1997) Progress in the authenticity assessment of vanilla. 1. Initiation of authenticity profiles. J Agric Food Chem 45:1752–1757. https://doi.org/10.1021/jf9606971

Kelly SD, Bateman AS (2010) Comparison of mineral concentrations in commercially grown organic and conventional crops—tomatoes (*Lycopersicon esculentum*) and lettuces (*Lactuca sativa*). Food Chem 119:738–745. https://doi.org/10.1016/j.foodchem.2009.07.022

Keppler F, Hamilton JTG (2008) Tracing the geographical origin of early potato tubers using stable hydrogen isotope ratios of methoxyl groups. Isotop Environ Health Studies 44:337–347. https://doi.org/10.1080/10256010802507383

Kim J, Jung Y, Song B, Bong Y-S, Ryu DH, Lee K-S, Hwang G-S (2013) Discrimination of cabbage (*Brassica rapa* ssp. *pekinensis*) cultivars grown in different geographical areas using ^{1}H NMR-based metabolomics. Food Chem 137:68–75. https://doi.org/10.1016/j.foodchem.2012.10.012

Kobayashi N, Tanoi K, Hirose A, Saito T, Noda A, Iwata N, Nakano A, Nakamura S, Nakanishi TM (2011) Analysis of the mineral composition of taro for determination of geographic origin. J Agric Food Chem 59:4412–4417. https://doi.org/10.1021/jf200264n

Kokkinofta R, Yiannopoulos S, Stylianou MA, Agapiou A (2020) Use of chemometrics for correlating carobs nutritional compositional values with geographic origin. Meta 10:62. https://doi.org/10.3390/metabo10020062

Krauß S, Vetter W (2019) Stable carbon and nitrogen isotope ratios of red bell pepper samples from Germany, The Netherlands, and Spain. J Agric Food Chem 67:4054–4063. https://doi.org/10.1021/acs.jafc.9b01631

Krejčová A, Návesník J, Jičínská J, Černohorský T (2016) An elemental analysis of conventionally, organically and self-grown carrots. Food Chem 192:242–249. https://doi.org/10.1016/j.foodchem.2015.07.008

Krueger DA, Krueger HW (1983) Carbon isotopes in vanillin and the detection of falsified natural vanillin. J Agric Food Feed Chem 31:1265–1268. https://doi.org/10.1021/jf00120a030

Krueger DA, Krueger HW (1985) Detection of fraudulent vanillin labeled with carbon-13 in the carbonyl carbon. J Agric Food Chem 33:323–325. https://doi.org/10.1021/jf00063a001

Lamprecht G, Pichlmayer F, Schmid ER (1994) Determination of the authenticity of vanilla extracts by stable isotope ratio analysis and component analysis by HPLC. J Agric Food Chem 42:1722–1727. https://doi.org/10.1021/jf00044a027

Laursen KH, Mihailova A, Kelly SD, Epov VN, Bérail S, Schjoerring JK, Donard OFX, Larsen EH, Pedentchouk N, Marca-Bell ED, Halekoh U, Olesen JE, Husted S (2013) Is it really organic?—Multi-isotopic analysis as a tool to discriminate between organic and conventional plants. Food Chem 141:2812–2820. https://doi.org/10.1016/j.foodchem.2013.05.068

Laursen KH, Schjoerring JK, Kelly SD, Husted S (2014) Authentication of organically grown plants—advantages and limitations of atomic spectroscopy for multi-element and stable isotope analysis. Trac-Trend Anal Chem 59:73–82. https://doi.org/10.1016/j.trac.2014.04.008

Laursen KH, Schjoerring JK, Olesen JE, Askegaard M, Helekoh U, Husted S (2011) Multielemental fingerprinting as a tool for authentication of organic wheat, barley, faba bean, and potato. J Agric Food Chem 59:4385–4396. https://doi.org/10.1021/jf104928r

LeRiche EL, Wang-Pruski G, Zheljazkov VD (2009) Distribution of elements in potato (*Solanum tuberosum* L.) tubers and their relationship to after-cooking darkening. HortScience 44:44:1866–44:1873. https://doi.org/10.21273/HORTSCI.44.7.1866

Liu X, Xue C, Wang Y, Li Z, Xue Y, Xu J (2012) The classification of sea cucumber (*Apostichopus japonicus*) according to region of origin using multi-element analysis and pattern recognition techniques. Food Contr 23:522–527. https://doi.org/10.1016/j.foodcont.2011.08.025

Lo Feudo G, Naccarato A, Sindona G, Tagarelli A (2010) Investigating the origin of tomatoes and triple concentrated tomato pastes through multielement determination by inductively coupled plasma mass spectrometry and statistical analysis. J Agric Food Chem 58:3801–3807. https://doi.org/10.1021/jf903868j

Lombardo S, Pandino G, Mauromicale G (2013) The influence of growing environment on the antioxidant and mineral content of "early" crop potato. J Food Compos Anal 32:28–35. https://doi.org/10.1016/j.jfca.2013.08.003

Longobardi F, Casiello G, Cortese M, Perini M, Camin F, Catucci L, Agostiano A (2015) Discrimination of geographical origin of lentils (*Lens culinaris* Medik.) using isotope ratio mass spectrometry combined with chemometrics. Food Chem 188:343–349. https://doi.org/10.1016/j.foodchem.2015.05.020

Longobardi F, Casiello G, Sacco D, Tedone L, Sacco A (2011) Characterisation of the geographical origin of Italian potatoes, based on stable isotope and volatile compound analyses. Food Chem 124:1708–1713. https://doi.org/10.1016/j.foodchem.2010.07.092

Magdas DA, Feher I, Dehelean A, Cristea G, Magdas TM, Puscas R, Marincaş O (2018) Isotopic and elemental markers for geographical origin and organically grown carrots discrimination. Food Chem 267:231–239. https://doi.org/10.1016/j.foodchem.2017.10.048

Mihailova A, Pedentchouk N, Kelly SD (2014) Stable isotope analysis of plant-derived nitrate—novel method for discrimination between organically and conventionally grown vegetables. Food Chem 154:238–245. https://doi.org/10.1016/j.foodchem.2014.01.020

Mir-Marqués A, Elvira-Sáez C, Cervera ML, Garrigues S, de la Guardia M (2016b) Authentication of protected designation of origin artichokes by spectroscopy methods. Food Contr 59:74–81. https://doi.org/10.1016/j.foodcont.2015.05.004

Mir-Marqués A, Martínez-García M, Garrigues S, Cervera ML, de la Guardia M (2016a) Green direct determination of mineral elements in artichokes by infrared spectroscopy and X-ray fluorescence. Food Chem 196:1023–1030. https://doi.org/10.1016/j.foodchem.2015.10.048

Mitic V, Stankov-Jovanovic V, Cvetkovic J, Dimitrijevic M, Ilic M, Nikolic-Mandic S (2015a) Application of multivariate statistical approach to identify element sources in parsley (*Petroselinum crispum*). Toxicol Environ Chem 97:754–765. doi:https://doi.org/10.1080/02772248.2015.1068315

Mitic VD, Stankov-Jovanovic VP, Tosic SB, Pavlovic AN, Cvetkovic JS, Dimitrijevic MV, Nikolic-Mandic SD (2015b) Chemometric approach to evaluate heavy metals' content in *Daucus carota* from different localities in Serbia. Hem Ind 69:643–650. https://doi.org/10.2298/HEMIND140705070M

Mohamed AE, Rashed MN, Mofty A (2003) Assessment of essential and toxic elements in some kinds of vegetables. Ecotoxicol Environ Saf 55:251–260. https://doi.org/10.1016/S0039-9140(98)00096-4

Mottese AF, Albergamo A, Bartolomeo G, Bua GD, Rando R, De Pasquale P, Saija E, Donato D, Dugo G (2018) Evaluation of fatty acids and inorganic elements by multivariate statistics for the traceability of the Sicilian *Capparis spinosa* L. J Food Compos Anal 72:66–74. https://doi.org/10.1016/j.jfca.2018.05.009

Moyib OK, Alashiri GO, Adejoye OD (2015) Chemometric dissimilarity in nutritive value of popularly consumed Nigerian brown and white common beans. Food Chem 166:576–584. https://doi.org/10.1016/j.foodchem.2014.06.069

Naccarato A, Furia E, Sindona G, Tagarelli A (2016) Multivariate class modeling techniques applied to multielement analysis for the verification of the geographical origin of chili pepper. Food Chem 206:217–222. https://doi.org/10.1016/j.foodchem.2016.03.072

Nakamura S, Suzuki T, Horita H, Nakano A (2012) Detection of falsely labeled Taro in Japan by elemental analysis: improvement of discrimination ability using a sampling plan. Food Sci Technol Res 18:723–733. https://doi.org/10.3136/fstr.18.723

Nassar AMK, Sabally K, Kubow S, Leclerc YN, Donnelly DJ (2012) Some Canadian-grown potato cultivars contribute to a substantial content of essential dietary minerals. J Agric Food Chem 60:4688–4696. https://doi.org/10.1021/jf204940t

Niu X, Mi S, Jin Q, Sang Y, Wang X (2022) Characterization and discrimination of two varieties of eggplants using multi-element and metabolomics profiles coupled with chemometrics. Food Res Inter Part A 162:111976. https://doi.org/10.1016/j.foodres.2022.111976

Novak V, Adler J, Husted S, Fromberg A, Laursen KH (2019) Authenticity testing of organically grown vegetables by stable isotope ratio analysis of oxygen in plant-derived sulphate. Food Chem 291:59–67. https://doi.org/10.1016/j.foodchem.2019.03.125

Ocvirk M, Nečemer M, Košir IJ (2019) The determination of the geographic origins of hops (*Humulus lupulus* L.) by multi-elemental fingerprinting. Food Chem 277:32–37. https://doi.org/10.1016/j.foodchem.2018.10.070

Ocvirk M, Ogrinc N, Košir IJ (2018) Determination of the geographical and botanical origin of hops (*Humulus lupulus* L.) using stable isotopes of C, N, and S. J Agric Food Chem 66:2021–2026. https://doi.org/10.1021/acs.jafc.7b06010

Oliveira AC, dos Santos VS, dos Santos DC, Carvalho RDS, Souza AS, Ferreira SLC (2014) Determination of the mineral composition of Caigua (*Cyclanthera pedata*) and evaluation using multivariate analysis. Food Chem 152:619–623. https://doi.org/10.1016/j.foodchem.2013.12.022

Opatić AM, Nečemer M, Budič B, Lojen S (2018a) Stable isotope analysis of major bioelements, multi-element profiling, and discriminant analysis for geographical origins of organically grown potato. J Food Compos Anal 71:17–24. https://doi.org/10.1016/j.jfca.2018.04.005

Opatić AM, Nečemer M, Kocman D, Lojen S (2017b) Geographical origin characterization of Slovenian organic garlic using stable isotope and elemental composition analyses. Acta Chim Slov 64:1048–1055. https://doi.org/10.17344/acsi.2017.3476

Opatić AM, Necemer M, Lojen S, Masten J, Zlati E, Sircelj H, Stopar D, Vidrih R (2018b) Determination of geographical origin of commercial tomato through analysis of stable isotopes, elemental composition and chemical markers. Food Contr 89:133–141. https://doi.org/10.1016/j.foodcont.2017.11.013

Opatić AM, Nečemer M, Lojen S, Vidrih R (2017a) Stable isotope ratio and elemental composition parameters in combination with discriminant analysis classification model to assign country of origin to commercial vegetables—a preliminary study. Food Contr 80:252–258. https://doi.org/10.1016/j.foodcont.2017.05.010

Ozturkoglu-Budak S, Aksahin I (2016) Multivariate characterization of fresh tomatoes and tomato-based products based on mineral contents including major trace elements and heavy metals. J Food Nutr Res 55:214–221

Padín PM, Peña RM, García S, Iglesias R, Barro S, Herrero C (2001) Characterization of Galician (N.W. Spain) quality brand potatoes: a comparison study of several pattern recognition techniques. Analyst 126:97–103

Papaefstathiou E, Agapiou A, Giannopoulos S, Kokkinofta R (2018) Nutritional characterization of carobs and traditional carob products. Food Sci Nutr 6:2151–2161. https://doi.org/10.1002/fsn3.776

Park JH, Choi S-H, Bong S-Y (2019) Geographical origin authentication of onions using stable isotope ratio and compositions of C, H, O, N, and S. Food Contr 101:121–125. https://doi.org/10.1016/j.foodcont.2019.02.023

Peña RM, García S, Iglesias R, Barro S, Herrero C (2001) Authentication of Galician (N.W. Spain) quality brand potatoes using metal analysis. Classical pattern recognition techniques *versus* a new vector quantization-based classification procedure. Analyst 126:2186–2193. https://doi.org/10.1039/b107114a

Pepi S, Chicca M, Telloli C, Di Roma A, Grisenti P, Tessari U, Vaccaro C (2019) Discrimination of geographical origin of hop (*Humulus lupulus* L.) using geochemical elements combined with statistical analysis. Environ Geochem Health 41:1559–1576. https://doi.org/10.1007/s10653-018-0232-7

Pepi S, Sardella A, Bonazza A, Vaccaro C (2018) Geochemical caper fingerprints as a tool for geographical origin identification. Environ Geochem Health 40:1385–1403. https://doi.org/10.1007/s10653-017-0063-y

Pérez-Rodríguez M, Gaiad JE, Hidalgo MJ, María V, Avanza MV, Roberto G, Pellerano RG (2019) Classification of cowpea beans using multielemental fingerprinting combined with supervised learning. Food Control 95:232–241. https://doi.org/10.1016/j.foodcont.2018.08.001

Perini M, Pianezze S, Strojnik L, Camin F (2019) C and H stable isotope ratio analysis using solid-phase microextraction and gas chromatography-isotope ratio mass spectrometry for vanillin authentication. J Chromatograph A 1595:168–173. https://doi.org/10.1016/j.chroma.2019.02.032

Pianezze S, Perini M, Bontempo L, Ziller L, D'Archivio AA (2019) Geographical discrimination of garlic (*Allium sativum* L.) based on stable isotope ratio analysis coupled with statistical methods: the Italian case study. Food Chem Toxicol 134:110862. https://doi.org/10.1016/j.fct.2019.110862

Pinto E, Almeida AA, Aguiar AARM, Ferreira IMPLVO (2014) Changes in macrominerals, trace elements and pigments content during lettuce (*Lactuca sativa* L.) growth: influence of soil composition. Food Chem 152:603–611. https://doi.org/10.1016/j.foodchem.2013.12.023

Potortì AG, Bua GD, Lo Turco V, Tekaya AB, Beltifa A, Ben Mansour HB, Dugo G, Di Bella G (2020) Major, minor and trace element concentrations in spices and aromatic herbs from Sicily (Italy) and Mahdia (Tunisia) by ICP-MS and multivariate analysis. Food Chem 313:126094. https://doi.org/10.1016/j.foodchem.2019.126094

Potortì AG, Lo Turco V, Di Bella G (2021) Chemometric analysis of elements content in Algerian spices and aromatic herbs. LWT—Food Sci Technol 138:110643. https://doi.org/10.1016/j.lwt.2020.110643

Potortì AG, Mottese AF, Fede MR, Sabatino G, Dugo G, Lo Turco V, Costa R, Caridi F, Di Bella M, Di Bella G (2022) Multielement and chemometric analysis for the traceability of the Pachino Protected Geographical Indication (PGI) cherry tomatoes. Food Chem 386:132746. https://doi.org/10.1016/j.foodchem.2022.132746

Prohens-Tomás J, Nuez F (2007a) Vegetables I: Asteraceae, Brassicaceae, Chenopodicaceae, and Cucurbitaceae. Springer, New York. ISBN-10: 0387722912; ISBN-13: 978-0387722917

Prohens-Tomás J, Nuez F (2007b) Vegetables II: Fabaceae, Liliaceae, Solanaceae, and Umbelliferae. Science. 0387741100, 9780387741109

Richter B, Gurk S, Wagner D, Bockmayr M, Fischer M (2019) Food authentication: Multi-elemental analysis of white asparagus for provenance discrimination. Food Chem 286:475–482. https://doi.org/10.1016/j.foodchem.2019.01.105

Rivero RC, Hernandez PS, Rodrıguez EMR, Martı JD, Romero CD (2003) Mineral concentrations in cultivars of potatoes. Food Chem 83:247–253. https://doi.org/10.1016/S0308-8146(03)00087-6

Rodríguez Galdón B, Rodríguez LH, Mesa DR, León HL, Pérez NL, Rodríguez Rodríguez EMR, Romero CD (2012) Differentiation of potato cultivars experimentally cultivated based on their chemical composition and by applying linear discriminant analysis. Food Chem 133:1241–1248. https://doi.org/10.1016/j.foodchem.2011.10.016

Rodríguez Galdon BR, Gonzalez RO, Rodrıguez ER, Romero CD (2008) Comparison of mineral and trace element contents in onion cultivars (*Allium cepa* L.). J Sci Food Agric 88:1554–1561. https://doi.org/10.1002/jsfa.3250

Romarís-Hortas V, García-Sartal C, Barciela-Alonso MC, Moreda-Piñeiro A, Bermejo-Barrera P (2010) Characterization of edible seaweed harvested on the Galician coast (northwestern Spain) using pattern recognition techniques and major and trace element data. J Agric Food Chem 58:1986–1992. https://doi.org/10.1021/jf903677y

Scharrer A, Mosandl A (2002) Progress in the authenticity assessment of vanilla. 2. Δ13CV-PDB correlations and methodical optimisation. Dtsch Lebensmitt Rundsch 98:117–121

Schipilliti L, Bonaccorsi IL, Mondello L (2017) Characterization of natural vanilla flavour in foodstuff by HS-SPME and GC-C-IRMS. Flavour Fragr J 32:85–91. https://doi.org/10.1002/ffj.3364

Sinkovič L, Nečemer M, Ogrinc N, Žnidarčič D, Stopar D, Vidrih R, Meglič V (2020) Parameters for discrimination between organic and conventional production: a case study for chicory plants (*Cichorium Intybus* L.). Food Chem Toxicol 136:111109. https://doi.org/10.1016/j.fct.2019.111109

Song B-Y, Ryu J-S, Shin HS, Lee K-S (2014) Determination of the source of bioavailable Sr using $^{87}Sr/^{86}Sr$ tracers: a case study of hot pepper and rice. J Agric Food Chem 62:9232–9238. https://doi.org/10.1021/jf503498r

Suárez M, Méndez E, Galdón B, Rodríguez E, Romero C (2011) Influence of agronomic variables on quality of tomato fruits. Agricultural Sci 2:424–431. https://doi.org/10.4236/as.2011.24054

Suárez MH, Pérez DM, Rodríguez-Rodríguez EM, Romero CD, Francisco Espinosa Borreguero FE, Galindo-Villardón P (2016) The compositional HJ-Biplot—a new approach to identifying the links among bioactive compounds of tomatoes. Int J Mol Sci 17:1828. https://doi.org/10.3390/ijms17111828

Suárez MH, Rodríguez EMR, Romero CD (2007) Mineral and trace element concentrations in cultivars of tomatoes. Food Chem 104:489–499. https://doi.org/10.1016/j.foodchem.2006.11.072

Subramanian NK, White PJ, Broadley MR, Ramsay G (2011) The three-dimensional distribution of minerals in potato tubers. Ann Bot 107:681–691. https://doi.org/10.1093/aob/mcr009

Swoboda S, Brunner M, Boulyga SF, Galler P, Horacek M, Prohaska T (2008) Identification of Marchfeld asparagus using Sr isotope ratio measurements by MC-ICP-MS. Anal Bioanal Chem 390:487–494. https://doi.org/10.1007/s00216-007-1582-7

Szefer P (2007) Chemometric techniques in analytical evaluation of food quality. In: Szefer P. Nriagu J (eds) Mineral components in foods. CRC Press. Taylor & Francis, FL, 69–121. doi:https://doi.org/10.1201/9781420003987

Szefer P, Grembecka M (2007) Mineral components in food crops, beverages, luxury food, spices, and dietary food. In: Szefer P. Nriagu J (eds) Mineral components in foods. CRC Press. Taylor & Francis, FL, p 231–322. doi:https://doi.org/10.1201/9781420003987

Tenailleau EJ, Lancelin P, Robins RJ, Akoka S (2004) Authentication of the origin of vanillin using quantitative natural abundance ^{13}C NMR. J Agric Food Chem 52:7782–7787. https://doi.org/10.1021/jf048847s

Trincherini PR, Baffi C, Barbero P, Pizzoglio E, Spalla S (2014) Precise determination of strontium isotope ratios by TIMS to authenticate tomato geographical origin. Food Chem 145:349–355. https://doi.org/10.1016/j.foodchem.2013.08.030

Vadalà R, Mottese AF, Bua GD, Salvo A, Mallamace D, Corsaro C, Vasi S, Giofrè SV, Alfa M, Cicero N, Dugo G (2016) Statistical analysis of mineral concentration for the geographic identification of garlic samples from Sicily (Italy). Tunisia Spain Foods 5:20. https://doi.org/10.3390/foods5010020

Varliklioz Er S, Eksi-Kocak H, Yetim H, Boyaci IH (2017) Novel spectroscopic method for determination and quantification of saffron adulteration. Food Anal Methods 10:1547–1555. https://doi.org/10.1007/s12161-016-0710-4

Wakefield J, Mccomb K, Ehtesham E, Van Hale R, Barr D, Hoogewerff J, Frew R (2019) Chemical profiling of saffron for authentication of origin. Food Contr 106:106699. https://doi.org/10.1016/j.foodcont.2019.06.025

Wilde AS, Frandsen HL, Fromberg A, Smedsgaard J, Greule M (2019) Isotopic characterization of vanillin ex glucose by GC-IRMS—new challenge for natural vanilla flavour authentication? Food Contr 106:106735. https://doi.org/10.1016/j.foodcont.2019.106735

Wilde AS, Hansen AMS, Fromberg A, Frandsen HL, Smedsgaard J (2020) Determination of $\delta^{13}C$ of vanillin in complex food matrices by HS-SPME-GC-C-IRMS. Flavour Fragr J 35:387–393. https://www.researchgate.net/publication/340505088

Yang L, Meng L, Gao H, Wang J, Zhao C, Guo M, He Y, Huang L (2021) Building a stable and accurate model for heavy metal detection in mulberry leaves based on a proposed analysis framework and laser-induced breakdown spectroscopy. Food Chem 338:127886. https://doi.org/10.1016/j.foodchem.2020.127886

Yuan Y, Hu G, Chen T, Zhao M, Zhang Y, Li Y, Xu X, Shao S, Zhu J, Wang Q, Rogers KM (2016) Improved discrimination for *Brassica* vegetables treated with agricultural fertilizers using a combined chemometric approach. J Agr Food Chem 64:5633–5643. https://doi.org/10.1021/acs.jafc.6b00453

Zampella M, Quétel CR, Paredes E, Asfaha DG, Vingiani S, Adamo P (2011) Soil properties, strontium isotopic signatures and multi-element profiles to authenticate the origin of vegetables from small-scale regions: illustration with early potatoes from southern Italy. Rapid Commun Mass Spectrom 25:2721–2731. https://doi.org/10.1002/rcm.5081

Zhang X, Liu Y, Li Y, Zhao X (2017) Identification of the geographical origins of sea cucumber (*Apostichopus japonicus*) in northern China by using stable isotope ratios and fatty acid profiles. Food Chem 218:269–276. https://doi.org/10.1016/j.foodchem.2016.08.083

Chapter 10
Mushrooms

Contents

10.1 Introduction

Mushrooms, which are low-calorie fruiting bodies found in forests, are rich sources of ergothioneine, glutathione, mineral elements (K, Se, Cu, Fe and P) and antioxidants, including Se, vitamin C and choline. Moreover, they contain vitamins such as vitamin D, riboflavin (B_2), folate (B_9), thiamine (B_1), pantothenic acid (B_5) and niacin (B_3). Wild-growing mushrooms are widely consumed as delicacies, especially in Central and Eastern European countries.

Fruiting bodies contain considerable amounts of structural polysaccharides and proteins, which are the main components of dry matter, in contrast to the lipid content, which is low. Chitin, mannitol and trehalose are characteristic carbohydrate components. The favourable nutritional profile of mushrooms, owing to their low caloric value, high proportion of indigestible fibres, specific β-glucans and antioxidative and flavour constituents, has attracted increasing attention from researchers and consumers. The ability of some species to accumulate several undesirable ingredients in edible mushrooms has also received special attention (Kalač 2009).

Supplementary Information The online version contains supplementary material available at https://doi.org/10.1007/978-3-031-99687-0_10.

P. Szefer, *Chemometrics in Food Control Based on Elemental Analysis*, https://doi.org/10.1007/978-3-031-99687-0_10

Many extensive studies have been performed by Cocchi et al. (2006) since the distributions of As, Cd, Pb, Hg and Se were investigated in 1194 samples of 60 common species of edible mushrooms collected mainly in the province of Reggio Emilia, Italy. The accumulation of As in the samples studied was generally modest, although species such as *Sarcosphaera eximia* may accumulate to a level of 1 µg/g d.w. The Cd contents of *Amanita caesarea*, *Boletus edulis* and *Boletus pinophilus* exceeded the maximum amount allowed. Moreover, the level of Cd in *Agaricus macrosporus* exceeded approximately 50 times the maximum weekly dose recommended by the WHO. With respect to Pb, the average content in all the samples was mostly below the maximum allowed concentration. High levels of Hg within the range of 0.005–0.010 µg/g d.w. were detected in *Agaricus bitorquis*, *Agaricus arvensis*, *Agaricus essettei*, *Agaricus albertii*, *B. pinophilus*, *Clitocybe geotropa* and *Macrolepiota rachodes*. Mushrooms in general, but species in the *B. edulis* group in particular, are rich in Se. The accumulation of toxic heavy metals may be species specific. Mushrooms, including some excellent edible species, such as *A. caesarea* and the *B. edulis* group, can accumulate high levels of heavy metals.

According to Kalač and Svoboda (2000) and Kalač (2010), some species are distinguished from others because they accumulate or even hyperaccumulate various elements. For example, mushrooms, mainly from the genera *Agaricus*, *Macrolepiota*, *Lepista* and *Calocybe*, accumulate considerable amounts of toxic heavy metals such as Cd and Hg even in unpolluted or mildly polluted regions. In the case of most mushroom species grown at unpolluted sites, their levels of, e.g., Al, Ba, Cd, Co, Cr, Cu, Fe, Hg, Mn, Ni, Pb, Se, Sb and Zn, can be considerably higher, even by an order of magnitude higher, than those of mushrooms grown in polluted areas; these findings were unexpected. The possibility of evaluating the toxicological risk or nutritional benefits has therefore been limited, and the consumption of the species that bioaccumulate unsafe levels of heavy metals should be restricted.

Some mushrooms have been heavily contaminated by radioactive fallout from the Chernobyl disaster in 1986. The radioactivity of ^{137}Cs and, to a lesser extent, ^{134}Cs was detected in the following years in some edible species. *Xerocomus* (Boletus) *badius*, *Xerocomus chrysenteron*, *Suillus variegatus*, *Rozites caperata* and *Hydnum repandum* are heavily accumulating mushrooms. Several environmental factors influence the activity concentrations of radionuclides, e.g., the rate of soil contamination with fallout, the horizon from which mycelium picks up nutrients, soil moisture and time from the accident. It has been reported that eating mushrooms and animals, such as deer, results in elevated levels of radionuclides in their tissues. The environmental consequences of the Chernobyl and Fukushima disasters concern, to a certain extent, the radioactivity status of the fruiting bodies of mushrooms. However, the radioactivity of cultivated mushrooms, in contrast to that of wild mushrooms, is negligible, and contamination can be considerably decreased by soaking or cooking dried or frozen mushroom slices (Kalač 2001, 2019).

Various modern analytical methodologies, including spectroscopy and mass spectrometry combined with multivariate chemometrics, have been widely used in the control of mushrooms for health safety. Therefore, such a technologically advanced approach has enabled qualitative and quantitative analysis and quality control of mushrooms in recent years (Liu et al. 2022).

Considering the substantial increase in the world's production of mushrooms, this will be accompanied by increased attention from producers, traders, consumers and authorities in the international food trade system to achieve safe, high-quality and authenticated products (El Sheikha and Hu 2018).

10.2 Chemometric Assessment of Mushrooms

10.2.1 Geographical Differentiation

10.2.1.1 European Countries

Kojta and Falandysz (2016) used ICP–OES and CV–AAS to study samples of Bay Bolete fruiting bodies from ten different Polish forest regions with respect to MEC combined with chemometric processing analysis (PCA and CA). The former, with the first three PCs explaining 71.86% of the total variance discriminated samples of mushroom fruiting bodies according to their geographical provenance. The most descriptive elements, K, Mg and Zn, were responsible for clear interspatial differentiation.

10.2.1.2 Asiatic Countries

Chen et al. (2021) combined ICP–MS with chemometrics to assess the origins of *Thelephora ganbajun* from four sites in Yunnan Province (China). PCA clearly differentiated mushrooms according to their location in four regions, with the first three PCs contributing 92.84% of the total variance. The most discriminative variables are as follows: Mg, Ca, Cr, Se, Cd, Fe, P, Al, As, Zn, K, Pb and Mn. The HCA also revealed four clusters related to the four origins of *T. ganbajun*. Like HCA and PCA, LDA was also an effective tool (Fig. 10.1), and this model provided an overall correct classification rate of 100% with an independent external sample set. They concluded that studying the multielement composition combined with multivariate chemometrics is a helpful approach for the identification of the geographical provenance of *T. ganbajun*.

Yao et al. (2019) applied a multimethodological approach to specify mushrooms from various sampling sites in Yunnan Province (China). The classification models were established on the basis of combined datasets with SVM. The combination of modern analytical and computational tools is a developing strategy resulting in better performance of classification models than when a single technique is used. The first five PCs, which explained 74.31% of the total variance, provided the main information of the *Boletaceae* mushroom samples and were used to combine mid-level data for further analysis. Owing to the use of combined data from three

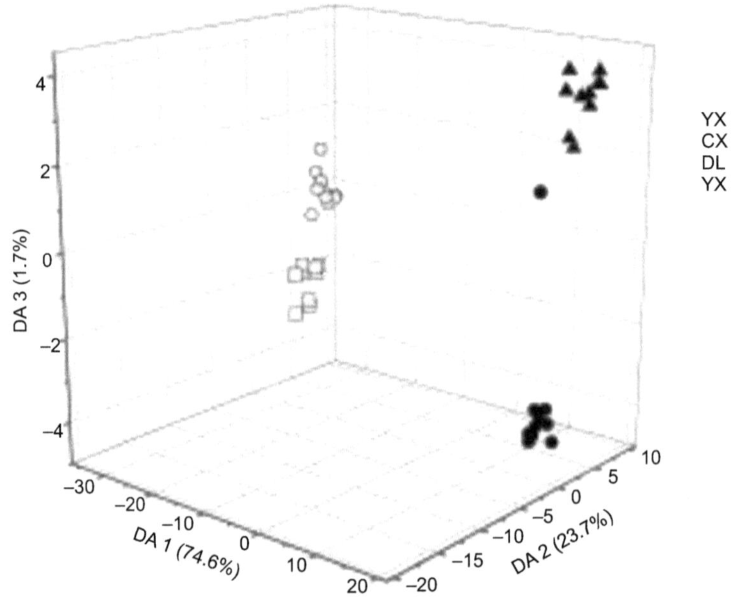

Fig. 10.1 Linear discriminant analysis plot of the elements in *T. ganbajun* taken from Yunnan Province, China. Kun Ming (KM); Chu Xiong (CX); Da Li (DL); Yu Xi (YX). (After Chen L-P et al., 2021. J Food Compos Anal 96:103699. https://doi.org/10.1016/j.jfca.2020.103699. With permission)

instrumental tools, i.e., FT-IR, UV spectroscopy and ICP–AES, it was possible to achieve an accuracy of 84.62% in the CV and an accuracy of 99.04% in the test set. It has been concluded that data fusion coupled with multivariate statistical analysis, such as SVM, may be a useful methodology for food traceability and quality control (Fig. 10.2). However, it should be kept in mind that the difference in species may present difficulties in tracing the geographical origins of *Boletaceae*.

According to Lim et al. (2012), coupling INAA with chemometric analysis (PCA and LDA) was useful for classifying *Phellinus* mushroom samples originating from various sites in three Asiatic countries (Cambodia, Korea and Vietnam). The PCA data indicated that the first three PCs, which explained 84.5% of the total variance, discriminated mushroom samples according to their geographical origin. LDA was also successful in achieving a correct classification rate of 100%.

10.2.1.3 Countries Across Multiple Continents

Chung et al. (2022) reported that EA–IRMS combined with chemometric processing analysis was applied to characterize the stable isotope profile of *Agaricus bisporus* mushrooms originating from three countries. The SIR distribution pattern of the mushroom significantly ($p < 0.05$) depended on the complete media used; in particular, high $\delta^{34}S$ values were detected in the Chinese complete media, low $\delta^{34}S$

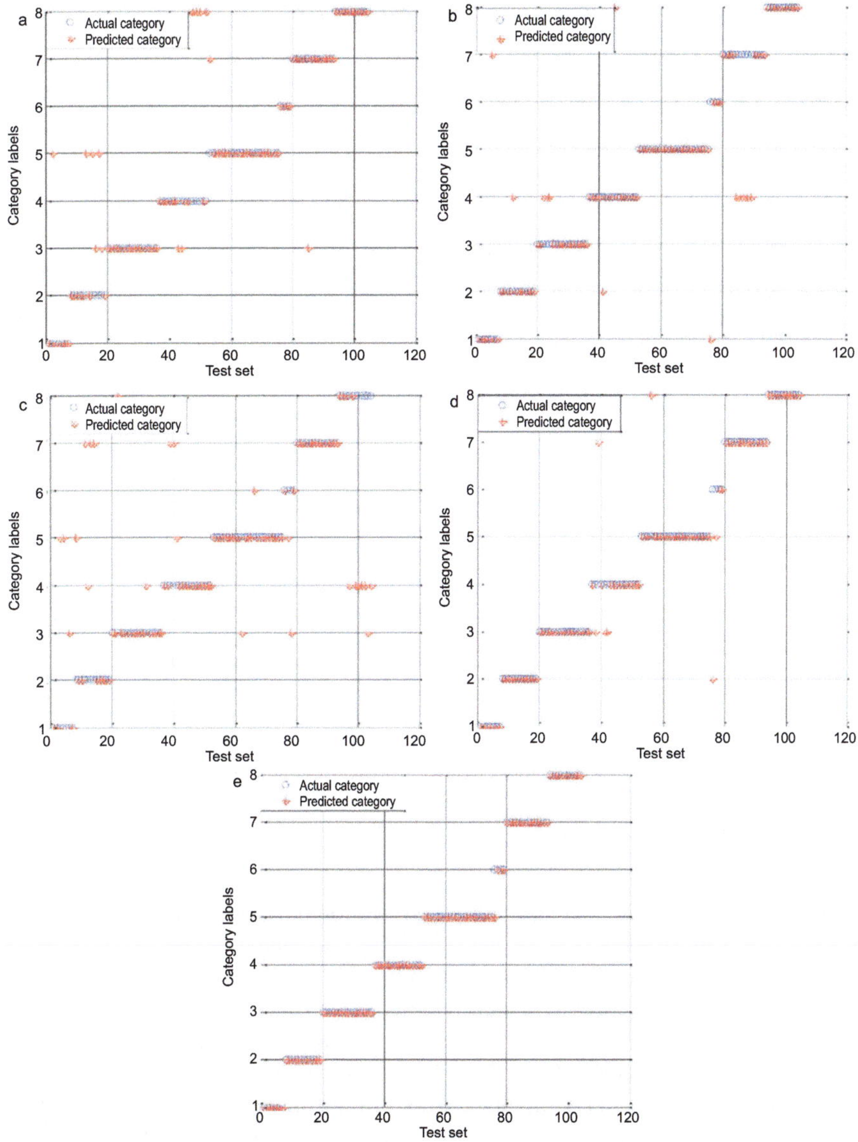

Fig. 10.2 Plot of the actual and predicted categories of test samples by SVM: (**a**) FT–IR spectra; (**b**) UV spectra; (**c**) ICP–AES; (**d**) combined data from FT–IR and UV; and (**e**) that from FT–IR, UV, and ICP–AES. (After Yao S et al., 2019. Int J Food Prop 22:414–426. https://doi.org/10.1080/10942912.2019.1588299. With permission)

values were detected in the Dutch complete media, and high $\delta^{15}N$ values were detected in the Korean complete media. In the case of LDA (Fig. 10.3), the overall classification accuracy was 87.3% for the original set and 85.8% for the cross-validated set. Compared with LDA, the SVM model, with an overall predictability

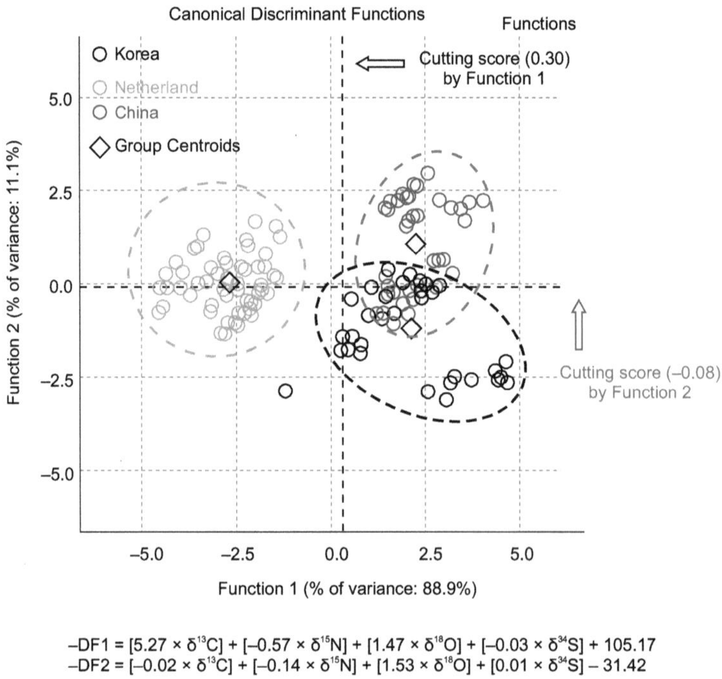

Fig. 10.3 $\delta^{13}C$, $\delta^{15}N$, $\delta^{18}O$ and $\delta^{34}S$-based geo-origin identification model and discriminant functions for *A. bisporus* grown in Korea using linear discriminant analysis (LDA). The cutting scores (0.30, −0.08) were calculated using the group size and centroid values. The dotted circles are for illustrative purposes only, not including statistical analysis. (After Chung IM et al., 2022. Food Chem 386:132820. https://doi.org/10.1016/j.foodchem.2022.132820. With permission)

of 100%, achieved better geographical classification of *A. bisporus* in the complete media. Descriptors such as $\delta^{15}N$ and $\delta^{34}S$ were the crucial feature weights in the SVM model. It was concluded that the combined tool allows the discrimination of mushroom samples according to their regional provenance.

Falandysz et al. (2017a) used ICP–MS to measure the elemental contents of samples from three species of *Cantharellus* mushrooms from various regions of two countries. The first two PCs explained 52% of the cumulative variance, indicating that V, As, Ag, Cd, Pb, Ba, U and Rb are responsible for the differentiation of Chinese chanterelles from Polish mushrooms. Both geogenic and anthropogenic Cd can bioaccumulate in chanterelles from the Świetokrzyskie Mountains in Poland, whereas geogenic sources can be significantly predominant in the background areas of Yunnan. The nutritional properties of Cu and Zn of mushrooms with respect to their physiological regulation was supported by data for *Cantharellus* mushrooms from both regions, although their geogenic source, and possibly an anthropogenic source, can also be attributed to the region of the Świetokrzyskie Mountains in Poland. The obtained results suggest a strong influence of the sampling site on the occurrence of the above eight elements in fruit bodies, which were present in relatively high levels in *Cantharellus* mushrooms from polymetallic soils of Yunnan in China. However, lower contents of Mn and Rb were observed compared with those in the fruit bodies from Poland.

10.2.2 Interregional and/or Interspecies Differentiation

Brzezicha–Cirocka et al. (2019) analysed 16 macro, micro- and trace elements quantified by FAAS and ICP–AES in approximately 1500 samples from 22 species of mushrooms collected from various regions in Poland and the area around Umeå. The FA technique, with the first two factors explaining 52.7–64.1% and 69.0% of the total variance, enabled the identification of samples with respect to five species of mushrooms originating from two regions in Poland. The most discriminative elements were K, Na, Mg, Ca, Fe, Zn, Cu, Mn and Cd (Fig. 10.4). PCA, with the first two factors accounting for between 52.7% and 64.1% of the total variance, achieved interregional differentiation in the case of three species, i.e., *C. cibarius*, *B. edulis* and *L. scabrum* (Fig. 10.5). It was concluded that the use of chemometric tools allows the interpretation of interspecies differentiation to be combined with geographical provenance.

Wang et al. (2019) coupled FT–IR with ICP–MS to categorize six species of porcini mushrooms from various sampling sites in China. The analytical data obtained were chemometrically processed successfully via PCA and sPLS–DA. Reliable data were obtained at the mushroom species level using PCA, with >81% cumulative variability for PC1 and PC2, as well as via sPLS–DA, with an accuracy rate greater than 92.1%. For example, estimated accuracy rates of 100 and 97.5% for interspecies differentiation were achieved for various species from Dali and Baoshan, respectively. Finally, the interspatial differentiation corresponded to the first two PCs, which explained 94% of the total variance with respect to the differentiation of three regional-scale producing areas, i.e., Central Yunnan, Western Yunnan and Northwestern Yunnan. This approach allows the discrimination of samples with respect to their origin, achieving a greater than 92% accuracy rate with FT–IR spectra. In summary, interspecies and interregional differentiation accuracies greater than 92.1 and 92.8%, respectively, were achieved.

Falandysz et al. (2008) used a combination of ICP–MS, ICP–OES, HG–AAS and CV–AAS with a chemometric tool. PCA, with the first three PCs explaining 55.6% of the total variance, was performed to characterize King Bolete samples collected from four sites in lowland and high mountain regions of Poland with different soil bedrock. The chemometric approach formed three clusters of object scores corresponding to Ba, Cd, Cs, Cr, Ga, Rb, Se, Sr and V; K and Mg; and Cu and Mo, respectively. The mushroom caps that originated from the lowland and mountain sites contained Ag, Ba, Co, Cr, Hg, K, Mg, Mn, Mo and Na at comparable levels, but the samples from the mountain areas were contained greater contents of Cd and Sb. In addition, King Bolete is enriched in Al, Pb and Rb in the metal ore-rich region of the Sudety Mountains. Interestingly, the mushrooms demonstrated

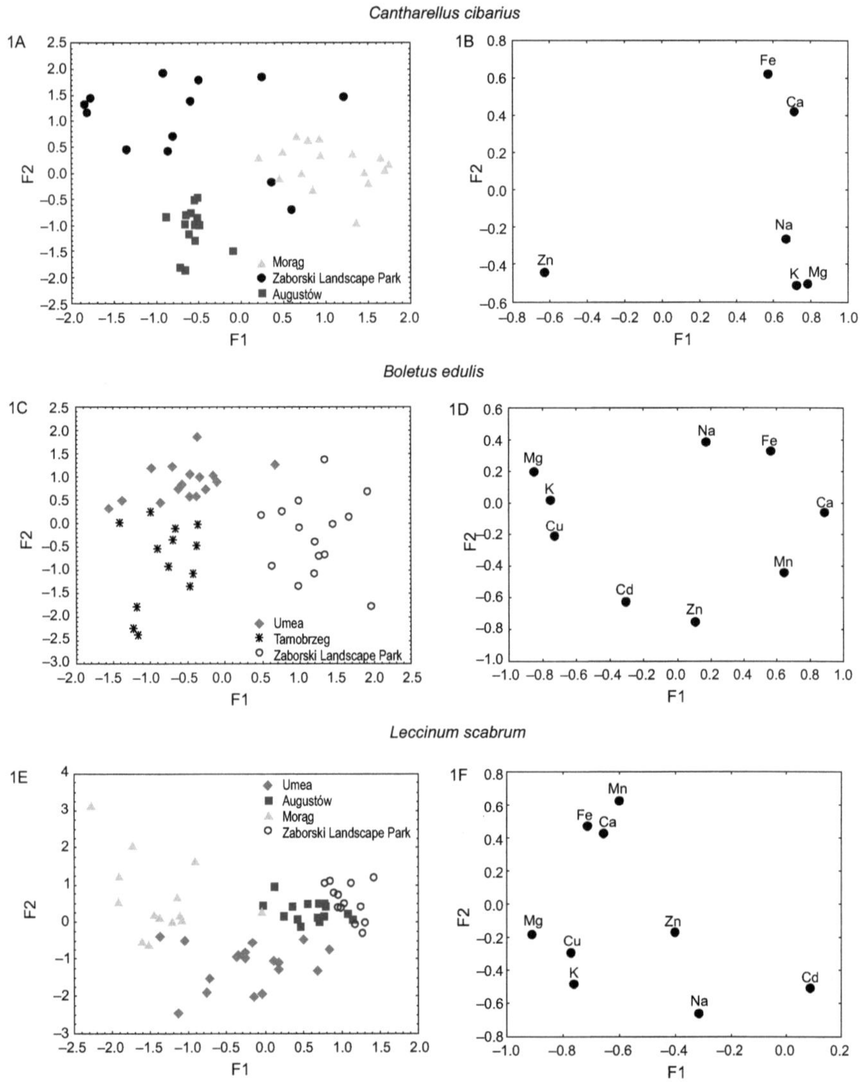

Fig. 10.4 A, C, E. Point scatterplots of Factors F1 and F2 for samples of *C. cibarius*, *B. edulis* and *L. scabrum* from Augustow, Morąg, Tarnobrzeg, the Zaborski Landscape Park and Umea. Fig. 1 B, D, F. Scatterplots of elemental loadings in the same mushroom samples as above. (After Brzezicha-Cirocka J et al., 2019. Ecotoxic Environ Saf 173, 353–365. https://doi.org/10.1016/j.ecoenv.2019.02.036. With permission)

high bioconcentration potential at the background sampling sites owing to their accumulation in fruiting bodies of high levels of toxic metals such as Cd, Pb and Hg.

According to Giannaccini et al. (2012), HG-ICP–OES was a useful tool for the study of topsoil and edible mushrooms, i.e., *Boletus edulis* and *Macrolepiota procera* harvested in green areas within the Italian province of Lucca, Tuscany. Using

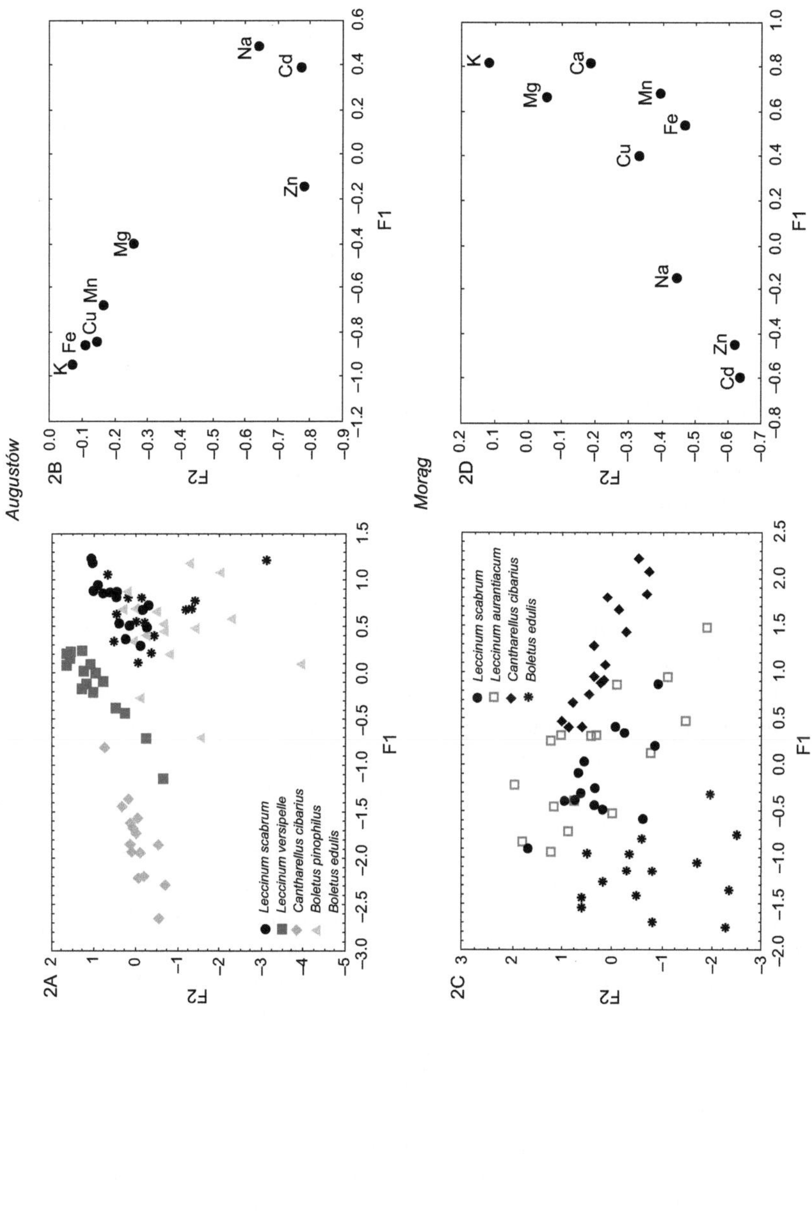

Fig. 10.5 A, C. Point scatterplots of Factors F1 and F2 for samples of *L. scabrum*, *L. aurantiacum*, *L. versipelle*, *C. cibarius*, *B. pinophilus* and *B. edulis* from Augustow and Morag. Fig. 2 B, D. Scatterplots of elemental loadings in the same mushroom samples as above. (After Brzezicha-Cirocka J et al., 2019. Ecotoxic Environ Saf 173, 353–365. https://doi.org/10.1016/j.ecoenv.2019.02.036. With permission)

CA, they could interpret two main clusters of soil element compositions corresponding to the most ecologically differentiated areas. Moreover, CA allowed the differentiation of groups for *M. procera* composition according to harvesting location, suggesting that this saprotrophic species is sensitive to the conditions of microhabitats. In conclusion, interspecies and interspatial differentiation using the elemental profile was achieved for *M. procera*.

Dimitrijevic et al. (2016) examined eleven species of wild-grown common edible mushrooms from the *Boletaceae* family from Serbia. Major, minor and trace elemental contents were determined for the mushrooms using ICP–OES and ICP–MS. The obtained concentration data were transformed via chemometric tools (PCA and HCA). It was possible to group chemometrically the mushroom samples into three statistically significant clusters. Moreover, PCA, with the first four PCs corresponding to 80.38% of the total variance, revealed the relationships among the analysed elements with respect to the differentiation of mushroom species. Overall, classification of the elemental profiles confirmed that mushrooms are a rich source of essential elements, confirming that they are nutritious.

10.2.3 Differentiation of Other Parameters

10.2.3.1 Interregional or Interspecies Differentiation Related to Parent Soil Geochemistry

According to Aloupi et al. (2012), five species of wild-grown common edible mushrooms from the *Boletaceae* family were investigated for the content of eight elements quantified by AAS. The heavy metal contents were processed chemometrically to characterize five species of wild edible mushrooms from the island of Lesvos (Greece). Both substrate geochemistry and species identity were studied, although not in a uniform way. For example, chemometrics demonstrated a species-specific interaction of the investigated mushrooms with different elements in species of the same genus. For example, *R. delica* had a relatively high affinity for Cu; *L. sanguifluus*, for Zn; *S. bellinii*, for Mn and Pb; and *L. semisanguifluus*, for Cr, Ni and Fe. Mushrooms that generally grow on serpentine soils contained significantly higher Cr and Ni levels than those from volcanic areas. The elevated Cd concentrations recorded in the mushrooms from the serpentine sites could also be attributed to increased levels of the metal in the soils. However, *R. delica* accumulated Cr and Ni proportionally from the soil substratum, but *S. bellinii* accumulated only Ni, whereas *L. sanguifluus* controlled metal levels because it was less affected by their levels in the soil. According to the CA dendrogram, only *L. deliciosus* and *L. sanguifluus* were similar, but all the remaining species formed single-species groups.

Gucia et al. (2012a) coupled ICP–OES and CV–AAS with chemometric processing to investigate fruiting bodies of parasol mushrooms (caps and stipes) and parent surface layers of soils from 11 Polish forest regions. The relationships between the concentrations of elements in mushrooms and associated soil substrata were presented and interpreted. The PC data allowed visualization for distinguishing samples with respect to their geographical origin, with the first three PCs, caps and stipes, accounting for 85.17 and 84.71% of the total variance, respectively. The CA dendrogram of the dataset for the mushroom caps clearly showed that the sampling sites could be divided into two major clusters with different specificities corresponding to the contents of the elements analysed, which means that the multivariate tools allowed the partial distinction of mushrooms in terms of their origin.

Chudzyński and Falandysz (2008), on the basis of the multielemental composition quantified by ICP–AES, investigated the caps and stipes of Larch Bolete with pooled samples of the underlying soil substrate originating from six Polish forest regions. PCA (with a VARIMAX rotation method) and CA were helpful for differentiating mushroom samples according to their geographical provenance. For example, according to the PCA data, PC1 (35%), PC2 (22%), PC3 (13%), PC4 (7%) and PC5 (6%) corresponded to elements Al, Ba, Cr, Fe and Sr; K and Mg; Ag and Cd; P and Zn; and Pb, respectively. With the use of these elements, the samples were differentiated according to their provenance. Importantly, the pattern of Pb distribution varied from that of the other elements analysed. It has been suggested that sources other than soil bedrock are responsible for the impact of local, regional or long-range atmospheric transport of Pb and ultimately deposition into forest ecosystems. As shown in Fig. 10.6, the dendrogram allows distinguishing two clusters, including mushroom samples related to the northern and southern parts of Poland. This means that the chemometric processing of the multielemental profile of the Larch Bolete caps allowed them to distinguish the mushroom samples according to their interspatial origin.

Kokkoris et al. (2019) used FAAS combined with chemometric processing analysis to classify six wild edible mushroom species from the Greater Athens area

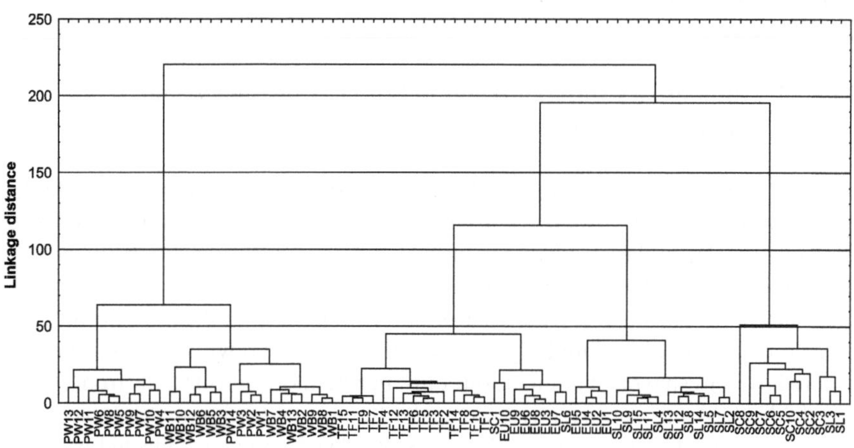

Fig. 10.6 Dendrogram of all the examined samples of the caps of the Larch Bolete fruiting bodies. (After Chudzyński K and Falandysz J. 2008. Chemosphere 73:1230–1239. https://doi.org/10.1016/j.chemosphere.2008.07.055. With permission)

(Greece) with their respective soil substrates (two depth layers). Significant relationships were observed between the bioavailable fractions of several heavy metals in soils and their respective levels in Greek *A. bisporus*, *C. comatus*, *P. candolleana* and *V. gloiocephalus*, in contrast to their total concentrations. The first two PCs accounted for 50.5% of the total variance, indicating that *A. bitorquis* and *A. gennadii* were differentiated from *C. comatus* and *P. candolleana* but not from *A. bisporus* and *V. gloiocephalus*. Overaccumulation of metals in fruit bodies was detected in the bioavailable fractions of Cu, Mn, Zn and Ni in the soil. The levels of Pb and Cu exceeded the maximum tolerable daily intake and RDA, respectively. ICP–OES was used to evaluate the chemical and multielemental profiles of mostly edible aboveground and wood-growing mushroom species collected near a heavily trafficked road. The aboveground species were distinguished by considerably higher levels of twelve elements, i.e., Ag, Al, As, B, Co, Cr, Cu, Fe, Ni, Pb, Se and Zn, than the wood-growing species. An inverse relationship was obtained for Au, Ba and Sr.

Drewnowska and Falandysz (2015) assayed common chanterelle mushrooms and associated topsoil samples from four regions of Poland with respect to their classification. Macro, micro- and trace elemental contents were quantified by ICP–OES and CV–AAS. PCA was applied to examine the interdependences between the elements in the fruiting bodies of *C. cibarius* and those in the soil substratum. The PCA data, with the first four PCs accounting for 96% of the total variation, indicated that the elemental profile variability in mushrooms is dependent on the geochemistry of the parent soil originating from all the examined locations. For example, the elemental content correlated with PC2 (Ag, Co and Ni) is responsible for the differentiation of mushrooms from Ciechocinek from other locations. PC3 (Hg and Mn) discriminated the soil samples from the Commune of Bukowiec and Augustowska Forest, whereas PC4 (Ca and Sr) distinguished the soil samples from the Commune of Bukowiec and Ciechocinek. The mushroom considerably accumulates Rb, K, P, Cu, Cd, Ag, Na and Zn from the associated soil, but Al, Ba, Cr, Fe and Pb do not have such a relationship.

10.2.3.2 Interspecies Differentiation Related to Cultivated/Wild Mushroom Variability

Sommer et al. (2021) applied EA–IRMS to investigate fresh and processed button mushrooms (*Agaricus bisporus*) from various sampling sites in four countries (Poland, The Netherlands, Germany and China). The $\delta^{15}N$ signatures allowed the differentiation of cultivated mushrooms from wild-growing samples since the former ones were considerably enriched in $\delta^{15}N$, indicating cultivation on different organic compost types. Similarly, all the studied fresh brown button mushrooms differed from fresh white button mushrooms. Furthermore, mushrooms from Poland and the Netherlands consistently presented different stable nitrogen isotopic signatures. A slight difference in the $\delta^{15}N$ values of related products collected in subsequent years revealed a high level of standardization of the individual cultivation processes over several years. It was concluded that analysis of the $\delta^{13}C$ and $\delta^{15}N$

signatures of cultivated button mushrooms could be important for long-term monitoring and target control.

Mleczek et al. (2021c) investigated edible wild-growing mushrooms from Polish forest regions, whereas cultivated mushroom species available in trade originating from 14 countries were investigated quantitatively by ICP–OES for their multielemental composition. The original data were processed using chemometric analysis, including one-way MANOVA, HCD and heatmaps. Notably, in the case of edible wild-growing and cultivated mushrooms, there was specific interspecies differentiation of the elements determined. Moreover, cultivated mushrooms generally contain higher elemental levels than wild mushroom species do, although both groups of mushrooms pose some risks to human health, especially if they are consumed relatively frequently. It was also reported that wild-type mushrooms had high Hg levels.

10.2.3.3 Interregional and/or Interspecies Differentiation Related to the Impact of Pollution

Malinowska et al. (2004) coupled FAAS and a spectrophotometric method with chemometric analysis to characterize samples of fruiting bodies (caps and stalks) of Bay Bolete and associated soil samples from 12 different Polish forest regions. Multivariate analysis, i.e., CA and FA, was used to differentiate mushrooms and the underlying soil substratum from Trojmiejski Landscape Park adjacent to the Tricity agglomeration. The most similar areas were even discriminated, e.g., Augustowska and Białowieska Forests as protected areas, which are far from industrial or urban regions. A FA with the first three Fs explained 75.8% of the total variance. Figure 10.7a shows that satisfactory discrimination between the cap samples from the various regions was achieved. Pb is also an identifier of mushroom samples originating from Trojmiejski Landscape Park, adjacent to the Tricity agglomeration (Fig. 10.7b). This toxic heavy metal is then widely emitted into the air because of burning car petrol containing tetraethyl lead. This means that *X. badius* is useful as an appropriate bioindicator for the environmental monitoring of heavy metal pollutants, especially Pb. FA, with the first three Fs values explaining 91.3% of the total variance, clearly differentiated samples of the mother soil corresponding to various forest regions (Fig. 10.8). The CA data demonstrated that it was also possible to discriminate the most similar forest areas with respect to the chemical composition of both the associated soil substratum and the biomass overgrown with *X. badius* in the explored area (Fig. 10.9). In addition to anthropogenic factors, natural factors such as the geochemical composition of bedrock, pH and granulometric structure of the soil, the genetic potential of fungi, ectomycorrhizal occurrence and the kind of undergrowth (mosses, lichens, ferns) may influence the contents of some metals in mushrooms.

Kojta et al. (2016) combined ICP–OES and Hg detection using CV–AAS with a multivariate chemometric tool to classify *Macrolepiota procera* (caps and stipes) and underlying soil samples regarding their mineral compositions that originated from upland regions of central Poland. The PCA, with the first two PCs accounting

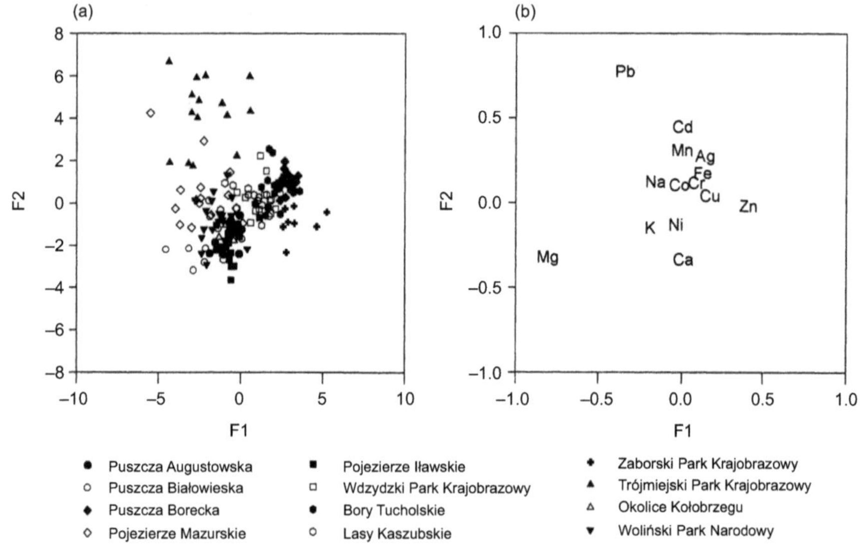

Fig. 10.7 (**a**) Scatterplot of the object scores of the first two discriminant functions for 166 samples of stalks; (**b**) locations of loadings for 14 metals in the stalks. (After Malinowska E et al., 2004. Food Chem 84:405–416. https://doi.org/10.1016/S0308-8146(03)00250-4. With permission)

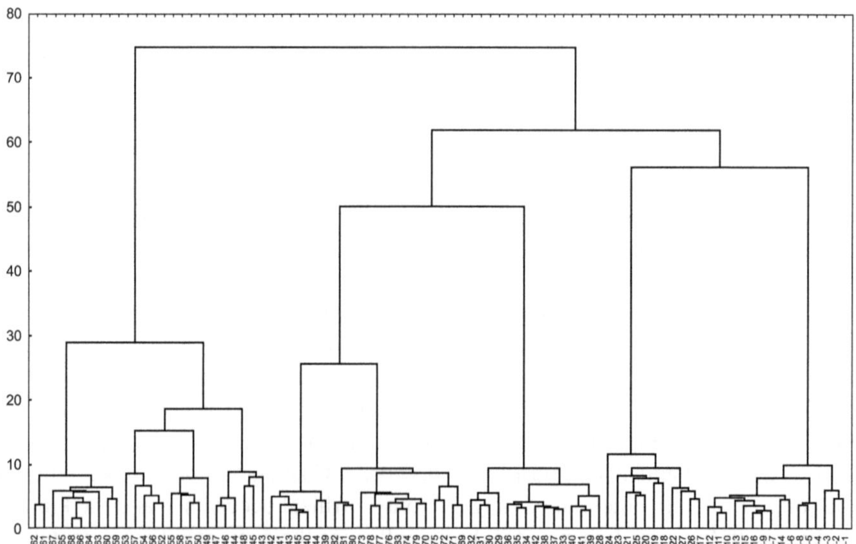

Fig. 10.8 Hierarchical dendrogram for 90 sampling sites (1–16 Augustowska Forest, 17–27 Bialowieska Forest, 28–42 Borecka Forest, 43–58 Adjacent area of Morag, 59–68 Ilawskie Lake District, 69–83 Wdzydzki Landscape Park, 139–145 Adjacent area of Kolobrzeg). (After Malinowska et al., 2004. Food Chem 84:405–416. https://doi.org/10.1016/S0308-8146(03)00250-4. With permission)

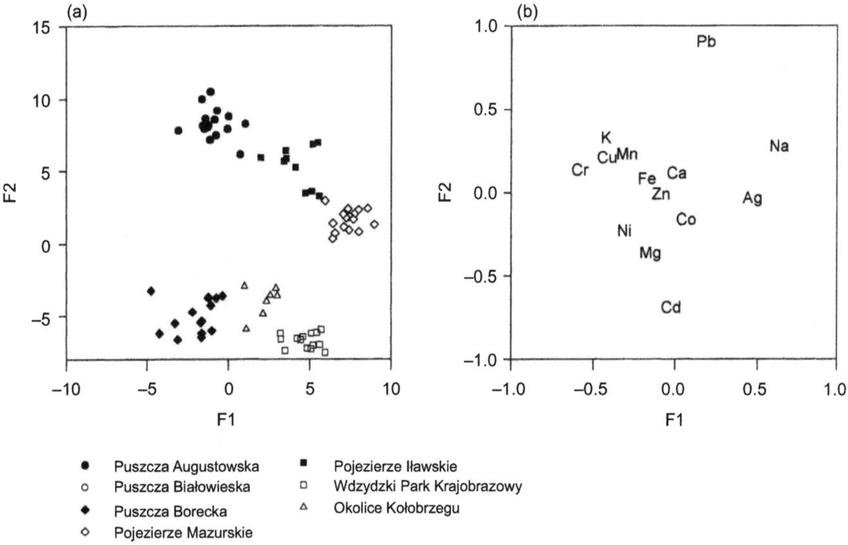

Fig. 10.9 (**a**) Scatterplot of object scores of the two first discriminant functions of 90 samples of soil; (**b**) locations of loadings for 14 metals in soil. (After Malinowska et al., 2004. Food Chem 84:405–416. https://doi.org/10.1016/S0308-8146(03)00250-4. With permission)

for 91.21% of the total variance, revealed interregional differences in the elemental profiles of the *Macrolepiota procera* mushrooms. For example, samples from the Poniatowa site were differentiated from groupings of samples corresponding to the Lubraniec site, and the Starachowickie Forest and Gostyńsko–Włocławskie Forest overlapped. Moreover, the CA model based on the cap and soil datasets discriminated *M. procera* from four stands in south-central Poland and parent soils collected from the same sites. According to the authors, occasional consumption (once a week) of *Macrolepiota procera* caps may be safe, whereas consumption more than once a week could exceed the provisionally allowed daily intake limits for humans with respect to toxic metals.

Falandysz et al. (2015) determined the content of Hg via the CV–AAS method in the fungi Boletus species from Latosols, Lateritic Red Earths, and Red and Yellow Earths in the Circum–Pacific Mercuriferous Belt of Southwestern China. Among the 21 species studied from 32 locations across Yunnan Province and two places in Sichuan Province, elevated Hg levels were detected in all samples from Yunnan Province, in contrast to the samples from Sichuan Province, which is located outside the Mercuriferous Belt. The caps of the fruiting bodies of ten species of *Boletus* were particularly distinguished, with higher contents of Hg (from 4.9 to 22 μg/g d.w.). The surface soil samples (0–10 cm) from all the sampling sites contained Hg in the range of 0.034 to 3.4 μg/g d.w., with Yunnan soil parent rock and fruiting bodies of *Boletus* spp. enriched in Hg, unlike the same species from Sichuan located outside the Mercuriferous Belt. This interregional differentiation can be explained by the fact that the Hg in the Yunnan samples mainly originated from geogenic sources rather than anthropogenic effects. Notably, for the first time, a highly elevated content of Hg was documented for several species of the edible fungi of the genus *Boletus* growing in Yunnan Province, China.

Liu et al. (2015) reported that several species of edible mushrooms from 16 regions of Yunnan Province in China were examined with respect to the toxic trace elements as determined by ICP–MS. The levels of As, Pb and Cd mostly exceeded the national standard values of China. In addition, As and Cd, in contrast to Pb, were distinguished by greater accumulation abilities, indicating greater health risk in the Yunnan community. It was concluded that heavy metal pollution in Yunnan wild edible mushrooms will create serious problems. Among the toxic elements, As and Cd in edible mushrooms in the area are the main sources of the risk, as they may cause severe health problems.

Mleczek et al. (2016a) used HPLC–HG–AAS to study edible fungal species (*Agaricus bisporus*, *Pleurotus ostreatus*, *Pleurotus eryngii* and *Hericium erinaceus*) grown separately on substrates containing different As species [As(III), As(V) and dimethylarsinic acid (DMA; each form individually) at various concentrations. The obtained data were processed by univariate chemometrics. To guarantee the status of cultured mushroom species and the safety of ending food products for human health, the contamination level of substrates with As compounds should be controlled. This scientific strategy results from the prerequisite that mushrooms accumulate various concentrations of different As species, which are regulated by their level in the substrate.

Falandysz et al. (2017b) examined the multielemental composition and associations between a group of 48 elements, including 16 REEs, determined by Q–ICP–MS in the fruiting bodies of *Macrolepiota procera* from 16 sites in the lowland areas in the northern and central regions of Poland. The correlation matrix obtained from a possible 48 × 16 data matrix was used to examine whether any associations existed between the 48 elements in the mushrooms from 16 sampling locations via PCA, in which eight PCs explained up to 93% of the variability with EVs ≥1. The absolute values of the correlation coefficient were above 0.72 (significance at $p < 0.05$) for 43 elements. With respect to the evaluation of toxicological risk or nutritional benefits from the point of view of the consumer, the levels of Cd, Hg and Pb in the caps of *M. procera* originated from the background; i.e., unpolluted areas could be considered elevated, whereas the irregular consumption of this mushroom was estimated to be safe. The chemometric data indicated that the relationships between numerous elements in *M. procera* revealed some features of a bioindicator species for anthropogenic Pb and for some geogenic metals.

Frankowska et al. (2010) quantified the MEC content by ICP–OES and CV–AAS to provide basic data, including the contents of toxic heavy metals such as Cd and Hg, for wild King Bolete mushrooms from the Płocka Dale area of Poland. In addition to the caps and stipes of the mushrooms, the surficial layers of the soil (0–10 cm) were also examined. In particular, Hg, Cu and Cd effectively accumulated in King Bolete in contrast to the other elements. To evaluate the potential risk due to the accumulation of Hg in the fruiting bodies of King Bolete, the RfD and

PTWI were calculated. The data implied that Hg levels in King Bolete from the Plocka Dale area could pose a health risk to consumers.

Ouzouni et al. (2009) analysed a combination of MECs with nutritive components quantified by FAAS and HG–AAS to examine ten wild edible mushroom species from West Macedonia and Epirus (Greece). PCA, with the first two PCs explaining 58.1% of the total variance, was a useful tool for the differentiation of essential elements with respect to interspecies mushroom classification. For example, *L. nuda, R. delica* and *B. aureus* contained the highest levels of Zn and Cu, whereas Co was considerably concentrated in *R. largentii.* The concentrations of both of these heavy metals were correlated negatively with Cr, whose content reached the highest level. The levels of toxic elements, i.e., Pb, Cd and As, indicated that the study areas are not polluted. According to the authors, protein, carbohydrate and essential element profiles suggested that the mushroom species studied could be used in well-balanced diets without posing any health risk.

Sarikurkcu et al. (2021) combined ICP–OES with chemometric analysis to categorize eight Russula species from Soguksu National Parkin in Ankara (Turkey). PCA, with the first four PCs explaining 86.9% of the total variation, as well as HCA allowed them to distinguish between the species; i.e., six clusters were selected: Cluster 1: *R. velenovskyi*; Cluster 2: *R. vinosa*; Cluster 3: *R. parazurea* and *R. delica*; Cluster 4: *R. cyanoxantha;* Cluster 5: *R. turci;* and Cluster 6: *R. risigallina* and *R. olivacea.* On the basis of the HRI estimate, the consumption of *R. risigallina* and *R. olivacea* may pose a risk of adverse effects on human health because the Cr level exceeds the high reference DIM land HRI values. Moreover, the consumption of *R. velenovskyi*, which had a high HRI value with respect to Cd, may also pose a risk to human health.

Giannaccini et al. (2012) combined HG–ICP–OES with multivariate analysis to characterize samples of *Boletus edulis* and *Macrolepiota procera* related to associated topsoil collected from five distinct green microhabitats inside Lucca Province (Italy). The CA results revealed that the concentration of elements in *B. edulis* was not dependent on the location of the microhabitat, whereas in the case of *M. procera*, it was suggested that metal uptake in this saprotrophic species is sensitive to the microhabitat. In addition, interspecies differentiation was observed, e.g., such significant variations were reflected by higher levels of Ni, Rb and Se in *B. edulis* or As, Pb and Cu in *M. procera.*

Sarikurkcu et al. (2020) examined wild edible mushrooms from six families from Turkish sampling locations for elemental composition quantified by ICP–OES. The health risk indices for these elements indicated that safe limits were exceeded for Cd in *L. nuda*, *L. decastes*, *M. exscissa*, *R. albonigra*, *R. delica* and *T. terreum.* Such elevated levels were noted for Ni in *A. auricula-judae* and *S. luteus* as well as for Fe in *A. auricula-judae* and *M. paedida.* The differences and similarities between mushroom species were quantified by PCA and HCA. The first two PCs, which explained 61.94% of the total variation, revealed that the mushroom samples of four species from three families (Agaricaceae, Suillaceae and Tricholomataceae) were clearly differentiated from the other samples along PC2 according to the contents of Al, Ca, Co, Cr, Mg, Mn, Fe, Ni and Pb (Fig. 10.10).

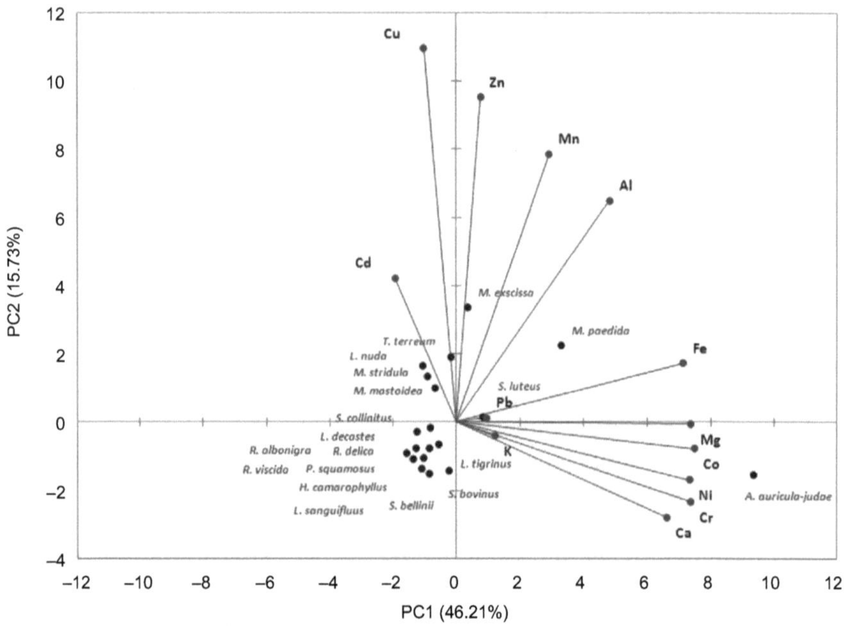

Fig. 10.10 Biplot of PC1 and PC2 of elemental analysis of wild edible mushroom species. (After Sarikurkcu C et al., 2020. Ecotoxicol Environ Saf 190:110058. https://doi.org/10.1016/j.ecoenv.2019.110058. With permission)

HCA grouped the mushroom samples into two clusters. The first cluster included only *A. auricula-judae* (Auriculariaceae family), which was clearly differentiated from the other studied mushrooms because it had the highest contents of Ca, Mg, Cr, Co, Fe and Ni and the lowest content of Cu. The second cluster contained the remaining species studied, including two subclusters; one of them consisted of *M. paedida,* but the second one was composed of the remaining mushroom species. Close positions of *S. luteus* and *L. tigrinus* (Suillaceae/Hygrophoraceae), *R. delica* and *L. decastes* (Russulaceae/Tricholomataceae), *S. belini* and *S. bovinus* (Suillaceae), *H. camarophyllus* and *S. collinitus* (Hygrophoraceae/Suillaceae), and *M. mastoidea* and *M. stridula* (Agaricaceae/Tricholomataceae) demonstrated similarities between these mushrooms with respect to elemental content.

Ivanić et al. (2021) combined HR–ICP–MS with chemometric processing to characterize 15 different edible wild mushroom species from the phylum Basidiomycota originating from the urban and forest areas of Croatia. The first two PCs, explaining 42.2% of the total variability, suggest that mushrooms of the same species or family from various regions have significantly different elemental levels, implying that different soil geochemical compositions and/or anthropogenic influences are the main factors determining their elemental contents. Notably, in the absence of possible anthropogenic impacts, soil geochemistry may be the major contributing factor determining these variations in the chemical composition of the mushroom. Elements such as As, Ni, Zn and macronutrients exhibit similar behaviours in certain mushroom species or families, suggesting the important influence of interspecies affinity for certain elements. However, As, Cd and Cu had the highest THQ values. The risk assessment revealed that 14 of the 19 mushrooms studied originated from both urban and forest areas and had HI values ˃1. This means that consumers of these mushrooms are exposed to a significant noncarcinogenic health risk due to high element levels.

Murati et al. (2019) used AAS to determine the contents of Ni, Cu, Zn, Fe, Mn, Cd and Pb in 15 species of wild mushrooms and soil samples from the thermal power plant Oslomej from 2012 to 2014. A significant correlation between the distance and levels of trace elements was obtained for ten mushroom species, particularly for Ni and Cu. The contents of elements such as Mn, Zn, Fe, Cd and Cu in fruiting bodies were primarily correlated with those in adjacent soils (total and extractable). In addition, CDA indicated the impact of the thermal power plant Oslomej on the trace elemental profile in *Boletus aestivalis*, *Russula cyanoxantha*, *Cantharellus cibarius* and *Hypholoma fasciculare*. Both the influence of pollution from the Oslomej thermal power plant and the soil composition affect the elemental composition of the fruiting bodies of mushrooms.

Nowakowski et al. (2021) determined the concentrations of Hg, Pb, Cd and As by ET–AAS and ICP–MS in the caps and stipes of seven species of edible wild mushrooms from the Podlaskie region and Northeast Poland, a region called the GLP. PCA, with the first three PCs explaining 99.5% of the total variance, was a useful tool for differentiating *L. perlatum* species from others because Pb is a discriminative variable. The highest levels of the four toxic elements were detected in *L. perlatum*. It was estimated that the caps and stipes of *L. perlatum* may pose a threat to human health related to Hg and As. Notably, the caps of both species, *L. perlatum* and *T. portentosum*, may pose health risks to consumers of these mushroom species.

Kojta et al. (2011) combined ICP–AES and CV–AAS to examine the MEC of Parasol mushrooms collected from two spatially related sites: the stands in the Augustowska Forest and the stands in the outskirts of Ełk town in the northeastern part of Poland. It has been reported that the Parasol mushroom effectively bioconcentrated Ag, Hg, Cu and K in fruiting bodies and moderately bioconcentrated Ca, Cd, Mg, Na, P, Rb and Zn. Caps of Parasol mushrooms from forested areas in northeastern Poland had various abilities to bioaccumulate the elements studied. Importantly, elevated levels of Cd and Pb in some samples may pose a risk of exceeding the permitted dose established in the EU for mushrooms. Additionally, consumption of these caps can lead to a greater tolerance limit for Hg and Cd, especially when they are consumed relatively frequently, i.e., more than once per week. Kojta et al. (2012) studied samples of edible Bay Bolete and underlying top soils from a forested mine dump at the site of a medieval gold and copper mine (near the town of Złotoryja in southern Poland) and at a reference forest site. The combination of ICP–OES and CV–AAS with multivariate techniques (CA and PCA) was useful for classifying mushroom samples. The Mann–Whitney *U* test revealed that mushrooms collected from the mine region accumulated higher levels of Ag and Pb in their flesh than did mushrooms from a reference site. PCA, with the first three

PCs corresponding to 72.2% of the total variance, was useful for differentiating Bay Bolete samples from Złotoryja and Tucholskie Forests with respect to their elemental profiles. It was concluded that elevated contents of both heavy metals in mushrooms were clearly associated with topsoil contamination by these elements.

Falandysz (2022) used Q–ICP–MS combined with PCA to characterize samples of *B. edulis* caps, stipes and whole fruiting bodies from various Polish regions ranging from the Baltic Sea south coast in the north to the Sudety and Tatra Mountains in the south. PCA, with the first four PCs explaining 88% of the total variance in the dataset, allowed the authors to differentiate mushrooms according to geographical origin. For example, samples from the Morąg site as well as Augustów Primeval Forest and Szczecinek were clearly differentiated from others. Moreover, samples originating from the Osowa Sudety Mountains and Augustów Primeval Forest sites were also distinguished. The applied approach allowed characterization of the elemental profile, including ΣREEs, of mushrooms collected across regions of Poland while considering the impact of natural or anthropogenic pollution.

Bosiacki et al. (2018) reported FAAS data for essential and toxic elements of various wild-growing *Agaricus bisporus* species from five regions (Poland) associated with CA processing. Owing to the use of this multivariate analysis technique, it was possible to differentiate the concentrations of essential elements (Zn, Cu, Mn and Fe) and pollutants (Cd, Cr, Pb and Ni) with respect to sampling sites as well as geographical regions. Among the 11 locations in Silesia, four locations contained mushrooms for which the level of Cd in fruiting bodies exceeded the permissible level.

Petkovšek and Pokorny (2013) extensively examined the contents of toxic metals such as Cd and Pb via ICP–AES in 699 samples of the fruiting bodies of 55 mushroom species, including 36 edible species, collected from 20,002,007 from two Slovenian regions subjected to anthropopression. Cd was present at relatively high levels at both sampling locations, especially near the former lead smelter, where 90% of the analysed species contained values exceeding tolerable values. The highest levels of Cd were detected in *Agaricus arvensis* and *Agaricus silvicola*, whereas the highest levels of Pb were detected in *Macrolepiota procera* and *Lycoperdon perlatum*. Furthermore, almost 70% of the analysed mushroom species near the lead smelter exceeded tolerable Pb levels. It was concluded that mushrooms from heavily polluted areas, such as near smelters, accumulate extremely high levels of metals and should therefore be excluded from human consumption.

Pankavec et al. (2019) investigated white button mushrooms processed under industrial conditions to determine the concentration of the elemental profile as quantified by ICP–DRC–MS and CV–AAS. Materials and methods concerning fungal materials originated from six Polish producers. The first five PCs explained 93% of the total variance. Some significant differences were detected in the concentration and composition of chemical elements in marinated white button mushrooms, depending on the producer. The first five PCs in the PCA accounted for 93% of the total variance for the total elemental content, and these elements could leak from mushrooms into the marinade at different rates from one side, whereas some of the elements could be absorbed (added to the vinegar or table salt as well as leaked from the vegetables if present) on the other side. Therefore, this chemometric technique could provide only incomplete information about the tendencies and detected the major differences in the technology or quality of marinades and/or containers used by producers. Moreover, conserved white button mushrooms do not have as many major essential elements or toxic metals such as Hg, As, Ag, Cd, Sb, Tl or Pb, as those detected in unprocessed mushrooms. It was also reported that the relatively higher content of Ag in some batches likely resulted from the quality of the substrate applied for mushroom cultivation, whereas elevated levels of Li, Rb, Cs, Cr, Al, U, V, As, and Mn as well as Ba and Sr may have been caused largely by the quality of the marinade.

Pająk et al. (2020) reported that although the Pb level in *Macrolepiota procera* seems to pose a threat to health, at present, knowledge regarding Pb bioaccessibility in mushrooms is quite limited. ICP–OES and ICP–AES combined with chemometric analysis were used to evaluate the degree of pollution of ten mushroom species belonging to three families in the northeastern part of the Upper Silesian Industrial Region (Poland). Multivariate analysis (MANOVA and CVA) revealed that mushrooms of the families *Boletaceae* and *Russulaceae* originating from heavily contaminated areas could be distinguished because they accumulate heavy metals more effectively than mushrooms of the family *Suillaceae*. Most of the studied mushrooms, especially *Imleria badia*, exhibited highly effective bioaccumulation of Cd. It was postulated that the levels of heavy metals in mushrooms originating from heavily polluted regions indicate a high potential health risk if the mushrooms are consumed. A spatial relationship was shown between the distance from the main metallic pollutant emitter in the studied area and the contents of heavy metals in the mushrooms.

Mleczek et al. (2016b) compared ten mostly edible aboveground mushroom species and ten wood-growing mushroom species originating from a heavily trafficked road in Poland with respect to their capacity to accumulate 26 trace elements in their fruiting bodies to examine mushroom differentiation. All the analyses were performed via ICP–OES. Compared with the wood-growing species, the aboveground species presented considerably greater levels of 12 elements, including Ag, As, Pb and Se, whereas an inverse relationship was obtained only for Au, Ba, and Sr. PCA, with the first two PCs explaining 38.66% of the total variance, revealed the following relationships for 20 tested mushroom species within both species groups: *Grifola frondosa*—Ag, As, Cu, Pb and Se; and *Laetiporus sulphureus*—Al, Co, Cr, Li, Ni and Mn. For the correlations between the mushroom species related to all the analysed elements jointly, the heatmap indicated the similarities and differences between the mushroom species. Among the studied mushroom species, the Pb level in *Macrolepiota procera* poses a health risk; however, according to current knowledge regarding Pb bioaccessibility from mushrooms, this threat is considerably reduced.

Zsigmond et al. (2019, 2020) studied the metal profiles of mushrooms and soil from six forests along an urbanization gradient and two city parks in Cluj-Napoca (Romania) via ICP–OES. Significant differences between the contents of some elements ($p < 0.05$) in the samples from the city compared with those in the samples from forests were observed. Moreover, a clear negative trend in the element accumulation capability of the mushrooms along the urbanization gradient was observed. It is postulated that the elevated carbonate content in the urban soil may cause the immobilization of metals in the soil. This specific mechanism can be explained by the presumption that the mushrooms studied have the potential to reduce Cd accumulation, indicating a positive environmental influence of urbanization in the case of the uptake of toxic metals by *R. cyanoxantha*. On this basis, the authors refute the hypothesis that urban soil pollution increases the element concentration in mushrooms.

10.2.3.4 Elemental Profile Assessment Related to Intermorphological Variability

The combination of ICP–MS and FT–IR with multivariate chemometric analysis (PCA and sPLS–DA) was used to categorize six species of porcini mushrooms, including caps and stipes, from various sampling sites in China. As shown in Fig. 10.11, the analytical data were successfully processed chemometrically. Many satisfactory results were obtained using discrimination analysis since accuracies >92.1 and 92.8% were achieved with respect to interspecies and interregional trends, respectively. For the differentiation between caps and stipes, they were classified with 96.7% accuracy. Polysaccharides significantly contribute to the identification of samples of different parts of mushrooms, i.e., caps and stipes. FT–IR spectroscopy data have been demonstrated to be sufficient for clearly classifying porcini mushroom samples, which may be useful for the quality assessment of edible mushrooms (Wang et al. 2019).

10.2.3.5 Assessment of the Elemental Profile Related to Annual/Seasonal Variability

Gucia et al. (2012b) examined variations and similarities in the elemental distribution patterns of the caps of Parasol mushrooms collected from the same two unpolluted sites in Poland between 2000 and 2003. The ICP–OES and CV–AAS data processed chemometrically demonstrated that elemental levels in mushrooms can fluctuate over time or over the lifespan of mycelia. Therefore, when the nutritional value of essential elements and the status of nonessential or toxic elements in Parasol's mushroom caps (and probably those of other edible mushroom species) are being assessed, the possible fluctuations in contents over time should be taken into account.

Chojnacka et al. (2013) found that the combination of ICP–AES and CV–AAS was an effective tool for examining the variability of MEC and the bioconcentration potential of yellow cracking bacteria collected from the same area over 3 years. Interannual differentiation of elements such as Al, Ba, Co, Hg, K, P, Pb, Sr and Zn in the fruiting bodies of the studied mushroom was achieved. This could be explained

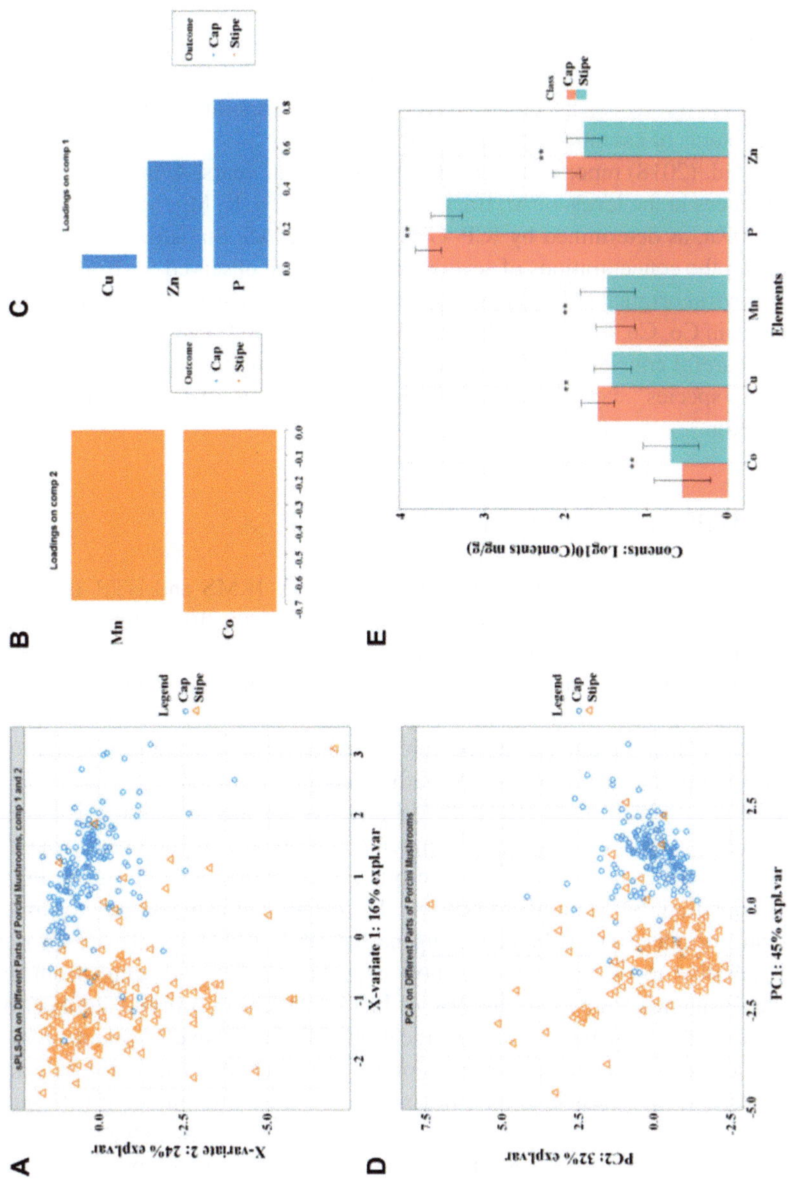

Fig. 10.11 Results of classification and important element contents in the caps and stipes of porcini mushrooms. (After Wang Y et al., 2019. J Food Sci 84:2112–2120. https://doi.org/10.1111/1750-3841.14715. With permission)

by unspecified differences in the availability of those elements from the soil substratum, as reflected by annual fluctuations.

Chudzyński et al. (2013) examined the MEC content by ICP–AES with USN/CFN in the flesh of *Suillus grevillei* mushrooms collected from the same site over two successive years. A differentiation between caps and stipes ($p < 0.05$) related to macro- and microelement profiles was achieved. In addition, interannual discrimination of the trace elemental contents of *Suillus grevillea* from the same site was performed. Therefore, it is necessary to consider such variability when assessing the nutritional status of essential elements in mushrooms.

Lipka et al. (2018) reported that over a period of four years, *A. muscaria* exhibited differences in the levels of Al, Ba, Cd, Ca, Co, Cu, Fe, K, Mg, Mn, Hg, Rb, Ag, Na, Sr and Zn, as determined by ICP–OES and CV–AAS. Statistically significant variations in the concentrations of several elements were observed for fruiting bodies. *A. muscaria* originated from the same area over the four-year period, indicating fluctuations in Co, Cu, Fe, K, Mn, Na and Zn in addition to Ca and Mg. This means that more careful consideration is needed when estimating the nutritional value of mushroom species caps and stipes in the case of detected interannual differentiation.

10.2.3.6 Verification of Authenticity/Adulteration

Krauß and Vetter (2020) used the combination of EA–IRMS and HTP–IRMS to assess the authenticity of six species of truffles originating from nine countries. Chinese truffles were clearly distinguished from all European samples by their significantly ($p < 0.05$) more negative δ^2H values. Samples of the second most expensive European species, *Tuber magnatum,* were more enriched in ^{15}N and more depleted in ^{13}C than the other samples were, resulting in significant differentiation. Therefore, SIRA is a useful tool for the authentication of truffles.

Segelke et al. (2020) used HR–ICP–MS and ICP–Q–MS to classify five commercially relevant species of truffles from ten countries to evaluate their elemental profiles. The original data obtained were processed chemometrically using PCA, *k*-NN and quadratic SVM multivariate tools. The classification models related to the species and origins were validated using nested cross validation. This approach made it possible to discriminate the most expensive *Tuber magnatum* from the remaining examined truffles. In the case of black truffles, an overall classification accuracy of 90.4% was attained, and falsification of the expensive *T. melanosporum* by *T. indicum* could be excluded. In terms of geographical provenance, for Italy and Spain, one-versus-all classification models were used to differentiate truffle samples from any other origin by 75.0 and 86.7%, respectively. By exclusively using the element ratios instead of the absolute element concentrations, the prediction was still possible according to an internal mathematical normalization scheme. The high potential of the methodology was achieved with respect to the element profile for parallel species and the authentication of the origin of truffles.

10.2.3.7 Assessment of Other Parameters

According to Stefanović et al. (2016), the contents of Cd, Cu, Pb and Zn measured via ICP–MS were significantly different in stipes than in caps, resulting in a discrimination between these parts of mushrooms originating from Serbia. PCA revealed that the metal content in various parts of the mushroom body corresponded to a five-PC model explaining 91.89% of the total variance. Using the PC2 and PC4 score values, the authors could partially discriminate two groups of samples related to caps and stipes. Additionally, the PCA data, with the first two PCs explaining approximately 85% of the total variance, formed three distinguished clusters of object scores corresponding to three geographical regions. Ba, Ni, Fe, Al, Cr and Sr had the greatest influence on the differentiation of samples, which was related to PC1, and Cd, Cu and Co were related to PC2. The HCA dendrogram confirmed that the first three phases of extraction corresponded to the most dominant forms of elements, as they were the most easily mobilized and accumulated by mushrooms from the soil substratum.

Cvetkovic et al. (2015) used the MEC content quantified by ICP–OES to study four wild edible mushroom species, i.e., scotch bonnet (*Marasmius oreades*), chanterelle (*Cantarellus cibarius*), porcini (*Boletus edulis*) and peppery milk cap (*Lactarius piperatus*), from unpolluted areas in Serbia. The mushrooms contained elevated levels of K. In addition, *M. oreades* was distinguished as having the highest level of almost all the elements, except for Cd, K, P and Pb. CA differentiated the elements into three well-shaped clusters differentiating toxic heavy metals such as Cd and Pb from the other analysed elements. PCA with the first two PCs explaining 95.51% of the total variance, confirmed the results obtained by CA. Notably, the studied mushrooms contain high levels of essential elements and are valuable sources of these nutrients.

Niedzielski et al. (2017) determined the MEC content, including REEs and PGEs, by ICP–OES in 12 mushroom species purchased between 2008 and 2016 from the Polish market; these mushrooms originated from both Poland (small-scale local production) and China (available in selected oriental or internet shops only). Both chemometric techniques, i.e., PCA (with the first two PCs explaining 42.49% of the total variance) and HCA, were used to process the obtained analytical data. Furthermore, Andrews curves demonstrated the transformation of multidimensional data into curves calculated for the mean concentration of elements in mushroom samples and for three randomly selected fruiting bodies of a particular mushroom species. This means that Andrews curves are a convenient tool for illustrating multidimensional data on food composition. The elemental profile was highly species-specific and did not follow a taxonomical pattern at the family level. The studied mushrooms were mostly distinguished by relatively high contents of PGEs, but the contents of toxic elements such as Al, As, Cd and Pb were within the safety limits established by the FAO/WHO.

Mleczek et al. (2021a, b) analysed various wild-growing mushroom species from selected sampling sites located in two macroregions, i.e., Lower and Upper Silesia and Wielkopolska Province (Poland). Chemometric techniques such as PCA and HCA were used to process the concentration data obtained via ICP–OES. According to Mleczek et al. (2021b), PCA, with the first two PCs accounting for 41.76% of the total variance, indicated interspecies differentiation with respect to their selective ability to accumulate one or more elements more effectively. For this reason, a heatmap was constructed, allowing the classification of all mushroom species with respect to the quantitative composition of 63 elements. For example, the following mushroom species were most similar: *M. esculenta* and *P. gilva*, *P. involutus* and *S. rugosoannulata*, *S. bovinus* and *T. equestre*. In addition, the heatmap enabled the selection of certain elements that were accumulated similarly by all the mushroom species. Different species may exhibited different levels of effectiveness in the accumulation different major or trace elements. For example, mushrooms belonging to the *Hymenochaetaceae* family were distinguished because they contain some of the most enriched elements, whereas mushrooms of the *Fomitopsidaceae* family have the lowest element levels (Mleczek et al. (2021a)).

Mleczek et al. (2021d) estimated toxicological risks and nutritional value of wild edible mushroom species with aspect of a half century monitoring study. ICP–MS data were processed chemometrically according to Niedzielski et al. (2023) to classify mushroom samples according to their genus and species. Species belonging to the *Lactarius* genus generally contained higher levels of major elements than *Russula*. The obtained results demonstrated a similarity in the elemental profile of the mushrooms within each individual genus. It was concluded that within each mushroom genus, a specific species having significantly greater element(s) levels suggests that both the mushroom species and genera may play important roles in regulating element accumulation from the soil to fruit bodies. Given that the studied mushrooms grew near trees, the presence of *Fagus sylvatica* may stimulate the accumulation of toxic elements in *Lactarius vellereus* fruiting bodies.

Mędyk et al. (2017) determined and examined the MEC content via ICP–MS in samples of edible mushrooms from four forest regions in the northern part of Poland. The first three PCs explained 87.64% of the total variance; however, no clear discrimination could be found in the levels of all the elements studied between the caps and stipes of fruiting bodies or interregional differentiation in the whole fruiting bodies of *S. imbricatus*. This finding could be explained by a lack of local or regional sources of emission/pollution, which could considerably the impact topsoil geochemistry and the studied mushrooms.

Siwulski et al. (2017) coupled ICP–OES data with chemometric analysis (HCA and a heatmap) to classify wood-growing mushroom species of the genus Pleurotus collected between 2013 and 2019 from 19 mushroom farms in Lower and Upper Silesia (Poland). Significant differences in element accumulation among various commercially cultivated species were achieved. The HCA dendrogram revealed three significantly diverse groups with similar abilities to accumulate elements within 7 years for the mushroom-producing regions. The species and strains formed three groups; i.e., the first group contained two units, the second group consisted of four units, and the third group also included four units. The content of most toxic elements was low, with the exception of Pb levels; therefore, Siwulski et al. (2017) recommended checking the elemental composition of the soil substrata used for the cultivation of these mushrooms.

According to Mirończuk-Chodakowska et al. (2019), 21 species of edible mushrooms collected from unpolluted areas of the eastern GLP territory were examined with respect to their Cu, Mn, Se and Zn contents to compare them with those of some popular species of cultivated mushrooms. The concentration data quantified by Q–ICP–MS were transformed by monovarietal statistics. Additionally, an assessment of the contribution of mushrooms to the daily intake of the studied essential elements was performed with consideration of their possible toxic impact associated with possibly excessive mushroom consumption from regions that are considered ecologically unpolluted. No toxicological risk related to mushroom consumption with the highest levels of Cu, Mn, Se and Zn (for the average adult) was detected.

Piatti et al. (2024) integrated ICP–MS and GC–MS with HS–SPME to assay the white truffle, the most expensive edible underground mushroom. The volatile and non-volatile components were identified and compared with the discriminators provided by sensory analysis. The non-volatile components were obtained as elemental profiles. The volatile profile mainly consisted of bis(methylthio)methane (78.72%) and other minor constituents related to seven odourant descriptors. Among the non-volatile components, the following deserve attention: K, P, S, Ca and Mg (97% of the elements investigated), which had higher concentrations than the other elements studied. It was concluded that this pioneering study was the first step, taking into consideration that truffles are still an unappreciated source of compounds with high commercial value.

10.3 Conclusions

In view of the considerable increase in the world's production of mushrooms, this increase will be accompanied by increased attention from producers, traders, consumers and authorities in the international food trade system to achieve a safe and high-quality product. It has been reported that mushrooms, which are mainly from the genera Agaricus, Macrolepiota, Lepista and Calocybe, can effectively accumulate toxic heavy metals even in unpolluted or mildly polluted regions. Therefore, in these circumstances, the possibility of evaluating toxicological risk or nutritional benefits has been limited, and the consumption of the species that bioaccumulate unsafe levels of heavy metals should be restricted. Among others, the assessment of the intermicrohabitat and interspecies differentiation was possible. The environmental consequences of the Chernobyl and Fukushima disasters involve, to a certain extent, to the radioactive status of mushrooms.

References

Aloupi M, Koutrotsios G, Koulousaris M, Kalogeropoulos N (2012) Trace metal contents in wild edible mushrooms growing on serpentine and volcanic soils on the Island of Lesvos. Greece Ecotox Environ Saf 78:184–194. https://doi.org/10.1016/j.ecoenv.2011.11.018

Bosiacki M, Siwulski M, Sobieralski K, Krzebietke S (2018) The content of selected heavy metals in fruiting bodies of *Agaricus bisporus* (Lange) Imbach. wild growing in Poland. J Elem 23:875–886. https://doi.org/10.5601/jelem.2017.22.3.1529

Brzezicha-Cirocka J, Grembecka M, Grochowska I, Falandysz J, Szefer P (2019) Elemental compositioon of selected species of mushrooms based on a chemometric evaluation. Ecotoxic Environ Saf 173:353–365. https://doi.org/10.1016/j.ecoenv.2019.02.036

Chen L-P, Zhu H-Y, Li Y-F, Zhang Y, Zhang W, Yang L-C, Yin H, Dong C-Y, Wang Y (2021) Combining multielement analysis and chemometrics to trace the geographical origin of *Thelephora ganbajun*. J Food Compos Anal 96:103699. https://doi.org/10.1016/j.jfca.2020.103699

Chojnacka A, Jarzyńska G, Lewandowska M, Nnorom IC, Falandysz J (2013) Trace elements of yellow-cracking bolete [*Xerocomus subtomentosus* (L.) Quél.] collected at the same site over three years. Fresenius Environ Bull 22:2707–2712

Chudzyński K, Falandysz J (2008) Multivariate analysis of elements content of Larch Bolete (*Suillus grevillei*) mushroom. Chemosphere 73:1230–1239. https://doi.org/10.1016/j.chemosphere.2008.07.055

Chudzyński L, Jarzyńska G, Falandysz J (2013) Cadmium, lead and some other trace elements in Larch Bolete mushrooms (*Suillus grevillei*) (Klotzsch) Sing., collected from the same site over two years. Food Addit Contam Part B 6:249–253. https://doi.org/10.1080/19393210.2013.807881

Chung I-M, Kim J-T, Moon H-S, Kim Y-J, Kim S-H (2022) Geographical origin discrimination of *Agaricus bisporus* produced by the complete medium: a pilot study in South Korea. Food Chem 386:132820. https://doi.org/10.1016/j.foodchem.2022.132820

Cocchi L, Vescovi L, Petrini LE, Petrin O (2006) Heavy metals in edible mushrooms in Italy. Food Chem 98:277–284. https://doi.org/10.1016/j.foodchem.2005.05.068

Cvetkovic JS, Mitic VD, Stankov-Jovanovic VP, Dimitrijevic MV, Nikolic-Mandic SD (2015) Elemental composition of wild edible mushrooms from Serbia. Anal Lett 48:2107–2121. https://doi.org/10.1080/00032719.2015.1010118

Dimitrijevic MV, Mitic VD, Cvetkovic JS, Stankov Jovanovic VP, Mutic JJ, Nikolic Mandic SD (2016) Update on element content profiles in eleven wild edible mushrooms from family *Boletaceae*. Eur Food Res Technol 242:1–10. https://doi.org/10.1007/s00217-015-2512-0

Drewnowska M, Falandysz J (2015) Investigation on mineral composition and accumulation by popular edible mushroom common chanterelle (*Cantharellus cibarius*). Ecotoxicol Environ Saf 113:9–17. https://doi.org/10.1016/j.ecoenv.2014.11.028

El Sheikha AF, Hu D-M (2018) How to trace the geographic origin of mushrooms? Trends Food Sci Technol 78:292–303. https://doi.org/10.1016/j.tifs.2018.06.008

Falandysz J (2022) Nutritional and other trace elements and their associations in raw King Bolete mushrooms, *Boletus edulis*. Int J Environ Res Public Health 19:417. https://doi.org/10.3390/ijerph19010417

Falandysz J, Kunito T, Kubota R, Bielawski L, Frankowska A, Tanabe S (2008) Multivariate characterization of elements accumulated in King Bolete *Boletus edulis* mushroom at lowland and high mountain regions. J Environ Sci Heaith Part A 43:1692–1699. https://doi.org/10.1080/10934520802330206

Falandysz J, Zhang J, Wang Y-Z, Saba M, Krasińska G, Wiejak A, Li T (2015) Evaluation of mercury contamination in fungi Boletus species from latosols, lateritic red earths, and red and yellow earths in the circum-Pacific mercuriferous belt of Southwestern China. PLoS One 10(11):e0143608. https://doi.org/10.1371/journal.pone.0143608

Falandysz J, Chudzińska M, Barałkiewicz D, Drewnowska M, Hanć A (2017a) Toxic elements and bio-metals in *Cantharellus* mushrooms from Poland and China. Environ Sci Pollut Res 24:11472–11482. https://doi.org/10.1007/s11356-017-8554-z

Falandysz J, Sapkota A, Dryżałowska A, Mędyk M, Feng X (2017b) Analysis of some metallic elements and metalloids composition and relationships in parasol mushroom *Macrolepiota procera*. Environ Sci Pollut Res 24:15528–15537. https://doi.org/10.1007/s11356-017-9136-9

Frankowska A, Ziółkowska J, Bielawski L, Falandysz J (2010) Profile and bioconcentration of minerals by King Bolete (*Boletus edulis*) from the Płocka Dale in Poland. Food Addit Contam Part B 3:1–6. https://doi.org/10.1080/19440040903505232

Gebrelibanos M, Megersa N, Taddesse AM (2016) Levels of essential and non-essential metals in edible mushrooms cultivated in Haramaya, Ethiopia. Inter J Food Contam 3:2. https://doi.org/10.1186/s40550-016-0025-7

Giannaccini G, Betti L, Palego L, Mascia G, Schmid L, Lanza M, Mela A, Fabbrini L, Biondi L, Lucacchini A (2012) The trace element content of top-soil and wild edible mushroom samples collected in Tuscany, Italy. Environ Monit Assess 184:7579–7595. https://doi.org/10.1007/s10661-012-2520-5

Gucia M, Jarzyńska G, Anna K, Kojta AK, Falandysz J (2012a) Temporal variability in 20 chemical elements content of parasol mushroom (*Macrolepiota procera*) collected from two sites over a few years. J Environ Sci Health, Part B 47:81–88. https://doi.org/10.1080/03601234.2012.611433

Gucia M, Jarzyńska G, Rafał E, Roszak M, Kojta AK, Osiej I, Falandysz J (2012b) Multivariate analysis of mineral constituents of edible Parasol Mushroom (*Macrolepiota procera*) and soils beneath fruiting bodies collected from northern Poland. Environ Sci Pollut Res 19:416–431. https://doi.org/10.1007/s11356-011-0574-5

Ivanić M, Turk MF, Tkalčec Z, Fiket Ž, Mešić A (2021) Distribution and origin of major, trace and rare earth elements in wild edible mushrooms: urban vs. forest areas. J Fungi 7:1068. https://doi.org/10.3390/jof7121068

Kalač P (2001) A review of edible mushroom radioactivity. Food Chem 75:29–35. https://doi.org/10.1016/S0308-8146(01)00171-6

Kalač P (2009) Chemical composition and nutritional value of European species of wild growing mushrooms: a review. Food Chem 113:9–16. https://doi.org/10.1016/j.foodchem.2008.07.077

Kalač P (2010) Trace element contents in European species of wild growing edible mushrooms: a review for the period 2000-2009. Food Chem 122:2–15. https://doi.org/10.1016/j.foodchem.2010.02.045

Kalač P (2019) Mineral composition and radioactivity of edible mushrooms. Academic Press. ISBN 10: 0128175656; ISBN 13: 9780128175651

Kalač P, Svoboda L (2000) A review of trace element concentrations in edible mushrooms. Food Chem 69:273–281. https://doi.org/10.1016/S0308-8146(99)00264-2

Kojta AK, Falandysz J (2016) Metallic elements (Ca, Hg, Fe, K, Mg, Mn, Na, Zn) in the fruiting bodies of *Boletus badius*. Food Chem 200:206–214. https://doi.org/10.1016/j.foodchem.2016.01.006

Kojta AK, Gucia M, Jarzyńska G, Lewandowska M, Zakrzewska A, Falandysz J, Zhang D (2011) Phosphorous and certain metals in Parasol Mushrooms (*Macrolepiota procera*) and soils from the Augustowska Forest and Ełk region in north-eastern Poland. Fresenius Environ Bull 20:3044–3052. https://www.researchgate.net/publication/299125959

Kojta AK, Jarzyńska G, Falandysz J (2012) Mineral composition and heavy metal accumulation capacity of Bay Bolete (*Xerocomus badius*) fruiting bodies collected near a former gold and copper mining area. J Geochem Explor 121:76–82. https://doi.org/10.1016/j.gexplo.2012.08.004

Kojta AK, Gucia M, Krasińska G, Saba M, Nnorom IC, Falandysz J (2016) Mineral constituents of edible field parasol (*Macrolepiota procera*) mushrooms and the underlying substrate from upland regions of Poland: bioconcentration potential, intake benefits, and toxicological risk. Pol J Environ Stud 25:2445–2460. https://doi.org/10.15244/pjoes/62997

Kokkoris V, Massas I, Polemis E, Koutrotsios G, Zervakis GI (2019) Accumulation of heavy metals by wild edible mushrooms with respect to soil substrates in the Athens metropolitan area (Greece). Sci Total Environ 685:280–296. https://doi.org/10.1016/j.scitotenv.2019.05.447

Krauß S, Vetter W (2020) Geographical and species differentiation of truffles (*Tuber* spp.) by means of stable isotope ratio analysis of light elements (H, C, and N). J Agric Food Chem 68:14386–14392. https://doi.org/10.1021/acs.jafc.0c01051

Lim JM, Lee JH, Sun GM, Moon JH, Chung YS, Kim KH (2012) The geographical origin and chemical composition in *phellinus* mushrooms measured by instrumental neutron activation analysis. J Radioanal Nucl Chem 291:451–455. https://doi.org/10.1007/s10967-011-1242-4

Lipka K, Saba M, Falandysz J (2018) Preferential accumulation of inorganic elements in *Amanita muscaria* from North-Eastern Poland. J Environ Sci Health Part A 53:968–974. https://doi.org/10.1080/10934529.2018.1470805

Liu B, Huang Q, Cai H, Guo X, Wang T, Gui M (2015) Study of heavy metal concentrations in wild edible mushrooms in Yunnan Province, China. Food Chem 188:294–300. https://doi.org/10.1016/j.foodchem.2015.05.010

Liu H, Liu H, Li J, Wang Y (2022) Review of recent modern analytical technology combined with chemometrics approach researches on mushroom discrimination and evaluation. Crit Rev Anal Chem 25:1–24. https://doi.org/10.1080/10408347.2022.2124839

Malinowska E, Szefer P, Falandysz J (2004) Metals bioaccumulation by bay bolete, *Xerocomus badius*, from selected sites in Poland. Food Chem 84:405–416. https://doi.org/10.1016/S0308-8146(03)00250-4

Mędyk M, Chudzińska M, Barałkiewicz D, Falandysz J (2017) Specific accumulation of cadmium and other trace elements in *Sarcodon imbricatus* using ICP-MS with a chemometric approach. J Environ Sci Health B 52:361–366. https://doi.org/10.1080/03601234.2017.1283145

Mirończuk-Chodakowska I, Socha K, Zujko ME, Terlikowska KM, Borawska MH, Witkowska AM (2019) Copper, manganese, selenium and zinc in wild-growing edible mushrooms from the eastern territory of "Green Lungs of Poland": nutritional and toxicological implications. Int J Environ Res Public Health 16:3614. https://doi.org/10.3390/ijerph16193614

Mleczek M, Niedzielski P, Siwulski M, Rzymski P, Gąsecka M, Goliński P, Kozak L, Kozubik T (2016a) Importance of low substrate arsenic content in mushroom cultivation and safety of final food product. Eur Food Res Technol 242:355–362. https://doi.org/10.1007/s00217-015-2545-4

Mleczek M, Niedzielski P, Kalač P, Budka A, Siwulski M, Gąsecka M, Rzymski P, Magdziak Z, Sobieralski K (2016b) Multielemental analysis of 20 mushroom species growing near a heavily trafficked road in Poland. Environ Sci Pollut Res 23:16280–16295. https://doi.org/10.1007/s11356-016-6760-8

Mleczek M, Gąsecka M, Budka A, Siwulski M, Mleczek P, Magdziak Z, Budzyńska S, Niedzielski P (2021a) Mineral composition of elements in wood-growing mushroom species collected from of two regions of Poland. Environ Sci Pollut Res 28:4430–4442. https://doi.org/10.1007/s11356-020-10788-y

Mleczek M, Budka A, Kalač P, Siwulski M, Niedzielski P (2021b) Family and species as determinants modulating mineral composition of selected wild-growing mushroom species. Environ Sci Pollut Res 28:389–404. https://doi.org/10.1007/s11356-020-10508-6

Mleczek M, Budka A, Siwulski M, Mleczek P, Budzyńska S, Proch J, Gąsecka M, Niedzielski P, Rzymski P (2021c) A comparison of toxic and essential elements in edible wild and cultivated mushroom species. Eur Food Res Technol 247:1249–1262. https://doi.org/10.1007/s00217-021-03706-0

Mleczek M, Siwulski M, Budka A, Mleczek P, Budzyńska S, Szostek M, Kuczyńska-Kippen N, Kalač P, Niedzielski P, Gąsecka M, Goliński P, Magdziak Z, Rzymski P (2021d) Toxicological risks and nutritional value of wild edible mushroom species—a half century monitoring study. Chemosphere 263:128095. https://doi.org/10.1016/j.chemosphere.2020.128095

Murati E, Hristovski S, Karadelev M, Melovski L (2019) The impact of thermal power plant Oslomej (Kichevo valley, Macedonia) on heavy metal contents (Ni, Cu, Zn, Fe, Mn, Pb, Cd) in fruiting bodies of 15 species of wild fungi. Air Qual Atmos Health 12:353–358. https://doi.org/10.1007/s11869-019-00665-0

Niedzielski P, Mleczek M, Budka A, Rzymski P, Siwulski M, Jasińska A, Gąsecka M, Budzyńska S (2017) A screening study of elemental composition in 12 marketable mushroom species accessible in Poland. Eur Food Res Technol 243:1759–1771. https://doi.org/10.1007/s00217-017-2881-7

Niedzielski P, Szostek M, Budka A, Budzyńska S, Siwulski M, Proch J, Kalač P, Mleczek M (2023) *Lactarius* and *Russula* mushroom genera—similarities/differences in mineral composition within the Russulaceae family. J Food Compos Anal 115:104970. https://doi.org/10.1016/j.jfca.2022.104970

Nowakowski P, Markiewicz-Żukowska R, Soroczynska J, Puścion-Jakubik A, Mielcarek K, Borawska MH, Socha K (2021) Evaluation of toxic element content and health risk assessment of edible wild mushrooms. J Food Compos Anal 96:103698. https://doi.org/10.1016/j.jfca.2020.103698

Ouzouni PK, Petridis D, Koller WD, Riganakos KA (2009) Nutritional value and metal content of wild edible mushrooms collected from West Macedonia and Epirus, Greece. Food Chem 115:1575–1580. https://doi.org/10.1016/j.foodchem.2009.02.014

Pająk M, Gąsiorek M, Jasik M, Halecki W, Otremba K, Pietrzykowski M (2020) Risk assessment of potential food chain threats from edible wild mushrooms collected in forest ecosystems with heavy metal pollution in Upper Silesia, Poland. Forests 11:1240. https://doi.org/10.3390/f11121240

Pankavec S, Hanć A, Barałkiewicz D, Dryżałowska A, Zhang J, Falandysz J (2019) Mineral constituents of conserved white button mushrooms: similarities and differences. Rocz Panstw Zakl Hig 70:15–25. https://doi.org/10.32394/rpzh.2019.0051

Petkovšek SAS, Pokorny B (2013) Lead and cadmium in mushrooms from the vicinity of two large emission sources in Slovenia. Sci Total Environ 443:944–954. https://doi.org/10.1016/j.scitotenv.2012.11.007

Piatti D, Marconi R, Caprioli G, Zannotti M, Giovannetti R, Sagratini G (2024) White Acqualagna truffle (*Tuber magnatum* Pico): evaluation of volatile and non-volatile profiles by GC-MS, sensory analyses and elemental composition by ICP-MS. Food Chem 439:138089. https://doi.org/10.1016/j.foodchem.2023.138089

Sarikurkcu C, Popović-Djordjević J, Solak MH (2020) Wild edible mushrooms from Mediterranean region: metal concentrations and health risk assessment. Ecotoxicol Environ Saf 190:110058. https://doi.org/10.1016/j.ecoenv.2019.110058

Sarikurkcu C, Akata I, Tepe B (2021) Metal concentration and health risk assessment of eight *Russula* mushrooms collected from Kizilcahamam-Ankara, Turkey. Environ Sci Pollut Res 28:15743–15754. https://doi.org/10.1007/s11356-020-11833-6

Segelke T, von Wuthenau K, Neitzke G, Müller M-S, Fischer M (2020) Food authentication: species and origin determination of truffles (*Tuber* spp.) by inductively coupled plasma mass spectrometry and chemometrics. J Agric Food Chem 68:14374–14385. https://doi.org/10.1021/acs.jafc.0c02334

Siwulski M, Mleczek M, Rzymski P, Budka A, Jasińska A, Niedzielski P, Kalač P, Gąsecka M, Budzyńska S, Mikołajczak P (2017) Screening the multi-element content of *Pleurotus* mushroom species using inductively coupled plasma optical emission spectrometer (ICP-OES). Food Anal Methods 10:487–496. https://doi.org/10.1007/s12161-016-0608-1

Sommer K, Krauß S, Vetter W (2021) Stable nitrogen and carbon isotope signatures ($\delta^{15}N$ and $\delta^{13}C$ values in ‰) of fresh and preserved button mushrooms (*Agaricus bisporus*) from the German market. Anal Lett 54:1401–1413. https://doi.org/10.1080/00032719.2020.1803346

Stefanović V, Trifković J, Mutić J, Tešić Ž (2016) Metal accumulation capacity of parasol mushroom (*Macrolepiota procera*) from Rasina region (Serbia). Environ Sci Pollut Res 23:13178–13190. https://doi.org/10.1007/s11356-016-6486-7

Wang Y, Li J, Liu H, Fan M, Wang Y (2019) Species and geographical origins discrimination of porcini mushrooms based on FT-IR spectroscopy and mineral elements combined with sparse partial least square-discriminant analysis. J Food Sci 84:2112–2120. https://doi.org/10.1111/1750-3841.14715

Yao S, Li J, Li T, Duan Z, Wang Y (2019) Geographical traceability of *Boletaceae* mushrooms using data fusion of FT-IR, UV, and ICP-AES combined with SVM. Int J Food Prop 22:414–426. https://doi.org/10.1080/10942912.2019.1588299

Zsigmond AR, Kántor I, May Z, Urák I, Heberger K (2019) Data on elemental composition of *Russula cyanoxantha* along an urbanization gradient in Cluj-Napoca (Romania). Data Brief 27:104572. https://doi.org/10.1016/j.dib.2019.104572

Zsigmond R, Kántor I, May Z, Urák I, Héberger K (2020) Elemental composition of *Russula cyanoxantha* along an urbanization gradient in Cluj-Napoca (Romania). Chemosphere 238:124566. https://doi.org/10.1016/j.chemosphere.2019.124566

Chapter 11
Fruits and Their Products

Contents

11.1 Introduction

Fruits contain many essential nutrients that people need. Fruits originate from plants and include high levels of carbohydrates and fibre—pectin. Depending on the variety of fruit, they can contain powerful antioxidant anthocyanins and isoflavones, mineral elements (K, Ca and Mg) and vitamins (vitamin C, folate, thiamin, riboflavin, niacin, vitamin B_6, folate and vitamin A). Appropriate analytical methods/techniques coupled with multivariate analysis are effective approaches for classifying fruit complexity with respect to geographical and botanical origin.

Badenes and Byrne (2002) presented information on applied plant breeding using current advances in the field, starting from the efficient application of genetic resources and ending with the impact of biotechnology. Important aspects of fruit breeding, such as trends, the marketing and protection of new varieties and the health benefits of fruits and new crops, were considered.

Supplementary Information The online version contains supplementary material available at https://doi.org/10.1007/978-3-031-99687-0_11.

P. Szefer, *Chemometrics in Food Control Based on Elemental Analysis*, https://doi.org/10.1007/978-3-031-99687-0_11

The dynamic progress of analytical methodologies for testing authenticity enables the application of multielemental and stable isotope analysis on the basis of atomic spectroscopy. This approach is especially useful for the identification of organically or conventionally cultivated edible plants, including fruits and their products. However, these techniques are considerably limited by the potential multiplicity of fertilization practices used for organic and conventional plant production. Consequently, analytical techniques integrated with chemometric tools are still intentional and are recommended for developing computational models that enable the accurate classification of agricultural resources with respect to geographic origin (Laursen et al. 2014).

According to such trends and recommendations, the combined use of various instrumental tools can improve the accuracy of classification or prediction of food identification, including for edible plants, in contrast to the application of a single analytical technique. Three levels of data fusion are known, and among them, low- and mid-level fusion approaches are the most commonly used. Data fusion is expected to not only contribute to global classification/prediction but also minimize the uncertainty of each individual result and allow more accurate prediction (Borràs et al. 2015).

A thorough knowledge of the chemical composition of fruit juice and access to dedicated databases are crucial requirements for the reliable selection of analytical tools and thorough interpretation of the obtained data. Moreover, the size of the fruit juice market has resulted in a limited number of qualified laboratories and relatively high control costs, which could be responsible for the low control level in markets (Rinke 2016).

Various analytical methodologies have been developed to trace and evaluate the authenticity of fruits and their products to detect falsification. For example, in the case of juices, the fraudulent addition of water and sugars, juice-to-juice adulteration, false labelling and the use of juice concentrates occur. The challenging tasks in the possible application of food fingerprinting tools for official food control include the establishment of consistent operating and standardization procedures and chemometric assessment as well as the development of appropriate validation protocols for computational models (Dasenaki and Thomaidis 2019).

Fruits and especially fruit juices are very popular worldwide. In this context, they have been the object of increased adulteration and fraudulent activities for economic profit. Since fruit juice quality is a key food quality and safety parameter, many analytical methods or techniques have been used in the assessment of fruit juice authenticity/adulteration. The targeted markers include SIR, MEC, phenolics, carbohydrates, organic acids, volatile components, DNA, amino acids, proteins and carotenoids. Untargeted techniques include ICP–MS, ICP–OES, L/GC–MS, NMR, FS and electronic sensors (Xu et al. 2022).

According to Drivelos and Georgiou (2012), an increasing number of scientific papers published in the early 2000s involved investigations of elemental contents and stable isotope ratios combined with strontium and lead isotope ratios as well as multielemental contents. This classification has been used to evaluate the interspatial and intervariety/cultivar differentiation of edible plants. The determination of trace, micro- and macroelement distributions in food products with different, sometimes unknown origins is important. Numerous analytical methods/techniques combined with multivariate analysis have been successfully used for the detection of adulteration of different food products, including fruits (Szefer 2007; Grembecka 2007; Grembecka and Szefer 2011, 2016).

11.2 Chemometric Assessment of Fruits and Their Products

11.2.1 Geographical Differentiation

11.2.1.1 European Countries

Fruit

Longobardi et al. (2015) combined EA–IRMS with EN of three varieties of sweet cherries from two Italian regions, i.e., Apulia and Emilia Romagna, to identify the geographical provenance of the two sample groups. In the case of the distribution pattern of sample scores in the subspace PC1 vs. PC2, no discrimination between samples corresponding to the two geographical regions was observed. Moreover, no single PC was sufficiently suitable to geographically differentiate the samples, resulting in only partial grouping of the object scores for each VES. However, the LDA results demonstrated that the EN–IRMS combination provided better CV prediction ability (92.3%) than the EN and IRMS methods, which were performed separately. To validate the results, they tested the models using an external set of samples with very satisfactory results.

11.2.1.2 Asiatic Countries

Fruit

According to Zhang et al. (2019), the use of GF–AAS with ICP–AES to study apple samples collected from five Chinese regions is an appropriate method associated with multivariate processing using PCA, LDA and BP–ANN. The first three PCs obtained by PCA explained 53.48% of the total variance, and the LDA and BP–ANN obtained 70.0 and 82.7% classification rates, respectively. It was postulated that apples could be classified on the basis of their elemental composition and constituted a basis for further exploration of the geographical provenance.

11.2.1.3 American Countries

Juice

Gaiad et al. (2016) reported that examination of the MEC content via ICP–MS allowed the study of lemon juice (prepared from three varieties) from four

Argentinean provinces. Chemometric models such as PCA, LDA, PLS–DA, k-NN, RF and SVM were useful for differentiating the juice samples. PCA, with the first two PCs accounting for 40.2% of the total variance, showed systematic differentiation of samples originating from the NE and NW regions. Moreover, SVM, as a nonlinear model, with the RBF kernel function was more effective than RF–k-NN, LDA and PLS–DA. SVM achieved an overall classification accuracy of 76.2%, followed by RF with 71%. This satisfactory discrimination likely resulted from the flexibility and ability of the algorithm for building a generalized model, even when it is used for small training groups.

11.2.1.4 Oceania Countries

Fruit

McComb et al. (2019) reported that the δ^2H data quantified by TC/EA–IRMS allowed the modelling (owing to the use of three models of LRA, MLR and kFCV) of a spatially representative set of blackberry leaf samples from different sampling sites on the north and south islands of New Zealand. The three models were constructed and compared for their accuracy and precision and with respect to samples assigned for their geographic origin. It was concluded that a δ^2H isoscape of a blackberry leaf is an example application for identifying the geographic provenance of plant materials in New Zealand. According to the authors, one of the models based on a precipitation isoscape was similar with respect to the accuracy and precision of assignment to the model built from the environmental parameters and provides a direction for determining valid isoscapes for future plant materials.

11.2.1.5 Countries Across Multiple Continents

Fruit

Lo Turco et al. (2020) used ICP–MS data to classify dried fig samples according to their geographical provenance. Using PCA (with the first two PCs explaining 82.85% of the total variance), it was possible to discriminate figs grown in three countries (Italy, Greece and Turkey) according to geographical origin (Fig. 11.1). It was also reported that good fig quality is related to essential trace element levels, highlighting the safety of figs.

Juice

Rummel et al. (2010) combined EA–IRMS with TIMS, providing SIR and $^{87}Sr/^{86}Sr$ values of authentic orange juices from different regions of 16 countries in Europe, America and Africa. LDA enabled the splitting of samples into three broad groups corresponding to different geographical regions, i.e., Mediterranean countries, South America and Middle America with Florida, with a correct classification of 88.5%. The most discriminatory loadings were δ^2H and $\delta^{87}Sr$. The discrimination of juice samples within European countries with respect to Greece, Israel, Italy and Spain was achieved with a correct classification of 75%.

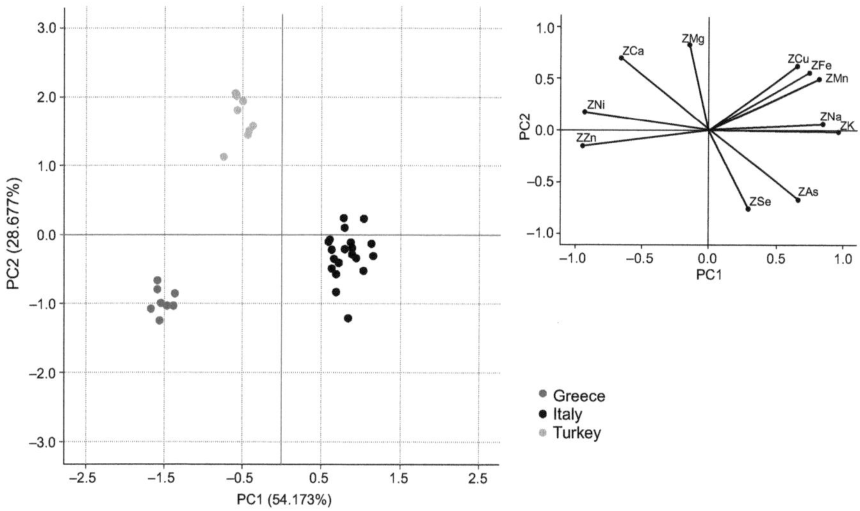

Fig. 11.1 2D score plots for 36 dried fig samples categorized according to geographical area. Insert: loading plot of minerals in the spaces defined by PC1 and PC2. (After Lo Turco V et al., 2020. J Food Compos Anal 90:103503. https://doi.org/10.1016/j.jfca.2020.103503. With permission)

11.2.2 Intercultivar/Brand Differentiation

Fruit

Mitić et al. (2012) integrated ICP–OES with a chemometric tool to process three sour cherry and three table grape cultivars from Serbia. PCA (with the first two PCs achieving 100% of the total variance separately for cherry and grape) was used to evaluate the distribution of elements. ANOVA revealed more or less significant differences in the mean content between the sour cherry and table grape cultivars. With as a benchmark, no health risks resulting from the consumption of fruits were reported.

Grembecka and Szefer (2013) used a combination of FAAS and spectrophotometric data with both FA and CA to examine 98 commercially available products of fresh fruits from 13 countries purchased from the local market. Owing to the use of the FA technique, with the first two factors (F) explaining 45.9% of the total variability, two types of classification were considered, i.e., based on the botanical type of fruit and on the fruit botanical family (Fig. 11.2). As demonstrated, several botanical families were well differentiated from each other. For example, P, Zn and Cr (described by higher values of F1) are descriptors of the *Ericaceae*, *Grossulariaceae*, *Lauraceae* and *Actinidiaceae* families, whereas *Cucurbitaceae* samples (with lower values of F1) identified by Na. *Ericacea*e samples (identified by the lowest F2) were differentiated by Mn and Fe, but *Lauracea*, *Musaceae*, *Vitaceae* and *Cucurbitacea*e (with higher F2) were described by Cu, Zn and Fe. It was also possible to classify fruits according to their type, i.e., accessory, berry, pip and stone fruits. Generally, CA results were in agreement with the FA data. Therefore, this method can also be

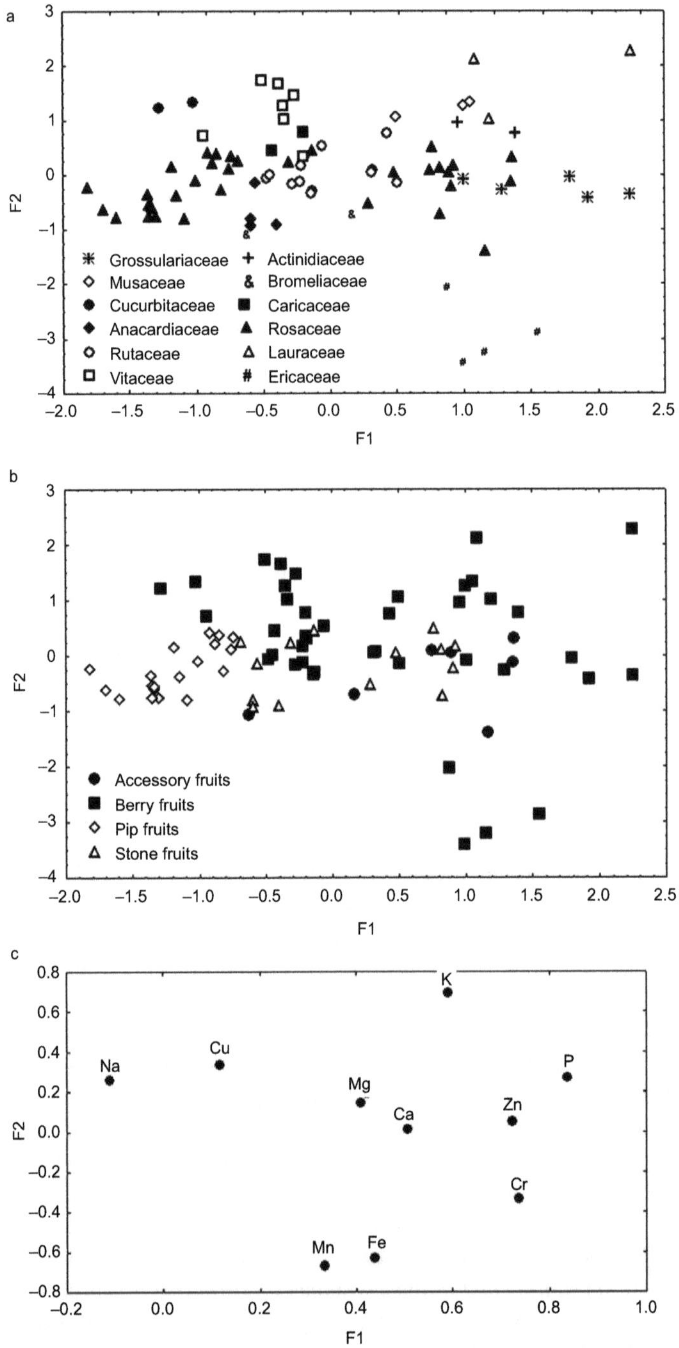

Fig. 11.2 (**a**) Scatterplot of object scores of the two discriminant functions of fresh fruit samples classified according to their botanical family. (**b**) Scatterplot of object scores of the two discriminant functions of fresh fruit samples classified according to their fruit type. (**c**) Scatterplot of loadings for 10 elements in fresh fruit samples. (After Grembecka M and Szefer P, 2013. Environ Monit Assess 185:9139–9160. https://link.springer.com/article/10.1007/s10661-013-3242-z. With permission)

a useful tool for distinguishing samples of various types as well as their botanical provenances. As shown in Fig. 11.3, the CA dendrogram is an effective tool for discriminating fruit samples belonging to the same botanical families, i.e., *Grossulariaceae*, *Actinidiaceae*, *Musaceae*, *Cucurbitaceae*, *Caricaceae*, *Rosaceae*, *Rutaceae*, *Lauraceae* and *Vitaceae* (Grembecka and Szefer 2013).

11.2.3 Interregional and/or Intercultivar Differentiation

Fruit

According to Mimmo et al. (2015), EA–IRMS of four apple varieties and their three sub-fractions (peel, pulp and seed) from orchards located in four districts of Italy was performed to determine the $\delta^{13}C$ and $\delta^{15}N$ contents. The obtained data were processed chemometrically (via PCA and LDA) to differentiate fruit samples according to their geographical origin and fruit cultivar. The first two PCs, which explained 78.6% of the total variance, partially discriminated the apple samples according to their origin. However, PCA, with contributions from PC1 (33.6%) and

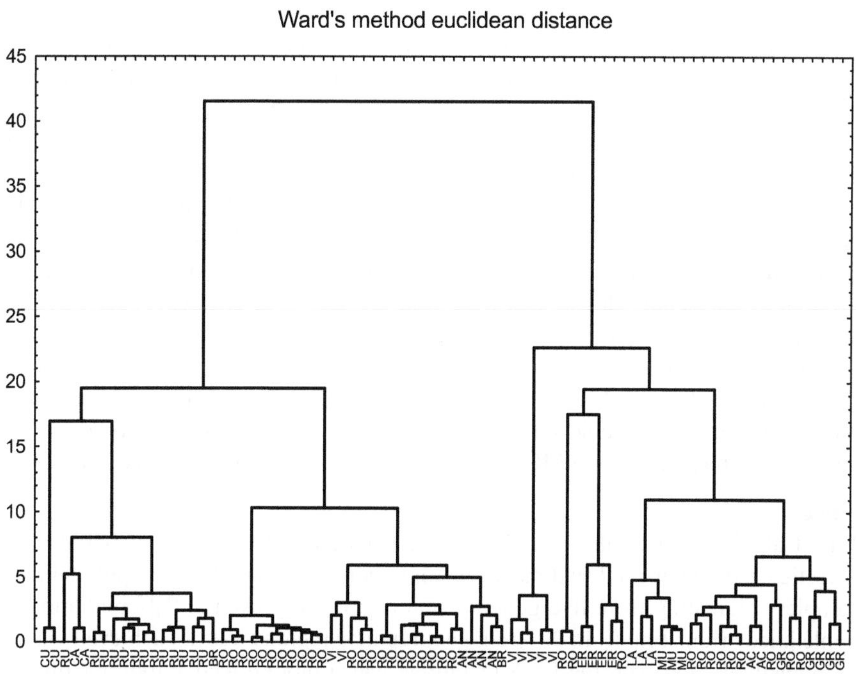

Fig. 11.3 Hierarchical dendrogram of fresh fruit samples (GR Grossulariaceae, AC Actinidiaceae, MU Musaceae, BR Bromeliaceae, CU Cucurbitaceae, CA Caricaceae, AN Anacardiaceae, RO Rosaceae, RU Rutaceae, LA Lauraceae, VI Vitaceae, ER Ericaceae). (After Grembecka M and Szefer P, 2013. Environ Monit Assess 185:9139–9160. https://link.springer.com/article/10.1007/s10661-013-3242-z. With permission)

PC2 (29.5%), revealed four distinct groups corresponding to the apple varieties only within each district. When LDA was used for all the apple subfractions, 99% of the samples were correctly identified with respect to their origin by cross-validation with the LOOCV. A distinct differentiation of apple varieties was also obtained; however, when each district was considered, the discriminations were different in each district. The discrimination of the samples was unclear, in agreement with the PCA results. It was concluded that the approach confirms the power of stable isotopes for differentiating the geographical provenance of apples grown in orchards located only a few hundred kilometres apart. The method could also distinguish various apple varieties, although only within small geographical areas.

Papapetros et al. (2018) studied the combination of ICP–OES, HPLC–RID and SPME–GC/MS (elemental profiles, glucose, fructose and semiquantified volatile compounds) for classifying four cultivars of sweet cherry originating from six regions of northern Greece. The analytical data were processed by MANOVA and LDA. The latter differentiated the studied samples according to their botanical and geographical origin (Skeena cultivar). The combination of volatile compounds–conventional properties–mineral components provided the highest correct prediction rate of 97.4% for botanical differentiation, whereas for interregional differentiation of the Skeena cultivar, conventional parameters and volatile compounds achieved 94.9% and 95.0% correct prediction rates, respectively.

Perez et al. (2006) combined ICP–AES with IRMS to categorize eight varieties of three kinds of fresh fruits (strawberry, blueberry and pear) originating from four countries. Models such as PCA, CDA, LDF, the QDF, ANN, GNN and HT were applied to transform the concentration data. A CDA frequency chart using the canonical variable is presented in Fig. 11.4. Each fruit type was analysed from two growing regions, i.e., Oregon vs. Mexico, Chile and Argentina. Excluding the LDF model (which has only a 60–80% success rate), the four other models, i.e., QDF, ANN, GNN and HT, enabled successful classification ranging from 70% to 100% depending on the commodity and model. According to the authors, although it was difficult to adequately distinguish fruit cultivars grown in different locations, preventing this inconvenience, it was possible to collect two varieties of strawberry and blueberry from adjacent plants (with the same soil, same environment and same agronomic practices). Therefore, owing to this strategy, an adequate evaluation of the variety effect was achieved. Interregional and intervarietal differentiation was achieved with respect to the MEC and SIR of strawberry, blueberry and pear.

Juice

Bat et al. (2016) used the combination of NMR, IRMS, TXRF and ICP–MS to investigate the elemental and stable isotopic compositions of three apple cultivar samples originating from five geographical regions in Slovenia. Differentiation of apple juice originating from different geographical regions was not achieved via the PCA method. However, the LDA technique was a more effective tool, enabling a clear differentiation of apple juice samples from the Mediterranean, Alpine and Pannonian regions from those originating from the Dinaric and sub-Mediterranean regions, with a total prediction ability of 83.9%. Moreover, LDA allowed the

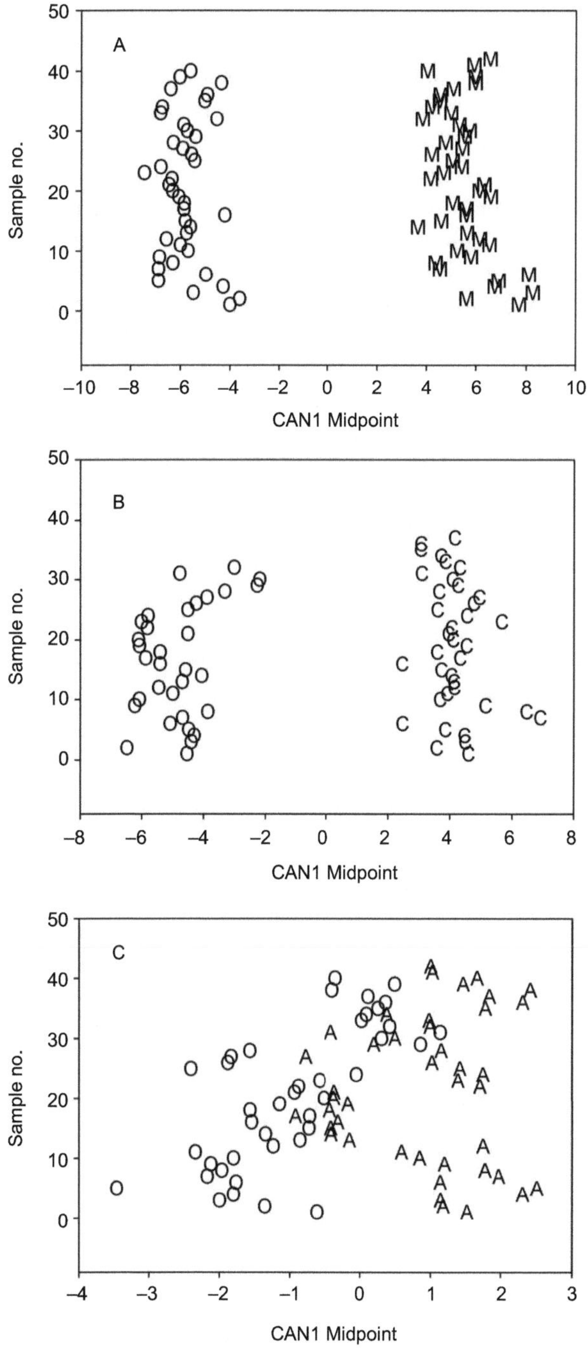

Fig. 11.4 CDA frequency chart using the canonical variable. All 10 available dimensions are utilized in this simplified visual representation of the differentiation between Oregon and Mexican strawberries (n = 80) (**a**), Oregon and Chilean blueberries (n = 68) (**b**) and Oregon and Argentine pears (n = 80) (**c**). (After Perez AL et al., 2006. J Agric Food Chem 54:4506–4516. https://doi.org/10.1021/jf0600455. With permission)

classification of the apple juices according to their three varieties, i.e., Idared, Golden Delicious and Topaz, achieving a total prediction ability of 75.8%.

According to Pellerano et al. (2008), the INAA of lemon juices (prepared from four apple varieties) from three regions of Northwest Argentina was integrated with chemometrics such as PCA and LDA. The first three PCs explained 65.5% of the total variance. However, partial differentiation between groups was achieved because only one sample corresponding to the cultivars farmed in the southern region with respect to the other varieties could be distinguished. In this case, the differentiation by these categories was not surprising since PCs are measures of total sample variation and do not explicitly consider differences between particular groups (locations–varieties). The LDA was useful for differentiating lemon juices, with a clear differentiation of three clusters related to the three production zones of the northwestern region of Argentina, i.e., the northern, middle and southern lemon-producing areas (Fig. 11.5). Owing to the LDA and cross-validation, the original grouped cases (93.2%) and cross-validated grouped cases (84.1%) were correctly classified. Moreover, 100.0% of the cross-validated grouped cases for the southern region were correctly classified.

Magdas et al. (2012) reported that the combination of IRMS–GC and ICP–MS was useful for characterizing the SIR and MEC profiles of organic apple juices (prepared from nine apple varieties) from four areas in Transylvania. The $\delta^{13}C$, $\delta^{18}O$ and $\delta^{2}H$ values indicated small differences, probably due to the environmental

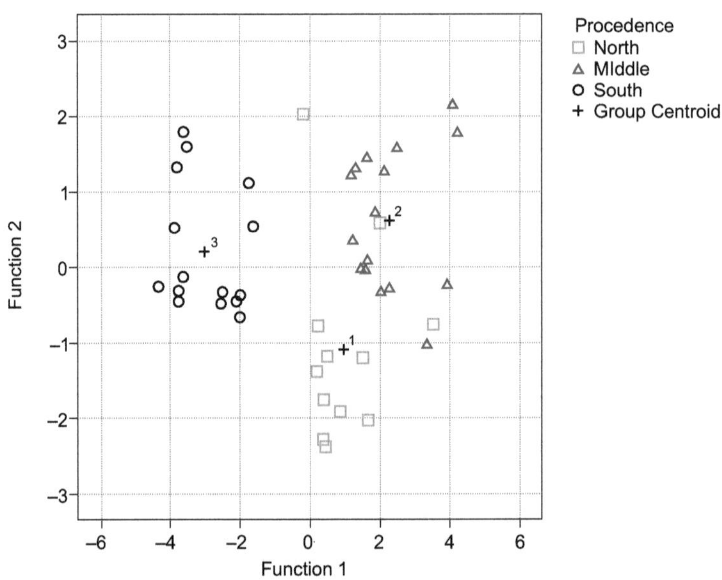

Fig. 11.5 LDA plot for 44 lemon juice samples based on the concentrations of 11 elements and using three groups as inputs a priori, that is, those corresponding to the lemon-producing areas. (After Pellerano RG et al., 2008. J Agric Food Chem 56:5222–5225. https://pubs.acs.org/doi/abs/10.1021/jf073555n. With permission)

conditions of the plants. No significant relationships between $\delta^{13}C$ levels and apple variety or geographical provenance were detected.

11.2.4 *Differentiation of Other Parameters*

11.2.4.1 Interregional Differentiation Related to Parent Soil Geochemistry

Wang et al. (2020) combined EA–IRMS and ICP–OES to characterize 563 banana (peel and pulp) samples from 15 Costa Rican farms in terms of their growing conditions, climates, topographies and soil compositions. In the case of the PCA data, the discrimination between the geographical origin (north and south) of banana, i.e., its pulp and peel as well as parent soil as substratum, was achieved, with the first two PCs explaining 36.9%, 40.8% and 71.7% of the total variance (Fig. 11.6). The greatest discrimination between the farms could be explained by the presence of volcanoes in the north in contrast to the south. The stable isotopes and elements Zn, Mg, Cu, Fe, Mn and Al are correlated with local soil composition and can be considered good indicators of geographical provenance when using multivariate analyses. The $\delta^{13}C$ and $\delta^{15}N$ values were influenced mainly by soil type, whereas rainfall and temperature were related more strongly to the $\delta^{18}O$ values. The elemental compositions of bananas were predominantly influenced by local rainfall and soil type.

Debbarma et al. (2021) separately applied ICP–MS data and physicochemical parameters of various accessions of Indian jackfruit samples combined with chemometric processing to differentiate samples according to their geographical origin. A significant relationship between only the Mo, Zn and Si contents in the fruit and associated soils was observed; however, the contents of soil Al, Cu, Mn, Ni, Rb and Xe were negatively correlated with the fruit contents. In the case of PCA, the first two PCs accounted for 39.9 (for fruit) and 55.1% (for soil) of the total variance. In the case of LDA of the fruit ionome of jackfruit related to geographical origin, the accessions were differentiated even at the 95% confidence level because all seven locations of accessions of jackfruit could be distinguished from each other. Using PCA and LDA, genomic variation and climatic differences were found to be important discriminating factors for fruit samples in addition to soil factors. Since all the fruits originating from each location were harvested at similar maturity stages, the differences in the ionome profiles of the fruits mainly corresponded to their geographical origins.

Yan et al. (2015) used MEA data to process pomelo samples and parent soil samples from four provinces of China via multivariate processing. PCA distinguished the pomelo and soil samples into four provinces, with the first two PCs explaining 69.66% and 67.06% of the total variance, respectively. Moreover, LDA was more effective for the classification of samples than PCA was. The former allowed for correct classification, with 91.30% and 83.40% for the pomelo and associated soil origins, respectively.

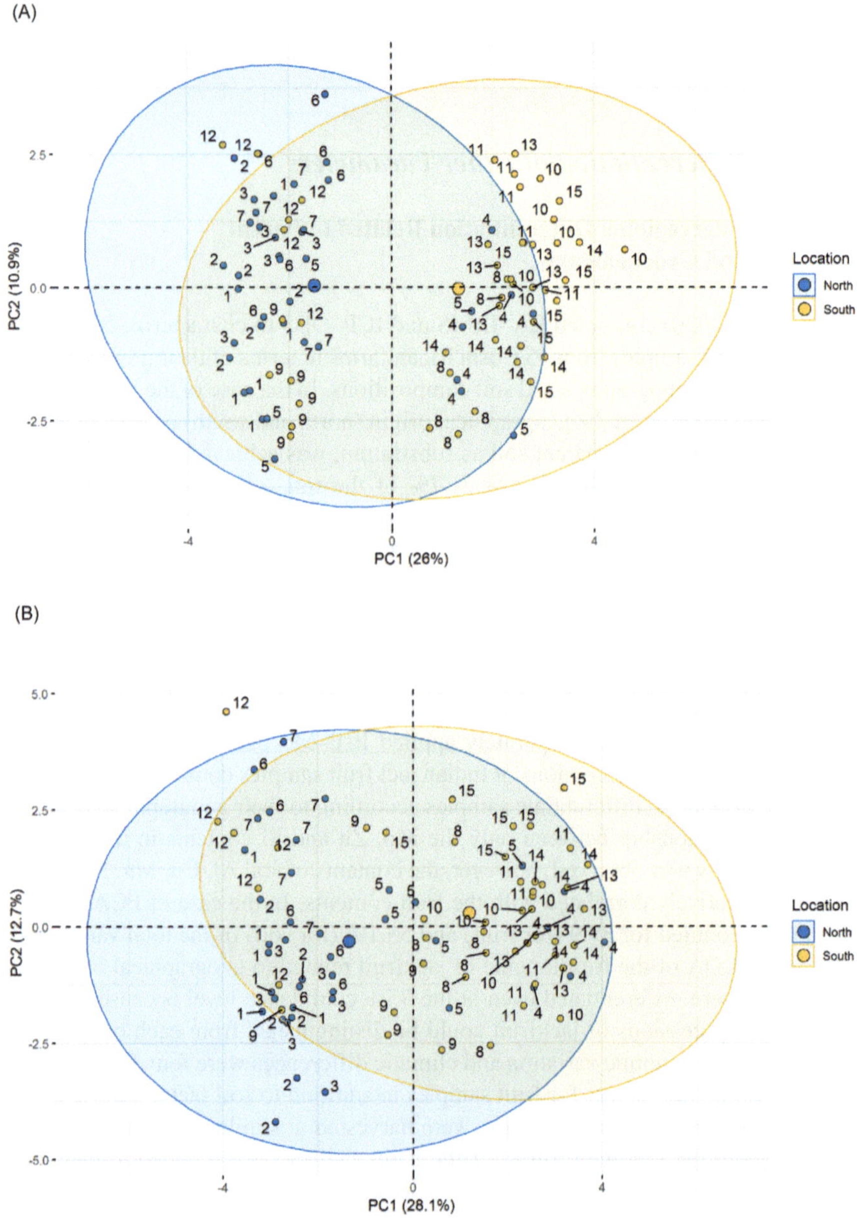

Fig. 11.6 Principal component analysis (PCA) plot of banana pulp (**a**), banana peel (**b**) and soil samples (**c**) collected from farms in Costa Rica. (After Wang Z et al., 2020. Food Res Inter 129:108882. https://doi.org/10.1016/j.foodres.2019.108882. With permission)

Simpkins et al. (2000a) separately used ICP–AES and ICP–MS in combination with chemometric processing analysis to assay Australian and Brazilian juices and peel extracts. PCA allowed the clear discrimination of Australian and Brazilian reconstituted concentrates, and peel extracts obtained from Australian and Brazilian juices were distinguished. The variables with the strongest contribution to this differentiation were the most important variables, such as Rb, Ba, Na, Ca, Sr and B. It was postulated that such a chemometric distinction between orange juices was caused by elemental differences in the associated soil and controlled by different types of rootstock.

11.2.4.2 Verifying Authenticity/Adulteration

In Bin et al. (2022), the combination of EA–IRMS and ICP–MS was used to group muskmelon samples originating from different administrative regions across Xinjiang Province (China). After the scope of the analysis was reduced from 13 administrative regions to six geographical production regions in Xinjiang, the analytical data obtained via LDA modelling were elaborated. Significantly improved classification accuracies were achieved, i.e., 100% for Altay, 91.9% for Hami and >79.1% for the other regions. This strategy could successfully be used to identify the origin of genuine Hami PGI muskmelons since the origin of mislabelling and fraudulent substitution of Hami PGI muskmelon products with non-PGI products has been frequently reported, which damages to their reputation and market advantage.

Hidalgo et al. (2023) assessed MEC data from MP–AES and ICP–MS to identify the origin of mandarin juices from three distinct citrus-producing zones in the northeast region of Argentina. PCA with the first seven PCs explaining 73.7% of the total variance revealed that PC1 and PC2 visualized the distribution of mandarin juice samples with respect to their geographical provenance. Among the elements analysed, Mg, Zn, Mo and Co (PC1) and Sr and Cr (PC2) were useful for differentiating the descriptors. In addition, supervised chemometric models, such as LDA, k-NN, SVM and RF, were applied. The overall classification rates ranged from 86.7% to 95.1% in the test subsets, with SVM emerging as the most proficient classifier. It was concluded that this methodology provides the ability to use mineral profiles to verify the authentic origin of mandarin juices, which prevents mislabelling, a form of fraud, contributing to consumer protection.

Pepi et al. (2019) combined XRF with ICP–MS to examine grape berries of the Cannonau cultivar in soil and leaf samples from two regions. The obtained data were integrated into a chemometric analysis, bearing in mind that this cultivar is used to produce the renowned Italian DOC wine "Cannonau" from Sardinia. The PLS–DA model yielded 100% sample validation, which clearly differentiated the two vineyards located in Pardu and Pelau. PLS–DA, which is based on data of indices of bioaccumulation and translocation coefficients, was used for differentiating the two vineyards on the basis of various concentrations of B, Ca, Cr, Fe, K, Mg, Na, Ni, Sr, V and Zr. Among these elements, B, Sr and Zr were significantly correlated with the soil vineyard and the leaves and grape berries from the two study areas. Therefore, these elements could be suitable for identifying possible geochemical fingerprints of *V. vinifera* grown in the Sardinia region, from which grapes are used to produce fine

Italian wines. It was possible to distinguish *V. vinifera* cv. Cannonau effectively with respect to geographical origin, thus confirming the cultivar authenticity.

Amenta et al. (2016) integrated several mod tools, such as FT–NIR, HPLC–PDA, ICP–OES and EA–IRMS, which were useful for investigating lemon fruits from several farms in the 'Limone di Siracusa PGI' production zone and outside the PGI area in Sicily and Campania (with two samples from Spain and Argentina). Chemometric techniques (ANOVA, PLS–DA, VIP and LDA) demonstrated satisfactory geographical discrimination potential. The PLS–DA model achieved high recognition (95%) and satisfactory predictive ability (85%), distinguishing between PGI and non-PGI lemon fruits.

Potorti et al. (2018) assayed the differentiation of MEC and SIR of fruit lemon samples originating from different regions of two countries by means of ICP–AES and IRMS methods. The combination of MEA, including lanthanide elements, together with multivariate processing (PCA and CDA) enabled the proper classification of the samples studied. The first two PCs, accounting for 73.65% of the total variance, identified two groups that were sufficiently distinguished; i.e., Turkish lemon samples were clearly differentiate from Sicilian samples. On the basis of CDA data, 100% of the total original dataset and the CV dataset of samples were correctly classified, indicating successful discrimination between PGI Interdonato lemon Messina (Italy) and non-PGI Turkish lemon pulps. It was concluded that Italian and Turkish lemon pulps cannot be treated as "rich in" or "sources of" analysed mineral components, although they can be considered safe because of their low levels of toxic elements.

According to Benabdelkamel et al. (2012), MEC was quantified via ICP–MS in Calabrian PGI Clementine, which originated from a specific region in southern Italy, and fruits were cultivated in Algeria, Tunisia and other locations in Spain. The concentration data obtained were processed chemometrically with PCA, S–LDA, SIMCA and PLS–DA to protect PGI clementines from non-PGI fruits. The scores and loading values of the first two PCs for the juice and peel samples are shown in Fig. 11.7a and Fig. 11.7b, respectively. In the case of juice samples, a differentiation

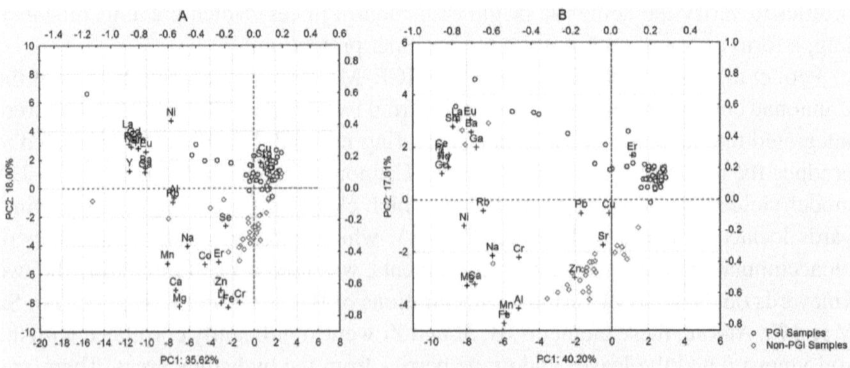

Fig. 11.7 Biplot of PC1 versus PC2 scores and loadings for clementine juice samples (**a**) and for clementine peel samples (**b**). (After Benabdelkamel H et al., 2012. J Agric Food Chem 60:3717–3726. https://doi.org/10.1021/jf2050075. With permission)

of objects corresponding to PGI and non-PGI samples on PC2 is well visible. In addition, with respect to the peel samples, differentiation of the groups related to the two production areas was principally observed for PC2. The CV procedure showed satisfactory values of prediction ability for the three models used, particularly S–LDA, with values of 96.6% for juice samples and 100% for peel samples, as with SIMCA, which showed 100% for both peel and juice samples.

Mir-Marqués et al. (2015) applied of the combination of ICP–OES and ICP–MS to examine the MEC of kaki fruit samples produced from different regions of Spain, including the PDO samples of "Kaki Ribera del Xúquer", Valencia. ANOVA indicated interspatial differences, therefore allowing the authentication of the PDO sample origin. It was suggested that the elemental profile of kaki fruits might be useful for verifying the authenticity of fruit with the PDO.

Bontempo et al. (2014) established SIR values on the basis of TC/EA–IRMS and SNIF–NMR in orange and lemon juice as well as orange concentrate juice from five citrus-producing regions of southern Italy. Variety did not affect the SIRs, whereas harvest year (from 2000 to 2012) and region of origin (five locations) affect the different ratios. PCA (with the first two PCs explaining 56.8% of the total variance) showed considerable overlap between the different groups of orange juices because the samples that originated exclusively from the southern regions of Italy presented very similar climate characteristics. In particular, it was postulated that the Italian range of values was estimated to be not completely in agreement with AIJN guidelines; this hypothesis was undergirded by the suspicion that these apparently genuine juices could not be truly authentic.

Cárnara et al. (1995) performed different chemical and physicochemical measurements to study pineapple juices, i.e., fresh juice (homemade), pineapple commercial juices and nectars. PCA, with the first two PCs accounting for 67.91% of the total variance, revealed that the organic acid, sugar and K levels and formol index are the most useful variables for quality control analysis of pineapple juices and nectar.

Antolovich et al. (2001) used IRMS to verify the authenticity of Australian commercial orange juices prepared from two cultivars. No significant seasonal variations in $\delta^{13}C$ content were observed. The variations in the combustion temperature in this method were insignificant. It was also concluded that SCIRA is a powerful method for identifying adulteration by C4 sugars in fruit juices.

According to Tormen et al. (2011), the MEC content determined via ICP–MS of several commercial types of 19 different juices from seven brands was integrated with chemometric analysis (PCA and HCA). For the former, the first three PCs represented 73.01% of the total variability of the data. The scatter plot scores for PC3 vs. PC1 showed that the grouping by brand was even better visualized than PC1 vs. PC2, especially in the case of the Rb and Na levels. Comparable results were achieved using the HCA technique. The Brazilian juice samples created clusters corresponding to five brands because of the parent soil characteristics and the different processing and storage conditions used by different brands, including water and additives. It was concluded that the samples were grouped with respect to brand, not by fruit, and the profile composition of the juice was dependent on the soil composition where the plant was grown as well as the industrial production and storage conditions applied by the factory.

Simpkins et al. (1999) processed Australian orange juices from major growing regions during a 5-year period via IRMS to test retail samples for authenticity. The $^{18}O/^{16}O$ ratios were useful for establishing a consistent pattern influencing cyclic seasonality (summer and wither) and regional variations in oranges growing in coastal areas. Concentrated orange juices, pulpwashes, peel extracts and other citrus types were also tested. The impact of some industry practices on $\delta^{18}O$ was also examined, as was its effectiveness in detecting the addition of water to authentic juices. A slight change to $\delta^{18}O$ can be an effect of storage, and these changes are unlikely to be significant in the case of authenticity testing.

Simpkins et al. (2000b) examined data for $\delta^{13}C$ in fresh and concentrated orange juices (from fruits: Trifoliata, Troyer and Carizzo citrange, sweet orange, Cleopatra mandarin, citrumelo and rough lemon), which represent production over a five-year period. The obtained data were used for adulteration testing. Peel extracts of Australian origin, Brazilian pulp washes and concentrates were also tested. The data for fresh juices showed regional differences in their isotopic abundances, which were related to the rootstock. This was helpful in the assessment of the authenticity of juices sold commercially with $\delta^{13}C$ values higher than 3 from the database. This means that the addition of cane sugar to juices could have undesirable impacts, even though it is a well-established practice.

Aguzzoni et al. (2021) aimed to apply a combination of TC/EA–CF–IRMS, ICP–MS and MC–ICP–M methods to characterize PDO and PGI apples cultivated in northern Italy and compare them with Italian apples without labels of geographical indications. LDA was used to determine the geographic origin of apples among three main districts of the South Tyrol cultivation area. The developed classification model was a useful tool, allowing complete recognition of all the samples according to their origin, with a satisfactory average accuracy of >96%. This approach is thus an effective tool for enhancing and protecting PDO and PGI Italian products.

Coelho et al. (2019) reported MEC data quantified by ICP–MS and ICP–OES for Rocha pears from the PDO and non-PDO production areas of Portugal. PCA identified six PCs with EVs above one explaining 74.7% of the total data variance; however, it allowed partial discrimination of pear samples according to their geographical origin. In contrast to PCA, CA revealed two distinct clusters corresponding to interregional differentiation. LDA confirmed that it was possible to recognize the geographical provenance of pear samples on the basis of their elemental distribution patterns. This predictive model, built and validated, discriminated samples with 100% accuracy, enabling the differentiation of pears from the PDO region and pears from non-PDO regions.

According to Matos-Reyes et al. (2013), ICP–OES data of Alicante Mountain cherry samples originating from five areas of Spain, including two PDO samples, were processed chemometrically to assess their ability to differentiate cherries according to their authenticity. LDA allowed complete discrimination of the edible samples from all five provinces. However, the cherry stones clearly differentiated between samples from the two areas and those originating from the three other provinces. Better discrimination results were achieved for the edible part of the cherry on the basis of the mineral composition, with 100% correctly classified samples; for cherry stones, 96.43% of the samples were correctly classified.

11.2.4.3 Differentiation Between Conventionally and Organically Grown Fruits

Camin et al. (2011) coupled TC/EA–IRMS with HPLC–IRD and HPLC–UV/Vis data to classify commercial and experimental fruits (oranges, clementines, strawberries and peaches) produced organically and conventionally. The obtained data were transformed by chemometric processing (ANOVA, HDS and CDA). Exclusively, strawberries were the most suitable for differentiating their geographic origin because more numerous and diverse samples were available. CDA enabled a clear differentiation of organic and conventional strawberries produced in each of the three Italian areas. It was also possible to correctly reclassify 98% of the samples via the cross-validation procedure. This means that reliable discrimination between the strawberry groups with respect to the agricultural regime and production site was achieved.

Forster et al. (2002) assayed two banana cultivars grown in Tenerife (Spain) and cv. Gran enana bananas from Ecuador with respect to MEC. The first two PCs, which explained 56.8% of the total variance, demonstrated that the bananas from Ecuador were clearly differentiated from those produced in Tenerife (Fig. 11.8). Differentiation of MEC content resulting from cultivation method (greenhouse and

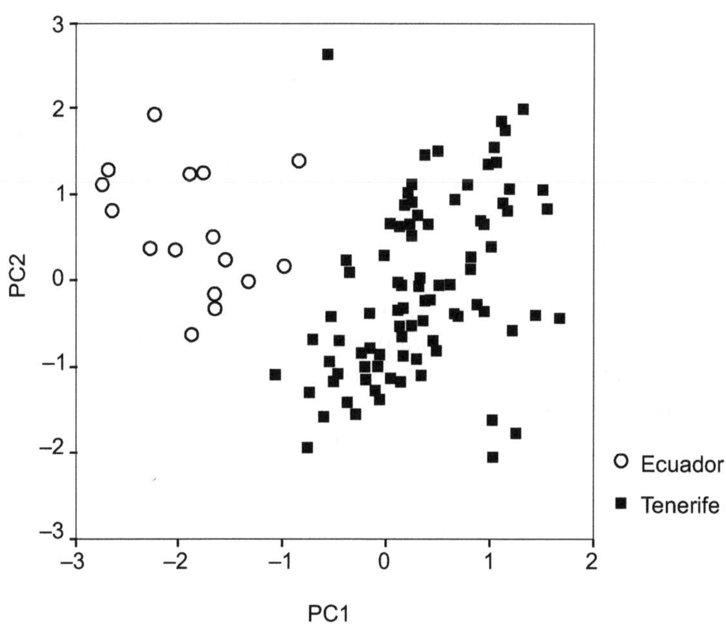

Fig. 11.8 Scores of the banana samples on axes representing the two PCs differentiating the precedence. (After Forster MP et al., 2002. J. Agric. Food Chem 50:6130–6135. https://doi.org/10.1021/jf0255578. With permission)

outdoors) and farming style (conventional and organic) in the samples from Tenerife was achieved. Moreover, the elemental content in the internal part of the banana was greater than that in the middle and external parts.

According to Borges et al. (2016), ICP–MS was applied to assay two classes of samples, i.e., ordinary and organic grape juices obtained from local Brazilian retail markets. The chemometric processing of the concentration data was performed with PCA and SIMCA. PCA using Ba, Ce, La, Mg, Na, P, Pb, Rb, Sn, Tl and V levels discriminated the two classes along PC1 and PC2, corresponding to 42% and 13% of the total variance, respectively. The SIMCA model could also recognize ordinary and organic grape juices, resulting in 82 and 95% prediction ability, respectively. The Coomans plot revealed that none of the constructed models admitted samples from the other category; thus, a specificity of 100% in all the cases was achieved.

Maione et al. (2016) grouped conventional and organic grape juice samples from different producers and used the MEC content determined by ICP–MS. Multivariate techniques such as SVM, CART and MLP–ANNs were effective at transforming the spectrometric data. The best models were as follows: the SVM model, with 89.18% accuracy, trained with K, Sn, P, Na, Sm and Nd, and the CART and MLP models, which achieved 86.48% accuracy when only Na was used. The elements mentioned above are the most important variables in the discrimination of Brazilian organic and conventional grape juice samples.

Bat et al. (2012) aimed to examine the combination of MEC and SIR values with selected chemical and physical parameters by means of IRMS, TXRF and HPLC—UV/VIS to differentiate varieties of Slovenian apples. The stable isotope ratios in sugar, pulp, protein and water were found to be useful variables in this respect. Intercultivar differentiation was achieved owing to the major influence of the $\delta^{13}C$ and $\delta^{15}N$ values of proteins and the $\delta^{18}O$ and $\delta^{2}H$ values of water. It was also possible to determine the geographical origin of samples on the basis of the $\delta^{18}O$ and $\delta^{2}H$ values in water and the contents of Rb and S in fruit juice. The most significant descriptors for distinguishing between organically cultivated fruits and conventionally cultivated fruits were the $\delta^{15}N$ values and antioxidant activity.

According to Hernández-Sánchez et al. (2012), FAAS data were useful for studying mango samples of uniform genetics (Lippens variety) cultivated on Gomera Island (Canary Islands) by conventional and organic farming. The PCA score plot generated via the first two or three PCs did not reveal any differences between the samples belonging to the two classes. Thus, supervised learning pattern recognition was used with Ca, Co, Cu, Fe, K, Mg, Mn, Na, Ni and Zn. LDA and SVM achieved classification performances of 73.2 and 93.1%, respectively, in terms of prediction ability. This means that the latter is the tool of choice for modelling nonlinear class boundaries between classes. Farming differences as well as soil differences may be reflected in the elemental content of the mangoes cultivated in these extensions.

11.2.4.4 Genetic Variability in the Mineral Content of Fresh Fruit Pulp

Devarajan et al. (2021) evaluated IMA with respect to nine elements in their FFP. The genetic variability of MEC was determined by ICP–OES, i.e., 4.7-fold for K and Mg to 111.1-fold for Ca. PCA, with the first two PCs representing 51.2% of the total variance, was unable to indicate any difference between the FPM compositions and the IMA of different ploidies/genomes. This means that the elements used as variables have no significant influence on the classification of IMA. However, both PCA and CA failed to group the IMA according to ploidy/genome/subgroups. Twenty commercial cultivars were placed in the top 10 positions on the basis of their MC. In addition to Ca and Mg, IMA was more abundant in terms of all micronutrients than the *Musa* gene pool.

Devarajan et al. (2024) demonstrated that the multielemental profile established by ICP–OES of Indian ripe banana combined with chemometric methods was an effective tool for distinguishing between banana genomic and subgenomic groups of 100 Indian Banana (Musa) accessions. The chemometric tool consisted of the PCA technique with a combination of PCA–LDA, PCA–SVM and PCA–ANN. The best results were obtained for the PCA–SVM model, which achieved the highest classification accuracy for the test/verification set prediction accuracy rate. The PCA–SVM model was the most useful for classification on the basis of set subgenomic and genomic group data, with accuracies of 83.7% and 100.0%, respectively. Moreover, multielemental fingerprint classification had the greatest effectiveness at the genomic group level compared with the subgenomic level.

According to Maseko et al. (2022), the combination of FAAS with multivariate analysis was used to categorize unripe South Africa banana flour from 33 banana varieties belonging to 4 genome groups and 11 subgenome groups. On the basis of PCA, with the first PCs contributing 54.2%, the genomic groups were not clearly differentiated. However, owing to the use of PCA–LDA, PCA–ANN and PCA–SVM models, the classification model verification set samples were reliably classified on the basis of the contributions of their genome groups with 100% accuracy and subgenome groups with 78.95–100% accuracy. It was concluded that multielemental fingerprinting combined with multivariate computational tools is an effective approach for classifying *Musa* genomic and subgenomic groups.

11.2.4.5 Interregional Differentiation Related to the Impact of Pollution

Dimitrijević et al. (2016) used ICP–OES to determine the contents of Cu, Zn, Pb, As, Cd and Ni in four parts of Serbian peach trees and topsoil from different locations in urban–industrial and rural areas located at various distances from the Bor copper smelter. PCA revealed that the first four PCs explained 86.3% of the total variance, allowing the identification of sources of heavy metals in the soil and particular parts of the peach trees. The first four PCs as well as HCA explained the contributions of metals from local anthropogenic activities (copper smelting). This means that peach tree parts, especially leaves, could be the best indicators of pollution originating from atmospheric deposition and could be suitable candidates for biomonitoring and bioremediation purposes.

Zinicovscaia et al. (2020) reported that integrated INAA and ENAA data from apple, plum and six varieties of grape and parent soil samples collected in four agricultural zones of the Republic of Moldova were processed chemometrically. The DA data enabled the classification of the studied fruits by type and region of origin, suggesting that MEC could be useful for differentiating between particular groups of fruits. On the basis of the assessment of anthropogenic contamination of the soil as a substratum, it was suggested that the analysed fruits are safe for consumption.

Bora et al. (2020) reported that MEC was determined via ICP–MS of Romanian samples of roots, canes and leaves of three varieties of grapevine with soil samples from Baia Mare Mining, the Smelting Complex and its surroundings. On the basis of ANOVA, LDA and CA, the levels of most elements in the plant parts were dependent on airborne pollution originating from nearby metallurgical activities, i.e., from the Cu smelter. The Ni content was affected mainly by geological factors. Since Cu and Zn are present at relatively high levels in all plants, soil, must and wine, except for grape samples from the Baia Mare and Baia Sprie areas, it has been concluded that the investigated grapevine cultivar effectively protects grapes from high contents of heavy metals, resulting in tolerance to high levels of Cu, Zn, Hg and As in other tissues, especially in root tissue.

Alagić et al. (2017) analysed trace elemental contents via ICP–OES to characterize different organs of Serbian peach and apple (roots, branches, leaves and fruits) in spatial soils. The contents of the studied elements were mostly within normal ranges, except in some cases for Cu, Zn and As. HCA revealed various specific features of the distribution of elements in both investigated systems, while one-way ANOVA revealed significant differences in the metal concentrations.

Alagić et al. (2015) combined trace element profiles with chemometric processing of different parts of grapevines and associated topsoil from selected zones near the Mining and Smelting Bor Complex (East Serbia). HCA confirmed that useful information was recorded for plant organs, namely, roots, stems, leaves and fruits, taken from five grapevines in the study area. The subsamples were bulked to obtain a representative sample for each region. In the case of atmospheric pollution, the most valuable data are found in unwashed leaves. It was suggested that the planting of grapevine cultivars could be recommended in all areas severely polluted with heavy metals. It was concluded that the studied grapevine presents highly effective strategies involved in tolerance to heavy-metal-induced stress. Consequently, there were generally no visible symptoms of phytotoxic effects of toxic elements, although grapevines often grow in soil and are highly polluted by Cu, Zn, Pb, As and Cd. Therefore, this authentic grapevine cultivar could be recommended as an effective candidate for phytostabilization.

According to Tošić et al. (2016), heavy metal contents were determined by ICP–OES for the roots, branches, leaves and fruits of apple trees and the parent spatial soil from polluted sites near the Mining and Smelting Bor Complex to obtain useful information about the environmental state of this region. Hierarchical CA and correlation analysis may suggest a better suitability of upper plant parts to reflect seasonal variations. In highly polluted areas, simple atmospheric deposition could be a significant source of metals in the upper parts. Therefore, this approach has been used for different biomonitoring studies involving both washed and unwashed plant parts. Notably, the dominant directions of winds in the study area during 2012 may have significantly influenced the obtained CA classification related to the characteristics of the polluted sites.

11.2.4.6 Identification of HLB-Infected Fruits

Rao et al. (2018) used LIBS to investigate Gannan navel oranges samples from Ganzhou, Jiangxi (China), which included healthy and HLB-infected fruits. The average accuracies obtained from the CWT–RF and PCA–RF models were very similar, reaching 97.45% and 97.89%, respectively, in the test set. These findings indicated that LIBS combined with a chemometric tool is a reliable approach for the identification of HLB-infected navel oranges.

Ranulfi et al. (2017) reported that LIBS combined with chemometric processing was useful for the effective study of MECs to identify nutritional changes caused by disease in Brazilian citrus leaves. PCA identified two main clusters corresponding to the HLB-symptomatic class and the other cluster related to the healthy and HLB-asymptomatic classes, which partially overlapped. Regression combined with PLSR provided a chemometric evaluation of the nutritional profile changes, with an accuracy of 73%, distinguishing the leaf samples from adult citrus trees according to three categories of leaves, i.e., healthy, HLB-symptomatic and HLB-asymptomatic.

Pereira et al. (2010) aimed to examine the effects of CLas on inoculated Brazilian citrus plants from a greenhouse using the mineral and organic signal profiles of leaves as a fingerprint. LIBS spectra related to major, macro- and microelements coupled with chemometric processing (PCA and SIMCA) were effective for assessing healthy leaves and CLas-inoculated leaves of citrus plants. Moreover, the evolution of the CLas on the inoculated plants was assayed via RT–qPCR amplification of bacterial sequences. PCA, with the first two PCs accounting for 95% of the explained variance for the first month, demonstrated two well-defined clusters, i.e., one for healthy plants and another for inoculated plants. The models were then applied to different inoculation times (from 1 to 8 months) (Fig. 11.9). The models effectively classified 82–97% of the diseased samples at the 95% significance level. It was concluded that LIBS combined with SIMCA was an effective tool for recognizing diseased samples from the first month of inoculation, with 97% correct predictions (classification model) over 4% of the current method of PCR.

According to Sankaran et al. (2015), LIBS spectra and ICP–OES data corresponding to citrus leaves with HLB anomalies (citrus canker) as well as nutrient deficiencies such as Fe, Mn, Fe, Mg and Zn were compared with those of healthy leaves. The PC scores derived from the preprocessed LIBS data were used during classification. Moreover, processing analytical data and their classification using two models, i.e., QDA and SVM, were achieved. The SVM-based classification of

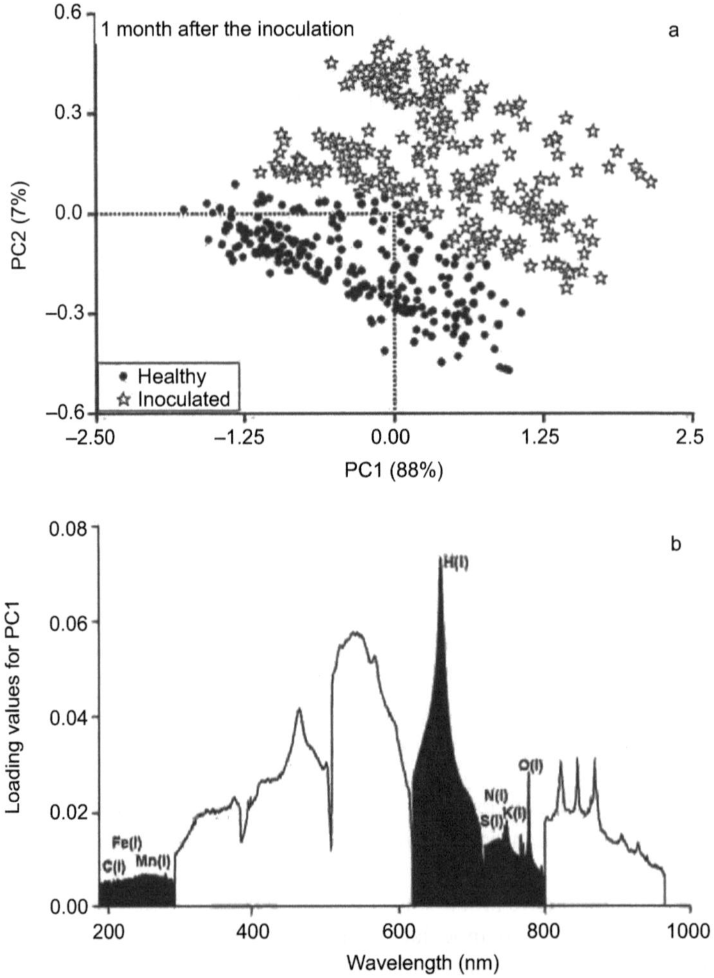

Fig. 11.9 Score (**a**) and loading plots (**b**) for the data matrix, 1 month after inoculation, containing 510 spectra and 1375 variables. (After Pereira FMV et al., 2010. Talanta 83:351–356. https://doi.org/10.1016/j.talanta.2010.09.021. With permission)

LIBS spectra resulted in 97.5% overall accuracy, with high average canker classification accuracy (96.5%). This result was slightly higher than the QDA-based classification accuracies. This means that a clear differentiation of healthy citrus leaves and the presence of other anomalies in citrus leaves could be observed. It was also possible to discriminate samples into different classes, i.e., healthy, HLB-infected, canker-infected, Mg-deficient, Mn-deficient and Zn-deficient.

11.2.4.7 Differentiation of Elemental Profiles Depending on Culinary Processing

de Souza et al. (2016) explored raw breadfruit samples acquired in several cities from Bahia state (Brazil) that were cooked using domestic methods such as boiling and microwave heating with digestion in acid media. The MEC was quantified via ICP–OES and GF–AAS. HCA and PCA, with the first three PCs explaining 81.96% of the total variance, demonstrated that both culinary processes resulted in a loss of nutrients, especially in the case of the samples cooked by boiling with a stove. It was concluded that the chemometric approach allows them to successfully distinguish between raw samples and both culinary processes.

11.2.4.8 Differentiation Related to Ripening Stages

The MEC has been determined via ICP–OES and combined with vitamin C and other physicochemical parameters of Atemoya fruit samples at different stages of ripening (green or ripe). PCA, with the first of three components describing 77.77% of the total variance, differentiated the samples according to the maturity stage, i.e., enabling a clear differentiation between green and ripe fruits. The variables Ba, Ca, Cu, K, Mg and P were effective descriptors contributing mostly to differentiating the two groups of fruits (dos Santos et al. 2016).

11.2.4.9 Assessment of the Extraction of Vitamin C from Two Tropical Fruits

The combination of numerous advanced methods, e.g., HPLC–DAD–MS, HPLC/ESI–MS/MS, NMR, SNIF–NMR, EA–IRMS and GC–C–IRMS, has been useful for examining independent samples of various origins: two sets of industrial samples (Sigma–Aldrich and Carlo Erba samples) and two sets of samples extracted from Brazilian camu-camu and acerola. A new procedure was described for the extraction of vitamin C from tropical fruits with high contents of ascorbic acid. Independent samples of ascorbic acid were of various origins, i.e., from natural and industrial sources. The PCA scatter plot revealed that the industrial and natural samples were differentiated, but the camu-camu and acerola samples were easily distinguishable from each other and from the industrial samples. Moreover, differentiating Sigma–Aldrich samples from Carlo Erba samples was not possible (Albertino et al. 2009).

11.3 Conclusions

As has been demonstrated, a properly chosen analytical methodology combined with multivariate processing is appropriate for verifying the authenticity of fruits and their products as well as adulterations/mislabelling, contributing to the determination of food safety. This procedure allows the classification of fruit assortment in view of geographical origin, intercultivar differentiation, the influence of environmental pollutants and culinary/technological food processing to determine the quality of products. It has also been possible to reliably recognize conventionally grown and organically farmed fruit crops. Owing to this approach, accurate results can enable the identification of falsely labelled products, thereby preserving brand reputations and protecting consumers from financial deceit.

References

Aguzzoni A, Bassi M, Pignotti E, Robatscher P, Scandellari F, Tirlerc W, Tagliavini M (2021) Multi-chemical analysis combined with chemometrics to characterize PDO and PGI Italian apples. J Sci Food Agric 101:5106–5115. https://doi.org/10.1002/jsfa.11156

Alagić SČ, Tošić SB, Dimitrijević MD, Antonijević MM, Nujkić MM (2015) Assessment of the quality of polluted areas based on the content of heavy metals in different organs of the grapevine (*Vitis vinifera*) cv Tamjanika. Environ Sci Pollut Res 22:7155–7175. https://doi.org/10.1007/s11356-014-3933-1

Alagić SČ, Tošić SB, Dimitrijević MD, Petrović JV, Medić DV (2017) Chemometric evaluation of trace metals in *Prunus persica* L. Batech and *Malus domestica* from Minićevo (Serbia). Food Chem 217:568–575. https://doi.org/10.1016/j.foodchem.2016.09.006

Albertino A, Barge A, Cravotto G, Luca Genzini L, Gobetto R, Vincenti M (2009) Natural origin of ascorbic acid: validation by ^{13}C NMR and IRMS. Food Chem 112:715–720. https://doi.org/10.1016/j.foodchem.2008.05.114

Amenta M, Fabroni S, Costa C, Rapisarda P (2016) Traceability of 'Limone di Siracusa PGI' by a multidisciplinary analytical and chemometric approach. Food Chem 211:734–740. https://doi.org/10.1016/j.foodchem.2016.05.119

Antolovich M, Li X, Robards K (2001) Detection of adulteration in Australian orange juices by stable carbon isotope ratio analysis (SCIRA). J Agric Food Chem 49:2623–2626. https://doi.org/10.1021/jf001384f

Badenes ML, Byrne DH (eds) (2002) Fruit breeding. Springer Science & Business Media, New York. ISBN: 1441907629, 9781441907622

Bat KB, Eler K, Mazej D, Vodopivec BM, Mulič I, Kump P, Ogrinc N (2016) Isotopic and elemental characterisation of Slovenian apple juice according to geographical origin: preliminary results. Food Chem 203:86–94. https://doi.org/10.1016/j.foodchem.2016.02.039

Bat KB, Vidrih R, Nečemer M, Vodopivec BM, Mulič I, Kump P, Ogrinc N (2012) Characterization of Slovenian apples with respect to their botanical and geographical origin and agricultural production practice. Food Technol Biotechnol 50:107–116. https://hrcak.srce.hr/79005

Benabdelkamel H, Di Donna L, Mazzotti F, Naccarato A, Sindona G, Tagarelli A, Taverna D (2012) Authenticity of PGI "Clementine of Calabria" by multielement fingerprint. J Agric Food Chem 60:3717–3726. https://doi.org/10.1021/jf2050075

Bin L, Cheng W, Zhi L, Weizhong H, Duoyong Z, Fang Y-y, Ying L, Zihong Z, Piao C, Wei L, Rogers KM (2022) Geographical origin traceability of muskmelon from Xinjiang province using stable isotopes and multi-elements with chemometrics. J Food Compos Anal 106:104320. https://doi.org/10.1016/j.jfca.2021.104320

Bontempo L, Caruso R, Fiorillo M, Gambino GL, Perini M, Simoni M, Traulo P, Wehrens R, Gaglianob G, Camin F (2014) Stable isotope ratios of H, C, N and O in Italian citrus juices. J Mass Spectrom 49:785–791. https://doi.org/10.1002/jms.3420

Bora FD, Bunea CI, Chira R, Bunea A (2020) Assessment of the quality of polluted areas in Northwest Romania based on the content of elements in different organs of grapevine (*Vitis vinifera* L.). Molecules 25:750. https://doi.org/10.3390/molecules25030750

Borges EM, Volmer DA, Brandelero E, Gelinski JMLN, Gallimberti M, Barbosa F Jr (2016) Monitoring the authenticity of organic grape juice via chemometric analysis of elemental data. Food Anal Methods 9:362–369. https://doi.org/10.1007/s12161-015-0191-x

Borràs E, Ferré J, Boqué R, Mestres M, Aceña L, Busto O (2015) Data fusion methodologies for food and beverage authentication and quality assessment—a review. Anal Chim Acta 891:1–14. https://doi.org/10.1016/j.aca.2015.04.042

Camin F, Perini M, Bontempo L, Fabroni S, Faedi W, Magnani S, Baruzzi G, Bonoli M, Tabilio MR, Musmeci S, Rossmann A, Kelly SD, Rapisarda P (2011) Potential isotopic and chemical markers for characterising organic fruits. Food Chem 125:1072–1082. https://doi.org/10.1016/j.foodchem.2010.09.081

Cárnara M, Díez C, Torija E (1995) Chemical characterization of pineapple juices and nectars. Principal components analysis. Food Chem 54:93–100. https://doi.org/10.1016/0308-8146(95)92668-A

Coelho I, Matos AS, Teixeira R, Nascimento A, Bordado J, Donard O, Castanheira I (2019) Combining multielement analysis and chemometrics to trace the geographical origin of Rocha pear. J Food Comp Anal 77:1–8. https://doi.org/10.1016/j.jfca.2018.12.005

Dasenaki ME, Thomaidis NS (2019) Quality and authenticity control of fruit juices - a review. Molecules 24:1014. https://doi.org/10.3390/molecules24061014

de Souza CT, Soares SAR, Queiroz AFS, dos Santos AMP, Ferreira SLC (2016) Determination and evaluation of the mineral composition of breadfruit (*Artocarpus altilis*) using multivariate analysis technique. Microchem J 128:84–88. https://doi.org/10.1016/j.microc.2016.04.001

Debbarma N, Manivannan S, Muddarsu VR, Umadevi P, Upadhyay S (2021) Ionome signatures discriminates the geographical origin of jackfruits (*Artocarpus heterophyllus* Lam.). Food Chem 339:127896. https://doi.org/10.1016/j.foodchem.2020.127896

Devarajan R, Dibakoane SR, Wokadala OC, Meiring B, Mlambo V, Kutu FR, Sibanyoni JJ, Jayaraman JK (2024) Genomic and subgenomic group discrimination between 100 Indian banana (Musa) accessions using ripe banana pulp multi-elemental fingerprints and chemometrics. J Food Compos Anal 131:106205. https://doi.org/10.1016/j.jfca.2024.106205

Devarajan R, Jayaraman JK, Somasundaram SM, Ragupathy S, Raman P, Sathiamoorthy K, Subbaraya U (2021) Genetic diversity in fresh fruit pulp mineral profile of 100 Indian Musa accessions. Food Chem 361:130080. https://doi.org/10.1016/j.foodchem.2021.130080

Dimitrijević MD, Nujkić MM, Alagić SČ, Milić SM, Tošić SB (2016) Heavy metal contamination of topsoil and parts of peach-tree growing at different distances from a smelting complex. Inter J Environ Sci Technol 13:615–630. https://doi.org/10.1007/s13762-015-0905-z

Dos Santos WNL, Sauthier MCS, Cavalcante DD, Benevides CMJ, Dias FS, Santos DCMB (2016) Mineral composition, nutritional properties, total phenolics and flavonoids compounds of the atemoya fruit (*Annona squamosa* L. x *Annona cherimola* Mill.) and evaluation using multivariate analysis techniques. Acad Bras Ciênc 88:1243–1252. https://doi.org/10.1590/0001-3765201620150537

Drivelos SA, Georgiou CA (2012) Multi-element and multi-isotope-ratio analysis to determine the geographical origin of foods in the European Union. TrAC Trends Anal Chem 40:38–51. https://doi.org/10.1016/j.trac.2012.08.003

Forster MP, Darias Martín RJ, Romero CD (2002) Statistical differentiation of bananas according to their mineral composition. J Agric Food Chem 50:6130–6135. https://doi.org/10.1021/jf0255578

Gaiad JE, Hidalgo MJ, Villafañe RN, Marchevsky EJ, Pellerano RG (2016) Tracing the geographical origin of Argentinean lemon juices based on trace element profiles using advanced chemometric techniques. Microchem J 129:243–248. https://doi.org/10.1016/j.microc.2016.07.002

Grembecka M (2007) Bromatological and chemometric assessment of plants food based on its mineral composition. Dissertation, Gdańsk Medical Academy

Grembecka M, Szefer P (2013) Comparative assessment of essential and heavy metals in fruits from different geographical origins. Environ Monit Assess 185:9139–9160. https://link.springer.com/article/10.1007/s10661-013-3242-z

Grembecka M, Szefer P (2011) Metals and metalloids in foods—essentiality, toxicity, applicability. In: Medina DA, Laine AM (eds) Food quality: control, analysis and consumer concerns. Nova Science Publishers, Inc., New York, pp 1–60. ISBN 10: 1611229170; ISBN 13: 9781611229172

Grembecka M, Szefer P (2016) Elemental trace analysis in studies of food products. In: Baranowska I (ed) Handbook of trace analysis—fundamentals and applications. Springer, pp 203–239. https://doi.org/10.1007/978-3-319-19614-5

Hernández-Sánchez C, Luis G, Moreno I, Cameán A, González AG, González-Weller D, Castilla A, Gutiérrez A, Rubio C, Hardisson A (2012) Differentiation of mangoes (*Magnifera indica* L.) conventional and organically cultivated according to their mineral content by using support vector machines. Talanta 97:325–330. https://doi.org/10.1016/j.talanta.2012.04.038

Hidalgo MJ, Gaiad JE, Goicoechea HC, Mendoza A, Pérez-Rodríguez M, Pellerano RG (2023) Geographical origin identification of mandarin fruits by analyzing fingerprint signatures based on multielemental composition. Food Chem X 20:101040. https://doi.org/10.1016/j.fochx.2023.101040

Laursen KH, Schjoerring JK, Kelly SD, Husted S (2014) Authentication of organically grown plants—advantages and limitations of atomic spectroscopy for multi-element and stable isotope analysis. TrAC Trends Anal Chem 59:73–82. https://doi.org/10.1016/j.trac.2014.04.008

Lo Turco V, Potortì AG, Tropea A, Dugo G, Di Bella G (2020) Element analysis of dried figs (*Ficus carica* L.) from the Mediterranean areas. J Food Compos Anal 90:103503. https://doi.org/10.1016/j.jfca.2020.103503

Longobardi F, Casiello G, Ventrella A, Mazzilli V, Nardelli A, Sacco D, Catucci L, Agostiano A (2015) Electronic nose and isotope ratio mass spectrometry in combination with chemometrics for the characterization of the geographical origin of Italian sweet cherries. Food Chem 170:90–96. https://doi.org/10.1016/j.foodchem.2014.08.057

Magdas DA, Dehelean A, Puscas R (2012) Isotopic and elemental determination in some Romanian apple fruit juices. Sci World J 878242:1–7. https://doi.org/10.1100/2012/878242

Maione C, de Paula ES, Gallimberti M, Batista BL, Campiglia AD, Barbosa F Jr, Barbosa RM (2016) Comparative study of data mining techniques for the authentication of organic grape juice based on ICP-MS analysis. Expert Syst Appl 49:60–73

Maseko KH, Regnier T, Anyasi TA, Du Plessis B, Da Silva LS, Kutu FR, Wokadala OC (2022) Discrimination of *Musa* banana genomic and sub-genomic groups based on multi-elemental fingerprints and chemometrics. J Food Compos Anal 106:104334. https://doi.org/10.1016/j.jfca.2021.104334

Matos-Reyes MN, Simonot J, López-Salazar O, Cervera ML, de la Guardia M (2013) Authentication of Alicante's Mountain cherries protected designation of origin by their mineral profile. Food Chem 141:2191–2197. https://doi.org/10.1016/j.foodchem.2013.04.097

McComb K, Sarker S, Hoogewerff J, Hayman A, Frew R (2019) A δ^2H isoscape of blackberry as an example application for determining the geographic origins of plant materials in New Zealand. PLoS One 14:e0226152. https://doi.org/10.1371/journal.pone.0226152

Mimmo T, Camin F, Bontempo L, Capici C, Tagliavini M, Cesco S, Scampicchio M (2015) Traceability of different apple varieties by multivariate analysis of isotope ratio mass spectrometry data. Rapid Commun Mass Spectrom 29:1984–1990. https://doi.org/10.1002/rcm.7306

Mir-Marqués A, Domingo A, Cervera ML, de la Guardia M (2015) Mineral profile of kaki fruits (*Diospyros kaki* L.). Food Chem 172:291–297. https://doi.org/10.1016/j.foodchem.2014.09.076

Mitić SS, Obradović MV, Mitić MN, Kostić DA, Pavlović AN, Tošić SB, Stojković MD (2012) Elemental composition of various sour cherry and table grape cultivars using inductively coupled plasma atomic emission spectrometry method (ICP-OES). Food Anal Methods 5:279–286. https://doi.org/10.1007/s12161-011-9232-2

Papapetros S, Louppis A, Kosma I, Kontakos S, Badeka A, Kontominas MG (2018) Characterization and differentiation of botanical and geographical origin of selected popular sweet cherry cultivars grown in Greece. J Food Compos Anal 72:48–56. https://doi.org/10.1016/j.jfca.2018.06.006

Pellerano RG, Mazza SS, Marigliano RA, Marchevsky EJ (2008) Multielement analysis of Argentinean lemon juices by instrumental neutronic activation analysis and their classification according to geographical origin. J Agric Food Chem 56:5222–5225. https://pubs.acs.org/doi/abs/10.1021/jf073555n

Pepi S, Chicca M, Piroddi G, Tassinari R, Vaccaro C (2019) Geographical origin of *Vitis vinifera* cv. Cannonau established by the index of bioaccumulation and translocation coefficients. Environ Monit Assess 191:436. https://doi.org/10.1007/s10661-019-7544-7

Pereira FMV, Milori DMBP, Venâncio AL, de Sá Tavares Russo M, Martins PK, Freitas-Astúa J (2010) Evaluation of the effects of *Candidatus Liberibacter asiaticus* on inoculated citrus plants using laser-induced breakdown spectroscopy (LIBS) and chemometrics tools. Talanta 83:351–356. https://doi.org/10.1016/j.talanta.2010.09.021

Perez AL, Smith BW, Anderson KA (2006) Stable isotope and trace element profiling combined with classification models to differentiate geographic growing origin for three fruits: effects of subregion and variety. J Agric Food Chem 54:4506–4516. https://doi.org/10.1021/jf0600455

Potorti AG, Di Bella G, Mottese AF, Bua GD, Fede MR, Sabatino G, Salvo A, Somma R, Dugo G, Lo Turco V (2018) Traceability of Protected Geographical Indication (PGI) Interdonato lemon pulps by chemometric analysis of the mineral composition. J Food Compos Anal 69:122–128. https://doi.org/10.1016/j.jfca.2018.03.001

Ranulfi AC, Romano RA, Magalhães AB, Ferreira EJ, Villas-Boas PR, Milori DMBP (2017) Evaluation of the nutritional changes caused by Huanglongbing (HLB) to citrus plants using laser- induced breakdown spectroscopy. Appl Spectrosc 71:1471–1480. https://doi.org/10.1177/0003702817701751

Rao G, Huang L, Liu M, Chen T, Chen J, Luo Z, Xu F, Xu X, Yao M (2018) Identification of Huanglongbing-infected navel oranges based on laser-induced breakdown spectroscopy combined with different chemometric methods. Appl Opt 57:8738–8742. https://doi.org/10.1364/AO.57.008738

Rinke P (2016) Tradition meets high tech for authenticity testing of fruit juices. In: Downey G (ed) Advances in food authenticity testing. Woodhead Publishing, pp 625–665. ISBN: 9780081002209

Rummel S, Hoelzl S, Horn P, Rossmann A, Schlicht C (2010) The combination of stable isotope abundance ratios of H, C, N and S with $^{87}Sr/^{86}Sr$ for geographical origin assignment of orange juices. Food Chem 118:890–900. https://doi.org/10.1016/j.foodchem.2008.05.115

Sankaran S, Ehsani R, Morgan KT (2015) Detection of anomalies in citrus leaves using laser-induced breakdown spectroscopy (LIBS). Appl Spectrosc 69:913–919. https://doi.org/10.1366/14-07767

Simpkins WA, Louie H, Wu M, Harrison M, Goldberg D (2000a) Trace elements in Australian orange juice and other products. Food Chem 71:423–433. https://doi.org/10.1016/S0308-8146(00)00150-3

Simpkins WA, Patel G, Harrison M, Goldberg D (2000b) Stable carbon isotope ratio analysis of Australian orange juices. Food Chem 70:385–390. https://doi.org/10.1016/S0308-8146(00)00086-8

Simpkins WA, Patel G, Collins P, Harrison M, Goldberg D (1999) Oxygen isotope ratios of juice water in Australian oranges and concentrates. J Agric Food Chem 47:2606–2612. https://doi.org/10.1021/jf981045a

Szefer P (2007) Chemometric techniques in analytical evaluation of food quality. In: Szefer P, Nriagu J (eds) Mineral components in foods. CRC Press. Taylor & Francis, Boca Raton, FL, pp 69–121. https://doi.org/10.1201/9781420003987

Tormen L, Torres DP, Dittert IM, Araújo RGO, Frescura VLA, Curtius AJ (2011) Rapid assessment of metal contamination in commercial fruit juices by inductively coupled mass spectrometry after a simple dilution. J Food Compos Anal 24:95–102. https://doi.org/10.1016/j.jfca.2010.06.004

Tošić S, Alagić S, Dimitrijević M, Pavlović A, Nujkić M (2016) Plant parts of the apple tree (*Malus* spp.) as possible indicators of heavy metal pollution. Ambio 45:501–512. https://doi.org/10.1007/s13280-015-0742-9

Wang Z, Erasmus SW, Dekker P, Guo B, Stoorvogel JJ, van Ruth SM (2020) Linking growing conditions to stable isotope ratios and elemental compositions of Costa Rican bananas (*Musa* spp.). Food Res Int 129:108882. https://doi.org/10.1016/j.foodres.2019.108882

Xu L, Xu Z, Liao X (2022) A review of fruit juice authenticity assessments: targeted and untargeted analyses. Crit Rev Food Sci Nutr 62:6081–6102. https://doi.org/10.1080/10408398.2021.1895713

Yan J, Liu J, Xiong Y, Qin W, Tang C (2015) Identification of the geographical origins of pomelos using multielement fingerprinting. J Food Sci 80:C228–C233. https://onlinelibrary.wiley.com/doi/full/10.1111/1750-3841.12746

Zhang J, Nie J, Kuang L, Shen Y, Zheng H, Zhang H, Farooq S, Asim S (2019) Geographical origin of Chinese apples based on multiple element analysis. J Sci Food Agric 99:6182–6190. https://onlinelibrary.wiley.com/doi/full/10.1002/jsfa.9890

Zinicovscaia I, Sturza R, Duliu O, Grozdov D, Gundorina S, Ghendov-Mosanu A, Duca G (2020) Major and trace elements in Moldavian orchard soil and fruits: assessment of anthropogenic contamination. Int J Environ Res Public Health 17:7112. https://doi.org/10.3390/ijerph17197112

Chapter 12
Tea, Its Products and Tea Infusions

Contents

12.1 Introduction

Tea plants are perennial evergreen plants; there are three species, viz. *Camellia sinensis*, *Camellia assamica* and *Camellia ambodgiensis*, which grow in more than 45 countries within the latitudinal range of 45°N to 34°S (though not in North America) (Wight 1962; Barua 2008; Karak and Bhagat 2010). Tea plants are cultivated in tropical and subtropical areas with adequate climatic conditions, such as rainfall, good drainage and acidic soils (Barua 2008; Karak and Bhagat 2010). The four main types of tea are black tea, green tea, oolong tea and brick tea; the differences between these types of teas are in their production techniques. Tea, which originated in China, is one of the most popular nonalcoholic beverages worldwide. The best measure of the popularity of this drink is the fact that approximately 18–20 billion cups of tea are consumed daily worldwide (Marcos et al. 1998; Karak and Bhagat 2010).

Supplementary Information The online version contains supplementary material available at https://doi.org/10.1007/978-3-031-99687-0_12.

P. Szefer, *Chemometrics in Food Control Based on Elemental Analysis*,
https://doi.org/10.1007/978-3-031-99687-0_12

According to Chung et al. (2003), the protective effects of tea against cancer have been reported in numerous studies using cell culture and animal models, although inconsistent results have been generated in epidemiological studies. The anticancer activity found in these studies has been attributed to tea by some researchers, but this attribution has been disputed by other researchers. The aim of the authors was to provide a better perspective on this contradiction by providing an overview of the literature concerning the results of laboratory studies of animals and humans, with an emphasis on animal tumour bioassays of the skin, lung, mammary glands and colon as well as the molecular and cellular mechanisms affected by tea. The authors also presented their own interesting point of view regarding future research directions in this area (Chung et al. (2003)).

It has been reported that tea infusions are responsible for reducing serum cholesterol, preventing LDL oxidation and decreasing the risk of cardiovascular disease. Polyphenols, including flavan-3-ols, phenolic acids and their derivatives, as well as flavonoids in tea beverages, belong to the most bioactive group of tea components and have antioxidative, antimutagenic and anticarcinogenic properties. Moreover, tea infusions contain other substances, e.g., fluoride, caffeine and biophile mineral components, which are beneficial to human health (Yao et al. 2004; Cabrera et al. 2003; Karak and Bhagat 2010; Zhong et al. 2020). However, trace elemental contents may have adverse effects on human health. For example, cumulative amounts of Al in tea infusions could be associated with Alzheimer's disease (Koch et al. 1988). Additionally, contamination with Cu and other trace elements, e.g., As, Cd, Cr, F, Pb, Mn and Ni, has been detected, especially in many South Asian countries, which is where tea is a traditional drink. Therefore, the quantification of major, minor and trace elements in tea is important because these elements have nutritional value and a protective effect, counteracting all potentially harmful influences. Although drinking tea is a popular habit in many countries, the possibly nonessential intake of trace elements, especially toxic heavy metals, from drinking tea could pose a health risk (Karak and Bhagat 2010).

The significant progress of analytical tools for testing authenticity is connected with the application of multielement and stable isotope analysis on the basis of atomic spectroscopy. This approach is especially effective for classifying agricultural resources, including tea plants, with respect to geographic and botanical origins. Consequently, analytical techniques integrated with chemometric processing are still intentional and are recommended for developing computational models that enable the evaluation of the authenticity/adulteration of agricultural resources, including tea plants (Efenberger-Szmechtyk et al. 2018).

According to such trends and recommendations, the use of multiple instrumental tools has improved the accuracy of classification or prediction of food identification, including for edible plants, in contrast to the application of a single analytical technique. Among the three levels of data fusion, low- and mid-level fusion approaches are the most commonly used. It is expected that data integration could not only contribute to the global classification/prediction ability but also be responsible for minimizing the uncertainty of each individual result. In addition, this methodology enables more accurate detection during prediction (Borràs et al. 2015).

The application of chemometrics in food chemistry has been discussed by Forina et al. (2009). Because the authenticity of food is an important aspect of food quality control, numerous analytical methods or techniques have been used in the classification of edible plants, including teas. The frequently used targeted markers include stable isotope ratios and multielement profiles. These analytical techniques include ICP–MS, ICP–OES, ICP–AES, FAAS, ET–AAS, HG–AAS, CV–AAS, EA–IRMS, LIBS, etc. The reliability and correctness of analytical data have been checked via quality assurance tests with appropriate certificate reference materials (CRMs) with known concentrations of analytes. Validated analytical data were then processed chemometrically using univariate, bivariate and other multivariate techniques, e.g., bar charts; histograms; one-way ANOVA; correlation or regression analysis; and other advanced and frequently applied multivariate techniques, such as PCA or FA, CA, HCA and supervised models, e.g., BP–ANN, LDA and PLS–DA. Numerous analytical methods/techniques combined with multivariate analysis have been successfully used for the detection of adulteration in different food products (Szefer 2007; Grembecka and Szefer 2011, 2016).

Recent research on the authentication of tea using a combination of instrumental and chemometric methods has been reviewed by Shuai et al. (2022). Instrumental methods, i.e., to enable analyses of stable isotope and mineral element contents, are suitable to identify the production region of tea since they are less affected by tea variety and processing methods. Chemometric analysis is a useful and effective method for authenticating tea. PCA and LDA are the most preferred techniques for elaborating numerous datasets and are effective tools for performing instrumental component analysis.

12.2 Chemometric Assessment of Tea

12.2.1 Geographical Differentiation

12.2.1.1 Asiatic Countries

Zhang et al. (2012a) used $\delta^{13}C$ and $\delta^{15}N$ values provided by EA–IRMS to classify tea samples from 12 provinces in China. However, one-way ANOVA data was found to be partially useful for classifying some teas. The interregional separation could not be fully accepted since it achieved satisfactory results for only four regions, not the eight other regions. According to the authors, it could be preferable to use additional variables such as mineral or organic component data.

Zhang et al. (2018a) examined tea samples from three Chinese growing regions via ICP–MS and ICP–AES methods coupled with computational tools (ANOVA, PCA, LDA and OPLS–DA). It was possible to clearly differentiate teas according to geographical origin. Visual inspection of the PCA plot revealed that the first three PCs explained 60.02% of the total variance. As shown in Figs. 12.1 and 12.2, the tea samples from the three regions were partially separated. The OPLS–DA model with

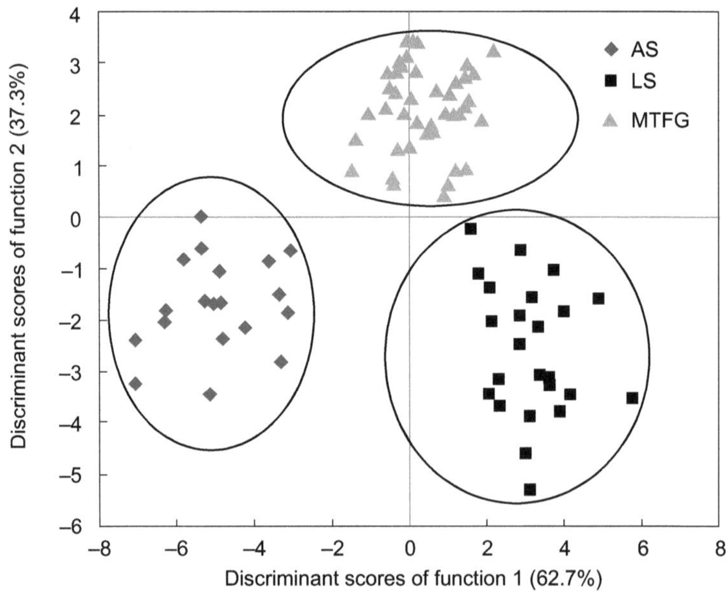

Fig. 12.1 Score scatter plot of the constructed standardized discriminant functions (function 1 and function 2) of the linear discriminant analysis (LDA) model based on the significantly different element concentrations. (After Zhang J et al., 2018a. Molecules 23:3013. https://doi.org/10.3390/molecules23113013. With permission)

a correct classification of 100% was more reliable than the LDA model (98.9%). It was concluded that the combination of multielemental fingerprinting and chemometrics is an effective approach for identifying the geographical origin of tea samples.

Zhang et al. (2020b) combined MEC data (determined by ICP–MS) with multivariate analysis processing (PCA, HCA and S-LDA) to study DYMJ tea samples from five original regions of Guizhou (China). A three-dimensional PCA scatter plot defined by the first three PCs contributing 52.42% of the total variance was unable to distinguish between the DYMJ and non-DYMJ regions of the samples. HCA outperformed the PCA model in the discrimination of the DYMJ teas; thus, this technique could clearly distinguish the subgroups in the noncore region. It was also demonstrated that the overall accuracy of the S-LDA model, amounting to 88.3%, was achieved by CV for all five regions, but more specifically, the predictive ability of the core region in the model of 96.0% was significantly greater than that of the noncore regions. It was concluded that S-LDA combined with ICP–MS might provide a more predictable classification model for the prediction of tea samples in the future.

Fu et al. (2021) coupled analytical methods (EA–IRMS, LC–MS/MS, MRM and TIC) with multivariate chemometric models (HCA, PCA and OPLS–DA) to study and classify partridge tea samples from four Chinese regions in Hainan (Wanning, Wenchang, Sanya and Baoting). Several parameters, such as 4 SIRs ($\delta^{15}N$, $\delta^{13}C$,

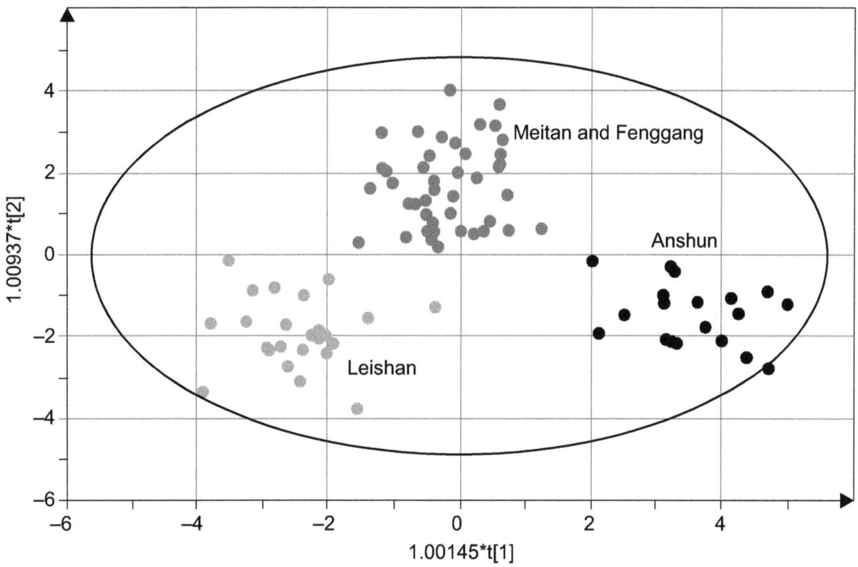

Fig. 12.2 Score scatter plot of principal components 1 and 2 of the orthogonal partial least squares discriminant analysis (OPLS–DA) results obtained from data on 28 multielement compositions of tea samples with three geographical origins (black represents Anshun, green represents Leishan, and red represents Meitan and Fenggang). The ellipse on the score plot represents the 95% confidence interval for Hotelling's T^2. (After Zhang J et al., 2018a. Molecules 23:3013. https://doi.org/10.3390/molecules23113013. With permission)

$\delta^{18}O$ and δ^2H) and 55 metabolites, including 49 polyphenols and 6 alkaloids, were used as potential approaches. The PCA 2D score plot for the SIRs and polyphenol and alkaloid contents indicated that the first two PCs accounting for 83.8% of the total variance achieved distinct geographical differentiation between the four groupings of samples. The OPLS–DA tool also provided many satisfying results.

Liu et al. (2020a) used ICP–MS and ICP–OES methods integrated with multivariate approaches (PCA, HCA, PLS–DA, BP–ANN and LDA) to provide improved discrimination of tea samples from six geographical origins. The methodology was used to demonstrate the differentiating approach using *Yongchuanxiuya* tea samples in an experimental test. The discriminating abilities of the two unsupervised learning methods, i.e., PCA (with the first three PCs explaining 69.8% of the total variance) and HCA, were less satisfactory than those of the three supervised learning methods, namely, PLS–DA, BP–ANN and LDA. Both the BP–ANN and LDA models achieved scores of 100% in recognition and prediction; ^{86}Sr and ^{112}Cd were the markers responsible for the successful classification of tea samples according to geographical origin. Under validation with the "blind" dataset, the prediction accuracies of the BP–ANN and LDA methods were greater than 90%. However, the LDA model showed the best performance and validity, with an accuracy of 100% for discriminating Chinese tea samples from six geographical origins.

Li et al. (2019) reported that a combination of ICP–MS, HPLC–DAD and HS–SPME/GC–MS instrumental techniques was an effective tool for studying PWT samples of a distinctive postfermentation variety of tea originating from four parts of the Pingwu region of Sichuan Province in China. PCA, an unsupervised method, was used to clearly distinguish tea samples with respect to their interspatial origin.

Lou et al. (2017) examined MEC combined with the SIR content using EA–IRMS, ICP–MS and FAAS to process oolong tea—Wuyi-Rock tea from nine Chinese producing areas and non-Wuyi-Rock tea from 11 nonproduction areas. The SVM non-linear model based on coupled data produced the best prediction accuracy of 0.9773. This means that analytical methods applying a classification model are effective and stable tools for the discrimination of geographical origin. Analytes such as δ^2H, δ^{18}O, Cs, Cu, Ca and Rb are recommended as significant descriptors for provenance designation.

Ye et al. (2017) studied white tea samples from three regions of China using ICP–OES and ICP–MS to determine their elemental contents. The LDA, SVM and KNN models provided successful differentiation of the elemental profile, with accuracies reaching 98.44, 95.31 and 100%, respectively. It was concluded that this approach provides a reference for white tea in further research studies.

Xiong et al. (2022) used a combination of TIMS, EA–IRMS, ICAP–Q and ICP–MS to determine 24 element contents and stable isotope ratios of ^{87}Sr/^{86}Sr, δ^{13}C, δ^{15}N, δ^2H and δ^{18}O in 379 samples of *D. officinale* from 11 sampling sites in six Chinese provinces. The ^{87}Sr/^{86}Sr, δ^{13}C and Ca data were significantly different at the regional scale, whereas the Fe data were significantly different at the local scale. To classify the samples studied, they used chemometric methods such as LDA, RF, SVM and KNN. The former two methods provided high accuracies for geographical origin authentication at both the regional (100%) and local (98.81%) scales. Among the most discriminative variables at the regional scale were Zn, Sr, δ^2H, Ba, Mn, δ^{15}N, Cu, ^{87}Sr/^{86}Sr, Nd and Rb, whereas Na, δ^{15}N, Cu, δ^2H, Zn, Ni, La, Rb, Nd and Pb were the most important at the local scale. The dataset obtained provides a reliable methodology that includes effective descriptors for the two regional scales within the framework of the authenticating process.

Chang et al. (2016) reported that combining the ^{87}Sr/^{86}Sr and δ^{11}B values with the MEC content via ICP–OES and MC–ICP–MS was a reliable approach for investigating tea leaf samples from four major tea plantations located in central and eastern Taiwan. Among the PCA and CA methods, the latter was used successfully to identify geographical tea provenance in Taiwan, with the exception of the Hualien gardens, possibly because of complicated local geological settings.

The aim of the study by Kanrar et al. (2022) was to estimate the profiles of 67 elements quantified by ICP–MS in 307 authentic Indian tea samples from eight tea-growing regions. Incomplete separation between tea samples with respect to their geographical origin was achieved. However, on the basis of the PCA data, a clear discrimination of the geographical origin of tea collected from eight regions of Northeast India was possible. Owing to the use of the OPLS–DA model, the tea samples from the eight tea areas could also be well separated from each other. The correct classification rate for the SVM, LDA and OPLS–DA models was 100%, and the confidence measure was estimated for the SVM and LDA models as 99.4 and 88.97%, respectively. Elements such as Cr, Mo, Rb, Mn and Ni were effective tracers for identifying the geographical origin of tea samples when OPLS–DA was used.

Barone et al. (2016) assayed MEC combined with individual congener PCB data provided by GF–AAS, HVG–AAS, FAAS and GC–MS with AIS and ECD. Although no statistical interspatial differentiation between Chinese and Indian teas was achieved, when the contents of individual PCB congeners, i.e., when the contents of PCBs 110, 114, 123 and 142, were considered, it was possible to discriminate black tea samples originating from India and China. For the green teas, no clear distinction between the concentrations of these four congeners was obtained. Because samples are contaminated at low levels, they do not pose any threat from a safety point of view. However, because tea beverages are frequently consumed, it has been emphasized that maximum safe levels should be provided, especially for toxic elements.

Brzezicha-Cirocka et al. (2016a) determined the concentration of MEC by FAAS of different types of green tea from three Asiatic countries. The FA data revealed the greatest degree of differentiation between the samples from China, Japan and India (Fig. 12.3). The first two F values cumulatively explain 69.28% of the total variance. CA also provided highly satisfying classification of samples, confirming the results provided by FA.

According to Pilgrim et al. (2010), ICP–MS and EA–IRMS were applied to quantify MEC and SIR in tea (*Camellia sinensis*) samples from different sampling sites in four Asiatic countries. The DA results confirmed the ability to assign samples to their growing region, although tea samples from Taiwan and China initially were unseparated. However, a clear discrimination of these two regions via the construction of a scatter plot of DF1/DF3 was successfully achieved using these three DFs, with 97.6% of the samples correctly classified. A separate dataset consisting exclusively of tea samples from the Nilgiri region of southern India was additionally processed by DA to assess intraspatial differentiation. A total of 97.4% of the CV cases were correctly classified, with the exception of tea samples from the Havukal and Kairbetta gardens, which overlapped. This could be caused by the gardens being close to each other, but after the samples clearly separated from each other were removed, it was possible to identify all the Nilgiri tea samples and their locations correctly into the specific production garden.

12.2.1.2 South American Countries

Marcelo et al. (2014) reported that with the use of ICP–MS and ICP–OES, it was possible to obtain MEA data that were transformed chemometrically. The obtained data were useful for characterizing the geographic provenance of commercial yerba mate samples from four countries in South America, i.e., Argentina, Brazil, Paraguay and Uruguay. The scores of the first three PCs, representing 68.21% of the total variance, incompletely discriminated the samples according to their interregional origin. However, among the supervised methods employed (kNN, SIMCA, PLS–DA and SVM–DA), the best geographical classification with 100% correctness was

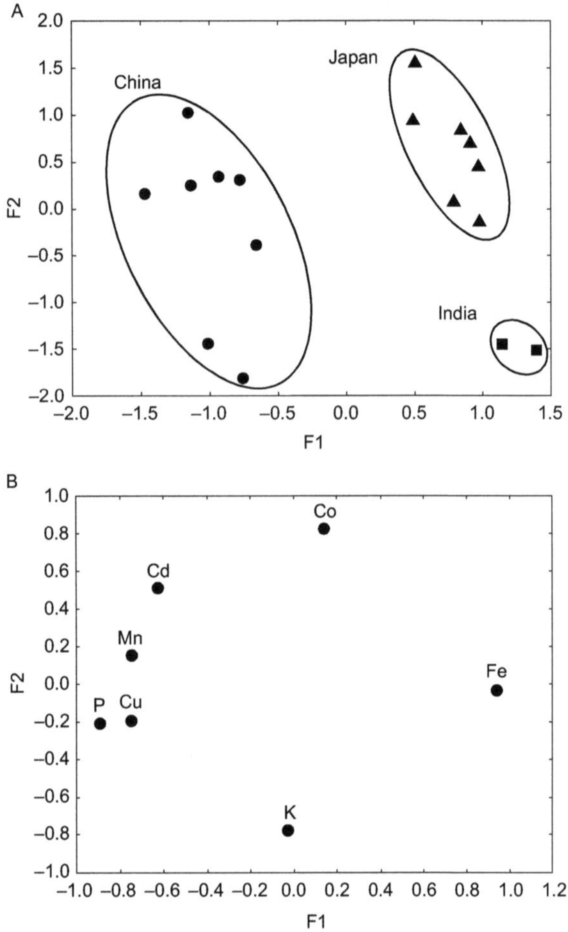

Fig. 12.3 (**a**) Scatterplot of the factors distinguishing green tea samples from various geographical origins. (**b**) Scatterplot of loadings for seven elements in green tea samples from various geographical origins. (After Brzezicha-Cirocka J et al., 2016a. Environ Monit Assess 188:183. https://doi.org/10.1007/s10661-016-5157-y. With permission)

achieved using SVM–DA after all the elements in the 54 tea samples were considered. Highly satisfactory results were also obtained by using 11 elements quantified via ICP–OES.

12.2.1.3 Countries Across Multiple Continents

Mbaye et al. (2013) differentiated 13 types of green tea samples using EDXRF to quantitatively establish MECs. The first three PCs, accounting for 99.29% of the total variance of the dataset, enabled the selection of the most discriminative elements, i.e., Ca, K, Mn and Fe. On the basis of the SVD, HCA and ANN multivariate techniques, the correctness of the classification of tea samples according to geographical provenance was satisfactory. The MLF–NN model was highly useful because it provided 100% correct classification for the tea varieties.

Cengiz et al. (2017) categorized tea samples from different locations in the Black Sea region (Turkey) and those imported from five other countries in Africa and Asia on the basis of SIR values provided by EA–IRMS. The classification of the Turkish tea samples and imported tea samples was achieved using PCA (with the first two PCs explained by 89.75% of the total variance) as well as HCA. It was concluded that, on the basis of the stable isotopic ratios used in the study, it was possible to distinguish the geographical provenance of Turkish tea from that of imported tea samples.

Moreda-Piñeiro et al. (2001) assessed the MEC data from ICP–AES and ICP–MS of Asian and African tea samples originating from 13 countries, including conventional blends. Data pretreatment was used for the three datasets before the application of multivariate models, i.e., PCA, CA, LDA and SIMCA. The half-range and central value transformation provided more appropriate classification of the tea samples than did classical normalization. Compared with LDA, SIMCA is a more adequate model; therefore, a proper classification of various classes/categories is achieved when a satisfactory SIMCA result is obtained.

According to Moreda-Piñeiro et al. (2003), ICP–AES and ICP–MS were used to classify tea samples originating from Africa, Asia and Oceania. PCA with the first three PCs explaining 78% of the total variance revealed that Cr, Cs, Fe, Ba, Ca, Rb, Sr, Pb and Al were the most descriptive variables of the tea samples. A discrimination between African and Asian teas was observed. Similar results were obtained after the CA technique was used (Fig. 12.4). The application of LDA enabled the correct classification of 100.0 and 94.4% of African and Asian tea samples, respectively, whereas for SIMCA, the correct classification of 100 and 91.79% of cases was achieved for the African and Asian groups, respectively. Moreover, LDA enabled 100% correct identification for tea samples from China, India and Sri Lanka

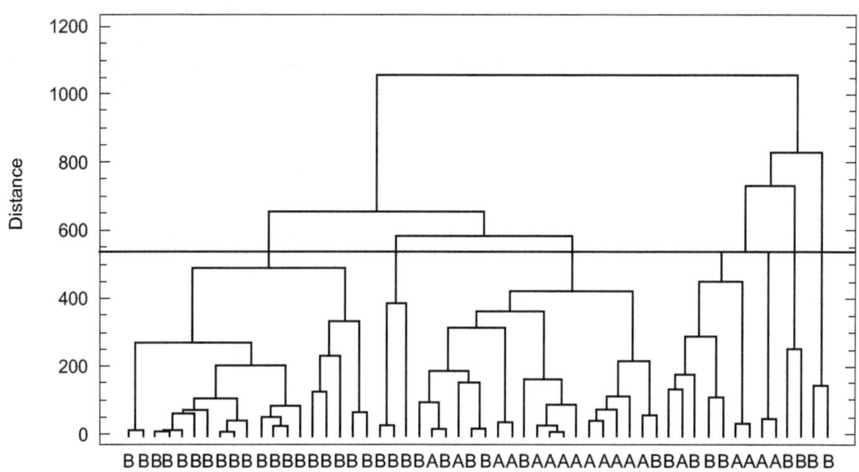

Fig. 12.4 Dendrogram of cluster analysis. (**a**) African tea samples; (**b**) Asian tea samples. (After Moreda-Piñeiro A et al., 2003. J Food Compos Anal 16:195–211. https://doi.org/10.1016/S0889-1575(02)00163-1. With permission)

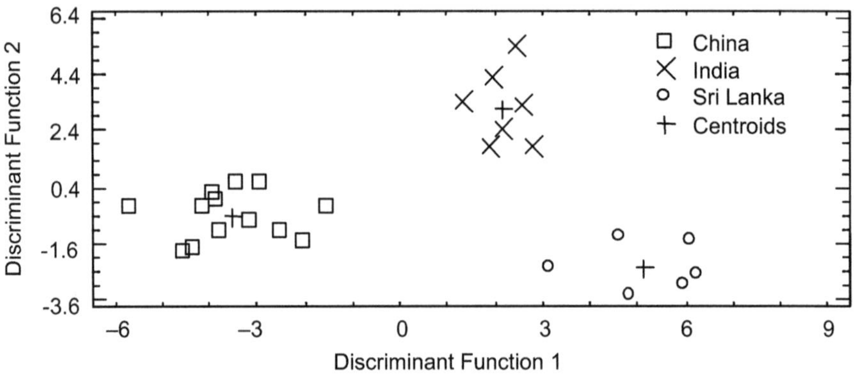

Fig. 12.5 Projections of tea samples from China, India and Sri Lanka in the space formed by the two discriminant functions after LDA. (After Moreda-Piñeiro et al., 2003. J Food Compos Anal 16:195–211. https://doi.org/10.1016/S0889-1575(02)00163-1. With permission)

(Fig. 12.5). However, a clear classification using SIMCA, i.e., a correct classification rate of 100%, was achieved only for the Chinese teas. In addition, the data obtained showed that it was possible to assign unknown samples and tea blends according to their country of origin (Moreda-Piñeiro et al. 2003).

12.2.2 Intercultivar/Type Differentiation

Chen et al. (2009) categorized eight cultivars of tea leaves and tea infusions by using ICP–AES data combined with chemometric transformation. For example, the first three PCs with many discriminative variables, i.e., Cu, P, Pb, Mg, Zn, Co and Fe, explained 67.8% of the total variance and clearly separated the eight Chinese tea cultivars. HCA data were in agreement with those obtained by PCA. The intercultivar differentiation was also correctly performed with 100% accuracy in total by LDA and BPNN analysis. This means that each cultivar presents a distinctive element fingerprint and that the MEC of teas can be a useful predictor for distinguishing all eight Chinese tea cultivars.

McKenzie et al. (2010) grouped Chinese tea leaf samples to differentiate the five tea varieties (white, green, black, oolong and pu-erh teas) that originated from China. ICP–AES combined with LDA and PNN was applied to construct classification models with overall classification performances of 81% and 97%, respectively. The latter model was built using Al, Ba, Ca, Cu, Fe, Mn, K, Sr, S and Zn as input variables.

According to Fernández et al. (2002), ICP–AES data were effective for characterizing several green and black tea beverages, infusions and instant and tea soft drinks. PCA data enabled them to only differentiate between the soft drinks and the remaining types of tea beverages. LDA was a more effective approach, allowing separation among four types of samples, i.e., infusions, soft drinks and two types of instant teas. Moreover, tea infusions are good sources of elements; this is especially the case for black teas, which are particularly good sources of Mn and K.

de Gois et al. (2016) used Mn, Ni, Rb, Sr and total phenolic compound data and antioxidant activity data obtained by SS–HR–CS ET AAS, ICP–MS, DPPH and ABTS radical scavenging methods to study six Brazilian species of stimulant plants. PCA revealed that the first two PCs explaining 72.17% of the total variance enabled successful distinction between species (catuaba, ginseng, ginger and mate), with the exception of coffee and guarana, indicating that these data could be used to discriminate between different stimulant plants.

Herrador and González (2001) used ICP–AES to quantify the MEC of commercial tea—black, green and oolong teas. With the PCA technique, unclear separation of the three types of tea was achieved. According to CA, nonfermented samples were mostly separated from fermented (fully or partially) samples. As shown in Fig. 12.6 and Fig. 12.7, complete classification of tea samples was achieved when supervised learning pattern recognition methods were used, i.e., LDA and BP–ANN. The latter was the most effective classification procedure, achieving a recall rate of 100% and a prediction ability of approximately 95% on average. These data are slightly better than those provided by LDA (the average recall rate and prediction ability were approximately 95% and 90%, respectively). It was possible to achieve the best differentiation of the three types of tea on the basis of Zn, Mn, Mg, Cu, Al, Ca, Ba and K as discriminative descriptors.

Paz-Rodríguez et al. (2015) applied HR–CS FAAS to investigate the infusions of black and green teas, rooibos and tea with seaweed. Preliminary classification using

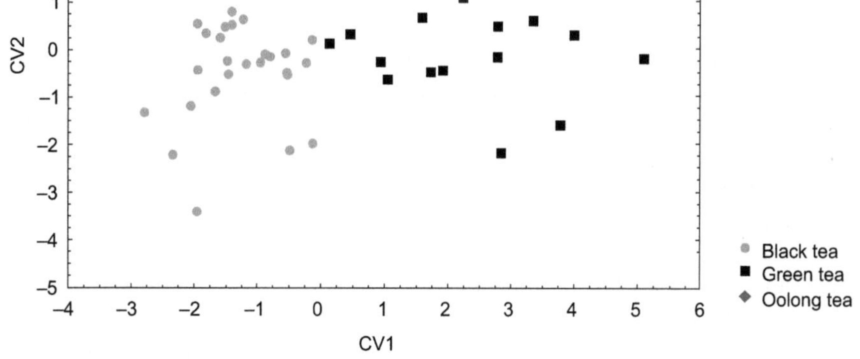

Fig. 12.6 Scatterplot of tea samples using CV1 and CV2 as axes. (After Herrador MA and González AG, 2001. Talanta 53:1249–1257. https://doi.org/10.1016/S0039-9140(00)00619-6. With permission)

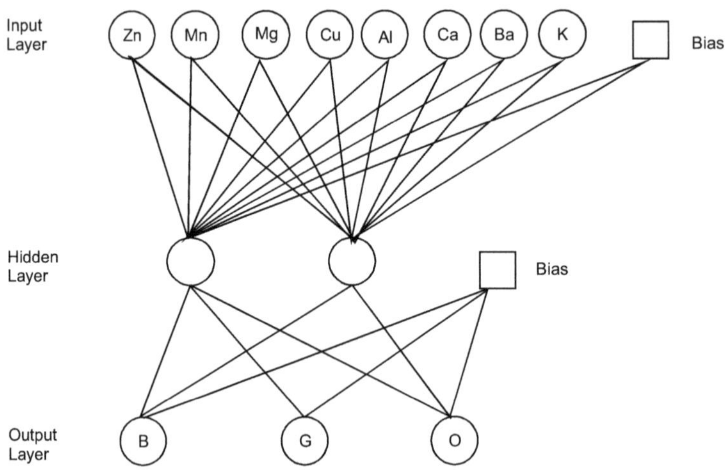

Fig. 12.7 Architecture scheme of the multilayer neural network used in this work (After Herrador and González, 2001. Talanta 53:1249–1257. https://doi.org/10.1016/S0039-9140(00)00619-6. With permission)

PCA and CA was successful for infusions made from rooibos and tea with seaweed, in contrast to black and green teas. With the former technique, the first three PCs explained 75.56% of the total variance, with mostly discriminative variables such as Cu, Mn, Zn, Ca and Na. The PCA results were consistent with those described for CA.

Brzezicha-Cirocka et al. (2016b) reported that MEC quantified by FAAS was applied to characterize Chinese dark (pu-erh), fruit tea leaves and their infusions. As shown in Fig. 12.8, the CA technique clearly distinguished tea samples in terms of tea quality and type with respect to the elemental profile composition of the analysed samples. For FA, two factors (F1 and F2) cumulatively explained 61.36% of the total variance. Elements such as Mg, Ca, K, Na, P, Mn, Cu, Fe, Zn, Co, Cr, Ni, Cd and Pb were good descriptors for the diversification of the analysed samples. It was concluded that F1 was the factor responsible for the differentiation of samples with respect to the tea type (pu-erh or fruit teas), whereas F2 provided additional F1 information concerning the studied samples. Therefore, both factors were responsible for the diversification within the group of fruit teas, i.e., pure fruit tea from black tea with fruit additions as well as pu-erh tea from the perspective of its confection (loose-form tea bags), which are usually associated with their quality. Multivariate techniques are efficient tools that are helpful in the differentiation of tea samples in terms of fermentation type, technological processing, quality, fermentation type and confection.

Brzezicha-Cirocka et al. (2016c) processed tea infusions in the context of chemical composition related to three parameters, i.e., $C_2O_4^{2-}$ (determined via the manganometric method), Mg and Ca. The oxalate content and Mg/Ca ratio values were processed chemometrically via the CA technique. Figure 12.9a presents a dendrogram, which was built of three main clusters containing particular kinds of tea samples. Compared with the remaining variables, it was possible to differentiate tea

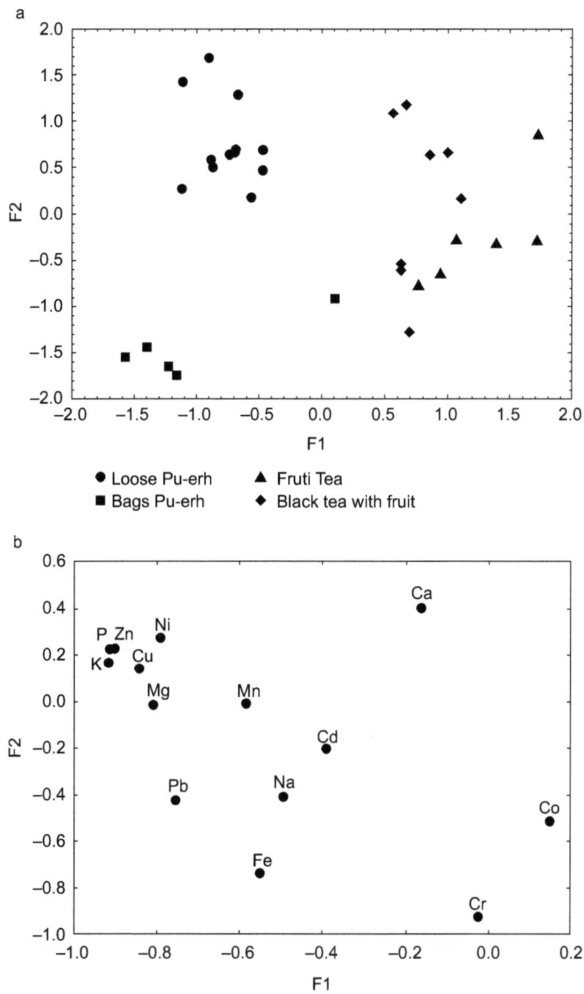

Fig. 12.8 (**a**) Scatterplot of object samples of two discriminant functions of all analysed tea samples. (**b**) Scatterplot of loadings for 14 elements in all analysed tea samples. (After Brzezicha-Cirocka J et al., 2016b. Biol Trace Elem Res 174:240–250. https://doi.org/10.1007/s12011-016-0669-4. With permission)

type with respect to oxalate, Mg and Ca contents. As shown in Fig. 12.9b, Mg is responsible for the differentiation of green tea. The black and dark tea samples are described by oxalate and Ca, respectively. The highest levels of oxalate were found in black and dark tea, and the lowest were found in green tea. The ratio of Mg to Ca in tea infusions was too low to protect against the formation of kidney stones; thus, it seems reasonable to use magnesium pharmaceuticals in the treatment of kidney stones. An alternative option for people with hyperoxaluria or a tendency to form kidney stones is to consume tea, especially a darker colour, in moderate quantities.

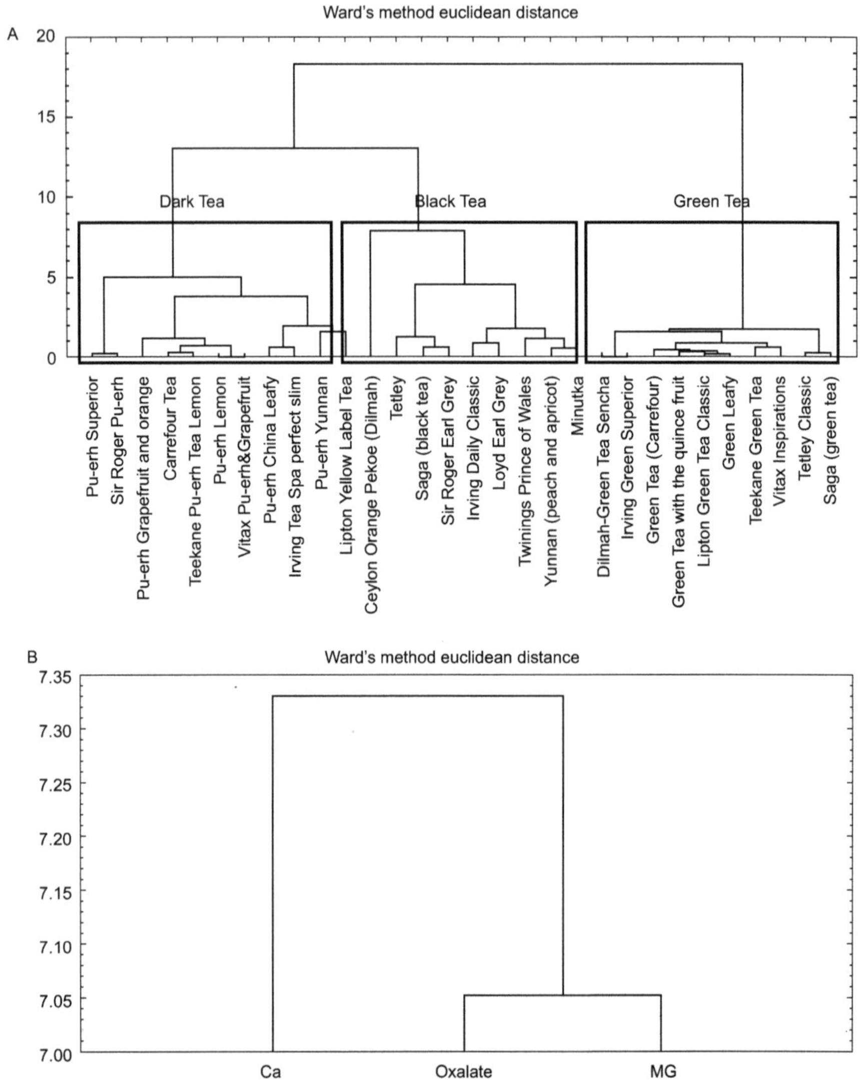

Fig. 12.9 (**a**) Hierarchical dendrogram for the analysed tea samples as objects from the perspective of the kind of tea. (**b**) Hierarchical dendrogram for the analysed elements in tea samples according to the type of tea. (After Brzezicha-Cirocka J et al., 2016c. Eur Food Res Technol 242:383–389. https://doi.org/10.1007/s00217-015-2548-1. With permission)

12.2.3 Interregional and Intercultivar/Type Differentiation

Attempts were made by Lim et al. (2021) to examine MEC by using EDXRF with the support of multivariate models to classify tea into five major geographical regions that produce most of the global tea. Only the first two PCs were used, with PC1 accounting for much of the explained variance of >95%. This model enabled the five geographical regions to be clearly separated since the tea samples from E Asia, S Asia and SE Asia overlapped with some samples originating from Africa and other countries. This means that Asian countries and non-Asian countries were successfully distinguished. In the case of Design I (E Asia, S Asia, SE Asia, Africa and other countries), models such as PLS–DA and SIMCA provided average overall correct classification rates between 54.4% (sensitivity) and 83.1% (specificity) and 70.6% (sensitivity) and 89.1% (specificity), respectively. The models also enable classification of tea type. For Design II, i.e., Asian and non-Asian countries, the highest average correct classification rates of 83.7 and 84.7% for SIMCA and PLS–DA, respectively, were achieved. It was concluded that building multivariate classifiers allows a broad classification with respect to geographical origin (Asian vs. non-Asian) as well as tea type with no sample preparation and at low cost of analysis. Notably, wholesalers and retailers require a rapid and reliable screening tool to support the provenance and type of tea.

Diniz et al. (2015) noted that MEC with other chemical parameters combined with computational techniques was an appropriate approach to classify tea samples from Argentina (black tea and green tea) and Sri Lanka (black tea), especially with their infusions. PCA was performed with the first two PCs employing the determination data for parameters such as moisture, total ash, caffeine, total polyphenols and fluoride, elements in tea leaves and elements in tea infusion datasets. For example, the elements present in tea infusions (Al, Mn, Mg, Ca, P and K) tend to separate among the three studied types of tea, with the first two PCs explaining 100% of the total variance. Among the other multivariate techniques, such as SIMCA, PLS–DA and SPA–LDA, the latter is the most reliable, providing, in some approaches, 100% discrimination, with the best classification results obtained when the elemental profile in tea infusions was used (Fig. 12.10). In the case of only three major elements, i.e., K, Al and Mg, it was possible to achieve a simpler, faster and less expensive discriminative methodology.

Brzezicha-Cirocka et al. (2017) reported that FAAS data processed chemometrically could be used to characterize black tea samples in terms of their geographical and botanical origin. FA, with the first two factors explaining 59.2% of the total variance, tended to group the tea samples relative to their origin from India, Ceylon, China and Kenya, with visualization of the analysed element distributions responsible for such a factorial distribution (Fig. 12.11). Moreover, FA (F1 and F2 accounting for 57.2% of the total variance) and CA discriminated between tea-type samples (Ceylon, Assam and Darjeeling) from Indian and Ceylon plantations by considering the element factorial profile, which revealed diversification of teas originating from various plantations within a single country. It was concluded that the recommended approach enables the differentiation of samples from the perspective of their intercountry and intertype differentiation and recognizes plantations within the common region.

Fernández-Cáceres et al. (2001) reported that ICP–AES integrated with a multivariate tool was applied for the study of green, black and instant teas from five countries. PCA was unable to clearly classify the tea samples according to their intertype differentiation. However, PCA enabled the selection of variables such as

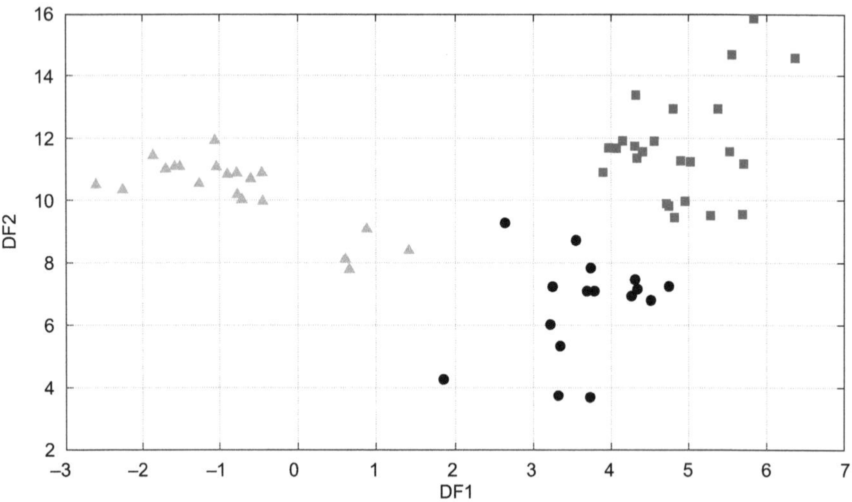

Fig. 12.10 Fisher's discriminant score plot obtained by using the parameters (K, Al and Mg) selected by SPA–LDA for the elements in the infusions. Argentinean black tea (■), Sri Lankan black tea (●) and Argentinean green tea (▲). (After Diniz PHGD et al., 2015. J Food Compos Anal 39:103–110. https://doi.org/10.1016/j.jfca.2014.11.012. With permission)

Fe, Ti, Na and Al, which had the greatest discriminating potential. Pattern recognition by LDA and ANN provided highly satisfactory results in the classification of tea varieties. Moreover, discrimination between Asian and African teas and between Chinese, Japanese, Indian and Sri Lankan teas was achieved (Fig. 12.12). On the basis of LDA, all the samples were correctly classified, representing a prediction ability of 93.5%. The MLP results were better than those obtained from LDA because of the prediction ability, with LOOCV accounting for 95.6% of the hits on average.

Marcos et al. (1998) used a combination of ICP–AES and ICP–MS with chemometric processing to classify teas from 10 Asian and African countries. On the basis of PCA data, modelling was performed with the second and third PCs because they were far more specific. A greatly improved plot was obtained since almost all of the teas belong to a regional grouping and even those not included in a grouping would seem to be logically located. Therefore, it was possible to discriminate between African and Asian teas as well as between Chinese teas and other Asian teas corresponding to three main groupings identifying certain tea types. It was also possible to differentiate between three main tea type groupings among Chinese, African and Indian samples together with Japanese and Bangladesh samples of only one type of tea. A similar structure was achieved via HCA. It was concluded that PCA modelling for the ICP–MS data was performed with PC2 and PC3 because of their far more specific characteristics; hence, the climatic conditions (combinations of factors such as rainfall, higher humidity and temperature) of the producing areas as well as the soil composition could be identified as influencing factors.

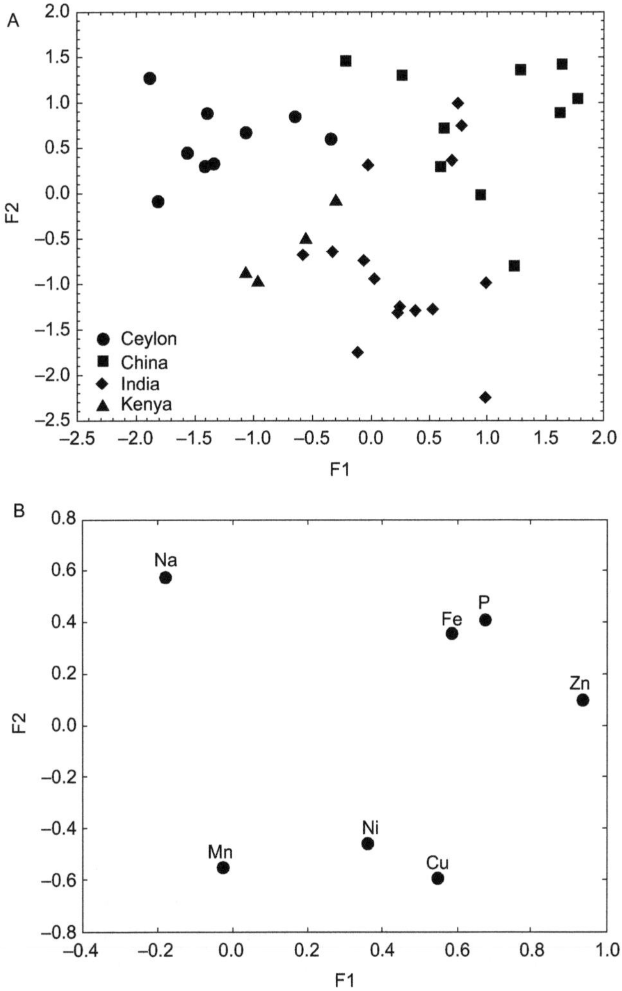

Fig. 12.11 (**a**) Scatter plot of samples from all tea samples from Ceylon, China, India and Kenya. (**b**) Scatter plot of loading for elements in all the analysed tea samples from Ceylon, China, India and Kenya. (After Brzezicha-Cirocka et al., 2017. Biol. Trace Elem Res 176:429–441. https://doi.org/10.1007/s12011-016-0849-2. With permission)

12.2.4 Differentiation of Other Parameters

12.2.4.1 Verifying Authenticity/Adulteration

According to Zhang et al. (2018b), ICP–MS was used to quantitatively examine traditional FHDC tea samples and other methods were integrated with chemometric analysis. The LDA data clearly separated the FHDC tea samples. The classification rate of tea samples between the production regions of Wuyi Mountain, Hangzhou and Phoenix Mountain reached 93.3%, with similarity degrees. A 96.6% predictive

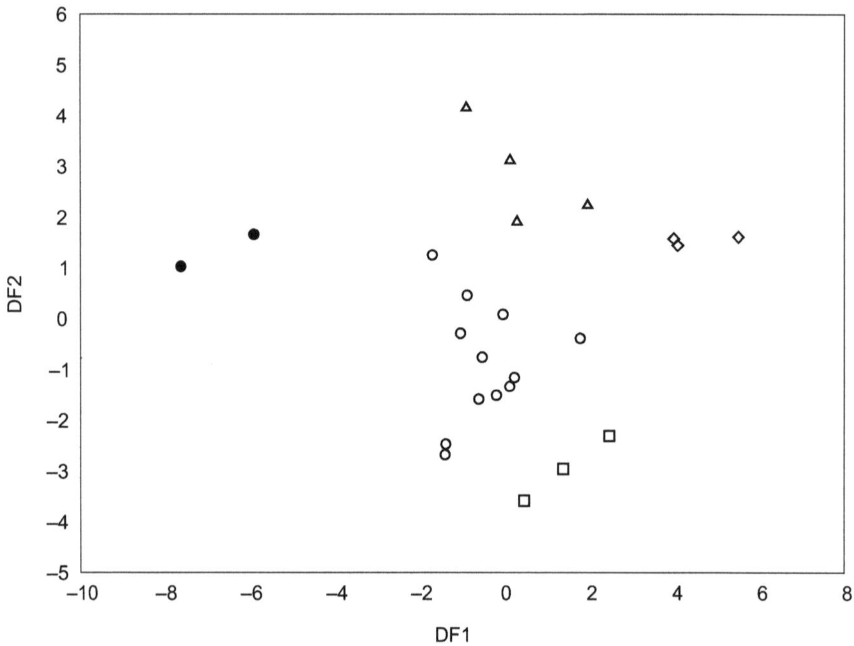

Fig. 12.12 Scatter plot of the two discriminant functions for tea samples with known geographical origins: China (o), Japan (□), India (◇), Sri Lanka (Δ) and Kenya (●). (After Fernández-Cáceres PL et al., 2001. J Agric Food Chem 49:4775–4779. https://doi.org/10.1021/jf0106143. With permission)

ability of the CV of the LDA model was achieved. This means that the applied methodology was useful for identifying the authenticity of FHDC.

Ma et al. (2016) assayed the Chinese green tea Dongting Biluochun (DTBLC) as a model to classify its geographic origin on the basis of multielement analysis, including REEs, via ICP–MS. PCA, CA and LDA were used to distinguish the BLC tea profiles from different geographical origins. LDA was a useful tool, providing high recognition (98.2%) and satisfactory predictive ability (96.4%). This means that the developed model offers a powerful and practical tool that is appropriate for fingerprinting unclassified BLC samples. In addition, it was possible to verify the authenticity of DTBLC that originated from different protection areas, and a classification rate as high as 95.2% was obtained. Notably, the methodology can also be effective for other green teas from nearby producing regions. It was concluded that this study provided a high classification rate, good sensitivity and a powerful tool to manage the quality control of green tea because of its significance for the tea industry.

Deng et al. (2020) reported that EA–IRMS and ICP–MS combined with multivariate models were effective tools for studying 623 green tea samples from different production regions around China from 2012 to 2016. The RF model achieved a high classification accuracy of 97.6% in discriminating XHLJ tea samples from those from other regions and correctly identified the tea samples from surrounding regions, with an accuracy of 97.9% (Fig. 12.13). Moreover, the discrimination of tea subsequently harvested in the following years showed good efficiency, with predictive accuracies >91%. The most important geographical discriminators were Rb, Mg, $\delta^{13}C$ and K, which enabled the identification of the geographical origin of tea. It was concluded that the RF model had greater classification accuracy than chemometric models such as PCA–LDA, PLS–DA, SIMCA, CART, ANN and SVM. Owing to the use of this methodology, it was possible to provide a new perspective for the application of predictive models that utilize historical data for geographical authentication of agricultural products with PDO.

Liu et al. (2019b) coupled EA–IRMS and ICP–MS data to classify green tea samples on the basis of their SIRs and elemental profiles. The results revealed that PCA and follow-up LDA were successful in processing the original obtained dataset. While PCA (with PC1 and PC2 explaining 42.14% of the variability) can generally classify the origins of teas from the different provinces, PCA (with PC1 and PC2 accounting for 32.22%) was unable to clearly separate Longjing green tea from Westlake regions and the surrounding regions in Zhejiang Province. Therefore, new LDA models were used, which were built on the first few PCs accounting for >90%. This supervised tool was effective for verifying the geographical origin of green tea from different provinces (Fig. 12.14). Consequently, PCA and follow-up LDA established the geographical origin, even in the case of zones localized in the same province, with discrimination accuracies higher than 92.3 and 87.8%, respectively. The methodology offers an accurate and effective verification tool for the

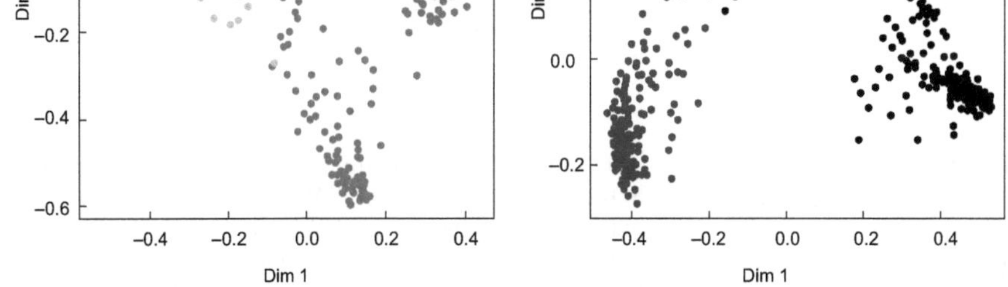

Fig. 12.13 Multidimensional scaling plots generated from a random forest model to discriminate Westlake green tea (XHLJ) from (**a**) other regions across China (NXHLJ) and (**b**) the surrounding Zhejiang region (SURD), with predictive accuracies of 97.6% and 97.9% (Table S3), respectively. Dim1 and Dim2 are the multidimensional scaling coordinates. Tea samples with similar geographical origins are indicated by closely spaced points, whereas tea samples with dissimilar geographical origins are represented by more dispersed points. (For interpretation of the references to colour in this figure legend, the reader is referred to the Web version of this article). (After Deng X et al., 2020. Food Contr 107:106807. https://doi.org/10.1016/j.foodcont.2019.106807. With permission)

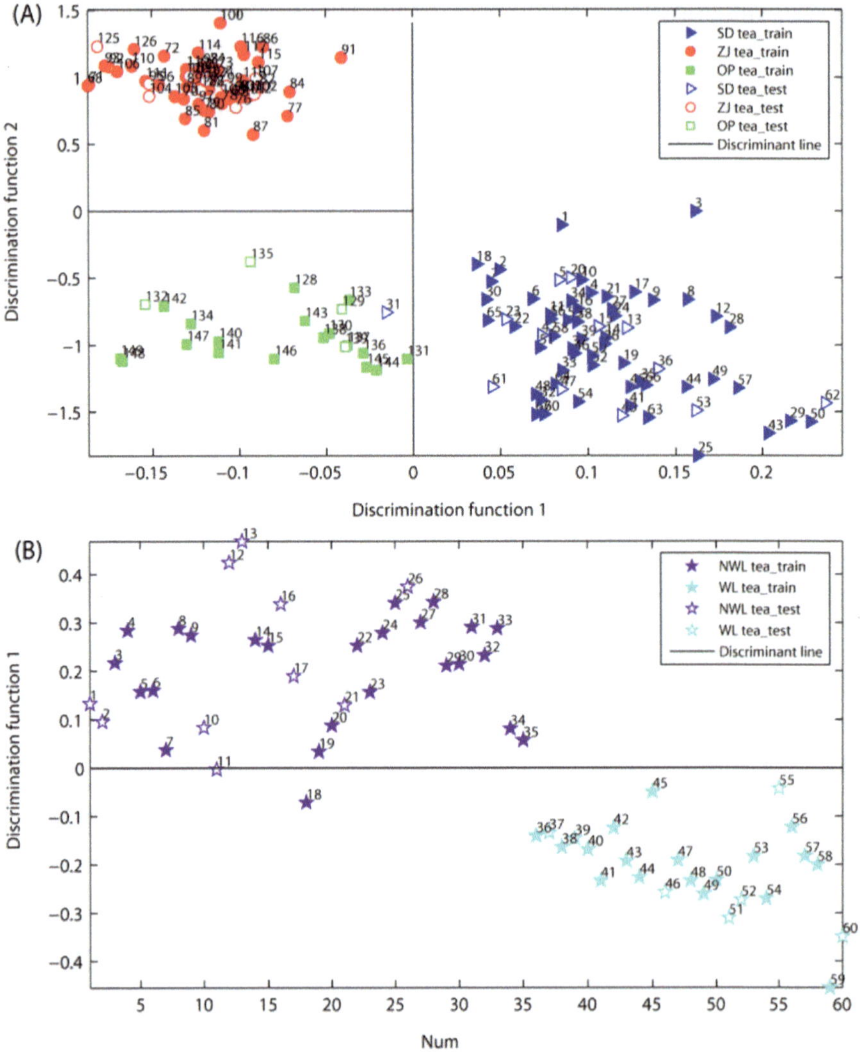

Fig. 12.14 LDA discrimination results of tea samples (**a**) from different provinces: Zhejiang Province (ZJ), Shandong Province (SD) and other provinces (OP); Westlake Longjing tea samples (**b**) from the Westlake region (WL) and the surrounding regions (NWL) in Zhejiang Province. The solid points represent training samples, and the hollow points represent testing samples. (After Liu H et al., 2019b. Rapid Commun Mass Spectrom 33:778–788. https://doi.org/10.1002/rcm.8405. With permission)

geographical provenance of Chinese green tea and is a useful approach to prevent fraudulent mislabelling of high-value green tea.

Liu et al. (2020c, d) integrated ICP–MS, ICP–OES and EA–IRMS to classify Xiangzhujing pu'erh tea samples in the harvesting seasons of 2014, 2015, 2016, 2017 and 2018. Chemometric methods are potentially powerful tools for authenticating Chinese tea. HCA and PCA revealed that tea from the five harvesting years slightly overlapped. However, using PCA based on 78 statistically significant variables enabled better discrimination of samples with respect to producing years. Supervised models such as the LDA, PLS–DA and BP–ANN algorithms were explored to authenticate pu'erh tea. The authentication performance of BP–ANN and LDA was better than that of PLS–DA, with a recognition ability of 100% and a prediction ability of 100% (Fig. 12.15.). Analytes such as Mn, ^{68}Zn and ^{203}Tl were the discriminative descriptors for the reliable verification of pu'erh tea authentication in different production years.

Liu et al. (2021) reported that ICP–MS and ICP–OES combined with four multivariate models were highly reliable potential tools for effective authenticity detection of Chinese Chongqing tuo tea with four formulations. PCA with the first two PCs accounting for 44.3% of the total variance did not clearly separate the four formulations of tea samples. It was demonstrated that 84 statistically significant elements and stable isotopes were useful as fingerprints for determining the authenticity of tuo tea with formulas. Both PLS–DA and LDA, with ^{114}Cd, ^{95}Mo, ^{85}Rb, Co, ^{133}Cs and P as the most important descriptors, were more discriminative models than HCA and PCA. Under validation by CV, the test and blind datasets as well as

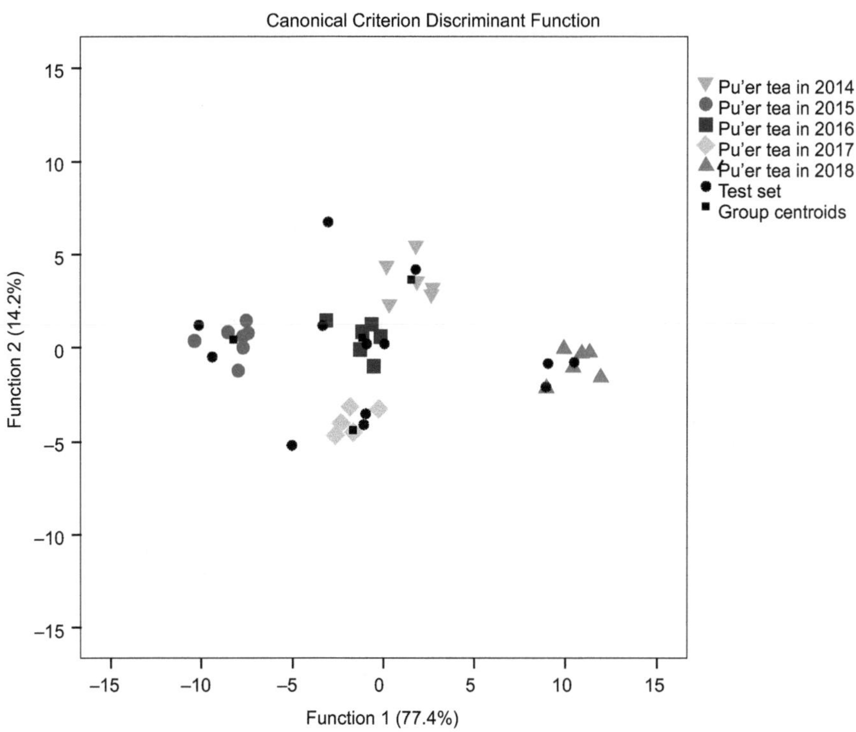

Fig. 12.15 Authentication of pu'erh tea in five production years (from 2014 to 2018) on the basis of the 18 multielemental stable isotope ratios ($p < 0.05$). (After Liu H et al., 2020c. J Food Drug Anal 28:248–260. https://doi.org/10.38212/2224-6614.1059. With permission)

the precisions of prediction of PLS–DA and LDA were > 90%, whereas LDA had the best prediction rate, with a precision of 100%. It was concluded that this approach is a promising for determination of the authenticity of teas worldwide.

Liu et al. (2020b) aimed to combine two instrumental techniques (EA–IRMS and ICP–MS MEA) with multivariate tools (HCA, PCA, PLS–DA, BP–ANN and LDA) as a convenient approach to verify the authenticity of Chinese Chongqing tuo tea according to different formulas. By using BP–ANN, it was possible to achieve 100% for the total correct authentication rates for model training and 91.2% for external validation. The most reliable results were achieved using LDA, since the total correct authentication rates for model training, cross-validation and external validation were 100%, 100% and 96.9%, respectively. Moreover, the δ^2H, $\delta^{18}O$, $^{98}Mo/^{95}Mo$, $^{96}Mo/^{95}Mo$ and $^{98}Mo/^{96}Mo$ ratios enabled the most successful authentication of Chinese tuo tea with respect to formulas.

Baskali-Bouregaa et al. (2020) used LIBS and ICP–HRMS to study tea samples consisting of leaves containing small pieces of fruit (lemon and red fruits) or flowers (jasmin tea). The elemental content data corresponding to the LIBS spectra were subjected to PCA, PLS–DA and FDA. The selected element lines were useful for differentiating most sample groups. Variables such as Ba, Ca, Mg, Sr, R and Na detected by the LIBS were important for classifying samples according to their geographical origin. Other elements, e.g., P, S, Fe and B, were also discriminative descriptors. PCA enabled the identification of the origin of four unknown tea samples, and the use of FDA or PLS–DA after PCA examination of the LIBS/ICP–MS data resulted in effective classification of the tea samples. It was concluded that a five-class model gave the best correct assignment (94.2%). In addition, this methodology enabled the origin of four unknown teas to be determined.

According to Chen et al. (2024), samples of pu-erh tea leaves at three ages and parent soils from three areas in Yunnan were studied on the basis of the MEC and SIR contents determined via ICP–MS and EA–IRMS, respectively. The manganese content was considerably influenced by the region and tree age interaction. Owing to the use of ANOVA, PCA (with the first four PCs accounting for 82.54% of the total variance) and SDA, 24 parameters were determined to be closely related to geographical origin. Moreover, 100.0% SDA of the original rates and cross-validation of the discriminant rates with only 6 variables, such as $\delta^{13}C$, $\delta^{15}N$, Mn, Mg, La and Tb, was possible. This combined methodology was an accurate and effective tool for establishing the geographic origin of pu-erh tea samples.

Zhu et al. (2023) developed a comprehensive elemental fingerprinting method to examine ripened pu-erh tea (RPT) using ICP–MS for accurate determination of 69 elements. The concentration data were transformed chemometrically (PCA and DA). The method was applied to assay the impact of major geographical origins and the pile fermentation process on the MEC of the RPT. In total, 35 and 31 elements significantly varied in pu'erh relative to the Lincang region and relative to the Xishuangbanna region ($p < 0.05$), respectively. DA was the most effective tool for discriminating the three producing regions. Moreover, predominant elemental accumulation was achieved during pile fermentation since 36 elements were highly concentrated in the RPT after pile fermentation. This study provides comprehensive data for the origin traceability of RPT and advances the understanding of the elemental enrichment mechanism during pile fermentation of RPT.

Zhang et al. (2020a) examined tea leaves and soils from three regions of Guizhou (China) with respect to their quantitative elemental composition via ICP–MS and ICP–AES. One-way ANOVA revealed a significant relationship ($P < 0.05$) between concentrations of 16 elements in tea leaves and parent soils related to interregional differentiation. Among the multivariate models, PCA and S-LDA were used; the former differentiated interspatially tea leaf samples owing to the use of 16 variables as discriminative descriptors. In addition, the S-LDA model provided a 100% differentiation rate and six effective markers (P, Sr, U, Pb, Cd and Cr) for the geographic traceability of tea leaves. However, the model was unable to offer an accurate discrimination rate of geographical origin by tea cultivars. Interregional differentiation of the elements in the studied tea was significantly related to that in associated soils and thus could be used to verify the geographical origin of the tea samples.

Zhang et al. (2012b) selected methods such as EA/IRMS and HT–RPLC/IRMS to classify naturally different caffeine samples, including four major types of tea leaves, coffee beans, maté leaves, guaraná powder, synthetic caffeine samples and tested drinks from different sources. This approach is a useful methodology for distinguishing two groups of samples according to caffeine $\delta^{13}C$ values for natural and synthetic caffeine. The proposed method is an effective approach for verifying the authenticity of caffeine-containing drinks.

12.2.4.2 Intervarietal and/or Interregional Differentiation Related to Seasonal Variability

Zhou et al. (2019) combined EA–IRMS with a chemometric tool to achieve a helpful approach for the classification of tea samples originating from four regions of China. The classification data of PCA, with the first two PCs explaining 88.6% of the total variance, were not satisfactory because four samples from Fujian Province were misclassified. Compared with the PLS–DA model, the OPLS–DA model was highly effective, with a 96.08% recognition rate, whereas the SLDA model achieved 100% correct classification. With respect to the seasonal variations in Fujian Province, the δD and $\delta^{18}O$ values were highly significant, indicating that rainfall had an important role in the isotope ratios in samples with seasonal variations. Both the PCA and HCA methods achieved 100% recognition rates for spring and autumn in the case of the Tieguanyin teas.

Huang et al. (2019) noted that ICP–OES was an appropriate method for studying Camelia tea leaves originating from Yunnan Province (China) from 2014–2016. PCA differentiated the tea samples with respect to premonsoon (spring) and monsoon (summer) periods. Significant differences were observed for the Fe, Ca, Mg, Mn, Al and Ba levels, which were greater in the monsoon season than in the premonsoon and postmonsoon seasons, whereas the Pb, K, Cu, Zn and Na concentrations were greater in the premonsoon and postmonsoon seasons. Although changes in elemental levels occur, extreme rainfall events do not affect tea leaf quality. The differences in micro- and macroelement levels at both elevations (1790 m and

1180 m) in Menghai County, Yunnan Province, for each harvest corresponded to the first two PCs, which explained 77% and 69% of the total variance, respectively. For example, the contents of Fe, Ca, Mg, Mn, Al and Ba were greater in June (monsoon) than in March (premonsoon) and September (postmonsoon).

Zhao and Zhao (2019) integrated ICP–MS with chemometric processing as an adequate tool to differentiate fresh tea leaves (closed buds and the associated two young leaves) from three major tea-growing regions in China in spring, summer and autumn of 2015 and 2016. LDA achieved overall discriminant accuracies (predicted, original) of 97.5% (2015) and 95.1% (2016), whereas in the case of the LOOCV, they were 95.1% (2015) and 87.7% (2016). The classification of tea leaves from 2 years achieved 100% accuracy. This means that the models have significant potential for discriminating the three production seasons (spring, summer, autumn) of tea samples, providing powerful technological support for the tea industry.

Ni et al. (2018) used three advanced instrumental tools, i.e., ICP–MS, ICP–AES and TC/EA–IRMS, integrated with chemometrics (PCA, LDA, PLS–DA and DT) to study 321 green tea samples from 11 different Chinese production areas. In addition, 107 blind samples were examined. In the case of PCA, the first two PCs accounted for 32% and 17% of the total variance. Two-way ANOVA revealed that there was significant difference between the sampling years. Interannual effects could be associated with climate control (precipitation and dry periods during tea tree growth) and elemental uptake. Under the validation by CV and the "blind" dataset, the prediction accuracies of the three methods, i.e., LDA, PLS–DA and DT, were greater than 70%. However, the DT method achieved the best performance, with an accuracy of 90%. Furthermore, for the discrimination of green tea, DT also presented the lowest error rate of 1.5%.

According to Zhao and Yang (2019), an integration of ICP—MS with chemometric processing was applied to study Chinese tea and the effects of geographical origin, harvest season, variety and their interactions. Geographical origin significantly influenced the variation in some REEs in tea leaves; hence, a significant relationship between the levels of REEs in tea leaves and the parent soils was observed. The S-LDA results were not satisfactory for identifying the geographical origin of tea leaves because the concentrations of most REEs are affected primarily by seasonal changes. However, the three varieties selected in this study had similar abilities to absorb REEs; hence, no significant impact on distinguishing geographical origin was observed. Therefore, only seasonal variations should be considered when REE fingerprinting is applied to verify tea authentication.

Zhao et al. (2017b) assayed samples of fresh tea leaves of three varieties from three regions during three harvest seasons. The levels of Mg, Ni, Rb, Sr, Cd, Cs, Ba and Pb measured via ICP–MS in tea samples were affected mainly by geographical provenance. Additionally, Al, Cu and U were significantly associated with variety, whereas Na, V, As, Se and Sn were highly related to the harvest season. The constructed discriminant models used for identifying tea geographical provenance are less affected by intercultivar and interseason variations, which is beneficial for verifying the authenticity of tea.

12.2.4.3 Interregional Differentiation and Other Parameters Related to Soil Geochemistry

Zhao et al. (2017a) classified tea leaf samples from three Chinese regions using ICP–MS combined with chemometrics. The samples used as calibration and validation sets to develop the model achieved a correct classification rate of 100%. One-way ANOVA with LDA or multiway ANOVA, RPA and SLDA were applied. A clear differentiation between tea related to parent topsoil and subsoil samples from three regions was attained. Na, Mg, Ca, Ni, Rb, Sr and Pb associated with parent soils are potential geographical identifiers of tea.

Lagad et al. (2013) reported that a combination of ICP–AES, MC–ICP–MS and EA–IRMS with PCA is a geographic tracing signature for Indian tea samples cultivated in four regions, i.e., Assam, Darjeeling, Munnar and Kangra. Variations in $\delta^{13}C$ were highly effective geographic indicators for the teas, and the $^{87}Sr/^{86}Sr$ values were statistically identical in the teas and the associated soil. PCA, with the first two PCs explaining 71% of the total variance, reliably differentiated samples according to four tea-producing regions.

According to Liu et al. (2019a), EA/IRMS was applied to study young (one bud with one leaf) and mature growth (older leaf) samples and parent soil samples from two depth layers. On the basis of chemometric evaluation, it was postulated that geographical traceability from a small growing region might be affected by possible interference factors. Significant differences were detected for the δ^2H and $\delta^{18}O$ values depending on leaf age. Therefore, leaf maturity effects should be considered since they can affect the precision and accuracy of models when the geographical origin, traceability and authenticity of tea are considered. Therefore, the influences of leaf age, species and soil depth on the authenticity and geographical origin of Chinese green tea should be considered.

Li et al. (2018) processed samples of tea leaves and soil from tea-growing areas by means of ICP–OES assisted by chemometric processing (PCA, CA and SIMCA). This approach enabled them to determine the corresponding relationships of elements between tea and parent soils. The results indicated the interspatial differentiation of mineral elements in tea and soil at the district scale. Elements such as Se, Mn, Cr and Ca were significantly correlated between tea and soil. CA and PCA (with the first two PCs explaining 90.0% of the total variance) clearly distinguished between tea from northern Lishui and that from central and southern Lishui (Fig. 12.16). This approach offers several simple, reliable and effective ways to determine the origin and regional distribution of tea in a small area on the basis of its elemental profile.

12.2.4.4 Elemental Profile of Tea Related to the Impact of Pollution

Milani et al. (2016) investigated Brazilian black, green, red oolong and white teas and their infusions from different locations via ICP–MS. PCA, with the first two PCs explaining 74.14% of the total variance, enabled the classification of tea leaves into three groups on the basis of only similarities in their trace elemental composition. Group 1, the red oolong tea samples, was assigned on the basis of the levels of

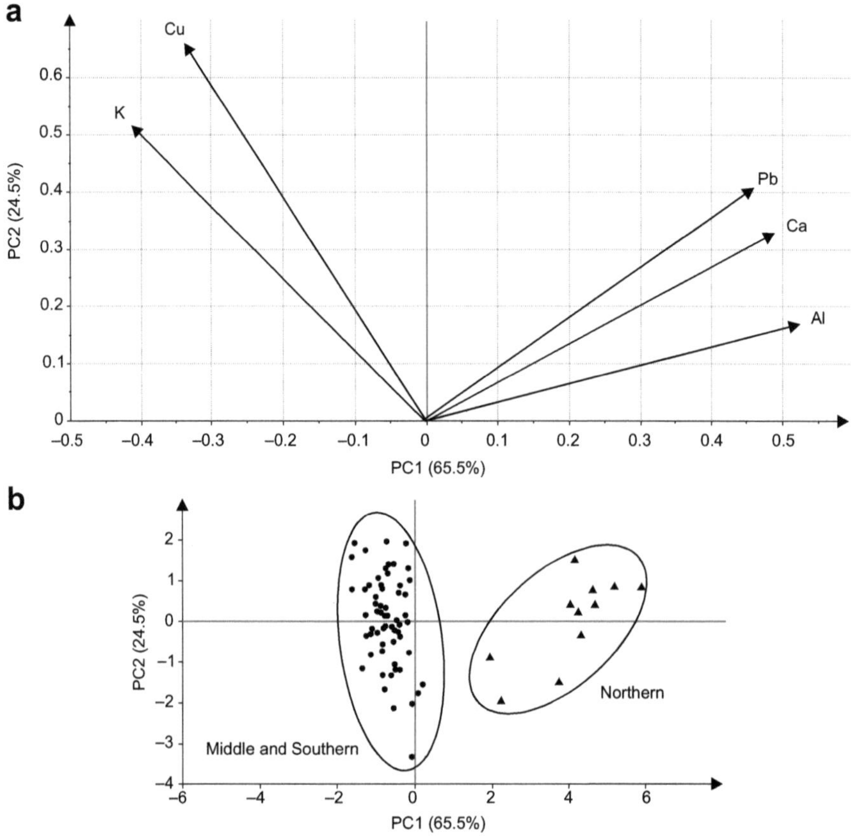

Fig. 12.16 PCA results: (**a**) loading scatter of five elements; (**b**) score scatter of 74 tea leaf samples. (After Li L et al., 2018. Food Contr 90:18–28. https://doi.org/10.1016/j.foodcont.2018.02.031. With permission)

As, Cd, Cu, Fe, Pb and Zn. Group 2, consisting of black and green tea samples, had low variability in their composition but was assigned on the basis of the Al, Ba and Mn contents. Group 3, white tea samples, could not be assigned on the basis of a predominance of any of the selected trace elements. Some tea leaf samples presented trace element levels above the maximum limits allowed by the Brazilian and MERCOSUL regulations, i.e., As (33% in red oolong tea) and Pb (67% and 100% in white and red oolong, respectively). However, the levels of the trace elements in the infusions appeared to be safe.

Idrees et al. (2020) applied GF–AAS to determine the elemental contents of the tea of 15 black tea brands from Pakistan. PCA explained 93.74% of the total variance with the first three PCs. Heavy metals such as Pb, Zn and Cd make a maximum contribution to F1, identifying some heavily contaminated tea brands. The HI values were below unity, which means that drinking tea infusions could pose no carcinogenic risk. The manganese and Cu levels were high, suggesting that the effects of these elements on human health should be studied more extensively.

Malinowska et al. (2008) examined the fluoride content by ISE in infusions of different types of commercially available teas originating from 13 countries. The concentration data were transformed chemometrically. HCA (Euclidian distance, single linkage) revealed two main clusters that were well identifiable. The white, oolong and pu-erh teas were grouped into one cluster, and the second cluster consisted of two subclusters that included all the black tea infusion samples and the green tea brews. It was concluded that consumers might be exposed to high levels of fluoride from drinking tea, especially black tea extracts, in large quantities. However, herbal infusions with the lowest fluoride content can be safely consumed by adults and children, supposing that intake from other sources does not exceed the SAI.

Motta et al. (2020) used ICP–MS and AAS to quantify the elemental profiles of washed and unwashed Yerba mate leaf samples from various Brazilian regions and processed them via PCA and LDA. The elemental concentrations of yerba mate leaves often follow the total soil concentration. On the basis of the LDA results (referring to the washed leaf levels of Mn, P, Co, Ni, Mg, Al, Fe, K, Na, Ca and Ba according to the sample region), no effective discrimination of the plant groups with 84% precision was achieved. However, when the discriminating objects were related to the soil parent material, clearer discrimination with 94% precision was obtained. This means that the yerba mate composition given above was dependent on the parent material of the soil, with high values of these elements in basalt soils, sedimentary soils and rhyolite/hyodacite soils. The washing process diminished the levels of Fe, Ti, As, Pb, Li, Mo and V, suggesting their atmospheric contributions/dust deposition origin. Therefore, washing yerba mate during industrial processing may alleviate a problem associated with some dust contamination, except for reducing the levels of some elements, e.g., Cd, below the current MAL.

Magri et al. (2021) quantified MEC via ICP–OES and GFAAS to process paired yerba mate leaves and soil samples from agroforestry and plantation systems in three South American countries. The first two PCs, which explained 68.4% of the total variance, indicated that the concentrations of bioavailable Cd, P, Cu, Fe, Mn and Zn were correlated in the soils. Moreover, positive relationships between available forms of Cd and Zn, as well as Pb and Fe, were observed. The legal limits of Cd and Pb affected the production of the yerba mate rather than its safety implications for consumers because of the estimated low intake of both heavy metals from yerba mate infusions.

Wong et al. (1998) aimed to investigate the eventual contamination of tea samples by trace metals, such as Al, Cu and Zn. Tea bushes from two tea plantations in Guangdong Province accumulated relatively high levels of Cu and Zn in young leaves but accumulated high levels of Al in old leaves. The greater content of Al in tea samples from Guangdong and Yunnan Provinces may be explained by higher rainfall and lower soil pH in these regions. Higher levels of trace metals were detected in black, oolong and pu-erh tea than in green tea. This contamination may be caused by the fermentation procedure, which green tea does not undergo. The transfer of Al, Cu and Zn from the soil to different parts of tea bushes was generally low, with the exception of Zn at the Lechang tea plantation, where greater transfer occurred from young leaves to tea products. These findings suggest the occurrence of metal contamination during tea processing. However, low concentrations of Cu and Zn and moderate concentrations of Al were detected in the tea liquor.

According to Li et al. (2020), MECs, including toxic and potentially toxic elements, i.e., As, Cr, Cd, Pb, Cu, Zn, Mn and Hg, of Chinese raw pu-erh tea infusions with different storage durations (from 2005–2029) were quantified via ICP–MS. PCA revealed that the first three PCs explained 75.7% of the total variance. PC1 had high load values of Cr, Cd, As, V, Th, Sb, Co, Lu, Eu, Tb, Sm, Ho, Nd, Sr, Rb, etc., whereas PC2 and PC3 had high load values of Cu, Hg, Pb, Ag and Mg and Zn, Mn, P, K, Se, S and Ni, respectively. It was postulated that both PC1 and PC3 represent the geological (soil) origin of heavy metals in tea leaves, but PC2 may correspond to elements derived from the atmosphere. There is no health risk when tea infusions are properly consumed by adults. The data demonstrate that the number of years of storage had no effect on the risk of exposure to toxic elements from tea infusions.

12.3 Conclusions

The determination of food authentication contributes to the assessment of food safety. To achieve this task, selected analytical methods/techniques integrated with computational tools were effectively applied to evaluate the quality of agricultural products with respect to adulteration/mislabelling. The thoughtful choice of such an approach for the study of tea plants allowed the evaluation of their authenticity. The integration of an analytical device with multivariate model processing enables fingerprints of the differentiation of edible plant resources, e.g., those related to interregional and intervarietal/brand differentiation. Moreover, variations were identified at annual or seasonal scales, and other differentiating parameters were characterized, i.e., the effects of chemical pollutants, technological processing and fermentation type recognition. The validated results make it possible to detect fraud and identify mislabelled products to preserve the brand reputations of tea brands and to protect consumers.

References

Barone G, Giacominelli-Stuffler R, Storelli MM (2016) Evaluation of trace metal and polychlorinated biphenyl levels in tea brands of different origin commercialized in Italy. Food Chem Toxicol 87:113–119. https://doi.org/10.1016/j.fct.2015.12.008

Barua DN (2008) Science and practice in tea culture. Tea Research Association, Calcutta/Jorhat

Baskali-Bouregaa N, Milliand M-L, Mauffrey S, Chabert E, Forrestier M, Gilon N (2020) Tea geographical origin explained by LIBS elemental profile combined to isotopic information. Talanta 211:120674. https://doi.org/10.1016/j.talanta.2019.120674

Borràs E, Ferré J, Boqué R, Mestres M, Aceña L, Busto O (2015) Data fusion methodologies for food and beverage authentication and quality assessment—a review. Anal Chim Acta 891:1–14. https://doi.org/10.1016/j.aca.2015.04.042

Brzezicha-Cirocka J, Grembecka M, Szefer P (2016a) Monitoring of essential and heavy metals green tea from different geographical origins. Environ Monit Assess 188:183. https://doi.org/10.1007/s10661-016-5157-y

Brzezicha-Cirocka J, Grembecka M, Szefer P (2016b) Analytical assessment of bio- and toxic elements distribution in Pu-erh and fruit teas in view of chemometric approach. Biol Trace Elem Res 174:240–250. https://doi.org/10.1007/s12011-016-0669-4

Brzezicha-Cirocka J, Grembecka M, Szefer P (2016c) Oxalate, magnesium and calcium content in selected kinds of tea: impact on human health. Eur Food Res Technol 242:383–389. https://doi.org/10.1007/s00217-015-2548-1

Brzezicha-Cirocka J, Grembecka M, Ciesielski T, Flaten TP, Szefer P (2017) Evaluation of macro- and microelement levels in black tea in view of its geographical origin. Biol Trace Elem Res 176:429–441. https://doi.org/10.1007/s12011-016-0849-2

Cabrera C, Gimenez R, Lopez MC (2003) Determination of tea components with antioxidant activity. J Agric Food Chem 51:4427–4435. https://doi.org/10.1021/jf0300801

Cengiz MF, Turan O, Ozdemir D, Albayrak Y, Fatih Perincek F, Kocabas H (2017) Geographical origin of imported and domestic teas (*Camellia sinensis*) from Turkey as determined by stable isotope signatures. Int J Food Prop 20:3234–3243. https://doi.org/10.1080/10942912.2017.1283327

Chang C-T, You C-F, Aggarwal SK, Chung C-H, Chao H-C, Liu H-C (2016) Boron and strontium isotope ratios and major/trace elements concentrations in tea leaves at four major tea growing gardens in Taiwan. Environ Geochem Health 38:737–748. https://doi.org/10.1007/s10653-015-9757-1

Chen Y, Yu M, Xu J, Chen X, Shi J (2009) Differentiation of eight tea (*Camellia sinensis*) cultivars in China by elemental fingerprint of their leaves. J Sci Food Agric 89:2350–2355. https://doi.org/10.1002/jsfa.3716

Chen M-M, Liao Q-H, Qian L-L, Zou H-D, Li Y-L, Song Y, Xia Y, Liu Y, Liu H-Y, Liu Z-L (2024) Effects of geographical origin and tree age on the stable isotopes and multi-elements of pu-erh tea. Foods 13:473. https://doi.org/10.3390/foods13030473

Chung FL, Schwartz J, Herzog CR, Yang Y-M (2003) Tea and cancer prevention: studies in animals and humans 1, 2. J Nutr 133:3268S–3274S. https://doi.org/10.1093/jn/133.10.3268S

de Gois JS, Almeida TS, de Andrade RM, Toaldo IM, Bordignon-Luiz MT, Borges DLG (2016) Direct solid analysis for the determination of Mn, Ni, Rb and Sr in powdered stimulant plants using high-resolution continuum source atomic absorption spectrometry followed by chemometric classification based on elemental composition, polyphenol content and antioxidant activity. Microchem J 124:283–289. https://doi.org/10.1016/j.microc.2015.08.020

Deng X, Liu Z, Zhan Y, Ni K, Zhang Y, Ma W, Shao S, Lv X, Yuan Y, Rogers KM (2020) Predictive geographical authentication of green tea with protected designation of origin using a random forest model. Food Contr 107:106807. https://doi.org/10.1016/j.foodcont.2019.106807

Diniz PHGD, Pistonesi MF, Alvarez MB, Fernández Band BS, Ugulino de Araújo MC (2015) Simplified tea classification based on a reduced chemical composition profile *via* successive projections algorithm linear discriminant analysis (SPA-LDA). J Food Compos Anal 39:103–110. https://doi.org/10.1016/j.jfca.2014.11.012

Efenberger-Szmechtyk M, Nowak A, Kregiel D (2018) Implementation of chemometrics in quality evaluation of food and beverages. Crit Rev Food Sci Nutr 58:1747–1766. https://doi.org/10.1080/10408398.2016.1276883

Fernández PL, Pablos F, Martin MJ, González AG (2002) Multi-element analysis of tea beverages by inductively coupled plasma atomic emission spectrometry. Food Chem 76:483–489. https://doi.org/10.1016/S0308-8146(01)00312-0

Fernández-Cáceres PL, Martín MJ, Pablos F, González AG (2001) Differentiation of tea (*Camellia sinensis*) varieties and their geographical origin according to their metal content. J Agric Food Chem 49:4775–4779. https://doi.org/10.1021/jf0106143

Forina M, Casale M, Oliveri P (2009) Application of chemometrics to food chemistry. In: Walczak B, Brown S (eds) Tauler R. Elsevier, Comprehensive chemometrics - Chemical and biochemical data analysis, pp 75–128. https://doi.org/10.1016/B978-044452701-1.00124-1

Fu J, Yu HD, Wu L, Zhang C, Yun YH, Zhang W (2021) Authentication of geographical origin in Hainan partridge tea (*Mallotus obongifolius*) by stable isotope and targeted metabolomics combined with chemometrics. Foods 10:2130. https://doi.org/10.3390/foods10092130

Grembecka M, Szefer P (2011) Metals and metalloids in foods—essentiality, toxicity, applicability. In: Medina DA, Laine AM (eds) Food quality: control, analysis and consumer concerns. Nova Science Publishers, Inc., New York, pp 1–60. ISBN 10: 1611229170; ISBN 13: 9781611229172

Grembecka M, Szefer P (2016) Elemental trace analysis in studies of food products. In: Baranowska I (ed) Handbook of trace analysis—fundamentals and applications. Springer, pp 203–239. https://doi.org/10.1007/978-3-319-19614-5

Herrador MA, González AG (2001) Pattern recognition procedures for differentiation of Green, Black and Oolong teas according to their metal content from inductively coupled plasma atomic emission spectrometry. Talanta 53:1249–1257. https://doi.org/10.1016/S0039-9140(00)00619-6

Huang H, Kfoury N, Orians CM, Griffin T, Ahmed S, Cash SB, Stepp JR, Xue D, Long C, Robbat A Jr (2019) 2014–2016 seasonal rainfall effects on metals in tea (*Camelia sinensis* (L.) Kuntze). Chemosphere 219:796–803. https://doi.org/10.1016/j.chemosphere.2018.12.075

Idrees M, Jan FA, Hussain S, Salam A (2020) Heavy metals level, health risk assessment associated with contamination of Black Tea; a case study from Khyber Pakhtunkhwa (KPK), Pakistan. Bio l Trace Elem Res 198:344–349. https://doi.org/10.1007/s12011-020-02059-1

Kanrar B, Kundu S, Khan P, Jain V (2022) Elemental profiling for discrimination of geographical origin of tea (*Camellia sinensis*) in north-east region of India by ICP-MS coupled with chemometric techniques. Food Chem Adv 1:100073. https://doi.org/10.1016/j.focha.2022.100073

Karak T, Bhagat RM (2010) Trace elements in tea leaves, made tea and tea infusion: a review. Food Res Int 43:2234–2252. https://doi.org/10.1016/j.foodres.2010.08.010

Koch KR, Pougnet MAB, de Villiers S, Monteagudo F (1988) Increased urinary excretion of Al after drinking tea. Nature 333:122. https://doi.org/10.1038/333122a0

Lagad RA, Alamelu D, Laskar AH, Rai VK, Singh SK, Aggarwal SK (2013) Isotope signature study of the tea samples produced at four different regions in India. Anal Methods 5:1604–1611. https://doi.org/10.1039/c3ay26142e

Li L, Wen B, Zhang X, Zhao Y, Duan Y, Song X, Ren S, Wang Y, Fang W, Zhu X (2018) Geographical origin traceability of tea based on multi-element spatial distribution and the relationship with soil in district scale. Food Contr 90:18–28. https://doi.org/10.1016/j.foodcont.2018.02.031

Li M-Y, Xiao Y, Zhong K, Bai J-R, Wu Y-P, Zhang J-Q, Gao H (2019) Characteristics and chemical compositions of Pingwu Fuzhuan brick-tea, a distinctive post-fermentation tea in Sichuan province of China. Int J Food Prop 22:878–889. https://doi.org/10.1080/10942912.2019.1614951

Li F, Lu Q, Li M, Yang X, Xiong C, Yang B (2020) Comparison and risk assessment for trace heavy metals in raw Pu-erh tea with different storage years. Biol Trace Elem Res 195:696–706. https://doi.org/10.1007/s12011-019-01886-1

Lim CM, Carey M, Williams PN, Koidis A (2021) Rapid classification of commercial teas according to their origin and type using elemental content with X-ray fluorescence (XRF) spectroscopy. Curr Res Food Sci 4:45–52. https://doi.org/10.1016/j.crfs.2021.02.002

Liu Z, Zhang Y, Zhang Y, Yang G, Shao S, Nie J, Yuan Y, Rogers KM (2019a) Influence of leaf age, species and soil depth on the authenticity and geographical origin assignment of green tea. Rapid Commun Mass Spectrom 33:625–634. https://doi.org/10.1002/rcm.8387

Liu Z, Yuan Y, Zhang Y, Shi Y, Hu G, Zhu J, Rogers KM (2019b) Geographical traceability of Chinese green tea using stable isotope and multi-element chemometrics. Rapid Commun Mass Spectrom 33:778–788. https://doi.org/10.1002/rcm.8405

Liu H-L, Zeng YT, Zhao X, Tong H-R (2020a) Improved geographical origin discrimination for tea using ICP-MS and ICP-OES techniques in combination with chemometric approach. J Sci Food Agric 100:3507–3516. https://doi.org/10.1002/jsfa.10392

Liu H, Zeng Y, Yan J, Huang R, Zhao X, Zheng X, Mo M, Tan S, Tong H (2020b) C N H O and mineral element stable isotope ratio analysis for authentication in tea. J Food Compos Anal 91:103513. https://doi.org/10.1016/j.jfca.2020.103513

Liu H, Zeng Y, Zhao X, Tong H (2020c) Chemometric authentication of Pu'er teas in terms of multielement stable isotope ratios analysis by EA-IRMS and ICP-MS. J Food Drug Anal 28:248–260. https://doi.org/10.38212/2224-6614.1059

Liu H-L, Zeng Y-T, Zhao X, Ye Y-L, Wang B, Tong H-R (2020d) Monitoring the authenticity of pu'er tea via chemometric analysis of multielements and stable isotopes. Food Res Int 136:109483. https://doi.org/10.1016/j.foodres.2020.109483

Liu H, Meng Q, Zhao X, Ye Y, Tong H (2021) Inductively coupled plasma mass spectrometry (ICP-MS) and inductively coupled plasma optical emission spectrometer (ICP-OES)-based discrimination for the authentication of tea. Food Contr 107735. https://doi.org/10.1016/j.foodcont.2020.107735

Lou Y-X, Fu X-S, Yu X-P, Ye Z-H, Cui H-F, Zhang Y-F (2017) Stable isotope ratio and elemental profile combined with Support Vector Machine for provenance discrimination of Oolong Tea (Wuyi-Rock Tea). J Anal Methods Chem 5454231:1–8. https://doi.org/10.1155/2017/5454231

Ma G, Zhang Y, Zhang J, Wang G, Chen L, Zhang M, Liu T, Lu C (2016) Determining the geographical origin of Chinese green tea by linear discriminant analysis of trace metals and rare earth elements: taking Dongting Biluochun as an example. Food Contr 59:714–720. https://doi.org/10.1016/j.foodcont.2015.06.037

Magri E, Valduga AT, Gonçalves IL, Barbosa JZ, de Oliveira RD, Menezes IMNR, de Andrade Nascimento P, Oliveira A, Corrêa RS, Vargas Motta ACV (2021) Cadmium and lead concentrations in yerba mate leaves from agroforestry and plantation systems: an international survey in South America. J Food Compos Anal 96:103702. https://doi.org/10.1016/j.jfca.2020.103702

Malinowska E, Inkielewicz I, Czarnowski W, Szefer P (2008) Assessment of fluoride concentration and daily intake by human from tea and herbal infusions. Food Chem Toxicol 46:1055–1061. https://doi.org/10.1016/j.fct.2007.10.039

Marcelo MCA, Martins CA, Pozebon D, Dressler VL, Ferrão MF (2014) Classification of yerba mate (*Ilex paraguariensis*) according to the country of origin based on element concentrations. Microchem J 117:164–171. https://doi.org/10.1016/j.microc.2014.06.027

Marcos A, Fischer A, Rea G, Hill SJ (1998) Preliminary study using trace element concentrations and a chemometrics approach to determine the geographical origin of tea. J Anal Atom Spectrom 13:521–525. https://doi.org/10.1039/A708658J

Mbaye M, Traoré A, Sadikh A, Wagué A (2013) Classification of tea consumed in Senegal using XRF techniques and chemometric based on their country of origin. African J Agric Res 8:5522–5529. https://doi.org/10.5897/AJAR2013.7512

McKenzie JS, Jurado JM, de Pablos F (2010) Characterisation of tea leaves according to their total mineral content by means of probabilistic neural networks. Food Chem 123:859–864. https://doi.org/10.1016/j.foodchem.2010.05.007

Milani RF, Marcelo A, Morgano MA, Cadore S (2016) Trace elements in *Camellia sinensis* marketed in southeastern Brazil: extraction from tea leaves to beverages and dietary exposure. LWT Food Sci Technol 68:491–498. https://doi.org/10.1016/j.lwt.2015.12.041

Moreda-Piñeiro A, Marcos A, Fisher A, Hill SJ (2001) Evaluation of the effect of data pretreatment procedures on classical pattern recognition and principal components analysis: a case study for the geographical classification of tea. J Environ Monit 3:352–360. https://doi.org/10.1039/b103658k

Moreda-Piñeiro A, Fisher A, Hill SJ (2003) The classification of tea according to region of origin using pattern recognition techniques and trace metal data. J Food Compos Anal 16:195–211. https://doi.org/10.1016/S0889-1575(02)00163-1

Motta ACV, Barbosa JZ, Magri E, Pedreira GQ, Santin D, Prior SA, Consalter R, Young SD, Broadley MR, Benedetti EL (2020) Elemental composition of yerba mate (*Ilex paraguariensis* A.St.-Hil.) under low input systems of southern Brazil. Sci Total Environ 736:139637. https://doi.org/10.1016/j.scitotenv.2020.139637

Ni K, Wang J, Zhang Q, Yi X, Ma L, Shi Y, Ruan J (2018) Multi-element composition and isotopic signatures for the geographical origin discrimination of green tea in China: a case study of Xihu Longjing. J Food Compos Anal 67:104–109. https://doi.org/10.1016/j.jfca.2018.01.005

Paz-Rodríguez B, Domínguez-González MR, Aboal-Somoza M, Bermejo-Barrera P (2015) Application of high resolution-continuum source flame atomic absorption spectrometry (HR-CS FAAS): determination of trace elements in tea and tisanes. Food Chem 170:492–500. https://doi.org/10.1016/j.foodchem.2014.08.003

Pilgrim TS, Watling RJ, Grice K (2010) Application of trace element and stable isotope signatures to determine the provenance of tea (*Camellia sinensis*) samples. Food Chem 118:921–926. https://doi.org/10.1016/j.foodchem.2008.08.077

Shuai M, Peng C, Niu H, Shao D, Hou R, Cai H (2022) Recent techniques for the authentication of the geographical origin of tea leaves from *Camellia sinensis*: a review. Food Chem 374:131713. https://doi.org/10.1016/j.foodchem.2021.131713

Szefer P (2007) Chemometric techniques in analytical evaluation of food quality. In: Szefer P, Nriagu J (eds) Mineral components in foods. CRC Press/Taylor & Francis, Boca Raton, FL, pp 69–121. https://doi.org/10.1201/9781420003987

Wight W (1962) Tea classification revised. Curr Sci 31:295–299

Wong MH, Zhang ZQ, Wong JWC, Lan CY (1998) Trace metal contents (Al, Cu and Zn) of tea: tea and soil from two tea plantations, and tea products from different provinces of China. Environ Geochem Health 20:87–94. https://doi.org/10.1023/A:1006545825302

Xiong F, Yuan Y, Li C, Lyu C, Wan X, Nie J, Li H, Yang J, Guo L (2022) Stable isotopic and elemental characteristics with chemometrics for the geographical origin authentication of *Dendrobium officinale* at two spatial scales. LWT Food Sci Technol 167:113871. https://doi.org/10.1016/j.lwt.2022.113871

Yao LH, Jiang YM, Shi J, Tomás-Barberán FA, Datta N, Singanusong R, Chen SS (2004) Flavonoids in food and their health benefits. Plant Foods Hum Nutr 59:113–122. https://doi.org/10.1007/s11130-004-0049-7

Ye X, Jin S, Wang D, Zhao F, Yu Y, Zheng D, Ye N (2017) Identification of the origin of based on mineral element content. Food Anal Meth 10:191–199. https://doi.org/10.1007/s12161-016-0568-5

Zhang L, Pan J, Zhu C (2012a) Determination of the geographical origin of Chinese teas based on stable carbon and nitrogen isotope ratios. J Zhejiang Univ Sci B 13:824–830. https://doi.org/10.1631/jzus.B1200046

Zhang L, Kujawinski DM, Federherr E, Schmidt TC, Jochmann MA (2012b) Caffeine in your drink: natural or synthetic? Anal Chem 84:2805–2810. https://doi.org/10.1021/ac203197d

Zhang J, Yang R, Chen R, Li YC, Peng Y, Liu C (2018a) Multielemental analysis associated with chemometric techniques for geographical origin discrimination of tea leaves (*Camelia sinensis*) in Guizhou Province, SW China. Molecules 23:3013. https://doi.org/10.3390/molecules23113013

Zhang X, Wu H, Huang X, Zhang C (2018b) Establishment of element fingerprints and application to geographical origin identification of Chinese Fenghuangdancong tea by ICP-MS. Food Sci Technol Res 24:599–608. https://doi.org/10.3136/fstr.24.599

Zhang J, Yang R, Li YC, Wen X, Peng Y, Ni X (2020a) Use of mineral multi-elemental analysis to authenticate geographical origin of different cultivars of tea in Guizhou, China. J Sci Food Agric 100:3046–3055. https://doi.org/10.1002/jsfa.10335

Zhang M, Huang C, Zhang J, Qin H, Ma G, Liu X, Yin J (2020b) Accurate discrimination of tea from multiple geographical regions by combining multi-elements with multivariate statistical analysis. Food Meas 14:3361–3370. https://doi.org/10.1007/s11694-020-00575-1

Zhao H, Yang Q (2019) The suitability of rare earth elements for geographical traceability of tea leaves. J Sci Food Agric 99:6509–6514. https://doi.org/10.1002/jsfa.9930

Zhao H, Zhao F (2019) The authenticity identification of teas (*Camellia sinensis* L.) of different seasons according to their multi-elemental fingerprints. Int J Food Sci Technol 54:249–255. https://doi.org/10.1111/ijfs.13935

Zhao H, Zhang S, Zhang Z (2017a) Relationship between multi-element composition in tea leaves and in provenance soils for geographical traceability. Food Contr 76:82–87. https://doi.org/10.1016/j.foodcont.2017.01.006

Zhao H, Yu C, Li M (2017b) Effects of geographical origin, variety, season and their interactions on minerals in tea for traceability. J Food Compos Anal 63:15–20. https://doi.org/10.1016/j.jfca.2017.07.030

Zhong J, Chen N, Huang S, Fan X, Zhang Y, Ren D, Yi L (2020) Chemical profiling and discrimination of green tea and Pu-erh raw tea based on UPLC-Q-Orbitrap-MS/MS and chemometrics. Food Chem 326:126760. https://doi.org/10.1016/j.foodchem.2020.126760

Zhou P, Li Z, Ouyang L, Gong X, Meng P, Dai M, Wang Z, Wang Y (2019) A multi-element stable isotope approach coupled with chemometrics for the determination of Tieguanyin tea geographical origin and harvest season. Anal Methods 11:346–352. https://doi.org/10.1039/C8AY02191K

Zhu J, Chen L, Chen Y, Rong Y, Jiang Y, Liu F, Zhou Q, Wei X, Yuan H, Zhang J, Li J (2023) Effect of geographical origins and pile-fermentation on the multi-element profiles of ripen Pu-erh tea revealed by comprehensive elemental fingerprinting. Food Contr 154:109978. https://doi.org/10.1016/j.foodcont.2023.109978

Chapter 13
Coffee and Its Products

Contents

13.1 Introduction

Coffee can be considered a drink with health-promoting properties, as shown by numerous authors. Its positive effect is due mainly to the presence of highly active antioxidative polyphenols. Coffee infusions is consumed frequently, the consumption of coffee can contribute to the intake of some dietary minerals. Consequently, coffee can enrich a diet with microelements that have antioxidant activity, i.e., Mn, Cu, Zn and Fe.

Postharvest and on-farm processing methods are currently recognized as particularly influential for bean chemistry and final cup quality. Although research concerning green coffee has been limited, multiple factors can play a role in determining green coffee chemistry, e.g., plant cultivation/fruit ripening issues and farming practices and postharvest storage conditions (Hall et al. 2022).

Elemental analysis of coffee using instrumental methods can be used to confirm the high quality and safety of raw coffee beans as well as various coffee by-products

Supplementary Information The online version contains supplementary material available at https://doi.org/10.1007/978-3-031-99687-0_13.

P. Szefer, *Chemometrics in Food Control Based on Elemental Analysis*,
https://doi.org/10.1007/978-3-031-99687-0_13

and final coffee products on the market. Typically, different analytical methods or techniques, usually combined with other analytical tools and computational processing, are used to determine the concentrations of various elements in green, roasted, ground or instant coffees and coffee infusions (Pohl et al. 2013).

According to Borràs et al. (2015), combining various analytical methods can improve the accuracy of classification or prediction of food identification, including edible plants, in contrast to the use of a single analytical technique. Low- and mid-level combined approaches are the most common. Moreover, it is expected that data integration could not only contribute to the global classification/prediction ability but also be useful for minimizing the uncertainty of each individual result. Moreover, this approach enables more accurate prediction.

To perform adequate analytical and chemometric assessments of original data, it is important to select and apply modern instrumental methodology coupled with computational processing that strictly follows the rules and research criteria. Dynamic progress in analytical and chemometric tools has been recently observed owing to new and advanced informative technologies. Owing to this opportunity, it makes it possible to obtain valuable information from an experimental database reliably. Therefore, multivariate analysis techniques are key statistical and mathematical approaches for extensively investigating databases and are helpful for simple and quick exploration. Among the frequently used analytical methods or techniques used in the analysis of coffee samples and other edible plants, some deserve special attention, e.g., AAS with specific techniques, such as FAAS, ET–AAS, CV–AAS and HG–AAS, as well as ICP–MS, ICP–OES, ICP–AES and IRMS. Other less frequently applied instrumental methods include XRF, NMR, AES, NAA, LA–ICP–MS, MC–ICP–MS, ICP–TOF–MS, ICP–SFMS, MC–TIMS, etc. The reliability and correctness of the concentration data have been checked via quality assurance tests involving the use of appropriate CRMs with known concentrations of analytes (Szefer 2007; Grembecka and Szefer 2011, 2016).

The validated analytical data are then processed chemometrically using univariate, bivariate and multivariate techniques. These advanced and frequently applied multivariate techniques include PCA or FA, CA, HCA, LDA, CDA or DA, DFA, PLS–DA, PLS, etc. In addition, less commonly used methods can be distinguished, such as QDA, kNN, GLM, SVM, MLP and NB (Szefer 2007; Marini 2013; Agriopoulou et al. 2023a, b).

13.2 Chemometric Assessment of Coffee

13.2.1 Geographical Differentiation

13.2.1.1 American Countries

Rodrigues et al. (2011a) used a combination of EA–IRMS, MC–ICP–SFMS and ICP–QMS to investigate green coffee samples from five Hawaiian coffee-producing regions. All obtained data were transformed by chemometric processing. One-way

ANOVA ($p < 0.05$) was a useful tool for achieving significant interregional differentiation among concentrations of elements such as Na, Mg, Al, Mn, Ga, Rb, Ba, Pb, Y, La, Ce, Pr, Sm, Nd, Eu, Dy and Gd. Moreover, Sr isotope analysis was shown to be an effective tool for determining the geographical origin of coffee beans. CDA enabled the discrimination of the different Hawaiian coffee-growing regions since three canonical components were achieved with coefficients >0.9 ($p < 0.05$). It was concluded that the significance of interspatial differentiation must be interpreted with caution. Moreover, more samples, including those from multiple years of harvest, are needed to construct a model.

13.2.1.2 African Countries

Habte et al. (2016) reported that a combination of ICP–MS, ICP–OES and DMA with chemometric tools (PCA, CA and LDA) could be used to evaluate the chemical composition of green coffee samples from 11 major provinces of Ethiopia. The most satisfactory results were achieved by LDA on the basis of the analysed elements in all the coffee samples, enabling their successful differentiation into provinces of origin. The samples introduced as ungrouped cases were 100% reliably differentiated interregionally (Fig. 13.1a), and 72.7% correct classification of samples was achieved when the entire dataset was considered. The samples from 2011 and 2012 were grouped together into their respective Illubabor coffee groups (Fig. 13.1b). The discriminative ability of PCA was not as effective as that of the LDA model. The LDA results were consistent with those of the ANOVA results for samples from the two production years. In summary, the LDA test successfully classified the coffee samples with respect to their interspatial differentiation.

13.2.1.3 Countries Across Multiple Continents

Muñiz-Valencia et al. (2013) classified samples of green coffee beans originating from three countries on the basis of MEC content determined by ICP–OES. The data processed by chemometric tools (PCA, LDA and SIMCA) were useful for assessing the interregional differentiation of elemental profiles. The Kruskal–Wallis test also revealed differences in the elemental contents of coffee samples from these three countries, although the PCA results, with the first three PCs accounting for 54.3%, 18.0% and 13.0%, respectively, revealed only trends, not significant relationships. Supervised pattern recognition techniques such as LDA and SIMCA enabled a clear classification of Mexican, Colombian and Brazilian coffee clusters, achieving high reliability, i.e., prediction abilities of 97% and 94%, respectively.

Akamine et al. (2010) examined samples of coffee beans originating from six countries by using a highly sensitive XRF. PCA with PC1 (44.5%) and PC2 (19.7%) was an appropriate tool for classifying the coffee beans according to their geographical origin. It was possible to successfully differentiate production areas, i.e., Brazil, Colombia, Guatemala, Vietnam, Indonesia and Tanzania. Importantly, roasted beans could be used according to the same criteria as their green beans. It

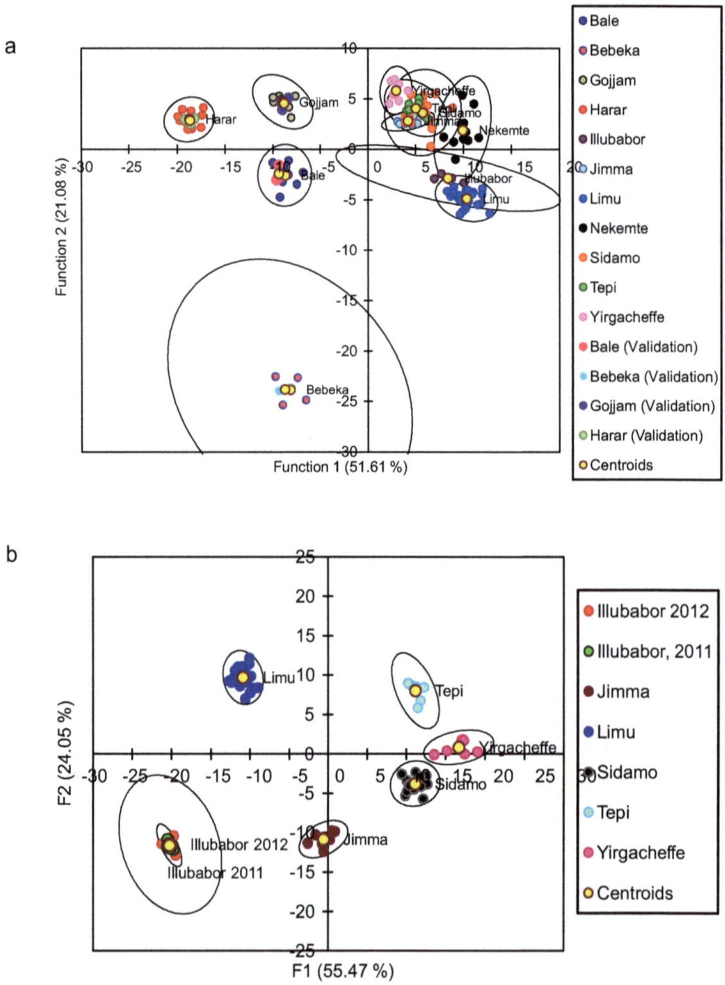

Fig. 13.1 (**a**). Linear discriminant analysis plot of the first two discriminant functions (Function 1 and Function 2; 72.69%) showing the cross-validation results on the basis of the concentrations of all the analysed elements in the coffee samples from Harar, Gojjam, Bale and Bebeka Provinces of Ethiopia; (**b**). Linear discriminant analysis plot of the first two discriminant functions (Function 1 and Function 2; 79.52%) showing the robustness of the technique where the samples from Illubabor 2011 and 2012 production years were grouped together. (After Habte G et al., 2016. Food Chem 212:512–520. https://doi.org/10.1016/j.foodchem.2016.05.178. With permission)

was concluded that the method used enabled the effective classification of samples originating from the six production areas of Asia, South/Central America and Africa.

Valentin and Watling (2013) combined ICP–MS and ICP–AES with multivariate analysis to categorize different types of coffee from 15 countries. LDA and PCA provided up to 96.9% correct classification of the coffee samples with respect to their continent of origin. It was possible to differentiate distinct groupings related to Asia, Africa, Oceania, Central America and South America (Fig. 13.2) as well as to

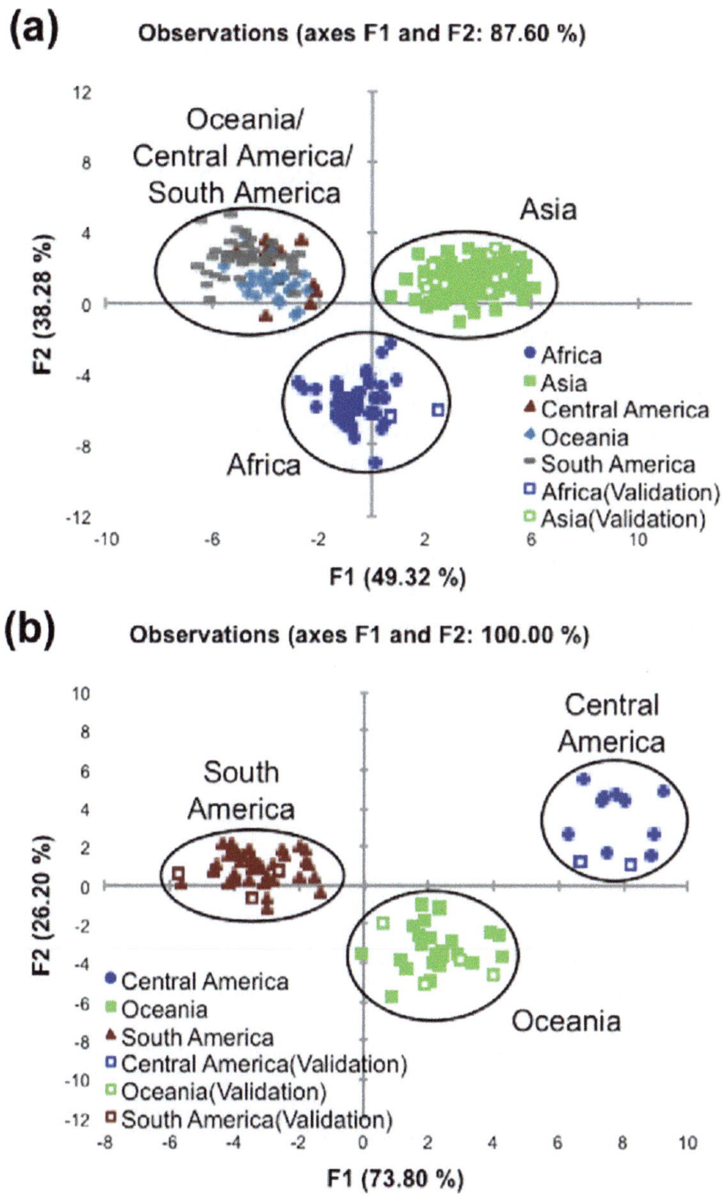

Fig. 13.2 (**a**) LDA plot of the first two discriminant functions (F1 and F2) showing the differentiation of coffees from Asia, Africa, Oceania and Central and South America on the basis of all the elemental concentrations; (**b**) LDA plot of the first two discriminant functions (F1 and F2) showing the differentiation of coffees from Oceania, and Central and South America on the basis of all the elemental concentrations. (After Valentin JL and Watling RJ, 2013. Food Chem 141:98–104. https://doi.org/10.1016/j.foodchem.2013.02.101. With permission)

the three African countries (Ethiopia, Kenya and Uganda) where the coffee samples originated. The five groups of coffees originating from India, Bali and Gajah Mountain were clearly distinguished. Moreover, natural groupings of Balinese coffees on the basis of the concentrations of Ca, Se, Rb, Sr and Ag were also achieved. In the case of samples originating from each continent and considered separately, up to 100% correct classification of coffee samples assigned to their countries and plantations of origin was achieved. The results also indicated that these classifications are considerably unaffected by the roasting process, degree of ripeness or harvest date.

Anderson and Smith (2002) used ICP–AES coupled with a chemometric tool to characterize the elemental profiles of coffee bean samples from seven countries in Asia, East Africa and Central/South America. PCA and CDA enabled the discrimination of samples originating from different countries (Figs. 13.3 and 13.4). Moreover, classification model data, i.e., ANN and DFA, were useful for differentiating sample groupings among the three major geographical regions of coffee. The use of both classification models successfully differentiated coffees relative to their subregional growing origin (70–86% successful classification).

Oliveira et al. (2015) integrated HR–CS–AAS and UV/VIS spectrophotometry to quantitatively assess the elemental profiles of roasted and ground coffee samples originating from major world-growing regions, i.e., 13 countries in Asia, Africa,

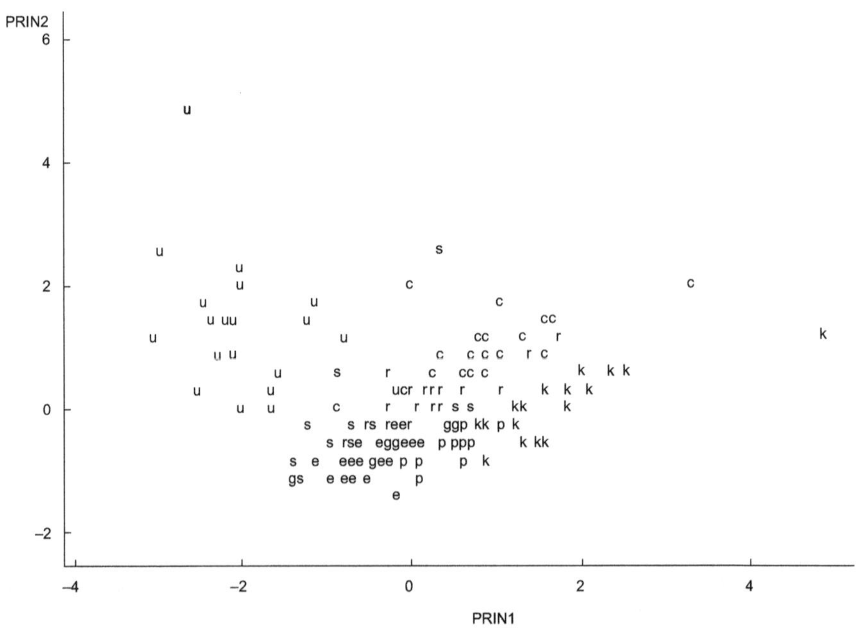

Fig. 13.3 Principal component 1 versus principal component two for the chemical profile of elements in roasted coffee beans from eight growing regions: r, Costa Rica; g, Guatemala; c, Colombia; p, Panama; s, Sulawesi; u, Sumatra; k, Kenya; e, Ethiopia. There are 41 observations that are hidden. (After Anderson KA and Smith BW, 2002. J Agric Food Chem 50:2068–2075. https://doi.org/10.1021/jf011056v. With permission)

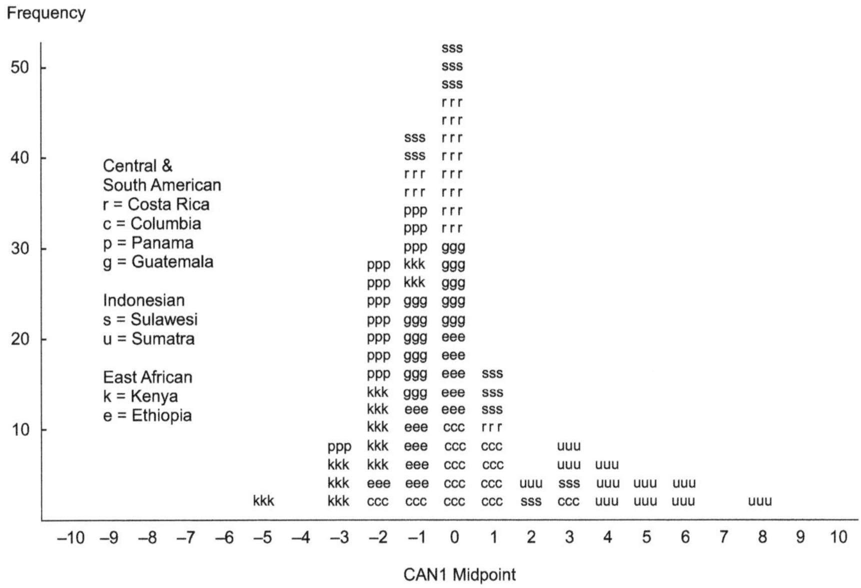

Fig. 13.4 Plot of the first canonical functions (country codes are given in the figure). This simplified visual representation demonstrates the differentiation of regions; any overlap viewed of this one-dimensional representation is not indicative of any intractable classification task. All 16 available dimensions are utilized. (After Anderson KA and Smith BW, 2002. J Agric Food Chem 50:2068–2075. https://doi.org/10.1021/jf011056v. With permission)

South America, Central America and Oceania. CDA enabled the differentiation of 11 sample groups, which included the major coffee producers, i.e., Brazil, Ethiopia, Colombia, India, Mexico, Honduras, Kenya and Cuba, except between Mussulo and Papua New Guinea as well as between Timor and Guatemala (Fig. 13.5). Two elements, i.e., Mn and Ca data, were reasonable descriptors that differentiated the coffees very well with respect to the various countries of origin. It was concluded that CDA is useful for differentiating coffees by identifying the chemical elements highly contributing to identifying inter- and intracontinental differences.

Carter et al. (2015) used advanced methods such as EA–IRMS, TC/EA–IRMS, ICP–MS, ICP–OES and ISQ GC–MS with chemometric tools. These combination methods are promising approaches for differentiating roasted and ground coffees (arabica and robusta) from 20 countries in Africa, Central/South America, Oceania and Asia. PCA was unable to provide clear discrimination between the regions of origin. SDA with jackknife tests using isotopic and elemental data provided up to 77% correct classification of coffee-growing regions. Samples from Africa and India were efficiently classified, whereas in the case of Central/South America, poor interspatial or intraspatial discrimination was achieved. On the basis of $\delta^{13}C_{caffeine}$ and $\delta^{13}C_{coffee}$ values, general trends were observed, with African samples being the most isotopically enriched and Central/South American samples being the most depleted and low-coffee-producing regions (Australia, India and Indonesia), forming discrete groupings within the data.

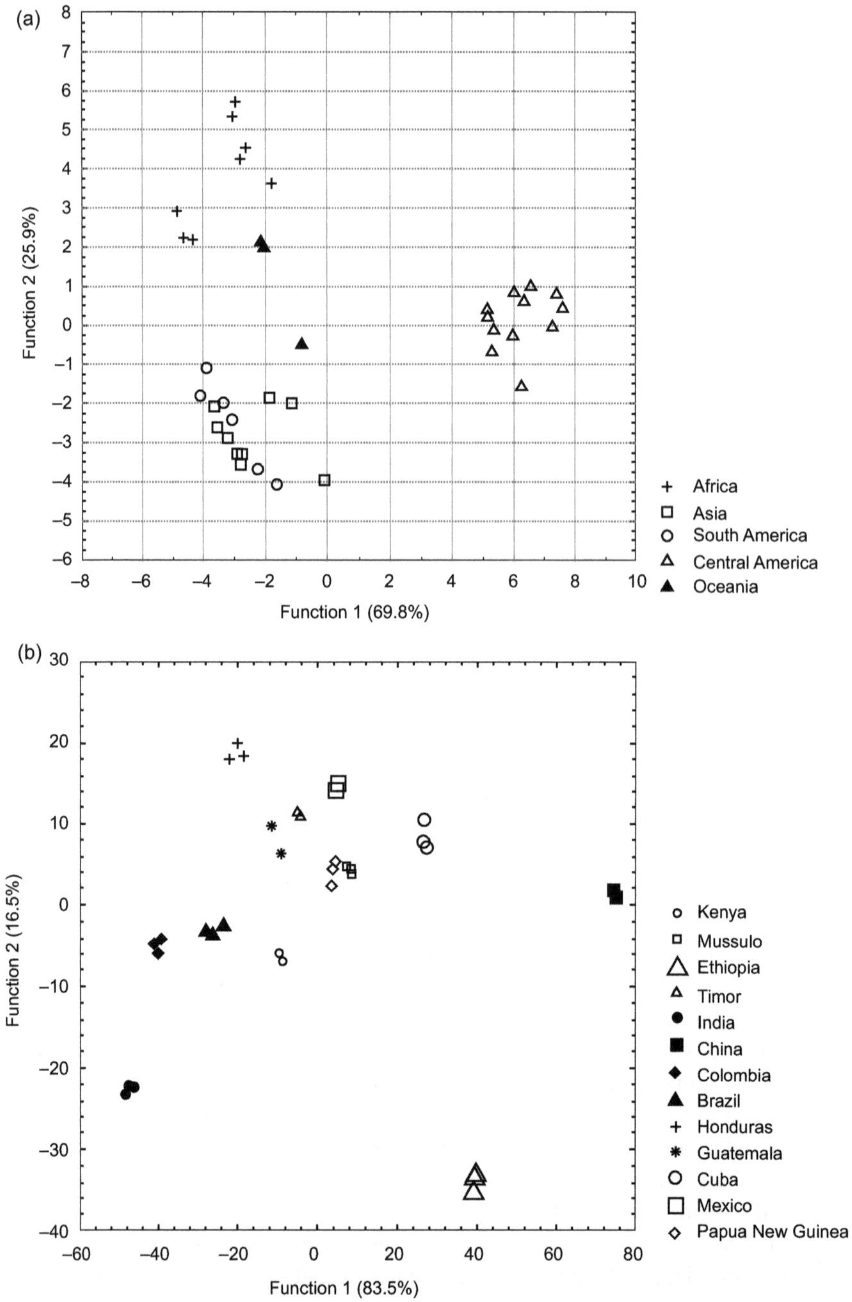

Fig. 13.5 Canonical analysis of mineral concentrations in pure-origin espresso coffees: (**a**) CDA scatter plot based on Ca, Mg, Na, K, P, Fe and Mn for intercontinental discrimination and (**b**) CDA scatter plot based on Mn and Ca for intercountry discrimination. (After Oliveira et al., 2015. Food Chem 177:330–338. https://doi.org/10.1016/j.foodchem.2015.01.061. With permission)

According to Weckerle et al. (2002), multiple combinations of advanced methods, such as EA–P–IRMS, have been developed. EA–C/P–IRMS, HPLC–UVD and ^{1}H NMR were useful for differentiating the δ^{13}C, δ^{2}H and δ^{18}O profiles of arabica green coffee bean samples from 16 countries in Central and South America, Africa, Indonesia and the Antilles. To evaluate the data chemometrically, they applied LDA and validation by classification and CART to the caffeine isotope data from African and American coffee samples. LDA resulted in an error rate for adaption of 5.1%, and for the CV, it was 10.3%. CART analysis, which is based on the δ^{18}O values, achieved an error rate for adaption of 5.1% and for CV of 15.4%. According to the authors, the chemometric data resulted in effective determination of geographical provenance.

Rodrigues et al. (2009) determined δ^{13}C, δ^{15}N and δ^{18}O by EA–IRMS in green coffee bean samples with 20 geographic origins across Central America, the Pacific, South America, Africa, Asia and Oceania. The data obtained were processed chemometrically, i.e., via the PCA technique, to discriminate coffee samples according to their provenance. The observed interspatial differences between the SIR and MEC of gourmet coffee may be the result of the influences of altitude and precipitation values associated with different geographical origins. In addition, the results of nonparametric measurements of the relationships between edaphic-climatic factors and the data studied were estimated to determine the pertinent factors for coffee origin differentiation by using the applied methodology.

Rodrigues et al. (2011b) applied a combination of MC–ICP–MS and EA–P–IRMS to determine the ^{87}Sr/^{86}Sr and δ^{18}O content ratios in samples of green coffee beans from 20 countries. A relationship between the isotopic profiles of the coffee and the environmental factors influencing the processes during coffee bean growth was found. PCA with the first two PCs explained 74.17% of the total variance. Coffee samples from UR Tanzania, East Timor, Hawaii, Rwanda and Kenya were in differentiate clusters; i.e., interregional differentiation of coffee samples related to their elemental profiles was possible. In addition, a relationship between the ^{87}Sr/^{86}Sr ratios of the green coffee beans and the parent rock and/or soil was observed. In the case of the combination of both δ^{18}O and ^{87}Sr/^{86}Sr, it was impossible to achieve differentiation among the 20 countries on the basis of single analysis, although it was possible to obtain a satisfying discrimination between selected origins or groups of origins. Effective differentiation was obtained for coffee samples from South America as well as those originating from islands. It was concluded that the applied methodology has the potential to verify the authenticity of coffees from all over the world. It was concluded that a conclusive geographical origin discrimination model for coffee needs further investigation to increase the sample number as well as to evaluate the influences on different parameters and how they are stable with time. Moreover, the use of specific compounds from the coffee bean for SIRA, i.e., caffeine and DNA, was postulated.

Vieira (2015) chemometrically elaborated the original concentration data obtained by Rodrigues et al. (2009, 2011a, b), i.e., SIR, with ^{87}Sr/^{86}Sr values of green coffee samples originating from growing regions within four countries. The δ^{18}O, δ^{15}N and strontium isotope ratios were identified as the most promising variables for the differentiation of green coffee beans. Moreover, $\delta^{34}S$ values were important in distinguishing between Hawaii Island and the other islands of the Hawaiian Archipelago. PCA, with the first three PCs accounting for 88% of the total variance, enabled the differentiation of two groups of samples, i.e., those from Hawaii Island and Kenya (mountainous areas) versus those from Brazil and the other islands of the Hawaiian Archipelago. Furthermore, a complete differentiation between African countries (Kenya and Ethiopia) and other producing regions was observed. According to CA, only 13% of all observations were assigned to a group other than its real geographical provenance. The Hawaiian Archipelago was differentiated into two groups: one with areas with low altitudes and the with areas with to medium altitudes, such as Hawaii Island.

Liu et al. (2014) combined HR–ICP–MS and MC–ICP–MS to quantify trace elements and $^{11}B/^{10}B$ and $^{87}Sr/^{86}Sr$ ratios to characterize green coffee beans originating from 14 countries in three major coffee-producing regions (Fig. 13.6). PCA, with the first three PCs accounting for approximately 75.7% of the total variance, revealed that the first two PCs significantly distinguished the geographic origin of the coffee samples (Fig. 13.7). This technique enabled the differentiation of coffee samples in Southeast Asia (Indonesia and Papua New Guinea), which presented the highest average contents of Rb, Sr and Ba, whereas American and African samples had intermediate levels. Coffee beans from Taiwan had the lowest levels of these three elements, and B, Fe and Zn (along PC2) were not significant descriptors differentiating geographical provenance. The Rb, Sr and Ba levels in the coffee beans were indicators for geographic determination, although the variability in such discrimination was not validated under the 95% confidence interval test because of the similar chemical compositions of the coffee samples. Therefore, only the isotope ratios of $^{11}B/^{10}B$ and $^{87}Sr/^{86}Sr$ could be considered potentially reliable indicators of the origin of coffee samples.

Serra et al. (2005) reported that the $\delta^{15}N$, $\delta^{13}C$ and $\delta^{11}B$ values processed chemometrically were promising datasets for differentiating green coffee samples originating from 19 countries, namely, Central America, South America, Africa, Oceania

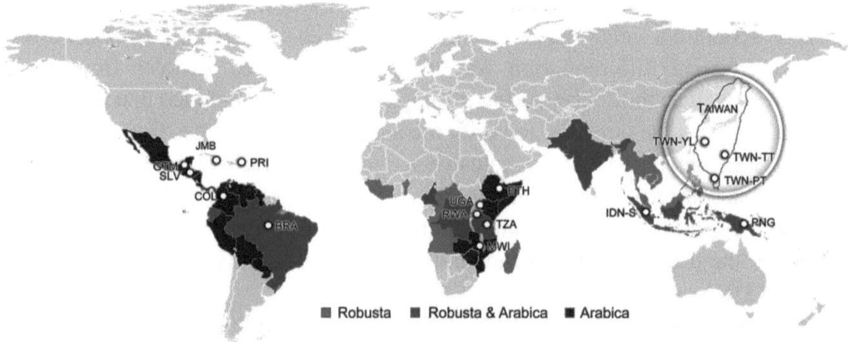

Fig. 13.6 Sampling locations among the major coffee-producing countries around the world. (Modified from Oestreich-Janzen (2010)). (After Liu H-C et al., 2014. Food Chem 142:439–445. https://doi.org/10.1016/j.foodchem.2013.07.082. With permission)

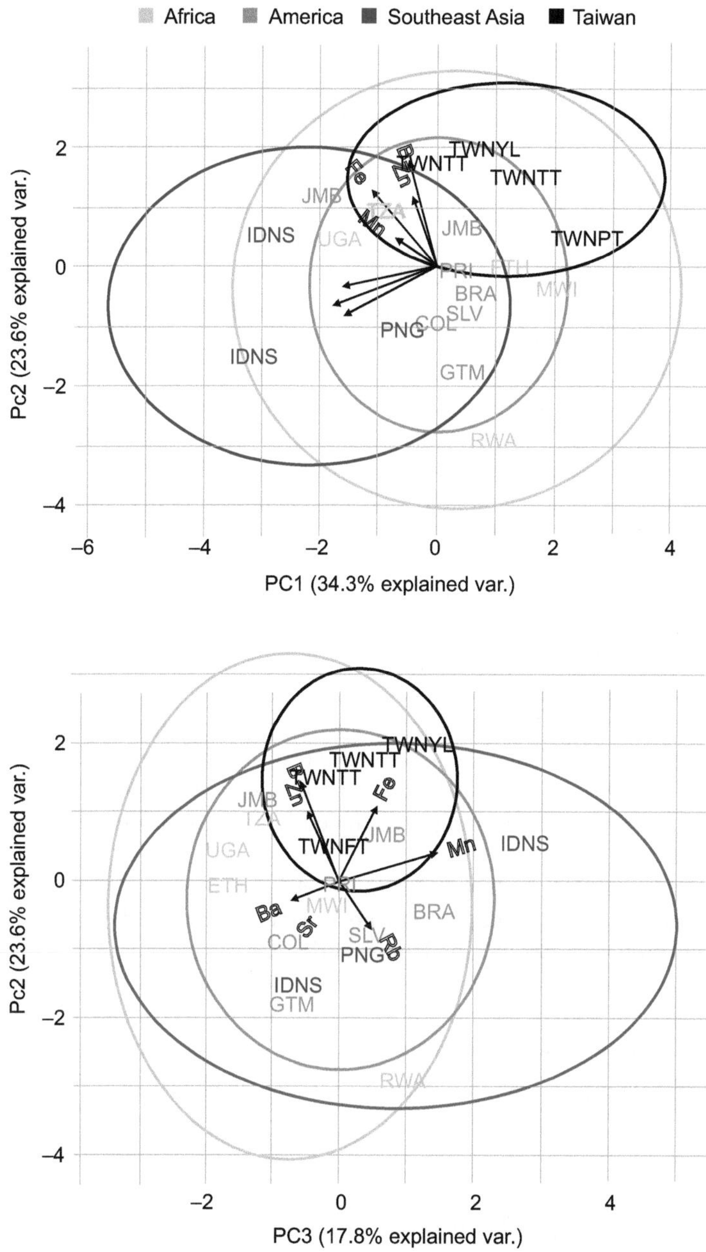

Fig. 13.7 Principal component analyses of the chemical compositions of coffee beans in major growth regions. The brackets represent the 95% confidence interval. (After Liu et al., 2014. Food Chem 142:439–445. https://doi.org/10.1016/j.foodchem.2013.07.082. With permission)

and Asia, with respect to their geographic characteristics. Because the first two PCs explained 78.0% of the total variance, good discrimination between African and American samples and African and Asian samples was achieved. Correct classification of continental origin was also obtained, with an overall success rate of 88.2% of the samples analysed. However, only 60% of the Asian samples were classified correctly, resulting from an insufficient number of samples used for analysis. Although PCA enabled discrimination among the three continental growing regions, i.e., Africa, Asia and America, intracontinental discrimination was not achieved. This significant limitation can be explained by the mismatch between the national and climatic boundaries. Therefore, small coffee-producing countries that are near each other have similar climatic conditions, making it impossible to differentiate samples from one country from samples from a nearby country. However, in the case of large countries with a significant variety of climatic conditions, such inter-spatial discrimination may be easier to achieve. Taking these limitations into account, the use of climatic data as an additional variable has been recommended for the construction of a chemometric model to improve its effectiveness.

Bitter et al. (2020), via ICP–MS, estimated the elemental composition of roasted arabica coffee beans originating from 21 countries located in South America, Central and North America, Africa, Asia and Oceania. The KDE evaluated the distributions of 1892 element content ratios for each of these countries with the combined profile of coffee samples from the other countries. Using this computational approach, it has been postulated that many of the world's coffee-producing regions can be successfully distinguished from other regions worldwide (Fig. 13.8). The data obtained for trace element ratios of roasted coffee are more predictable because some potential complications, such as variations caused by degradation during storage or any changes during roasting, are avoided in the identification of the region of origin of the coffee.

Krivan et al. (1993) combined INAA, GF–AAS, FAAS and CHN-o-Rapid elemental analyzers to determine 20 elements in green coffees of the Arabica species produced in crop year 1987/88 in eight countries. On the basis of the data obtained, an attempt was made to verify the provenance of the green coffee. Mn is a promising indicator, since in 84% of the cases in which the elemental concentration was compared, the Mn content was significantly different between eight coffee samples of different geographic origins. Moreover, C, Co, Cs, Na and Rb also showed similar indicator abilities.

13.2.2 Intervarietal/Type Differentiation

Zhang et al. (2018) integrated LIBS with chemometric processing of original data to identify four cultivars of arabica and robusta coffee from Yunnan and Hainan Provinces in China. Six chemometric tools, WT, PCA, PLSR, PLS–DA, RBFNN and SVM, were used to construct classification models. To select the spectral variables contributing significantly to the recognition of coffee varieties, they used PCA

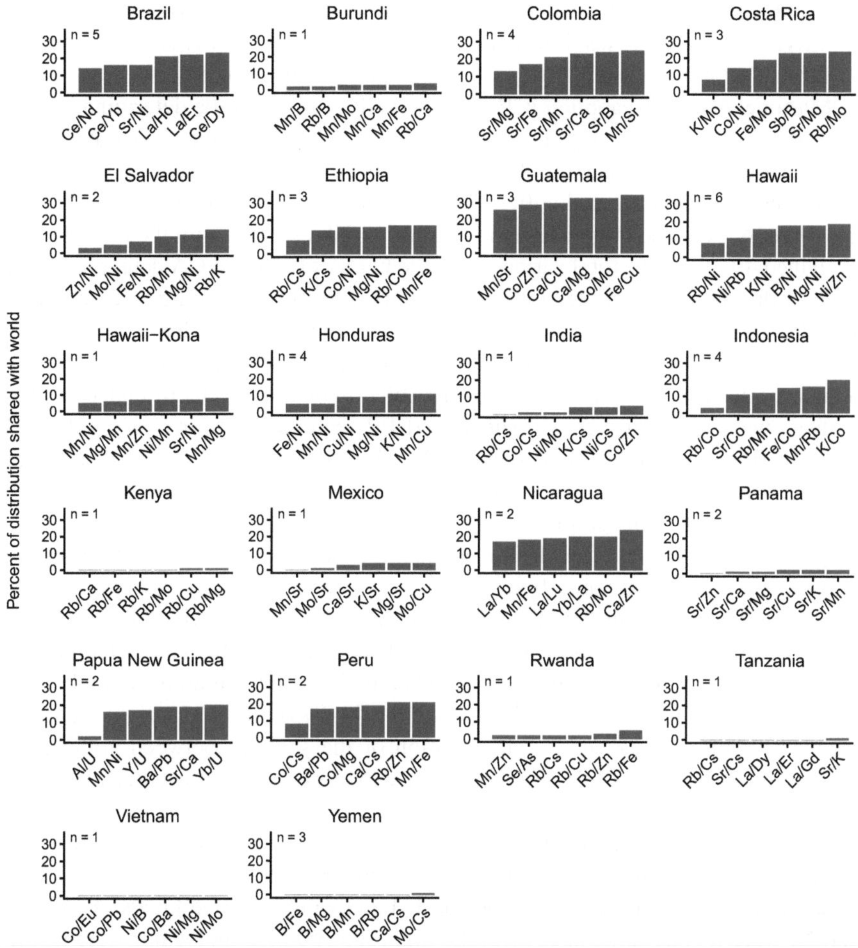

Fig. 13.8 Percent of the kernel density estimate of each ratio for samples from each country that overlaps with the kernel density estimate of coffees from other countries. For each country, the six element ratios with the lowest percent overlap are shown. Countries that have a lower percentage of their distribution shared with other countries are more easily distinguishable than countries with a higher percentage of their distribution shared with other countries. The ratios with no visible bar height have a shared distribution of less than 0.5%. (After Bitter NQ et al., 2020. Food Chem 320:126602. https://doi.org/10.1016/j.foodchem.2020.126602. With permission)

with the first three PCs explaining 81.17%, 9.90% and 2.76%, respectively. However, the sample scores representing different varieties overlapped, making it impossible to differentiate them. However, PCA loadings were used to select the spectral variables, enabling them to distinguish coffee varieties. Consequently, 20 wavelength variables were selected. Moreover, the PLS–DA, RBFNN and SVM models provided adequate results, whereas in the case of the latter two models, more accurate data were obtained with a classification accuracy of over 80% in the prediction set for both the full spectra and the selected variables. Overall, the results revealed that the combination of LIBS with chemometric tools was an appropriate approach for identifying coffee varieties. It has been recommended that further studies should be performed with more samples to produce robust classification models.

Martín et al. (1996) applied ICP–AES and HPLC–UV for the determination and selection of K, Na, Ca, Mg, Fe, Mg and antioxidants (caffeine, aqueous extract, amino acids, polyphenols and 5-HMF) to classify different types of coffee into four categories and decaffeinated coffee. Chemometric models such as PCA, CA, LDA and SIMCA were used to process the original variables. PCA with the first two PCs explained 49.3% of the total variance. On the basis of PCA and CA data, it was possible to differentiate samples into two clusters, i.e., corresponding to commercial samples of nontorrefacto coffees and commercial samples that contain some (50–70%) or 100% torrefacto coffee. However, the three classes of coffees, i.e., torrefacto coffees, nontorrefacto coffees and blends of nontorrefacto and torrefacto coffees, were distinguished via supervised PR techniques. Moreover, with LDA using HMF content and SIMCA using HMF, Mg and Mn achieved good discrimination between the coffee classes, i.e., from 90.0% to 100% of the three correctly classified categories of samples.

According to Martín et al. (1998), elemental data from ICP–AES were processed chemometrically for green coffee samples originating from 13 Central American, African and Asiatic countries. The P and Cu levels were greater for robusta, whereas arabica coffees were distinguished by a greater content of Mn. The differentiation of the two green coffee varieties was achieved by PCA and hierarchical agglomerative CA. PCA, with the first two PCs explaining 49.7% of the total variance, enabled them to obtain a satisfactory differentiation of the samples between arabica and robusta coffees (Fig. 13.9). P, Ca, Mn and Mg were the most discriminating variables.

Pohl et al. (2016) integrated FAAS, FAES and ICP–OES with multivariate analysis (PCA and LDA) to develop an analytical scheme for physicochemical fractionation of elemental profiles of ground roasted and soluble coffee brews. Al, Cu, Fe and Ni were mostly associated with the highest >100 kDa molecular weight fraction, but Co, K, Mg and Na were associated with the lowest <5 kDa fraction. Consequently, these elements effectively differentiated the brews of ground coffees and soluble coffees on the basis of LDA for both metal fraction concentrations.

Filho et al. (2007) determined the contents of several essential elements by FAAS with FES in Brazilian green and roasted coffee samples. Compared with green coffee, roasted coffees had higher levels of Na, K, Ca, Mg and Fe. Both PCA (with the first three PCs accounting for 65.7%) and HCA were effective tools for differentiating between green and roasted coffee samples because Na, K, Ca, Cu, Mg and Fe were used as discriminative variables.

de Gois et al. (2016) used Mn, Ni, Rb, Sr and total phenolic compound data and antioxidant activity data obtained by SS–HR–CS ET AAS, ICP–MS, DPPH and ABTS radical scavenging methods to study six Brazilian species of stimulant plants. PCA revealed that the first two PCs explaining 72.17% of the total variance enabled the successful distinction between species (catuaba, ginseng, ginger and mate), with

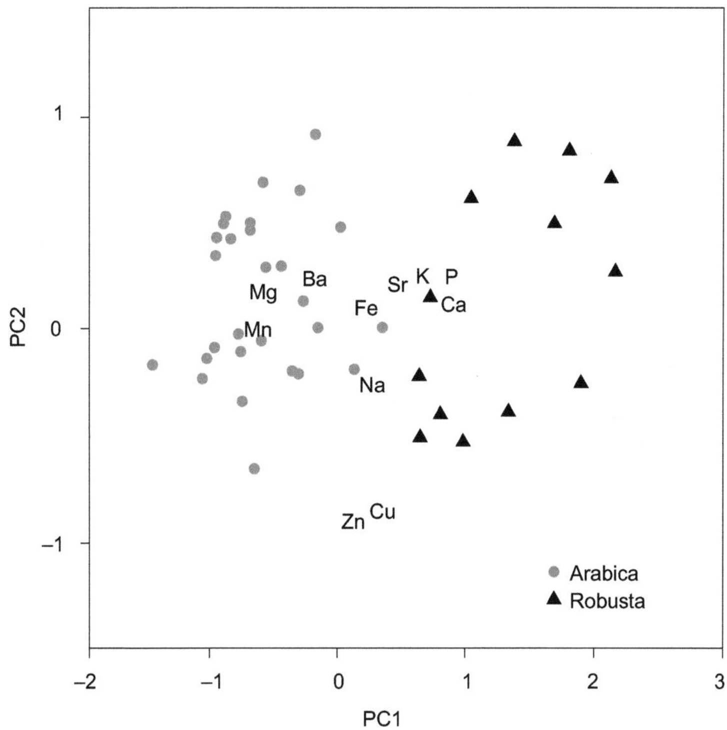

Fig. 13.9 PCA biplot for the first PCs. (After Martín MJ et al., 1998. Anal Chim Acta 358:177–183. https://doi.org/10.1016/S0003-2670(97)00610-7. With permission)

the exception of coffee and guarana, indicating that these data could be used to discriminate between different stimulant plants.

13.2.3 Interregional and Interspecies Differentiation

Voica et al. (2016) characterized the elemental composition of ground and instant coffee from bought in nine countries by means of ICP–MS. ANOVA highlighted elements distinguishing coffee on the basis of geographical origin or distinguishing ground from instant coffee. For PCA, the first three PCs explained 57.08% of the total variance, enabling the classification of some elements into appropriate categories. Co, Na, Mg and K were characteristic variables for soluble coffee, whereas Mn, Zn and Ca were characteristic variables for ground coffee.

Santato et al. (2012) classified green arabica and robusta coffee beans grown at different sampling sites in 16 countries from 4 continent macroareas, i.e., South and Central America, Africa and Asia. ICP–MS and EA–IRMS in combination with chemometric tools were useful for the interdifferentiation of coffee samples. SDA using $\delta^{13}C$, $\delta^{15}N$, $\delta^{18}O$, Li, Mg, P, K, Mn, Co, Cu, Se, Y, Mo, Cd, La and Ce contributed to 83% of the variability. This model correctly assigned 98% of the green coffee beans to the appropriate four continental macroareas. By performing CDA on the basis of the most significant elements, such as P, K, Mn, Cu and Mo, it was possible to perform a reclassification test, which correctly assigned 95% of the arabica and robusta green coffee beans. This means that the combined methodology is a useful approach for correctly verifying the authenticity of these two coffee varieties.

Bertrand et al. (2008) evaluated the chemical composition (macro- and microelements, chlorogenic acids and fatty acid content) of samples of *C. arabica* genotypes (four introgressed lines and Caturra) and two field replications from three growing Colombian regions. The combination of HPLC–UV, GC–FI and ICP–AES with chemometric tools was useful for classifying the studied samples. The first three PCs explained approximately 75% of the total variance. Factorial scores of PCs corresponding to EV > 1 were applied to calculate the DF models. Chemical elements have significant differentiating potential, contributing to the correct classification of coffee samples according to their origin. The effectiveness of these variables for reliable identification of the geographical provenance of coffee samples likely corresponds to the fact that the elemental profile of coffee reflects the composition of the soil in which the plants grow, which is highly dependent on the coffee-producing location. However, the elemental composition was not useful for discriminating coffee varieties. However, the usefulness of fatty acids resulted from distinguishing both varieties and environments, with satisfactory correct classifications of 79% and 90%, respectively.

Mehari et al. (2016) assayed green coffee bean samples (washed and unwashed samples), which represent six varieties from east, west and south Ethiopia. The contents of macro- and microelements were quantified by CP–OES and FAAS methods. PCA, with the first two PCs accounting for 57% of the total variance, clearly differentiated the coffee samples from the three regions. Variables such as P and Mn as well as Cu and Fe were the most effective contributors to PC1 and PC2, respectively. The application of LDA supplied models with 92% of the CV samples correctly classified for the production region. Elements such as P, Mn, S, Cu and Fe had the best discriminatory abilities. A less effective discrimination of six types of coffee was achieved, with a prediction ability of 88%, whereas 84% was obtained through the CV. An even lower prediction ability of 79% was achieved for this intervarietal discrimination model when it was obtained at the regional level.

Fernandes et al. (2005) examined coffee and sugar-cane spirit (cachaça) samples by ICP–OES. Pattern recognition techniques, i.e., PCA and CA, were used to classify samples according to their origin and production mode. Although it was impossible to verify the provenance of the coffee or the production mode (organic or conventional) of cachaça, partial interregional differentiation of cachaça across three regions (northeast, central and southern) was achieved. The coffee samples were also not fully distinguished with respect to the production mode. However, a clear distinction between industrial and homemade cachaças or between instantly soluble and roasted coffees was obtained. Notably, in the case of the cachaça samples, the KNN classification model with four nearest neighbours provided a success rate of 84.6%, which means that the samples were correctly classified.

13.2.4 Differentiation of Other Parameters

13.2.4.1 Verification of Authenticity/Adulteration

Martı́n et al. (1999) differentiated between green coffee samples belonging to the arabica and robusta varieties from ten countries in Central and South America, Africa and Asia. On the basis of the elemental data, PCA, CA and PLS enabled discrimination between arabica and robusta coffees relative to their geographical provenances. As shown by the PCA data, the first two PCs, which explained 61.0% of the total variance, enabled a clear differentiation of the Arabica samples from the robusta coffee samples owing to the significant contributions of P, Cu and Mn as discriminative descriptors. The best results were obtained using PLS calibration, which enabled them to estimate the authenticity of three commercial ground roasted coffee samples of 100% arabica, 100% robusta and a blend with 25% robusta (according to the label clam), with highly satisfactory prediction abilities of 94.0%, 103.7% and 25.3%, respectively.

Sezer et al. (2018) used LIBS elemental profiles to study coffee originating from 21 countries. They aimed to identify the adulteration of ground roasted coffee arabica with wheat, corn and chickpea samples and to distinguish these samples. The first two PCs, which represent 98.27% of the total variance, clearly differentiated seven clusters. This chemometric technique involves blending two wheat varieties to obtain a general wheat matrix, and a PLS study was performed using this mixture for the remainder of the investigation. Consequently, a good prediction ability was achieved via PLS, and the low RMSEC and RMSEP values indicated that this approach could be applied for the identification and quantification of adulterants. This means that LIBS combined with chemometric techniques can be used as a routine technique for determining adulteration in ground roasted coffee and identifying its authenticity.

According to de de Oliveira et al. (2023), a widely used tool to verify authenticity is the application of one-class classifiers (OCCs), and this approach could have the potential to detect adulterations in specialty coffees with low-quality coffee. Therefore, the aim of their study was to use multielement analysis quantified by ICP–MS. The next step was to assess the efficiency of OCC (DD–SIMCA, OCRF and OCPLS). Using this approach, authentic specialty coffee samples were adulterated with low-quality coffee in 25–75% (w/w) proportions. The OCRF and OCPLS models had satisfactory sensitivity, specificity and reliability rates. Moreover, the PTEs, i.e., ^{208}Pb, ^{114}Cd, ^{75}As and ^{53}Cr, were distinguished as the variables mostly responsible for differentiation between authentic and adulterated samples in the OCPLS and OCRF models. Among the one-class classifiers, the OCPLS method was the best option for detecting adulterations with low-quality coffee in specialty coffees, which stood out for its higher specificity (SPE = 100%) and reliability rate (RLR = 94.3%). The classified specialty of coffees was examined for the first time using REEs and PTEs combined with the abovementioned three one-class classifiers. This effect was possibly due to the selection of coffees even when the adulterant originated from the same region.

Worku et al. (2019) coupled ICP–MS, ICP–OES and WD–XRF with chemometric tools for the characterization of elemental and stable isotope profiles of green arabica coffee samples from 45 districts covering four producing regions in the cropping seasons of 2010–2011, 2014–2015 and 2015–2016. LDA clearly discriminated the three major coffee regions (Harar, southeastern and southwestern) on the basis of the multielement data obtained by WD-XRF, with $\delta^{13}C$ values achieving an overall classification accuracy of 88%. A very similar percentage was obtained for the ICP data (approximately 89%). Moreover, the clear differentiation of Harar coffee from the other regions (Fig. 13.10) was in agreement with earlier published studies concerning the geographic provenance of Ethiopian coffee at local (Habte et al. 2016; Mehari et al. 2016) and country scales (Valentin and Watling 2013) via multielement data. Harar coffee may be more traceable via multielement quantification with or without the use of $\delta^{13}C$, $\delta^{15}N$ and $\delta^{18}O$ than coffees originating from other regions in Ethiopia. It was concluded that XRF, as a more rapid, easier and less expensive analysis method compared with ICP, is preferable to the multielement approach in that it can be used to authenticate or identify the geographic provenance of Ethiopian coffee and can be applied to help protect consumers financially. It was also postulated that the methodology requires further standardization; hence, a CRM for coffee should be developed.

Endaye et al. (2020) integrated ICP–OES with multivariate techniques (PCA and LDA) to classify Ethiopian green coffee beans collected from four major cultivation zones. PCA was used to explore the groupings of samples and the discriminatory ability of the elements. The first two PCs, which explained 83.7% of the total variance, differentiated coffee samples according to their production origins. Among the studied elements, K, Mg and Na were key chemical predictors of the production zones of the coffee samples. LDA supplied a model to classify the samples with an accuracy of 94.2% and a prediction ability of 93.4%. In conclusion, the elemental profile of green coffee beans can be applied as a chemical descriptor in the authentication of coffee originating from the Amhara region.

Zhang et al. (2012) examined naturally different caffeine samples, including coffee beans, tea leaves, guaraná powder, maté leaves, synthetic caffeine samples and tested drinks from different sources in Asia, East Africa, South America, Central America and Oceania. EA/IRMS and HT–RPLC/IRMS were useful analytical methods for distinguishing two groups of caffeine $\delta^{13}C$ values for natural and synthetic caffeine. The proposed methods were effective approaches for controlling the authenticity of caffeine-containing drinks.

Arana et al. (2016) quantified $\delta^{13}C$ and seven compounds, including caffeine, α-tocopherol, β-tocopherol, stigmasterol, β-sitosterol, 2-furanmethanol and palmitic acid (C16:0), in roasted coffees from three countries in South America, i.e., Colombia, Brazil and Peru. Alternative analytical techniques, such as GC–C–IRMS

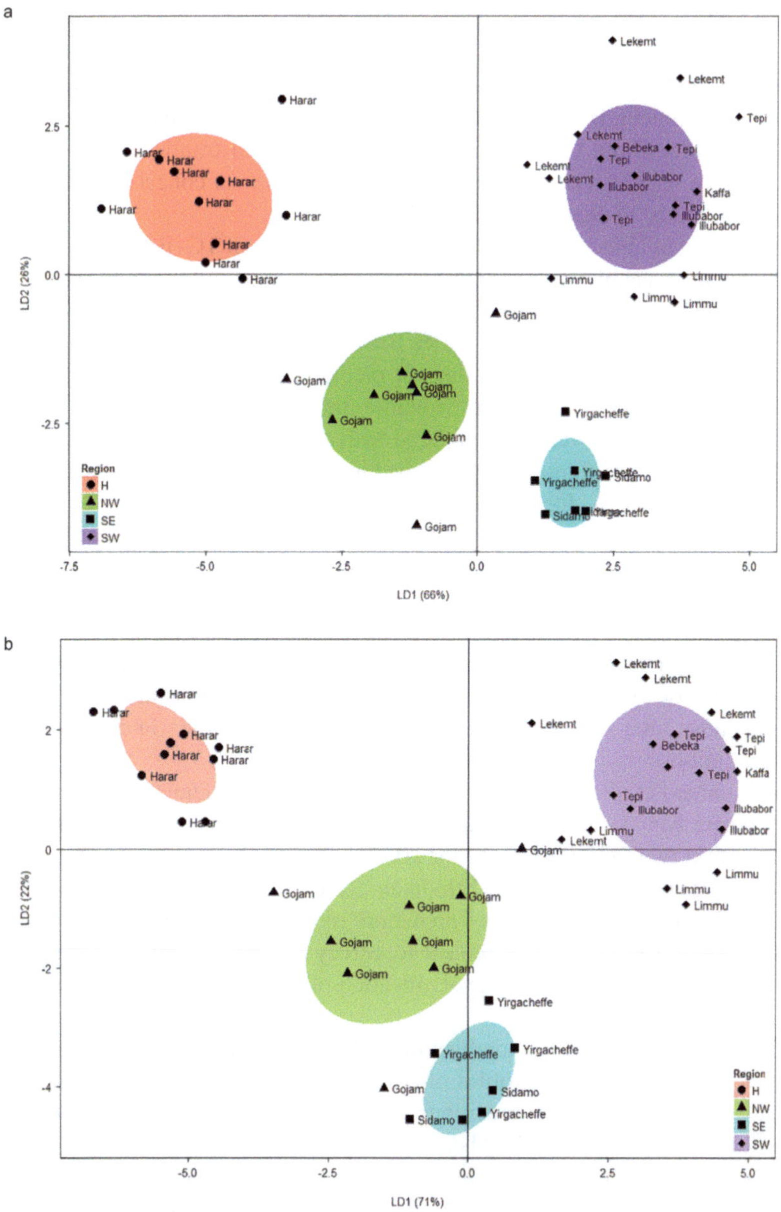

Fig. 13.10 (**a**) LDA plot of the first two discriminant functions (LD1 and LD2; 92%) showing the differentiation of coffee samples originating from Harar (H, red) and northwestern (NW, green), southeastern (SE, blue) and southwestern (SW, purple) coffee regions of Ethiopia, which is based on the multielement data obtained by WD–XRF; (**b**) LDA plot of the first two discriminant functions (LD1 and LD2; 93%) showing the differentiation of coffee samples originating from Harar (H, red) and northwestern (NW, green), southeastern (SE, blue) and southwestern (SW, purple) coffee regions of Ethiopia, which is based on the multielement data obtained by WD–XRF and the $\delta^{13}C$ value. (After Worku M et al., 2019. Food Chem.290:295–307. https://doi.org/10.1016/j.foodchem.2019.03.135. With permission)

and GC–MS, are promising alternatives to ^{1}H-NMR. Among the chemometric models, PCA, SIMCA and PLS–DA validated by seven-fold CV were applied. NMR fingerprinting combined with chemometric processing enabled the observation of numerous compounds (over a thousand variables), most of which are essential for clearly distinguishing Colombian coffee from potential frauds. It was concluded that positive results were obtained despite investigations concerning high-profile coffees produced in nearby bordering countries.

According to Driscoll et al. (2020), the combination of TC–EA/IRMS with computational tools enables geographical characterization of coffee samples originating from 21 countries in South America, Central America, North America, Africa, Asia and Oceania. The $\delta^{18}O$ values of coffee bean water ($\delta^{18}O_{sw}$) were empirically modelled as a function of the $\delta^{18}O$ values of precipitation, relative humidity and temperature using MLR. Coffees grown within the same country or region frequently have comparable $\delta^{18}O$ values. Compared with coffees from other areas worldwide, African coffees were significantly enriched in $\delta^{18}O$. The interfarm value differentiation approached 7‰ for samples grown in the USA (Hawaii), El Salvador, Guatemala and Colombia, indicating that highly significant interregional variations in coffee cellulose $\delta^{18}O$ values may be observed within some countries. The $\delta^{18}O$ value of coffee bean α-cellulose can effectively discriminate between different regions of provenance and therefore could be useful for determining the authenticity of coffees that may be fraudulently labelled. Using this model of the $\delta^{18}O$ values of coffee cellulose combined with other parameters that are representative of the origin will achieve the most precise, accurate and reliable interspatial coffee differentiation.

Mohammed et al. (2019) used ICP–OES to study the elemental composition of green coffee bean samples from different producing regions in two countries. PCA, with the first two PCs accounting for 100% of the total variance, significantly distinguished Yemeni coffee from Ethiopian coffee. This finding was confirmed by HAC. Calcium and Na were key predictors of coffee-producing countries. It has been suggested to use this approach to certify the geographical origin of high-quality Yemeni coffee; hence, the certification of its origin will make it possible to reduce fraud.

Grembecka et al. (2007) combined FAAS and a spectrophotometric method to characterize the elemental profiles of two types of coffee, i.e., instant and ground coffees, including arabica and robusta, which originated from 18 countries in America, Africa, Asia, the Caribbean and Oceania. On the basis of the FA results, F1 (46.4%) and F2 (22.7%) results, as well as the CA results, ground coffee infusions were generally differentiated from dry ground coffee and instant coffee samples (Figs. 13.11 and 13.12). As shown in Fig. 13.13, arabica coffee samples were clearly differentiated from robusta samples. Manganese was the best descriptor for the identification of arabica coffee, whereas P identified robusta samples. The FA two-dimensional scatterplot shows that expensive coffee brands are generally located near arabica coffee brands, whereas less expensive brands correspond to robusta brands. The chemometric results enabled the evaluation of the authenticity of retail coffee according to the declared content of the labelled product.

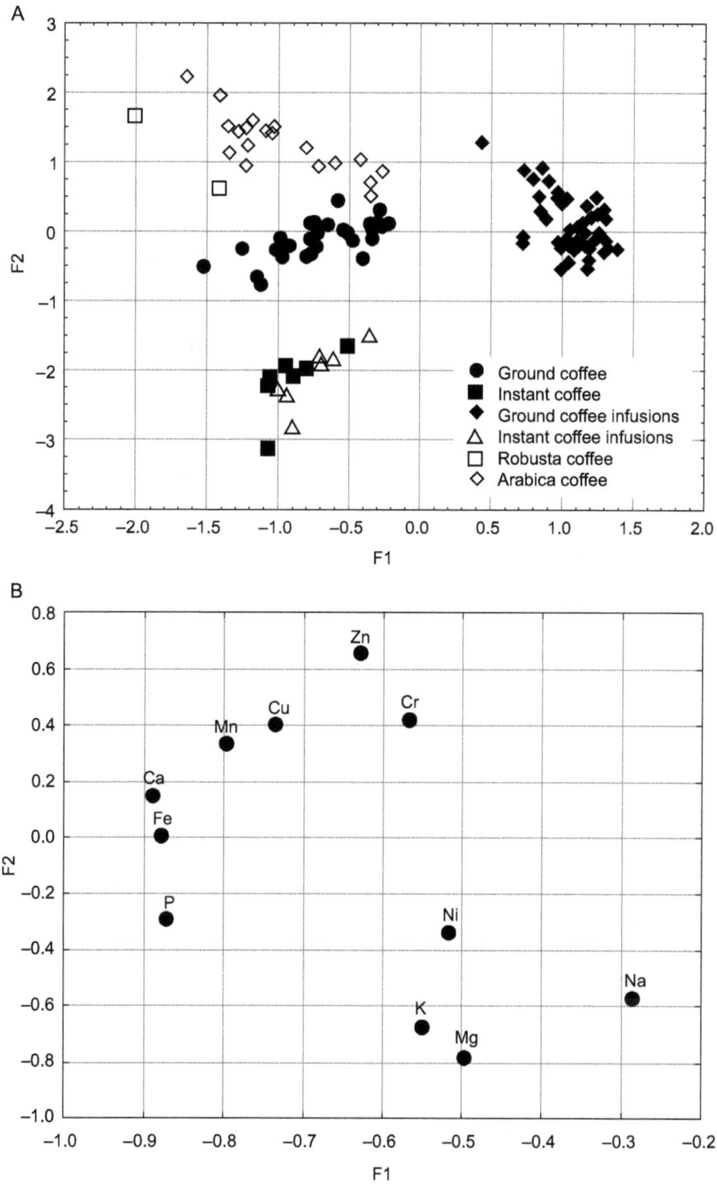

Fig. 13.11 (**a**) Scatterplot of the object scores of the two discriminant functions of 120 coffee samples. (**b**) Scatterplot of loadings for 11 elements in 120 coffee samples. (After Grembecka M et al., 2007. Sci Total Environ 383:59–69. https://doi.org/10.1016/j.scitotenv.2007.04.031. With permission)

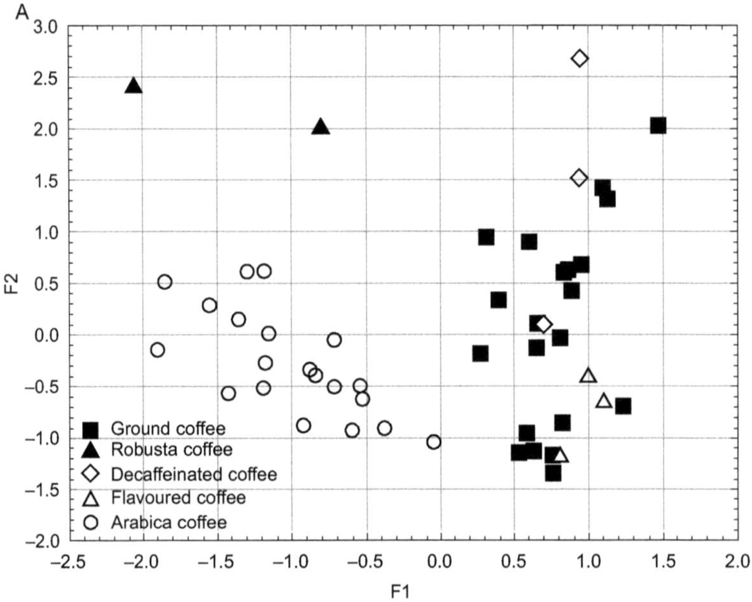

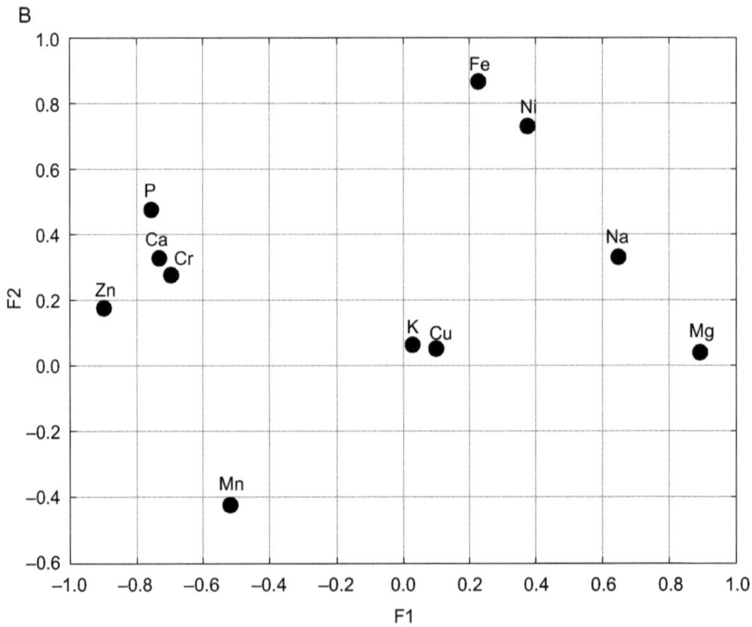

Fig. 13.12 (**a**) Scatterplot of the object scores of the two discriminant functions of the ground coffee samples. (**b**) Scatterplot of loadings for different elements in the ground. (After Grembecka M et al., 2007. https://doi.org/10.1016/j.scitotenv.2007.04.031. With permission)

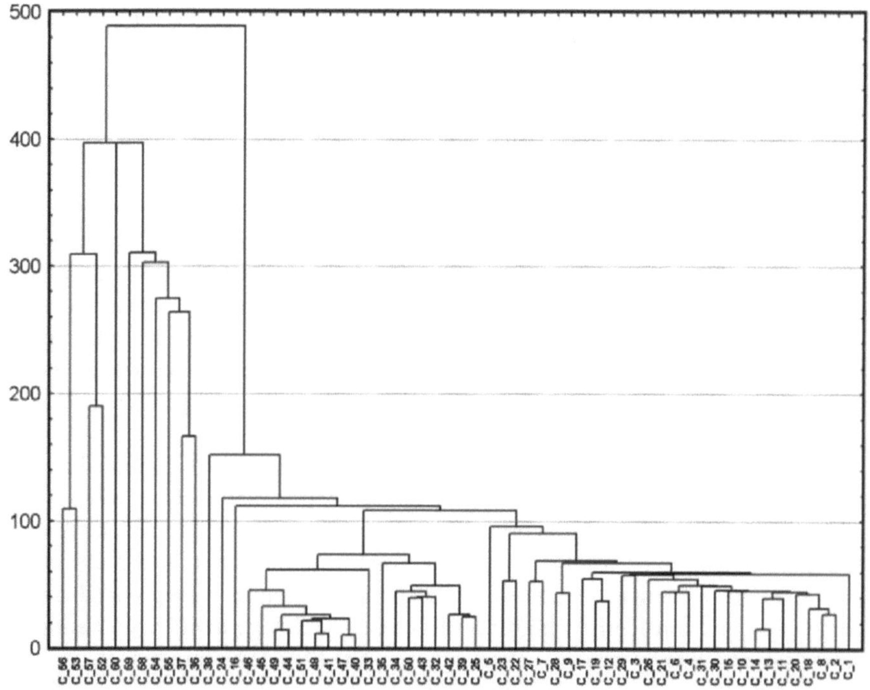

Fig. 13.13 Hierarchical dendrogram for 60 ground and instant coffee samples (1–23 ground coffee, 24–25 robusta coffee, 26–28 decaffeinated coffee, 29–31 flavoured coffee, 32–51 arabica coffee, 52–60 instant coffee). (After Grembecka M et al., 2007. https://doi.org/10.1016/j.scitotenv.2007.04.031. With permission)

13.2.4.2 Differentiation Between Conventional and Organic Cultivation Systems

Peng et al. (2019) integrated EA–IRMS with chemometric processing using supervised models (LDA, *k*-NN and SVM) to identify stable isotope profiles of Brazilian coffees in five producing regions of organic and conventional systems. Organic coffees (certified and containing the seal of authenticity) had relatively high values of $\delta^{15}N$ ($p = 0.0006$), and *k*-NN achieved the most reliable discrimination between organically and conventionally cultivated coffee beans, with 87% and 67% accuracy, respectively. Although the classification of coffees from different geographical origins is rather complex, the use of SVM and LDA enabled the correct classification of coffees produced in São Paulo with 75% accuracy and in Minas Gerais with 71% accuracy, respectively. Notably, both of these producers are the most important producers in terms of quality and planted area.

Barbosa et al. (2014) used ICP–MS to investigate certified organic coffee and nonorganic coffee from three Brazilian states. The data mining algorithms, i.e., MLP, SVM and NB, were useful for classifying the coffee samples. Many accurate results (96.3%) for these three models were obtained when all the variables analysed were considered. However, a better result, i.e., > 98.2% accuracy for NB, was achieved when only eight elements, namely, Zn, Se, Ba, U, Tl, Th, Dy and Mo, selected by the algorithm were applied. The use of trace element contents processed by mining techniques was concluded to be a powerful approach for reliably evaluating the authenticity of organic coffee samples.

Dos Santos et al. (2009) studied samples of coffee cultivated during the process of conversion for organic agriculture in southwestern Bahia (Brazil). The contents of essential and toxic elements in coffee and parent soils from two transitional organic farms were quantified via FAAS. Although the coffee samples had relatively high levels of Cd, Zn and Cu, these levels did not yet exceed the limits specified by the Brazilian Food Legislation. Owing to the use of computational methods, the addition of high amounts of organic matter in the flowering tree period may affect the efficiency of elemental bioavailability in soils.

According to Fiamegos et al. (2021), EDX–RF was used to determine different elements in organic and conventionally produced coffee, i.e., arabica, robusta and mixtures of arabica and robusta originating from 15 American, African and Asiatic countries. The original results were reprocessed using univariate and multivariate computational techniques. PCA revealed that organic samples were grouped almost completely, and considerably higher levels of Cl and K were found in conventional coffee samples than in organic coffee samples. The PLS–DA score plot clearly differentiated the two clusters, which corresponded to organic and conventional coffees.

13.2.4.3 Evaluation of the Nutritional Quality of Coffee in Relation to Other Parameters

dos Santos and de Oliveira (2001) coupled ICP–AES with chemometric analysis to assay soluble Brazilian coffee in the context of nutritional quality acceptable for human consumption. The application of PCA and AHC enabled the identification of samples with different elemental profiles, reflecting possible differences between the processes of industrial production and/or factors influencing the cultivation of the coffee plant, including the type of soil, the use of fertilizers with different chemical compositions and ambient conditions. It was postulated that the values obtained were generally acceptable for human consumption from both nutritional and toxic points of view, except for one sample containing a Cr level exceeding a limit above that recommended by the Brazilian Food Legislation. This elevated value could be associated with contamination from the metal container.

Cloete et al. (2019) examined roasted organic coffee beans from four countries with respect to the physicochemical profiles determined by XRM and PIXE methods combined with multidimensional analysis. These methods are promising tools for assessing nutritional content and qualitative characteristics. LDA found no strong relationship between the elemental data and production region, whereas a heatmap coupled with HCA revealed that the physicochemical properties of soil and plants might influence interspatial elemental designation. The observed variation in the elemental profile between coffee beans characterizes both the nutritional and organoleptic qualities of beans in specific regions of Ethiopia, Colombia, Honduras and Mexico.

Árvay et al. (2019) examined arabica green and roasted coffee beans of various roasting degrees and their infusions originating from five countries. The contents of mineral elements in the samples were analysed via ICP–OES. As a result of the chemometric analysis performed, significant differences in the elemental profile were observed between the green coffee beans and the different roasting degrees. It was also reported that the degree of roasting influences the levels of elements in resulting infusions. Moreover, on the basis of the PTWI value, the consumption of coffee does not pose a health risk to consumers.

13.2.4.4 Elaboration of Improved Methodology

Szymczycha-Madeja et al. (2015) proposed a simplified alternative to wet digestion of roasted ground coffee and a preparation procedure for instant coffee samples, which was developed and validated for the quantification of MECs via ICP–OES and FAAS. The recommended procedure was fast and simple, requiring minimal use of reagents, and was satisfactory in many of validation parameters, such as the LOD, accuracy and precision. Notably, a preliminary classification of 18 ground and instant coffee samples was an effective approach owing to the coupling of elemental concentrations with PCA and CA.

13.3 Conclusions

The evaluation of food authentication methods contributes to the assessment of food safety. Selected analytical methods/techniques coupled with computational tools can be helpful in the evaluation of coffee in terms of its authenticity or adulteration/mislabelling to achieve this task. The proper selection of such a methodology for the classification of coffee beans has enabled the evaluation of the degree of influence of various environmental and technological/culinary parameters on coffee quality. The integration of an analytical device with multivariate models enables the fingerprints of edible plant resources, including coffees, to be differentiated. Therefore, it is possible to characterize coffee quality in terms of its interregional, intervarietal/brand or annual/seasonal variations as well as to discriminate between conventional and organic coffees. Validated results make it possible to identify fraud and detect product mislabelling to preserve the brand reputations of original products and protect consumers from being taken advantage of financially.

References

Agriopoulou S, Tarapoulouzi M, Varzakas T (2023a) Introduction. In: Agriopoulou S, Tarapoulouzi M, Varzakas T (eds) Chemometrics and authenticity of foods of plant origin. CRC Press/Taylor & Francis Groups, Boca Raton, FL, pp 1–8. https://doi.org/10.1201/9781003266167

Agriopoulou S, Tarapoulouzi M, Varzakas T (eds) (2023b) Chemometrics and authenticity of foods of plant origin. CRC Press/Taylor & Francis Groups, Boca Raton, FL. https://doi.org/10.1201/9781003266167

Akamine T, Otaka A, Hokura A, Ito Y, Nakai I (2010) Determination of trace elements in coffee beans by XRF spectrometer equipped with polarization optics and its application to identification of their production area. Bunseki Kagaku 59:863–871. https://doi.org/10.2116/bunsekikagaku.59.863

Anderson KA, Smith BW (2002) Chemical profiling to differentiate geographic growing origins of coffee. J Agric Food Chem 50:2068–2075. https://doi.org/10.1021/jf011056v

Arana VA, Medina J, Esseiva P, Pazos D, Wist J (2016) Classification of coffee beans by GC-C-IRMS, GC-MS, and ^{1}H-NMR. J Anal Methods Chem 8564584:1–11. https://doi.org/10.1155/2016/8564584

Árvay J, Šnirc M, Hauptvogl M, Bilčíková J, Bobková A, Demková L, Hudáček M, Hrstková M, Lošák T, Král M, Kováčik A, Štefániková J (2019) Concentration of micro- and macro-elements in green and roasted coffee: influence of roasting degree and risk assessment for the consumers. Biol Trace Elem Res 190:226–233. https://doi.org/10.1007/s12011-018-1519-3

Barbosa RM, Batista BL, Varrique RM, Coelho VA, Campiglia AD, Barbosa F Jr (2014) The use of advanced chemometric techniques and trace element levels for controlling the authenticity of organic coffee. Food Res Int 61:246–251. https://doi.org/10.1016/j.foodres.2013.07.060

Bertrand B, Villarreal D, Laffargue A, Posada H, Lashermes P, Dussert S (2008) Comparison of the effectiveness of fatty acids, chlorogenic acids, and elements for the chemometric discrimination of coffee (*Coffea arabica* L.) varieties and growing origins. J Agric Food Chem 56:2273–2280. https://doi.org/10.1021/jf073314f

Bitter NQ, Fernandez DP, Driscoll AW, Howa JD, Ehleringer JR (2020) Distinguishing the region-of-origin of roasted coffee beans with trace element ratios. Food Chem 320:126602. https://doi.org/10.1016/j.foodchem.2020.126602

Borràs E, Ferré J, Boqué R, Mestres M, Aceña L, Busto O (2015) Data fusion methodologies for food and beverage authentication and quality assessment—a review. Anal Chim Acta 891:1–14. https://doi.org/10.1016/j.aca.2015.04.042

Carter JF, Yates HSA, Tinggi U (2015) Isotopic and elemental composition of roasted coffee as a guide to authenticity and origin. J Agric Food Chem 63:5771–5779. https://doi.org/10.1021/acs.jafc.5b01526

Cloete KJ, Šmit Ž, Minnis-Ndimba R, Vavpetič P, du Plessis A, le Roux SG, Pelicon P (2019) Physico-elemental analysis of roasted organic coffee beans from Ethiopia, Colombia, Honduras, and Mexico using X-ray micro-computed tomography and external beam particle induced X-ray emission. Food Chem X 2:100032. https://doi.org/10.1016/j.fochx.2019.100032

de Gois JS, Almeida TS, de Andrade RM, Toaldo IM, Bordignon-Luiz MT, Borges DLG (2016) Direct solid analysis for the determination of Mn, Ni, Rb and Sr in powdered stimulant plants using high-resolution continuum source atomic absorption spectrometry followed by chemometric classification based on elemental composition, polyphenol content and antioxidant activity. Microchem J 124:283–289. https://doi.org/10.1016/j.microc.2015.08.020

de Oliveira CT, Botelho JR, Nascimento MHC, Krause M, Weitzel Dias Carneiro MT, Coelho Ferreira D, Filgueiras PR, de Oliveira Souza M (2023) A one-class classification approach for authentication of specialty coffees by inductively coupled plasma mass spectroscopy (ICP-MS). Food Chem 138268. https://doi.org/10.1016/j.foodchem.2023.138268

dos Santos ÉJ, de Oliveira E (2001) Determination of mineral nutrients and toxic elements in Brazilian soluble coffee by ICP-AES. J Food Compos Anal 14:523–531. https://doi.org/10.1006/jfca.2001.1012

dos Santos JS, dos Santos MLP, Conti MM, dos Santos SN, de Oliveira E (2009) Evaluation of some metals in Brazilian coffees cultivated during the process of conversion from conventional to organic agriculture. Food Chem 115:1405–1410. https://doi.org/10.1016/j.foodchem.2009.01.069

Driscoll AW, Howa JD, Bitter NQ, Ehleringer JR (2020) A predictive spatial model for roasted coffee using oxygen isotopes of α-cellulose. Rapid Commun Mass Spectrom 34:e8626. https://doi.org/10.1002/rcm.8626

Endaye M, Atlabachew M, Mehari B, Alemayehu M, Mengistu DA, Kerisew B (2020) Combining multi-element analysis with statistical modeling for tracing the origin of green coffee beans from Amhara region, Ethiopia. Biol Trace Elem Res 195:669–678. https://doi.org/10.1007/s12011-019-01866-5

Fernandes AP, Santos MC, Lemos SG, Ferreira MMC, Nogueira ARA, Nóbrega JA (2005) Pattern recognition applied to mineral characterization of Brazilian coffees and sugar-cane spirits. Spectrochim Acta Part B 60:717–724. https://doi.org/10.1016/j.sab.2005.02.013

Fiamegos Y, Papoci S, Dumitrascu C, Ghidotti M, Zdiniakova T, Ulberth F, de la Calle Guntinas MB (2021) Are the elemental fingerprints of organic and conventional food different? ED-XRF as screening technique. J Food Compos Anal 99:103854. https://doi.org/10.1016/j.jfca.2021.103854

Filho VRA, Polito WL, Neto JAG (2007) Comparative studies of the sample decomposition of green and roasted coffee for determination of nutrients and data exploratory analysis. J Braz Chem Soc 18:47–53. https://doi.org/10.1590/S0103-50532007000100005

Grembecka M, Szefer P (2011) Metals and metalloids in foods—essentiality, toxicity, applicability. In: Medina DA, Laine AM (eds) Food quality: control, analysis and consumer concerns. Nova Science Publishers, Inc., New York, pp 1–60. ISBN 10: 1611229170; ISBN 13: 9781611229172

Grembecka M, Szefer P (2016) Elemental trace analysis in studies of food products. In: Baranowska I (ed) Handbook of trace analysis—fundamentals and applications. Springer, pp 203–239. https://doi.org/10.1007/978-3-319-19614-5

Grembecka M, Malinowska E, Szefer P (2007) Differentiation of market coffee and its infusions in view of their mineral composition. Sci Total Environ 383:59–69. https://doi.org/10.1016/j.scitotenv.2007.04.031

Habte G, Hwang IM, Kim JS, Hong JH, Hong YS, Choi JY, Nho EY, Jamila N, Khan N, Kim KS (2016) Elemental profiling and geographical differentiation of Ethiopian coffee samples through inductively coupled plasma-optical emission spectroscopy (ICP-OES), ICP-mass spectrometry (ICP-MS) and direct mercury analyzer (DMA). Food Chem 212:512–520. https://doi.org/10.1016/j.foodchem.2016.05.178

Hall RD, Trevisan F, de Vos RCH (2022) Coffee berry and green bean chemistry—opportunities for improving cup quality and crop circularity. Food Res Int 151:110825. https://doi.org/10.1016/j.foodres.2021.110825

Krivan V, Barth P, Morales AF (1993) Multielement analysis of green coffee and its possible use for the determination of origin. Mikrochim Acta 110:217–236. https://doi.org/10.1007/BF01245106

Liu H-C, You C-F, Chen C-Y, Liu Y-C, Chung M-T (2014) Geographic determination of coffee beans using multi-element analysis and isotope ratios of boron and strontium. Food Chem 142:439–445. https://doi.org/10.1016/j.foodchem.2013.07.082

Marini M (2013) Chemometrics in food chemistry. Elsevier. ISBN: 9780444595287; https://www.elsevier.com/books/chemometrics-in-food-chemistry/marini/978-0-444-59528-7

Martín MJ, Pablos F, González AG (1996) Application of pattern recognition to the discrimination of roasted coffees. Anal Chim Acta 320:191–197. https://doi.org/10.1016/0003-2670(95)00550-1

Martín MJ, Pablos F, González AG (1998) Characterization of green coffee varieties according to their metal content. Anal Chim Acta 358:177–183. https://doi.org/10.1016/S0003-2670(97)00610-7

Martín MJ, Pablos F, González AG (1999) Characterization of *arabica* and *robusta* roasted coffee varieties and mixture resolution according to their metal content. Food Chem 66:365–370. https://doi.org/10.1016/S0308-8146(99)00092-8

Mehari B, Redi-Abshiro M, Chandravanshi BS, Combrinck S, McCrindle R (2016) Characterization of the cultivation region of Ethiopian coffee by elemental analysis. Anal Lett 49:2474–2489. https://doi.org/10.1080/00032719.2016.1151023

Mohammed F, Guillaume D, Dowman S, Abdulwali N (2019) An easy way to discriminate Yemeni against Ethiopian coffee. Microchem J 145:173–179. https://doi.org/10.1016/j.microc.2018.10.039

Muñiz-Valencia R, Jurado JM, Ceballos-Magaña SG, Alcázar A, Reyes J (2013) Geographical differentiation of green coffees according to their metal content by means of supervised pattern recognition techniques. Food Anal Methods 6:1271–1277. https://doi.org/10.1007/s12161-012-9538-8

Oestreich-Janzen S (2010) Chemistry of coffee. CAFEA GmbH, Elsevier Ltd., Hamburg, Germany, 3:1085-1096. https://doi.org/10.1016/B978-008045382-8.00708-5

Oliveira M, Ramos S, Delerue-Matos C, Morais S (2015) Espresso beverages of pure origin coffee: mineral characterization, contribution for mineral intake and geographical discrimination. Food Chem 177:330–338. https://doi.org/10.1016/j.foodchem.2015.01.061

Peng C-Y, Zhang Y-L, Song W, Cai H-M, Wang Y, Granato D (2019) Characterization of Brazilian coffee based on isotope ratio mass spectrometry ($\delta^{13}C$, $\delta^{18}O$, $\delta^{2}H$, and $\delta^{15}N$) and supervised chemometrics. Food Chem 297:124963. https://doi.org/10.1016/j.foodchem.2019.124963

Pohl P, Stelmach E, Welna M, Szymczycha-Madeja A (2013) Determination of the elemental composition of coffee using instrumental methods. Food Anal Methods 6:598–613. https://doi.org/10.1007/s12161-012-9467-6

Pohl P, Szymczycha-Madeja A, Stelmach E, Welna M (2016) Differentiation of roasted and soluble coffees through physical fractionation of selected essential and nonessential metals in their brews and exploratory data analysis. Talanta 160:686–693. https://doi.org/10.1016/j.talanta.2016.08.020

Rodrigues CI, Maia R, Miranda M, Ribeirinho M, Nogueira JMF, Máguas C (2009) Stable isotope analysis for green coffee bean: a possible method for geographic origin discrimination. J Food Compos Anal 22:463–471. https://doi.org/10.1016/j.jfca.2008.06.010

Rodrigues C, Brunner M, Steiman S, Bowen GJ, Nogueira JMF, Gautz L, Prohaska T, Máguas C (2011a) Isotopes as tracers of the Hawaiian coffee-producing regions. J Agric Food Chem 59:10239–10246. https://doi.org/10.1021/jf200788p

Rodrigues C, Máguas C, Prohaska T (2011b) Strontium and oxygen isotope fingerprinting of green coffee beans and its potential to proof authenticity of coffee. Eur Food Res Technol 232:361–373. https://doi.org/10.1007/s00217-010-1362-z

Santato A, Bertoldi D, Perini M, Camin F, Larcher R (2012) Using elemental profiles and stable isotopes to trace the origin of green coffee beans on the global market. J Mass Spectrom 47:1132–1140. https://doi.org/10.1002/jms.3018

Serra F, Claude G, Guillou CG, Reniero F, Ballarin L, Cantagallo MI, Wieser M, Iyer SS, Héberger K, Vanhaecke F (2005) Determination of the geographical origin of green coffee by principal component analysis of carbon, nitrogen and boron stable isotope ratios. Rapid Commun Mass Spectrom 19:2111–2115. https://doi.org/10.1002/rcm.2034

Sezer B, Apaydin H, Bilge G, Boyaci IH (2018) *Coffee arabica* adulteration: detection of wheat, corn and chickpea. Food Chem 264:142–148. https://doi.org/10.1016/j.foodchem.2018.05.037

Szefer P (2007) Chemometric techniques in analytical evaluation of food quality. In: Szefer P. Nriagu J (eds) Mineral components in foods. CRC Press. Taylor & Francis, Boca Raton, FL 69–121. https://doi.org/10.1201/9781420003987

Szymczycha-Madeja A, Welna M, Pohl P (2015) Simplified multi-element analysis of ground and instant coffees by ICP-OES and FAAS. Food Addit Contam Part A 32:1488–1500. https://doi.org/10.1080/19440049.2015.1067928

Valentin JL, Watling RJ (2013) Provenance establishment of coffee using solution ICP-MS and ICP-AES. Food Chem 141:98–104. https://doi.org/10.1016/j.foodchem.2013.02.101

Vieira JFA (2015) Classification of green coffee beans according to the geographical origin: a statistical approach. Analysis of the producing regions of Hawaii, Brazil, Ethiopia and Kenya. Agri Food Sci. Extended abstract. Corpus ID: 208540097

Voica C, Feher I, Iordache AM, Cristea G, Dehelean A, Magdas DA, Mirel V (2016) Multielemental analysis of coffee by inductively coupled plasma-mass spectrometry. Anal Lett 49:2627–2643. https://doi.org/10.1080/00032719.2015.1116003

Weckerle B, Richling E, Heinrich S, Schreier P (2002) Origin assessment of green coffee (*Coffea arabica*) by multi-element stable isotope analysis of caffeine. Anal Bioanal Chem 374:886–890. https://doi.org/10.1007/s00216-002-1560-z

Worku M, Upadhayay HR, Latruwe K, Taylor A, Blake W, Vanhaecke F, Duchateau L, Boeckx P (2019) Differentiating the geographical origin of Ethiopian coffee using XRF- and ICP-based multi-element and stable isotope profiling. Food Chem 290:295–307. https://doi.org/10.1016/j.foodchem.2019.03.135

Zhang L, Kujawinski DM, Federherr E, Schmidt TC, Jochmann MA (2012) Caffeine in your drink: natural or synthetic? Anal Chem 84:2805–2810. https://doi.org/10.1021/ac203197d

Zhang C, Shen T, Liu F, He Y (2018) Identification of coffee varieties using laser-induced breakdown spectroscopy and chemometrics. Sensors 18:95. https://doi.org/10.3390/s18010095

Chapter 14
Cocoa, Its Products and Sweets

Contents

14.1 Introduction

Cocoa is enriched in antioxidants such as flavonoids and procyanidins, to which antiaging properties are assigned. This product also contains high levels of epicatechin, which is helpful for cardiovascular health and is a rich source of essential mineral elements, including K, Mg, Ca, P, Cu, Zn, B, Mn and Fe.

Cocoa beans are the seeds of the tree *Theobroma cacao* L., the raw material that is used for chocolate production. The properties of this plant product depend mostly on the cocoa origin and composition, and the manufacturing procedure determines the specific sensory properties of the final product. The excessive global consumption of cocoa products, which are important sources of organoleptic value, and their

Supplementary Information The online version contains supplementary material available at https://doi.org/10.1007/978-3-031-99687-0_14.

P. Szefer, *Chemometrics in Food Control Based on Elemental Analysis*,
https://doi.org/10.1007/978-3-031-99687-0_14

chemical composition, i.e., dietary polyphenols with well-known health benefits, are unique. Therefore, increasing attention has been given to fingerprinting the geographical provenance of cocoa and the authentication/adulteration of chocolate to guarantee product quality. However, the reliable production of high-quality cocoa products is still a serious challenge in the cocoa sector. Progress with respect to the authenticity and traceability of cocoa products is still needed; hence, different issues need to be resolved by chocolate producers. A review by Perez et al. (2022) provided an update on the progress towards the authenticity, traceability and sustainability of cocoa products, issues that chocolate producers still need to resolve.

Cocoa bean flavour is a key parameter for the acceptability of cocoa beans and cocoa products such as chocolate. This sensory feature depends on the bean genotype, postharvest treatments, such as pulp preconditioning, fermentation and drying, industrial processes, such as roasting, the type of soil and the age of the cocoa tree (Kongor et al. 2016). Cocoa bean fermentation and drying change the chemical profiles of plant material through enzymatic degradation of the stored proteins into amino acids and short-chain oligopeptides, while polysaccharides are transformed into monosaccharides, i.e., glucose and fructose. The breakdown products react with each other during the roasting process to flavour volatiles. Polyphenols are also enzymatically oxidized during fermentation and drying, intensifying the taste of cocoa beans. The increasing requirements for the sustainable production of high-quality cocoa beans have led to a greater understanding of the factors contributing to shaping flavour characteristics, which implies significant commercial implications (Kongor et al. 2016).

To improve the understanding of the origin, classification and population differentiation within the species *Theobroma cacao* L. germplasm, researchers have genotyped 1241 accessions covering a large geographic sample with 106 microsatellite markers of the cacao germplasm into 10 genetic clusters, in contrast to the two genetic groups traditionally recognized within *T. cacao*, which were identified via Bayesian statistics. As a result, a new classification of the cacao germplasm that will enhance its management was proposed (Motamayor et al. 2008).

With respect to the pollution issue concerning cocoa bean and its products, cocoa and chocolate can contain Cd and lead Pb from natural and anthropogenic sources (Abt and Robin 2020). Relatively high levels of Bi and As have been found in the cocoa bean shells but not in the cocoa bean or chocolate. This difference may be attributed to environmental contamination; however, the origin of other elements is associated with the manufacturing processes of cocoa and chocolate products. Children are known to be frequent consumers of chocolate; hence, they may be at risk of exceeding the daily limit of Pb (Yanus et al. 2014).

Another issue, in contrast to the abovementioned positive aspects of cacao and its products, is the problem of authenticating natural sweeteners. For example, honeys and maple syrups could be adulterated in practice by cheaper, commercially available sugar syrups with similar compositions (Banerjee et al. 2015; Saksangawong et al. 2021).

14.2 Chemometric Assessment of Cocoa and its Products

14.2.1 Geographical Differentiation

14.2.1.1 American Countries

Nimalaratne et al. (2020) studied the physicochemical profiles of maple syrup samples from North America while considering the seasonal compositional changes related to Nova Scotia maple syrup. Interregional differences in physicochemical composition, i.e., syrup pH, colour and Brix, were found to be strongly dependent on the sampling location. For example, among Canadian provinces, some interspatial differences in elemental profiles were found in three Nova Scotia provinces. Moreover, analysis of the seasonal compositional variations in Nova Scotia maple syrup revealed significant changes in syrup colour, mineral element content and phenolic content throughout the season. According to the authors, the data obtained could be used to assess the potential health benefits of maple syrup as a functional food ingredient.

14.2.1.2 Countries Across Multiple Continents

Dark chocolate samples with unroasted cocoa beans from two countries were investigated by ICP–AES, ICP–MS and GC–MS to determine the MEC content and fatty acid profile. Compared with those in Ghanaian cocoa beans, significantly higher values of P, Fe, Zn, Na and Se were detected in Ecuadorian cocoa beans, and lower levels of Mn were detected. The FA profile also differed depending on geographical provenance, while in the case of chocolates, only carbohydrate and fat contents varied significantly due to the effect of origin, but an insignificant effect was observed for processing conditions. However, no significant variations in the Ca, Mg, K or Cu contents between the two types of beans were observed (Torres-Moreno et al. 2015).

Lee (2020) used IRMS combined with PCA (with the first two PCs explaining 90% of the total variance) to estimate the SIR profile for the characterization of cocoa beans and chocolates from six countries within Asia, Africa, Central America and Oceania. Some interspatial differences reflected the geographical and climatic factors of the countries because the Solomon Islands and Samoa are island countries in the Pacific Ocean, whereas Ghana, Nicaragua and Vietnam belong to countries within a larger continental setting; the latter is a long coastal country along the Pacific Ocean. Therefore, PCA differentiated samples belonging to Vietnam scores, which can be explained by the fact that the cocoa beans originated from different regions within the country. In addition, the chocolate samples had positive PC1 scores, whereas the samples from Ghana had negative PC1 scores.

Saito et al. (2021) combined EDXRF with chemometric analysis to characterize the elemental profiles of cacao beans from different cocoa bean growing regions in ten countries. PCA, with the first two PCs accounting for 68.8% of the total variance, allowed the differentiation of cacao samples between African, Asiatic, Caribbean and South American countries. Elements such as Mn, Fe, Ni, Cu, Zn, Rb and Sr were the discriminative variables for recognizing geographical provenance.

According to Perini et al. (2016), TC/EA–IRMS was used to investigate samples of fermented and dried cocoa beans originating from 14 principal cocoa-producing countries in Asia, West Africa, East Africa, South America and Central America. They found that CDA differentiated samples originating from various countries (Fig. 14.1). This model was validated with all the samples and for all analyses. Approximately 84% of the samples (ranging from 81% to 88% according to the considered model) were correctly classified with respect to their geographical region.

Bertoldi et al. (2016) differentiated the elemental profiles of cocoa beans and commercial dark chocolates originating from 22 countries in East and West Africa, Asia and Central and South America. ICP–MS data processed by PCA, CDA and SDA were helpful for classifying the cocoa beans. The score plot of the first two PCs explaining 56% of the total variance indicated an initial discrimination between cocoa beans from the five macroareas of origin. The four canonical variables explained 100% of the observed variability, with 100% correct reclassification of cocoa samples in the five macroareas of origin (Fig. 14.2). The parameters with larger standardized coefficients for the first three canonical variables were Y, Fe, U, Li, Rb, Tl, Ga, Th, Ni, Be, Ba, Mn, Cr and Sr. This cross-validated model achieved

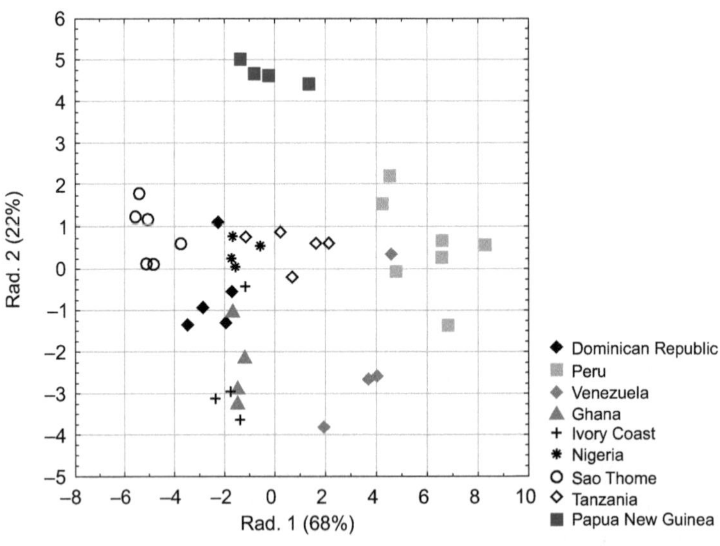

Fig. 14.1 Scatter plot of the first two canonical variables in stepwise discriminant analysis (cumulative variance explained: 81%) on the basis of the isotopic compositions of cocoa beans of different geographical origins. (After Perini M et al., 2016. J Mass Spectrom 51:684–689. https://doi.org/10.1002/jms.3833. With permission)

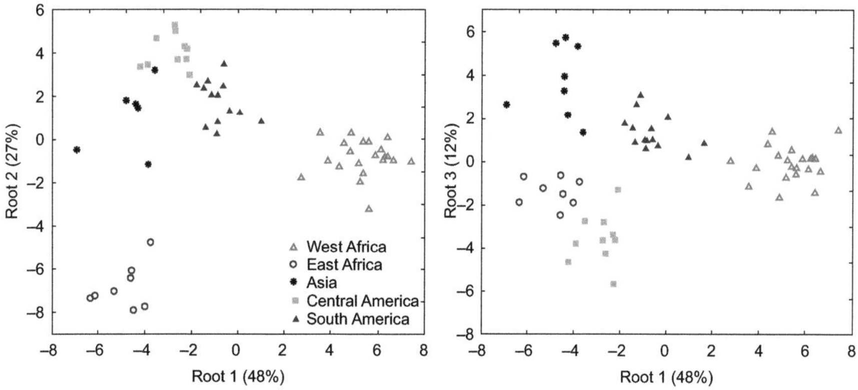

Fig. 14.2 Score plot of the first three discriminant functions for discriminating cocoa beans in different macroareas. (After Bertoldi D et al., 2016. Food Contr 65:46–53. https://doi.org/10.1016/j.foodcont.2016.01.013. With permission)

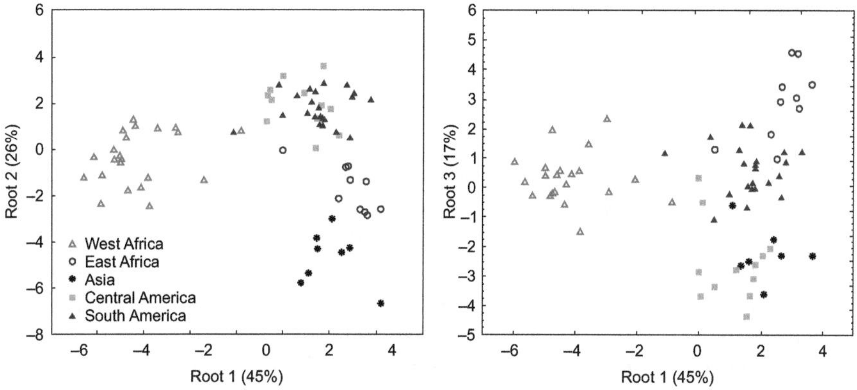

Fig. 14.3 Score plot of discriminant functions 1, 2 and 3 to discriminate cocoa beans and commercial dark chocolate of different macroareas. (After Bertoldi D et al., 2016. Food Contr 65:46–53. https://doi.org/10.1016/j.foodcont.2016.01.013. With permission)

a satisfactory effect corresponding to >85% correct reclassification. In the case of tracing the geographic origin of commercial dark chocolate, this approach yielded worse results, with a 77% correct reclassification (Fig. 14.3).

14.2.2 Intervarietal/Type Differentiation

Pedro et al. (2006) aimed to classify different brands of chocolate-flavoured beverages from São Paulo State (Brazil). Macro- and microelements were determined via ICP–OES. CA data enabled the establishment of similarities among the different brands. The pattern recognition of hierarchical grouping of CA made it possible to classify 13 brands into five groups depending on their mineral profile. For example,

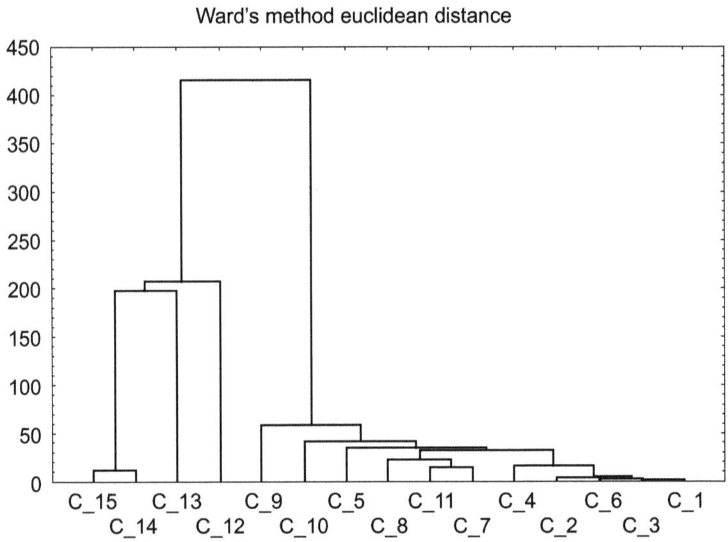

Fig. 14.4 Hierarchical dendrogram of analysed samples of sugar and other confectionery products as objects. (After Grembecka M and Szefer P, 2012. Food Anal Methods 5:250–259. https://doi.org/10.1007/s12161-011-9234-0. With permission)

one of the groups corresponded to micronutrients such as Cr, Mo, Cu, Se, Mn, Zn and Fe, indicating 100% similarity. The second group of macroelements, i.e., Mg, S, Na, Ca and P, also showed approximately 100% similarity among its components, and the similarity between the two groups was 95%. Potassium was present in the beverages at the highest level, located separately in the CA dendrogram, showing only 75% similarity in relation to the others.

Grembecka and Szefer (2012) applied FAAS and a spectrophotometric method to differentiate sugar samples according to product type. The CA data were considered for beet, cane sugar, maple syrup and molasses. As shown in Fig. 14.4, the dendrogram consists of two main clusters. The first contains samples represented by other confectionery products, including molasses and maple syrup, whereas the latter corresponds to sugar samples. It was also possible to discriminate two subclusters in the sugar sample. The first subcluster represents refined beet sugar, whereas the second represents natural cane sugar samples.

14.2.3 Interregional and Interspecies Differentiation

EA–IRMS data obtained by Diomande et al. (2015) were used for their chemometric transformation by PCA and PLS–DA. It is possible that the $\delta^{13}C$ and $\delta^{15}N$ signatures obtained from cocoa beans and their different submatrices can be useful for differentiating their macroareas of provenance. Moreover, the $\delta^{13}C$ and $\delta^{15}N$ values enable discrimination between cocoa beans of different varieties. The $\delta^{15}N$ values of proteins, cotyledons, defatted cocoa, shells and cocoa beans allowed the differentiation of cocoa samples from Africa, America, Asia and Oceania. An SPCA with the first two components showed variability from 85% on the score plot corresponding to Africa and the other countries. Moreover, the $\delta^{15}N$ values of theobromine and defatted cocoa and the $\delta^{13}C$ values of proteins, defatted cocoa and lipids could discriminate between cocoa beans from three African towns on the Ivory Coast (Fig. 14.5). As shown in Fig. 14.6, both stable isotope values were effective

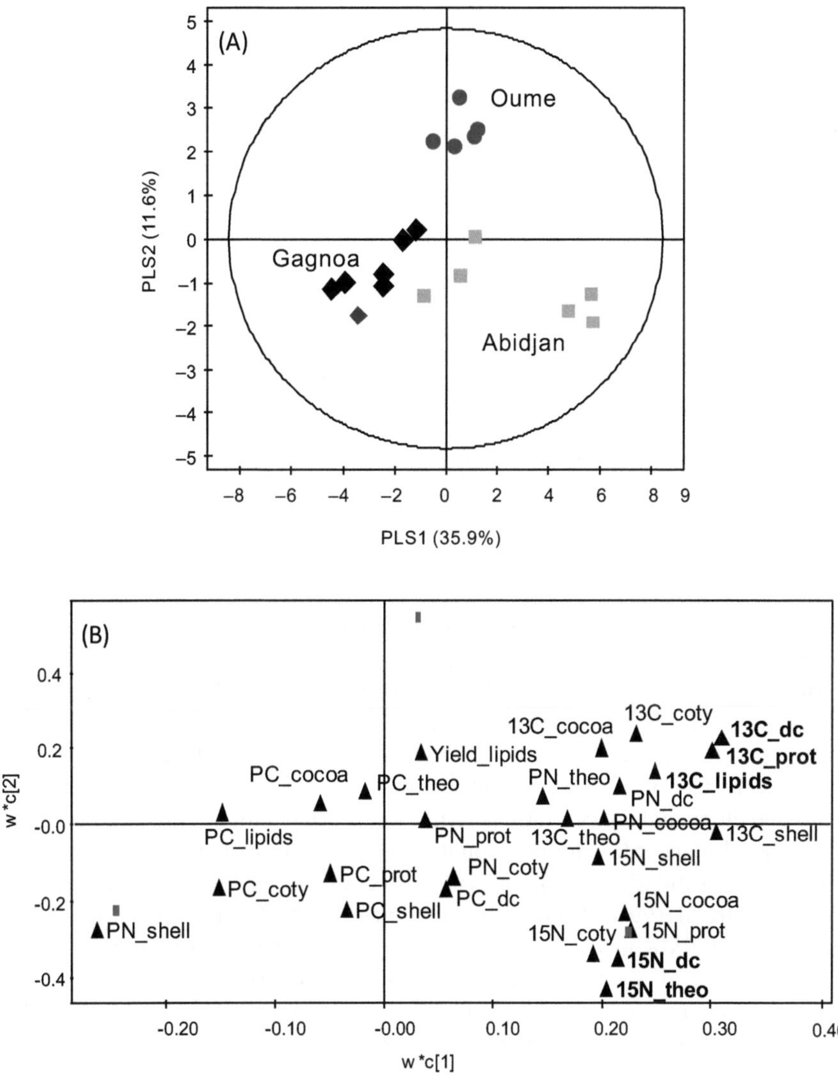

Fig. 14.5 Statistical analyses of the isotopic data of the three cities in the Ivory Coast: Abidjan, Gagnoa and Oumé: (**a**) Score plot of the PLS–DA along the two principal PLS axes and (**b**) loading plot of the PLS–DA. The principal discriminating metabolites are the $\delta^{15}N$ values of theobromine and defatted cocoa. Statistical parameters for validation: Axes = 2; $Q^2 = 0.63$; R^2 X = 0.48; R^2 Y = 0.77; CV-ANOVA = 6.10^{-4}. (After Diomande D et al., 2015. Food Chem 188:576–582. https://doi.org/10.1016/j.foodchem.2015.05.040. With permission)

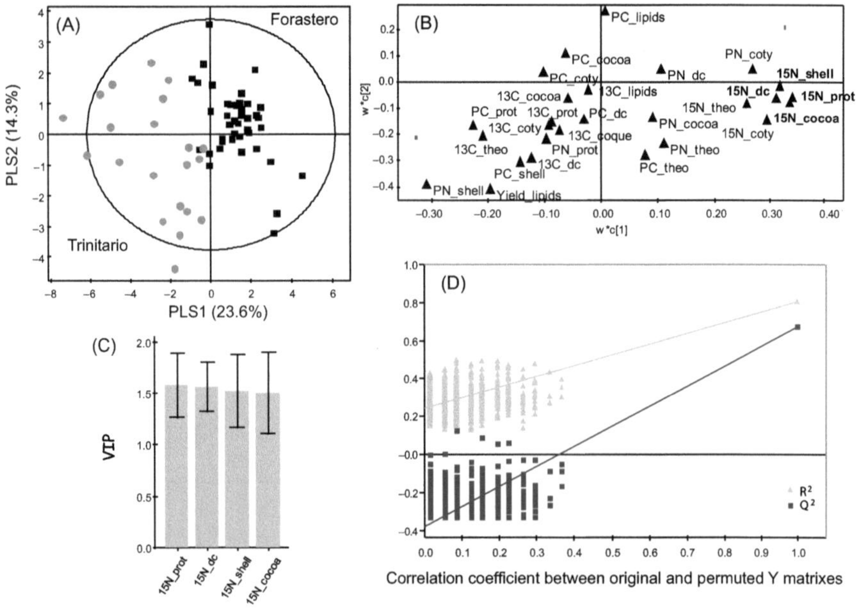

Fig. 14.6 Statistical analyses of the isotopic data of the cocoa, Forastero and Trinitario varieties: (**a**) Score plot of the PLS–DA along the two principal PLS axes. (**b**) Loading plot of the PLS–DA. The principal discriminating metabolites are the $\delta^{15}N$ values of shells, proteins, cocoa beans and defatted cocoa. (**c**) VIPs of the discriminating parameters and (**d**) results from the response permutation test (999 permutations). Statistical parameters for validation: Axes = 3; $Q^2 = 0.67$; R^2 X = 0.50; R^2 Y = 0.80; CV-ANOVA = $3.9.10^{-10}$. (After Diomande D et al., 2015. Food Chem 188:576–582. https://doi.org/10.1016/j.foodchem.2015.05.040. With permission)

markers for distinguishing between cocoa samples from two varieties, i.e., Forastero and Trinitario.

Bindereif et al. (2019) reported that EA–IRMS and ^{1}H NMR integrated with multivariate models were appropriate approaches for assessing the SIR profile in authentic cocoa beans from 20 countries in Africa, Central, South and North America, the Caribbean region and Oceania. In particular, the δ^2H, $\delta^{15}N$ and $\delta^{18}O$ levels and elemental concentrations reliably contributed to the classification of the cocoa samples according to their country of origin. For example, the LDA model enabled successful differentiation, with correct classification of samples originating from Brazil (99%), Peru (100%), Ecuador (88%) and Colombia (88%). Moreover, ^{1}H NMR made it possible to identify intervarietal differences using a small subgroup of samples, i.e., four samples of the Criollo cultivar exclusively grown in one specific region of Peru. Further research should be conducted to confirm the obtained results by improving the methodology due to the relatively low number of samples studied as well as because of a lack of detailed metadata for some samples.

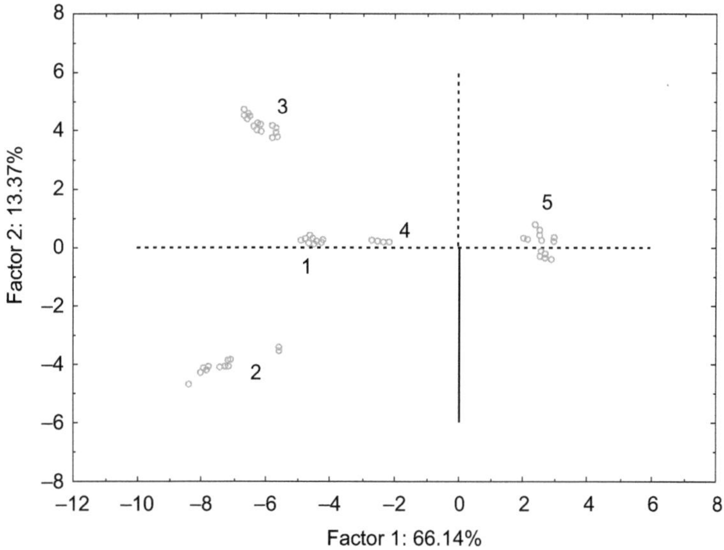

Fig. 14.7 PCA results from using 22 elements and 14 sugar samples. Groups 1–4 represent four raw cane sugar samples from four different geographic locations: Louisiana, USA (1), Costa Rica (2), Argentina (3) and Swaziland (4). Group 5 represents ten refined sugar samples: IAEA-C6, Ultrex sucrose and eight beet sugar samples from eight different geographic locations; see Fig. 4 for locations. Factor 1 was most highly correlated with Al, Ca, Cr, Cu, Fe, Ni and V; Factor 2 was most highly correlated with Cs, Pb and Zn. (After Rodushkin I et al., 2011. J Food Compos Anal 24:70–78. https://doi.org/10.1016/j.jfca.2010.05.005. With permission)

Rodushkin et al. (2011) used multiple advanced methods, such as TIMS, ICP–OES, ICP–SFMS, ICP–QMS, MC–ICP–MS and IRMS, combined with PCA processing. It was possible to detect raw sugars with much higher levels of most elements than in refined sugars, whether of cane or beet origin. $\delta^{13}C$ was useful for clearly distinguishing between pure samples of beet and cane sugars. As shown in Fig. 14.7, PCA allowed the discrimination of sugar samples of different geographic origins along the F1 and F2 axes, explaining 79.51% of the total variance. Four raw cane sugar samples were distinguished between four locations, i.e., Louisiana (USA), Costa Rica, Argentina and Swaziland. However, the fifth unseparated score group of ten refined sugar samples was processed by PCA, and the first two factors explained 62.91% of the total variance (Fig. 14.8). It was then possible to distinguish refined sugar samples between IAEA-C6 and Ultrex sucrose (both of cane origin) as well as 8 beet sugar samples originating from eight different geographic areas, i.e., Moldova, Poland, Idaho (USA), the Netherlands, France, Germany, Hungary and Michigan (USA). Although the statistical significance of the discriminative analysis was not noted, it was recommended that greater numbers of analytical measurements be used to improve the validation parameters, such as precision. Thus, it would be possible to increase the likelihood of differentiating groups.

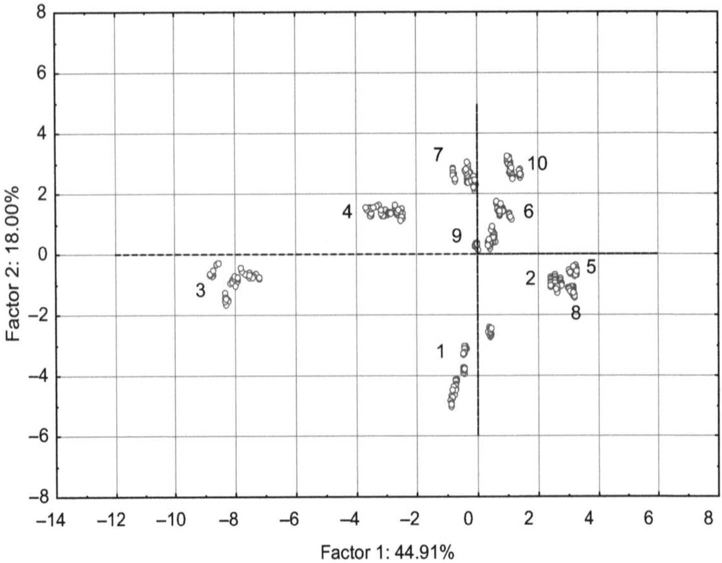

Fig. 14.8 PCA results from using 22 elements and ten refined sugar samples (raw cane sugars excluded). Group 1 represents the IAEA-C6 sample (cane origin). Group 2 represents Ultrex sucrose (cane origin). Groups 3–10 represent eight refined beet sugar samples from eight different geographic locations: Moldova (3), Poland (4), Idaho, USA (5), the Netherlands (6), France (7), Germany (8), Hungary (9) and Michigan, USA (10). Factor 1 was most highly correlated with Al, Ca, Cs, Fe, Mn and Ni; Factor 2 was most highly correlated with Ba, K, Mg, Na and Pb. (After Rodushkin I et al., 2011. J Food Compos Anal 24:70–78. https://doi.org/10.1016/j.jfca.2010.05.005. With permission)

14.2.4 Differentiation of Other Parameters

14.2.4.1 Verification of Authenticity/Adulteration

Shittu and Badmus (2009) integrated FAAS data with chemometric processing to assess labelled product information and retail prices of Nigerian-made powdered cocoa beverages. The results obtained allowed them to determine the relationships among retail price, the mineral element profile and the practices of powdered cocoa beverage manufacturers. The products generally had high levels of Ca and P. The extreme locations of Pb and P in the clustering of the HCA tree dendrogram indicated that the elements were present as contaminants and as nutritional supplements, respectively. PCA and HCA successfully differentiated the beverage brands into two main groups consisting mainly of unregistered and registered products with GMP certificates. It was concluded that retail prices and Ca and P levels are reliable indices of GMP in powdered cocoa beverages.

Because coconut sugar may become an object of fraudulent manipulation, Zdiniakova and de la Calle (2020) combined ED–XRF with PCA and SIMCA to assess the elemental profiles of coconut, cane and beetroot sugars. PCA clearly differentiated two sample groupings related to the coconut and cane. SIMCA generally classified samples as coconut sugar, corresponding to a specificity of 72.7%, and the remaining samples could not be classified as coconut sugar or cane sugar.

Banerjee et al. (2015) used triple combination analysis via ICP–OES, ICP–MS and TC/EA–IRMS of natural and artificial sweeteners, including syrups, honeys and sugars, to evaluate their potential as parameters for food authentication. For example, the addition of 10% corn syrup to pure maple syrup could be detected using molar ratios of Na/(Na + K) and Na/(Na + Ca) for maple syrup. However, adulteration of honeys and maple syrups of >15% by corn syrup or cane sugar could be detected via the $\delta^{13}C$ carbon isotope values of C3 plants. This approach was useful in the case of authentic honeys and maple syrups (25.1 ± 1.1‰) relative to the composition of C4 plants for corn syrups and cane sugars (−11.4 ± 0.8‰) as major adulterant agents.

Grembecka and Szefer (2012) applied FAAS and a spectrophotometric method to classify cocoa and chocolate samples. According to FA (with the first two Fs explaining up to 89.5% of the total variance), higher values of F1 corresponded to instant cocoa samples with the greatest contents of Na, whereas lower values corresponded to cocoa powder containing high levels of Mg, K, Zn, Cu, Fe and Mn. Moreover, CA and FA (with the first two Fs accounting for 81.9% of the total variance) were useful for distinguishing two types of chocolate samples with different cocoa contents, i.e., dark chocolate and milk samples (Figs. 14.9 and 14.10). The methodology used might be helpful for detecting fraud or confirming the authenticity of a product.

Papaefstathiou et al. (2018) applied ICP–OES, HPLC–RID and other methods to quantify essential elements and select organic compounds, including nutritional compounds, in traditional carob products. PCA was performed on 46 samples in four preformed groups, and PLS–DA was applied to validate the previous data on the impact of the production process on the composition. PCA clearly differentiated the samples into four groups, although the samples of the carob creamy products showed wide variation. PLS–DA analysis differentiated the carob products according to their type and discriminative components, highlighting their particular nutritional value. The data obtained were compared with the pulp of Cypriot carob cultivars. Only 60% of the products studied had a label indicating their nutritional value, and the majority (75%) were consistent with those used for labelling. Carobs, which are functional foods, are low in fat content and high in dietary fibres, and they may be a source of mineral components. Moreover, carob products partially satisfy health and nutritional requirements as expected. This research was the first to determine the chemical composition of Cyprus carob cultivars.

Bononi et al. (2015) used LC–IRMS to evaluate the $\delta^{13}C$ of vanillin in chocolate bars (dark chocolate, milk chocolate, white chocolate) and chocolate snack foods (chocolate snacks, chocolate nutmeg) and industrial flavourings. The obtained data allowed them to discriminate between natural vanillin from vanilla beans and vanillin from other sources (mixtures from beans, synthesis or biotechnology). Moreover, the flavour of vanilla from industrial quality control processes was also analysed. Only 30% of products were marked as "vanilla" on the packaging label in accordance with the requirements of EU Regulation 1334/2008. Moreover, all of the

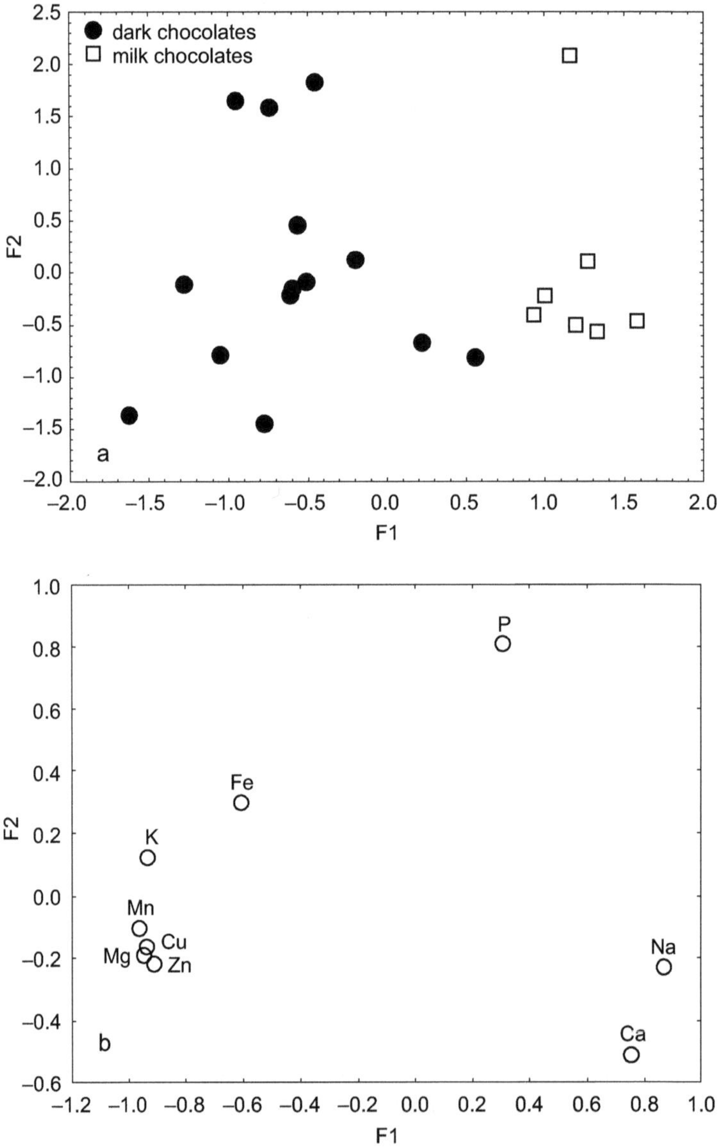

Fig. 14.9 (**a**) Scatterplot of the object scores of the two discriminant functions of chocolate samples. (**b**) Scatterplot of loadings for nine elements in chocolate samples (After Grembecka M and Szefer P, 2012. Food Anal Methods 5:250–259. https://doi.org/10.1007/s12161-011-9234-0. With permission)

samples that were not declared "vanilla" or "natural" on the label also complied with the same EU regulations. The data obtained was useful after statistical processing.

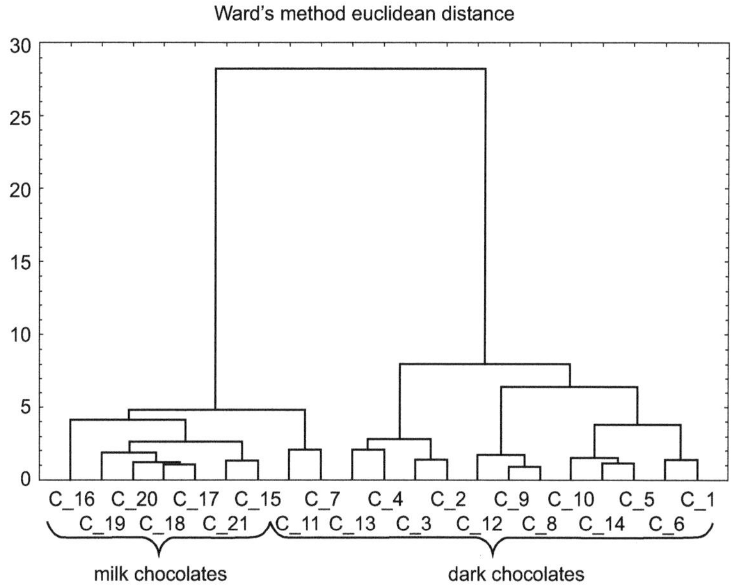

Fig. 14.10 Hierarchical dendrogram for the analysed samples of chocolate (C1–C14, dark chocolates; C2–C4 and C13, chocolates with a cocoa content >70%; C7 and C11, chocolates with a cocoa content of at least 45%; C15–C21, milk chocolates). (After Grembecka M and Szefer P, 2012. Food Anal Methods 5:250–259. https://doi.org/10.1007/s12161-011-9234-0. With permission)

14.2.4.2 Differentiation Between Organic and Conventional Origins and Related to Other Parameters

Fernandes et al. (2005) determined MECs via ICP–OES in coffee and sugar cane spirit (cachaça) samples. The original variables were transformed by multivariate models such as PCA and CA to differentiate samples in relation to their geographical origin and production mode (industrial or homemade and organically or conventionally produced). Some interregional discrimination for cachaça was related to three regions (northeast, central and south). For coffee samples with respect to production mode, PCA, with the first three PCs accounting for 73% of the total variability, clearly differentiated samples of cachaça industrially produced from homemade samples.

Junior et al. (2018) applied ICP–MS to evaluate the elemental profile of Brazilian chocolate samples produced from cocoa beans grown in different cultivation systems. The first two PCs, which explained 57.2% of the total variance, differentiated between organic chocolate samples and conventional chocolate samples. LDA was a more reliable model than PCA, allowing them to distinguish the samples of the two types of chocolate (Fig. 14.11) with an 83% probability of correct prediction. Comparable levels of Zn and Mg or even lower levels of Fe were found in organic chocolates than in conventional chocolates, which contradicts the current opinion that organic food is rich in essential elements.

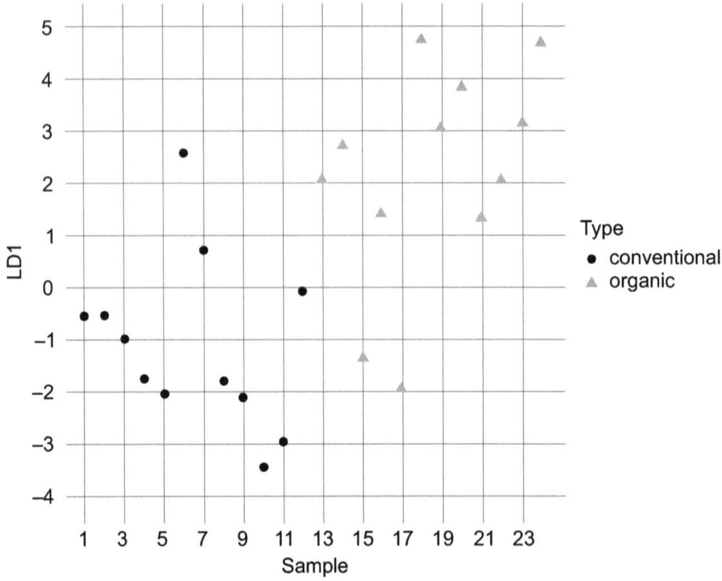

Fig. 14.11 Chocolate samples and their values obtained from the generated discriminant function. (After Junior ACM et al., 2018. J Chemom 32:e3036. https://doi.org/10.1002/cem.3036. With permission)

Fiamegos et al. (2021) integrated ED–XRF with PCA to characterize the mineral composition of chocolate, although the specific geographical origin of the cocoa used was not assigned for all samples, and only reference to geographical origin was generally declared non-EU. When chocolates with a cocoa concentration ranging from 7075% were exclusively used, two clearly differentiated sample groupings corresponding to organic and conventional chocolates were obtained. The elements whose contents significantly differed in both types of chocolates were Fe, Mn and Rb.

Barbosa et al. (2015) used ICP–MS with the NB MLA and RF MLA tools to classify certified Brazilian organic sugarcane and nonorganic sugarcane samples on the basis of the MEC database. It was possible to assess the differentiation of the studied samples. Using all the variables, i.e., 32 elements, resulted in accurate results (>90%); however, better accuracy was achieved, i.e., 95.4% for NB and RF when only eight elements (Rb, U, Al, Sr, Dy, Nb, Ta and Mo) chosen by a feature selection algorithm were utilized. The chemometric approach was useful for classifying Brazilian organic and nonorganic sugarcane samples.

According to Fiamegos et al. (2021), despite a reduced number of samples, the combination of both noncentrifugal cane sugar (NCS) and brown sugars with cane sugars from Antilles and Latin America established two clearly distinguished clusters. These samples corresponded to samples of organic and conventional origin; the "rapadura" and "panela" samples were distinguished from brown sugar samples but were generally within their respective organic/conventional clusters. NCSs containing higher levels of Ca, Cl, Cr, Cu, I, Fe, Mg, Mn, P, K, Se, Na and Zn than brown sugars are also called "rapadura" and "panela" and are obtained by evaporation of cane juice, whereas "brown sugar" is obtained by mixing refined saccharose with molasses samples.

14.2.4.3 Differentiation of Various Steps of the Manufacturing Process

Kruszewski and Obiedziński (2018) applied ICP–AES to distinguish the elemental profiles of raw material samples and trace changes in mass during all the steps of the chocolate manufacturing process by three producers. Tracking of mineral composition changes in masses during the production steps at one of the three companies was successfully determined via PCA. The first two PCs, accounting for 93.17% of the total variance, allowed them to clearly differentiate two groups of samples with respect to the geographical provenance of cocoa beans from the Dominican Republic and Ecuador. Moreover, PCA and CA allowed a clear discrimination between the elemental compositions of the research material taken from the production lines of the two companies. These findings indicated that dark chocolate could be considered an important dietary source of Fe as well as a good supplement of Mg, P and Zn.

Škrbić et al. (2010) used a combination of FAAS and ET–AAS with computational analysis (PCA and CA) to evaluate sugar beet and its by-products, i.e., sugar beet, extracted sugar beet pulp, dried sugar beet pulp, molasses and white sugar as well as the soil substratum. On the basis of the PCA biplot, with the first two PCs explaining 83.0% of the total variance, the following groups were obtained: Group I consisted of sugar beet, dried sugar beet and extracted sugar beet pulps, group II consisted of molasses samples and group III included soil and sugar samples. The three groups presented the highest levels of Cu and Mg (PC1) and intermediate levels of Na and K (PC2). The occurrence of the investigated elements in sugar was comparable throughout the entire study, in contrast to fluctuations in the metal contents in sugar beet and its byproducts, indicating the effects of the purification processes. Furthermore, graphical presentation of the correlation coefficients demonstrated a relationship between element levels in sugar beet and soil originating from the fields where beets were grown.

Awadallah et al. (1995) assessed the elemental profiles of sugarcane plants and crude juice samples collected from cultivated areas in five cities where the main Egyptian sugar industry factories are located. Methods such as INAA, ICP–AES and flameless AAS were integrated with CA to obtain helpful differentiating methodologies. Samples from each progressive industrial step, i.e., syrup, molasses and raw and refined sugar, were collected. The CA data showed that the products of both the Kom Ombo and Armant factories were similar but significantly different from those of the other two companies (due to their different textures, structures and different soil types), as reflected by the composition of the juice and the processing steps used in these two areas. Among the 18 elements analysed, the relatively higher levels of Pb and As exceeding the allowed levels in the final products than in the juice could be attributed to the contamination of the containers by soil particles, which can easily be avoided.

Acierno et al. (2020) examined fermented and dried cocoa bean samples and chocolate samples industrially manufactured from beans originating from seven different countries. Notably, the chain from cocoa beans to chocolate is a long and complex process. Therefore, the aim of this study was to examine the retention and loss of compositional signatures from cocoa beans to chocolate. The volatile compounds, MECs and SIRs of cocoa beans of different origins and the corresponding chocolates were determined by HS–PTR–MS, ICP–MS and EA–IRMS, respectively. The volatile parameters provided the most important information on the origin and main processing features of the raw cocoa beans in the chocolates. The MEC and SIR characteristics were more recognizable in terms of the cocoa content and added ingredients. It was found that Fe, Cr and Cd could be useful discriminators. Moreover, VOCs were the most effective descriptive variables. The combined approach provided the ability to track the technological processing of cocoa beans to the final product as chocolates. It was concluded that the methodology provides a scientific basis for cacao quality certification programmes.

Gohara et al. (2013) studied the addition of chia and azuki flour to gluten-free chocolate cake using FAAS and the AOAC method. The results obtained by ANOVA for each of the responses studied in duplicate in the 2^2 full factorial designs were interpreted. PCA with the first two PCs explaining 94.94% of the total variance distinguished the samples with a higher content of chia in PC1, but PC2 discriminated those formulations with the highest level of azuki from the others. Moreover, the factors associated with the greatest contents of chia and azuki flour were significant, and their increase resulted in positive Cu, Fe, K and Mg responses. The quadrant analysis indicated that the Ca, Mg, Mn and Zn contents were responsible for the discrimination of the cakes with the greatest contents of chia and azuki. ANOVA revealed that the increase in elemental content was greater with the addition of 20% of both flours.

14.2.4.4 Assessment of Differences Related to Pollution Impact/Elemental Impurities

Manton (2010) coupled Pb/Nd ratios with model Sm–Nd ages, and the isotope ratios of Pb and Sr provided reliable tools to indicate, among other factors, which samples were contaminated. Moreover, model ages indicated where the cocoa powder was grown. For example, African samples are considered uncontaminated, but Asiatic samples contained 50% anthropogenic Pb from the Australian provenance. It was also concluded that no measurable Pb contamination occurred during the transport of beans and the manufacture of chocolate. It was also shown that cocoa shells accumulate metals from soil as substrates and that the effectiveness of this process depends on drying time. For this reason, in the case of slowly drying shells, more Pb may be taken up from uncontaminated soil than from rapidly drying shells from contaminated substrata.

Vanderschueren et al. (2019) found that ICP–MS with multivariate analysis was an effective approach for identifying the power of elemental fingerprinting for tracing the origin of chocolate from different manufacturers in 10 countries (originating from 27 producing countries). The CART effectively classified chocolate samples according to their cocoa origin, with an overall misclassification rate of 23%, although it was impossible to correctly classify Asia Pacific samples. It was revealed that 12% of single-origin chocolates, generally from South America, exceeded the Cd limit established for trade in the European market. These anomalies can be explained by the geogenic origin of Cd in the South American ecosystem. Furthermore, in the case of Pb, six samples from Ecuador, India, Peru and Trinidad exceeded the limit of the Codex Alimentarius for edible fats. It has been suggested that Pb levels are unrelated to raw cacao material but could be a consequence of postharvest processing.

Guedes and Pereira (2018) and Andrade et al. (2018) combined LIBS with chemometric processing to identify elemental impurities leached from Brazilian raw sugarcane. The detection of even low levels of solid impurities originating from soil particles and green and dry/brown sugarcane leaves in raw sugarcane is relevant for improving sugar mill production performance. These impurities directly influence the quality of sugarcane juice by affecting the entire sugar manufacturing process. Guedes and Pereira (2018) also reported that LIBS data combined with PCA, PLS2–DA and kNN models were assigned to impurity ranges for raw sugar manufacturing processes and need to be characterized from 0 to 5 wt%, as the desired material, and 8 to 10 wt%, as the undesired material, i.e., designated as ranges 1 and 2, respectively. PCA is an effective tool for verifying the slight tendency of differentiation for the mean-centred data in the first two PCs, explaining 75% of the total variance (Fig. 14.12a). As shown in Fig. 14.12b, the most relevant chemical elements corresponded to the impurity ranges with the use of loading values based on 634 variables. Owing to chemical elements, the models effectively classified the

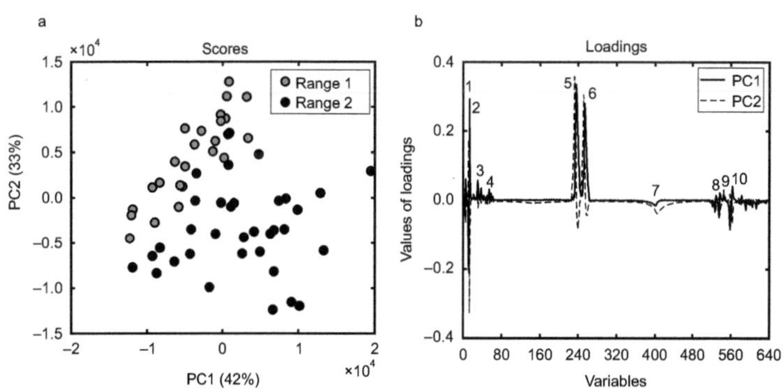

Fig. 14.12 Scores (**a**) and loadings from principal component analysis (PCA) performed with 634 selected variables using mean-centred data (**b**). For samples in ranges 1 (0–5 wt%, green circles) and 2 (8–10 wt%, red circles), each number represents the loading values for the following emission lines (in nanometres): 1 and 2, C I, 193.09 and 247.86; 3, both Mg II, 279.55 and 280.27; 4, both Ca II, 393.37 and 396.85; 5 and 6, Na I, 589.00 and 589.59, 7, H I, 656.30; 8, both N I, 744.23 and 746.83; 9, K I, 766.49; and 10, O I, 777.42. I and II are atomic and ionic emission lines, respectively. (After Guedes WN et al., 2018. Microchem J 143:331–336. https://doi.org/10.1016/j.microc.2018.08.030. With permission)

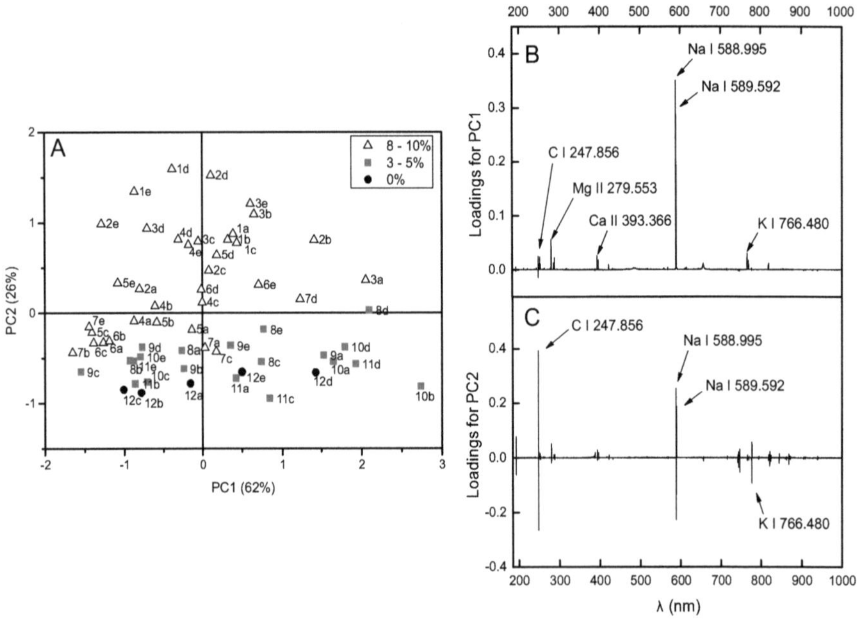

Fig. 14.13 Scores (**a**) and loadings for PC1 (**b**) and PC2 (**c**) calculated using the mean spectra standardized with the C I 247.86 nm line. (After Andrade DF et al., 2018. Microchem J 137:443–448. https://doi.org/10.1016/j.microc.2017.12.005. With permission)

results with 97% accuracy for both ranges. The classification results indicated potential for future applications in sugarcane refineries. To address the data generated by LIBS, Andrade et al. (2018) tested 12 different standardization strategies for the dependent variables. As shown in Fig. 14.13, LIBS emission lines and standardization modes coupled with PCA were good options for identifying impurities in Brazilian sugarcane. After standardization, it was possible to visualize a narrow range of impurities between 3% and 10 wt%. Chemical elements with positive loadings, i.e., Ca and Mg, were present in the samples with the highest impurity range.

14.2.4.5 Elaboration of Improved Methodology

Lo Dico et al. (2018) validated the method for the quantification of Cd, Pb, As, Cr, Sb, Se and V extracted from cocoa powder and chocolate matrices, with efficient recovery (90–110%). The elemental profile of the study samples was established via ICP–MS in accordance with the conditions recommended by the EC Commission Regulation. The proposed method is an appropriate tool for obtaining results in accordance with the performance criteria and the EU requirements for method validation for use in official food controls. The obtained data may exclude chocolate contamination due to manufacturing processes. All the chocolate samples achieved a Cd content under the limits established by Commission Regulation (EC) Reg.

488/20114. Significant differences were observed between the cocoa powder and chocolate samples.

Souza et al. (2019) quantified essential elements in sugarcane juice by ICP–OES. Satisfactory validation parameters of the developed method were obtained. The original results were processed by descriptive statistics, including LRA, etc. The proposed analytical method was a good alternative for the simultaneous determination of biophile elements in sugarcane juice via the introduction of slurries and detection via ICP–OES.

This novel and simple pretreatment ICP–OES method is simple to perform, fast, precise and accurate and may be applied as an alternative for the routine analysis of elements in sugarcane juice samples.

Nunes et al. (2010) aimed to combine LIBS and ICP–OES with univariate calibration and chemometric methods. Consequently, PLSR and iPLS were used for the analysis of pellets of sugarcane leaves for the quantification of macro- and micronutrients by choosing calibration and validation samples with similar matrix compositions. The significant relationship between the LIBS and ICP–OES results could be a promising indication of better fitness for plant analysis and essential element diagnosis purposes.

de Carvalho et al. (2015) reported that direct quantification of the nutrient profile of plant materials, including sugarcane material, by femtosecond LIBS with ns-LIBS and ICP–OES was performed. Although the values predicted by ns-LIBS multivariate modelling were in better agreement with reference to mass fractions than those predicted by univariate functions, fs-LIBS showed greater potential for the determination of essential elements in plant materials because of its susceptibility to undesirable effects on the chemical composition of the matrices.

14.3 Conclusions

Food authentication contributes to guaranteeing food safety. To achieve this goal, appropriate analytical methods/techniques integrated with chemometric processing can estimate the authenticity or adulteration of cocoa, its products and sugar and sweet products. The reasonable choice of such an approach for classifying cocoa beans, sugars and their products enables the assessment of various environmental and technological parameters that affect the quality of plant material. A combination of multielemental and/or stable isotope analysis with multivariate elaboration enables categorization of samples with respect to their interspatial and intervarietal variation or identification of conventional and organic cocoa beans and sugarcane/sugar beetroot and their byproducts. Validated results make it possible to identify fraud and detect product mislabelling to preserve brand reputations, protecting the consumer from fraud.

References

Abt E, Robin LP (2020) Perspective on cadmium and lead in cocoa and chocolate. J Agric Food Chem 68:13008–13015. https://doi.org/10.1021/acs.jafc.9b08295

Acierno V, de Jonge L, van Ruth S (2020) Sniffing out cocoa bean traits that persist in chocolates by PTR-MS, ICP-MS and IR-MS. Food Res Inter 133:109212. https://doi.org/10.1016/j.foodres.2020.109212

Andrade DF, Guedes WN, Pereira FMV (2018) Detection of chemical elements related to impurities leached from raw sugarcane: use of laser-induced breakdown spectroscopy (LIBS) and chemometrics. Microchem J 137:443–448. https://doi.org/10.1016/j.microc.2017.12.005

Awadallah RM, Ismail SS, Mohamed AE (1995) Application of multi-element clustering techniques of five Egyptian industrial sugar products. J Radioanal Nucl Chem 196:377–385. https://doi.org/10.1007/BF02038058

Banerjee S, Kyser TK, Vuletich A, Leduc E (2015) Elemental and stable isotopic study of sweeteners and edible oils: constraints on food authentication. J Food Compos Anal 42:98–116. https://doi.org/10.1016/j.jfca.2015.03.011

Barbosa RM, Batista BL, Barião CV, Varrique RM, Coelho VA, Campiglia AD, Barbosa F Jr (2015) A simple and practical control of the authenticity of organic sugarcane samples based on the use of machine-learning algorithms and trace elements determination by inductively coupled plasma mass spectrometry. Food Chem 184:154–159. https://doi.org/10.1016/j.foodchem.2015.02.146

Bertoldi D, Barbero A, Camin F, Caligiani A, Larcher R (2016) Multielemental fingerprinting and geographic traceability of *Theobroma cacao* beans and cocoa products. Food Contr 65:46–53. https://doi.org/10.1016/j.foodcont.2016.01.013

Bindereif SG, Brauer F, Schubert J-M, Schwarzinger S, Gebauer G (2019) Complementary use of ^{1}H NMR and multi-element IRMS in association with chemometrics enables effective origin analysis of cocoa beans (*Theobroma cacao* L.). Food Chem 299:125105. https://doi.org/10.1016/j.foodchem.2019.125105

Bononi M, Quaglia G, Tateo F (2015) Easy extraction method to evaluate $\delta^{13}C$ vanillin by liquid chromatography–isotopic ratio mass spectrometry in chocolate bars and chocolate snack foods. J Agric Food Chem 63:4777–4781. https://doi.org/10.1021/acs.jafc.5b02136

de Carvalho GGA, Moros J, Santos D Jr, Krug FJ, Laserna JJ (2015) Direct determination of the nutrient profile in plant materials by femtosecond laser-induced breakdown spectroscopy. Anal Chim Acta 876:26–38. https://doi.org/10.1016/j.aca.2015.03.018

Diomande D, Antheaume I, Leroux M, Lalande J, Balayssac S, Remaud GS, Tea I (2015) Multi-element, multi-compound isotope profiling as a means to distinguish the geographical and varietal origin of fermented cocoa (*Theobroma cacao* L.) beans. Food Chem 188:576–582. https://doi.org/10.1016/j.foodchem.2015.05.040

Fernandes AP, Santos MC, Lemos SG, Ferreira MMC, Nogueira ARA, Nóbrega JA (2005) Pattern recognition applied to mineral characterization of Brazilian coffees and sugar-cane spirits. Spectrochim Acta Part B At Spectrosc 60:717–724. https://doi.org/10.1016/j.sab.2005.02.013

Fiamegos Y, Papoci S, Dumitrascu C, Ghidotti M, Zdiniakova T, Ulberth F, de la Calle Guntinas MB (2021) Are the elemental fingerprints of organic and conventional food different? ED-XRF as screening technique. J Food Compos Anal 99:103854. https://doi.org/10.1016/j.jfca.2021.103854

Gohara AK, Souza AHP, Rodrigues AC, Stroher GL, Gomes STM, Souza NE, Visentainer JV, Matsushita M (2013) Chemometric methods applied to the mineral content increase in chocolate cakes containing chia and azuki. J Braz Chem Soc 24:771–776. https://doi.org/10.5935/0103-5053.20130101

Grembecka M, Szefer P (2012) Differentiation of confectionery products based on mineral composition. Food Anal Methods 5:250–259. https://doi.org/10.1007/s12161-011-9234-0

Guedes WN, Pereira FMV (2018) Classifying impurity ranges in raw sugarcane using laser-induced breakdown spectroscopy (LIBS) and sum fusion across a tuning parameter window. Microchem J 143:331–336. https://doi.org/10.1016/j.microc.2018.08.030

Guerra MBB, Adame A, de Almeida E, Brasil MAS, Schaefer CEGR, Krug FJ (2018) In situ determination of K, Ca, S and Si in fresh sugar cane leaves by handheld energy dispersive X-Ray Fluorescence Spectrometry. J Braz Chem Soc 29:1086–1093. https://www.researchgate.net/publication/321792174

Junior ACM, Maione C, Barbosa RM, Gallimberti M, Paulelli ACC, Segura FR, Souza VCO, Batista BL, Barbosa F (2018) Elemental fingerprint profiling with multivariate data analysis to classify organic chocolate samples. J Chemom 32:e3036. https://doi.org/10.1002/cem.3036

Kongor JE, Hinneh M, Van de Walle D, Afoakwa EO, Boeckx P, Dewettinck K (2016) Factors influencing quality variation in cocoa (*Theobroma cacao*) bean flavour profile—a review. Food Res Int 82:44–52. https://doi.org/10.1016/j.foodres.2016.01.012

Kruszewski B, Obiedziński MW (2018) Multivariate analysis of essential elements in raw cocoa and processed chocolate mass materials from three different manufacturers. LWT 98:113–123. https://doi.org/10.1016/j.lwt.2018.08.030

Lee T (2020) Determining the geographical origin of cocoa beans in chocolate using stable isotope ratios. Dissertation, University of Otago, New Zealand

Lo Dico GM, Galvano F, Dugo G, D'Ascenzi C, Macaluso A, Vella A, Giangrosso G, Cammileri G, Ferrantelli V (2018) Toxic metal levels in cocoa powder and chocolate by ICP-MS method after microwave-assisted digestion. Food Chem 245:1163–1168. https://doi.org/10.1016/j.foodchem.2017.11.052

Manton WI (2010) Determination of the provenance of cocoa by soil Protolith ages and assessment of anthropogenic lead contamination by Pb/Nd and lead isotope ratios. J Agric Food Chem 58:713–721. https://doi.org/10.1021/jf9029207

Motamayor JC, Lachenaud P, da Silva e Mota JW, Loor R, Kuhn DN, Brown JS, Schnell RJ (2008) Geographic and genetic population differentiation of the Amazonian chocolate tree (*Theobroma cacao* L). PLoS One 3:e3311. https://doi.org/10.1371/journal.pone.0003311

Nimalaratne C, Blackburn J, Lada RR (2020) A comparative physicochemical analysis of maple (*Acer saccharum* Marsh.) syrup produced in North America with special emphasis on seasonal changes in Nova Scotia maple syrup composition. J Food Compos Anal 92:103573. https://doi.org/10.1016/j.jfca.2020.103573

Nunes LC, Braga JWB, Trevizan LC, de Souza PF, de Carvalho GGA, Junior DS, Poppi RJ, Krug FJ (2010) Optimization and validation of a LIBS method for the determination of macro and micronutrients in sugar cane leaves. J Anal At Spectrom 25:1453–1460. https://doi.org/10.1039/c003620j

Papaefstathiou E, Agapiou A, Giannopoulos S, Kokkinofta R (2018) Nutritional characterization of carobs and traditional carob products. Food Sci Nutr 6:2151–2161. https://doi.org/10.1002/fsn3.776

Pedro NAR, De Oliveira E, Cadore S (2006) Study of the mineral content of chocolate flavoured beverages. Food Chem 95:94–100. https://doi.org/10.1016/j.foodchem.2004.12.021

Perez M, Lopez-Yerena A, Vallverdú-Queralt A (2022) Traceability, authenticity and sustainability of cocoa and chocolate products: a challenge for the chocolate industry. Crit Rev Food Sci Nutr 62:475–489. https://doi.org/10.1080/10408398.2020.1819769

Perini M, Bontempo L, Ziller L, Barbero A, Caligiani A, Camin F (2016) Stable isotope composition of cocoa beans of different geographical origin. J Mass Spectrom 51:684–689. https://doi.org/10.1002/jms.3833

Rodushkin I, Baxter DC, Engström E, Hoogewerff J, Horn P, Papesch W, Watling J, Latkoczy C, van der Peijl G, Berends-Montero S, Ehleringer J, Zdanowicz V (2011) Elemental and isotopic characterization of cane and beet sugars. J Food Compos Anal 24:70–78. https://doi.org/10.1016/j.jfca.2010.05.005

Saito N, Hokura A, Inoue K, Ono M (2021) Determination of trace elements in cacao nibs and cacao mass by X-ray fluorescence analysis and its application to identification of their production areas. Adv X-Ray Chem Anal, Japan 52:243–256

Saksangawong C, Trevanich A, Wongwattanakul M, Chio-Srichan S, Sukon P (2021) Analysis of raw honey and syrup concentration in adulterated honey by Fourier transform infrared spectroscopy. Int J Appl Eng Res 16:127–134. http://www.ripublication.com

Shittu TA, Badmus BA (2009) Statistical correlations between mineral element composition, product information and retail price of powdered cocoa beverages in Nigeria. J Food Compos Anal 22:212–217. https://doi.org/10.1016/j.jfca.2008.10.006

Škrbić B, Đurišić-Mladenović N, Mačvanin N (2010) Determination of metal contents in sugar beet (*Beta vulgaris*) and its products: empirical and chemometrical approach. Food Sci Technol Res 16:123–134. https://doi.org/10.3136/fstr.16.123

Souza SO, Costa SSL, Brum BCT, Santos SH, Garcia CAB, Araujo RGO (2019) Determination of nutrients in sugarcane juice using slurry sampling and detection by ICP OES. Food Chem 273:57–63. https://doi.org/10.1016/j.foodchem.2018.03.060

Torres-Moreno M, Torrescasana E, Salas-Salvadó J, Blanch C (2015) Nutritional composition and fatty acids profile in cocoa beans and chocolates with different geographical origin and processing conditions. Food Chem 166:125–132. https://doi.org/10.1016/j.foodchem.2014.05.141

Vanderschueren R, Montalvo D, De Ketelaere B, Delcour JA, Smolders E (2019) The elemental composition of chocolates is related to cacao content and origin: a multi-element fingerprinting analysis of single origin chocolates. J Food Compos Anal 83:103277. https://doi.org/10.1016/j.jfca.2019.103277

Yanus RL, Sela H, Borojovich EJC, Zakon Y, Saphier M, Nikolski A, Gutflais E, Lorber A, Karpas Z (2014) Trace elements in cocoa solids and chocolate: an ICPMS study. Talanta 119:1–4. https://doi.org/10.1016/j.talanta.2013.10.048

Zdiniakova T, de la Calle MB (2020) Feasibility study about the use of element proiles determined by ED-XRF as screening method to authenticate coconut sugar commercially available. Eur Food Res Technol 246:2101–2109. https://doi.org/10.1007/s00217-020-03559-z

Chapter 15
Mineral and Tap Water

Contents

15.1 Introduction

Water has considerable effects on health since proper hydration is essential to maintain body water equilibrium. The Food and Drug Administration (FDA, USA) regulates bottled water as a food and defines natural mineral water as "water containing not less than 250 parts per million (ppm) of total dissolved solids (TDS), coming from a source at one or more bore holes or springs, originating from a geologically and physically protected underground water source" (Code of Federal Regulations, Title 21, Part 165—Beverages; Quattrini et al. 2016). The consumption of bottled mineral water has recently considerably increased worldwide. By 2014, China accounted for more than 15% of the global bottled water volume (Rodwan Jr 2014), and by 2017, the world population was expected to consume approximately 391 billion litres of bottled water (Statistica 2009). According to Statistica (2016), China accounted for approximately 17% of the global bottled water volume sales in 2015.

Supplementary Information The online version contains supplementary material available at https://doi.org/10.1007/978-3-031-99687-0_15.

P. Szefer, *Chemometrics in Food Control Based on Elemental Analysis*,
https://doi.org/10.1007/978-3-031-99687-0_15

Physical and chemical characterization is useful for classifying different mineral waters by the analysis of the main parameters (European Legislation (2009/54/EC Directive)). According to the mineral elemental distribution pattern, natural mineral waters are categorized as bicarbonate mineral waters, sulphate mineral waters, chloride mineral waters, calcic mineral waters, magnesium mineral waters, fluorurate mineral waters, ferrous mineral waters and sodium-rich mineral waters. Since the quality of natural mineral waters can be affected by plasticizers and endocrine disruptors, several studies have explored their properties and performance under different physiological and pathological conditions (Quattrini et al. 2016).

For bottled mineral water, the release of chemicals from bottles and uptake by the water can occur. According to Bosnir et al. (2003), plasticizers, such as DEHP, are widely present in PET bottles. PET is a chemically inactive material, but according to some in vitro studies, storage conditions (exposure to sunlight, high temperature) may contribute to the release of chemicals from bottles into the water (Nawrocki et al. 2002; Biscardi et al. 2003).

According to Cohen et al. (2022), bottled and packaged water, frequently accompanied by branding and marketing, is promoted as healthier and safer than alternative drinking water sources. However, in China and in other LMICs as well as in HICs, this may not always be true. As before, concerns related to impacts from contaminated bottled water consumption in both LMICs and HICs have inspired more research on this issue.

It has been possible to detect fraud and identify water mislabelling to preserve the brand reputations of original products. Analytical methods/techniques such as ICP–MS, ICP–OES and ICP–AES have been most often used to determine cationic species in water, whereas IC enables the quantification of anionic species of elements. To improve the efficiency of the obtained data, analytical tools have been integrated with chemometrics. Owing to this methodical coupling, a more advanced approach has been achieved for the assessment of the authentication/adulteration of drinking/mineral water. For example, such combinations have been used in the quantification of multiple molecular profiles and other parameters in water samples, such as the use of ICP–QMS, ICP–AES, IC and AFS; ICP–OES, ICP–QMS, HG–ICP–MS and IC; and ICP–QMS, ICP–AES, IC and AFS. Notably, the IRMS method is less frequently included in multiple advanced instrumental tools, such as ICP–MS and IC or SIMS. Other rare systems include TIMS and MC–ICP–MS; IRIS; ICP–OES; and IC or INAA with PIXE. Chemometric techniques have been used to transform original data, resulting in effective classification of water samples with respect to their chemical composition. Owing to this processing, it has been possible to distinguish samples in terms of interregional or intertype/brand differentiation of drinking and mineral water samples.

15.2 Chemometric Assessment of Mineral and Drinking Waters

15.2.1 Geographical Differentiation

15.2.1.1 European Countries

Sipos et al. (2012) combined ICP–AES, IC–CD, ET and sensory evaluation with multivariate analysis (PCA, HCA and LDA) to assess the elemental composition of different Hungarian water classes (commercially available, noncarbonated spring water and tap water). No discrimination between different water brands from the same geographical origin was achieved, as expected. However, Cl^-, SO_4^{2-} and Mg^{2+} with two ET sensors were useful for building a much more appropriate model. Owing to this approach, a clear discrimination between water samples originating from three geographical regions was observed (Fig. 15.1). The first two LDA canonical variables related to the combined results of ET and element levels are shown in Fig. 15.2. The three groups of "Albertirsa K-131" and "Albertirsa K-125", "Visegrádi K7" and "Miskolc B-108" were distributed properly. In summary, LDA correctly classified the studied waters, achieving 90.8%, i.e., a better result was achieved than in the classification with the sensory evaluation.

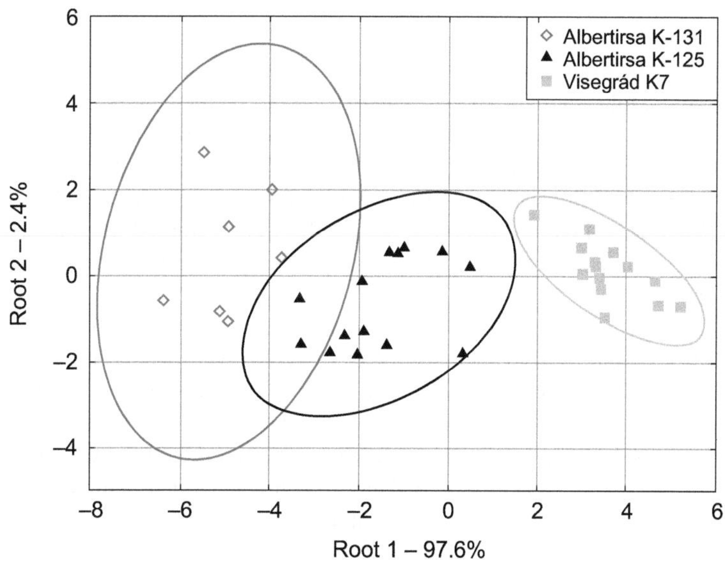

Fig. 15.1 LDA results for different water samples from the first sensory session; confidence ellipses: 90%. (After Sipos L et al., 2012. Food Chem 135:2947–2953. https://doi.org/10.1016/j.foodchem.2012.06.021. With permission)

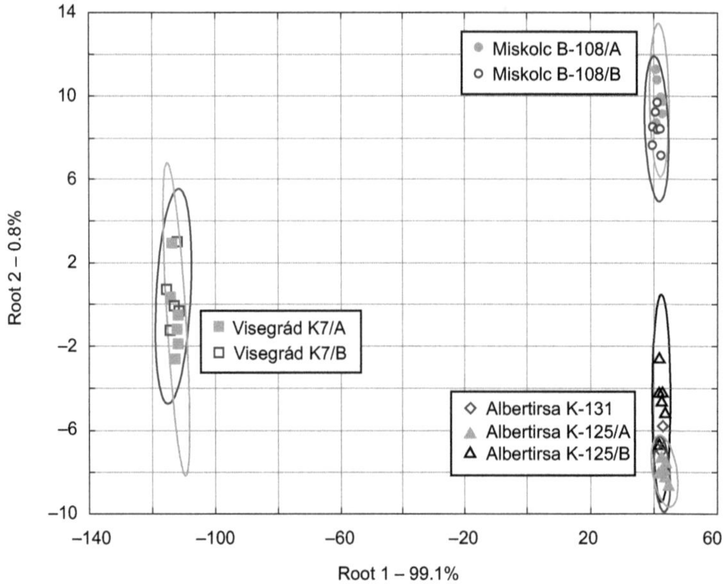

Fig. 15.2 LDA for the electronic tongue and analytical measurements; confidence ellipses: 95%. (After Sipos L et al., 2012. Food Chem 135:2947–2953. https://doi.org/10.1016/j.foodchem.2012.06.021. With permission)

According to Dotsika et al. (2010), IRMS is a helpful method for assessing the δ^2H and $\delta^{18}O$ distribution patterns in bottled and spring waters originating from different spring/well regions (Greece). The isotope ratio data were effective as tracers for identifying the provenance of bottled waters. The isotopic composition of bottled water and natural spring water were comparable, indicating that the production process may not influence the isotopic composition of the waters. Therefore, no technological modification was observed for the waters studied. In conclusion, the applied approach can be used for the authentication and identification of the origin of bottled waters.

Gutiérrez-Reguera et al. (2018) integrated ICP–AES, ICP–MS, IC–CD and standard analytical methods with PCA, SLDA, ANN and MLP–ANN as advanced approaches to classify commercial bottled mineral waters from five locations in Spain. Variables such as DE, Ca, Mg, K, Na and Ti were not sufficiently effective to discriminate all the sample origins, as shown by the first three PCs, which explained 47.67% of the total variance. SLDA data discriminated water samples originating from different mountainous regions. An iterative model (ANN3) was obtained by combining ANN1 and ANN2, including SLDA selection of the variables. Moreover, ANN achieved better results, with the overall performance of ANN2 reaching 93% and 98% sensitivity and specificity, respectively. It was concluded that the proposed model distinguished water samples from Cordillera Costero–Catalana from the other mountainous regions and therefore enabled the classification of the remaining groups. This model achieved an average classification ability of 94%.

Raco et al. (2013) reported that IRMS, SIMS and standard analytical methods were helpful for identifying the distribution of stable isotopes in samples of bottled water originating from spring/well locations across Italy. All the aquifers were grouped into five classes, i.e., carbonate rocks, sandstone, igneous rocks, crystalline rocks and alluvium aquifers. The δ^2H and $\delta^{18}O$ values of the bottled water samples were within a range of values typical of meteoric waters obtained from northern, central and southern Italy, suggesting that the stable isotope ratios in the water protected the record of the water source origin. This technique can be used not only to determine the natural origin of bottled water but also to identify differences between natural and processed water. However, no such technological modification was observed in the case of the studied waters.

Tanasković et al. (2012) reported that hydrochemical and radiological data (obtained by GRS, IC, FAAS and standard analytical methods) combined with multivariate analysis (PCA and HCA) of Serbian spa waters can be considered reliable tools. PCA, with the first four PCs accounting for 74.2% of the total variance, enabled them to distinguish the groundwater sampling sites into three distinct clusters related mainly to their geotectonic units. The first cluster corresponded to the Serbo–Macedonian massif (in metamorphic and volcanic areas); the second cluster, which was mixed, was attributable mainly to the Vardar zone from volcanogenic masses and waters from the Serbo–Macedonian massif, including waters emerging from the karstic region in the geotectonic unit Carpatho–Balkans; and the third cluster belonged to the Pannonian Basin, with the water sample from Vrdnik spa as an outlier. The total number of correct classifications was 83.3% for the predefined geotectonic units. It was concluded that HCA is a less effective technique than PCA since the former uses less information; i.e., it exclusively uses distance.

15.2.1.2 Countries Across Multiple Continents

Grošelj et al. (2010) aimed to establish a database of major ions via chemical processing to verify the geological provenance of mineral water samples originating from eight European countries. Four models were built using two geological classification schemes, tested for RA and validated by LOOCV. Using CP–ANNs, it was possible to predict the source geology of mineral waters successfully via a lithological classification based on four classes, i.e., magmatic rocks, metamorphic rocks, biogenic–chemical sediments and classic sediments. The optimal model achieved correct predictions of 85 and 65% for RA and LOOCV, respectively, enabling correct prediction of the geology of the mineral water samples. It was concluded that, in view of the obtained data, it is possible to verify the geographical origin of bottled waters with respect to a specific geology or lithology.

Voerkelius et al. (2010) coupled TIMS and MC–ICP–MS with computational tools to characterize 656 samples of bottled and raw water from 13 European countries. The obtained results were used to provide interspatial prediction of $^{87}Sr/^{86}Sr$ ratios by combining the measured data with GIS based on a geological map of Europe. They found a satisfactory relationship of the strontium isotope ratios in the natural mineral water samples with the surface geology. Integrating the measured data with GIS based on a geological map enabled the prediction of qualitative inter-spatial information of Sr isotope values. This "prediction" map considers large-scale geographical regional differentiation, e.g., the $^{87}Sr/^{86}Sr$ values obtained for honey and wheat originating from specific sample regions were significantly related to those of natural mineral water, as predicted by the map. On the basis of complex computational and graphical attempts, the map may provide overlapping values for the specific units within large-scale regional differences. According to the authors, this viable approach is highly suitable for the authentication of the geographical origin of food because it enables assessment of the provenance of foods without the urgent need for geographical authentication, which may not always be achievable.

Hoogewerff et al. (2019) applied bioavailable $^{87}Sr/^{86}Sr$ isotope ratios for approximately 1000 soil samples collected by the GEMAS consortium from grazing and agricultural soils in Europe. The measured or modelled data can be used as reference data for the overall distribution of $^{87}Sr/^{86}Sr$ in Europe or elsewhere.

15.2.2 Intertype/Brand Differentiation

Versari et al. (2002) characterized Italian water sample profiles with respect to MEC data processed by PCA, HCA and LDA. PCA, with the first two PCs explaining 66% of the total variance, was not particularly useful. HCA confirmed the data obtained by PCA, making it possible to separate the Italian bottled waters into six clusters related generally to the TDS content. However, an insignificant interregional relationship was achieved. For training the classification suggested by HCA, LDA enabled the classification of water samples into three groups according to the TDS value, i.e., low-mineral, oligomineral and mineral water. A large variation in the chemical composition of the different brands of water samples studied was observed. The levels of some macroelements, e.g., Na, Cl, SO_4 and F, in several mineral waters exceeded the maximum concentration enabled for natural waters by Italian regulations.

Lee et al. (2019) noted that ICP–AES data combined with physicochemical parameters were useful for classifying groundwater samples collected directly from the main production wells drilled in 60 manufacturing plants. The geochemical properties of groundwater for bottled waters in South Korea differ in terms of lithology. Groundwater types are commonly classified into three types, e.g., Ca–HCO_3, Ca–Na–HCO_3 and Ca–Mg–HCO_3, depending on the bedrock. The correlation matrix demonstrated that the groundwater chemistry was controlled by water–rock interactions. The relationships between groundwater compositions and bedrock geology in Korea were observed.

Oyebog et al. (2012) determined MEC and SIR using ICP–MS, IC, IRMS and standard analytical methods to assess the physicochemical profiles of water brands originating from plants located across the southwestern and littoral regions of Cameroon. According to Piper's diagram, five water types could be classified, i.e.,

Ca–HCO_3, Ca–Mg–HCO_3, Ca–Na–Mg–HCO_3, Na–HCO_3 and Na–Mg–HCO_3. One of the three R-Mode FA associations showed high loadings corresponding to NO_3^- and Pb along with As, PO_4^{3-} and Zn. The latter likely indicated anthropogenic contributions in the volcano–sedimentary aquifers of the study region. Q-Mode HCA distinguished four major groups among the bottled water samples. The δD and δO^{18} values revealed the meteoric origin of the bottled waters and other groundwaters in the explored area.

Lourenço et al. (2010) processed chemical MEC data to classify samples of natural mineral and spring (noncarbonated, carbonated) bottled waters. PCA was used, with the first three PCs explaining 81.93% of the total variance. PC1 is associated with the highest positive values of loadings, i.e., HCO_3^-, Na^+, Ca^{2+}, Mg^{2+} and K^+, contributing significantly to mineralization. PC2 differentiated SO_4–Cl-type waters from another type of water, which consisted of SiO_2 and F^-, whereas PC3, which was the pollution index associated with the highest positive value of the NO_3 parameter, was related mainly to agricultural activities. Therefore, this method enabled the classification of Portuguese samples into three hydrochemical groups according to their evaporative origin: high-mineralization waters, low-mineralization waters and medium-mineralization waters.

Petrović et al. (2010) and Petrović Pantić et al. (2015) used ICP–QMS, ICP–AES, IC, AFS and photometric and electrochemical methods to quantify the MEC of mineral and thermal waters. The direct relationships between the hydrochemical composition of water and complex geological parameters, in which the formation and movement of water have occurred throughout geological history, was revealed. Mineral waters of Serbia had significantly different TDS values than prevailing HCO_3-type waters. For the cation composition, two major groups were defined with dominant contents of Na + K and Ca (i.e., Ca + Mg), with the exception of the Mivela region, which was dominated by Mg. The Q-mode CA enables the classification of thermal groundwater samples. The dendrogram depicts the similarities/dissimilarities among the collected samples. All of the thermal groundwater samples were of the HCO_3–Na to SO_4–Na type (Petrović Pantić et al. 2015).

Van der Aa (2003) processed 291 mineral waters from 41 countries in Europe, Asia, Africa, North America, Central America, South America and Australia. Nine to 20% of the samples exceeded the Dutch drinking water standards for Cl^-, Ca, Mg, K, Na, SO_4^{2-} and F^-. It was possible to perform an intercomparison assessment of the van Wirdum diagram and Stuyfzand water classification with respect to hardness, fresh-salt classes and dominant cations and anions as well as for European mineral waters, as indicated by country. The fresh/salt subdivision according to the Stuyfzand method was characterized; i.e., the oligohaline waters were in the lithocline–atmocline water zone, and the fresh-brackish, brackish and salt-brackish classes were in the transitional area between the lithocline and thalassocline waters. Moreover, the Stuyfzand hardness classes could also be recognized in groups that increase from the atmocline to the thalassocline water. The dominant group of mineral waters according to Stuyfzand ($CaHCO_3$) was located, as expected, around the lithocline type of water in the van Wirdum system. This also applies to the $MgHCO_3$ and $CaSO_4$ waters. In the case of the characteristics of the mineral water dataset, the content of dissolved compounds, hardness and chloride content were the most distinctive criteria. As shown in Fig. 15.3, the majority of the mineral European waters in a van Wirdum diagram belonged to the lithocline, calcium-rich type of water. Dutch tap water is not very rich in Ca. Dutch, Spanish and Belgian mineral waters are atmocline–lithocline types, whereas German mineral waters are primarily characterized by being much saltier and predominantly connected with the lithocline–thalassocline zone. French mineral waters are highly diverse, whereas Italian, Austrian and Swiss mineral waters are generally categorized as the lithocline type.

Platikanov et al. (2013) estimated the MEC and sensory data by ICP–OES, IC and FPA of Spanish bottled mineral water and tap water. PCA revealed that the water samples were rich in HCO_3^-, SO_4^{2-}, Ca^{2+} and Mg^{2+}. Moreover, it was possible to discriminate between bottled and tap water samples owing to the detection of minor but still considerable contributions of residual chlorine. The application of chemometrics to physicochemical and sensory data in properly designed experiments is useful for the quantification of taste features of bottled mineral and tap water samples.

Brenčič and Vreča (2007) used IRMS to classify bottled water (58 brands) and 16 replicates (sparkling, still and flavoured waters) that were purchased from a Slovene market and that were produced in Slovenia and three other European countries. The stable isotopic ($\delta^{13}C$) profiles of bottled sparkling water (with natural and artificial addition of CO_2) and still water from the Slovenian market produced in Slovenia, Croatia, Austria and Italy were evaluated. It was possible to differentiate natural and artificial sparkling waters as well as various origins of still waters. The

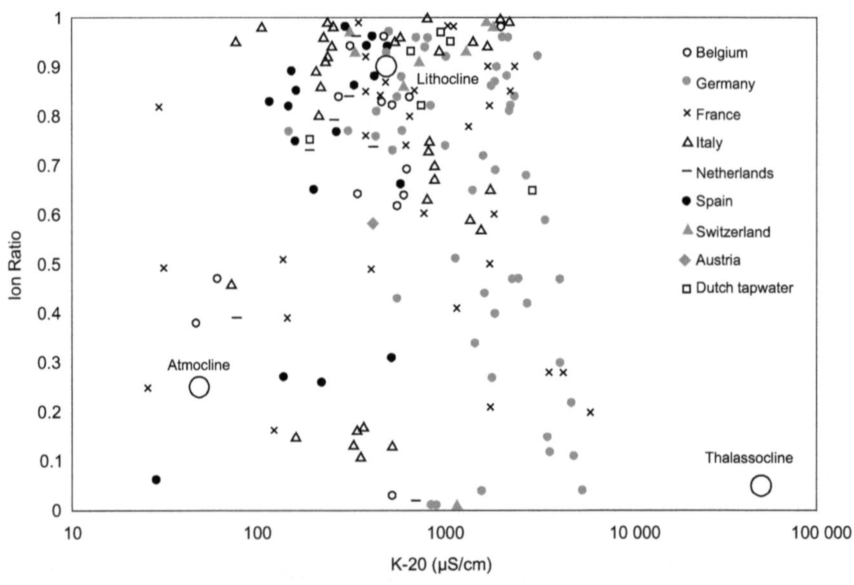

Fig. 15.3 Van Wirdum diagram for European mineral waters, indicated per country. (After Van der Aa M, 2003. Env Geol 44:554–563. https://doi.org/10.1007/s00254-003-0791-4. With permission)

various origins of still waters as well as the analysis of bottled waters enabled the discrimination of three types of water, i.e., water originating from deeper aquifers, aquifers with recharge in mountainous areas and aquifers in equilibrium with soil CO_2.

Mouhoumed et al. (2020) integrated MEC with chemometrics to classify different brands of local and imported bottled mineral water. The bottled water was classified into eight major African water types according to the Piper diagram, i.e., Ca–Mg–SO_4, Mg–Ca–Na–HCO_3–Cl, Ca–Mg–SO_4–Cl, Ca–SO_4, Ca–HCO_3–SO_4 and Ca–Mg–SO_4–HCO_3, as well as the most frequently observed types, such as Mg–Ca–Cl–HCO_3 and Ca–Mg–HCO_3. PCA revealed that the first two PCs explained 68.67% of the total variance. PC1 was related to high positive contributions of Mg, Ca, SO_4 and HCO_3; hence, it was clearly dependent on the geological composition of the substrate, which was located mostly in association with carbonate rocks. This component characterizes hardness (Ca and HCO_3 as major dissolved species in limestone aquifers and Mg as a part of magnesium calcite or dolomite). PC2 is loaded with negative values of K, Na and Cl, indicating dissolution of the evaporite rocks. Furthermore, HCA confirmed the PCA results.

Ghasemi et al. (2013) combined ICP–OES with multivariate tools (PCA, HCA, PLS–DA, LS–SVM and SIMCA), which were effective dual approaches for classifying drinking water and 30 brands of Iranian mineral water according to the different brands. PCA provided the first two PCs, accounting for 50.09% of the total variance. PC1 showed positive loadings for Li, Na, K, Sr, Mg, Si and Zn, i.e., representing water saltiness, whereas PC2 was loaded with high positive values of Ca and Ba and negative values of B connected with water hardness. Eight classes could be differentiated according to their mineral content. The performance of three supervised pattern recognition techniques, i.e., PSL–DA, SIMCA and SVM, in the classification of mineral water samples resulted in correct classification rates of 97, 100 and 100%, respectively. These models have the ability to discriminate different brands of mineral water with high accuracy and can differentiate tap water samples from classified mineral water samples.

Bodor et al. (2021) classified the chemical profile of bottled natural mineral waters available on the Romanian market. The analytical data were processed chemically. HCA was performed by analysing parameters such as HCO_3^-, Ca^{2+}, Mg^{2+} and Na^+. It was possible to discriminate two clusters and five subclusters. The first one had three mineral elements and HCO_3^- (1.1-low, 1.3-medium and 1.2-high), and the average Ca^{2+}/Mg^{2+} ratios varied between 3.61 and 5.71. Cluster 2 consisted of mineral water samples with very high mineral element and HCO_3^- contents. The differences in the chemical profiles of the bottled waters were related mainly to the geological structure of the hydromineral basin, postvolcanic activity and residence time.

Carstea et al. (2016) determined the MEC and fluorescence emission spectra via ICP–OES, ICP–MS, IC–CD, FS and standard analytical methods integrated with chemometric processing. Despite the small number of samples, it seems promising to characterize Romanian still mineral water and tap water. FA (with the first four factors explaining 73.32% of the total variance) and the correlation data matrix revealed great variability between the bottled mineral water samples. Both computational techniques were limited by the small sample size and the low variability of particular elements. However, it was possible to recognize the relationships between components and the main contributing factors and to determine the characteristics of the water samples. However, in the case of tap water, the chemical profiles were relatively similar for samples originating from the same city.

Khan et al. (2016) quantified the contents of BrO_3^-, NO_3^- and NO_2^- by UPLC–ESI/MS in metropolitan water from three areas and commercial bottled water from well or unknown sources after filtration. The levels of BrO_3^- and NO_3^- in bottled waters and metropolitan water were, in many cases, above the recommended levels. PCA, with the first two PCs explaining 99% of the total variance, enabled differentiation between desalinated water, well water and desalinated water blended with well water in Saudi Arabia. However, classification of the metropolitan drinking water samples was not achieved in terms of interspatial origin.

Kermanshah et al. (2010) studied the major elemental composition via multivariate analysis of Iranian bottled water and divided it into 11 classes; most waters were Ca–Mg–HCO_3-type. As shown from the PC plots (with the first three PCs explaining 81.4% of the total variance), one mineral water sample and one drinking water sample were clearly distinguished from the remaining samples. HCA supported this finding, distinguishing four major groups consisting of mineral and drinking waters of different brands.

15.2.3 Interregional and Intertype/Brand Differentiation

Bong et al. (2009) combined ICP–AES, IC and IRMS methods with multivariate analysis as an effective approach for evaluating different bottled waters originating from South Korea and 8 other countries. The PCA data, with the first two PCs explaining 56.7% of the total variance, were described by Ca, Mg, Na and HCO_3^- (F1) and Cl^-, δ^2H with $\delta^{18}O$ (F2). The four former descriptors mainly corresponded to TH and dissolution of carbonate, whereas the three latter descriptors clearly distinguished marine water from other types of bottled water. It was possible to distinguish marine water completely from other water types (Fig. 15.4). Similarly, sparkling water was much better differentiated, although a clear differentiation between natural and artificial CO_2 water was not possible. It was concluded that combined methods, including MEC and SIR with computational tools, are better for distinguishing water types and characterizing the geographical provenance of bottled water than one of these methods alone.

Kim et al. (2013) demonstrated that higher values of $^{87}Sr/^{86}Sr$ ratios established by ICP–MS in desalinated marine waters than in seawater indicate that they are the result of posttreatment processing after desalination. In contrast, the strontium isotope ratios of still and sparkling waters clearly clustered according to their bedrock origins. The relationships between $^{87}Sr/^{86}Sr$ and other geochemical parameters are useful for tracing the geographical provenance of bottled water, even revealing

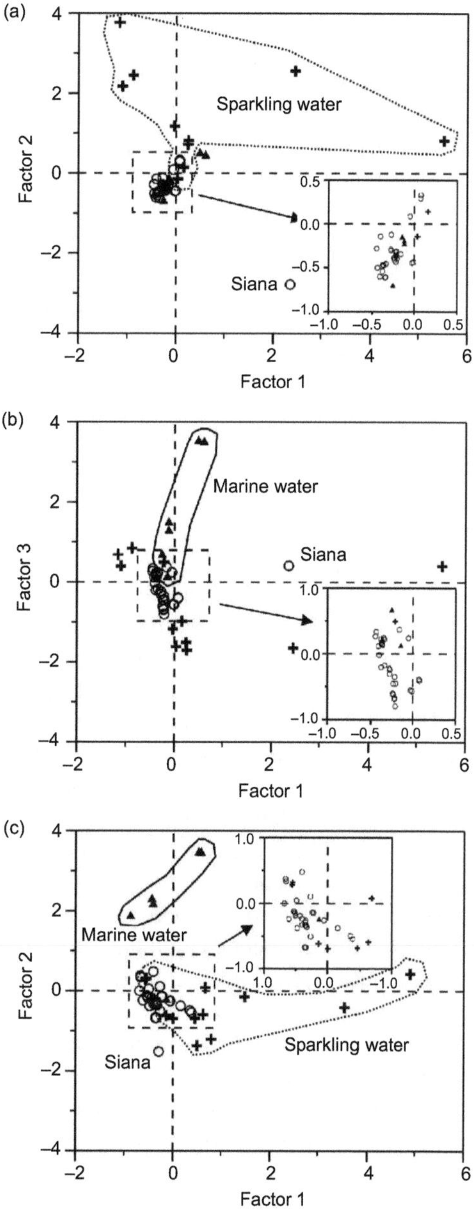

Fig. 15.4 Plots of PCA factor 1 vs. factor 2 scores (**a**) and factor 1 vs. factor 3 scores (**b**) using a factor analysis for major elements only and factor 1 vs. factor 2 scores (**c**) using a factor analysis for a combined dataset including stable isotopic compositions, explaining the grouping of various brands of bottled water. The symbols used are as follows: O still water, + sparking water, ▲marine water. (After Bong YS et al., 2009. Anal Chim Acta 631:189–195. https://doi.org/10.1016/j.aca.2008.10.032. With permission)

slight differences between bedrock types. In addition, Korean bottled water is typically characterized by a negative correlation between $^{87}Sr/^{86}Sr$ ratios and $\delta^{18}O$ values, which is specific to South Korea. Kim et al. (2012) reported that the combination of ICP–OES, IC and IRMS data of mineral waters was used to characterize water origin and types on the basis of MEC and SIR distribution patterns. Chemical changes in desalinated marine water resulting from desalination were identified. The elemental compositions of still and sparkling waters were quite similar, whereas desalinated marine water was clearly distinguished by its high levels of Ca, Mg, B and Cl. Furthermore, the $\delta^{18}O$ and $\delta^{2}H$ data of desalinated marine water were significantly higher than those of still and sparkling waters. It was concluded that a combination of elemental and stable isotopic tracers is a helpful tool for identifying the provenance of bottled water, including their intertype differentiation, and verifying elemental and isotopic modifications during desalination processes (Kim et al. 2012, 2013).

Kraic et al. (2008) applied HR–ICP–MS to determine MEC for the study of five types of water samples, i.e., tap, mineral, mineral carbonated, spring and spring carbonated waters from four European countries. On the basis of the ANOVA results, tap, spring and mineral water classes were mutually discriminated only by ^{121}Sb and ^{123}Sb. PCA with the first two PCs accounted for 37.7% of the total variance. High positive PC1 values corresponded to Sr, Li, B, Ni, Co, As, Sb, Sn, Tl, Be and Cr, whereas negative values of PC1 and large positive values of PC2 were related to Cd, Zn, Cu, Bi and Fe. To evaluate the provenance of water samples, ANOVA showed that only Al enables them to distinguish quantitatively the three countries of origin, i.e., Croatia, Slovenia and the Czech Republic. The LDA model effectively differentiated the three classes of water samples, i.e., tap, mineral and spring water samples, as demonstrated by the 100% performance obtained in the SPSS output for both the discriminant model and the LOOCV.

Lee et al. (2014) interpreted LIBS spectra related to MEC to investigate their discriminating power in 10 kinds of edible sea salts originating from six countries in Asia, Europe, North America and Oceania. The computational approach enabled three groups of elements representing dissolved ions in seawater (K, Mg and Li) associated with calcified particles (Ca and Sr) and those included in soils containing sea salts (Al, Si, Ti and Fe) to be distinguished. The seven models were classified on the basis of PCA and PLS–DA with different types of spectral information. It was suggested that elements within the same group are associated with the same source for sea salts. The confusion matrix of Models 2 and 3 for seven PCs is shown in Fig. 15.5. The results obtained can be used to develop a reliable classification methodology for sea salts via LIBS and other elemental analysis techniques.

According to Redondo and Yélamos (2005), IRMS was used to identify the origin of CO_2 in Spanish sparkling bottled waters. A single analysis of $\delta^{13}C$ distinguished the carbon dioxide in the two types of sparkling bottled waters, i.e., originating from a natural source or adding exogenous CO_2 of industrial origin. The method has two limitations; namely, it is not possible to detect sparkling water saturated with the original gas from the source or with industrial carbon dioxide produced from marine limestone.

Peh et al. (2010) used ICP–QMS, ICP–OES, IC and standard analytical methods to classify bottled water samples originating from two parts of Croatia, i.e., Dinaric and Pannonian samples. Spring waters from the two investigated regions were significantly differentiated with the values of EC and TDS as well as the levels of K, Mg, SiO_2, Al, As, Cd, Ba and Zr. Most of these factors are present at relatively high

(a) Model 2 (K I, Ca II, Al I)

Response (columns) / Truth (rows)

Truth \ Response	Jeung-Do	Haenam	Taean	Laizhou	Hainandao	Kochi	Camargue	Guerande	Mexico	New Zealand
Jeung-Do	25	43.75	0	6.25	25	0	0	0	0	0
Haenam	0	93.75	0	0	0	6.25	0	0	0	0
Taean	0	0	100	0	0	0	0	0	0	0
Laizhou	0	0	0	100	0	0	0	0	0	0
Hainandao	0	0	0	0	100	0	0	0	0	0
Kochi	0	0	0	0	0	100	0	0	0	0
Camargue	0	0	0	0	0	0	100	0	0	0
Guerande	0	0	0	0	0	0	0	100	0	0
Mexico	0	0	0	0	0	0	0	0	100	0
New Zealand	0	0	0	0	0	0	6.25	0	0	93.75

(b) Model 3 (K I, Ca II)

Response (columns) / Truth (rows)

Truth \ Response	Jeung-Do	Haenam	Taean	Laizhou	Hainandao	Kochi	Camargue	Guerande	Mexico	New Zealand
Jeung-Do	12.5	43.75	0	0	43.75	0	0	0	0	0
Haenam	0	93.75	0	0	0	6.25	0	0	0	0
Taean	0	0	100	0	0	0	0	0	0	0
Laizhou	0	0	0	100	0	0	0	0	0	0
Hainandao	0	0	0	0	100	0	0	0	0	0
Kochi	0	0	6.25	0	0	93.75	0	0	0	0
Camargue	0	0	0	0	0	0	100	0	0	0
Guerande	0	0	0	0	0	0	0	75	25	0
Mexico	0	0	0	0	0	0	0	0	93.75	6.25
New Zealand	0	0	0	0	0	0	6.25	0	0	93.75

Fig. 15.5 The confusion matrix of Model 2 developed with the K I, Ca II and Al I lines (**a**) and that of Model 3 developed with the K I and Ca II lines (**b**). For both models, seven PCs (PC1–PC7) are included. (After Lee Y et al., 2014. Spectrochim Acta B: At Spectrosc 101:57–67. https://doi.org/10.1016/j.sab.2014.07.012. With permission)

levels in Pannonian spring waters, except for Al, Cd and Zr, which occur at relatively high concentrations in Dinaric samples. These two parts of Croatia differ considerably with respect to their water types; i.e., there is a lack of mineral water types in the Dinaric region, whereas in the Pannonian region, even the spring waters are more mineralized than their Dinaric counterparts. In general, natural waters from Croatia are HCO_3^--type waters, with rationally balanced Ca–Mg associations in the spring water, and the dominant Na–K combination is attributed to mineral water types.

15.2.4 *Differentiation of Other Parameters*

15.2.4.1 Verification of Authenticity/Adulteration Related to Analytical Quality Control

Birke et al. (2010a) determined 71 chemical parameters by ICP–QMS, ICP–AES, IC, AFS and standard analytical methods in 1785 European marketed bottled water samples purchased at supermarkets throughout 40 European countries and in tap water from 29 European countries. According to the authors, general analytical problems occurred because a large number of elements were analysed and no suitable reference materials (standards) were available; hence, there was a lack of certified values for all these elements. Notably, the analytical results obtained in the BGR laboratories were compared with those given on the bottle labels. In the case of most bottled water samples with concentrations higher than usual, they were reanalysed to eliminate any possibility of mistakes during analysis. According to the authors, "the high quality dataset for bottled water in the European Groundwater Geochemistry atlas (EGS 2010) provides a first impression of the natural concentrations and variations of the determined chemical elements and additional parameters in groundwater at the European scale".

Fugedi et al. (2010) used ICP–OES, ICP–QMS, IC and standard analytical methods to assess the elemental composition of bottled mineral water samples originating from different water-producing wells in Hungary. Bottles with water such as sparkling (blue cap), still (pink cap) and medium (green cap) were collected. Moreover, information on bottle materials, i.e., plastic (PET) or glass bottles, bottle colour and plastic or metal caps, was considered and included in the dataset. The use of FA (with the first two factors explaining 40% of the total variance) was unsuccessful because the factor weight was too small due to the strong heterogeneity and low sample size. Other computational techniques are useful for comparable evaluations of the obtained results with those reported on bottle labels by producers. Technological processes seriously alter the original water mineral profile; hence, bottled water samples, in contrast to original water originating from natural sources, are not useful for the interregional mapping of groundwater quality. The obtained data were related to two informative sources of data, i.e., compared with archived hydrochemical data from the producing wells and with respect to the chemical composition reported on bottle labels.

Demetriades (2010a) demonstrated that MEC data are useful for estimating the elemental profiles of bottled waters from different locations in Greece. Analytical measurement uncertainty in a probabilistic scheme was used to assess compliance of bottled water with drinking water standards. R-CA data revealed two large clusters, i.e., "lithological" and "mineralization", corresponding to element groups, i.e., Al to Mo and As to Sb, respectively. This difference could be explained by the impact of both of these formations on the mineral composition of Hellenic bottled water.

Demetriades (2010b) processed different brands of bottled waters from different locations in Greece via ICP–AES, ICP–QMS and IC instrumental methods. The R-CA dendrogram presents two clusters related to the “lithological” and “mineralization” classification of the majority of elements. The chemical profiles of the source aquifers of the bottled water samples were relatively stable over time because there was a significant relationship between the results obtained by Demetriades (2010a) and those reported for bottle labels dating back to 1998. This agreement is in accordance with a requirement in the Natural Mineral Water Directive (EU 2009; Annex I. I.3, p. 51). It was concluded that the quality of European bottled water is considered good, with the exception of water samples containing levels of Mn, B, Ba, As, Fe, Ni, Se and Al that are definitely above the estimated respective UTL and hence exceed the corresponding legislative standard values established by European Union directives. It was postulated that national regulatory bodies should verify these results and develop an efficient evaluation mechanism to manage limits by means of the methodology proposed by Demetriades (2010a).

According to Brenčič and Vreča (2010), SIRs such as δ^2H, $\delta^{18}O$ and $\delta^{13}C$ were established in domestic and foreign bottled waters randomly collected from the Slovene market. The components of the normal finite mixture model of $\delta^{13}C_{DIC}$ represent four groups of bottled water, i.e., a group of artificial sparkling bottled water, two groups of still and flavoured bottled water and a group of bottled water originating from deeper aquifers. It was concluded that this model indicated that $\delta^{13}C_{DIC}$ for bottled water may be useful as a highly effective discriminator for the categorization of bottled water and therefore could be an authentication tool.

Methods such as ICP–QMS, AFS, IC and other standard methods (Brenčič et al. 2010) were applied to compare two datasets from two time sampling campaigns via Kolmogorov–Smirnov independent two-sample tests. The obtained data for the two bottled water samples were in agreement with the drinking water and mineral water standards, except for discrepancies observed only for B and Ni in highly mineralized waters. The obtained analytical results were in general agreement with the values reported on the labels. Slovenian bottled water is predominantly controlled by a $CaCO_3$–CO_2–H_2O association, where Na and Cl^- as well as SO_4^{2-} are present as the major components in different combinations. The analyses of the labels on the bottle packaging indicated that the analytical results obtained in the study were in general agreement with the values stated on the product’s packaging label.

Montgomery et al. (2006) studied spring and mineral waters from England, Scotland and Wales on the basis of the $^{87}Sr/^{86}Sr$ composition quantified by MC–ICPMS. Using this approach, it was possible to characterize the relationship with the aquifer rocks and broadly reflect the age and lithology of their host aquifers. It was postulated that strontium isotopic signatures could be used to fingerprint and recognize mineral waters to reduce fraud. According to the authors, this is the first reported spatial profile of bioavailable $^{87}Sr/^{86}Sr$ in Britain, which can be used in modern, historical and archaeological studies.

Souza et al. (2011) used ICP–OES data processed by multivariate analysis (PCA, HCA, k-NN and SIMCA) as a promising combined tool to assess the authenticity and adulteration of commercial mineral water samples from different Brazilian aquifers. For example, PCA (with the first two PCs explaining 98% of the total variance) was performed by applying Pareto scaling, in contrast to autoscaling, and achieved a clearer preliminary differentiation of mineral water samples corresponding to their geographical origin (Fig. 15.6). Similar results were obtained using HCA. It was also found that prediction based on autoscaling with SIMCA provides better results, but k-NN yields more satisfactory results when the data are

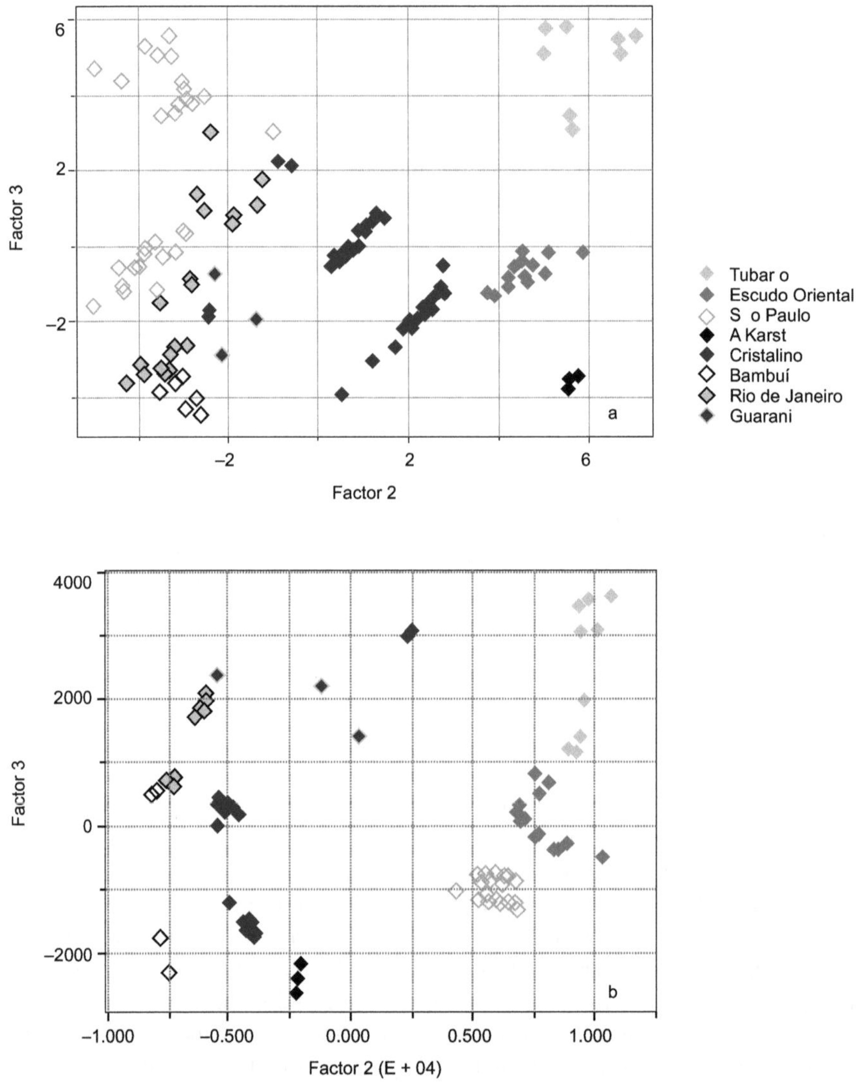

Fig. 15.6 Score plot for PCA performed on individual lines measured under robust conditions by applying autoscaling (**a**) and Pareto scaling (**b**). blue ◆ Tubarão; dark purple ◆ Escudo Oriental; green ◆ São Paulo; black ◆ Karst; ◆ violet Cristalino; red ◆ Bambuí; dark olive ◆ Rio de Janeiro; dark orange ◆ Guarani. (After Souza AL et al., 2011. J Anal At Spectrom 26:852–860. https://doi.org/10.1039/c0ja00071j. With permission)

preprocessed with Pareto scaling. Moreover, both supervised methods are characterized by satisfactory efficiency, and equivalent global classification rates amount to approximately 100% correct prediction, depending on the type of preprocessing or transformation. In general, some combinations of preprocessing and transformation methods insignificantly improve the classification rate. It has been postulated that ICP–OES fingerprinting combined with multivariate analysis is a useful tool with the potential to rapidly recognize authenticity and adulteration of mineral water samples as well as other assortments of samples directly related to their mineral composition.

15.2.4.2 Elemental Profile Influenced by Different Types of Packaging Materials Related to Other Parameters

Birke et al. (2010b) used a combination of ICP–QMS, ICP–AES, IC, AFS and standard analytical methods with computational tools, which were useful approaches for evaluating the elemental profiles of German water samples from 502 wells/brands in glass bottles (clear glass, green glass, bluish glass and greenish glass), PET bottles and tetrapacks. The glass bottles had metal, PE, crown and ceramic caps. PCA (R-mode) was applied with 18 factors, accounting for 72.8% of the total variance. The criterion for factor extraction was an eigenvalue >1. This technique differentiated most of the elements in the bottled waters associated with their origin from different well regions. For example, Figs. 15.7, 15.8 and 15.9 show geochemical maps of the factor scores for Factors 1, 3 and 8 in German bottled water. Most of the variables were assigned to geological sources, e.g., Triassic sediments, especially Cenozoic volcanic rocks, salt beds in highly fractured regions and the crystalline basement. Moreover, the different bottle materials and colours were associated with significant levels of Pb, Al, Zr, La and Sb, which can be leached from glass and PET bottles.

Misund et al. (1999) integrated ICP–MS, ICP–AES and IC to evaluate the elemental profiles of mineral waters from 15 European countries and Georgia (transcontinental country in Eastern Europe and West Asia). Differences in the contents of some elements detected between countries/regions are related to the geological environment. An influence of bottle type on the mineral elemental profile was also observed. For example, most mineral waters from Germany and Russia sold in glass bottles had significantly high median concentrations of Pb. Elements such as Pb, U, Zr, Li, K and Na are also present at a much higher median level in water stored in glass bottles, whereas most mineral waters from other countries are sold in plastic bottles (blank, green or blue). This means that there is a clear differentiation between glass and plastic bottles corresponding strictly to the different bottle types, colours and origins.

Felipe-Sotelo et al. (2015) integrated ICP–MS and ICP–OES with standard analytical methods as a reliable approach to assess the elemental composition of bottled waters from England and seven other European countries. PCA (with the first two PCs representing 36.0% of the total variance) enabled the discrimination of two

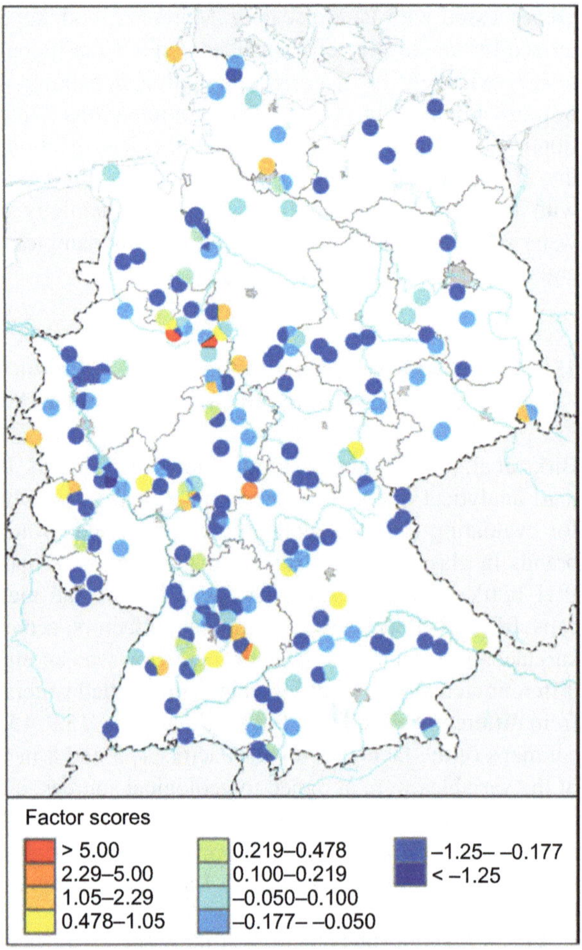

Fig. 15.7 Geochemical map of the factor scores for factor 1: Gd–Tb–Sm–Dy–Nd–Y– Pr–Ho–La–Er–Tm–(Al–Ce) in German bottled water. (After Birke M et al., 2010b. J Geochem Explor 107:245–271. https://doi.org/10.1016/j.gexplo.2010.06.002. With permission)

groups of samples, although it was impossible to achieve a complete differentiation of bottled waters according to geographical provenance. Importantly, British water samples were effectively differentiated from continental bottled water samples. Compared with the PCA technique, the CA technique provided worse results. The results of one-way multivariate ANOVA (one-way MANOVA) agreed with those obtained for four-way ANOVA with respect to the country of origin and type of water as the main factors affecting the water chemistry. The concentrations of Ca, Co, Mo, Pb, Sb and U were largely responsible for the significant interregional differentiation. Four factors were considered to affect the chemical distribution pattern of the samples, i.e., country of origin, geological characteristics, type of water and container material. For example, the geology of a water source significantly influences the Mo level. In terms of the influence of different types of packaging material on the elemental profile of the studied water samples, the waters in PET bottles had relatively high levels of Sb, whereas the greatest content of Al was detected in

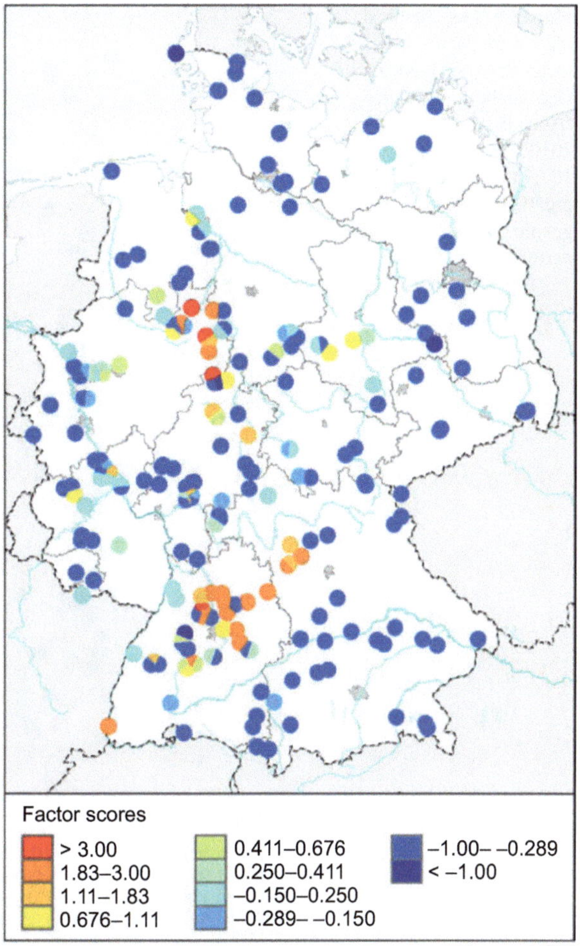

Fig. 15.8 Geochemical map of the factor scores for factor 3: SO_4–Ca–Sr–Te–EC–Mg–Ta–(Mo–B) in German bottled water (After Birke M et al., 2010b. J Geochem Explor 107:245–271. https://doi.org/10.1016/j.gexplo.2010.06.002. With permission)

waters bottled in glass. Additionally, elevated levels of As and TOC were present in the waters in PET bottles. To confirm this finding, further research should continue with the utilization of samples from the same origin and placed in two types of containers.

15.2.4.3 Interregional Differentiation Related to Geological Composition and the Corrosion Effect of the Distribution System

Dinelli et al. (2012a) aimed to combine ICP–QMS, ICP–OES, AFS, IC and standard analytical methods with computational processing to characterize the elemental patterns of tap water samples in private or public distribution systems throughout Italy. The impacts of bedrock geology and/or aquifer composition were detected for some elements, e.g., As, F, Rb and V, whereas other elements, e.g., Br and I, clearly

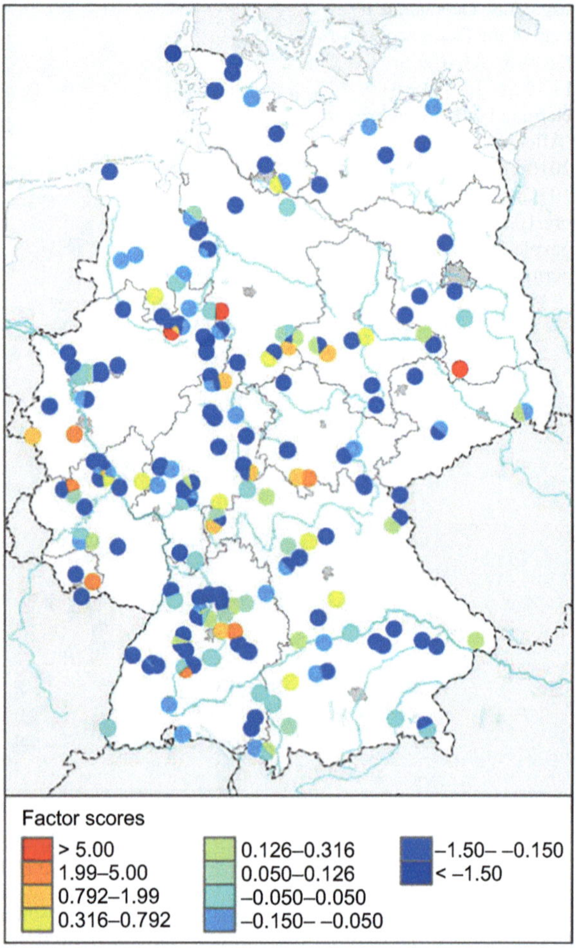

Fig. 15.9 Geochemical map of the factor scores for factor 8: Co–Cd–Ni–Zn in German bottled water. (After Birke M et al., 2010b. J Geochem Explor 107:245–271. https://doi.org/10.1016/j.gexplo.2010.06.002. With permission)

indicated that mixing with deep formation water or sea-spray contributions occurred. Other elements were identifiers of geographic provenance, e.g., B in Tuscany and Se in Emilia–Romagna. The R-mode FA technique was useful for processing the analytical data.

This tool confirms the distribution of different groups of elements reflecting geological patterns, e.g., contributions from carbonate dissolution, interactions with volcanic rocks and geographic origins and contributions from seawater (Fig. 15.10). The potential effects of corrosion, including the preferential release of Pb, Zn, Cd, Cu, Mn, Sn and Fe into the environment, are related to interactions with the distribution system pipelines. Dinelli et al. (2012b) investigated tap and bottled mineral waters originating from 20 administrative regions of Italy via ICP–QMS, ICP–OES, AFS, IC and standard analytical methods to create a valuable elemental database. Al, Cd, Cu, Fe, Ni, Pb and Zn were distinguished by their high levels in tap water, whereas Be, Cs, Sb, Sn and Tl were significantly enriched in bottled mineral water.

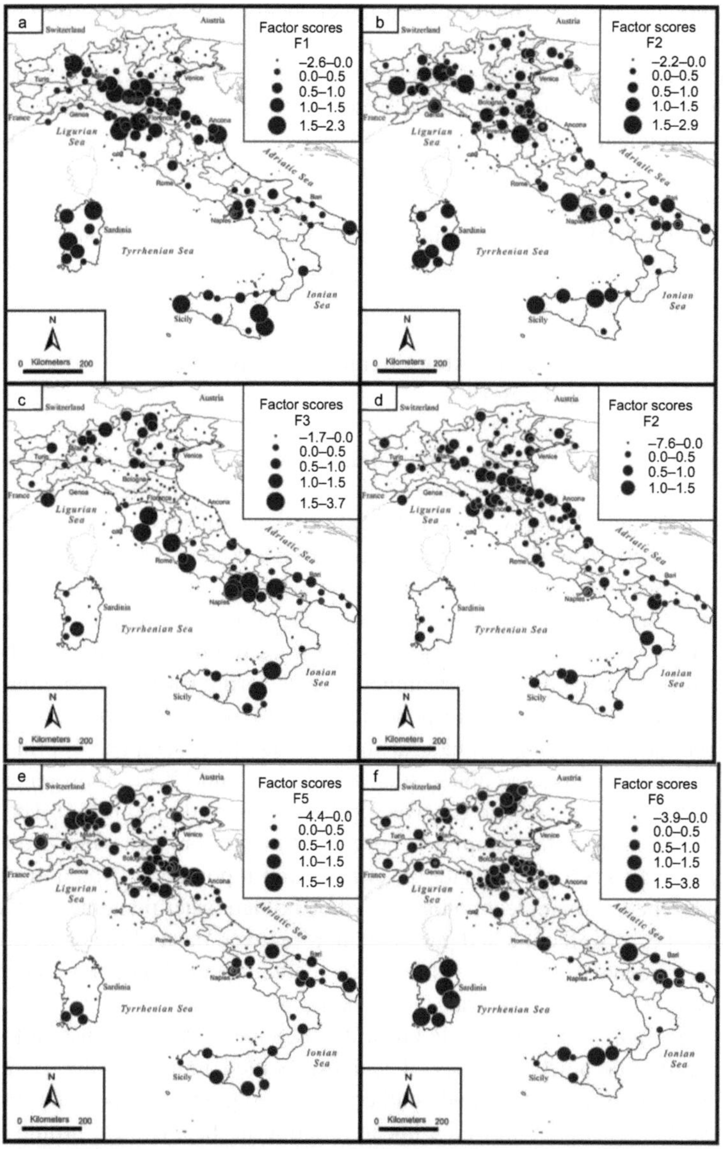

	F1		F2		F3		F4		F5		F6
Na	0.940	Pb	0.818	Rb	0.836	Ca	0.906	Se	0.754	Nd	0.629
Cl^-	0.884	Zn	0.810	As	0.774	Sr	0.850	U	0.718	Al	0.619
B	0.844	Cd	0.795	Cs	0.763	Mg	0.806	SO_4^{2-}	0.687	Y	0.610
Br^-	0.811	Cu	0.697	V	0.662	Ba	0.733	Sb	0.579	Fe	0.541
I	0.785	Mn	0.601	F^-	0.599	HCO_3^-	0.494	Ni	0.440	*V*	*-0.353*
K	0.604	Sn	0.536	Mo	0.572			*Si*	*-0.268*	*HCO_3^-*	*-0.425*
Li	0.577	Fe	0.492	K	0.549					*Cr*	*-0.554*
Si	0.573	Nd	0.419	Tl	0.481						
Co	0.549	Tl	0.407								
HCO_3^-	0.466										
% variance	24.6		13.8		9.1		6.8		5.8		5.5

Fig. 15.10 Factor score distributions of the five-factor solution by R-mode factor analysis. (After Dinelli E et al., 2012a. J Geochem Explor 112:54–75. https://doi.org/10.1016/j.gexplo.2011.07.009. With permission)

Such differences can be explained for some elements, e.g., the first group is essentially exposed to corrosion processes in the distribution system, showing a random interspatial distribution, which is unrelated to geological or geochemical characteristics. However, Sb in bottled mineral water is released from PET containers. For some other elements not included in the above two groups, e.g., Cr and V, it can be suspected that their nonnatural origin cannot be excluded because of pipeline corrosion of the distributing systems or container leaching. It was concluded that the geographical distribution of the elements provides similar information as well as the significance of geology, groundwater conditions, etc.

15.2.4.4 Differentiation of Water Chemistry Modified by Various Parameters

According to Güler (2007), an analytical database of 34 physicochemical parameters was compiled from the production licences issued to bottlers by the Turkish Ministry of Health. Notably, the MEC and other parameters seem to be reliable tools for characterizing different brands of Turkish bottled waters. The HCA and PCA techniques enabled the determination the natural groupings among the brands, providing information on the composition of the waters originating from various sources. HCA extracted four major groups that were significantly distinguished and related to different brands of natural mineral water, spring water and drinking water. PCA, with the first two PCs corresponding to 66.0% of the total variance, enabled the differentiation of samples of natural mineral water from samples of spring water. The most discriminative variables, such as Ca, Mg and HCO_3, represented total hardness (PC1), and Na, Cl and K corresponded to water saltiness (PC2). The chemical composition of water samples is dependent mostly on various environmental parameters, e.g., geologic formations, climate and topography, which are further modified by the treatment technologies used by bottled water manufacturers during production, i.e., disinfection and purification. Moreover, most of the obtained water quality constituents were distinguished by their significantly higher levels in natural mineral waters than in the natural spring and bottled waters.

Lee et al. (2021) noted that ICP–OES, IC, IRIS and TIMS were useful tools for classifying two commercial types of Chinese bottled mineral water from wells located in the northern (BSS) and southern (BDS) regions near potential volcanic areas. BSS bottled water showed a linear variation in the chemical profile against Cl^-, whereas the BDS water reflected sudden changes in this respect. The observed variation in the case of the former was likely due to pumping the aquifer at different depths, whereas the change in the latter was due to water–rock interactions with bedrock of various mineral compositions. It was concluded that the chemical distribution pattern of the bottled water produced from groundwater in the volcanic area might be a useful tool for monitoring the geochemical environmental changes in the groundwater aquifer.

Flem et al. (2015) collected a dataset (with the use of ICP–QMS, ICP–AES, IC and standard analytical methods) of 579 tap water samples from 29 European countries. The samples were analysed in a single laboratory, and the obtained results were evaluated statistically in terms of the spatial and national distributions of the determined parameters. The profile of most parameters is dependent on different artificial and natural factors, i.e., land use, distribution networks, water sources and treatments, geographical locations and geology. In the case of most parameters, a level of 99% compliance or better was obtained with internationally accepted norms; e.g., NO_3^- compatibility could reach 100%. The distributions of alkalinity, Ca, Mg, Sr and Li are dependent on the water source and the geological variability between base-poor crystalline rock terrains and carbonate-rich sedimentary rock. Scandinavian tap water is distinguished by the highest levels of most of the REEs, possibly reflecting their geological availability and mobility in low-pH raw water sources. The profiles of F, U and As are also related to geological origin, i.e., groundwater versus surface water controls. However, the Hungarian samples contained high As levels, probably reflecting the As-rich reducing groundwaters of the Pannonian basin. Most Estonian tap water indicates a distinct hydrochemical environment, i.e., Palaeozoic near-coastal aquifers with deep, reducing or saline groundwater, perhaps affected by marine intrusions enriched with Ba, B, Br, Cl, Eu, F, I, Li, K, Mn and Na.

15.2.4.5 Assessment of Geogenic and Pollution Impacts

Dinelli et al. (2010) used ICP–QMS, ICP–AES, IC, MEA and standard analytical methods to investigate 1785 samples of bottled mineral water from supermarkets throughout Europe, including Italian waters bottled at 126 sites. The application of R-Mode FA enabled the discrimination of individual elements related to lithology or other surface enrichment compartments. For example, waters draining through volcanic rocks are enriched in As, B, Br^-, Cl^-, Cs, I, K, Li, Na, NO_3^-, PO_4^{3-}, Rb, Sc, SiO_2, Sr, Te, Ti and V up to 3 orders of magnitude higher than those in waters draining through other lithologies. Factor 4 of R-mode FA demonstrated that determinants such as NO_3, V, As, PO_4^{3-} and especially Se can have geogenic and anthropogenic contributions to some aquifers in volcano-sedimentary or silicoclastic deposits as well as in intensively cultivated areas. It was concluded that the need for better knowledge of the chemical composition of increasingly popular mineral waters is a key issue for establishing their quality, especially for elements that are not regularly monitored.

According to Takem et al. (2015), physicochemical parameters provided by ICP–AES, ISE and standard analytical methods were useful for characterizing Cameroonian groundwater, streams and rainwater samples collected seasonally from 2006–2008 and from 2010–2013. PCA data (with the first four PCs explaining 89.51% of the total variance) in combination with hydrogeochemical studies indicated that four main factors are responsible for groundwater chemistry and acidity, i.e., acid atmospheric deposition, anthropogenic activities (industrial effluent discharges and acid spills), chemical weathering and coastal atmospheric deposition/cation exchange. It was postulated that the continuous acid inputs into acidic soils weakened their buffering capacity, which resulted in the acidification of shallow groundwater. This leads to increased mobilization of toxic elements such as Al, Pb, Cu, Zn and Mn as well as their possible harmful effects on health.

Djimi et al. (2021) reported that a combination of physicochemical parameters with multivariate processing was used to classify samples of deep and shallow groundwater and surface water from five watersheds in Cameroon. PCA with seven and six PCs accounted for 77.65% (northern region) and 72.24% (Adamawa region) of the total variance, respectively. PCA/FA revealed that potential pollution sources, such as iron-bearing minerals (northern region), Al-rich soils (Adamawa region) and surface runoff, appear to be of natural origin, with significant contributions from Fe^{2+}, Al^{3+}, NH_4^+, NO_3^-, K^+, EC and turbidity outside the CDWS and WHO norms. However, HCA distinguished sample origins into three clusters, i.e., low-, moderate- and high-pollution areas in the northern region and unpolluted, low- and moderate-pollution areas in the Adamawa region (Fig. 15.11).

In studies by Birke et al. (2009, 2010c), EGS (2010) and Schnug et al. (2008), the uranium concentration in stream and mineral water and bottled and tap water originating from various regions of Germany and nine other European countries was quantified via ICP–QMS. PCA (R-mode) data were used to study the relationships among the 71 inorganic parameters measured in the samples. The multivariate approach enabled them to discriminate the geographic distribution patterns of element associations in the studied bottled water. It was possible to identify factors that not only cause geochemical anomalies but also influence the element enrichment and migration of all European bottled water. PCA of stream water presented two factors with high factor loads of U. The distribution of the factor NO_3^-–Se–Cu–U in the bottled water in Germany (Fig. 15.12) was closely related to the distribution of the factor U–(Se–Mg) for surface water, confirming the impact of fertilizer application in agriculture. The interspatial variation in U content in bottled and tap water is influenced mainly by lithological and anthropogenic (old mining districts, phosphate fertilizers and industrial emissions) factors. Such anomalous scores were

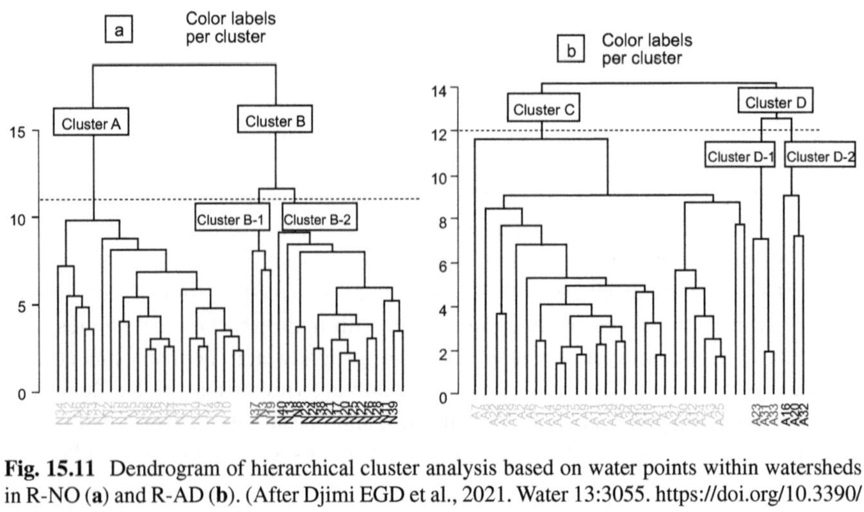

Fig. 15.11 Dendrogram of hierarchical cluster analysis based on water points within watersheds in R-NO (**a**) and R-AD (**b**). (After Djimi EGD et al., 2021. Water 13:3055. https://doi.org/10.3390/w13213055. With permission)

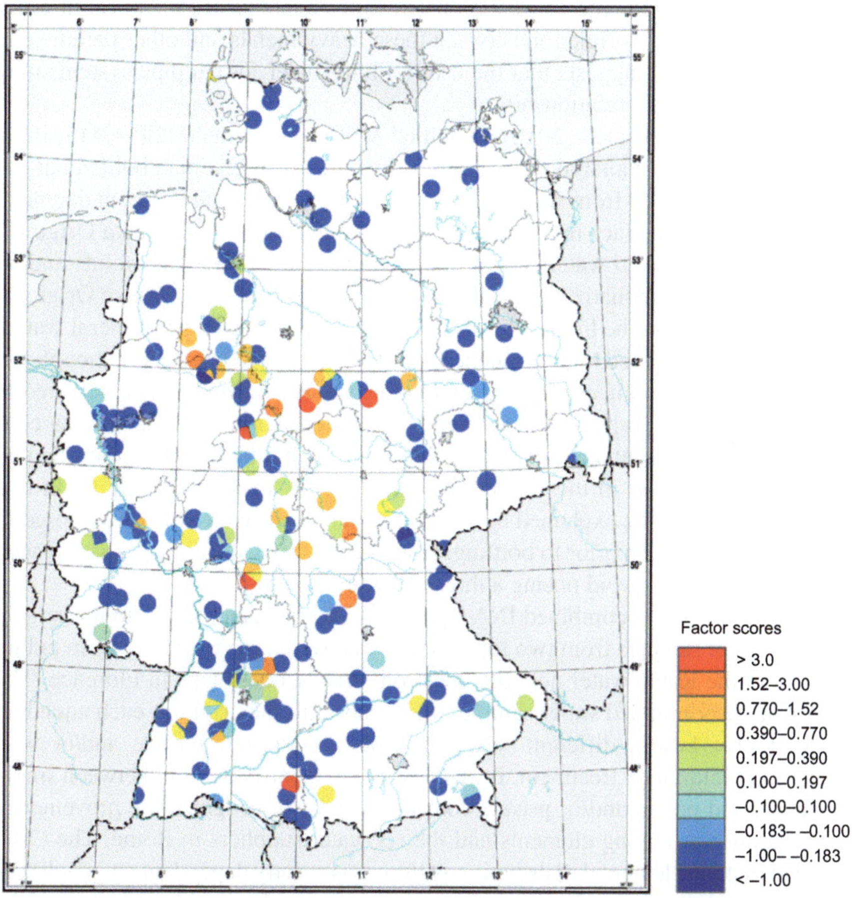

Fig. 15.12 Distribution of the factor values for the factor NO_3^-–Se–Cu–U in German bottled water. (After Birke M et al., 2010c. J Geochem Explor 107:272–282. https://doi.org/10.1016/j.gexplo.2010.04.003. With permission)

observed not only in Germany but also in the other six European countries. When PCA was used, anomalous scores of the factor U–SO_4^{2-}–Mg–EC–Na were observed for bottled water in the Czech Republic, Slovakia and Hungary as well as Lithuania, Belarus and Finland. The geology of the bedrock of a source is the most important factor affecting the uranium content of bottled water. Moreover, anthropogenic influences were postulated for all three water types, i.e., surface, bottled and tap water.

Banks et al. (2015) used instrumental methods such as ICP–QMS, ICP–AES, IC and standard analytical methods to process 579 tap water samples from 29 European countries. For most parameters, a level of 99% compliance or better was obtained according to internationally accepted norms. For NO_3, compliance seems to be 100%, corresponding to the strong emphasis in the EU to control this parameter. Moreover, As and Na deserve special attention since among the parameters with the highest rates of noncompliance, As and Na had noncompliance rates of 1% and 0.6%, respectively, in EU member states.

There is also a moderate correlation between the contents of Pb and Cu in water samples. The lack of correlation between these heavy metals and other parameters, such as pH and EC, suggests that the use of Pb in solder for Cu pipe systems may be responsible for this relationship.

Bityukova and Petersell (2010) quantified MEC by integrated ICP–QMS, ICP–AES, IC and standard analytical methods in five brands of Estonian bottled natural mineral water pumped from wells drilled into various aquifers at different depths in sequences of sedimentary rocks and tap water from Tallinn. The Värska Originaal and Värska mineralized waters had a high content of total dissolved solids and an acidic pH due to carbonatization of the bottled waters. Moreover, Värska Originaal was distinguished by the highest levels of Cl^- and NH_4^+ ions. Other mineral waters were weakly mineralized and influenced by HCO_3^- and SO_4^{2-} ions, although the lowest levels of the main constituents, including Ca, Fe, K, Mg and Na, occurred in Saku mineral water, which was purified by filtration before bottling. It was concluded that the differentiation of chemical elements in bottled water is controlled by the host rock geology in the studied area. The higher levels of Ni and Cr in this mineral water could be explained by its contamination since the water was stored in a stainless steel cistern prior to bottling. However, in 2010, the cistern was replaced with a plastic one to avoid posing a threat to public health.

Avino et al. (2011) combined INAA and PIXE with chemometrics to investigate drinking water samples from two Italian cities that were drawn directly from public supplies in the Rome water network and from private house taps in Florence. The CDA data discriminated water samples with respect to the origin of each aqueduct (hydrographic basin), different water provenances (spring, surface and/or well water) and preliminary treatment. It was possible to identify the split between single aqueducts and public and/or private consumers, i.e., to recognize the provenience and relationships among elements and the six water suppliers in Rome. The CDA revealed that the elemental distribution pattern was strictly dependent on geochemical factors. Finally, the application of chemometrics enabled the identification of the apportionments from single aqueducts to public and/or private consumers. In general, the quality of bottled water in Europe is good, with the exception of some brands that contained element levels above the estimated UTL and that exceeded the legislative standard values established by EU directives, i.e., in the case of Mn, B, Ba, As, Fe, Ni, Se and Al. According to the authors, exceptionally high levels of Fe and Al may be considered critical in drinking water originating from Rome and Florence, respectively.

15.2.4.6 Evaluation of Chemical Composition Related to Historical Records

Bodiš et al. (2010) evaluate chemical composition stability of mineral waters using both historical records and the most recent data, i.e., related to the database from 1965–2009 for Slovak water samples. The basic factors influencing mineral water stability in the Western Carpathians geological setting were characterized. The criteria for the temporal stability of the chemical distribution patterns of particular natural mineral water resources in Slovakia were also evaluated. Moreover, to improve the overall knowledge of the stability of mineral water resources, a comparative study of the chemical profiles of natural and bottled mineral waters originating from the Slovak market was performed. Both historical chemical analysis records and modern water quality monitoring information were interpreted. Historical records can provide a precious background, but their main drawback is a lack of information as well as the short association of the analysed components and analytical methods applied; hence, they can be assessed merely descriptively. The choice of the appropriate statistical criterion was tested on time series of various sources of Slovak mineral waters.

15.2.4.7 Water Samples in the Context of Legal Limits and Standards

Bertoldi et al. (2011) performed an extensive survey of MEC with other parameters quantified by ICP–MS, IC, ISE and standard analytical methods on 571 samples of mineral bottled water originating from 23 European countries. The purpose of this study was not to evaluate geographical differences in water mineral profiles but to report extensive up-to-date information on the element levels in water samples in the context of the legal limits available on the European market. Approximately 2.8% of the samples exceeded at least one parameter for the most restrictive limits set by various European countries for Se, NO_2, Mn, Ni, Ba, F and NO_3, and approximately 9% of the samples contained B, NO_3 or NO_2 levels above the legal limit established in European countries.

Cicchella et al. (2010) applied a combination of ICP–QMS, ICP–OES, AFS, IC and standard analytical methods with computational processing to estimate the elemental profiles of noncarbonated, artificially carbonated and naturally occurring CO_2 contents of 158 brands of mineral water. The levels of some elements were higher than the Italian legal limit established for infant consumption; e.g., Be reached surprisingly high levels in drinking water, as did F^-. It was also estimated that 5% of the water samples contained NO_2^-, which violates Italian mineral water standards, and 10% exceeds the Italian legal limit set for NO_3^- in the case of the mineral water for infant consumption. This high NO_3 level may be a result of the fertilization of agricultural fields. Such information should be reported on the label. Moreover, packaging can release some trace elements into bottled water. For example, bottled waters stored in PET are contaminated with Sb originating from their containers, whereas glass packaging can increase the levels of some elements (Al, Co, Cr, Cu, Fe, Pb, Sn, Zr and REEs) in bottled mineral water. However, the release of these elements from packing occurs at rates considered negligible for consumers.

Cidu et al. (2011) used instrumental tools such as ICP–OES, ICP–QMS, HG-ICP–MS, IC and standard analytical methods to assess the quality of two types of water, i.e., bottled and tap waters. On the basis of the Italian and WHO guidelines for drinking water, the obtained results indicated significant agreement between toxic elements such as Cr, Cd, Hg and Pb, since in 20% of the bottled water samples, one or more elements could be found at levels exceeding the Italian regulations, such as Cl^-, SO_4^{2-}, NO_3^- (limit for infants), F^- and As, as well as the WHO guidelines, such as B and U. The overall quality of the studied bottled waters was not always higher than that of municipal tap water. There is a need to update the current guidelines for drinking water (including bottled waters) on the basis of assessing the toxicity related to long-term exposure to toxic and harmful trace elements. In the case of mineral waters with excess levels of toxic elements, reporting the maximum daily uptake, referring to the lowest allowable health risk exposure, has been recommended. According to the authors, the obtained results indicate that worries about the quality of municipal drinking water are frequently unjustified, although moderate use of specific bottled waters should be recommended. Owing to such an approach, a reduced consumption of bottled waters would also help decrease the impact of plastic packaging on the environment, especially in Italy, where the recycling of plastic materials is very low at present.

Baba et al. (2008) used ICP–AES techniques with standard analytical methods to analyse the chemical composition of bottled mineral water purchased randomly throughout Turkey. The original variables were processed by HCA, and five groups were observed via this multivariate tool. The results revealed a wide range of chemical profiles of the mineral waters that were characterized. This elemental distribution pattern of the water samples observed in the regions may be caused by the geological environment as well as the fact that the majority of bottled mineral waters exceed the pH limit of Turkish drinking water standards. Except for Ba and Mn, the concentrations of heavy metals did not exceed the normative values established by the US EPA in Turkey.

15.3 Conclusions

The combination of multielemental analysis with computational processing is a useful methodology for classifying the studied brands of mineral water. Chemometric assessment of the analytical data has enabled the discrimination of the elemental profile of water controlled by rock geology and modified by contamination released into water stored in cisterns prior to bottling. The influence of technological processes on the elemental composition of the original water should also be considered. It has also been possible to characterize the elemental distribution pattern of water resulting from pipeline corrosion of distributing systems or leaching from containers. Notably, interregional differentiation could be influenced mainly by lithological and anthropogenic factors and additionally related to water type and container material. Notably, analytical quality control of the mineral profile of water samples with reference to bottle labels is needed. It has also been estimated that in some cases, water samples contain exceeding levels of mineral components in the context of European or national legal limits. Furthermore, interregional and intertype/brand differentiation related to various parameters have been assessed.

References

Avino P, Capannesi G, Rosada A (2011) Ultra-trace nutritional and toxicological elements in Rome and Florence drinking waters determined by Instrumental Neutron Activation Analysis. Microchem J 97:144–153. https://doi.org/10.1016/j.microc.2010.08.007

Baba A, Ereeş FS, Hıçsönmez Ü, Çam S, Özdılek HG (2008) An assessment of the quality of various bottled mineral water marketed in Turkey. Environ Monitor Assess 139:277–285. https://doi.org/10.1007/s10661-007-9833-9

Banks D, Birke M, Flem B, Reimann C (2015) Inorganic chemical quality of European tap-water: 1. Distribution of parameters and regulatory compliance. Appl Geochem 59:200–210. https://doi.org/10.1016/j.apgeochem.2014.10.016

Bertoldi D, Bontempo L, Larcher R, Nicolini G, Voerkelius S, Lorenz GD, Ueckermann H, Froeschl H, Baxter MJ, Hoogewerff J, Brereton P (2011) Survey of the chemical composition of 571 European bottled mineral waters. J Food Compos Anal 24:376–385. https://doi.org/10.1016/j.jfca.2010.07.005

Birke M, Rauch U, Lorenz H (2009) Uranium in stream and mineral water of the Federal Republic of Germany. Environ Geochem Health 31:693–706. https://doi.org/10.1007/s10653-009-9247-4

Birke M, Reimann C, Demetriades A, Rauch U, Lorenz H, Harazim B, Glatte W (2010a) Determination of major and trace elements in European bottled mineral water—Analytical methods. J Geochem Explor 107:217–226. https://doi.org/10.1016/j.gexplo.2010.05.005

Birke M, Rauch U, Harazim B, Lorenz H, Glatte W (2010b) Major and trace elements in German bottled water, their regional distribution, and accordance with national and international standards. J Geochem Explor 107:245–271. https://doi.org/10.1016/j.gexplo.2010.06.002

Birke M, Rauch U, Lorenz H, Kringel R (2010c) Distribution of uranium in German bottled and tap water. J Geochem Explor 107:272–282. https://doi.org/10.1016/j.gexplo.2010.04.003

Biscardi D, Monarca S, De Fusco R, Senatore F, Poli P, Buschini A, Rossi C, Zani C (2003) Evaluation of the migration of mutagens/carcinogens from PET bottles into mineral water by *Tradescantia*/micronuclei test, Comet assay on leukocytes and GC/MS. Sci Total Environ 302:101–108. https://doi.org/10.1016/S0048-9697(02)00349-2

Bityukova L, Petersell V (2010) Chemical composition of bottled mineral waters in Estonia. J Geochem Explor 107:238–244. https://doi.org/10.1016/j.gexplo.2010.07.006

Bodiš D, Kordik J, Slaninka I, Malík P, Liščák P, Panák D, Božíková J, Marcin D (2010) Mineral waters in Slovakia — evaluation of chemical composition stability using both historical records and the most recent data. J Geochem Explor 107:382–390. https://doi.org/10.1016/j.gexplo.2010.06.009

Bodor K, Bodor Z, Szép A, Szép R (2021) Classification and hierarchical cluster analysis of principal Romanian bottled mineral waters. J Food Compos Anal 100:103903. https://doi.org/10.1016/j.jfca.2021.103903

Bong Y-S, Ryu J-S, Lee K-S (2009) Characterizing the origins of bottled water on the South Korean market using chemical and isotopic compositions. Anal Chim Acta 631:189–195. https://doi.org/10.1016/j.aca.2008.10.032

Bosnir J, Puntarić D, Skes I, Klarić M, Simić S, Zorić I (2003) Migration of phthalates from plastic products to model solutions. Coll Antropol 27(Suppl 1):23–30

Brenčič M, Vreča P (2007) Isotopic composition of dissolved inorganic carbon in bottled waters on the Slovene market. Food Chem 101:1516–1525. https://doi.org/10.1016/j.foodchem.2006.04.003

Brenčič M, Vreča P (2010) The use of a finite mixture distribution model in bottled water characterisation and authentication with stable hydrogen, oxygen and carbon isotopes — Case study from Slovenia. J Geochem Explor 107:391–399. https://doi.org/10.1016/j.gexplo.2010.08.006

Brenčič M, Ferjan T, Gosar M (2010) Geochemical survey of Slovenian bottled waters. J Geochem Explor 107:400–409. https://doi.org/10.1016/j.gexplo.2010.09.007

Carstea EM, Levei EA, Hoaghia M-A, Savastru R (2016) Quality assessment of Romanian bottled mineral water and tap water. Environ Monit Assess 188:521. https://doi.org/10.1007/s10661-016-5531-9

Cicchella D, Albanese S, De Vivo B, Dinelli E, Giaccio L, Lima A, Valera P (2010) Trace elements and ions in Italian bottled mineral waters: Identification of anomalous values and human health related effects. J Geochem Explor 107:336–349. https://doi.org/10.1016/j.gexplo.2010.04.004

Cidu R, Frau F, Tore P (2011) Drinking water quality: comparing inorganic components in bottled water and Italian tap water. J Food Compos Anal 24:184–193. https://doi.org/10.1016/j.jfca.2010.08.005

Cohen A, Cui J, Song Q, Xia Q, Huang J, Yan X, Guo Y, Sun Y, Colford JM Jr, Ray I (2022) Bottled water quality and associated health outcomes: a systematic review and meta-analysis of 20 years of published data from China. Environ Res Lett 17:013003. https://doi.org/10.1088/1748-9326/ac2f65

Demetriades A (2010a) Use of measurement uncertainty in a probabilistic scheme to assess compliance of bottled water with drinking water standards. J Geochem Explor 107:410–422. https://doi.org/10.1016/j.gexplo.2010.11.001

Demetriades A (2010b) General ground water geochemistry of Hellas using bottled water samples. J Geochem Explor 107:283–298. https://doi.org/10.1016/j.gexplo.2010.10.012

Dinelli E, Lima A, De Vivo B, Albanese S, Cicchella D, Valera P (2010) Hydrogeochemical analysis on Italian bottled mineral waters: effects of geology. J Geochem Explor 107:317–335. https://doi.org/10.1016/j.gexplo.2010.06.004

Dinelli E, Lima A, Albanese S, Birke M, Cicchella D, Giaccio L, Valera P, De Vivo B (2012a) Major and trace elements in tap water from Italy. J Geochem Explor 112:54–75. https://doi.org/10.1016/j.gexplo.2011.07.009

Dinelli E, Lima A, Albanese S, Birke M, Cicchella D, Giaccio L, Valera P, De Vivo B (2012b) Comparative study between bottled mineral and tap water in Italy. J Geochem Explor 112:368–389. https://doi.org/10.1016/j.gexplo.2011.11.002

Djimi EGD, Abia ALK, Belibi PDB, Soh PT, Che RN, Ghogomu JN, Ketcha JM (2021) Multivariate statistical and hydrochemical analysis of drinking water resources in Northern Cameroon watersheds. Water 13:3055. https://doi.org/10.3390/w13213055

Dotsika E, Poutoukis D, Raco B, Psomiadis D (2010) Stable isotope composition of Hellenic bottled waters. J Geochem Explor 107:299–304. https://doi.org/10.1016/j.gexplo.2010.07.002

EGS (EuroGeoSurveys) Geochemistry Expert Group (2010) European groundwater geochemistry (EGG): bottled water. Gebr. Borntraeger Science Publishers, Stuttgart

Felipe-Sotelo M, Henshall-Bell ER, Evans NDM, Read D (2015) Comparison of the chemical composition of British and Continental European bottled waters by multivariate analysis. J Food Compos Anal 39:33–42. https://doi.org/10.1016/j.jfca.2014.10.014

Flem B, Reimann C, Birke M, Banks D, Filzmoser P, Frengstad B (2015) Inorganic chemical quality of European tap-water: 2. Geographical distribution. Appl Geochem 59:211–224. https://doi.org/10.1016/j.apgeochem.2015.01.016

Fugedi U, Kuti L, Jordan G, Kerek B (2010) Investigation of the hydrogeochemistry of some bottled mineral waters in Hungary. J Geochem Explor 107:305–316. https://doi.org/10.1016/j.gexplo.2010.10.011

Ghasemi JB, Zolfonoun E, Khosrokhavar R (2013) Linear and nonlinear multivariate classification of Iranian bottled mineral waters according to their elemental content determined by ICP-OES. J Sci Islam Repub Iran 24:15–22

Grošelj N, van der Veer G, Tušar M, Vračko M, Novič M (2010) Verification of the geological origin of bottled mineral water using artificial neural networks. Food Chem 118:941–947. https://doi.org/10.1016/j.foodchem.2008.11.085

Güler C (2007) Characterization of Turkish bottled waters using pattern recognition methods. Chemometr Intell Lab Syst 86:86–94. https://doi.org/10.1016/j.chemolab.2006.08.009

Gutiérrez-Reguera F, Jurado JM, Montoya-Mayor R, Ternero-Rodríguez M (2018) Geographical classification of Spanish bottled mineral waters by means of iterative models based on linear discriminant analysis and artificial neural networks. Neural Comput Applic 29:459–468. https://doi.org/10.1007/s00521-016-2459-5

Hoogewerff JA, Reimann C, Ueckermann H, Frei R, Frei KM, van Aswegen T, Stirling C, Reid M, Clayton A, Ladenberger A (2019) Bioavailable $^{87}Sr/^{86}Sr$ in European soils: a baseline for provenancing studies. Sci Total Environ 672:1033–1044. https://doi.org/10.1016/j.scitotenv.2019.03.387

Kermanshah KY, Tabaraki R, Karimi H, Nikorazm M, Abbasi S (2010) Classification of Iranian bottled waters as indicated by manufacturer's labellings. Food Chem 120:1218–1223. https://doi.org/10.1016/j.foodchem.2009.11.067

Khan MR, Wabaidur SM, Alothman ZA, Busquets R, Naushad M (2016) Method for the fast determination of bromate, nitrate and nitrite by ultra-performance liquid chromatography–mass spectrometry and their monitoring in Saudi Arabian drinking water with chemometric data treatment. Talanta 152:513–520. https://doi.org/10.1016/j.talanta.2016.02.036

Kim G-E, Ryu J-S, Shin W-J, Bong Y-S, Lee K-S, Choi M-S (2012) Chemical and isotopic compositions of bottled waters sold in Korea: chemical enrichment and isotopic fractionation by desalination. Rapid Commun Mass Spectrom 26:25–31. https://doi.org/10.1002/rcm.5292

Kim G-E, Shin W-J, Ryu J-S, Choi M-S, Lee K-S (2013) Identification of the origin and water type of various Korean bottled waters using strontium isotopes. J Geochem Explor 132:1–5. https://doi.org/10.1016/j.gexplo.2013.03.002

Kraic F, Mocák J, Fiket Ż, Kniewald G (2008) ICP MS analysis and classification of potable, spring, and mineral waters. Chem Pap 62:445–450. https://doi.org/10.2478/s11696-008-0063-6

Lee Y, Ham K-S, Han S-H, Yoo J, Jeong S (2014) Revealing discriminating power of the elements in edible sea salts: line-intensity correlation analysis from laser-induced plasma emission spectra. Spectrochim Acta B At Spectrosc 101:57–67. https://doi.org/10.1016/j.sab.2014.07.012

Lee BD, Oh YH, Cho BW, Yun U, Choo CO (2019) Hydrochemical properties of groundwater used for Korea bottled waters in relation to geology. Water 11:1043. https://doi.org/10.3390/w11051043

Lee S-G, Koh D-C, Ha K, Ko K-S, Lee YS, Jung Y-Y, Cheng Z, Chen S-S (2021) Geochemical implication of chemical composition of mineral water (bottled water) produced near Mt. Baekdu (Changbai), Northeast China. Water 13:2191. https://doi.org/10.3390/w13162191

Lourenço C, Ribeiro L, Cruz J (2010) Classification of natural mineral and spring bottled waters of Portugal using Principal Component Analysis. J Geochem Explor 107:362–372. https://doi.org/10.1016/j.gexplo.2010.08.001

Misund A, Frengstad B, Siewers U, Reimann C (1999) Variation of 66 elements in European bottled mineral waters. Sci Total Environ 243–244:21–41. https://doi.org/10.1016/S0048-9697(99)00307-1

Montgomery J, Evans JA, Wildman G (2006) $^{87}Sr/^{86}Sr$ isotope composition of bottled British mineral waters for environmental and forensic purposes. Appl Geochem 21:1626–1634. https://doi.org/10.1016/j.apgeochem.2006.07.002

Mouhoumed EI, Abdillahi MM, Elmi YM, Doubad CO, Dirieh EA-N (2020) Classification of various bottled mineral waters marketed in Djibouti. WJET 8:720–738. https://doi.org/10.4236/wjet.2020.84052

Nawrocki J, Dabrowska A, Borcz A (2002) Investigation of carbonyl compounds in bottled waters from Poland. Water Res 36:4893–4901. https://doi.org/10.1016/S0043-1354(02)00201-4

Oyebog SA, Ako AA, Nkeng GE, Suh EC (2012) Hydrogeochemical characteristics of some Cameroon bottled waters, investigated by multivariate statistical analyses. J Geochem Explor 112:118–130. https://doi.org/10.1016/j.gexplo.2011.08.003

Peh Z, Šorša A, Halamić J (2010) Composition and variation of major and trace elements in Croatian bottled waters. J Geochem Explor 107:227–237. https://doi.org/10.1016/j.gexplo.2010.02.002

Petrović Pantić T, Birke M, Petrović B, Nikolov J, Dragišić V, Živanović V (2015) Hydrogeochemistry of thermal groundwaters in the Serbian crystalline core region. J Geochem Explor 159:101–114. https://doi.org/10.1016/j.gexplo.2015.08.009

Petrović T, Zlokolica-Mandić M, Veljković N, Vidojević D (2010) Hydrogeological conditions for the forming and quality of mineral waters in Serbia. J Geochem Explor 107:373–381. https://doi.org/10.1016/j.gexplo.2010.07.009

Platikanov S, Garcia V, Fonseca I, Rullán E, Devesa R, Tauler R (2013) Influence of minerals on the taste of bottled and tap water: a chemometric approach. Water Res 47:693–704. https://doi.org/10.1016/j.watres.2012.10.040

Quattrini S, Pampaloni B, Brandi ML (2016) Natural mineral waters: chemical characteristics and health effects. Clin Cases Miner Bone Metab 13:173–180. https://doi.org/10.11138/ccmbm/2016.13.3.173

Raco B, Dotsika E, Cerrina-Feroni A, Battaglini R, Poutoukis D (2013) Stable isotope composition of Italian bottled waters. J Geochem Explor 124:203–211. https://doi.org/10.1016/j.gexplo.2012.10.003

Redondo R, Yélamos JG (2005) Determination of CO_2 origin (natural or industrial) in sparkling bottled waters by $^{13}C/^{12}C$ isotope ratio analysis. Food Chem 92:507–514. https://doi.org/10.1016/j.foodchem.2004.08.018

Rodwan JG Jr (2014) "Bottled water: REINVIGORATION". U.S. and International Developments and Statistics. BWR-July Aug-2015-Issue BMC 2014-Bottled-Water-Statistics-Article.pdf

Schnug E, Birke M, Costa N, Knolle F, Fleckstein J, Panten K, Lilienthal H, Hahneklaus S (2008) Uranium in German mineral and tap waters. In: De Kok LJ, Schnug E (eds) Loads and fate of fertilizer-derived uranium. Backhuys Publishers, Leiden, pp 91–110. ISBN: 978-3-8236-1546-0

Sipos L, Kovács Z, Sági-Kiss V, Csiki T, Kókai Z, Fekete A, Héberger K (2012) Discrimination of mineral waters by electronic tongue, sensory evaluation and chemical analysis. Food Chem 135:2947–2953. https://doi.org/10.1016/j.foodchem.2012.06.021

Souza AL, Lemos SG, Naozuka J, Correia PRM, Oliveira PV (2011) Exploring the emission intensities of ICP OES aided by chemometrics in the geographical discrimination of mineral waters. J Anal At Spectrom 26:852–860. https://doi.org/10.1039/c0ja00071j

Takem GE, Kuitcha D, Ako AA, Mafany GT, Takounjou-Fouepe A, Ndjama J, Ntchancho R, Ateba BH, Chandrasekharam D, Ayonghe SN (2015) Acidification of shallow groundwater in the unconfined sandy aquifer of the city of Douala, Cameroon, Western Africa: implications for groundwater quality and use. Environ Earth Sci 74:6831–6846. https://doi.org/10.1007/s12665-015-4681-3

Tanasković I, Golobocanin D, Miljević N (2012) Multivariate statistical analysis of hydrochemical and radiological data of Serbian spa waters. J Geochem Explor 112:226–234. https://doi.org/10.1016/j.gexplo.2011.08.014

van der Aa M (2003) Classification of mineral water types and comparison with drinking water standards. Env Geol 44:554–563. https://doi.org/10.1007/s00254-003-0791-4

Versari A, Parpinello GP, Galassi S (2002) Chemometric survey of Italian bottled mineral waters by means of their labeled physico-chemical and chemical composition. J Food Compos Anal 15:251–264. https://doi.org/10.1006/jfca.2002.1058

Voerkelius S, Lorenz GD, Rummel S, Quétel CR, Heiss G, Baxter M, Brach-Papa C, Deters-Itzelsberger P, Hoelzl S, Hoogewerff J, Ponzevera E, Van Bocxstaele M, Ueckermann H (2010) Strontium isotopic signatures of natural mineral waters, the reference to a simple geological map and its potential for authentication of food. Food Chem 118:933–940. https://doi.org/10.1016/j.foodchem.2009.04.125

Chapter 16
Wines and Other Alcoholic Beverages

Contents

16.1 Introduction

Wine authenticity is the control of geographical provenance and is becoming an increasingly serious challenge in the current situation of the global wine market. Single parameters are often insufficient to determine product identity in relation to wine labelling. It has been shown that such control is improved by the application of multivariate chemometric methods based on a combination of stable isotopic data and other chemical components (Schlesier et al. 2009).

Multielemental analysis in the context of the European project "Establishing of a wine data bank for analytical parameters from Third Countries" has been performed to apply multivariate classification techniques for identification of the country of origin of wines. It was then possible to evaluate whether models built for authentic samples can be used to discriminate commercial samples (Smeyers-Verbeke et al. 2009).

Supplementary Information The online version contains supplementary material available at https://doi.org/10.1007/978-3-031-99687-0_16.

P. Szefer, *Chemometrics in Food Control Based on Elemental Analysis*,
https://doi.org/10.1007/978-3-031-99687-0_16

Multivariate discrimination and classification methods have been applied to characterize and identify the geographical origin of wines (Römisch et al. 2009).

Wines of declared geographic origin should be produced, processed and prepared in a selected region using traditional procedures. Geographical indications guaranteeing the protection of Italian wines include DOC (Denominazione di Origine Controllata) and DOCG (Denominazione di Origine Controllata e Garantita) (European Commission, European Union (EU) Agricultural Product. Quality Policy, Specific EU Quality Schemes—Guaranteeing Quality; Council Regulation (EC) No 510/2006 of 20 March 2006 on the protection of geographical indications and designations of origin for agricultural products and foodstuffs).

The key task is accurately and precisely determining the analytes responsible for distinguishing different types of wines according to their region of origin. For this purpose, the official method of isotope-ratio mass spectrometry (IRMS) has been established on the basis of the ratio of stable isotopes of essential elements, such as H, C and O (SIR), of wine components. The δ^2H, δ^{13}C and δ^{18}O values are dependent on different parameters, e.g., climatic factors as well as the photosynthetic cycle of the plant [6]. Therefore, they can be used to determine which wine-producing region a wine sample is from and detect adulteration, such as the addition of cane sugar, beet sugar or water. Moreover, the δ^2H and δ^{18}O values of plants are affected by their content in the local groundwater (Clark and Fritz 1997; van der Veer et al. 2009), as well as by the climatic conditions of the preharvest vintage, resulting in evapotranspiration in plants (Rossmann et al. 1999). Furthermore, the δ^{13}C ratio in plant material is influenced by different environmental and physiological parameters (O'Leary 1995; Scheidegger et al. 2000; Farquhar et al. 1982).

In addition to the analysis of SIR in wine samples by IRMS, multielemental compositions are frequently quantified using analytical methods/techniques, e.g., AAS (FAAS, ET (GF)–AAS, HG–AAS and CV–AAS), ICP–MS, ICP–AES and ICP–OES. To improve the efficiency of the obtained data, a combination of analytical tools are applied, e.g., SIRA with ICP–MS or SIRA with ICP–AES. Owing to this combined methodology, it was possible to achieve a more advanced assessment of the authentication/adulteration of wine and its products.

The validated analytical data were then processed chemometrically using univariate, bivariate and multivariate techniques.

Recent chemometric methods/techniques have been used to process original analytical data, enabling the assessment of wine quality with respect to its wine-producing origin. The application of multivariate analysis, in contrast to univariate analysis, can lead to large improvements, resulting in an increase in the number of applications of multivariate chemometrics in food authentication and traceability using SIR (e.g., Angerosa et al. 1999; Camin et al. 2007; Schellenberg et al. 2010; Bontempo et al. 2011; Camin et al. 2012), primarily in combination with other analytical approaches, assessing geographical origin (Ogrinc et al. 2001; Gremaud et al. 2004; Capron et al. 2007a) and fraud detection (Wachter et al. 2009).

Among the frequently applied multivariate techniques, PCA, FA, CA, HCA are more frequently used, and LDA, CDA, DA, PLS–DA and PLS are less frequently used.

16.2 Chemometric Assessment of Wine and Other Alcoholic Beverages

16.2.1 Geographical Differentiation

16.2.1.1 European Countries

Wine

Brescia et al. (2002) combined ICP–AES, HPIC, HPICE, NMR and routine official methods to process MEC and organic acid data with ^{1}H-NMR data processed by chemometric analysis (PCA, HCA and DA) to classify Italian red wines from three provinces. PCA, with the first two PCs explaining 40% of the total variance, was a useful technique for distinguishing wine samples between three geographical provinces (Central, Southern and Northern Apulia). DA was applied to the first nine PCs to characterize the red wines. They found that 100% of the training set samples were correctly assigned, and the validation set achieved 100% correctness. This means that heavy metals and amino acids are highly descriptive variables, enabling the differentiation of red wine samples originating from the central, southern and northern Apulian regions. The obtained data were significantly comparable to the HCA dendrogram data as well as the NMR dataset, indicating highly effective interspatial differentiation.

Galgano et al. (2008) used ICP–Q–MS data processed by multivariate chemometric tools such as LSD, CA, DA and CVA to characterize 18 varieties of commercial red wines (classified as DOC) originating from three regions in southern Italy. Among all the computational techniques, CVA (the first CV1 and CV2 accounting for 68.91% and 31.05% of the total variance, respectively) was the most effective approach for distinguishing wines on the basis of their geographical origin. The differentiation of samples was possible owing to descriptive variables such as Ag, B, Ca, Cd, Eu, Fe, Ga, Mn, Th, V and Zr for Basilicata; Al, B, Fe, Mg, Mn, Ti, Tl, Sc and Zn for Calabria; and Ba, Eu, K, I, Rb and Tl for the Campania region. On the basis of the additional use of CVA, REEs were effective identifiers of the interregional differentiation of the wine samples, with the first two CVs accounting for 99.94%, indicating the reliability of the tool for authenticity assessment.

Di Martino et al. (2013) classified samples of white and red wines originating from different municipalities of Italy during four vintages (2009, 2010, 2011 and 2012) on the basis of ICP–MS measurements. Among all the elements analysed, Sr, Ba and Rb were considered specific markers. The LDA technique was a reliable approach for differentiating the wines produced in the two municipalities, independent of the type of wine; i.e., an accuracy of 100% was achieved for the categorization of wines.

Perez Trujillo et al. (2011) estimated MEC data via FAAS and AES of 299 wine samples originating from seven regions of Canary Islands. With the first two PCs, PCA explained 40.60%of the total variance. As shown in Fig. 16.1, the samples from Fuerteventura are isolated the other samples, but the samples from La Gomera

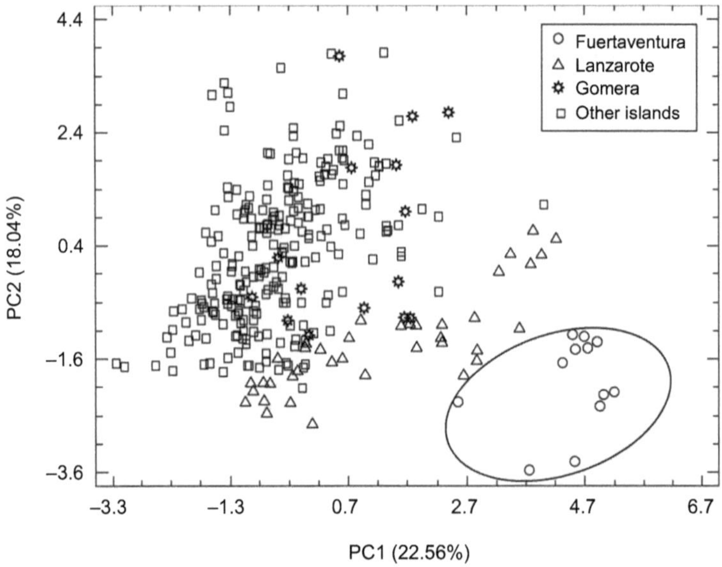

Fig. 16.1 Scores of the 299 samples of Canary Island wines on the first two principal components (40.6% of the total variance) taking into account the 11 features. Fuerteventura samples inside the circle. (After Pérez Trujillo JP et al. 2011. CyTA – J Food 9:135–140. https://doi.org/10.1080/19476337.2010.491582. With permission)

are dispersed in the central part of the plot. Among the the elements studied, PC1 significantly distinguished Li and Na, especially Fuerteventura wines and this differentiation may be explained by the effects of the marine spray.

Díaz et al. (2003) reported that FAAS and FAES data were transformed chemometrically to evaluate the MEC of red wines from four locations in the Canary Islands. Compared with the PCA data with the first five PCs (76.4% of the total variance), the application of Kohonen SOMs improved the ability to distinguish Canary wines with respect to interspatial differentiation. Similar results were achieved using LDA and BP–ANN. In the case of the former (Li, Sr, Mg, Mn, Ca, K, Fe and Zn), it was possible to achieve 94.0% recognition ability and 90.4% prediction ability. BP–ANN (Li, Sr, Mg and Mn) enabled the classification of wines according to geographical origin, with 95.2% and 94.0%, respectively.

Sun et al. (1997) aimed to estimate the elemental composition of wine samples collected from six German regions via ICP–OES. The obtained data were transformed by ANN, CA, PCA, BDA and FDA tools. Both PCA and CA were inadequate approaches for clearly distinguishing wine samples. On the basis of FDA, it was possible to classify wine samples with respect to category and geographical origin. The average discrimination rates for training and the corresponding prediction rates were 95.3% and 91.8%, respectively. ANN demonstrated a better prediction of 100% with the jackknife LOO procedure, achieving satisfactory classification of wine samples belonging to six regional categories. Both the BDA and FDA methods could also be applied with slightly lower distinguishing rates but shorter computation times than the ANN.

Thiel et al. (2004) investigated wine samples and grape-growing soils from four growing areas in Germany using ICP–MS and ICP–OES data. Both PCA and CA partially discriminated the wine samples relative to their production region origin. Among the three techniques of multivariate analysis, DA enabled them to distinguish the four groups of wine samples corresponding to four regions of origin. Elements such as As, B, Be, Cs, Li, Mg, Pb, Si, Sn, Sr, Ti, W and Y were the most indicative variables. The discrimination model enabled the application of a prediction rate with a CV of 88.6%.

Kallithraka et al. (2001) classified red and white wines of 15 varieties from various regions of Greece by employing both instrumental methods, i.e., ICP, GC and HPLC–DAD and sensory analysis. The application of PCA achieved satisfactory classifications for exclusively Greek red wines with respect to their geographical origin. Greek wines were clearly distinguished into two or three groups of wines from the Aegean/Ionean islands, North Greece and South Greece. However, the authors recommended further work with the use of many samples to make a detailed map of Greek wines in terms of their geographical origin and variety.

Dinca et al. (2016) analysed MEC combined with SIRA by means of EA–IRMS, CF–IRMS and ICP–MS of red and white wine samples from four Romanian wine-producing regions. Chemometric models (LDA, S-LDA and CDA) were useful for processing the original variable data. On the basis of SLDA, highly discriminative variables, i.e., Mn, Ni, Cu, Zn, Pb, V and $\delta^{18}O$, enabled the effective differentiation of wines by producing region, with total correct classification rates of 90.75% with a CV rate of 95.20% for the 2010 harvest and 98.7% with a CV rate of 94.07% for the 2009 harvest.

The aim of the study by Dembroszky et al. (2020) was to characterize the elemental profile of noncommercial traditional wines from three cultural regions of Eastern Europe (Romania) and to identify important chemical factors differentiating the wines in these producing regions. The ICP–OES method was a useful tool for analysing 21 elements. The wines contained low levels of elemental pollutants. Interregional differentiation was achieved via multivariate processing of the original dataset. PCA with the first three PCs explained 71.7% of the total variance. Among the analysed variables, Mg, Sr and Fe were the discriminative descriptors. LDA, which was based on these elements, exhibited 88.9% recognition ability and 87.0% prediction ability and was based on ten-fold cross validation (prediction abilities of three groups). The KNN method, like LDA, which was based on the same three elements, had 100% efficiency in both the recognition and prediction of the groups of red wines. It was concluded that the noncommercial wines from traditional cultural areas in Eastern Europe were of good quality in terms of their elemental profile. According to the authors, such monitoring may reveal significant changes in the properties of these wines because rural areas undergo significant social–ecological changes.

Šelih et al. (2014) considered numerous groups of elements quantified via ICP–MS and ICP–OES of wines originating from three Slovenian regions, including seven subregions. Categorization of white wines using PCA (with the first three PCs explaining less than 50% of the variance) did not achieve satisfactory differentiation of the wine samples, but the CPANN model clearly discriminated wine samples with respect to the main wine-growing regions, although some overlap was noted. Moreover, some partial regional categorization could be achieved for wines originating from Primorska. The top map of the selected CPANN model showed three incorrect predictions in each of three sets, i.e., 52/55, 9/12 and 9/12 of the correct predictions for TR, TE and validation test (VA), respectively. Validation success rates of 94.5% (TR) and 75% (TE) were achieved.

Klarić et al. (2011) evaluated the elemental distribution patterns of samples of commercial blackberry wine from three subregions of Croatia. The analytical data provided by the FAAS, FAES and GF–AAS methods were processed by ANOVA, PCA and LDA. The PCs (with the first seven PCs explaining 99.99% of the total variance) in the original dataset were used by LDA. This supervised learning model demonstrated satisfactory discrimination of wine samples corresponding to three subregions, with a sensitivity and specificity ranging between 80.00% and 83.33% and between 71.43% and 100.00%, respectively.

Brescia et al. (2003) reported that analysis of the MEC with three organic acids from wines from wine-growing regions in Italy and Slovenia enabled the assessment of their chemical distribution profile. Highly satisfying discrimination between wines from the two countries was achieved owing to the use of ICP–OES, NMR and HPICE–CD in combination with PCA and HCA. The multivariate classification of wines according to geographical origin was obtained with a nearly 100% degree of achievement. The models obtained from the application of RDA to the first five PC scores via both methodologies yielded 100% HPICE and ICP–OES results, as did the ^{1}H NMR dataset.

Karasinski et al. (2018) reported that MECs and different non-volatile organic compounds were quantified by QICP–MS and GC–TOF–MS in samples of red wine from four European countries. PCA, with the first two PCs describing 76% of the total data variation, discriminated the samples according to their regional origin. Moldovan products generally contained higher levels of elements than products from other countries did. Hungarian and Moldovan wines had relatively high contents of V, As, Al and Fe, whereas Be was related to Hungarian products. Wine samples from Poland were associated with Cd and Pb, but Bulgarian wines were identified by Tl, Ba, Rb and Mo descriptors. Moreover, U and Mn described wines from Bulgaria and Moldova. According to the authors, chemometric data analysis revealed that geographical factors considerably affected the metabolic profiles.

Beer

The contents of selected cations, anions and total polyphenols provided by the combination of ICP–AES, single-column IC/CD and UV–VIS spectrometry have been processed chemometrically to classify beer samples according to geographical origin. BS–LDA, which is based on the levels of Fe, K, P, PO_4^{3-} and total polyphenols, clearly differentiated the studied beer from the three production geographical countries, i.e., Germany, Portugal and Spain. Consistent results were obtained by the SVM model, which presents an overall prediction ability exceeding 99%, with an overall sensitivity and specificity of 99.3% and 99.5%, respectively (Alcázar et al. 2012a).

16.2.1.2 Asiatic Countries

Wine

Fan et al. (2018) explored nine grape varieties of red and white wines from different Chinese wine-producing regions. A triple method, i.e., ICP–MS, ICP–OES and IRMS, was an effective approach for determining the MEC and SIR profiles of wines. The supervised PLS–DA and SVM models distinguished wine samples originating from three regions, achieving average correct classification rates for the TR of 98% and TS of 95% and 97%, respectively. Manganese, K and Co obtained from the PLS–DA model were the most descriptive elements in the classification of Chinese wines with respect to their geographical origin. Furthermore, the average rate of correct classification for both joint models was >95%. Therefore, the combined data from the PLS–DA and SVM methods could be used to reliably verify the geographical origins of wine samples. The combination of $\delta^{18}O$ and elemental profiling with multistep multivariate analysis is a promising approach for the verification of the three geographical origins of Chinese wines.

Samples of rice wines originating from three regions of China were investigated by Shen et al. (2013) in terms of ICP–MS data combined with ANOVA, PCA, LDA and PLS–DA tools. The first two PCs clearly explained 45.14% of the total variance but preliminarily differentiated samples between Guyelongshan and Kuaijishan; therefore, LDA and PLS–DA were used to achieve improved classification. The samples were then distinguished from the three wineries with major contributions of Cu, Na, Mn, Mg, K, Mo, Ca and Ni as the most discriminative elements. LDA and PLS–DA achieved correct classification rates of 100% and 97.4%, respectively, in the prediction.

16.2.1.3 African Countries

Wine

Coetzee et al. (2014), on the basis of investigations of red and white wine samples from five wards in the Stellenbosch area in the Western Cape Wine Region (Africa), suggested that classification of wines is possible using reliable ICP–MS data transformed by a chemometric tool, i.e., CA and DA. PCA was exclusively used to select appropriate elements as discriminative variables (Fig. 16.2). DA differentiated the 17 regions into three groups corresponding to the results obtained by CA. Elements such as B, Ba, Cs, Cu, Mg, Rb, Sr, Tl and Zn were useful as suitable descriptors responsible for such interspatial differentiation. The three clusters identified by applying CA and DA corresponded to three particular wards and a set of soil types, namely, the first cluster included the Blaauwklippen/Helderberg wards, which included five estates; the second cluster included the Stellenbosch West ward, which included three estates; and the third cluster included the Simonsberg ward, which also included three estates (Fig. 16.3). All three clusters were correctly classified, corresponding to 95–100% of the original grouped cases (estates).

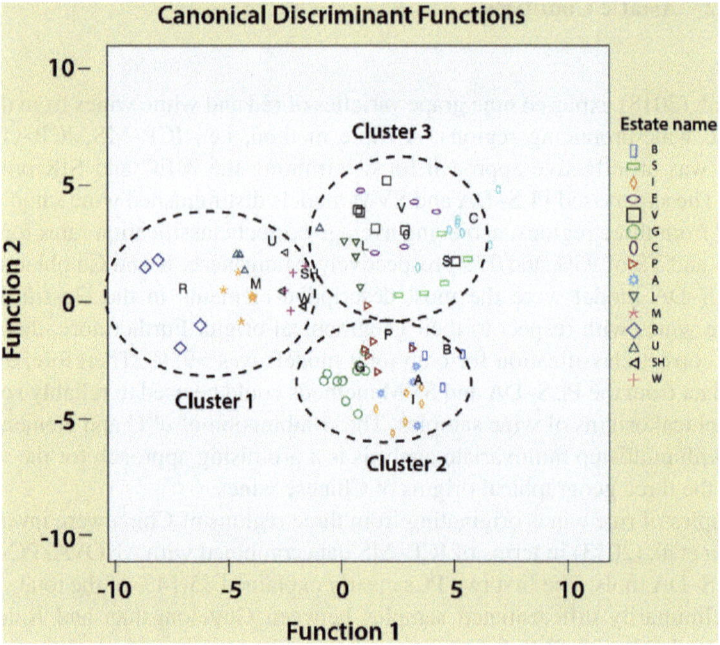

Fig. 16.2 Canonical discriminant function plot for red wines from 17 estates showing estates grouped into three clusters. (After Coetzee PP et al., 2014. Food Chem 164:485–492. https://doi.org/10.1016/j.foodchem.2014.05.027. With permission)

16.2.1.4 American Countries

Wine

Azcarate et al. (2015) explored white wine samples of the main varieties from four wine-growing regions of Argentina to evaluate the distribution pattern of ICP–MS data. PCA, with the first two PCs explaining 95.95% of the total variance and based on Ba, As, Pb, Mo and Co as descriptors differentiated the samples into four groups according to the evaluated regions. LDA included the same five variables used in PCA and made it possible to clearly differentiate four Argentinean wine-growing regions (Fig. 16.4). Discrimination rates exceeding 96% for the prediction and validation datasets were achieved. Espinoza Cruz et al. (2020) used MIP–OES data from samples of red wine from different regions in Mexico to identify their geographical provenance. PCA, with the first two PCs accounting for 74% of the total data variance, enabled the differentiation of three winery regions, i.e., Guanajuato, Valle de Guadalupe and Aguascalientes, corresponding to the most descriptive variables, namely, Rb–K–Ca–Sr, Na–Cu–Mn and Fe–Al, respectively; however, a clear clustering of variables was not achieved. According to the authors, to improve the discrimination model, additional samples possibly collected from other Mexican regions must be investigated.

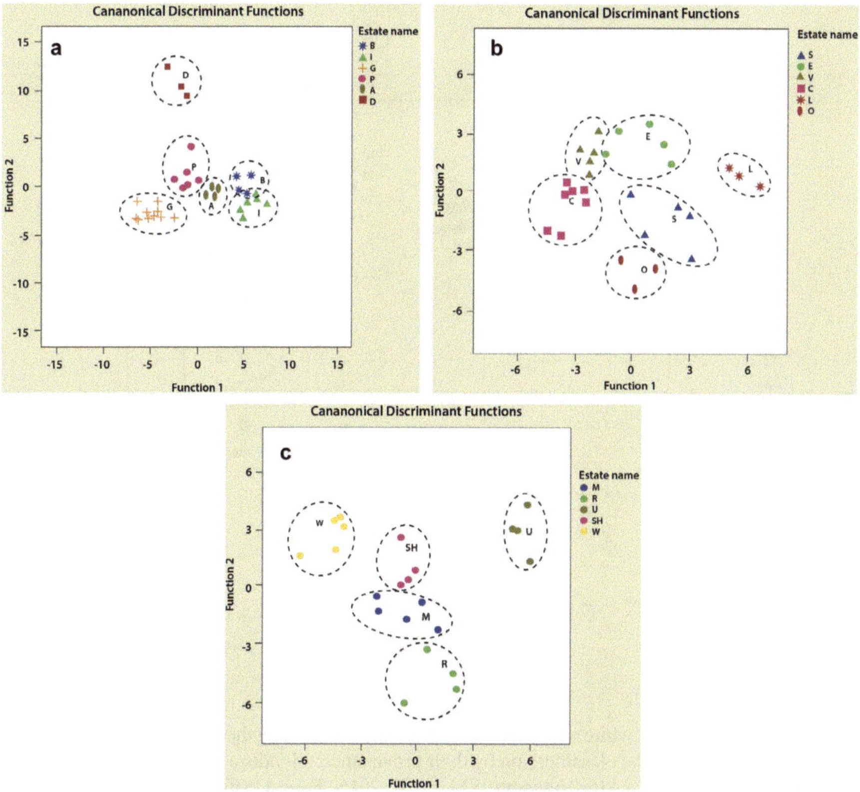

Fig. 16.3 Canonical discriminant function plot for red wines from 17 estates showing estates reported in (**a**) Cluster 1, (**b**) Cluster 2 and (**c**) Cluster 3. (After Coetzee PP et al., 2014. Food Chem 164:485–492. https://doi.org/10.1016/j.foodchem.2014.05.027. With permission)

Boschetti et al. (2013) characterized wine samples from eight regions within four macroregions of Brazil via combined HR–CS, FAAS and ICP–OES methods. PCA, with the three PCs explaining 85.75% of the total variance, showed that among all eight sample groups, six were clearly distinguished from each other except for two regions that had very similar mineral profiles. The HCA data also clearly distinguished different groups with high similarity indices, confirming the obtained PCA data. The most effective descriptors, such as K, Mn, Rb and Sr, were the discriminators for the geographical origin of the wines.

Saavedra et al. (2011) used a combination of MEC with nonacylated anthocyanin content (determined by HPLC/UV–VIS, ICP–OES, UV–VIS spectroscopy and HG–AFS) to classify wine samples from different geographical locations of six wineries of two valleys (Chile). The obtained data were transformed by means of multivariate tools, i.e., PCA, LDA, SIMCA and PLS–DA. PCA, with the first two PCs explaining 91.7% of the total variance, enabled the differentiation of three

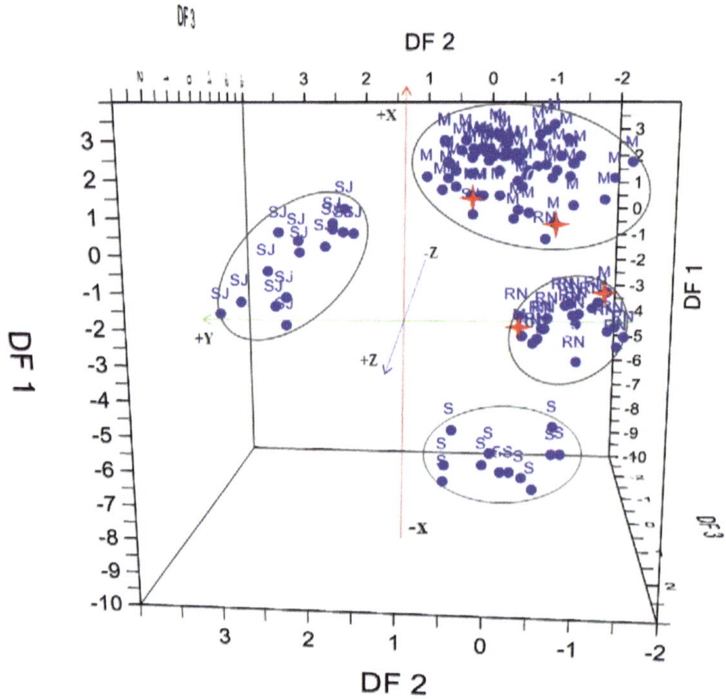

Fig. 16.4 Distribution of white wine samples in the space defined by the first three canonical functions of LDA, depicting the classification by their provenance: Mendoza (M), RíoNegro (RN), San Juan (SJ), and Salta (S). (After Azcarate SM et al., 2015. Food Contr 57:268–274. https://doi.org/10.1016/j.foodcont.2015.04.025. With permission)

sample groups corresponding to two subzones in Casablanca Valley and a third one related to Loncomilla Valley. For such interspatial differentiation, four nonacylated anthocyanins, Na, Mg, Mn, As and Cu, are responsible. It was concluded that on the basis of the LDA, SIMCA and PLS–DA results, these three regional groups were more clearly differentiated and correctly classified. According to the authors, owing to the simplicity and powerful effect of the methodology used, SIMCA is recommended for further routine exploration.

Numerous chemical elements were analysed by Bentlin et al. (2011) in red wines from four wine-producing countries in South America, i.e., Argentina, Brazil, Chile and Uruguay. PCA, with the first two PCs explaining 80.26% of the total variance, indicated an interdependent ability to differentiate geographical origin. Comparable results were obtained by CA and CD. As shown in Fig. 16.5, CD clearly discriminated the wine samples according to their country of origin. The wines could be classified according to their country of origin, regardless of the type of grape with discriminant elements such as Tl, U, Li and Rb and Mg.

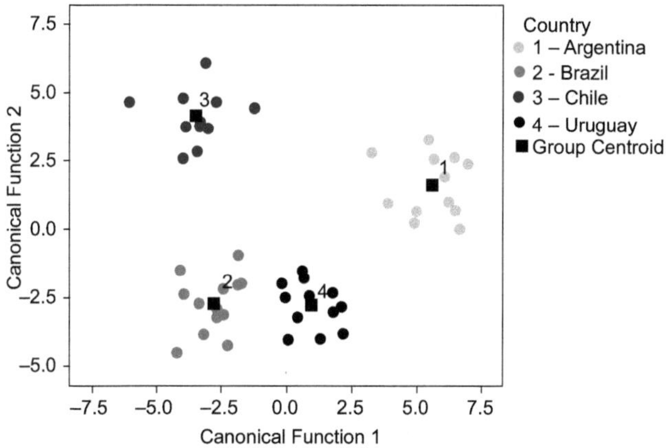

Fig. 16.5 Scatter plot of the unstandardized canonical discriminant functions 1 and 2 describing the countries of red wines. (49) origin. (After Bentlin FRS et al., 2011. J Braz Chem Soc 22:327–336. https://doi.org/10.1590/S0103-50532011000200019. With permission)

16.2.1.5 Oceanic Countries

Wine

ICP–MS data were used to characterize red wine samples from different major wine-growing regions in New Zealand. At a broad regional scale, it was possible to classify samples by LDA using Rb, Sr, Ni, Fe, Mn and Pb, with a classification accuracy of 94.4%, which was achieved when regions were combined into three major regions, i.e., North Island, Nelson and other South Island regions. However, at the finer regional scale, i.e., for seven regions, an overall accuracy of 80.8% was obtained for the LOOCV when parameters such as Sr, Rb, Ni, Co, Pb, Mn, Cd, Ga and Cs were used. In conclusion, despite the slight influence of grape variety on the mineral profile, it was possible to clearly distinguish the geographic provenance of the wines studied (Angus et al. 2006).

16.2.1.6 Countries Across Multiple Continents

Wine

Coetzee and Vanhaecke (2005) reported $^{11}B/^{10}B$ ratios determined by ICP–QMS in wines from different producing regions in three countries. The average $^{11}B/^{10}B$ ratios in red wines from selected regions of South Africa, France and Italy showed significant variability, ranging from 0.5% to 1.5% in the respective groups. This means that the isotope ratio is an advantageous tool for the identification of inter-spatial origin.

A combination of MEC, SIR and organic compound data was obtained by Capron et al. (2007a) for wine samples from four countries. The original data were processed chemometrically via various multivariate tools, i.e., PCA, CART, PLS–DA, UVE–PLS and PLS–UVE–SEL. PCA revealed that the first three PCs, accounting for 46% and 43% of the obtained data, corresponded to authentic and commercial wines, respectively. It was possible to discriminate between South African, Romanian and Czech samples, while the Hungarian group overlapped with the Czech group and, to a lesser extent, with the Romanian cluster. However, when PLS–DA was used, three models were developed, i.e., Hungary vs. Czech Republic + Romania, Czech Republic vs. Hungary + Romania and Romania vs. Hungary + Czech Republic. The models were used for authentic data as well as commercial data. Moreover, classical PLS and UVE–PLS models were compared. Although such PLS discriminating models with the use of a complete dataset of parameters have achieved very successful distinctions almost all the time, the use of UVE–PLS followed by UVE–PLS–SEL has been recommended, especially for commercial samples. Thus, very significant classification rates can be obtained on the basis of only a few parameters, i.e., stable isotopic ratios or trace element contents.

According to Römisch et al. (2009), IRMS, HR–ICP–MS, ^{2}H-NMR, HPLC, GC and wet chemical and enzymatic analyses were used to classify 1188 wine samples. Using CART, samples of South African wines were clearly differentiated from those of East European wines because only one isotopic parameter was used, although less satisfactory results for the differentiation of East European wines were achieved. Tools such as RDA, PLS–DA and PLS–UVE result in discriminant models, enabling the correct classification of wines from East European countries with rates between 88% and 100%. Comparisons of RDA and PLS–DA, as well as RDA models, include somewhat fewer variables than PLS–DA models do, since PLS–DA is limited to two-group comparisons, i.e., one country against the other countries. The discrimination of wines on the basis of preferred Models (M1) was achieved.

According to Capron et al. (2007b), the discrimination of South African wines is uncomplicated for both 1188 authentic and commercial samples, in contrast to the classification of authentic wines originating from three Eastern European countries, i.e., Hungary, the Czech Republic and Romania. However, this inconvenience can be overcome by using kernel-SVM models based on only six parameters, i.e., V, Cr, Cd, As, U and δ^{2}H, with a correct classification rate of generally $\geq$95% (Capron et al. (2007b)).

Forina et al. (2009) noted that a dataset consisting of 1188 wine samples from four countries was analysed for 63 chemical variables processed by UNEQ, SIMCA, MRM, PF and CV tools. The studies were performed within the European project WINES-DB "establishing of a wine databank for analytical parameters from third countries". Models of South African wines can be achieved via the use of only 2–3 variables (stable isotope ratios). However, commercial white wines from Eastern European countries at 3 years of vintage for each country required approximately 25 variables with efficiency that was not highly acceptable, especially in the case of Czech wines. Consequently, South African and European wines were modelled separately using both the UNEQ and MRM models, which achieved 100% forced model efficiency and excellent CV sensitivity of >99% with only ethanol$(D/H)_1$, ethanol$(D/H)_2$ and Li variables. In the case of geographical classification of wines, the efficiency achieved with UNEQ modelling was the most effective technique when the CV efficiency was considered. Moreover, the most reliable results were obtained by SIMCA and MRM, although MRM was more effective than SIMCA.

Serapinas et al. (2008) reported that the combination of ICP–MS data with chemometric processing analysis (ANOVA and PCA) was a helpful approach for classifying different wines originating from seven countries. PCA enabled them to discriminate between wine samples from Spain, Bulgaria and Hungary (Fig. 16.6); Chile and the USA (California); Italy and France; and Chile, Spain and France (Fig. 16.7). It was possible to select three or four descriptors, e.g., Li, Rb, Sr and Zn, which are often useful for effective discrimination between countries at the 0.9 probability level.

Sass-Kiss et al. (2008) performed analytical measurements of MEC, amine and organic acid contents by FAAS, ET–AAS, HPLC–FD and FP of Tokaji aszú wines and foreign botrytised wines from nine countries using multivariate chemometric techniques, i.e., PCA and LDA. The first two PCs explained 51.1% of the total variance, indicating that the differentiation of Tokaji aszú wines and foreign wines was insignificant when variables such as Ca, Cd, Cu, Fe, K, Mg, Mn and Zn were used,

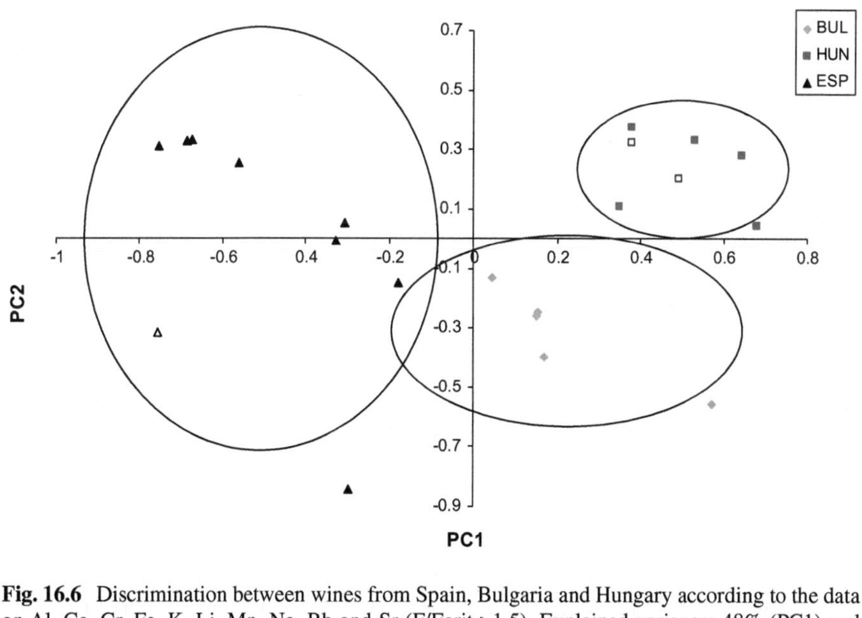

Fig. 16.6 Discrimination between wines from Spain, Bulgaria and Hungary according to the data on Al, Ca, Cr, Fe, K, Li, Mn, Na, Rb and Sr (F/Fcrit >1.5). Explained variance: 48% (PC1) and 23% (PC2). The absolute loadings of K to the principal components were 0.4 and 0.7, those of Al were 0.5 and 0.4, and those of the other elements were less than 0.4. (After Serapinas P et al., 2008. Food Chem 107:1652–1660. https://doi.org/10.1016/j.foodchem.2007.09.003. With permission)

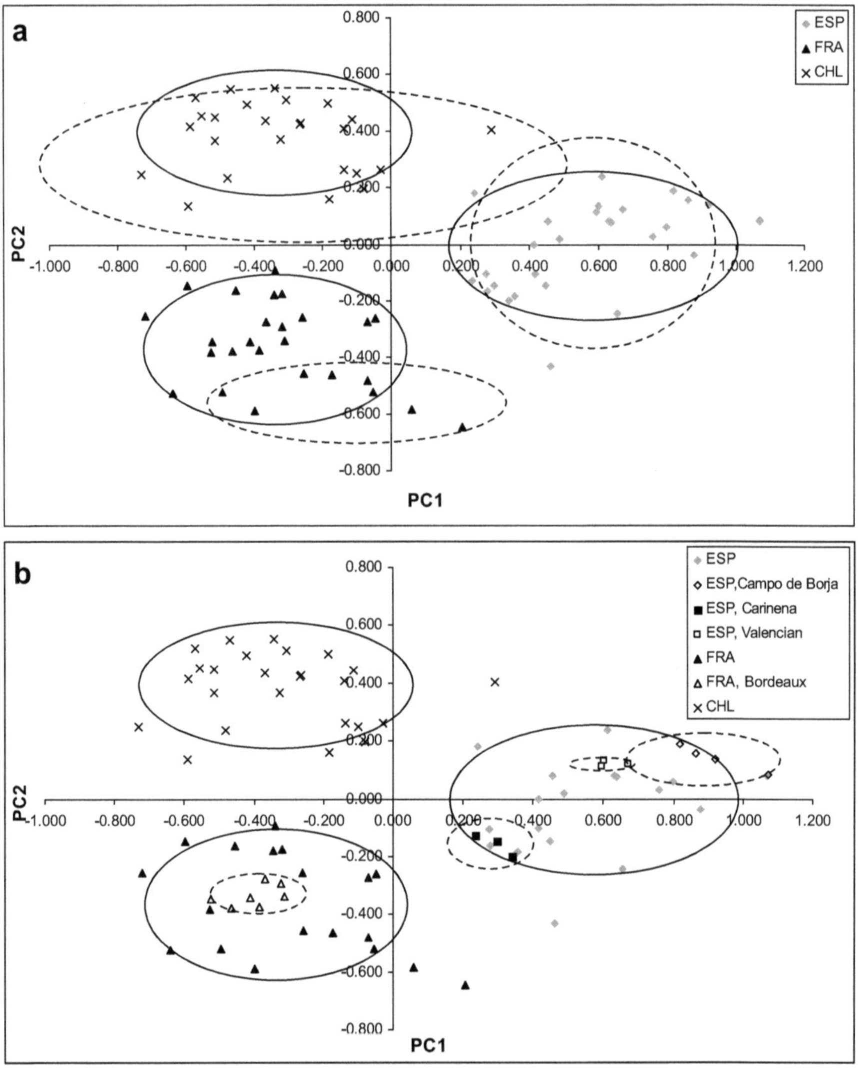

Fig. 16.7 Two principal component patterns for wines from Chile, Spain and France (solid line) measured in both experiments (Li, Rb, Sr and Zn, F/Fcrit >11) and some regions (dotted line). Explained variance: 59% (PC1) and 27% (PC2). The absolute loadings of the elements to the principal components ranged from 0.1 to 0.8. (After Serapinas P et al., 2008. Food Chem 107:1652–1660. https://doi.org/10.1016/j.foodchem.2007.09.003. With permission)

although a clear trend towards differentiation between Hungarian and foreign wine samples was achieved. The contents of acids and elements can support the amine analysis results in the identification of wine origin.

Other Alcoholic Beverages

Styburski et al. (2018) determined the K, Ca, Cl and P contents of various types of bottled beer imported to Poland from 11 countries in Europe, Asia and North

America via EDXRF. The significant differences in the Ca, K and Cl levels in the studied beers depended on their country of origin. The highest level of Ca was detected in German and Armenian beers, whereas high levels of Cl were detected in Ukrainian and Portuguese beers. There were no statistically significant differences in P content between most of the countries of origin. According to the authors, the differences in the elemental profiles of various types of beer may be a result of the type of water, the resources utilized for beer production, the kind of yeast and the technological methods used during production.

Bong et al. (2016) combined MEC data with SIR data provided by IRMS, ICP–AES and ICP–MS to estimate the chemical profiles of bottled beers available in the South Korean market originating from 27 countries. Elemental and isotopic compositions enabled the clear discrimination of diverse beers according to their geographical provenance. The PCA data with the first two factors explained 77.30% of the total variance and clearly differentiated the samples according to geographical origin on a continental scale, i.e., Europe, Asia, America and Oceania (Fig. 16.8). Moreover, the LDA results showed that 98.9% of the samples were effectively discriminated according to origin. Elements such as K, Ca, Mg, P, Fe, S, Li, Cr, Ni and Sr, especially $\delta^{18}O$ and $^{87}Sr/^{86}Sr$, were significant factors for distinguishing beers in terms of their interregional distributions. Water, as one of the main components of beer, significantly influences the isotopic composition of beer because isotope ratios directly reflect both geographical and geological characteristics.

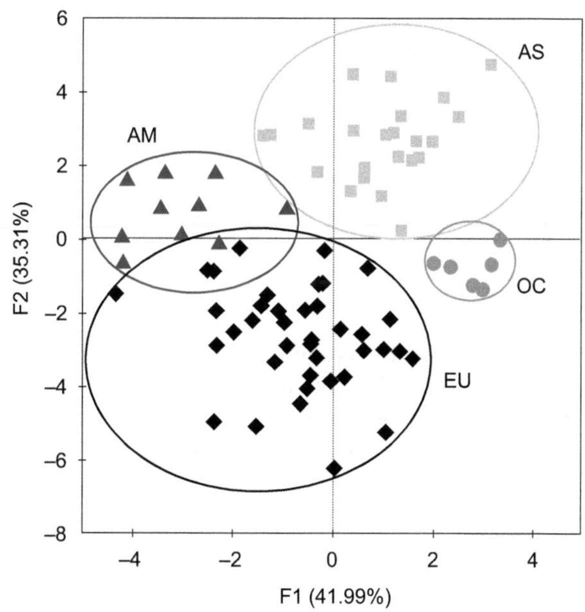

Fig. 16.8 Discriminant scatter plot showing relatively good discrimination of the geographical origin of beers according to their continent of production. Note the clear discrimination among AS, EU and OC. (After Bong YS et al., 2016. Food Contr 60:378–381. https://doi.org/10.1016/j.foodcont.2015.08.017. With permission)

16.2.2 Intervarietal/Type Differentiation

Wine

Pořízka et al. (2018) explored three popular varieties of Moravian white wines from two wine-producing regions in the Czech Republic for the quantification of numerous groups of elements determined by ICP–MS and ICP–OES to classify wines with respect to grapevine variety. PCA (with the first two PCs corresponding to 65.66% of the total variance) and CDA, which are based on variables such as Al, Sn, Gd, Tb, Tm/Yb, Yb/Lu, Mo/Sn and Mn/Cr, enabled the classification of three known varietal wines. CDA clearly demonstrated varietal differentiation. This technique provided the distribution of wines into varietal groups at a success rate of 95.83% (discrimination efficiency) and a predictive ability of the model (finally tested by CV) with a classification effectiveness for unknown samples of 70.83%.

Ruggieri et al. (2020) used the ICP–OES method to analyse MEC coupled with chemometric processing, which is a useful approach for classifying three cultivars of Italian wine. LDA and PLS–LDA were applied to establish parameters for the discrimination of wines. PLS–LDA provided the best performance in contrast to stepwise LDA. The proposed model achieved a total classification ability in calibration of 92% for the three wine cultivars, and with the external validation set, the classification ability in prediction was 80%. The characterization of white wine based on a combination of elemental profiles and chemometric tools may be an effective approach for the intercultivar differentiation of the investigated samples.

Moret et al. (1994), using different instrumental techniques, such as FES, AAS, LLE–CGC and GC–MS, including classical analytical procedures, studied five types of Venetian white wines from two regions of Italy between 1977 and 1984 with respect to Na, K, Ca, Mg, Li, Cl, P, N and 19 other organic compounds as analytical variables. Interannual differentiation was considered. Univariate and multivariate tools were used to classify the wine varieties that achieved satisfactory CVA results, with an overall correct classification of approximately 90%. Many discriminative descriptors, such as cis-3-hexen-1-01, l-hexanol, K, N compounds and total phenols, were identified. However, PCA and its derived method, SIMCA, provided less effective differentiation of the group structure.

Other Alcoholic Beverages

According to Zambrzycka-Szelewa et al. (2020), HR CS and FAAS data enabled the evaluation of elemental profiles, including other parameters such as antioxidant activity, in different beer types and brands in Poland. Multivariate techniques such as FAMD and HCPC were helpful tools for identifying the relationships among the type, colour, refractive index, antioxidant activity, extraction, alcohol content, essential elemental content and total phenolic content. For example, HCPC enabled the clustering of beer samples into five main groups, establishing the similarities/dissimilarities of the elements studied. The achieved sample discrimination enabled the identification of different types and brands of beer. It was concluded that the developed method was a powerful approach for the direct analysis of beers to assess their quality and safety.

Alcázar et al. (2002) integrated ICP–AES data with multivariate preprocessing analysis to classify three types of beer, i.e., lager, dark and low alcoholic content beer. PCA with the first two PCs explaining 36.53% of the total variance enabled the observation of the trends in discrimination of the three types of beer studied. To recognize the observed trends, supervised learning methods such as LDA and ANNs were used. The most discriminative results were achieved when the PR method based on BP–MLP was used, with an average prediction ability of approximately 92%.

Alcázar et al. (2012b) measured the contents of Cl^-, PO_4^{3-} and $SO_4^{2=}$ with total polyphenols and amino acids by IC/CD, including classical analytical procedures, in three types of blonde beers. The results obtained were processed by PCA and LDA. The first three PCs, accounting for 61.6% of the total variance, incompletely differentiated three types of blonde beers. However, LDA was a more reliable approach than PCA. The constructed LDA model clearly discriminated lager, pilsner and low-alcohol beers with satisfactory overall prediction ability, i.e., a sensitivity of 92.7% and a specificity of 96.3%.

Using ICP–MS, Alexa et al. (2018) assessed the elemental profiles of four types of beers produced by different breweries in five European countries. A combination of MEA with chemometric processing by ANOVA, LSD and LDA indicated that barley and wheat beers could be differentiated according to their Mn and Cu contents. According to the authors, to separate all the groups, further investigations with more samples are needed. LDA enabled the differentiation of four types and subtypes of beer, i.e., pale barley, dark barley, pale wheat and dark wheat beers, to be verified. Overall, 75.0% of the cross-validated grouped cases were correctly classified.

Bellido-Milla et al. (2000), on the basis of FAAS, FAES and MAS, interpreted the MEC data obtained from molecular absorption spectra subjected to chemometric processing to classify four types of beer. ANOVA enabled the differentiation of canned and bottled beers. LDA was a useful tool for correctly classifying each type of beer, i.e., lager, stout, wheat beer and ale, although in the case of stout and ale, some overlap was observed (Fig. 16.9). Moreover, LDA satisfactorily classified the beer samples relative to their raw materials, i.e., barley, barley and other cereals and wheat (Fig. 16.10).

Sakellari et al. (2017) reported that DPASV and ICP–MS methods provided MEC data with Cu complexing ligands (LT) and other parameters of six types of beer bottled in glass or PET bottles commercially available locally in Greece. Importantly, the physicochemical properties of the organic material, surface-active substances and catalytically active compounds were measured for the first time in beers by electrochemical techniques. The LT values of the beers analysed were not exceeded by the total Cu level, which means that the Cu content was complex. The PCA data indicated that the first two PCs explained up to 63.3% of the total variance. Statistically insignificant differences between the elemental contents of particular beer types were detected. However, ale-type beers were likely responsible for the dispersion of object scores with respect to lager and pilsner scores. This was

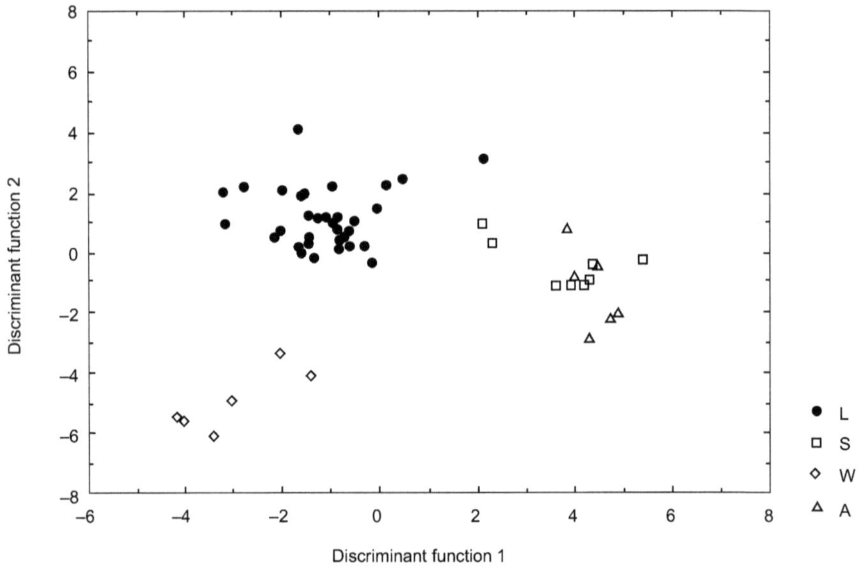

Fig. 16.9 Plot of the first discriminant functions vs. the second discriminant functions for each type of beer: L, lager; S, stout; W, wheat beer; A, ale. (After Bellido-Milla D et al., 2000. Spectrochim Acta Part B: At Spectrosc 55:855–864. https://doi.org/10.1016/S0584-8547(00)00164-6. With permission)

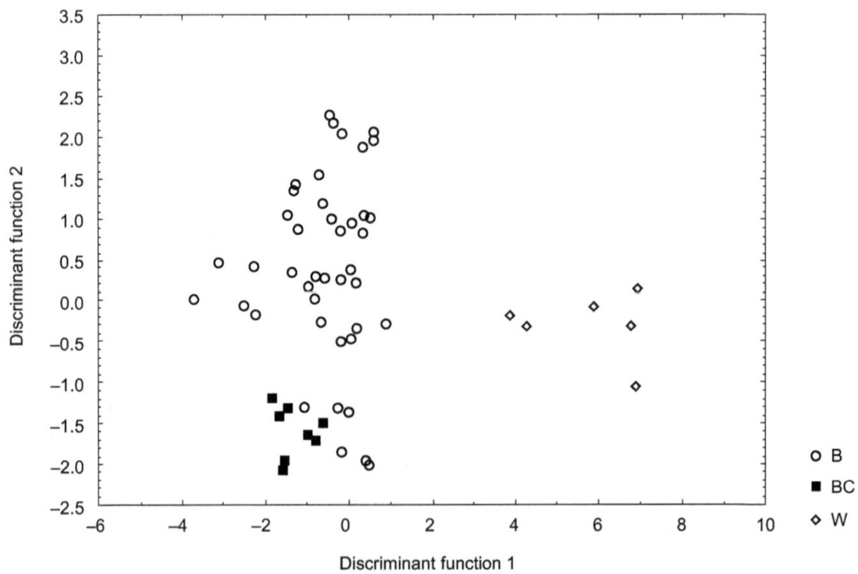

Fig. 16.10 Plot of the first discriminant functions vs. the second for raw materials: B, barley; BC, barley plus cereals; W, wheat. (After Bellido-Milla D et al., 2000. Spectrochim Acta Part B: At Spectrosc 55:855–864. https://doi.org/10.1016/S0584-8547(00)00164-6. With permission)

probably due to the expansive variety of ale beers consisting of brown, pale ale, India pale ale and golden Scotch and therefore to the ingredients used.

Mahmood et al. (2012) determined the MEC content via ICP–MS in commercial bottled and canned beer samples via PCA and ROC analysis. PCA, with the first three PCs explaining 52.1% of the total variance, discriminated the B27 and B38 samples with relatively high loading values of As, Cs and U, which were well differentiated from the other beers. Importantly, these object samples originated from the same company (different brands, however) and country of origin. In addition, the AUC ROC curve shows that the identification of some errors may be expected (AUC ROC <1), and the correlation coefficient acts as an excellent classifier metric for the beers. This means that all beer samples were discriminated via this methodology. ROC analysis enabled the correct identification of individual beer samples by the magnitude of their correlation coefficient relative to the other beers with a low inaccurate positive rate. It was concluded that the elemental fingerprint methodology is suitable for sample discrimination.

Cardoso et al. (2004) used a combination of ICP–AES, GC–MS, GC–FID and HPLC–UV–vis/PDA methods to quantify the content of MECs and selected different organic compounds in samples of cachaça and rum from nine countries in South America, North America, the Caribbean Islands and Europe. PCA, with the first three PCs accounting for 77.7% of the total variance, enabled a clear differentiation between cachaça and rum based on the use of protocatechuic acid, propanol, isobutanol, isopentanol, Cu, Mn and Mg as effective descriptors. Analyses of these seven chemical descriptors by PCA, HCA, KNN and PLS–DA were highly appropriate for distinguishing cachaça and rum. KNN using three nearest neighbours and one-factor model PLS–DA could distinguish these two types of products with 100% certainty. This procedure can be recommended as a routine methodology for verification of authenticity/adulteration.

16.2.3 Interregional and Intertype/Brand Differentiation

Wine

According to Baxter et al. (1997), different types of wine samples from different regions in two European countries, England and Spain, were investigated by FI–ICP–MS. The obtained analytical data were processed chemometrically using CDA. Complete interspatial differentiation was possible owing to the use of seven elements as variables. Classification success rates of 94.7%, 100% and 86.4% for Spanish wine groups originating from the three producing regions were achieved. Furthermore, a clear differentiation of wines originating from England and Spain was achieved by using 48 elements with GenStat via a plot of the squared distance of each individual white wine from the mean of the English group against the squared distance of each individual white wine from the mean of the Spanish group. They found that 95% of the data were classified correctly. Moreover, a clear classification of Spanish wines according to their type, i.e., rose, white and red, was also possible by means of SPSS.

Šperková and Suchánek (2005) processed chemometrically (ANOVA, MANOVA, PCA and DA) multielemental data determined by ICP–MS and ICP–OES from wine samples from the four most important Bohemian wine regions (Czech Republic) to confirm their authenticity. PCA with the first three PCs explained more than 60% and more than 80% of the total variance for white and red wines, respectively. DA clearly differentiated the wine samples into two types. The most descriptive variables were as follows: Al, Ba, Ca, Co, K, Li, Mg, Mn, Mo, Rb, Sr, V, Sr/Ba, Sr/Ca and Sr/Mg. Correct classifications of 100% and 97.4% were achieved for red and white wines, respectively.

Tian et al. (2017) used LIBS and ICP–OES data to classify red or white wines from five Bordeaux production regions (France). The recorded spectra were then analysed with PCA and RF tools for the differentiation of wines according to their production regions. As shown in Fig. 16.11, PCA (with the first two PCs corresponding to 56.2% of the total variance) was useful for differentiating the white and red wines. Moreover, this unsupervised technique with the first two PCs (44.6% and 14.1%, respectively) partially discriminated wines from three wines instead of five Bordeaux regions differentiated from California wines (Fig. 16.12). Therefore, a supervised classification model based on RF was developed. By optimizing the model parameters, satisfactory classification results were achieved in the best cases, and the classification accuracy was 100% for all the tested wines. Five French producing regions were clearly differentiated from California producing regions as outliers with respect to the tested French wines. It was concluded that LIBS combined

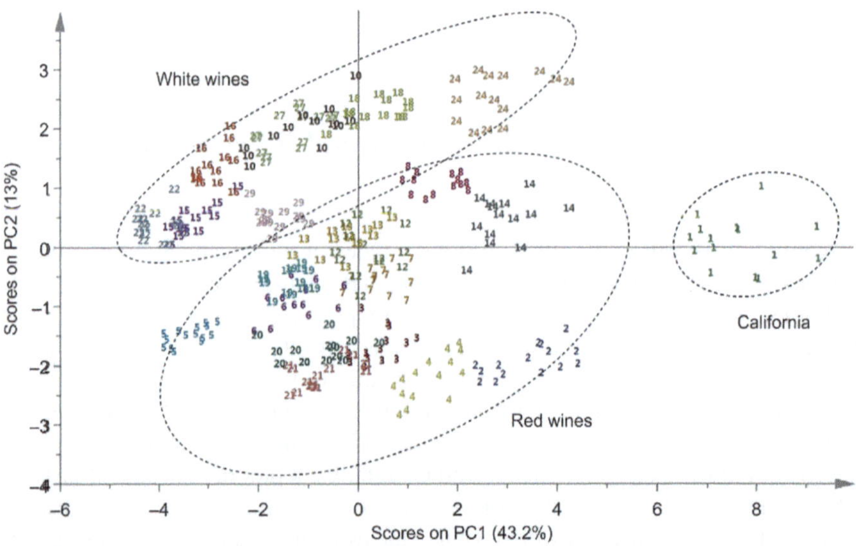

Fig. 16.11 PCA score plot of red and white wines with a plasma temperature of approximately 8000 K. (After Tian et al., 2017. Spectrochim Acta B 135:91–101. https://doi.org/10.1016/j.sab.2017.07.003. With permission)

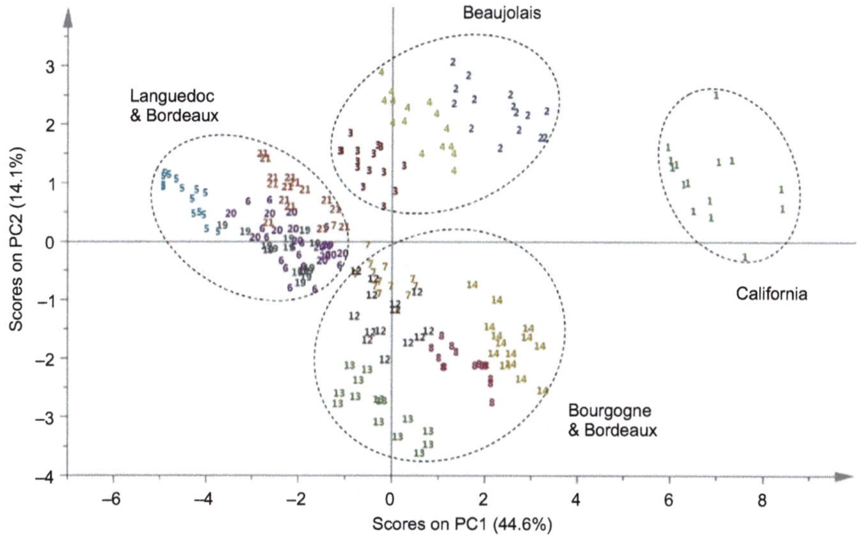

Fig. 16.12 PCA score plot of red wines from the five regions. (After Tian et al., 2017. Spectrochim Acta B 135:91–101. https://doi.org/10.1016/j.sab.2017.07.003. With permission)

with a chemometric tool enables the classification of wines according to their inter-spatial and intertype differentiation.

Ivanova-Petropulos et al. (2013) characterized samples of white, rose and red wines from different wine regions of Macedonia with respect to numerous groups of major, minor and trace elements quantified by ICP–OES and ICP–MS. FA, with the first two factors explaining 46.36% of the total variance, clearly differentiated the samples according to wine type, i.e., white vs. red. For example, REEs, Ce, La, Mo, Zr, V, Be, Al and Pb. B, P, Mg and Mn were highly discriminative descriptors for the red wines located in the positive part of F2. However, Ba, Ca, Cu, P, Na and S were the elements responsible for the differentiation of the wines according to type (white vs. red) and geographical origin. FA and CA were able to distinguish the wines in terms of their interspatial and intertype variability.

Košir et al. (2001) assayed natural and enriched wines originating from three wine-producing regions in Slovenia with respect to $\delta^{13}C$ and $\delta^{2}H$ data, including ^{2}H NMR spectra determined by IRMS and SNIF–NMR methods. PCA based on the first two PCs, accounting for 88.61% of the total variance, enabled them to obtain a satisfactory differentiation of samples originating from the coastal region and those originating from two continental regions. However, PCA, which provided a scatterplot of PC1 vs. PC2 (93.82%), demonstrated good discrimination between natural and enriched wines from the coastal region. Unfortunately, insufficient satisfactory discrimination between wine samples from the two continental regions (Sava, Drava) was obtained because of insignificant climatic differentiation between these two continental regions. However, better differentiation between natural wines and wines enriched with beet sugar from the Drava region was possible owing to the use of the KANN method, although in all cases, CA had poorer efficiency than PCA and KANN did.

Kokkinofta et al. (2014) combined SNIF–NMR, IRMS and ICPS to investigate wines from three Cyprus regions. The original obtained results were processed chemometrically via PCA, CA and RDA. It was reported that analytes such as H, C, O, $\delta^{18}O$, $\delta^{2}H$, Ni and Pb were the discriminative variables. PCA based on SIR achieved correct discrimination and enabled partial differentiation of the wine samples from the three regions, with the exception of the other regions. The RDA technique provided cross-validated data with 80.4% and 98.0% correct recognition for white and red wines, respectively.

Kokkinofta et al. (2017) demonstrated that the original variables provided by combining SNIF–NMR, IRMS and ICP–AES for Cypriot wines originating from two regions, i.e., Pafos and Limassol, were processed chemometrically using SIMCA, OPLS–DA, PLS and CV–ANOVA. PCA, with the first two PCs explaining 56.0% of the total variance, confirmed the importance of grape variety and geographical origin. It was possible to differentiate the Pafos region (attributable to Cu, Fe and $\delta^{13}C$) from the Limassol region varieties (corresponding to $(D/H)_{II}$, R and $\delta^{18}O$). Using the OPLS–DA model, all the studied wine varieties were successfully differentiated. For example, it was possible to obtain a clear discrimination between the Shiraz and Maratheftiko varieties. Moreover, distinguishing between indigenous and foreign wine varieties was also confirmed. The OPLS–DA models provided very high values related to their accuracy (>94%), sensitivity (>90%) and specificity (>88%), indicating their predictability.

According to Marengo and Aceto (2003), the multielement content of red wines from five Italian production zones was determined via ICP–MS using PCA and SLDA tools. The latter contained five variables, i.e., Mn, Mg, Mo, Si and Ti, with the NER% of this final model estimated to be 70.6%. Scatter plots of the three most discriminative elements, i.e., Mo, Ti and Mg, distinguished between the five classes of wines. LDA based on these metals enabled the differentiation of the five classes of wines with highly variable sensitivity and specificity ranging between 45.5% and 84.6% and 38.5% and 91.7%, respectively. However, three of four misclassifications were caused by Nebbiolo d'Alba, which were also poorly characterized by PCA. This confusion could be explained by the fact that the region of production of this wine type is located exactly in the middle of the other regions. Therefore, it is reasonable that both the low sensitivity and specificity for this class are unsatisfactory. The methodology applied was able to satisfactorily discriminate two classes of high-quality wines from three less fine wine classes.

Pérez-Trujillo et al. (2003) used ICP–MS to quantify all the lanthanides, enabling the differentiation of red wines from white and rosé wines, although the REEs were not suitable for classifying wines according to the island of origin. FA, with the first two F values accounting for 36.0% of the total variance, differentiated the three types of wine studied, whereas F3, F4 and F5 (explaining 30.8% of the total variance) enabled the grouping of wines according to geographical origin. Moreover, good recognition and prediction abilities were achieved when both LDA (96.7%

and 95.4%, respectively) and BP–ANN (100% and 99.3%, respectively) were used to distinguish wines with respect to their geographical provenance.

Pérez-Trujillo et al. (2002) systematized white, rosé and red wines from four Canarian Islands using ICP–MS data. It was possible to classify the studied wine samples using the LSD as a criterion of comparison at the 95% confidence level. Elements such as Te, Pt and Ti were distinguished by significant variations in their levels among the three types of wine. However, the identification of intervintage variations was possible owing to specific elements such as Te, Pt, Au, Co, Zn, Re, Tl and Sb. This result could be explained by the effects of climate or vine handling on the elemental contents of wine. Intertype discrimination was also observed, which could have resulted from the differences in the elaboration processes. Furthermore, interspatial variability was detected for four4 Canarian Islands, which may be attributable to specific geochemical features of the soil on the various islands.

Kruzlicova et al. (2013) processed nine wine brand samples from three wine-producing regions in Croatia to determine their elemental profiles by HR–ICP–MS. Chemometric processing of the original data was performed using ANOVA, PCA, CA and LDA. PCA, with the first two PCs explaining 44.8% of the total variance, enabled partial differentiation of the wine samples, in contrast to the LDA model, which achieved highly satisfactory discrimination. The classification of the wines studied was attained with a success rate of approximately 100% using two various classification criteria, i.e., the variety of wine and region of origin.

Martin et al. (2012) investigated red and white wines from different regions of grape varieties and vintages with respect to MEC content assigned to spectrometric analysis, i.e., ICP–MS and ICP–AES. The chemometric evaluation of the data indicated that, for red and white wines, the vintage and grape variety had little influence on the elemental profile of the wine, although significant discrimination was noted between red and white wines. Using LDA, a clear differentiation was also found between the three regions of different states, i.e., Coonawarra (South Australia), the Yarra Valley (Victoria) and the great south (Western Australia), with 100% of the estimation samples and 100% of the validation samples classified correctly (Fig. 16.13). It was also possible to distinguish between regions from the same state, for example, Chardonnay wines from four regions, with 94% of the estimation samples and 75% of the validation samples classified correctly (Fig. 16.14). However, given the close proximity of several regions within South Australia, some difficulties arose when attempting to classify samples with respect to their interregional origin. In general, it was possible to discriminate between wines originating from different Australian regions, and the potential of using the LDA technique for accurate provenance establishment has been sufficiently well documented.

Gajek et al. (2021b) applied MEC data provided by ICP–TOF–MS, ICP–OES and CV–AAS to categorize the elemental profiles of different varieties of wines originating from 22 countries. HCA enabled the selection of five variable clusters. Multivariate chemometric assessment demonstrated that the achieved grouping should correspond more to the country of origin than to the grape variety. For example, wines from Poland were among the most characteristic elemental fingerprints

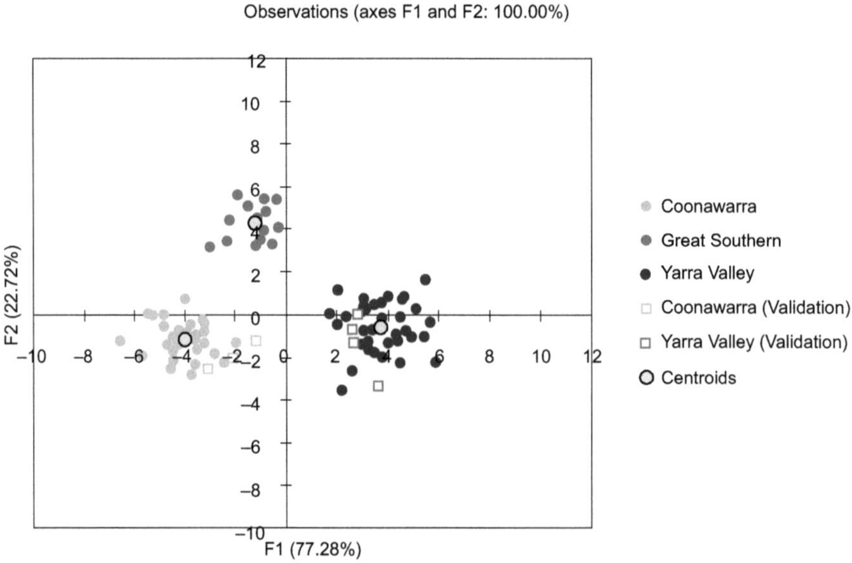

Fig. 16.13 Discriminant plot generated using forward stepwise LDA detailing the separation of data associated with red wines from Coonawarra, the Great Southern and the Yarra Valley. (After Martin EA et al., 2012. Food Chem 133:1081–1089. https://doi.org/10.1016/j.foodchem.2012.02.013. With permission)

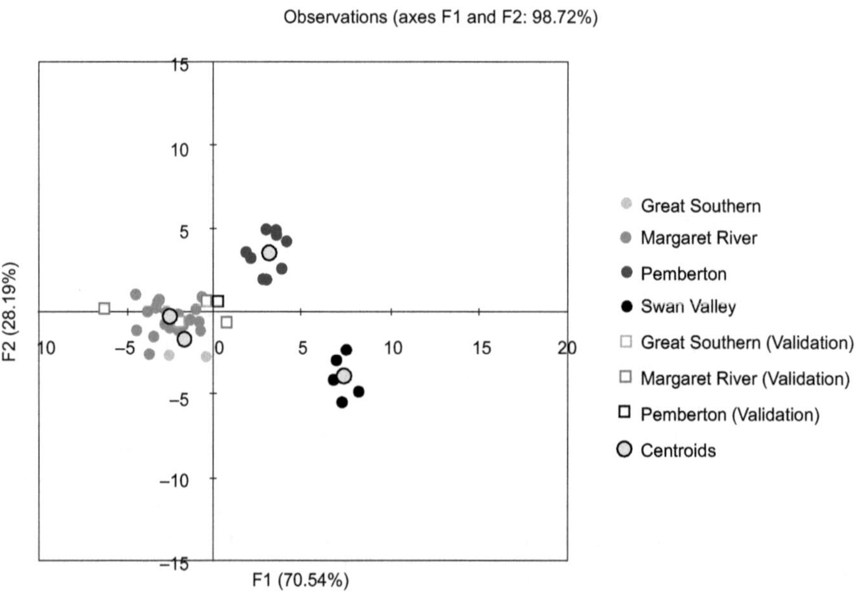

Fig. 16.14 Discriminant plot generated using forward stepwise LDA detailing the separation of data associated with Chardonnay samples from the Margaret River, the Swan Valley, Pemberton and the Great Southern. (After Martin EA et al., 2012. Food Chem 133:1081–1089. https://doi.org/10.1016/j.foodchem.2012.02.013. With permission)

because of relatively low levels of many elements, whereas samples from the USA and Australia could be differentiated from the remaining studied wines.

Rodrigues et al. (2011) noted that ICP–MS enabled the assessment of the elemental profiles of white and red wines originating from four Portuguese Appellations. The FA technique with F1 (lithophile elements) and F2 (fertilization process) differentiated wines with respect to their regional origin and type. Owing to the use of DA, it was possible to clearly separate wine samples originating from the four regions, achieving a highly satisfactory recognition percentage of 98.2% for the classification of wine samples with respect to appellation of origin. The most significant absolute correlations between functions and variables were obtained, namely, with Mg, Mn and Ba for DF1 and with P, B and Zn for DF2.

Pérez-Trujillo et al. (2011) used spectroanalytical procedures such as FAAS and FAES for the determination of macro- and micronutrients to characterize samples of liquor and table wines from Madeira and Azores (Portugal). Chemometric processing (ANOVA, PCA and LDA) of the original variables was a reliable approach for segregating wine samples. PCA, with the first two PCs accounting for 65.02% of the total variance, significantly differentiated between the wines originating from both Azores archipelagos, i.e., Madeira liquor wines from the Pico Islands and those from the Terceira Islands. Liquor wines of Madeira were described according to the "stufagem" process and therefore tend to be differentiated from table wines, i.e., both from the Azores and from Madeira. However, PC2 tended to discriminate the samples between Terceira and Pico. LDA using archipelago (Azores and Madeira) as a factor and Na, Ca, Cu, Zn, Mn, Sr and Li enabled the authors to achieve statistically significant discriminant functions with 90.6% and 89.0% recognition ability and prediction ability, respectively. In the case of stepwise LDA with the island as a factor (relative to the Azores samples), elements such as Na, K, Ca, Mg, Fe, Cu, Zn, Sr, Li and Rb enabled the classification of samples with 100% and 88.9% recognition ability and prediction ability, respectively.

Gremaud et al. (2004) reported that the combination of analytical tools, i.e., ^{1}H NMR, IRMS, ICP–OES, ICP-MS and FT–IR, was a promising approach for discriminating authentic and commercial red and white wine samples from different Swiss locations. LDA enabled a clear differentiation of three groups of white wines (Valais, Romandie, Alemanique+Graubunden) and four groups of red wines (Valais, Tessin, Alemanique+Graubunden, Romandie) on the basis of a combination of $\delta^{18}O$, $\delta^{2}H$ and multielement data. The Tessin group was split into two subgroups corresponding to different harvest periods owing to the inclusion of FT–IR data.

Griboff et al. (2021) classified white and red wines from Austria and Argentina on the basis of EA–IRMS and ICP–MS quantifying MEC and SIR data. LDA provided clear intercountry and intertype differentiation. Elements such as Sr, Li, V, Pb, B, Mn, Co, Rb, As, Na, Mg, Zn and $\delta^{18}O$ were highly descriptive indicators with the potential to differentiate wines according to their production origin, whereas Sr, Li, Na, $\delta^{13}C$, $\delta^{18}O$, Ca, B, Fe, Mn, V, Mg, Co and Zn enabled the discrimination of wines in terms of both origin and type. LDA enabled the classification of 100% of the wines according to their origin and 98.4% of the samples with respect to whether the wine was red or white.

Potortí et al. (2017) reported that ICP–OES and ICP–MS data enabled the assessment of the elemental profiles of Sicilian wines originating from two grape cultivars. PCA, with the first two PCs explaining 56.46% of the total variance, clearly differentiated the two cultivars of *V. vinifera*, and the interspatial origin of the Italian wines was determined. The samples from Nero d'Avola had relatively high levels of Zn, As, Ni, Cr and Cd, whereas the K, Mg, Cu and Sb contents corresponded to those in the Syrah region. Reliable results were obtained when LDA was used as a suitable model to discriminate the samples according to the botanical and geographical origins of the wines. For the CV results, 98.6% of the cases were correctly classified, which means that this method can be useful for guaranteeing wine authenticity.

Geană et al. (2014) reported that the chemical distribution patterns of white and red wines from vineyards located in 2 Romanian wine regions were characterized via the integration of ICP–MS, FAAS and RP–HPLC (HPLC–PDA). On the basis of the PCA data, with the first three PCs explaining 62% of the total variance, classifying wines with respect to the two wine-producing regions and specifying the Dragasani region as two subregions, i.e., northern and southern vineyards, was possible. The most descriptive variables responsible for interregional differentiation included Be, Ba, Na, Mg, Cr, Sr and Li combined with phenolic compounds, i.e., (+)-catechin, (−)-epicatechin, *p*-coumaric acid, ferulic acid and resveratrol. The distributions of the object samples and loadings enabled the classification of the wine samples. Furthermore, intervarietal grape discrimination was achieved using PCA, which explained 74% of the total variance on the basis of six elements and three organic markers.

According to Temerdashev et al. (2024), the combination of ICP–OES and ICP–MS with multidimensional analysis of original elemental data was performed to determine the geographical origin of three varieties of dry wines, i.e., Chardonnay, Riesling and Muscat, grown in four regions of the Krasnodar Territory, Russia. Relationships between the elemental compositions of the soils, grapes and wines were found, which was useful for distinguishing the areas of grape growth and wine varieties with high accuracy. Despite the close location of the regions, wine intervarietal differentiation was achieved on the basis of the mean concentrations of the elements in the studied samples. The differences in the concentrations of Al, Ba, Ca and Rb in the wines were related to wine variety, whereas the concentrations of Al, Ba, Rb, Fe, Li and Sr were linked with the region of grape growth. The quality of the constructed predictive models was recognized by the percentage of correctly identified wine samples. The neural networks achieved the highest classification accuracy (100% on the test and training samples) among the machine learning methods, such as SVM (98.69%) and GDA (94.77%). The C&RT (97.39%), CHAID (98.69%), boosted trees (98.69%) and random forest (98.69%) methods also achieved satisfactory overall accuracies. It was concluded that data mining machine learning methods based on numerous dimensional data integrated with ICP spectrometric analysis were effective tools for determining the authenticity of wines in terms of their geographical origin, component composition and variety.

Coetzee et al. (2005) reported that the combination of ICP–MS data with computational processing was a useful approach for categorizing wines from three important wine-producing regions in Africa. The first two PCs, which explained 68.0% of the total variance, clearly distinguished between the three regions and between the red and white wines within each region. The SDA procedure also correctly classified wines from the three regions that achieved 100% success with or without CV. Red and white wines included in the same dataset could be classified correctly according to the region of origin. Moreover, PDA enabled classification in three steps, with each step classifying a wine as originating from a certain region or not. The discriminant functions were based on the selected elements for each of the three regions. Owing to this procedure, the classification of the wines into one of the groups was 100% successful.

Leder et al. (2015) processed white and red wine samples from two wine-producing regions in Croatia for the quantification of MEC by ICP–OES to estimate their regional origin. PCA (with the first two PCs representing 41.5% of the total variance) was used to separate white from red wines. Furthermore, the first two PCs, representing 46.7% of the total variance, revealed that K, Ca, Mg, Na and Co were related to red wines originating from the continental region, whereas Cd, Ni, Cr, Zn, Al and As discriminated red wines originating from the coastal region. Differentiations of white wines in terms of their origin were obtained on the basis of Cd, Fe, Cr, K, Mg, Ca, Pb and Co for the continental region, whereas Na, Sn, Ni, Al, Cu and Zn were distinctive variables for the coastal region.

Leder et al. (2021) aimed to classify microvinified and commercial wine samples from continental and coastal winegrowing areas as well as from three viticultural zones in Croatia. SIR and MEC were determined by EA–IRMS and ICP–OES procedures, including different routine analytical methods. The first three PCs, which represented 41.0% of the total variance, revealed that partial differentiation was possible with respect to the viticulture zone. GDA achieved the highest effectiveness for the differentiation of wines from the continental and coastal zones of Croatia, with a correct classification of 97.9% for all investigated samples in the CV matrix. Moreover, in the case of the microvinified and commercial wines, the correct classification was obtained for 100.0% and 84.8% of the samples, respectively. The results revealed that the relationships between the isotopic ratios and levels of different elements combined with an appropriate chemometric model are powerful for the discrimination of wines produced in different Croatian winegrowing areas.

According to Conde et al. (2002), white, rosé and red wines from five regions of Tenerife Island were assayed using the FAAS and AES methods. Variations in elemental content with respect to vintage, type of wine and designation of origin were identified. PCA discriminated wines in terms of their type and designation of provenance. PC1 (24%) corresponded to the type of wine, PC2 (17.6%) corresponded to the soil and marine spray, PC3 (11.8%) corresponded to the equipment used in the wine-making process, and PC4 (10%) corresponded to the soil composition. SLDA-supported PCA data yielded approximately 80% correct classification of samples relative to their specification of interregional origin.

Other Alcoholic Beverages

Carter et al. (2015a) performed extensive studies of approximately 160 brands of bottled and canned beers originating from 42 countries using IRMS–TC/EA, ICP–OES, IC and FIA data processed by CDA. According to the authors, a statistically significant relationship, "consistent with natural waters but offset from the Global Meteoric Water Line, was observed between the $\delta\ ^{2}H/\delta/\delta\ ^{18}O$ composition of the beers". A general trend towards stable isotopic depletion was observed moving from Africa to Australasia, Asia, South America, Europe and North America. In the opinion of the authors, the spread of data assigned to each group was not surprising because "the continents comprise large landmasses with significant variations in geography". The correctness of the classification related to beer type and country of origin was reliable, with DFs dominated by $\delta^{2}H$. For example, it was possible to distinguish Australasian and European beer samples effectively, in contrast to African and Asian samples. Discrimination between various types of beer as well as different continents of manufacture was achieved, with a satisfactory degree of correctness of 73%. Importantly, on the basis of the isoscape, it may be possible to evaluate the geographical origin of samples when genuine comparative samples are not available.

Other Alcoholic Beverages

Adam et al. (2002) analysed trace elements by GF–AAS to differentiate between 6- and 20-year-old Scotch whiskies distilled in different regions. CA was used to process chemometrically the concentration data obtained; however, no characteristic metal fingerprints were identified for the different geographical regions. In the case of the second set of samples, Cu was exclusively analysed, and highly significant differentiation was achieved, suggesting that its simple analysis could be used as a singular test to verify interregional whiskey production. Therefore, the Cu level could be useful as a criterion to distinguish blended or grain Scotch from malt whisky (Adam et al. 2002).

16.2.4 Differentiation of Other Parameters

16.2.4.1 Discrimination of Conventional and Organic Farming

Wine

Longobardi et al. (2017) used SIR data obtained via EA–IRMS to classify samples according to their agronomic practices (conventional versus organic farming) and the origin of the table grape. PCA with the first two PCs (75.1% of the total variance) enabled only partial grouping of objects with respect to the geographical origin of the samples. Therefore, a GDA model was applied, achieving mean recognition and prediction abilities of 93.5% and 92.2%, respectively. Moreover, the single CV prediction rates were 94.6% for the Iberian samples and 90.0% for the Italian samples.

Wine Vinegars

Guerrero et al. (1997) investigated genuine wine vinegars with the D.O. trademark from Spain for elemental component levels determined by FAAS, FES and HG–AAS to distinguish conventional (quick) vinegars from traditional (slow) vinegars. PCA, with the first two PCs accounting for 46.6% of the total variance, was unable to completely discriminate class samples. However, on the basis of Mg, Na, Mn and Fe, models such as KNN, LDA and BPANN achieved the most satisfactory performance in terms of both recall and prediction ability (approximately 100%), whereas SIMCA was unable to achieve differentiation of class envelopes.

16.2.4.2 Classification of Alcoholic Beverages According to Price and Other Parameters

Other Alcoholic Beverages

Brooks et al. (2002) integrated $\delta^{13}C$ data with chemometric processing analysis (ANOVA and GLM) to characterize a wide range of beer styles originating from six regions worldwide, i.e., Europe, Brazil, Mexico, Canada, the United States and the Pacific Rim. The beer samples were classified according to price, style, brewery and location of purchase. Although the use of C4 carbon (corn, cane sugar, etc.) was not assigned to the analysed beers from any particular region worldwide, its application was ubiquitous in established beer styles, especially the common US-style lagers and ales brewed by large breweries in the United States, Canada, Brazil, Mexico and the Pacific Rim. European-style lagers, e.g., import pilsners from Germany and the Netherlands, and specialty ales, e.g., porters and stouts, fashioned after the English style of ale brewing, contain insignificant C4 carbon levels. European beers with mostly C3 plant carbon differ from U.S. and Canadian beers with only either C3 or C3–C4 mixtures, whereas Brazilian, Mexican and Pacific Rim beers were identified to be mostly C3–C4 mixtures. A simple GLM was used to indicate the tendency in price on the basis of $\delta^{13}C$ values. When the percentage of C4 carbon in beer was normalized to a standard volume of beer, the C4 carbon content increased as beer prices decreased. Beer prices are significantly dependent on the cost of workload, processing, wrapping and marketing in addition to the ingredient cost.

Mardegan et al. (2013) coupled $\delta^{13}C$ data with computational processing, i.e., GLM and ANOVA, to classify 77 brands of beer, including 49 brands produced in Brazil and 28 produced in Europe, South and North America and China. European beers are distinctly differentiated from Asiatic, South American and North American beers. The duality of beer isotopic signatures refers to beers composed of a notable mixture of C3- and C4 sources with higher $\delta^{13}C$ values, which originate from large breweries compared with those composed only of C3-cereals produced in artisanal breweries. Moreover, there is a strong relationship between beer $\delta^{13}C$ values and their final cost per litre, either on the basis of considering all samples analysed or only beers produced in Brazil. Notably, although ingredient expenses represent only a small share of the total beer cost, a strong inverse relationship between price and C4 content has been found.

16.2.4.3 Assessment of the Impact of Pollution Related to Wine Manufacturing Practices/Soil Composition

Wine

Kment et al. (2005) differentiated the elemental compositions of wines and parent soils originating from six viticultural areas with respect to their geographical origin. On the basis of the combination of q-ICP–MS, FAAS and traditional spectrophotometric methods with multivariate analysis (CA, PCA and FA), PCA (with the first two PCs explaining 35%) and FA (with F1 and F2 accounting for 28%) were useful techniques for the differentiation of Czech wines with respect to elements derived from vineyard soil of lithogenic origin (Mg, K, Mn, Cs, Sr, Ba and Cr) or human activity (Cu, As, Cd, Pb and V). The grouping of wine samples generally did not correspond to the clustering of soil samples; therefore, it was suggested that the elemental fingerprints of wines reflect not only the soil chemistry but also the impact of pollution and wine manufacturing practices used by individual producers.

Almeida and Vasconcelos (2003) assayed Portuguese wines and their precursors, including provenance soil, and their possibilities as identifiers of wine origin by means of ICP–MS, FAAS and ET–AAS. Most of the elements had similar or even lower levels in the wine than in the respective grape juice prepared in the laboratory, possibly owing to precipitation or coprecipitation with suspended particles during fermentation and/or wine ageing. Significant relationships were also detected between the elemental content of the wine and the respective grape juice for the two types of wines as well as between that in the wine and the corresponding soil of the grape-growing region. Two studied vinification processes were characterized, i.e., one modern one used for producing table wine and one old-fashioned one used to produce fortified wine. These processes may influence the multielemental composition of wine. Relatively low levels, much lower than the legal threshold limit values, were established for either toxic elements such as Cd, Cr, Ni and Pb or those potentially affecting the organoleptic properties of wine, such as Al, Fe, Zn and Cu. However, relative contamination during winemaking steps such as grape pressing, fermentation and/or fining of wines was detected for Cd, Cr, Cu, Fe, Ni, Pb, V and Zn in the fortified wine and for Al, Cr, Fe, Ni, P and V in the table wine.

According to Blesić et al. (2017), the contents of seven metals were quantified by the ET–AAS and FAAS techniques in six varieties of white and red wines originating from four viticultural localities in Herzegovina. Some wines contained levels of Cu, Cr, Ni and Pb that exceeded the maximum acceptable limits. Moreover, white wines from the Stolac locality were distinguished by relatively high contents of Cu, Pb and Zn. However, on the basis of PCA, with the first three PCs accounting for 82.1%, it was not possible to achieve a strong relationship between the contents of the studied metals in wine samples and producing regions. The metal profile of the wines studied may be influenced by different viticultural and oenological practices rather than parent soil composition and other environmental conditions in wine-producing regions. According to the authors, further extended research should focus on potential sources of pollutants in Herzegovinian wines, covering the scope of the entire production chain from the vineyard to the final product. Notably, although the levels of metals in wines were mostly below their regulated maximum acceptable limits, unacceptably high levels of Cu, Cr, Ni and Pb were detected in a few wine samples.

Other Alcoholic Beverages

Redan et al. (2019) used two analytical methods, ICP–MS/MS and LC–ICP–MS, to determine the levels of toxic elements, i.e., Pb, Cd and inorganic species of As, in lager and ale-styles of commercial beer from 11 countries in Europe, the United States, Caribbean countries and Australia to determine whether filter aids, i.e., DE, used in the case of filtration methods for this alcoholic beverage contain elevated amounts of these elements. The experiments indicated that the elemental profile of DE might be a source of toxic element levels, especially As(V) in beer filtrates.

Wyrzykowska et al. (2001) applied ICP–MS data to classify bottled and canned Polish beers. PCA demonstrated that the first two PCs explaining 51.45% of the total variance enabled the identification of elements correlated with PC1 (As–Hg–Tl–Bi–Sn–Th–Hg and U–Sb) and PC2 (Fe–Zn). The PC1 descriptors could be related to industrial and other anthropogenic sources. Some similarities in the above elemental profiles of beers suggested the accumulation of anthropogenic elements on plant (barley) and soil surfaces, most likely through airborne particle deposition. The Fe–Zn clusters linked with PC2 could correspond to their provenance from other sources, i.e., the widely used zinc coating on steel and other ferrous substrates for corrosion protection.

Markovski et al. (2018) reported that the MEC quantified by ICP–OES was useful for evaluating the elemental profiles of beers from 32 countries that are commercially available on the US market, i.e., those originating from Europe, Asia, Africa, North America, South America, Caribbean countries and Australia. In general, there was no significant differentiation between the elemental profiles of imported and domestic beers sold on the market. The impact of the geographic origin of the beers and grains used in their production on the presence of various elements, including contaminants, was studied. Some specific toxic elements, such as As, were present in both imported and domestic beers at levels exceeding the national primary drinking water standard MCLs. According to the authors, not only may water quality be a factor responsible for the elevated levels of concerning elements in beers, but also more likely, the grain ingredients are more important contributors to the observed trends.

16.2.4.4 Elemental Profile Related to Different Types of Packaging Material

Other Alcoholic Beverages

According to Voica et al. (2015), a combination of ICP–MS and EA–IRMS data was applied to characterize the chemical compositions of different blonde and dark beer samples bottled in various forms available in Romania. The analysed beers were characterized by the presence of different plant types used in brewing, i.e., C3,

C3–C4 mixtures and C4, depending on the producer. Furthermore, the CA data classified all the beer samples into three clusters corresponding to small, moderate and high Al levels. Importantly, beers are manufactured by the same producer but are bottled in different types of packaging, such as glass, dose or PET. Elements such as Mg, K and P were significantly different among the three types of packaging material; however, surprisingly, no significant interpackaging differentiation with respect to Al was achieved. Moreover, Al, when used as a part of beverage packaging (dose), is frequently covered by a polymeric film; hence, its level of migration is considerably limited.

Rodrigo et al. (2017) aimed to use ICP–MS data to differentiate the elemental compositions of commercial beer samples that retailed in the United Kingdom and originated from 10 countries. The first two PCs, which explained 75% of the total variance, incompletely differentiated beer styles with respect to their country of origin. The computational tool differentiated some types of beer; i.e., ANOVA revealed higher levels of Sr, Mo and Pb in stout/porter beer as well as higher levels of Al, Mn, Ni and Cd in lambic beer. Moreover, the levels of As, Mg, Na and V in beer were significantly influenced by container type, i.e., metal barrel, bottle and can. For the country of origin, higher levels of Se were observed for beer brewed in the United States, possibly because the greater content of this element in US cereal grains corresponded to prevailing soil geochemical characteristics. Moreover, high Ni and Fe levels were obtained for Belgian beers, but high Cs contents were obtained for Italian beers. The DA data for beer style had a prediction ability greater than 81%, whereas the data for beer origin country had a lower prediction ability of 76%, which means that only 76% of the beer samples were classified into the correct group. This means that beer style has a more significant effect on the elemental profile of beer than does geographic origin or container type.

Gajek et al. (2021a) confirmed that the chemical profile of ciders quantified by ICP–OES, ICP–MS and CVAAS depends on the mineral fingerprint of the ingredients used for apple varieties. Moreover, the technology and equipment applied by the producer also affect the type of packaging used. The PCA data, with the first two factors explaining 50.66% of the total variance, revealed significant differences in the contents of Ag, Al, B, Bi, Co, Cr, Cu, Fe, K, Li, Mg, Na, Ni, Ti and Zn with respect to the type of cider origin, i.e., commercial or homemade. Furthermore, in the case of different packages (cans or bottles within one brand of commercial cider), significant differentiation was achieved because of the presence of many discriminative elements, such as Cu, Mn and Na. Moreover, the elements were present in levels that were lower than the allowable limits.

16.2.4.5 Production Chain Related to the Variability of Different Parameters

Wine

Durante et al. (2013) integrated FAAS, ICP–Q–MS and MC–ICP–MS to evaluate the elemental and isotopic compositions of vine branch and grape juice samples from Modena (Italy). The Sr/Rb differentiation procedure was optimized on the basis of the synergistic use of experimental design techniques. It was possible to achieve maximum Sr recovery in contrast to Rb recovery. Multivariate analysis (PCA and MLR) was successfully used, e.g., PCA was a useful tool, with the first two PCs explaining 66.55% of the total variance. Research on the variability of the $^{87}Sr/^{86}Sr$ ratio along the wine production chain (soil–vine branch–grape juice) indicated that the differentiation of samples of grape juice/vine branches, depending on their provenance, is hopefully possible.

Braschi et al. (2018) aimed to trace the $^{87}Sr/^{86}Sr$ ratio from vine to wine by means of TIMS to process grapes from three adjacent vines harvested as Sangiovese bunches. The material was examined under controlled laboratory conditions through microvinification to obtain three experimental wines for isotope analyses over a period of four harvest years. The uptake of Sr from grapevine roots to its final product is a time-independent process. Consequently, the $^{87}Sr/^{86}Sr$ ratio of wines can be safely examined as a significant parameter to characterize the terroir typicity of the wine, even at a regional scale, i.e., in relation to a single vine. The experimental results revealed that the biological activity of the vine was unable to change the original $^{87}Sr/^{86}Sr$ profile originating from the bioavailable fraction of the parent soil. Therefore, the Sr isotopic ratio of the wine is an unadulterated feature of the terroir.

As reported by del Mar Castiñeira et al. (2004a), the entire winemaking process was investigated, from the must to ready wine. Two winemaking processes were compared, i.e., the addition of clarifying agents (bentonites) before and after fermentation. The multielemental group was quantified via ICP–MS. Notably, the element contents of Li, B, Mg, Ca, Rb, Cs and Pb did not considerably change throughout the winemaking process irrespective of the time of bentonite addition. However, the quantitative profile of some elements was susceptible to change during winemaking processes. For example, the contents of Cu and Ba, as well as those of Al, Y and Zr, changed as a result of the fermentation process, with clarification (addition of bentonites) and/or fining, respectively. These elements are therefore robust elements for studying the origin of German white wines. Moreover, del Mar Castiñeira et al. (2004b) characterized the elemental composition of white wines from four wine-growing regions by using SF–ICP–MS combined with supervised pattern recognition techniques, i.e., QDA and DTs. QDA enabled the discrimination of four regions with 83% accuracy using Li, B, Mg, Fe, Zn, Sr, Cs and Pb and the prediction ability for classifying new samples was 76%. However, DTs were a more effective tool since they were able to classify samples with an accuracy of 84% for only four elements, i.e., Li, Zn, Mg and Sr and had a prediction ability of 74% in the identification of unknown samples. Therefore, the latter is recommended for the classification of new wine samples from these areas of German wine production.

Fabani et al. (2009) integrated FAAS data with computational processing tools, i.e., KNN, SLDA and CCA, to assess the impact of vine variety on the absorption of metals from soil. SLDA revealed that descriptors such as Na, Mn, Mg, Cu, Ca, K and Fe were able to distinguish among the four juice varieties. Comparable results were achieved by the application of KNN, enabling 100% discrimination among the four wine varieties. Elements such as Mn, Mg, Na, Fe, Zn, K, Ca and Cr were the most discriminative variables for distinguishing between different wine varieties. Moreover, CCA demonstrated that juices and wines from the same vineyard can be clearly differentiated in terms of their variety, and a good correlation between grape juice and wine and their associated soil was achieved. Among the elements analysed, Mg could be used as a marker of wine provenance, whereas Mn could be used as an indicator of intervarietal differentiation associated with plant physiology.

Shimizu et al. (2020) used ICP–MS and ICP–AES to study the effects of various winemaking techniques, including fining, ageing with oak tips, maceration with grape skins, chaptalization and acidification as well as yeast nutrient addition for alcohol fermentation, on the contents of Li, B, Na, Mg, Si, P, S, K, Ca, Mn, Co, Ni, Ga, Rb, Sr, Mo, Ba and Pb. Among the various winemaking techniques, maceration with grape skin and fining treatment with bentonites had the most significant effects on element concentrations, i.e., considerably changing the levels of most of the elements analysed. However, PCA, with the first two PCs explaining 78.7% of the total variance, enabled clear discrimination of the wine samples with respect to the grapes used in the experiments rather than the winemaking technique. It was concluded that some winemaking techniques significantly altered the contents of some elements in wine; however, the key factor affecting wine chemical composition is likely its geographical provenance.

According to Mihucz et al. (2006), the REE profiles of young wine samples from the Dealurile Moldovei viticulture area (Romania) were classified via DF–ICP–MS. The red wine samples were generally more concentrated in terms of REEs than the white wine samples were. The REE contents of the four bentonites (Gelbenton, Evergel, BW200 and Tükrös) determined after their fusion to wine were greater by three orders of magnitude than those of the unprocessed wine samples. After a simulated wine purification process was performed with these three bentonite samples, the REE levels of the clarified wine samples increased differently for red and white wines. On the basis of HCA evaluation using REE concentrations as descriptors, it was suggested that the adequate selection of bentonite could enable the use of REEs as identifiers for distinguishing the type of grapes as well as the origin of wine.

Aceto et al. (2013) combined ICP–MS data with chemometric data (PCA and CA) to process *Moscato d'Asti* musts from 25 Italian geographical provinces. However, despite the proximity to the neighbourhood of the sampling sites, it was possible to distinguish four zones for both historical and physical reasons. The first two PCs accounted for approximately 59% of the total variance. The effects of blending, rectification and clarification were also assessed by reprojecting these samples onto the PCA. The first two PCs explained approximately 90% of the total variance. Moreover, the REE fingerprint remains unaltered with respect to passage from the soil via grapes to the must, whereas fractionation takes place when wine is clarified with bentonites. This means that traceability is possible from the soil to the must since the REE profile is maintained from the soil to the grape and must, but blending, rectification and clarification result in a change in the distribution of REEs, probably because of the release of metals from bentonite.

Petrini et al. (2015) used TIMS and XRD focused on must and soil samples. Glera vineyards from the Prosecco wine district in northern Italy were characterized with respect to the $^{87}Sr/^{86}Sr$ isotope ratios of the musts and grape components (skin, seeds, must, stem) combined with their values obtained for the labile fractions of the soils. A statistically significant relationship between the soil labile fraction and must was noted. A significant range of isotope ratios for musts from the different ten vineyards was a good marker of must origin despite the relatively limited extent of the investigated geographic area.

16.2.4.6 Interregional or Intervarietal Differentiation Related to Soil

Wine

Geană et al. (2013) coupled ICP–MS data with chemometric processing to differentiate white and red wine samples and associated soil samples from the three most important Romanian wine regions. On the basis of the PCA data, with the first three PCs explaining 83% of the total variance, the wines from the Valea Calugareasca and Murfatlar regions were clearly distinguished, whereas differentiating the wines from Moldova was more difficult. Mn, Cr, Sr, Ag and Co were the indicators for discriminating between wine and soil samples from the three wine-producing regions.

Fabani et al. (2010) employed FAAS and MEA to assess the chemical composition of samples of wine- and grape-growing soils from major wine-producing regions of Argentina. On the basis of LDA, clear differentiation was achieved among wine samples originating from three provinces (Fig. 16.15). LDA correctly classified 100% of the wines studied using only seven elements, K, Fe, Ca, Cr, Mg, Zn and Mn. Moreover, a significant relationship was obtained between the contents of elements in wine and soil, with an overall correlation of 86%. Notably, the correlation calculated from the CCA was compatible with the correct classification achieved by LDA using Ca, Fe and Cr. In addition to the prevailing influence of the chemistry of vineyard soil on the metal contents in wines, the remaining 14% could be attributed to other factors, e.g., climate, agricultural practices, pollution, grape maturity at harvest and wine-making techniques.

Zinicovscaia et al. (2017) noted that macro, minor- and trace element contents were determined by ENAA in red and white wines as well as associated soil samples from two corresponding major wine-producing regions in Moldavia. Subsequent DA enabled the differentiation of all wine samples according to their type, i.e., red or white, as well as their vineyard origin. Moreover, the correlation between the contents of the elements studied in wine and soil from grapes was better revealed utilizing TF, especially for all 19 elements, with K having the highest TF.

Taylor et al. (2003) processed samples of wines and soil from two major wine-producing regions in Canada, and their mineral compositions were determined via ICP–MS and XRF. A DA function classified two wine-producing regions, i.e., Niagara and Ontario, with a 100% success rate using U, V, Al, Sb, Co, Zn, Sr, Rb,

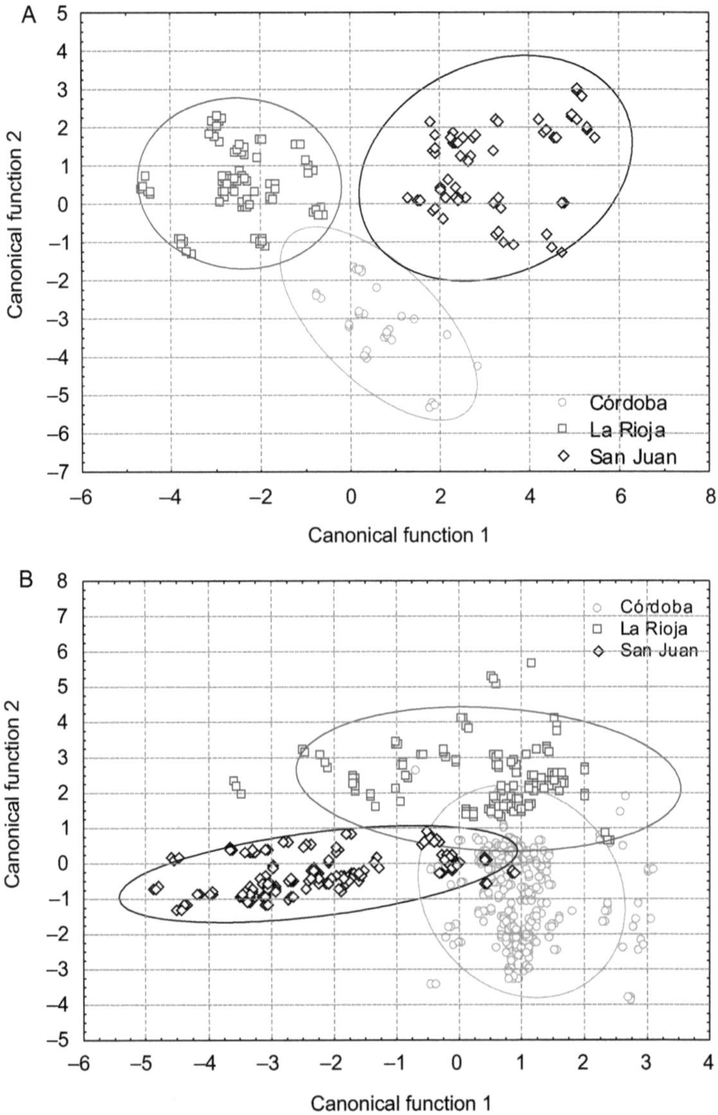

Fig. 16.15 Distribution of wines (**a**) and (**b**) pseudototal elements in soil from Córdoba, La Rioja and San Juan in the plane defined by the first two canonical functions of LDA. (After Fabani MP et al., 2010. Food Chem 119:372–379. https://doi.org/10.1016/j.foodchem.2009.05.085. With permission)

Mo and Mn. Furthermore, with respect to the relationship between Sr in wine and soil, the levels of Sr were greater in both soil and wine originating from the same region. The difference in the mean Sr contents in the wine samples demonstrated a possible regional influence on the soil composition in addition to other factors affecting its wine concentration.

Di Paola-Naranjo et al. (2011) utilized multiple instrumental methods, i.e., HPLC–ESI–MS/MS, Q–ICP–MS, FAAS, TIMS and EA–IRMS, to classify Argentinean red wines from three producing regions with respect to their soil composition. The application of backwards SDA enabled 100% discrimination between wines from three regions on the basis of descriptive variables, i.e., trans-resveratrol, kaempferol, (+)-catechin, B, Na, Mg, Ca, Mn, Co, Ni, Cu, Rb, Sr, Ba, La, Pb, Ca/Sr, $\delta^{13}C$ and $^{87}Sr/^{86}Sr$. It was also possible to correctly classify 100% of the wines analysed from the three varieties, owing to significant variables such as ferulic acid, kaempferol, (+)-catechin, Li, Mg, Al, K, Ca, Co, Cu, Zn, Rb, Cd, La, Ce, Lu, Pb, U, Ca/Sr, $\delta^{13}C$ and $^{87}Sr/^{86}Sr$. The configuration of the different geographical regions reflects the consensus between the wines and soils studied, which explains 100% of the variability between samples and geographical origins and was well discriminated because of the use of MEC and isotopes from the soil and wine samples. In addition to LDA, GPA and CCA indicated a significant relationship between the elements present in wine and soil during grape growth.

Vorster et al. (2010) integrated QICP–MS with ANOVA and CDA to characterize the wines and soils of four major South African wine-producing regions. The $^{11}B/^{10}B$ isotope ratio was a useful approach for discriminating wines with respect to their regional origin. Moreover, the B isotope ratios together with the elemental concentrations of selected indicator elements, as independent variables, were helpful in LDA and enabled the classification of wine according to geographical provenance. A good correlation between the B isotope ratios in the wine and parent soils was observed. However, the $^{87}Sr/^{86}Sr$ ratio showed limited potential for recognition of provenance in terms of wine-producing region since only one region could be differentiated from the others.

16.2.4.7 Interregional and/or Intervarietal Differentiation Related to the Interseasonal/Vintage or Ageing Time of Wine

Wine

Paneque et al. (2010) analysed macro- and microelements via FAAS and FAES of white wines from two wine-producing regions in Andalusian Province of Córdoba in Spain. Univariate techniques enabled the differentiation of wine samples according to their origin from the two regions on the basis of Na, Mg, Mn, Fe and Zn as the most descriptive elements. Among the chemometric techniques (PCA, CA, LDA and LOOCV), LDA was a useful tool for satisfactory recognition and prediction abilities of 93.6%, corresponding to four categories, i.e., young and aged wines from the two regions. The classification of wine samples attributable to both provenance and ageing on the basis of the content of 10 elements was achieved. This means that unsupervised and supervised pattern recognition methods were able to distinguish Cordoba wine samples in terms of their provenance and their type of elaboration, with or without ageing.

According to Agazzi et al. (2018), investigations of the evolution of MEC with polyphenols and anthocyanins profiles quantified by MP–AES and HPLC–DAD were performed during the ageing of Malbec red wines from different regions of Mendoza (Spain) and California. The chemometric analyses (two-way ANOVA, PCA and LDA) were useful for fingerprinting the effects of ageing time and region on the chemical composition of wine. On the basis of PCA, with the first two PCs accounting for 58.4% of the total variance, a reliable distinction of wine samples originating from Spain and the United States was achieved. Spanish Mendoza wines had high relative contents of gallic acid, syringic acid, Na, Sr and vitisin A, but California wines had high relative levels of t-caftaric acid, p-coumaric acid, catechin, U1, K, Rb, Mg and myricetin (Fig. 16.16). LDA demonstrated significant differentiation between three Spanish departments as well as California wine departments. It was possible, after 5 years of ageing, to differentiate wines with respect to their four regions of provenance. According to the authors, the results obtained enabled decision-making regarding the storage of Malbec wines in the future.

Owing to the use of the ICP–MS method, Feher et al. (2019) differentiated wine samples of two varieties from three important wine areas in Romania within five vintage years. This investigation enabled the classification of wine samples according to different criteria. The multivariate processing approach was useful for characterizing and classifying white wines according to geographical origin, vintage or specific variety. Multivariate techniques, including multi-instrumental chemometric methodologies such as FPCA, FPCA–LDA, FDHAC, HCA, PCA, FPCA, PCA–LDA and PCA/FPCA–LDA, were applied to process the obtained analytical data. In the case of FPCA, the first two PCs explained 64.53% of the total variance, indicating a satisfactory differentiation of samples according to geographic area and variety. The most effective classification of wines according to different criteria, such as geographical area, vintage or variety, was achieved by applying PCA/FPCA–LDA. In all the cases, the classification rate was estimated to be 100%. According to the authors, the new fuzzy clustering procedure was presented and applied for the first time to distinguish wine samples with respect to their elemental pattern distribution.

Geană et al. (2016b) processed five authentic varieties of young and aged red wines with respect to the assessment of SIR combined with several anthocyanins and organic acids established by a multi-instrumental methodology such as CF–IRMS, HPLC/UV–VIS, ^{1}H NMR, ^{13}C NMR and MA-E-AS315–11-ANCYAN. LDA revealed highly satisfactory classification of the wines, both in terms of variety and year of production. A reliable 85.7% classification of Romanian wines according to grape variety was achieved, and a 71.1% classification of Romanian wines according to vintage year was also possible.

According to Monakhova et al. (2014), 1383 samples of wines from 37 grape varieties from vintages 2005 to 2010 from 13 growing regions were investigated. Spectroscopic methods such as ^{1}H NMR, SNIF–NMR and IRMS coupled with multidimensional analysis (multiway (n-way) ANOVA, CLV, LDA, PLS–DA, FDA, ICA, PCA, (MILCA), LOOCV, MB–PLS–DA, CPCA-W MB and ComDim MB) were effective approaches for achieving clear discrimination when two datasets

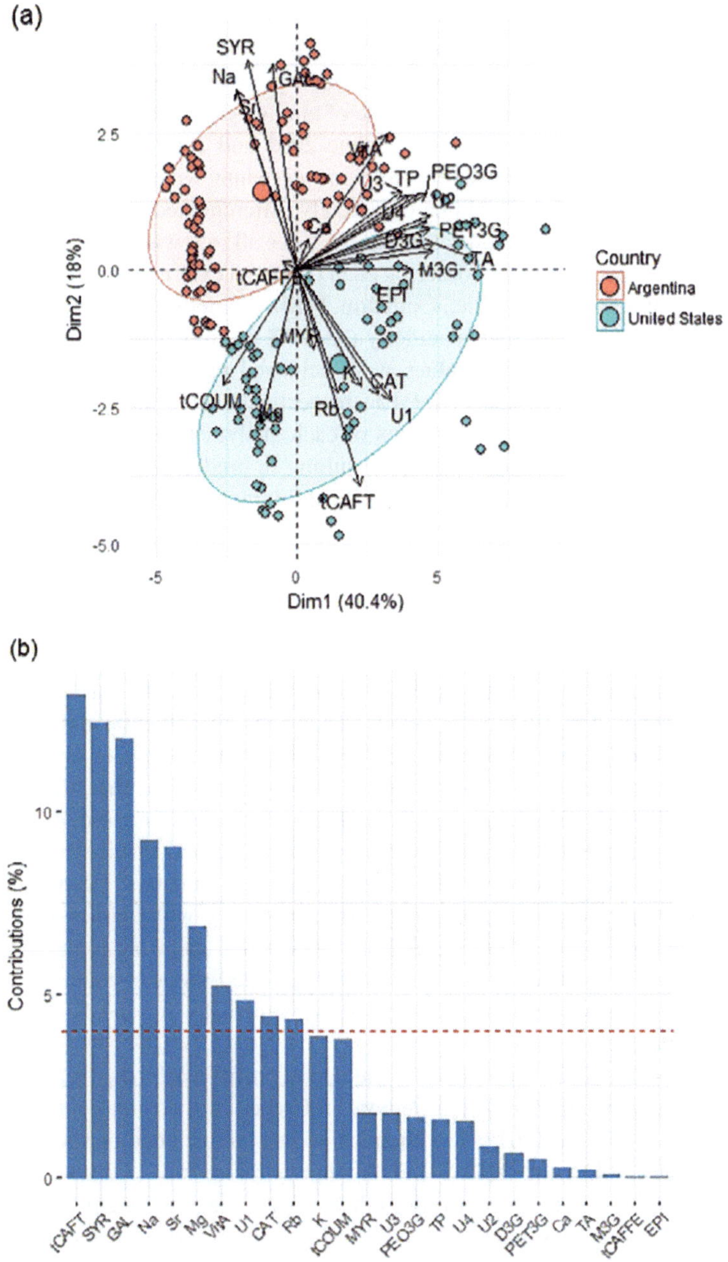

Fig. 16.16 (**a**) PCA biplot of all the analysed parameters in Mendoza and California wines from the two ageing times: t0 and t5; ellipses and colour represent the classification in accordance with the ageing time. (**b**) The contribution of variables to the first component; the red dashed line on the graph indicates the expected average contribution. (After Agazzi FM et al., 2018. Food Chem 269:103–110. https://doi.org/10.1016/j.foodchem.2018.06.142. With permission)

were considered, e.g., when one of the multiblock methods of PCA was used. PCA, with the first two PCs accounting for 77.5% of the total variance, incompletely differentiated wine samples corresponding to their producing regions in Germany, in contrast to the ANOVA results. In the case of vintage year, sufficient PCA differentiation (PC1 vs. PC3) was achieved for the 2005 and 2009 groups, although no significant discrimination between red wine or white wine grape varieties was obtained. All four clusters could be completely differentiated when CPCA-W MB was used. This finding is somewhat surprising since all four wine production regions are very closely located, even bordering each other. An extremely similar effect was achieved by another MB tool, i.e., ComDim. When all methods were used, satisfactory sensitivity (88–96%) and specificity (85–90%) results were obtained for geographical origin. Comparable values were achieved in the case of vintage year, corresponding to 89–97% and 87–94%, respectively. This means that both MB methods could integrate the advantages of each analytical technique, leading to a much more effective combined model, resulting in prediction of the geographical origin up to 100%. This combination significantly improves the accuracy of verifying food authenticity.

Smeyers-Verbeke et al. (2009) reported that 63 chemical parameters, e.g., SIs and MEC data, were used to classify commercial and authentic wines originating from new European member states and from overseas countries. The original variables were processed by chemometric tools (Boxplots, ANOVA, F test, Kruskal–Wallis test, multiway or multifactor ANOVA and PCA). The variability of the obtained data could resulted from the difference between authentic and commercial samples for each country, followed by the differentiation between white and red wines. Comparative analysis of the various vintages indicated that, depending on the country, several variables changed over the 3-year period assessed. According to the authors, considerable variation observed for the authentic wines could be the effect of small-scale production affecting more wine parameters. Despite this variability, authentic and commercial South African red and white wines were clearly distinguished from wines from wines from other countries owing to the use of isotopic ratios such as δ^2H or $\delta^{18}O$. Differentiation of the European countries was significantly less visible because of overlapping, especially for the Hungarian and Czech samples.

Pasvanka et al. (2021) classified the elemental profile established by ICP–MS of white and red wines from different regions in Greece for two consecutive vinification years, i.e., 2017 and 2018. The chemical tools used included one-way ANOVA, HCA, DA, OPLS–DA and O2PLS–DA. As shown in Fig. 16.17, OPLS–DA enabled the classification of wines according to their variety, i.e., they were clearly distinguished, with accuracies of 97.78% and 96.67% for both harvest years, respectively. OPLS–DA explained 100% of the classification of the samples with satisfactory goodness of fit and predictability for the discrimination of the samples with respect to the six regions of production (Fig. 16.18). Most of the elements studied were present in higher contents in wines from vintage 2018 than in those from harvest 2017, except for Na and La, which were present at lower levels in the same harvest year.

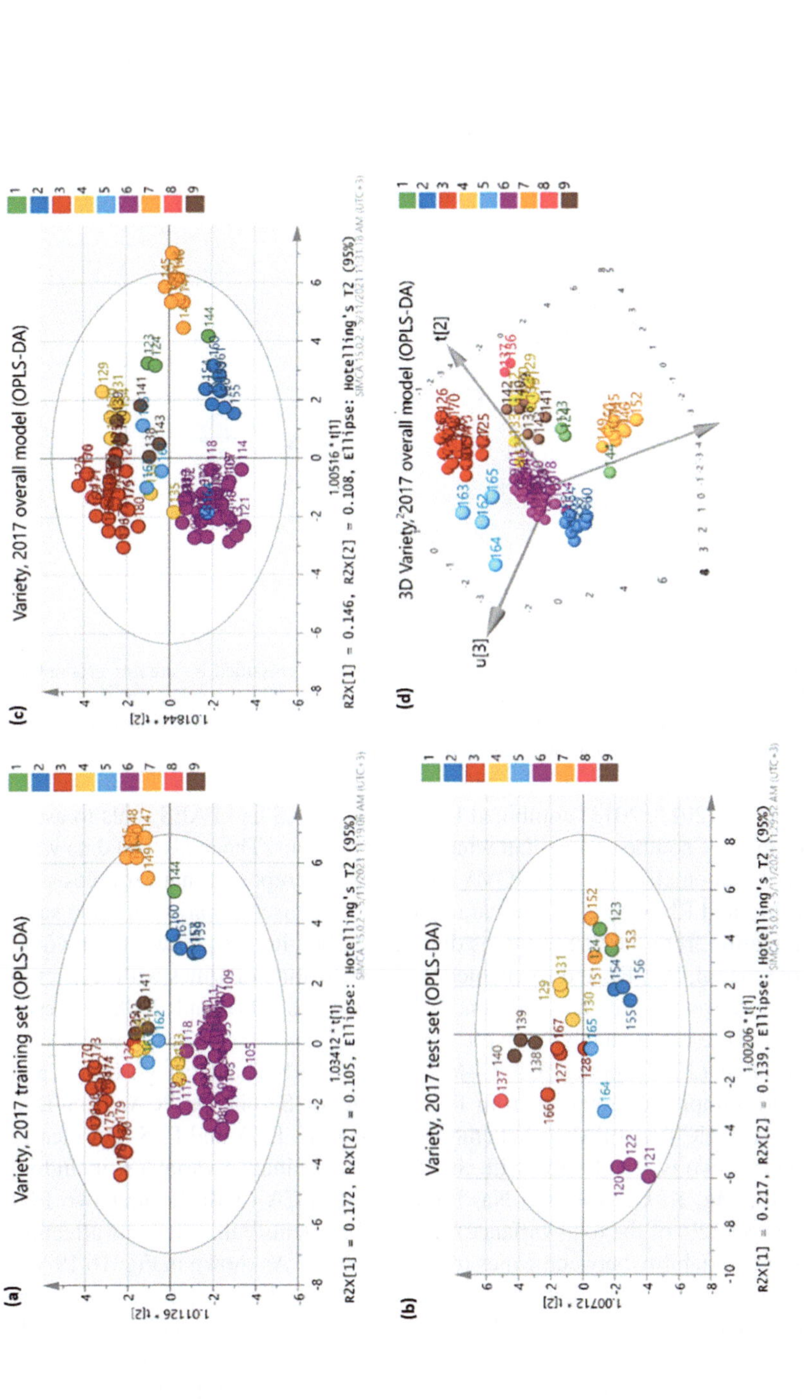

Fig. 16.17 Score scatterplots for the wine samples from 2017 classified according to variety (i.e., 1: Syrah, 2: Agiorgitiko, 3: Asyrtiko, 4: Malagouzia, 5: Muscat, 6: Moschofilero, 7: Xinomavro, 8: Rodoti and 9: Savatiano), where **(a)** training set, **(b)** test set, **(c)** overall 2D model and **(d)** overall model (3D). (After Pasvanka K et al., 2021. Separations 8:119. https://doi.org/10.3390/separations8080119. With permission)

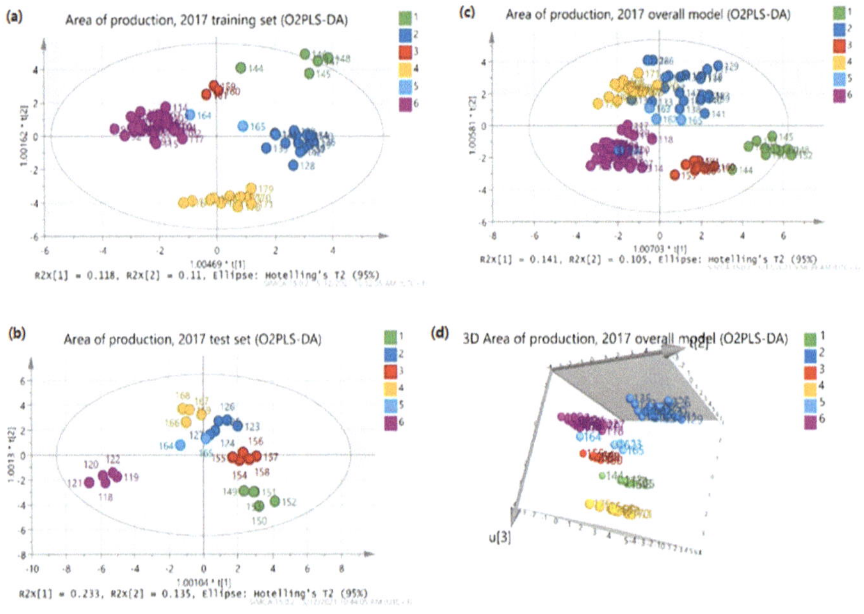

Fig. 16.18 Score scatterplots for the wine samples from 2017 classified by area of production (i.e., 1: Naousa, 2: Attika, 3: Nemea, 4: Santorini, 5: Samos and 6: Arkadia), where (**a**) training set, (**b**) test set, (**c**) overall 2D model and (**d**) overall. (After Pasvanka K et al., 2021. Separations 8:119. https://doi.org/10.3390/separations8080119. With permission)

Dutra et al. (2011, 2013) combined EA–IRMS, FAAS and FAES tools to assay wines from three southern Brazilian wine-making regions. The analytical data were processed chemometrically by ANOVA and DA. It was reported that variables such as $\delta^{18}O$, Mg and Rb were effective parameters for the discrimination of wine samples originating from the three regions during the 2007 harvest. However, the $\delta^{13}C$ values of ethanol, Mg, Rb and Li enabled the classification of wines with respect to variety. Moreover, significant differences between the 2007 and 2008 harvests, regardless of the variety analysed, were observed.

Đurđić et al. (2017) applied ICP–Q–MS and ICP–OES data to categorize red and white wine samples originating from four regions of Serbia via PCA, PLS–DA, RMSEC, RMSECV and RMSEP chemometric models. PCA and PLS–DA clearly differentiated two types of wine with the most discriminative variables, including Be, Al, Rb, Mg, K, Cu, Mn and Na. Furthermore, PCA (with the first two PCs explaining 39.72% of the total variance) was used to identify the elements affecting the seasonal variability between wines from one winery. As shown in Fig. 16.19, the model distinguished three distinctive groups corresponding to vintages with the highest impact from descriptors such as Cd, Pb, As, Sb, V, Na, K and Zn. This difference could result from the different soil types in these three Serbian regions. In addition, PCA demonstrated that seasonal variation insignificantly affected the elemental composition of red wines.

Paneque et al. (2010) studied white wines originating from 2 wine-producing regions of the Andalusian Province of Córdoba in Spain for MEC quantification by

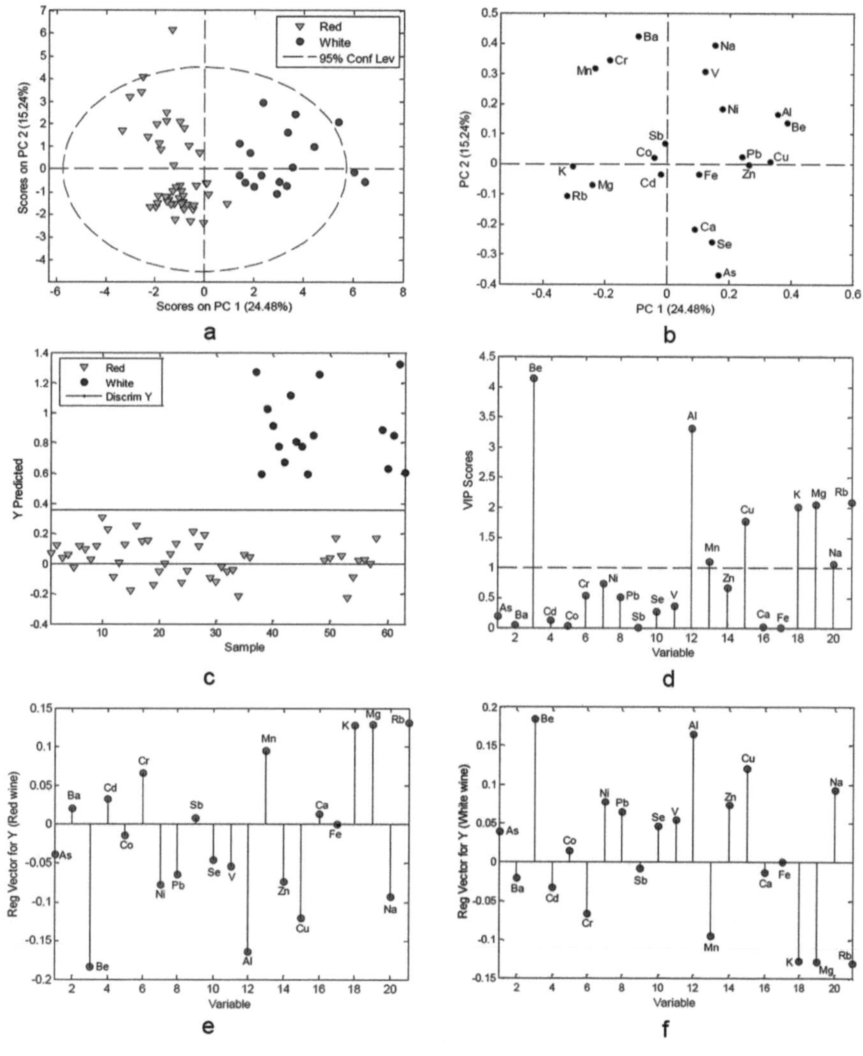

Fig. 16.19 Classification models for wine. (**a**) PCA score plot for red and white wine samples, (**b**) PCA loading plot for red and white wine samples, (**c**) PLS–DA score plot, (**d**) VIP scores for the PLS–DA model, (**e**) regression coefficients of the PLS–DA model for red wine samples and (**f**) regression coefficients of the PLS–DA model for white wine samples. (After Đurđić S et al., 2017. RSC Adv 7:2151–2162. https://doi.org/10.1039/c6ra25105f. With permission)

FAAS and FAES. Among the chemometric techniques used, univariate techniques enabled the differentiation of wine samples according to their origin from the two regions on the basis of Na, Mg, Mn, Fe and Zn, which were the most descriptive elements. Among the multivariate techniques (PCA, CA and LDA), LDA was also a useful tool for satisfactory recognition and prediction abilities corresponding to the four categories, i.e., young and aged wines from the two regions. Classification of the wine samples attributable to both provenance and ageing on the basis of the content of 10 elements showed a total recognition ability of 93.6%, whereas in the case of validated results, the total prediction ability was 76.2%. This means that unsupervised and supervised pattern recognition methods could distinguish Cordoba wine samples in terms of their provenance and their type of elaboration, with or without ageing.

Studies by Gimenez-Miralles et al. (1999) involved the analysis of SIR and NMR spectra via the use of ^{13}C IRMS and ^{2}H [^{1}H] NMR tools to characterize red wines from three regions of DO Valencia over three vintage years. Significant interspatial and interannual differences in the $\delta^{2}H$ and $\delta^{13}C$ ratios were identified, possibly corresponding to environmental and ecophysiological factors. The LDA model was highly discriminative with respect to both stable isotopes, resulting in reasonable differentiation of wine samples even at a microregional scale with respect to their geographic provenance and grapevine genotypic features. The data obtained for the 1991–1993 period were useful for identifying the most important red wine-producing regions of Valencia and, therefore, for quality control with respect to regional origin evaluation.

González and Peña-Méndez (2000) established an elemental profile with other physicochemical parameters in musts and wines from the Tacoronte–Acentejo PDO area (Spain (Canary Islands)) obtained via an artisanal winemaking procedure. Univariate and multivariate data analyses were used to discriminate between two harvest years for both musts and wines. Chemometric techniques, including DA, PCA, biplot analysis and FA, enabled the characterization of the studied wines. PCA, with the first two PCs explaining 45.2% of the total variance, enables them to clearly distinguish between must and wine samples. Although differentiation between musts and wines was obtained, no significant differentiation was achieved with respect to the vintage year. However, owing to the use of biplot analysis and FA methods, it was possible to select many discriminative descriptors for distinguishing between musts and wines, as well as to classify these products on the basis of their vintage year and category.

16.2.4.8 Verification of Authenticity/Adulteration

Wine

According to Barbaste et al. (2002), the integration of ICP–MS with multivariate processing analysis was helpful for categorizing wines (white, rose, red and sweet wines) belonging to eight DOs for three vintage years from four Canary Islands. CA was used with all the variables, and five classification SIMCA models were used with the elements together or differentiated into boxed groups, i.e., Pb isotope ratios, REEs and other elements. Therefore, it was concluded that more adequate information is available for grouping of variables “other metals”, which is helpful for identifying the origin of wines from the Canary Islands. Owing to the use of CA, it was possible to demonstrate the similarity between the SIMCA boxes of the model and define the distance between models corresponding to sensitivity and specificity for the eight DOs with the five groups of variables considered.

Frías et al. (2001) combined FAAS data with chemometric processing, enabling the classification of sweet wines from two DO islands of the Canary Islands. PCA, with the first two PCs explaining 71% of the total variability, revealed that the five variables explained 84% of the variability among wine samples, with K, Na, Ca, Mg and Cu originating from two DO islands. This result was confirmed by CA and SLDA; however, in the latter case, Na and Mg were adequate to achieve 100% classification and 100% prediction in each category.

Pérez-Magariño et al. (2002) studied commercial rosé wines originating from three Spanish PDOs and from two consecutive vintages. The MEC and different phenolic compounds were determined by AAS and different traditional methods, respectively. ANOVA, LSD and SLDA tools were used to process the original obtained analytical data. The latter clearly differentiated the different PDO wines relative to all three regions with 12 selected parameters, including ethanol and Ca, as the most descriptive variables. The evaluation of the model was performed by CV, which achieved a total percentage of correct classification and a global prediction of 98.8% and 97.3%, respectively. The results were considered satisfactory and acceptable; therefore, the selected variables were utilized to differentiate wines according to geographical origin.

García-Rodríguez et al. (2011) quantified MEC by ICP–AES in red wines from two important DO wine-producing Spanish regions, i.e., "Ribera del Guadiana" (RG) and "Méntrida" (M). Both classes of certified wines were clearly distinguished owing to LDA as well as PCA, with the first three PCs explaining 47.3% of the total variance. The former methods of selecting As, Cd, Cu, Li, Pb and Sr achieved 97.6% and 88.3% recognition ability for RG and M red wines, respectively, with a significant squared Mahalanobis distance.

Geană et al. (2017) performed analyses of the MEC combined with $^{87}Sr/^{86}Sr$ ratios determined by ICP–MS, GF–AAS, FAAS and Q–ICP–MS to characterize red wines with denominations of PDOs and GIs from different wine regions of Romania. On the basis of the LDA data, among the considered variables, Ga, Al and $^{87}Sr/^{86}Sr$ were the most important identifiers for interspatial differentiation of the wines, with 100% predicted membership. The samples were grouped into three separate clusters attributable to region of origin, distinguishing mainly the wines between the Dealurile Vrancei and Terasele Dunării regions (along DF1), whereas discrimination of the wines originating from the two neighbouring regions, i.e., Cuza Vodă and Însurăţei, was achieved along DF2.

Jos et al. (2004), on the basis of ICP–AES, GF–AAS and HG–AAS data, investigated two kinds of Spanish and French sparkling wines with the D.O. trademark, i.e., cava and champagne of various brands. When pattern supervised learning recognition techniques such as LDA and SIMCA were applied, the two classes of samples, i.e., brut cava and brut champagne, were highly discriminative with variables such as Zn, Sr, Pb, Na, Cu, Ni, As, P and Cd. The SIMCA procedures achieved both SENS and SPEC values of 100%, and consequently, the probabilities of false-negatives and false-positives were negligible.

Charlton et al. (2010) reported that 593 authentic red and white wine samples collected from four countries were analysed with respect to the quantification of 63 chemical parameters, including mainly mineral components. The original chemical parameters were transformed into numerous multidimensional models, such as DPLS, UVE–DPLS, CART, NFS, CART—NFS and CART–DPLS. For example, all 43 descriptive variables selected by CART were potentially significant for differentiating wines with respect to their vintages, regardless of their geographical origin. Moreover, all 36 discriminate variables selected by CART were the most useful for distinguishing the grape varieties of the wines studied, regardless of their geographical origin. Notably, these parameters could be used to evaluate the authenticity of wines.

Perini and Camin (2013) analysed the $\delta^{18}O$ of ethanol to improve detection of the watering of wine and to determine the origin of ethanol. The authors processed fermented samples, i.e., authentic wine samples from all over Italy, spirits from fruit and cereals and chemically synthesized ethanol with the combined use of EA–IRMS, TC/EA–IRMS and SNIF–NMR with LRA. The $\delta^{18}O$ of ethanol was strongly related to the $\delta^{18}O$ of fermentation water and could be used as an appropriate internal reference. The method was useful in improving the detection of the watering of wine and establishing the provenance of ethanol (from grapes, other fruits or synthesis), except for the addition of cane or beet sugar to wine. According to the authors, the methodology was useful for improving the detection of illegal watering of wine.

Dordevic et al. (2012) reported results based on 5220 DOC and DOCG wine samples, which originated from 17 Italian regions, from the Italian Wine Databank from 2000 to 2010. A combination of RMS and NMR data (δ^2H, $\delta^{18}O$ and $\delta^{13}C$) with univariate (*t* test) and multivariate (MD) methods, including mixture modelling, was used to increase the sensitivity and specificity, leading to improvements in the authenticity of wines. It was concluded that the multivariate method was more effective than the univariate method, and in the absence of a normal distribution of results, additional improvements based on mixture modelling could be achieved. According to the authors, taking into consideration the diversity in climatic and geographic conditions related to wine-growing areas, the use of mixture modelling tools should lead to further improvements.

According to Pepi et al. (2019), ICP–MS and XRF are advanced approaches used to estimate the mineral distribution pattern of the *Vitis vinifera* cultivar Cannonau used to produce the renowned Italian-controlled DOC wine "Cannonau" from Sardinia. Two Cannonau vineyards were studied to determine the relationship between the geochemical composition of vineyard soil and the chemical profiles of leaves and grape berries. The PLS–DA technique differentiated the "Cannonau" cultivar according to geographical origin. Elements such as B, Sr and Zr may be useful determinants as geochemical markers because they can be used to establish a reliable relationship between the soil vineyard and the leaves and grape berries from the two study areas.

According to Epova et al. (2020), Pb contents and isotope ratios were determined by MC–ICP–MS and Q–ICP–MS to assess their ability to authentically and geographically determine authentic Bordeaux wines from prestigious châteaux wines and wines that were labelled as from Bordeaux but whose label was dubious. The total Pb levels in Bordeaux wines have considerably decreased over the last 50 years, with a significant shift in isotopic signatures towards geogenic values related to the general trend of European environmental lead monitoring. This may be explained by the positive consequences of the ban on leaded gasoline in Europe. Moreover, the Pb isotopic signatures of the allegedly Bordeaux wines of dubious origin were significantly different from those of authentic wines. This finding was supported by the use of three-isotope mixing systems to discriminate between the geogenic and the anthropogenic Pb isotope data characterizing European and Asian origins. According to the authors, this approach did not enable very clear differentiation between authentic and counterfeit ones.

Durante et al. (2015) reported that interpretation of $^{87}Sr/^{86}Sr$ data established by MC–ICP–MS and ICP–Q–MS enabled the classification of samples of *Lambrusco* PDO wine obtained from three areas of the district of Modena. Generally, no significant interspatial variability was achieved due to the proximity of the wine production zones and the similarities of the geo-, litho- and pedological features of the zones, which are predominantly affected by the Po River Basin. It was postulated that further studies are needed to develop a powerful geographically attributable model that considers, e.g., the isotopic variation in soil and the characteristics of groundwater and wine branches. Moreover, there was no significant variability within wines of the same variety across years. According to the authors, such temporal invariance of the $^{87}Sr/^{86}Sr$ ratio highlights its powerful geographical provenance and offers wide potential for effectively verifying the authenticity of the food tested.

Moreover, Durante et al. (2016) used a combination of advanced MC–ICP–MS, SNIF–NMR and EA–IRMS methods to analyse the light and heavy element isotope ratios or exclusively the $^{87}Sr/^{86}Sr$ ratios of soil from grapes to wine originating from different wine-producing regions. A decreasing trend in the $\delta^{87}Sr$ value was observed from the grape juices to the final products, which could be due to the winemaking process. Notably, the $\delta^{15}N$ values for grapes corresponded to the variability of soil, which could be used as a marker for future geographical traceability of wine. The obtained $\delta^{87}Sr$ data were used to construct strontium isotopic maps to confirm the Lambrusco PDO wine provenance as well as other products of the Modena district. A strong relationship was observed between the $^{87}Sr/^{86}Sr$ isotope ratios of soils and vine branches on a large scale, demonstrating that vine branches can be useful as an optimal sampling approach to support geographical traceability (Durante et al. 2018).

Rebolo et al. (2000) employed FAAS, AES, GC/FID and HPLC/DAD methods with multivariate models such as CA, PCA, LDA, KNN and SIMCA to classify Galician red wines collected from Ribeira Sacra CBO (RS) and non-Ribeira Sacra CBO (N-RS). The contents of mineral elements, 16 volatile compounds and phenolic compounds were processed chemometrically to differentiate the studied samples. For example, PCA (with the first three PCs amounting to 40.06% of the total variance) discriminated wines of Ribeira Sacra (RS) origin from non-Ribeira Sacra

(N-RS) wines. Owing to the reduction of 35 variables to Fe, Li, Rb, delphinidin and epicatechin, classification methods such as LDA, kNN and SIMCA achieved essentially the same or even better results than did the use of the complete database. LDA enabled correct recognition and prediction abilities between 90% and 100% for the two wine classes, i.e., RS and N-RS. LDA and kNN had the best recognition and prediction abilities for RS samples, whereas SIMCA provided a powerful ability for N-RS wines. These studies confirmed the authenticity of Galician CBO wines from the RS area.

According to Iglesias et al. (2007), both ICP–AES and ICP–MS were useful for studying wine samples from the DO Empordà–Costa Brava (Catalonia). PCA, with the first two PCs accounting for 49.6% of the total variance, enabled them to differentiate DO Empordà–Costa Brava wines from the wines of other DOs studied, i.e., from three other wine-producing regions, i.e., Penedès, Somontano and Rioja. The Sr and Ba levels could be used to differentiate the two DO groups. Furthermore, a DA function based on both variables discriminated the two groups, achieving a 100% classification rate. At the level of the LOOCV, all of the Empordà–Costa Brava samples were satisfactorily classified.

Pérez-Álvarez et al. (2019) aimed to quantify MEC via ICP–MS to test wine authenticity, namely, to estimate the elemental composition of white and red wines collected from AOC Rioja, with D.O.Ca., the highest category in Spanish wine regulation. DA was an effective tool for clearly discriminating samples with respect to white and red grape varieties, with a correct classification of 98.5% of the analysed wine samples. Reliable differentiation of the Na and Cs contents between aged and young wines was also obtained, with 100% accuracy. With respect to the influence of the addition of SO_2 to the wines, owing to the contributions of most descriptive elements, such as Cs, Mg, Cu, Pb and Zn, the samples were distinguished according to their addition (function 1) and according to whether the wines were stored in oak barrels (function 2), with 100% of the studied samples being correctly classified. Spatial differentiation, with contributions from the most descriptive elements, such as Cs, Pb and As, was possible for the wines from the three studied soil types, with correct classification of 100% for the investigated samples.

Gonzálvez et al. (2009) applied ICP–OES in combination with multidimensional tools, such as CA, DA, PCA, HCA and CARTs, to identify the origins of different red wine samples from four protected designations of origin (PDOs). Multivariate analysis methods were useful tools for classifying wine samples according to their elemental profile. The HCA and PCA results clearly indicated that Utiel–Requena and Jumilla red wines were distinguished from the wines of the other PDOs in their vicinity, although wines from Valencia and Yecla PDOs were not distinguished on the basis of the abovementioned methodical solution. For the usefulness of CART, wines from Utiel–Requena and Jumilla could also be effectively authenticated with the use of only the contents of Li and Mg, respectively. Additionally, wine samples from the Valencia PDO subregions could also be differentiated with an acceptable resubstitution rate. The DA model was also able to separate all the studied Spanish PDO samples, achieving accurate results by means of the resubstitution method of 100%.

According to Marini et al. (2006), the use of two chemometric class-modelling techniques, such as SIMCA and UNEQ, enabled effective processing of concentration data for Cu, Zn, Pb, SO_2 and 31 other indices related to Italian wine samples from seven Italian CDOs. When only four indices (Cu, Zn, total antocyans and SO_2) were included in the UNEQ models, it was possible to achieve a high percentage of the test samples accepted by the corresponding category models built using samples from different vintages. When these four variables were applied to build the corresponding SIMCA models, the results obtained were comparable and even significantly better than those of the SIMCA models when all the variables were used, with all the test samples recognized by their corresponding category model. Both techniques were the most effective in identifying the levels of Cu, Zn and Pb in SIMCA, as well as the levels of Cu and Zn in UNEQ, as the most discriminative variables. These indices are mostly related to the soil composition and therefore strictly correspond to the geographical provenance of the samples. Owing to the use of the two chemometric class-modelling techniques, it was possible to authenticate the origin of Italian wines from different CDOs.

Shen et al. (2012) reported that instrumental tools, such as HPLC–UV, HPAEC–PAD, ICP–MS, and other official methods enabled the determination of the MEC, sugars and lactic acid contents via chemometric techniques to characterize semidry rice wines from three well-known wineries in China. Multivariate analysis was applied to predict and classify samples of different ageing statuses at 3, 9, 15, 21 and 33 months. Systematic differences between samples were revealed by two-way ANOVA and PCA. The first two PCs, accounting for 43.58% of the total variance, provided unsatisfactory results because PCA was unable to clearly discriminate the rice wine samples. However, LDA based on 16 selected variables was more useful since it enabled the discrimination of wine samples, achieving 88.5% accuracy in terms of the LOOCV. The results demonstrated that the chemical parameters enabled the authentication of the ageing status of rice wine.

Martinelli et al. (2003) used EA–IRMS to process the wine and the CO_2 bubbles produced during the second fermentation for several sparkling wines originating from 10 countries around the world, bearing in mind that such analyses were performed to assess the addition of sugar obtained from C4 plants (sugarcane or corn). The average $\delta^{13}C$ values of Brazilian sparkling wines were significantly different from those of sparkling wines produced in other parts of South America, i.e., Argentina and Chile, as well as Europe, i.e., France, Germany, Italy, Portugal and Spain. However, no statistical differentiation was achieved for sparkling wines produced in the United States or Australia. Such intercountry discrimination related to $\delta^{13}C$ levels in sparkling wines may be explained by the addition of C4 sugar during the production of some sparkling wines in Australia, Brazil and the United States. Similar trends were observed in the case of the $\delta^{13}C$ composition of the CO_2 bubbles. For example, the $\delta^{13}C$ values obtained for Brazilian and Argentinean sparkling wines suggest that the source of carbon for the second fermentation was sugarcane. However, in the case of sparkling wines produced in Chile and Europe, a different sugar (most likely sugar beet) was used mainly in the second fermentation. It was concluded that the $\delta^{13}C$ values of sparkling wine and CO_2 bubbles could frequently be useful in identifying sugar sources.

Cabañero et al. (2007) determined $\delta^{13}C$ isotope ratios to investigate the authenticity of commercial sparkling drink samples from various origins (Spain, Portugal, Italy, France and the United Kingdom). The $\delta^{13}C$ (cava) values were quantified when C3 sugar addition was performed for a second alcoholic fermentation. The addition of exogenous CO_2 to sparkling water and wine as well as the addition of C4 sugar to "aguja" wine could also be detected. According to the authors, the measurement of CO_2 enables verification of whether products are compatible with the declaration on the label.

Geană et al. (2016a) applied the combination of CF–IRMS, HPLC–PDA, HPLC–UV–VIS and MA-E-AS315-11-ANCYAN according to OIV to determine the SIR with other variables, e.g., antioxidants, synthetic sweeteners and red dyes in commercial sweet and medium sweet red wines as well as authentic red wines from EU countries and non-EU countries, e.g., Romania, Bulgaria and the Republic of Moldova. LDA differentiated 100% of the table wines according to category, i.e., good table wines, suspect table wines and adulterated table wines. The profile of anthocyanins, an indicator of the red colour of wine, was assessed in table wines and compared to that of authentic wines obtained by microvinification of *Vitis vinifera* to determine their authenticity. Chemometric interpretation of the data enabled the discrimination of the table wines in terms of quality. Among the investigated table wines, 28% were adulterated, and 17% were suspected of adulteration, reflecting a disturbing increase in fraudulent practices in the wine industry, such as the addition of natural or artificial sugars, water and colouring agents.

According to Dugo et al. (2005), the use of IC and ASCP methods enabled the determination of selected anions and cations in Marsala wines produced in the D.O.C. zone of Marsala (Italy), which were also classified according to their age, colour and sugar content. ANOVA revealed that ageing in wooden barrels significantly affected the levels of Cd^{2+}, Cl^-, NO_3^-, Pb^{2+}, SO_4^{2-} and Zn^{2+}. The results of PCA and, largely, those for LDA indicated a clear differentiation of the four groups representing various ages of the wine. Although MLRA provided a strong positive relationship between wine age and the studied variables, when LRA was used, no significant correlation was detected between the sugar content of wines and the element concentration.

Raco et al. (2015) determined SIR data by SIMS in Italian wine with ground water and grape berries. The data interpretation revealed the ability of $\delta^{18}O$ in wine and grape berries, especially in the case of its combination with δ^2H, to determine the geographical provenance of wine. It was concluded that this method is an effective tool for verifying compliance with national and international regulations via correct labelling of geographical origins, identification and lack of adulteration.

Ogrinc et al. (2001) assayed samples of selected wines from three growing regions of Slovenia over 3 years using a combination of IRMS and SNIF–NMR. In the case of the $\delta^{13}C$ data, a clear PCA separation of wine samples (with the first two

PCs explaining 91.1%) originating from the coastal region and those from two continental regions was possible, although separation between samples from the Drava and Sava regions was not observed. Comparable good discrimination of samples was obtained by LDA. However, complete separation between the wines from the Sava and Drava regions was achieved when all SNIF–NMR combined with $\delta^{13}C$, $\delta^{18}O$ and δD data were included as a dataset on which PCA and LDA calculations were performed. Moreover, this study explained the factors causing seasonal differences. For example, in the case of PCA and LDA, which are based on δD and $\delta^{13}C$ data, separation between wines from 1996 and 1997 was possible. This methodology limits adulteration, such as chaptalization and/or watering.

Marchionni et al. (2016), on the basis of TIMS data, reported that no isotopic variations were detected during winemaking (in the case of musts and wines), which means that the $^{87}Sr/^{86}Sr$ signature of wines is strongly related to the bioavailable fraction of the soil rather than to its bulk. Therefore, Sr isotopes can be useful as robust geochemical tools for fingerprinting the geographic authenticity and provenance of wine.

According to Tescione et al. (2018), grapes inherit their Sr isotopic composition only from the geologic substratum of the vineyard of production, and it does not depend on the variety of the cultivar. It has also been shown that for four harvest years, the $^{87}Sr/^{86}Sr$ values remained constant, supporting the idea that climatic variations do not impact on these values. This means that Sr isotopes could be useful tools for evaluating the geographical provenance of grapes and may be used to label wine production.

Samples of Canarian dry and sweet DO wines from three regions were studied for the contents of 11 elements processed chemometrically by one-way ANOVA, CA, LDA, ANN and BP–ANN (Frías et al. (2002) or PCA and SIMCA (Frías et al. (2003). MEA was used to investigate dry and sweet wines corresponding to the three DOs. CA enabled them to obtain two main clusters that were well distinguished, corresponding to samples from Lanzarote and those divided into the remaining groups attributable to La Palma and El Hierro. CA and SOM differentiated wine samples with respect to their geographical origin and ripening state. Figure 16.20 shows a dendrogram built with the variables using Ward's method and the Euclidean squared distance. PCA, with the first two PCs accounting for 57.52% of the total variance, provided comparable results with CA data. PC1 is related to the region of wine origin, whereas PC2 is associated with grape maturity. The wine samples were clearly distinguished according to the three Canarian Islands. A clear discrimination of wines originating from the three islands was achieved via the use of LDA, which showed 100% recognition ability and 95.6% prediction ability. The SIMCA index showed a highly acceptable degree of sensitivity (97.8%) and specificity (100%). It was concluded that BP–ANN provided satisfactory 100% recognition and 100% prediction ability, and the slightly less effective LDA and SIMCA were useful tools for the recognition of interregional discrimination of wine samples. Most of the potential discrimination ability was due to Sr, Zn and Li between El Hierro and La Palma, Li and Rb between El Hierro and Lanzarote and Rb between La Palma and Lanzarote.

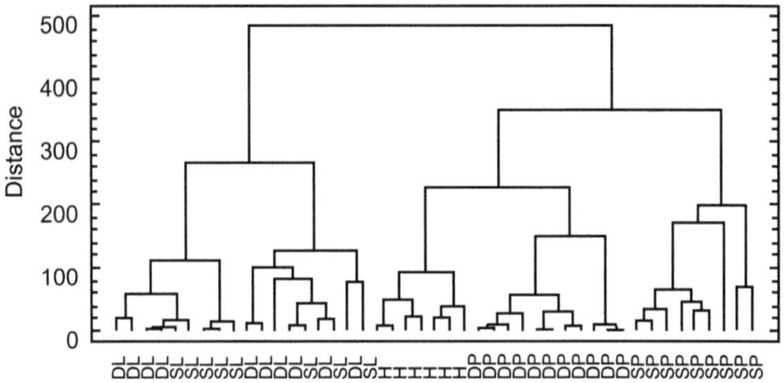

Fig. 16.20 Dendrogram built with the variables using Ward's method and the Euclidean squared distance. Codification: SL = sweet wines from Lanzarote; DL = dry wines from Lanzarote; DH = dry wines from El Hierro; DP = dry wines from La Palma; SP = sweet wines from La Palma. (After Frías S et al. 2002. Food/Nahrung 46:370–375. https://doi.org/10.1002/1521-3803(20020901)46:5<370::AID-FOOD370>3.0.CO;2-F. With permission)

Paneque et al. (2009) used AES and AAS to study "Oloroso" sherry wine samples of different DOs from three wine-producing regions in Spain. CA indicated that Andalusian Oloroso wines were approximately grouped with respect to the denomination of origin, whereas the best result was achieved by LDA. Classification of the samples from the three DO regions using Na, Mg, Fe, Al and Sr was attained, with a total recognition ability of 100% and a total prediction ability of 93% for LDA and LOOCV, respectively.

According to Latorre et al. (1994), the elemental patterns of Galician white wines from Rlas Baixas (RB) and non-Rlas Baixas (NRB) were analysed via AAS and AES. Both CA and PCA clearly discriminated Spanish wine samples from two regions, i.e., RB and NRB. The three-dimensional scatter plot of the PCA objects, with the first three PCs accounting for 72.0% of the total variability, demonstrated that Li and Rb, as dominant descriptors, correlated significantly with PC1 and PC2, respectively. LDA for all the elements correctly classified all the samples, with a highly satisfactory 100% recognition ability for the two classes. It was also found that KNN and SIMCA for Li and Rb achieved a correct recognition ability above 95% and a correct prediction ability above 91%. It was concluded that samples from NRB brands could be correctly classified, enabling for the detection of consumer fraud, because the probability of classification of an NRB trademark as a genuine RB does not exist in practice.

Moncayo et al. (2016) used LIBS to analyse samples of red wine from different PDOs as well as foreign red wines and table wines from Spain, Germany, France and Italy. Owing to its capacity to model complex nonlinear input–target relationships, the ANN enabled the recognition of the PDO pattern and correct classification of all wines tested, maximizing inter-PDO differences and minimizing intra-PDO variability. All the validation procedures achieved successful results. A high sensitivity and generalization ability were possible, with averages of 99.2% and 98.6% correct classification, respectively, with a robustness of 100%. It was postulated that the approach applied could provide in situ classification results, advancing adulteration control and developing quality control of agronutritional products.

According to Calderone and Guillou (2008), the analysis of δ^2H, $\delta^{13}C$ and $\delta^{18}O$ was performed by TC/EA–IRMS to assess watering in beverages such as wine and fruit juices from eight European countries. A real case of multi-isotopic and multi-component analysis of wine (^{13}C measured on the distilled ethanol from wines, ^{18}O determined on the water of the wine) for detection of adulteration was also examined and interpreted. The results of the stable isotopic ratio measurements were useful for detecting the illegal watering of beverages, including wines.

Other Alcoholic Beverages

Vinegars

Bearing in mind that the production of wine vinegar by fermenting dried grapes and rehydrating with tap water is illegal according to European Regulations, Camin et al. (2013) analysed 14 Italian groups of authentic samples (wine, raw vinegar, the respective diluted vinegar and the tap water used for dilution from different regions) and 92 dubious wine vinegars marketed in Italy and produced in Greece. It was proven that $\delta^{18}O$ analysis of beverage water, which is officially used to detect the watering of wine and rehydration of concentrated fruit juice, could also be applied to vinegar for the detection of this type of fraud. These production chains, from wine to the corresponding raw vinegar and diluted vinegar, were examined, as was the official European wine databank established according to EU Reg. 555/2008. The $\delta^{18}O$ limits were designated for raw and diluted vinegars produced from fresh grapes. Values lower than the normative values indicate a significant addition of water to a starting matrix with a sugar content significantly greater than that of fresh grapes. In this context, more than 60% of the 92 dubious wine vinegars imported to the Italian market were not authentic. According to the authors, the suspected samples could most likely be obtained by diluting dried grapes with water.

Perini et al. (2014) aimed to study SIR and 33 volatile compounds in the production chains considered, e.g., wine, the corresponding wine vinegar and grape must. A combination of EA–IRMS, SNIF–NMR and GC–MS/MS methods with chemometric tools such as ANOVA, HSD Tukey, Kruskall–Wallis and U-Mann–Whitney tests was a useful approach. The quantification of δ^2H and $\delta^{13}C$ in ethanol and acetic acid and $\delta^{18}O$ in water was performed to determine whether these compounds could be used for the ingredients of ABM to verify their authenticity. The standard deviations of repeatability and reproducibility were comparable in wine vinegar and ABM and overall lower than those reported in the official methods. In addition, the impurities in solutions extracted from wine vinegar and ABM were verified and quantified, the validation parameters of the methods used for the ingredients of AMB were specified, and the parameters were compared with those reported in the CEN and OIV standards. The variation in δ^2H and $\delta^{13}C$ between wine ethanol and acetic acid in wine vinegar and ABM was assessed to determine whether the wine isotope databank might also be useful as a reference for δ^2H and $\delta^{13}C$ of acetic acid for monitoring the authenticity of vinegar and ABM.

Benito et al. (1999) determined the levels of Cl, Ca, Cu, Fe, K and Na combined with selected organic compounds in wines from the CDO Rioja and Jerez vinegars via chemometric analysis, i.e., CA, PCA, SIMCA, GINN and MLF–ANN. According to the authors, all the analyses were performed in the Estación Enológica de Haro, the official laboratory that deals with the quality and origin certificates relating to the Qualified Denomination of Origin Rioja, using official methods as a part of a cooperation at national and international scales. Satisfactory sensitivity and specificity were obtained for SIMCA and GINN. For example, the former provided highly acceptable values corresponding to sensitivity (84.9%) and specificity (88.9%) for Rioja vinegars, whereas these parameters were 100% and 94.4%, respectively, for Jerez vinegars. The predictions by CV with SIMCA and GINN were also satisfactory, achieving 92.9% and 88.7% (median of 10 training steps), respectively.

Paneque et al. (2017) used ICP–OES to assay wine vinegars of three PDOs with different ageing durations from Andalusia. Concerning the effect of geographical origin on young and aged vinegars, LDA with selected variables such as B, Ca, Pb and S obtained 100% recognition ability, and external validation results achieved good discrimination among PDOs, with a prediction ability of 80%. An identical result was obtained with SVM analysis. For the effect of ageing time on the elemental profile within each PDO, good differentiation was obtained by CA between young and aged vinegars, with 93.3% of the aged vinegars in the first sample group and 84.6% of young samples in the other group. Using variables such as S, Al, Mn, Zn, Ca, K, Na and Fe, forward SLDA achieved a total recognition ability of 92.86% when classifying three PDO vinegars as young and aged vinegars. The results obtained in the external validation were 86.7% and 90% for the test set in LDA and SVM, respectively. The classification models were obtained by LDA and SVM, with satisfactory prediction abilities, i.e., 73% and 80%, respectively.

Cider

García-Ruiz et al. (2007) quantified MEC by a combination of MC–ICP–MS, ICP–AES and ORS–ICP–MS to assess the chemical composition of cider samples from different locations in four European countries, including Asturian samples (Spain) mostly produced under the PDO. PCA, with PC1 and PC2 accounting for 53.2% of the total variance, was unable to clearly distinguish the cider samples with regard to their geographic origin. For example, Asturian ciders, i.e., PDOs and non-PDOs, were undifferentiated and similar to ciders originating from Basque country. However, LDA was the most effective technique for correctly classifying 100% of cider samples originating from different Spanish regions, as well as other countries, i.e., France, England and Switzerland. Elements such as Na, Mg, Al, K, Ca, Ti, V, Mn, Zn, As, Rb, Sr, Mo, Ba and the $^{87}Sr/^{86}Sr$ isotope ratio were the most discriminative variables.

According to Carter et al. (2015b), multi-instrumental methodologies such as TC/EA–IRMS, ICP–OES, ICP–MS, compact IC, HPLC–RID MEA and SIRA with computational tools were reliable approaches for evaluating the mineral and stable isotope ratio profiles of European and Australasian ciders produced in eight countries. The loadings for PC1 and PC2 were mainly described by the K and S variables, respectively. The CDA was also determined by the levels of K, P, Ca and Mg but not by S. This could be explained by the fact that the majority of S had a source from an additive rather than from a natural component of the fruit juice. Therefore, in contrast to PCA data, CDA, which is based exclusively on elemental data and especially on their combination with isotopic data, achieved a clear distinction between Australian, New Zealand and EU countries. The most discriminative elements were Sr, Ba, Mn and Cl. In summary, it was possible to designate the ciders according to their country of provenance with 100% correct classification. The graphical presentation of data in the form of a geospatial system enabled much clearer identification of the country of origin. Given the inconsistency in the isotopic composition of cider samples with the claimed country of origin, such methodology can require verification of the authenticity of cider samples without requiring reference samples.

Gaillard et al. (2013) reported that authentic ciders originating from various French production areas and commercial ciders were analysed via MultiFlow-IRMS to determine the $\delta^{13}C$ values of cider CO_2 bubbles. Although most of the samples analysed presented a gas $\delta^{13}C$ value in the expected range, some ciders presented $\delta^{13}C$ values outside the 3σ limit, reflecting carbonation by technical CO_2. In the case of the latter samples, exogenous carbonation must be investigated. Some cider samples were suspected of exogenous carbonation because the $\delta^{13}C$ value was significantly lower. Therefore, this approach was intended to control natural carbonation (coming from in situ sugar fermentation in the bottle) or exogenous carbonation of CO_2 bubbles, making possible more effective control in agreement with labelling to protect quality products from adulteration. It was concluded that the methodology is useful for cider CO_2 measurements as well as for any sparkling beverage after its simplicity and reliability are considered.

Singerling et al. (2018) processed hard ciders (15 of different brands) from the United States on the basis of the $\delta^{13}C$ values to identify specifically added C4 sugars (corn syrup and cane sugar) in opposition to the main ingredient of hard ciders, mostly apple juice, which is derived from C3 plants. Most imported ciders studied from the United Kingdom and Ireland did not indicate the presence of added C4-based sweeteners, whereas US ciders generally had elevated $\delta^{13}C$ values, confirming the presence of C_4-based sugar additives. It was also reported that the labelled ingredients in ciders are inconsistent with the carbon isotope ratios. Moreover, the $\delta^{13}C$ values were not correlated with the sugar content in the cider (according to the nutritional information label), likely reflecting varying fermentation times.

Cristea et al. (2019) combined dual inlet IRMS with ICP–MS to characterize the MEC and SIR profiles of apple and pear cider from Romania to evaluate the addition of cane sugar or corn glucose syrup. Apple and pear ciders, fruit or other ingredients were added to the apple cider to provide flavour. Moreover, apple with cinnamon, strawberry, cranberry, elderberry, blackberry, kiwi with lime, fructul pasiunii and sour cherry, fruit or other ingredients were added to the pear cider to provide flavour: vodka, cherries, cola, linden, forest fruits and passion fruit. The stepwise LDA method was applied to limit the number of redundant variables used as markers for fruit differentiation, i.e., In, Cu, Al, K, Fe and $\delta^{18}O$. Consequently, it was possible to clearly discriminate cider samples with respect to the variety of fruit achieving percentages for initial and cross-validation tests of 96.9%. In this case, the K level is one of the strongest predictors since it is also added by producers in different forms as preservatives and additives. The percentages of initial and cross-validation classification were estimated to be 100% and 95.5%, respectively. Chemometric assessment of the elemental and stable isotope distribution patterns of ciders might be useful for the authentication of fruit for ciders and cider producers.

Whiskey, Rum and Others

Parker et al. (1998) analysed the $\delta^{13}C$% values of volatile congeners in whisky samples by integrated IRMS, GC–FID and GC–C–IRMS methods. RDA was used to compare the carbon isotope profiles of the whiskies. Owing to this approach, it was possible to distinguish other whisky samples from the Scotch whisky blend on the basis of the $\delta^{13}C$ values. It was also demonstrated that this approach can be used to establish the authenticity of the brands of whiskey. Two other whisky samples differed from the whisky samples on the basis of their $\delta^{13}C$ values.

Meier-Augenstein et al. (2012) established both $\delta^{2}H$ and $\delta^{18}O$ data via the TC/EA–IRMS method to characterize neat Scottish whiskies originating from 11 distilleries from different locations and the local source of water used and during whisky production. Statistical evaluation of the data obtained enabled them to differentiate between authentic and counterfeit whiskies. Bulk $\delta^{2}H$ and $\delta^{18}O$ isotope analyses of neat spirit have the potential to quickly recognize counterfeit Scotch whisky originating from different regions.

Kokkinofta et al. (2003) aimed to establish the authenticity of the traditional Cypriot spirit zivania. Both CDA and CBT models, which process ICPS data for alcoholic beverages (40–55 alcoholic degree) from 17 countries, revealed the grouping of the analysed samples with respect to their elemental contents. Notably, the ability of Cu to discriminate zivania from other samples could be attributed to the use of copper distillation vessels. CDA and CBT analyses demonstrated the discrimination between the analysed samples, according to their elemental profile corresponding to their geographic origin. Furthermore, CDA and CBT revealed that Mg, Zn and Cu were able to differentiate zivania from other spirits with similar degrees of alcoholism. Moreover, both of these chemometric tools enabled very accurate classification because only six variables sufficiently predicted the origin of each distillation. On the basis of CDA, the overall classification was estimated to be very good at 93.75%. It was concluded that this discrimination in the elemental profile between the alcoholic beverages studied was related to the geological and meteorological conditions specific to the island of Cyprus.

Kokkinofta and Theocharis (2005) presented the ICPS, HPLC–RID, GC–FID and ^{1}H NMR analysis results for alcoholic beverages, including zivania spirit from Cyprus, samples of eau de vie and other distillations from Greece and 16 other countries. Twenty-six variables, i.e., the contents of Fe, Cu, Mg, Ca, K, Na, six alcohols, acetaldehyde, ethyl acetate, ethyl lactate, furfural, six sugars and other physicochemical parameters, were obtained and processed by a chemometric tool. With respect to the chemometric procedures (PCA, CA, RDA and CART), acceptable results were achieved using PCA with the first three PCs accounting for 56.3%

of the total variance. A clear differentiation of red zivanias (two groups: sweet and dry), eau de vie samples and Greek distillates was achieved. Zivanias were well identified except for the sample from northern Cyprus (Yeni Raki and the double-distilled sample from Kaminaria). The samples with origins outside Cyprus and Greece could not be discriminated from each other using this technique. Furthermore, CA showed less significant efficiency in emphasizing the group structure. The total correct recognition ability was much better with the use of the RDA and CART techniques, but the prediction for zivania was higher owing to the use of RDA. The total recognition ability was 97.6%, and the cross-validated error enabled for correct prediction, with 92.8% for zivanias and 100% for red zivanias. These results are well prognostic for the application of chemometrics to classify spirit samples and wines.

According to Ballabio et al. (2007), the GC–FID and ICPS methods were used to assay zivania and other alcoholic beverages originating from Cyprus, Greece, Italy, Russia and the Czech Republic. Profiles of Fe, Mg, Cu, Ca, K, Zn and other physicochemical parameters were quantified and characterized. PCA, with the first two PCs explaining 42.5% of the total variance, enabled the differentiation of zivania and red zivania samples, although it was impossible to clearly separate all the groups of studied samples. CP–ANN enabled the classification of zivania with acceptable predictive and fitting performances, despite the difference between the zivania and the red zivania samples. The model provided error rates in prediction and in fitting of 8.6% and 0.5%, respectively. Counter-propagation ANN data were analysed via PCA to select several discriminative variables for use in the classification model. PCA and Kohonen mapping were used to differentiate between red zivania and other zivania samples on the basis of Ca, Mg, K, furfural, sucrose, Brix and the RI. Variables such as Cu, Mg and Ca have particular significance since they are related to the production area or to the ingredients used, such as cinnamon and the other extracts added to red zivania samples. Copper is a parameter that is particularly related to Cyprus soil and is one of the discriminative variables in the classification model.

Cameán et al. (2001) evaluated the elemental profile determined by spectroscopic methods such as FAAS, FAES and GF–AAS of genuine Sherry and Penede brandies for their effective differentiation. PCA, with the first four PCs explaining approximately 87% of the data variance, was an ineffective technique since it was impossible to achieve a linear differentiation of classes on the basis of their origin. Therefore, to assess this preliminary study, unsupervised pattern recognition methods were applied, i.e., MPN–BP was performed. This model enabled the suitable discrimination of Sherry and Penede samples via the use of Fe, Mg, Cu and Ca as descriptors, leading to 93% prediction ability.

According to Ceballos-Magaña et al. (2009), differentiation between four types of tequila and mezcal on the basis of their elemental composition established by ICP–OES was achieved. Chemometric techniques such as PCA, LDA and PNN were applied to characterize particular types of tequila and mezcal. With respect to PCA, with the first three PCs explaining 60.9% of the total variance, no clear differentiation between the samples was attained, although some trends were observed. However, complete differentiation of mezcal from silver and extra-aged tequilas was achieved, although no complete discrimination between gold and silver tequilas was obtained. Moreover, the aged and silver classes overlapped, and some extra-aged samples were considered silver. However, the LDA model could differentiate completely mezcal from extra-aged and silver tequilas (Fig. 16.21). In addition, gold and extra-aged could be differentiated. Using the PNN, a specificity value of 100% for classification success was achieved for silver, gold, extra-aged tequila and mescal, but in the case of aged tequila, 90% of the samples were successfully classified. Elements such as Na, K, Ca, S, Mg, Fe, Sr, Cu and Zn were the most discriminant variables.

Langasco et al. (2018) determined volatile compounds and MECs via multi-instrumental methodologies such as GC–MS, ICP–MS and GC–FID in traditional *Filuferru* as an ancient spirit from Sardinia, which is usually obtained from the distillation of wine or grape marc. PCA, with the first two PCs explaining 73% of the total variance, enabled the discrimination of volatile compounds in the two groups. The significant difference in composition indicated the artisanal nature of these spirits. However, there were doubts, such as the high content of low-boiling-point compounds and acetic acid as well as the high level of Cu that can be released into the distillate from the distillation apparatus. According to the authors, careful quality control should be applied in the future to establish a regulation for valuable traditional production.

According to Shand et al. (2017), the MEC contents established by diverse methodologies, such as ICP–MS, ICP–OES, TXRF, FTIR and CPA, of Scottish whiskies originating from four producing regions were processed chemometrically (PCA, CA and LDA) to evaluate the mineral composition. PCA with the first

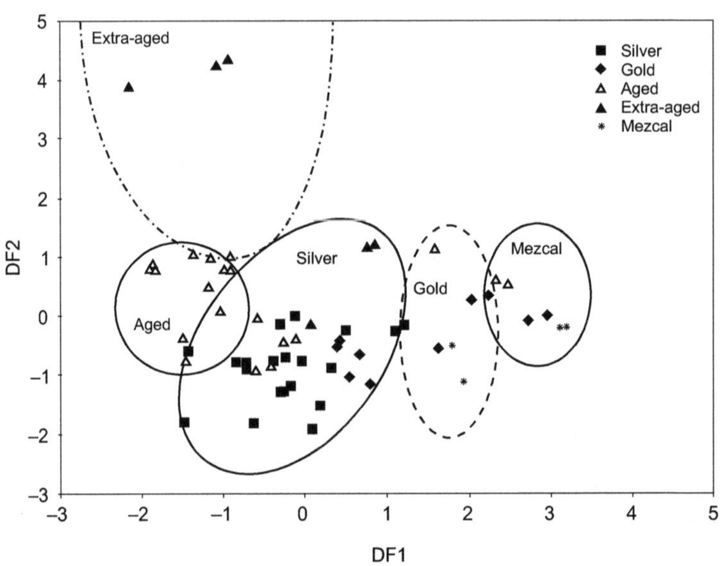

Fig. 16.21 LDA. Scatter plot of the samples in the DF1 and DF2 planes. (After Ceballos-Magaña SG et al., 2009. J Agric Food Chem 57:1372–1376. https://doi.org/10.1021/jf803626k. With permission)

three PCs corresponding to 79.6% of the total variance enables them to partially differentiate the whiskies by region or type and clearly separate the counterfeit samples from the other samples (Fig. 16.22). CA showed the grouping of the five counterfeit samples that were not completely distinguished and had similarity to

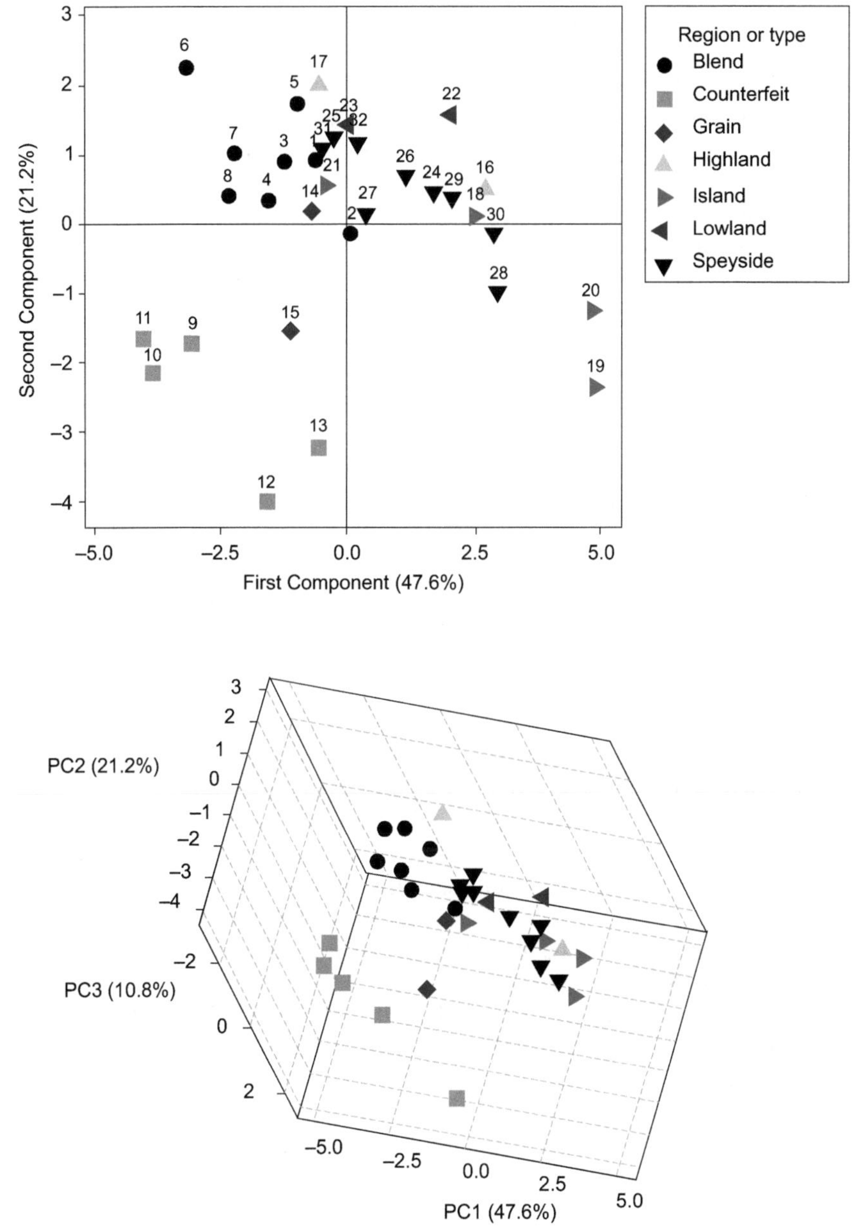

Fig. 16.22 Principal component analysis score plots prepared from log-transformed trace element concentration data for 32 whisky samples. (After Shand CA et al., 2017. Anal Chim Acta 976:14–24. https://doi.org/10.1016/j.aca.2017.04.041. With permission)

some of the blends and to the unripe grain whisky. However, PCA and LDA enabled the correct classification of adulterated whisky samples, although it should be mentioned that there were some limitations due to the relatively small number of counterfeit whisky samples identified compared with the total number of samples.

Gajek et al. (2022) studied two types of whisky originating from 11 countries and used MEC quantified by combining ICP–MS, ICP–OES and CV–AAS methods. The obtained results processed chemometrically were useful for performing a comparative analysis of elemental distribution patterns with respect to selected parameters, i.e., country of origin, type of whisky (single malt or blended) and age of the products. Cu clearly discriminated single malt and blended types of whisky. PCA with the first two PCs explained 43.3%, 43.5% and 46.1% of the total variance for the USA, Ireland and Poland, respectively. This technique did not show any clustering according to brand or production region, although simultaneous overlap of two parameters, such as age and type, was observed. Therefore, only samples with the same ageing period (3 years) were included in the newly tested dataset, which consisted of blended and single malt whisky samples. This approach enabled them to conclude that P and Zn could be related to the age parameter since the influence of this factor was theoretically eliminated because of the rejection of samples that had matured for longer than 3 years. Moreover, investigations of samples from the United State and Ireland demonstrated that the samples represented by the same product but originating from independent bottles (e.g., JB, JDG and BUS brands) presented similar elemental profiles. In contrast, comparative analysis of Scotch whisky with respect to ageing time revealed that the longer the alcohol was aged (i.e., stayed in the barrel), the higher the levels of Cu and Mn.

Flores et al. (2009) reported that ICP–MS and HPLC–DAD were used to quantify MEC, five furanic compounds and syringaldehyde in commercial Mexican tequila samples. The 40 brands studied had low contents of these elements, following current Mexican regulations, except for Cu. The concentration of this element was detected in 100% of the tequila, which exceeded the maximum permissible value. This finding can be explained by the fact that Cu or its alloys are commonly used to construct pot stills used in distillation, which could be responsible for relatively high Cu levels in all the tequila samples. PCA revealed that 100% of mixed and aged tequila samples were differentiated by the first three PCs, accounting for 94% of the total variance. The most discriminative variables were 2-FAL, 5-MFA, Cu, Hg, Th, Mo, SIR and 2-FAC.

Diverse methodologies, such as ICP–OES, FAAS, UV–VIS, GC–FID, LC–DAD, LC–FD and HPLC–UV/DAD, were used by Sampaio et al. (2008) to determine macro, micro- and trace element contents as well as 42 volatile compounds, polyphenols, aldehydes, ketones and other organic parameters in Cuban and non-Cuban rums from 15 countries. Among the chemometric analysis data, such as PCA, PLS–DA and LDA data, the former, with the first three PCs corresponding to 70.4% of the total variance, selected descriptive variables, i.e., isoamyl alcohol, n-propyl alcohol, Cu, Fe, furfuraldehyde, benzaldehyde, epicatechin and vanillin.

The LDA model enabled the discrimination of the original Cuban rums from the others with 88.2% accuracy.

Iglesias Rodríguez et al. (2010) investigated Spanish samples of traditional Galician orujo alcoholic distillates with and without CBO distillates with respect to eight elements determined by ET–AAS and ICP–OES. Moreover, three kinds of samples were analysed, i.e., samples produced with alambique and alquitara (two traditional copper-made distillation devices employed in Galicia for the production of spirits) or industrial distillation apparatuses (manufactured in refineries with industrial distillation apparatuses, i.e., stainless steel devices and glass columns). Many advanced and numerous multivariate tools, such as PCA, CA, LDA, KNN, SIMCA, UNEQ and ANN (MLF–NN, SVM, LVQ and PNN), were useful for classifying distillates. No direct relationship between the distillation procedure and the mineral element content in the orujo samples was detected. However, the most acceptable results were provided by the PNN, in which the correct recognition abilities for CBO classes and without CBO classes were 98.6% and 98.0%, respectively; the prediction ability results were 87.7% and 86.1%, respectively. This means that the applied approach enabled the development of a classification procedure to authenticate Galician CBO orujo samples.

According to Hernández-Caraballo et al. (2003), the contents of Cu, Fe and Zn were determined by FAAS in samples of Cocuy de Penca firewater from different producers from the Lara and Falco'n states (Venezuela). Among the three chemometric tools applied, i.e., LDA, QDA and various ANNs, the latter was the most effective tool, achieving highly satisfactory overall estimation and prediction capabilities of 96.5% and 97.0%, respectively. The results enabled the identification of the geographic origin of the manufacturers and the presence or absence of sugar in the product.

Suto and Kawashima (2019) used LC/IRMS to quantify $\delta^{13}C$ of bulk sake, ethanol and glucose in sake samples from three regions of Japan. Criteria for the use of $\delta^{13}C_{eth}$ and $\delta^{13}C_{glu}$ to check brewers' alcohol and sugar contents were developed. There was a strong relationship between the percentage of added brewers' alcohol and the $\delta^{13}C$ values. According to the authors, the price of sake was found to be inversely related to the percentage of brewers' alcohol in the sake; hence, it was concluded that $\delta^{13}C_{eth}$ and $\delta^{13}C_{glu}$ are useful diagnostics for identifying either brewers' alcohol or sugar, respectively.

Otim et al. (2019) used ICP–MS to determine MEC data combined with a multivariate tool (CA and PCA) to characterize, identify and authenticate African *Lira-Lira* spirits. In an elaborated methodology, the elemental profile was effective for differentiating chemometrically unrecorded spirits named locally as Lira-Lira from other spirits because of their Cu contents and for detecting their purchase origin. It was possible to identify similar sources of pollutants in 13 Ugandan spirits, i.e., Se–As–Ni, Sn–Cd, Pb–Cr, Tl–Cr, Pb–Ni or Cu–Ag. It was concluded that the methodology was useful for differentiating authentic spirits from other spirits sold in Uganda.

16.2.4.9 Assessment of Potential Markers Related to Different Parameters

Wine

Marchionni et al. (2013), via TIMS, analysed the Sr isotopes of bottled wines from four geographical regions in Italy. The $^{87}Sr/^{86}Sr$ composition was compared with that of rocks from the substrata (i.e., rocks) of their vineyards. In addition, wines from the same winemaker but different vintage years were analysed to verify the consistency of the $^{87}Sr/^{86}Sr$ ratio with time. A large variation was observed among wines from four geographical areas, i.e., Tuscany, Latium, Campania and Basilicata, supporting the relationship with the geological substratum of the production area. $^{87}Sr/^{86}Sr$ was also quantified in wines, must grape juices, soils and rocks from six vineyards to determine the ability of this variable as a potential marker of interspatial origin.

Catarino et al. (2018) aimed to explain whether the elemental profile established by ICP–MS, followed by chemical processing, might be useful for discriminating wines according to geographical origin, winemaking process and year of production, with consideration of the effects of the parent soil. The results obtained enabled the authors to assume that winemaking processes and vintage year should be considered when elemental composition is used as a tool for Portuguese wine traceability. PCA (with the first two PCs explaining 69.5% of the total variance) as well as DA revealed that Li, Mn, Sr and REEs significantly differentiated between the four various vineyards concerning the two vintages. Moreover, chemometric analysis demonstrated that the elemental composition pattern was transferred through the soil–wine system, and differences observed for soils were reflected in grape musts and wines. According to the authors, recognition of the evolution of the elemental pattern composition from soil to wine, and in what way it is influenced by the climatic year, is required to realize traceability aims.

16.2.4.10 Elaboration of Improved Methodology

Cabañero et al. (2008) aimed to clarify new procedures for wine ethanol $\delta^{13}C$ determination using EA–IRMS and HPLC/IRMS. This combined tool was developed to improve the isotopic methodology used to investigate wine authenticity. Importantly, the developed methods improved the currently available methodologies for wine adulteration.

Moreover, the study performed by Cabañero et al. (2010) also revealed a novel procedure for the simultaneous classification of wine glycerol and ethanol $\delta^{13}C$ using LC–IRMS, which was a useful tool enabling $\delta^{13}C$ quantification of both analytes in the same run directly from a liquid sample with no previous glycerol or ethanol isolation. Owing to this developed approach, it was possible to overcome the technical limitations resulting from complex sample treatment. Moreover, this approach improved this methodology in terms of its simplicity and speed.

Cozzolino et al. (2008) applied VIS and NIRS to determine the contents of macro- and micronutrients in seven varieties of white and red wine samples from Australia. The contents of Ca, Mg, Na, K, P, S, Fe, Mn and B in the wine samples were also investigated via ICP–MS. The samples were scanned in transmission mode using a scanned monochromator. The spectra were pretreated by second derivative and SNV prior to developing calibration models using PLS regression with CV. The highest cross-validation R^2 coefficients of determination and the lowest errors of SECV were obtained for Ca, Fe and K. Some elements studied in the wine could be measured by VIS–NIRS.

Cozzolino et al. (2011) used NIR spectroscopy to quantify essential elements and EC in red grape homogenates from two vintages in Australia. A wide range of varieties and regions were studied by ICP–OES for the levels of Ca, K, Mg, P, S, Fe and Mn and scanned in reflectance with an NIR tool. The spectra were preprocessed via MSC before developing the calibration models via PLS regression and CV. Cross-validation R^2 and SECV values were obtained for the elements studied and EC. Mg, S and EC in grape berries could be determined by NIR spectroscopy.

Thomas and Jamin (2009) presented the results of a collaborative study demonstrating the ability of laboratories to perform food isotopic testing. SNIF–NMR and IRMS were used to provide complementary information to characterize the botanical origin of acetic acid and to detect adulterations of vinegar from cider using synthetic acetic acid. Statistical calculations were performed according to the ISO Standard 5725 and the IUPAC protocol via the Cochran and Grubbs tests. On the basis of the satisfactory interlaboratory precision and accuracy demonstrated by a spiking experiment, the authors recommended the adoption of standard operating procedures, including δ^2H, $\delta^{13}C$ and $\delta^{18}O$ determinations, as official methods for controlling the authenticity of vinegar.

16.3 Conclusions

Single parameters are usually not adequate to identify a product with respect to wine labelling. It has been reported that such control is improved by the use of multivariate chemometric tools based on a combination of stable isotopic data and other multielemental components. Recent chemometric methods/techniques have been used to process original analytical data, enabling the differentiation of wine quality according to wine-producing origin. Moreover, it is possible to verify the geographical origin of wine to protect brand reputations and assess intertype or brand differentiation related to other parameters, e.g., winemaking processes and vintage year.

References

Aceto M, Robotti E, Oddone M, Baldizzone M, Bonifacino G, Bezzo G, Di Stefano R, Gosetti F, Mazzucco E, Manfredi M, Marengo E (2013) A traceability study on the Moscato wine chain. Food Chem 138:1914–1922. https://doi.org/10.1016/j.foodchem.2012.11.019

Adam T, Duthie E, Feldmann J (2002) Investigations into the use of copper and other metals as indicators for the authenticity of Scotch whiskies. J Inst Brew 108:459–464. https://doi.org/10.1002/j.2050-0416.2002.tb00576.x

Agazzi FM, Nelson J, Tanabe CK, Doyle C, Boulton RB, Buscema F (2018) Aging of Malbec wines from Mendoza and California: evolution of phenolic and elemental composition. Food Chem 269:103–110. https://doi.org/10.1016/j.foodchem.2018.06.142

Alcázar A, Pablos F, Martín MJ, González AG (2002) Multivariate characterisation of beers according to their mineral content. Talanta 57:45–52. https://doi.org/10.1016/S0039-9140(01)00670-1

Alcázar Á, Jurado JM, Palacios-Morillo A, de Pablos F, Martín MJ (2012a) Recognition of the geographical origin of beer based on support vector machines applied to chemical descriptors. Food Contr 23:258–262. https://doi.org/10.1016/j.foodcont.2011.07.029

Alcázar Á, Jurado JM, Palacios-Morillo A, de Pablos F, Martín MJ (2012b) Differentiation of blonde beers according to chemical quality indicators by means of pattern recognition techniques. Food Anal Methods 5:795–799. https://doi.org/10.1007/s12161-011-9313-2

Alexa L, Kántor A, Kovács B, Czipa N (2018) Determination of micro and trace elements of commercial beers. J Microbiol Biotech Food Sci 7:432–436. https://doi.org/10.15414/jmbfs.2018.7.4.432-436

Almeida CMR, Vasconcelos MTSD (2003) Multielement composition of wines and their precursors including provenance soil and their potentialities as fingerprints of wine origin. J Agric Food Chem 51:4788–4798. https://doi.org/10.1021/jf034145b

Angerosa F, Breas O, Contento S, Guillou C, Reniero F, Sada E (1999) Application of stable isotope ratio analysis to the characterization of the geographical origin of olive oils. J Agric Food Chem 47:1013–1017. https://doi.org/10.1021/jf9809129

Angus NS, O'Keeffe TJ, Stuart KR, Miskelly GM (2006) Regional classification of New Zealand red wines using inductively-coupled plasma-mass spectrometry (ICP-MS). Aust J Grape Wine Res 12:170–176. https://doi.org/10.1111/j.1755-0238.2006.tb00057.x

Azcarate SM, Martinez LD, Savio M, Camiña JM, Gil RA (2015) Classification of monovarietal Argentinean white wines by their elemental profile. Food Contr 57:268–274. https://doi.org/10.1016/j.foodcont.2015.04.025

Ballabio D, Kokkinofta R, Todeschini R, Theocharis CR (2007) Characterization of the traditional Cypriot spirit Zivania by means of Counterpropagation Artificial Neural Networks. Chemometr Intell Lab Syst 87:52–58. https://doi.org/10.1016/j.chemolab.2006.09.002

Barbaste M, Medina B, Sarabia L, Ortiz MC, Pérez-Trujillo JP (2002) Analysis and comparison of SIMCA models for denominations of origin of wines from the Canary Islands (Spain) builds by means of their trace and ultratrace metals content. Anal Chim Acta 472:161–174. https://doi.org/10.1016/S0003-2670(02)00979-0

Baxter MJ, Crews HM, Dennis MJ, Goodall I, Anderson D (1997) The determination of the authenticity of wine from its trace element composition. Food Chem 60:443–450. https://doi.org/10.1016/S0308-8146(96)00365-2

Bellido-Milla D, Moreno-Perez JM, Hernández-Artiga MP (2000) Differentiation and classification of beers with flame atomic spectrometry and molecular absorption spectrometry and sample preparation assisted by microwaves. Spectrochim Acta Part B: At Spectrosc 55:855–864. https://doi.org/10.1016/S0584-8547(00)00164-6

Benito MJ, Ortiz MC, Sanchez MS, Sarabia LA, Iniguez M (1999) Typification of vinegars from Jerez and Rioja using classical chemometric techniques and neural network methods. Analyst 124:547–552. https://doi.org/10.1039/a809635j

Bentlin FRS, Pulgati FH, Dressler VL, Pozebon D (2011) Elemental analysis of wines from South America and their classification according to country. J Braz Chem Soc 22:327–336. https://doi.org/10.1590/S0103-50532011000200019

Blesić M, Drmać M, Batinić K, Spaho N, Smajić Murtić M, Zele M (2017) Levels of selected metals in wines from different Herzegovinian viticultural localities. Croat J Food Sci Technol 9:1–10. https://doi.org/10.17508/CJFST.2017.9.1.01

Bong Y-S, Ryu J-S, Choi S-H, La M-R, Lee K-S (2016) Investigation of the geographical provenance of the beer available in South Korea using multielements and isotopes. Food Contr 60:378–381. https://doi.org/10.1016/j.foodcont.2015.08.017

Bontempo L, Camin F, Manzocco L, Nicolini G, Wehrens R, Ziller L, Larcher R (2011) Traceability along the production chain of Italian tomato products on the basis of stable isotopes and mineral composition. Rapid Commun Mass Spectrom 25:899–909. https://doi.org/10.1002/rcm.4935

Boschetti W, Rampazzo RT, Dessuy MB, Vale MGR, de Oliveira RA, Hertz P, Manfroi V, Celso PG, Ferrão MF (2013) Detection of the origin of Brazilian wines based on the determination of only four elements using high-resolution continuum source flame AAS. Talanta 111:147–155. https://doi.org/10.1016/j.talanta.2013.02.060

Braschi E, Marchionni S, Priori S, Casalini M, Tommasini S, Natarelli L, Buccianti A, Bucelli P, Costantini EAC, Conticelli S (2018) Tracing the $^{87}Sr/^{86}Sr$ from rocks and soils to vine and wine: an experimental study on geologic and pedologic characterisation of vineyards using radiogenic isotope of heavy elements. Sci Total Environ 628–629:1317–1327. https://doi.org/10.1016/j.scitotenv.2018.02.069

Brescia MA, Caldarola V, De Giglio A, Benedetti D, Fanizzi FP, Sacco A (2002) Characterization of the geographical origin of Italian red wines based on traditional and nuclear magnetic resonance spectrometric determinations. Anal Chim Acta 458:177–186. https://doi.org/10.1016/S0003-2670(01)01532-X

Brescia MA, Kosir IJ, Caldarola V, Kidric J, Sacco A (2003) Chemometric classification of Apulian and Slovenian wines using ^{1}H NMR and ICP-OES together with HPICE data. J Agric Food Chem 51:21–26. https://doi.org/10.1021/jf0206015

Brooks JR, Buchmann N, Phillips S, Ehleringer B, Evans RD, Lott M, Martinelli LA, Pockman WT, Sandquist D, Sparks JP, Sperry L, Williams D, Ehleringer JR (2002) Heavy and light beer: a carbon isotope approach to detect C_4 carbon in beers of different origins, styles, and prices. J Agric Food Chem 50:6413–6418. https://doi.org/10.1021/jf020594k

Cabañero AI, San-Hipolito T, Ruperez M (2007) GasBench/isotope ratio mass spectrometry: a carbon isotope approach to detect exogenous CO_2 in sparkling drinks. Rapid Commun Mass Spectrom 21:3323–3328. https://doi.org/10.1002/rcm.3212

Cabañero AI, Recio JL, Rupérez M (2008) Isotope ratio mass spectrometry coupled to liquid and gas chromatography for wine ethanol characterization. Rapid Commun Mass Spectr 22:3111–3118. https://doi.org/10.1002/rcm.3711

Cabañero AI, Recio JL, Ruperez M (2010) Simultaneous stable′ carbon isotopic analysis of wine glycerol and ethanol by liquid chromatography coupled to isotope ratio mass spectrometry. J Agric Food Chem 58:722–728. https://doi.org/10.1021/jf9029095

Calderone G, Guillou C (2008) Analysis of isotopic ratios for the detection of illegal watering of beverages. Food Chem 106:1399–1405. https://doi.org/10.1016/j.foodchem.2007.01.080

Cameán AM, Moreno I, López-Artíguez M, Repetto M, González AG (2001) Differentiation of Spanish brandies according to their metal content. Talanta 54:53–59. https://doi.org/10.1016/S0039-9140(00)00623-8

Camin F, Bontempo L, Heinrich K, Horacek M, Kelly SD, Schlicht C, Thomas F, Monahan FJ, Hoogewerff J, Roßmann A (2007) Multi-element (H,C,N,S) stable isotope characteristics of lamb meat from different European regions. Anal Bioanal Chem 389:309–320. https://doi.org/10.1007/s00216-007-1302-3

Camin F, Wehrens R, Bertoldi D, Bontempo L, Ziller L, Perini M, Nicolini G, Nocetti M, Larcher R (2012) H, C, N and S stable isotopes and mineral profiles to objectively guarantee the authenticity of grated hard cheeses. Anal Chim Acta 711:54–59. https://doi.org/10.1016/j.aca.2011.10.047

Camin F, Bontempo L, Perini M, Tonon A, Breas O, Guillou C, Moreno-Rojas JM, Gagliano G (2013) Control of wine vinegar authenticity through $\delta^{18}O$ analysis. Food Contr 29:107–111. https://doi.org/10.1016/j.foodcont.2012.05.055

Capron X, Smeyers-Verbeke J, Massart DL (2007a) Multivariate determination of the geographical origin of wines from four different countries. Food Chem 101:1585–1597. https://doi.org/10.1016/j.foodchem.2006.04.019

Capron X, Massart DL, Smeyers-Verbeke J (2007b) Multivariate authentication of the geographical origin of wines: a kernel SVM approach. Eur Food Res Technol 225:559–568. https://doi.org/10.1007/s00217-006-0454-2

Cardoso DR, Andrade-Sobrinho LG, Leite-Neto AF, Reche RV, Isique WD, Ferreira MMC, Lima-Neto BS, Franco DW (2004) Comparison between Cachaça and Rum using pattern recognition methods. J Agric Food Chem 52:3429–3433. https://doi.org/10.1021/jf035262+

Carter JF, Yates HSA, Tinggi U (2015a) A global survey of the stable isotope and chemical compositions of bottled and canned beers as a guide to authenticity. Sci Justice 55:18–26. https://doi.org/10.1016/j.scijus.2014.05.002

Carter JF, Yates HSA, Tinggi U (2015b) Stable isotope and chemical compositions of European and Australasian ciders as a guide to authenticity. J Agric Food Chem 63:975–982. https://doi.org/10.1021/jf5030054

Catarino S, Madeira M, Monteiro F, Caldeira I, Bruno de Sousa R, Curvelo-Garcia A (2018) Mineral composition through soil-wine system of Portuguese vineyards and its potential for wine traceability. Beverages 4:85. https://doi.org/10.3390/beverages4040085

Ceballos-Magaña SG, Jurado JM, Martín MJ, Pablos F (2009) Quantitation of twelve metals in tequila and mezcal spirits as authenticity parameters. J Agric Food Chem 57:1372–1376. https://doi.org/10.1021/jf803626k

Charlton AJ, Wrobel MS, Stanimirova I, Daszykowski M, Grundy HH, Walczak B (2010) Multivariate discrimination of wines with respect to their grape varieties and vintages. Eur Food Res Technol 231:733–743. https://doi.org/10.1007/s00217-010-1299-2

Clark I, Fritz P (1997) Environmental isotopes in hydrogeology. CRC Press/Lewis Publishers, Boca Raton, FL. ISBN 1-56670-249-6

Coetzee PP, Vanhaecke F (2005) Classifying wine according to geographical origin via quadrupole-based ICP-mass spectrometry measurements of boron isotope ratios. Anal Bioanal Chem 383:977–984. https://doi.org/10.1007/s00216-005-0093-7

Coetzee PP, Steffens FE, Eiselen RJ, Augustyn OP, Balcaen L, Vanhaecke F (2005) Multi-element analysis of South African wines by ICP-MS and their classification according to geographical origin. J Agric Food Chem 53:5060–5066. https://doi.org/10.1021/jf048268n

Coetzee PP, van Jaarsveld FP, Vanhaecke F (2014) Intraregional classification of wine via ICP-MS elemental fingerprinting. Food Chem 164:485–492. https://doi.org/10.1016/j.foodchem.2014.05.027

Conde JE, Estévez D, Rodríguez Bencomo JJ, García Montelongo FJ, Pérez-Trujillo JP (2002) Characterization of bottled wines from the Tenerife Island (Spain) by their metal ion concentration. Italian J Food Sci 14:375–387

Cozzolino D, Kwiatkowski MJ, Dambergs RG, Cynkar WU, Janik LJ, Skouroumounis G, Gishen M (2008) Analysis of elements in wine using near infrared spectroscopy and partial least squares regression. Talanta 74:711–716. https://doi.org/10.1016/j.talanta.2007.06.045

Cozzolino D, Cynkar W, Shahn N, Smith P (2011) Quantitative analysis of minerals and electric conductivity of red grape homogenates by near infrared reflectance spectroscopy. Comput Electron Agric 77:81–85. https://doi.org/10.1016/j.compag.2011.03.011

Cristea G, Voica C, Feher I, Radu S, Magdas DA (2019) Isotopic and elemental characterization of cider commercialized on Romanian market. Anal Lett 52:139–149. https://doi.org/10.1080/00032719.2018.1434189

del Mar Castiñeira M, Brandt R, Jakubowski N, Andersson JT (2004a) Changes of the metal composition in German white wines through the winemaking process. A study of 63 elements by inductively coupled plasma–mass spectrometry. J Agric Food Chem 52:2953–2961. https://doi.org/10.1021/jf035119g

del Mar Castiñeira M, Feldmann I, Jakubowski N, Andersson JT (2004b) Classification of German white wines with certified brand of origin by multielement quantitation and pattern recognition techniques. J Agric Food Chem 52:2962–2974. https://doi.org/10.1021/jf035120f

Dembroszky XO, May Z, Hartel T, Zsigmond A-R (2020) Elemental profile of non-commercial wines in changing traditional rural regions from eastern Europe. Environ Eng Manag J 19:625–634. http://www.eemj.eu

Di Martino M, Domenico C, Di Giacomo F, Civitarese C, Cichelli A (2013) ICP-MS analysis for the characterization of the origins of wines. Agro Food Indus Hi Tech 24:30–34

Di Paola-Naranjo RD, Baroni MV, Podio NS, Rubinstein HR, Fabani MP, Badini RG, Inga M, Ostera HA, Cagnoni M, Gallegos E, Gautier E, Peral-García P, Hoogewerff J, Wunderlin DA (2011) Fingerprints for main varieties of Argentinean wines: terroir differentiation by inorganic, organic, and stable isotopic analyses coupled to chemometrics. J Agric Food Chem 59:7854–7865. https://doi.org/10.1021/jf2007419

Díaz C, Conde JE, Esté vez D, Pérez Olivero SJ, Pérez Trujillo JP (2003) Application of multivariate analysis and Artificial Neural Networks for the differentiation of red wines from the Canary Islands according to the Island of origin. J Agric Food Chem 51:4303–4307. https://doi.org/10.1021/jf0343581

Dinca OR, Ionete RE, Costinel D, Geană IE, Popescu R, Stefanescu I, Radu GL (2016) Regional and vintage discrimination of Romanian wines based on elemental and isotopic fingerprinting. Food Anal Methods 9:2406–2417. https://doi.org/10.1007/s12161-016-0404-y

Dordevic N, Wehrens R, Postma GJ, Buydens LMC, Camin F (2012) Statistical methods for improving verification of claims of origin for Italian wines based on stable isotope ratios. Anal Chim Acta 757:19–25. https://doi.org/10.1016/j.aca.2012.10.046

Dugo G, La Pera L, Pellicanó TM, Di Bella G, D'Imperio M (2005) Determination of some inorganic anions and heavy metals in D.O.C. Golden and Amber Marsala wines: statistical study of the influence of ageing period, colour and sugar content. Food Chem 91:355–363. https://doi.org/10.1016/j.foodchem.2004.09.001

Durante C, Baschieri C, Bertacchini L, Cocchi M, Sighinolfi S, Silvestri M, Marchetti A (2013) Geographical traceability based on $^{87}Sr/^{86}Sr$ indicator: a first approach for PDO Lambrusco wines from Modena. Food Chem 141:2779–2787. https://doi.org/10.1016/j.foodchem.2013.05.108

Durante C, Baschieri C, Bertacchini L, Bertelli D, Cocchi M, Marchetti A, Manzini D, Papotti G, Sighinolfi S (2015) An analytical approach to Sr isotope ratio determination in Lambrusco wines for geographical traceability purposes. Food Chem 173:557–563. https://doi.org/10.1016/j.foodchem.2014.10.086

Durante C, Bertacchini L, Bontempo L, Camin F, Manzini D, Lambertini P, Marchetti A, Paolini M (2016) From soil to grape and wine: variation of light and heavy elements isotope ratios. Food Chem 210:648–659. https://doi.org/10.1016/j.foodchem.2016.04.108

Durante C, Bertacchini L, Cocchi M, Manzini D, Marchetti A, Rossi MC, Sighinolfi S, Tassi L (2018) Development of $^{87}Sr/^{86}Sr$ maps as targeted strategy to support wine quality. Food Chem 255:139–146. https://doi.org/10.1016/j.foodchem.2018.02.084

Đurđić S, Pantelić M, Trifković J, Vukojević V, Natić M, Tešić Ž, Mutić J (2017) Elemental composition as a tool for the assessment of type, seasonal variability, and geographical origin of wine and its contribution to daily elemental intake. RSC Adv 7:2151–2162. https://doi.org/10.1039/c6ra25105f

Dutra SV, Adami L, Marcon AR, Carnieli GJ, Roani CA, Spinelli FR, Leonardelli S, Ducatti C, Moreira MZ, Vanderlinde R (2011) Determination of the geographical origin of Brazilian wines by isotope and mineral analysis. Anal Bioanal Chem 401:1571–1576. https://doi.org/10.1007/s00216-011-5181-2

Dutra SV, Adami L, Marcon AR, Carnieli GJ, Roani CA, Spinelli FR, Leonardelli S, Vanderlinde R (2013) Characterization of wines according the geographical origin by analysis of isotopes and minerals and the influence of harvest on the isotope values. Food Chem 141:2148–2153. https://doi.org/10.1016/j.foodchem.2013.04.106

Epova EN, Bérail S, Séby F, Barre JPG, Vacchina V, Médina B, Sarthou L, Donard OFX (2020) Potential of lead elemental and isotopic signatures for authenticity and geographical origin of Bordeaux wines. Food Chem 303:125277. https://doi.org/10.1016/j.foodchem.2019.125277

Espinoza Cruz TL, Guerrero Esperanza M, Wrobel K, Yanez Barrientos E, Acevedo Aguilar FJ, Wrobel K (2020) Determination of major and minor elements in Mexican red wines by microwave-induced plasma optical emission spectrometry, evaluating different calibration methods and exploring potential of the obtained data in the assessment of wine provenance. Spectrochim Acta B At Spectrosc 164:105754. https://doi.org/10.1016/j.sab.2019.105754

Fabani MP, Toro ME, Vázquez F, Díaz MP, Wunderlin DA (2009) Differential absorption of metals from soil to diverse vine varieties from the valley of Tulum (Argentina): consequences to evaluate wine provenance. J Agric Food Chem 57:7409–7416. https://doi.org/10.1021/jf901572k

Fabani MP, Arrúa RC, Vázquez F, Diaz MP, Baroni MV, Wunderlin DA (2010) Evaluation of elemental profile coupled to chemometrics to assess the geographical origin of Argentinean wines. Food Chem 119:372–379. https://doi.org/10.1016/j.foodchem.2009.05.085

Fan S, Zhong Q, Gao H, Wang D, Li G, Huang Z (2018) Elemental profile and oxygen isotope ratio ($\delta^{18}O$) for verifying the geographical origin of Chinese wines. J Food Drug Anal 26:1033–1044. https://doi.org/10.1016/j.jfda.2017.12.009

Farquhar GD, O'Leary MH, Berry JA (1982) On the relationship between carbon isotope discrimination and the intercellular carbon dioxide concentration in leaves. Aust J Plant Physiol 9:121–137. https://doi.org/10.1071/PP9820121

Feher I, Magdas DA, Dehelean A, Sârbu C (2019) Characterization and classification of wines according to geographical origin, vintage and specific variety based on elemental content: a new chemometric approach. J Food Sci Technol 56:5225–5233. https://doi.org/10.1007/s13197-019-03991-4

Flores CR, Figueroa JAL, Wrobel K, Wrobel K (2009) ICP-MS multi-element profiles and HPLC determination of furanic compounds in commercial tequila. Eur Food Res Technol 228:951–958. https://doi.org/10.1007/s00217-009-1010-7

Forina M, Oliveri P, Jäger H, Römisch U, Smeyers-Verbeke J (2009) Class modeling techniques in the control of the geographical origin of wines. Chemometr Intell Lab Syst 99:127–137. https://doi.org/10.1016/j.chemolab.2009.08.002

Frías S, Pérez Trujillo J, Peña E, Conde JE (2001) Classification and differentiation of bottled sweet wines of Canary Islands (Spain) by their metallic content. Eur Food Res Technol 213:145–149. https://doi.org/10.1007/s002170100344

Frías S, Conde JE, Rodríguez MA, Dohnal V, Pérez-Trujillo JP (2002) Metallic content of wines from the Canary Islands (Spain). Application of artificial neural networks to the data analysis. Food/Nahrung 46:370–375. https://doi.org/10.1002/1521-3803(20020901)46:5<370::AID-FOOD370>3.0.CO;2-F

Frías S, Conde JE, Rodríguez-Bencomo JJ, García-Montelongo F, Pérez-Trujillo JP (2003) Classification of commercial wines from the Canary Islands (Spain) by chemometric techniques using metallic contents. Talanta 59:335–344. https://doi.org/10.1016/S0039-9140(02)00524-6

Gaillard L, Guyon F, Marie-Hélène Salagoïty M-H, Médina B (2013) Authenticity of carbon dioxide bubbles in French ciders through multiflow-isotope ratio mass spectrometry measurements. Food Chem 141:2103–2107. https://doi.org/10.1016/j.foodchem.2013.04.112

Gajek M, Pawlaczyk A, Wysocki P, Szynkowska-Jozwik MI (2021a) Elemental characterization of ciders and other low-percentage alcoholic beverages available on the Polish market. Molecules 26:2186. https://doi.org/10.3390/molecules26082186

Gajek M, Pawlaczyk A, Szynkowska-Jozwik MI (2021b) Multi-elemental analysis of wine samples in relation to their type, origin, and grape variety. Molecules 26:214. https://doi.org/10.3390/molecules26010214

Gajek M, Pawlaczyk A, Jóźwik K, Szynkowska-Jóźwik MI (2022) The elemental fingerprints of different types of whisky as determined by ICP-OES and ICP-MS techniques in relation to their type, age, and origin. Foods 11:1616. https://doi.org/10.3390/foods11111616

Galgano F, Favati F, Caruso M, Scarpa T, Palma A (2008) Analysis of trace elements in southern Italian wines and their classification according to provenance. LWT 41:1808–1815. https://doi.org/10.1016/j.lwt.2008.01.015

García-Rodríguez G, Hernandez-Moreno D, Soler F, Pérez-López M (2011) Characterization of "Ribera del Guadiana" and "Méntrida" Spanish red wines by chemometric techniques based on their mineral contents. J Food Nutr Res 50:41–49

García-Ruiz S, Moldovan M, Fortunato G, Wunderli S, Alonso JIG (2007) Evaluation of strontium isotope abundance ratios in combination with multi-elemental analysis as a possible tool to study the geographical origin of ciders. Anal Chim Acta 590:55–66. https://doi.org/10.1016/j.aca.2007.03.016

Geană I, Iordache A, Ionete R, Marinescu A, Ranca A, Culea M (2013) Geographical origin identification of Romanian wines by ICP-MS elemental analysis. Food Chem 138:1125–1134. https://doi.org/10.1016/j.foodchem.2012.11.104

Geană EI, Marinescu A, Iordache AM, Sandru C, Ionete RE, Bala C (2014) Differentiation of Romanian wines on geographical origin and wine variety by elemental composition and phenolic components. Food Anal Meth 7:2064–2074. https://doi.org/10.1007/s12161-014-9846-2

Geană EI, Popescu R, Costinel D, Dinca OR, Stefanescu I, Ionete RE, Bala C (2016a) Verifying the red wines adulteration through isotopic and chromatographic investigations coupled with multivariate statistic interpretation of the data. Food Contr 62:1–9. https://doi.org/10.1016/j.foodcont.2015.10.003

Geană EI, Popescu R, Costinel D, Dinca OR, Ionete RE, Stefanescu I, Artem V, Bala C (2016b) Classification of red wines using suitable markers coupled with multivariate statistic analysis. Food Chem 192:1015–1024. https://doi.org/10.1016/j.foodchem.2015.07.112

Geană E-I, Sandru C, Stanciu V, Ionete RE (2017) Elemental profile and $^{87}Sr/^{86}Sr$ isotope ratio as fingerprints for geographical traceability of wines: an approach on Romanian wines. Food Anal Methods 10:63–73. https://doi.org/10.1007/s12161-016-0550-2

Gimenez-Miralles JE, Salazar DM, Solana I (1999) Regional origin assignment of red wines from Valencia (Spain) by H^2 NMR and C^{13} IRMS stable isotope analysis of fermentative ethanol. J Agric Food Chem 47:2645–2652. https://doi.org/10.1021/jf9811727

González G, Peña-Méndez EM (2000) Multivariate data analysis in classification of must and wine from chemical measurements. Eur Res Food Technol 212:100–107. https://doi.org/10.1007/s002170000207

Gonzálvez A, Llorens A, Cervera ML, Armenta S, de la Guardia M (2009) Elemental fingerprint of wines from the protected designation of origin Valencia. Food Chem 112:26–34. https://doi.org/10.1016/j.foodchem.2008.05.043

Gremaud G, Quaile S, Piantini U, Pfammatter E, Corvi C (2004) Characterization of Swiss vineyards using isotopic data in combination with trace elements and classical parameters. Eur Food Res Technol 219:97–104. https://doi.org/10.1007/s00217-004-0919-0

Griboff J, Horacek M, Wunderlin DA, Monferrán MV (2021) Differentiation between Argentine and Austrian red and white wines based on isotopic and multi-elemental composition. Front Sustain Food Syst 5:657412. https://doi.org/10.3389/fsufs.2021.657412

Guerrero MI, Herce-Pagliai C, Cameán AM, Troncoso AM, González AG (1997) Multivariate characterization of wine vinegars from the south of Spain according to their metallic content. Talanta 45:379–386. https://doi.org/10.1016/S0039-9140(97)00139-2

Hernández-Caraballo EA, Avila-Gómez RM, Capote T, Rivas F, Pérez AG (2003) Classification of Venezuelan spirituous beverages by means of discriminant analysis and artificial neural networks based on their Zn, Cu and Fe concentrations. Talanta 60:1259–1267. https://doi.org/10.1016/S0039-9140(03)00226-1

Iglesias Rodríguez R, Fernández Delgado M, Barciela García J, Peña Crecente RM, García Martín S, Herrero Latorre C (2010) Comparison of several chemometric techniques for the classification of orujo distillate alcoholic samples from Galicia (northwest Spain) according to their certified brand of origin. Anal Bioanal Chem 397:2603–2614. https://doi.org/10.1007/s00216-010-3822-5

Iglesias M, Besalú E, Anticó E (2007) Internal standardization–atomic spectrometry and geographical pattern recognition techniques for the multielement analysis and classification of Catalonian red wines. J Agric Food Chem 55:219–225. https://doi.org/10.1021/jf0629585

Ivanova-Petropulos V, Wiltsche H, Stafilov T, Stefova M, Motter H, Lankmayr E (2013) Multielement analysis of Macedonian wines by inductively coupled plasma–mass spectrometry (ICP–MS) and inductively coupled plasma–optical emission spectrometry (ICP–OES) for their classification. Maced J Chem Chem Eng 32:265–281. https://doi.org/10.20450/mjcce.2013.447

Jos A, Moreno I, González AG, Repetto G, Cameán AM (2004) Differentiation of sparkling wines (cava and champagne) according to their mineral content. Talanta 63:377–382. https://doi.org/10.1016/j.talanta.2003.11.015

Kallithraka S, Arvanitoyannis IS, Kefalas P, El-Zajouli A, Soufleros E, Psarra E (2001) Instrumental and sensory analysis of Greek wines; implementation of principal component analysis (PCA) for classification according to geographical origin. Food Chem 73:501–514. https://doi.org/10.1016/S0308-8146(00)00327-7

Karasinski J, Torres Elguera JC, Gonzalez Ibarra AA, Wrobel K, Bulska E, Wrobel K (2018) Comparative evaluation of red wine from various European regions using mass spectrometry tools. Anal Lett 51:2645–2659. https://doi.org/10.1080/00032719.2018.1442472

Klarić DA, Klarić I, Velić D, Dragojević IV (2011) Evaluation of mineral and heavy metal contents in Croatian blackberry wines. Czech J Food Sci 29:260–267. https://doi.org/10.17221/132/2009-CJFS

Kment P, Mihaljevič M, Ettler V, Šebek O, Strnad L, Rohlová L (2005) Differentiation of Czech wines using multi-element composition—a comparison with vineyard soil. Food Chem 91:157–165. https://doi.org/10.1016/j.foodchem.2004.06.010

Kokkinofta RI, Theocharis CR (2005) Chemometric characterization of the Cypriot Spirit "Zivania". J Agric Food Chem 53:5067–5073. https://doi.org/10.1021/jf0478814

Kokkinofta R, Petrakis PV, Mavromoustakos T, Theocharis CR (2003) Authenticity of the traditional Cypriot Spirit "Zivania" on the basis of metal content using a combination of coupled plasma spectroscopy and statistical analysis. J Agric Food Chem 51:6233–6239. https://doi.org/10.1021/jf034476o

Kokkinofta R, Economidou N, Tzioni E, Damianou K, Poulli K, Savvidou C, Louka C, Kanari P (2014) Studies on the authenticity of local wines by spectroscopic and chemometric analysis. J Chem Chem Eng 8:101–107

Kokkinofta R, Fotakis C, Zervou M, Zoumpoulakis P, Savvidou C, Poulli K, Louka C, Economidou N, Tzioni E, Damianou K, Loupasaki S, Kefalas P (2017) Isotopic and elemental authenticity markers: a case study on Cypriot wines. Food Anal Methods 10:3902–3913. https://doi.org/10.1007/s12161-017-0959-2

Košir IJ, Kocjančič M, Ogrinc N, Kidrič J (2001) Use of SNIF-NMR and IRMS in combination with chemometric methods for the determination of chaptalisation and geographical origin of wines (the example of Slovenian wines). Anal Chim Acta 429:195–206. https://doi.org/10.1016/S0003-2670(00)01301-5

Kruzlicova D, Fiket Ž, Kniewald G (2013) Classification of Croatian wine varieties using multivariate analysis of data obtained by high resolution ICP-MS analysis. Food Res Int 54:621–626. https://doi.org/10.1016/j.foodres.2013.07.053

Langasco I, Caredda M, Sanna G, Panzanelli A, Pilo MI, Spano N, Petretto G, Urgeghe PP (2018) Chemical characterization of craft *Filuferru* Spirit from Sardinia, Italy. Beverages 4:62. https://doi.org/10.3390/beverages4030062

Latorre MJ, Garcia-Jares C, Medina B, Herrero C (1994) Pattern recognition analysis applied to classification of wines from Galicia (Northwestern Spain) with certified brand of origin. J Agric Food Chem 42:1451–1455. https://doi.org/10.1021/jf00043a012

Leder R, Kubanović V, Petric IV, Vahcić N, Banović M (2015) Chemometric prediction of the geographical origin of Croatian wines through their elemental profiles. J Food Nutr Res 54:229–238

Leder R, Petric IV, Jusup J, Banovic M (2021) Geographical discrimination of Croatian wines by stable isotope ratios and multielemental composition analysis. Front Nutr 8:625613. https://doi.org/10.3389/fnut.2021.625613

Longobardi F, Casiello G, Centonze V, Catucci L, Agostiano A (2017) Isotope ratio mass spectrometry in combination with chemometrics for characterization of geographical origin and agronomic practices of table grape. J Sci Food Agric 97:3173–3180. https://doi.org/10.1002/jsfa.8161

Mahmood N, Petraco N, He Y (2012) Elemental fingerprint profile of beer samples constructed using 14 elements determined by inductively coupled plasma–mass spectrometry (ICP-MS): multivariation analysis and potential application to forensic sample comparison. Anal Bioanal Chem 402:861–869. https://doi.org/10.1007/s00216-011-5452-y

Marchionni S, Braschi E, Tommasini S, Bollati A, Cifelli F, Mulinacci N, Mattei M, Conticelli S (2013) High-precision $^{87}Sr/^{86}Sr$ analyses in wines and their use as a geological fingerprint for tracing geographic provenance. J Agric Food Chem 61:6822–6831. https://doi.org/10.1021/jf4012592

Marchionni S, Buccianti A, Bollati A, Braschi E, Cifelli F, Molin P, Parotto M, Mattei M, Tommasini S, Conticelli S (2016) Conservation of $^{87}Sr/^{86}Sr$ isotopic ratios during the winemaking processes of 'red' wines to validate their use as geographic tracer. Food Chem 190:777–785. https://doi.org/10.1016/j.foodchem.2015.06.026

Mardegan SF, Andrade TMB, de Sousa Neto ER, de Castro Vasconcellos EB, Martins LFB, Mendonça TG, Martinelli LA (2013) Stable carbon isotopic composition of Brazilian beers - a comparison between large- and small-scale breweries. J Food Compos Anal 29:52–57. https://doi.org/10.1016/j.jfca.2012.10.004

Marengo E, Aceto M (2003) Statistical investigation of the differences in the distribution of metals in Nebbiolo-based wines. Food Chem 81:621–630. https://doi.org/10.1016/S0308-8146(02)00564-2

Marini F, Bucci R, Magrì AL, MagrI AD (2006) Authentication of Italian CDO wines by class-modeling techniques. Chemometr Intell Lab Syst 84:164–171. https://doi.org/10.1016/j.chemolab.2006.04.017

Markovski J, Markovski M, Knežević B, Hristovski KD (2018) Metals in select beers commercially available in the US: unmonitored sources of concerning exposure. Maced J Chem Chem Eng 37:159–172. https://doi.org/10.20450/mjcce.2018.1501

Martin AE, Watling RJ, Lee GS (2012) The multi-element determination and regional discrimination of Australian wines. Food Chem 133:1081–1089. https://doi.org/10.1016/j.foodchem.2012.02.013

Martinelli LA, Moreira MZ, Ometto JPHB, Alcarde AR, Rizzon LA, Stange E, Ehleringer JR (2003) Stable carbon isotopic composition of the wine and CO_2 bubbles of sparkling wines: detecting C_4 sugar additions. J Agric Food Chem 51:2625–2631. https://doi.org/10.1021/jf026088c

Meier-Augenstein W, Kemp HF, Hardie SML (2012) Detection of counterfeit scotch whisky by ^{2}H and ^{18}O stable isotope analysis. Food Chem 133:1070–1074. https://doi.org/10.1016/j.foodchem.2012.01.084

Mihucz VG, Done CJ, Tatár E, Virág I, Záray G, Baiulescu EG (2006) Influence of different bentonites on the rare earth element concentrations of clarified Romanian wines. Talanta 70:984–990. https://doi.org/10.1016/j.talanta.2006.05.080

Monakhova YB, Godelmann R, Hermann A, Kuballa T, Cannet C, Schäfer H, Spraul M, Rutledge DN (2014) Synergistic effect of the simultaneous chemometric analysis of ^{1}H NMR spectroscopic and stable isotope (SNIF-NMR, ^{18}O, ^{13}C) data: application to wine analysis. Anal Chim Acta 833:29–39. https://doi.org/10.1016/j.aca.2014.05.005

Moncayo S, Rosales JD, Izquierdo-Hornillos R, Anzano J, Caceres JO (2016) Classification of red wine based on its protected designation of origin (PDO) using Laser-induced Breakdown Spectroscopy (LIBS). Talanta 158:185–191. https://doi.org/10.1016/j.talanta.2016.05.059

Moret I, Scarponi G, Cescon P (1994) Chemometric characterization and classification of five Venetian white wines. J Agric Food Chem 42:1143–1153. https://doi.org/10.1021/jf00041a019

O'Leary MH (1995) Environmental effects on carbon isotope fractionation in terrestrial plant. In: Wada E, Yoneyama T, Mingawa M, Ando T, Fry BD (eds) Stable isotopes in the biosphere. Kyoto University Press, Kyoto, pp 78–91. ISBN: 9784876980222

Ogrinc N, Košir IJ, Kocjančič M, Kidrič J (2001) Determination of authenticy, regional origin, and vintage of Slovenian wines using a combination of IRMS and SNIF-NMR analyses. J Agric Food Chem 49:1432–1440. https://doi.org/10.1021/jf000911s

Otim EO, Chen IR, Otim O (2019) Applying multivariate analysis to characterize *waragi* spirits from Acoli, Uganda, by their metal contents. Heliyon 5:e01417. https://doi.org/10.1016/j.heliyon.2019.e01417

Paneque P, Álvarez-Sotomayor MT, Gómez IA (2009) Metal contents in "oloroso" sherry wines and their classification according to provenance. Food Chem 117:302–305. https://doi.org/10.1016/j.foodchem.2009.04.006

Paneque P, Álvarez-Sotomayor MT, Clavijo A, Gómez IA (2010) Metal content in southern Spain wines and their classification according to origin and ageing. Microchem J 94:175–179. https://doi.org/10.1016/j.microc.2009.10.017

Paneque P, Morales ML, Burgos P, Ponce L, Callejon RM (2017) Elemental characterisation of Andalusian wine vinegars with protected designation of origin by ICP-OES and chemometric approach. Food Contr 75:203–210. https://doi.org/10.1016/j.foodcont.2016.12.006

Parker IG, Kelly SD, Sharman M, Dennis MJ, Howie D (1998) Investigation into the use of carbon isotope ratios ($^{13}C/^{12}C$) of Scotch whisky congeners to establish brand authenticity using gas chromatography-combustion-isotope ratio mass spectrometry. Food Chem 63:423–428. https://doi.org/10.1016/S0308-8146(98)00010-7

Pasvanka K, Kostakis M, Tarapoulouzi M, Nisianakis P, Thomaidis NS, Proestos C (2021) ICP–MS analysis of multi-elemental profile of Greek wines and their classification according to variety, area and year of production. Separations 8:119. https://doi.org/10.3390/separations8080119

Pepi S, Chicca M, Piroddi G, Tassinari R, Vaccaro C (2019) Geographical origin of *Vitis vinifera* cv. Cannonau established by the index of bioaccumulation and translocation coefficients. Environ Monit Assess 191:436. https://doi.org/10.1007/s10661-019-7544-7

Pérez-Álvarez EP, Garcia R, Barrulas P, Dias C, Cabrita MJ, Garde-Cerdán T (2019) Classification of wines according to several factors by ICP-MS multi-element analysis. Food Chem 270:273–280. https://doi.org/10.1016/j.foodchem.2018.07.087

Pérez-Magariño S, Ortega-Heras M, González-San José ML (2002) Multivariate classification of rosé wines from different Spanish protected designations of origin. Anal Chim Acta 458:187–190. https://doi.org/10.1016/S0003-2670(01)01500-8

Perez-Trujillo J-P, Barbaste M, Medina B (2002) Contents of trace and ultratrace elements in wines from the Canary Islands (Spain) as determined by ICP-MS. J Wine Res 13:243–256. https://doi.org/10.1080/0957126022000046529

Pérez-Trujillo J-P, Barbaste M, Medina B (2003) Chemometric study of bottled wines with denomination of origin from the Canary Islands (Spain) based on ultra-trace elemental content determined by ICP-MS. Anal Lett 36:679–697. https://doi.org/10.1081/AL-120018257

Pérez-Trujillo JP, Pérez Pont ML, Conde González JE (2011) Content of mineral ions in wines from Canary Islands (Spain). CyTA – J Food 9:135–140. https://doi.org/10.1080/19476337.2010.491582

Perini M, Camin F (2013) $\delta^{18}O$ of ethanol in wine and spirits for authentication purposes. J Food Sci 78:C839–C844. https://doi.org/10.1111/1750-3841.12143

Perini M, Paolini M, Simoni M, Bontempo L, Vrhovsek U, Sacco M, Thomas F, Jamin E, Hermann A, Camin F (2014) Stable isotope ratio analysis for verifying the authenticity of balsamic and wine vinegar. J Agric Food Chem 62:8197–8203. https://doi.org/10.1021/jf5013538

Petrini R, Sansone L, Slejko FF, Buccianti A, Marcuzzo P, Tomasi D (2015) The $^{87}Sr/^{86}Sr$ strontium isotopic systematics applied to Glera vineyards: a tracer for the geographical origin of the prosecco. Food Chem 170:138–144. https://doi.org/10.1016/j.foodchem.2014.08.051

Pořízka J, Diviš P, Dvořák M (2018) Elemental analysis as a tool for classification of Czech white wines with respect to grapevine varieties. J Elem 23:709–727. https://doi.org/10.5601/jelem.2017.22.4.1379

Potortí AG, Lo Turco V, Saitta M, Bua GD, Tropea A, Dugo G, Di Bella G (2017) Chemometric analysis of minerals and trace elements in Sicilian wines from two different grape cultivars. Nat Prod Res 31:1000–1005. https://doi.org/10.1080/14786419.2016.1261341

Podunavac-Kuzmanović SO, Jevrić LR, Kovačević SZ, Marošanović BV, Lojović MM (2015) Chemometric approach to determination of the authenticity of different types of Serbian and Macedonian brandy. Glasnik hemičara, tehnologa i ekologa Republike Srpske 11:11–17. https://doi.org/10.7251/ghte1511011k

Raco B, Dotsika E, Poutoukis D, Battaglini R, ChantzI P (2015) O–H–C isotope ratio determination in wine in order to be used as a fingerprint of its regional origin. Food Chem 168:588–594. https://doi.org/10.1016/j.foodchem.2014.07.043

Rebolo S, Peña RM, Latorre MJ, Garcia S, Botana AM, Herrero C (2000) Characterisation of Galician (NW Spain) Ribeira Sacra wines using pattern recognition analysis. Anal Chim Acta 417:211–220. https://doi.org/10.1016/S0003-2670(00)00929-6

Redan BW, Jablonski JE, Halverson C, Jaganathan J, Mabud MA, Jackson LS (2019) Factors affecting transfer of the heavy metals arsenic, lead, and cadmium from diatomaceous-earth filter dids to alcoholic beverages during laboratory-scale filtration. J Agric Food Chem 67:2670–2678. https://doi.org/10.1021/acs.jafc.8b06062

Rodrigo S, Young SD, Talaverano MI, Broadley MR (2017) The influence of style and origin on mineral composition of beers retailing in the UK. Eur Food Res Technol 243:931–939. https://doi.org/10.1007/s00217-016-2805-y

Rodrigues SM, Otero M, Alves AA, Coimbra J, Coimbra MA, Pereira E, Duarte AC (2011) Elemental analysis for categorization of wines and authentication of their certified brand of origin. J Food Compos Anal 24:548–562. https://doi.org/10.1016/j.jfca.2010.12.003

Römisch U, Jäger H, Capron X, Lanteri S, Forina M, Smeyers-Verbeke J (2009) Characterization and determination of the geographical origin of wines. Part III: multivariate discrimination and classification methods. Eur Food Res Technol 230:31–45. https://doi.org/10.1007/s00217-009-1141-x

Rossmann A, Reniero F, Moussa I, Schmidt H-L, Versini G, Merle MH (1999) Stable oxygen isotope content of water of EU data-bank wines from Italy, France and Germany. Z Lebensm Unters Forsch 208:400–407. https://doi.org/10.1007/s002170050437

Ruggieri F, D'Archivio AA, Foschi M, Maggi MA (2020) Multivariate optimization of an analytical method for the analysis of Abruzzo white wines by ICP OES. Anal Methods 12:2772–2778. https://doi.org/10.1039/D0AY00478B

Saavedra J, Fuentealba C, Yáñez L, Bravo M, Quiroz W, Lukacsy G, Carot JM (2011) Chemometric approaches for the zoning of Pinot Noir wines from the Casablanca valley, Chile. Food Chem 127:1842–1847. https://doi.org/10.1016/j.foodchem.2011.01.132

Sakellari A, Karavoltsos S, Plavšić M, Bempi E, Papantonopoulou G, Dassenakis M, Kalogeropoulos N (2017) Copper complexing properties, trace metal content and organic matter physico-chemical characterization of Greek beers. Microchem J 135:66–73. https://doi.org/10.1016/j.microc.2017.07.010

Sampaio OM, Reche RV, Franco DW (2008) Chemical profile of rums as a function of their origin. The use of chemometric techniques for their identification. J Agric Food Chem 56:1661–1668. https://doi.org/10.1021/jf0726841

Sass-Kiss A, Kiss J, Havadi B, Adányi N (2008) Multivariate statistical analysis of botrytised wines of different origin. Food Chem 110:742–750. https://doi.org/10.1016/j.foodchem.2008.02.059

Scheidegger Y, Saurer M, Bahn M, Siegwolf R (2000) Linking stable oxygen and carbon isotopes with stomatal conductance and photosynthetic capacity: a conceptual model. Oecologia 125:350–357. https://doi.org/10.1007/s004420000466

Schellenberg A, Chmielus S, Schlicht C, Camin F, Perini M, Bontempo L, Heinrich K, Kelly SD, Roßmann A, Thomas F, Jamin E, Horacek M (2010) Multielement stable isotope ratios (H, C, N, S) of honey from different European regions. Food Chem 121:770–777. https://doi.org/10.1016/j.foodchem.2009.12.082

Schlesier K, Fauhl-Hassek C, Forina M, Cotea V, Kocsi E, Schoula R, van Jaarsveld F, Wittkowski R (2009) Characterisation and determination of the geographical origin of wines. Part I: overview. Eur Food Res Technol 230:1–13. https://doi.org/10.1007/s00217-009-1140-y

Šelih VS, Šala M, Drgan V (2014) Multi-element analysis of wines by ICP-MS and ICP-OES and their classification according to geographical origin in Slovenia. Food Chem 153:414–423. https://doi.org/10.1016/j.foodchem.2013.12.081

Serapinas P, Venskutonis PR, Aninkevičius V, Ežerinskis Ž, Galdikas A, Juzikienė V (2008) Step by step approach to multi-element data analysis in testing the provenance of wines. Food Chem 107:1652–1660. https://doi.org/10.1016/j.foodchem.2007.09.003

Shand CA, Wendler R, Dawson L, Yates K, Stephenson H (2017) Multivariate analysis of Scotch whisky by total reflection x-ray fluorescence and chemometric methods: a potential tool in the identification of counterfeits. Anal Chim Acta 976:14–24. https://doi.org/10.1016/j.aca.2017.04.041

Shen F, Li F, Liu D, Xu H, Ying Y, Li B (2012) Ageing status characterization of Chinese rice wines using chemical descriptors combined with multivariate data analysis. Food Contr 25:458–463. https://doi.org/10.1016/j.foodcont.2011.11.019

Shen F, Wu J, Ying Y, Li B, Jiang T (2013) Differentiation of Chinese rice wines from different wineries based on mineral elemental fingerprinting. Food Chem 141:4026–4030. https://doi.org/10.1016/j.foodchem.2013.06.119

Shimizu H, Akamatsu F, Kamada A, Koyama K, Iwashita K, Goto-Yamamoto N (2020) Variation in the mineral composition of wine produced using different winemaking techniques. J Biosci Bioeng 130:166–172. https://doi.org/10.1016/j.jbiosc.2020.03.012

Singerling SA, Burkemper LK, Sharp ZD (2018) Measurement of adjuncts in hard ciders obtainable in the United States using carbon isotopes. J Agric Food Chem 66:11422–11425. https://doi.org/10.1021/acs.jafc.8b05415

Smeyers-Verbeke J, Jäger H, Lanteri S, Brereton P, Jamin E, Fauhl-Hassek C, Forina M, Römisch U (2009) Characterization and determination of the geographical origin of wines. Part II: descriptive and inductive univariate statistics. Eur Food Res Technol 230:15–29. https://doi.org/10.1007/s00217-009-1142-9

Šperková J, Suchánek M (2005) Multivariate classification of wines from different Bohemian regions (Czech Republic). Food Chem 93:659–663. https://doi.org/10.1016/j.foodchem.2004.10.044

Styburski D, Janda K, Baranowska-Bosiacka I, Łukomska A, Dec K, Goschorska M, Michalkiewicz B, Ziętek P, Gutowska I (2018) Beer as a potential source of macroelements in a diet: the analysis of calcium, chlorine, potassium, and phosphorus content in a popular low-alcoholic drink. Eur Food Res Technol 244:1853–1860. https://doi.org/10.1007/s00217-018-3098-0

Sun L-X, Danzer K, Thiel G (1997) Classification of wine samples by means of artificial neural networks and discrimination analytical methods. Fresenius J Anal Chem 359:143–149. https://doi.org/10.1007/s002160050551

Suto M, Kawashima H (2019) Compound specific carbon isotope analysis in sake by LC/IRMS and brewers' alcohol proportion. Sci Rep 9:17635. https://doi.org/10.1038/s41598-019-54162-6

Taylor VF, Longerich HP, Greenough JD (2003) Multielement analysis of Canadian wines by inductively coupled plasma mass spectrometry (ICP-MS) and multivariate statistics. J Agric Food Chem 51:856–860. https://doi.org/10.1021/jf025761v

Temerdashev Z, Khalafyan A, Abakumov A, Bolshov M, Akin'shina V, Kaunova A (2024) Authentication of selected white wines by geographical origin using ICP spectrometric and chemometric analysis. Heliyon 10:e29607. https://doi.org/10.1016/j.heliyon.2024.e29607

Tescione I, Marchionni S, Casalini M, Vignozzi N, Mattei M, Conticelli S (2018) $^{87}Sr/^{86}Sr$ isotopes in grapes of different cultivars: a geochemical tool for geographic traceability of agriculture products. Food Chem 258:374–380. https://doi.org/10.1016/j.foodchem.2018.03.083

Thiel G, Geisler G, Blechschmidt I, Danzer K (2004) Determination of trace elements in wines and classification according to their provenance. Anal Bioanal Chem 378:1630–1636. https://doi.org/10.1007/s00216-003-2444-6

Thomas F, Jamin E (2009) ^{2}H NMR and ^{13}C-IRMS analyses of acetic acid from vinegar, ^{18}O-IRMS analysis of water in vinegar: international collaborative study report. Anal Chim Acta 649:98–105. https://doi.org/10.1016/j.aca.2009.07.014

Tian Y, Yan C, Zhang T, Tang H, Li H, Yu J, Bernard J, Chen L, Martin S, Delepine-Gilon N, Bockova J, Veis P, Chen Y, Yu J (2017) Classification of wines according to their production regions with the contained trace elements using laser-induced breakdown spectroscopy. Spectrochim Acta B 135:91–101. https://doi.org/10.1016/j.sab.2017.07.003

van der Veer G, Voerkelius S, Lorentz G, Heiss G, Hoogewerff JA (2009) Spatial interpolation of the deuterium and oxygen-18 composition of global precipitation using temperature as ancillary variable. J Geochem Explor 101:175–184. https://doi.org/10.1016/j.gexplo.2008.06.008

Voica C, Magdas D-A, Feher I (2015) Metal content and stable isotope determination in some commercial beers from Romanian markets. J Chem 192032:1–10. https://doi.org/10.1155/2015/192032

Vorster C, Greeff L, Coetzee PP (2010) The determination of $^{11}B/^{10}B$ and $^{87}Sr/^{86}Sr$ isotope ratios by quadrupole-based ICP-MS for the fingerprinting of south African wine: research article. S Afr J Chem 63:207–214. https://hdl.handle.net/10520/EJC24493

Wachter H, Christoph N, Seifert S (2009) Verifying authenticity of wine by Mahalanobis distance and hypothesis testing of stable isotope pattern—a case study using the EU Wine Databank. Mitt Klosterneubg 59:237–249

Wyrzykowska B, Szymczyk K, Ichichashi H, Falandysz J, Skwarzec B, Yamasaki S-I (2001) Application of ICP sector field MS and principal component analysis for studying interdependences among 23 trace elements in polish beers. J Agric Food Chem 49:3425–3431. https://doi.org/10.1021/jf010184g

Zambrzycka-Szelewa E, Nalewajko-Sieliwoniuk E, Zaremba M, Bajguz A, Godlewska-Żyłkiewicz B (2020) The mineral profile of polish beers by fast sequential multielement HR CS FAAS analysis and its correlation with total phenolic content and antioxidant activity by chemometric methods. Molecules 25:3402. https://doi.org/10.3390/molecules25153402

Zinicovscaia I, Duliu OG, Culicov OA, Sturza R, Bilici C, Gundorina S (2017) Geographical origin identification of Moldavian wines by Neutron Activation Analysis. Food Anal Methods 10:3523–3530. https://doi.org/10.1007/s12161-017-0913-3

Summary

This book presents the chemometric assessment of analytical dataset to determine authenticity/adulteration of the animal and plant food origin, such as meat, fish, seafood and its products, milk and dairy products, honey, cereals and grains, olive oils and oilseeds, vegetables, legume seeds and spices, mushrooms, fruits including fruit juices, tea and its infusions, coffee and its products, cocoa and its products, sugar and its products, mineral and drinking water, and wines and other alcoholic beverages. The state-of-the-art analytical methods/techniques coupled with appropriate chemometrics appear to be useful approach to determine food authenticity. It is demonstrated that this methodology is a necessary and powerful tool for the food quality control based on the multielement analysis. The other main topics concern traceability, that is, the possibility of verifying the geographical, animal/botanical, and/or productive origin of a foodstuff; hence the use of chemometric techniques is not only recommended but also essential: Since the food is characterized by a complex matrix, the strategy to study its elemental profile should be multivariate. The reported data are extensively tabulated and graphically presented. This book will be useful for food chemists, quality control professionals, nutrition and public health educators, analytical chemists, food researchers needing to use chemometric techniques in practice, food manufacturers, environmental scientists and students, and practicing professions in food chemistry and food safety.

P. Szefer, *Chemometrics in Food Control Based on Elemental Analysis*,
https://doi.org/10.1007/978-3-031-99687-0

Index

P. Szefer, *Chemometrics in Food Control Based on Elemental Analysis*, https://doi.org/10.1007/978-3-031-99687-0

MIX
Papier aus verantwortungsvollen Quellen
Paper from responsible sources
FSC® C105338

If you have any concerns about our products,
you can contact us on
ProductSafety@springernature.com

In case Publisher is established outside the EU,
the EU authorized representative is:
Springer Nature Customer Service Center GmbH
Europaplatz 3, 69115 Heidelberg, Germany

Printed by Libri Plureos GmbH
in Hamburg, Germany